D1058099

Case Studies in
Reliability and Maintenance

WILEY SERIES IN PROBABILITY AND STATISTICS
Established in Part by WALTER A. SHEWHART and SAMUEL S. WILKS

A complete list of the titles in this series appears at the end of this volume.

Case Studies in Reliability and Maintenance

Edited by

WALLACE R. BLISCHKE

D. N. PRABHAKAR MURTHY

A JOHN WILEY & SONS PUBLICATION

This text is printed on acid-free paper. ♾

Published by John Wiley & Sons, Inc., Hoboken, New Jersey.
Published simultaneously in Canada.

Library of Congress Cataloging-in-Publication Data

Case studies in reliability and maintenance / edited by Wallace R. Blischke, D. N. Prabhakar Murthy.
p. cm. — (Wiley series in probability and statistics)
Includes bibliographical references and index.
ISBN 978-0-471-41373-8
1. Reliability (Engineering)—Case studies. 2. Maintainability (Engineering)—Case studies. I. Blischke, W. R., 1934– II. Murthy, D. N. P. III. Series.

TA169 .C37 2003
620'.00452—dc21 2002191015

To our wives, Carol and Jayashree,
for their patience and loving support

Contents

Contributors

Kim M. Aaron
Jet Propulsion Laboratory
California Institute of Technology
Pasadena, California

Kakuro Amasaka
School of Science and Engineering
Aoyama Gakuin University
Tokyo, Japan

Hari P. Annadi
Ford Motor Company
Dearborn, Michigan

Wallace R. Blischke
5401 Katherine Avenue
Sherman Oaks, California

Barry Boehm
Henry Salvatori Hall
University of Southern California
Los Angeles, California

Aarnout C. Brombacher
Section Product and Process Quality
Faculty Technology Management
Eindhoven University of Technology
Eindhoven, The Netherlands

Michael Bulmer
Department of Mathematics
The University of Queensland
St. Lucia, Australia

Chun Kin Chan
Bell Laboratories
Lucent Technologies
Holmdel, New Jersey

Anthony H. Christer
Centre for Operational Research and Applied Statistics
The University of Salford
Manchester, England

Sunita Chulani
IBM Research
San Jose, California

Roger M. Cooke
Department of Mathematics and Informatics
Delft University of Technology
Delft, The Netherlands

John Crocker
Data Systems & Solutions
Bristol, United Kingdom

Yuling Cui
Novellus Systems, Inc.
San Jose, California

Abhijit Dasgupta
CALCE Electronics Products and Systems Center
University of Maryland
College Park, Maryland

Donald H. Ebbeler, Jr.
Jet Propulsion Laboratory
California Institute of Technology
Pasadena, California

John Eccleston
Department of Mathematics
The University of Queensland
St. Lucia, Australia

Elsayed A. Elsayed
Department of Industrial Engineering
Rutgers University
Piscataway, New Jersey

Luis A. Escobar
Department of Experimental Statistics
Louisiana State University
Baton Rouge, Louisiana

Malcolm J. Faddy
School of Mathematics and Statistics
The University of Birmingham
Birmingham, United Kingdom

George Fox
Jet Propulsion Laboratory
California Institute of Technology
Pasadena, California

Hal Gurgenci
Department of Mechanical Engineering
The University of Queensland
St. Lucia, Australia

Per-Erik Hagmark
Tampere University of Technology
Laboratory of Machine Design
Tampere, Finland

Nicholas A. J. Hastings
Samford, Australia

Guan Yue Hong
Institute of Information and Mathematical Sciences
Massey University at Albany
Auckland, New Zealand

Antoine Huijben
Philips Nederland Medical Systems
Eindhoven, The Netherlands

Bermawi P. Iskandar
Department Teknik Industri

Institut Teknologi Bandung
Bandung, Indonesia

Eric Jager
Gasunie Research
Groningen, The Netherlands

Xisheng Jia
Department of Management Engineering
Shijazhuang Mechanical Engineering College
Shijazhuang, People's Republic of China

Chung Wai Kong
Department of Operations Research
School of Engineering and Applied Science
The George Washington University
Washington, DC

U. Dinesh Kumar
Operations Management Group
Indian Institute of Management
Calcutta, India

D. Lewandowski
Department of Mathematics and Informatics
Delft University of Technology
Delft, The Netherlands

Steven Luitjens
Philips Research
Eindhoven, The Netherlands

Karl D. Majeske
The University of Michigan Business School
Ann Arbor, Michigan

William Q. Meeker
Department of Statistics
Iowa State University
Ames, Iowa

D. N. Prabhakar Murthy
Department of Mechanical Engineering
The University of Queensland
St. Lucia, Australia

Soon Huat Ong
National Semiconductor
Manufacturer Singapore Pte Ltd.
Toa Payoh, Singapore

Shunji Osaki
Department of Industrial and Systems Engineering
Hiroshima University
Higashi-Hiroshima, Japan

Michael Osterman
CALCE Electronics Products and Systems Center
University of Maryland
College Park, Maryland

Valia T. Petkova
Section Product and Process Quality
Faculty Technology Management
Eindhoven University of Technology
Eindhoven, The Netherlands

Mark D. Riches
Ford Motor Company
Dearborn, Michigan

Anthony T. Rivers
Viatec Research
Raleigh, North Carolina

Peter C. Sander
Section Product and Process Quality
Faculty Technology Management
Eindhoven University of Technology
Eindhoven, The Netherlands

Nozer D. Singpurwalla
Department of Operations Research
School of Engineering and Applied Science
The George Washington University
Washington, DC

Karen A. Slijkhuis
Bouwdienst Rijkswaterstaat
Utrecht, The Netherlands

Thomas Stadterman
U.S. Army Material Systems Analysis Activity
APG, Maryland

Bert M. Steece
Marshall School of Business
University of Southern California
Los Angeles, California

Loon Ching Tang
Department of Industrial and Systems Engineering
National University of Singapore
Singapore

Michael Tortorella
Department of Industrial Engineering
Rutgers University
New Brunswick, New Jersey

Luis M. Toscano
C/Cambrila 29-1
Madrid, Spain

Peter G. A. Townson
The Centre for Mining Technology and Equipment
Department of Mechanical Engineering
The University of Queensland
St. Lucia, Australia

Seppo Virtanen
Tampere University of Technology
Laboratory of Machine Design
Tampere, Finland

Mladen A. Vouk
Department of Computer Science
North Carolina State University
Raleigh, North Carolina

W. John Walker
Jet Propulsion Laboratory
California Institute of Technology
Pasadena, California

Richard J. Wilson
Department of Mathematics
The University of Queensland
St Lucia, Australia

Gerry M. Winter
Facet Consulting Engineers
Ascot, Australia

Claes Wohlin
Department of Software Engineering and Computer Science
Blekinge Institute of Technology
Ronneby, Sweden

Linda C. Wolstenholme
Department of Actuarial Science and Statistics
City University
London, United Kingdom

Min Xie
Department of Industrial and Systems Engineering
National University of Singapore
Singapore

Stephen A. Zayac
1610 Christian Hills Drive
Rochester, Michigan

Gilles C. Zwingelstein
Department of System Engineering
Institut Universitaire Professionnalisé
The University of Paris, France

Preface

For both the manufacturer and the purchaser, reliability is one of the most important characteristics defining the quality of a product or system. High reliability is achieved through design efforts, choice of materials and other inputs, production, quality assurance efforts, proper maintenance, and many related decisions and activities, all of which add to the costs of production, purchase, and product ownership.

On the other hand, lack of reliability can also lead to significant costs. Loss of revenue due to grounding of a wide-body commercial aircraft may be substantial. In the case of breakdown of a large production facility, revenue loss can amount to millions of dollars per day. Failure of a spacecraft can result in the loss of tens or hundreds of millions of dollars. Failure of a medical device can result in injury or death of a patient and can have serious legal implications. A recall by an automobile manufacturer can cost tens of millions of dollars.

Efforts at achieving high reliability involve many disciplines—engineering, mathematics, materials science, operations analysis, statistics, computer science, and so forth. A scientific approach to reliability theory and methods began during and subsequent to World War II. Since then, the discipline has developed rapidly and much literature addresses this area, including more than a hundred books on general reliability, many more on specific issues (design and testing, materials science, and so forth), scores of journals, and numerous conferences each year.

The notion of a book on case studies originated during the preparation of our previous book in this area, *Reliability: Modeling, Prediction, and Optimization* (Wiley, 2000). That book included an early chapter devoted to brief descriptions of reliability applications, along with associated data sets taken from the engineering, reliability, and statistical literature—in essence, the input for small case studies. These were used throughout the book both to motivate and to illustrate the reliability and statistical methodologies presented. In developing this approach, it occurred to us that a book of fully developed case studies in reliability would be a most useful addition to the literature, for the same reasons. As we began work on the casebook project, this soon expanded to include maintenance and maintainability as well.

Although there are many case studies on reliability, maintainability, and related areas scattered in the literature, cases in these areas had not yet not been collected into a single book. In attempting to compile such a collection, we contacted a large number of practitioners, reliability engineers, statisticians, and others, soliciting both ideas and potential contributions. Our perception of the need for such a book was strongly reinforced by the enthusiastic response we obtained.

In our contacts and in the cases themselves, we intended to achieve broad coverage, both geographically and in the types of applications included. The book that resulted includes 26 cases by a total of 60 authors from 14 countries in North America, Europe, Asia, and Australia. Applications include aerospace, automotive, mining, electronics, power plants, dikes, computer software, weapons, photocopiers, industrial furnaces, granite building cladding, chemicals, and aircraft engines, with analyses done at every stage from product design to post-sale operations and service. Reliability, maintenance, and statistical techniques illustrated include modeling, reliability assessment and prediction, simulation, testing, failure analysis, FMEA, use of expert judgment, preventive maintenance, statistical process control, regression analysis, reliability growth modeling and analysis, repair policy, availability analysis, and many others.

All cases include some mathematical modeling, and some include a fairly thorough treatment of the mathematical foundations of the techniques employed in the case. Mathematical derivations, however, are put in appendices, and many references are given so that the reader may pursue the theory further, if desired. The intent is to provide the reader with an understanding of the methodology, including its strengths and limitations, as well as to illustrate its application.

As the listing of applications makes clear, reliability is important in the context of a very broad range of areas, including physical structures, biological systems, intellectual entities, and services. Correspondingly, much of the reliability research carried out to date has been very discipline-specific, with relatively little interaction between the disciplines. In addition, as is often the case, there is a gap between theory and applications.

A key objective of this book is to help to bridge the gaps between disciplines and between theory and its application. It is hoped that the spread and diversity of the cases and of the backgrounds and disciplines of the authors will help to engineer and construct the first bridge. The second bridge is, as always, more difficult to construct. The connection that is given in each case between the mathematical formulation and the application of the theory to solve a specific problem should help in this regard.

These issues are discussed in more detail in Chapter 1, which provides an introduction and overview of the book. In this chapter, we look briefly at some of the basic concepts of reliability, maintenance, maintainability and quality, and at some of the important practical issues in these areas, many of which are the focus on one or more of the cases that follow.

The format of the cases is generally as follows. Each begins with a discussion of the context, issues, and specific problems to be addressed, emphasizing both the reliability and/or maintenance aspects and the practical implications of the project.

Relevant tools and techniques for solving the problems at hand are discussed along with associated theory, assumptions, models, and related issues. Data issues are addressed and data, if any, are given, analyzed and discussed. The results are interpreted in the context of the original objectives of the project, and conclusions and recommendations are given.

We note that, while not all cases are organized in strictly this fashion, all include these basic elements. Also, as is inevitable in a contributed book, writing styles vary. The editors have attempted to minimize the variability in content, style, and level of the presentations, attempting to achieve a coherent set of case studies.

Each case study concludes with a set of exercises. These include numerical problems for solution, verification of results, extensions of the results given, conceptual questions, and occasional group projects. The intent is to provide both exercises for classroom use and aids for the practitioner desiring to expand on the results, techniques, or theory presented in the case.

The organization of the book is based on the life cycle of a product or system. In this context, we define the life cycle to be the sequence of stages beginning with product design and proceeding through development and testing, production, postsale service, and maintenance. The cases are grouped in this framework in accordance with the main thrust of the case. Most of the cases deal with other issues as well, and this is recognized in tabulations of the cases in several dimensions. A listing of the cases and details of their classification are given in Chapter 1.

The intended audience for the casebook is practitioners, researchers, and graduate students or advanced undergraduates. Because of the wide coverage of techniques and applications, practitioners in reliability engineering, maintenance, maintainability, general engineering, statistics, operations, operations research, and related disciplines should find relevant illustrations in nearly any area of interest. It is our hope that this will help in broadening the readers' knowledge of these important areas and in suggesting further applications. We hope that theorists will find motivation for research among the many unresolved problems and issues cited by the contributors. Finally, we hope that all of this will promote further interaction between theorists and practitioners and among the many disciplines represented in the cases.

The editors hereby express their sincere thanks to all of the contributors of case studies. Each case is unique and was written expressly for this book. Without exception, the cases required a great deal of work on the part of the contributors (and then more work at the request of the editors), and these efforts are greatly appreciated.

Thanks are also due to John Wiley & Sons and to our editors, Steven Quigley and Heather Haselkorn, for their unfailing support of the project, helpful suggestions, and enthusiasm. Finally, we wish to thank the reviewers and our many colleagues for their helpful comments and encouragement.

WALLACE R BLISCHKE
D. N. PRABHAKAR MURTHY

Sherman Oaks, California
Brisbane, Australia
October, 2002

CHAPTER 1

Introduction and Overview

Wallace R. Blischke and D. N. Prabhakar Murthy

1.1 INTRODUCTION

The world of the twenty-first century is a highly complex world in many respects—socially, politically, economically, technologically, and so forth. This complexity is reflected in many of the products and systems we manufacture and use. Buyers' expectations with regard to the ability of a product or system to perform its intended function, in basic terms its *reliability,* are usually high and are becoming ever higher. Except for cheap, throw-away items, this is pretty much true across the board. Simple consumer items such as small electronics, small appliances, and mechanical devices are expected to work without fail. For more complex items such as PCs, communications devices, and even automobiles, most consumers will tolerate very few failures. For some very complex systems such as nuclear power plants and rocket propulsion systems, failures can be disastrous and very high reliabilities are required.

How can such high reliabilities be achieved? There are many factors that affect reliability, and there are many issues that must be addressed. These include engineering design, materials, manufacturing, operations, and maintenance. Addressing the issues requires information, modeling, analysis, testing, data, additional analysis, often additional testing, reengineering when necessary, and so forth. The objectives of a reliability study may be understanding the failure phenomena, estimating and predicting reliability, optimization, and many others. Two other related issues important in the context of product reliability are maintenance and maintainability. Maintenance deals with actions to reduce the likelihood of failure and to restore a failed item to an operational state. Maintainability deals with maintenance issues at the product design stage.

This book is intended to provide a sample of reliability studies done in real-life situations—that is, actual applications with (usually) real data, addressing real problems, and arriving at realistic solutions that can be implemented. The number of possible applications in which reliability analysis may be applied is huge; the sample given here is necessarily small. The range of applications and methodolo-

Case Studies in Reliability and Maintenance, Edited by W. R. Blischke and D. N. P. Murthy.

ISBN 0-471-41373-9

gies applied (and in some cases, developed), however, should serve to illustrate the broad applicability of the techniques, their power and limitations, and the kinds of results that a practitioner may expect to obtain.

1.1.1 Objectives of the Book

Reliability, maintenance, and maintainability (henceforth denoted R&M) are important in nearly every endeavor that deals with engineered and manufactured goods. Thus the number of areas of application is very large, and the number of specific applications is virtually endless. The cases presented in this book illustrate the range of applications. Reliability applications include consumer goods (a motorcycle, an automobile, a DVD player), commercial goods (a photocopier), software, infrastructure (an underground gas pipeline, a system of dikes), aerospace (a space interferometer, a missile system), construction (cladding for buildings), and so forth. Maintenance applications include plant maintenance, aircraft engines, and mining equipment, to name a few.

In addition, reliability overlaps many disciplines, and both researchers and practitioners come from many backgrounds. As a result, reliability and maintenance are considered from different perspectives, some relatively narrow and some quite broad and inclusive. Disciplines involved in the study of R&M issues include all fields of engineering (aerospace, automotive, civil, electrical, mechanical, etc.), materials science, probability, statistics, operations analysis, marketing, management, and so forth, as well as application specific disciplines.

There is a vast literature on reliability, maintainability, and maintenance, including many thousands of articles and a large number of books. Most early books on reliability dealt mainly with the theoretical aspects, focusing on concepts, tools, and techniques. With few exceptions, the examples given in these and many later books are often contrived or based on simulated data, or both. Furthermore, when real data are used, there is typically very little discussion of its origin, on the one hand, or of the broader implications of the results (e.g., engineering design, business management, economic), on the other. As is often the case, therefore, there exists a considerable gap between theory and applications.

One of the key objectives of this book is to attempt to bridge this gap. The intent is to present a number of cases dealing with a range of different problems, with each case highlighting the integration of the different issues involved, the use of appropriate models, data collection where appropriate, data analysis and interpretation, and so on. The cases will serve to illustrate the breadth of applications, as well as some of the many approaches taken to modeling, analyzing, estimating, predicting, and improving reliability, maintainability, and maintenance. In so doing, we hope that the practitioner will obtain some insight, knowledge, and understanding of the application of modern techniques for solving real-world R&M problems. In addition, it is hoped that these case studies may serve to motivate reliability researchers to develop new models and techniques, resulting in valuable contributions to reliability theory as well.

1.1.2 Chapter Outline

The purpose of this chapter is to provide an introduction to the case studies. Prior to discussing the cases, we look at some aspects of reliability, maintenance, maintainability, and quality. This is done in Sections 1.2 through 1.8, which deal with background material intended to provide a very brief review of R&M and to establish a framework for organization and discussion of the cases. The reader familiar with the theory and methodology of R&M may wish to skip these sections and proceed to Section 1.9.

The outline of the chapter is as follows: In Section 1.2, we look more carefully at definitions of terms and at the role and importance of R&M. A short history of R&M is given in Section 1.3, and a brief description of some typical applications is presented in Section 1.4. In Section 1.5, we discuss the concept of a product life cycle, which will serve as a basis for classification of the cases. Section 1.6 is devoted to a discussion of some of the methodologies of reliability analysis. Data-related issues in R&M are discussed in Section 1.7. The different types of problems typically encountered in R&M in the different stages of a product life cycle are discussed in Section 1.8. Finally, the classification of cases by application, life-cycle stage, and R&M issues dealt with is given in Section 1.9.

1.2 RELIABILITY, MAINTENANCE, MAINTAINABILITY, AND QUALITY

1.2.1 Reliability

The reliability of a product or system conveys the concept of dependability, successful operation or performance, and the absence of failure. Unreliability conveys the opposite. More precisely, we define reliability and related concepts as follows:

The ***reliability*** of a product (system) is the probability that the item will perform its intended function throughout a specified time period when operated in a normal (or stated) environment.

Reliability theory deals with the interdisciplinary use of probability, statistics, and stochastic modeling, combined with engineering insights into the design and the scientific understanding of the failure mechanisms, to study the various aspects of reliability. As such, it encompasses issues such as (i) reliability modeling, (ii) reliability analysis and optimization, (iii) reliability engineering, (iv) reliability science, (v) reliability technology, and (vi) reliability management.

Reliability modeling deals with model building to obtain solutions to problems in predicting, estimating, and optimizing the survival or performance of an unreliable system, the impact of unreliability, and actions to mitigate this impact.

Reliability analysis can be divided into two broad categories: (i) qualitative and (ii) quantitative. The former is intended to verify the various failure modes and causes that contribute to the unreliability of a product or system. The latter uses real failure data (obtained, for example, from a test program or from field operations) in conjunction with suitable mathematical models to produce quantitative estimates of product or system reliability.

Reliability engineering deals with the design and construction of systems and products, taking into account the unreliability of its parts and components. It also includes testing and programs to improve reliability. Good engineering results in a more reliable end product.

Reliability science is concerned with the properties of materials and the causes for deterioration leading to part and component failures. It also deals with the effect of manufacturing processes (e.g., casting, annealing, assembly) on the reliability of the part or component produced.

Reliability management deals with the various management issues in the context of managing the design, manufacture, and/or operation and maintenance of reliable products and systems. Here the emphasis is on the business viewpoint, because unreliability has consequences in cost, time wasted, and, in certain cases, the welfare of an individual or even the security of a nation.

Some of the items on this list involve many other issues, including prediction, assessment, optimization, and related topics. These are defined as follows:

Reliability prediction deals basically with the use of models, past history regarding similar products, engineering judgment, and so forth, in an attempt to predict the reliability of a product at the design stage. The process may be updated in later stages as well, in an effort to predict ultimate reliability.

Reliability assessment is concerned with the estimation of reliability based on actual data, which may be test data, operational data, and so forth. It involves system modeling, goodness-of-fit to probability distributions, and related analyses.

Reliability optimization covers many areas and is concerned with achieving suitable trade-offs between different competing objectives such as performance, cost, and so on.

Reliability test design deals with methods of obtaining valid, reliable, and accurate data, and doing so in an efficient and effective manner.

Reliability data analysis deals with estimation of parameters, selection of distributions, and many of the aspects discussed above.

For a further discussion of the various aspects of reliability, see Lloyd and Lipow (1962), Ireson and Coombs (1988), and Blischke and Murthy (2000).

1.2.2 Maintenance

The performance of a product or system depends not only on its design and operation, but also on the servicing and maintenance of the item during its operational lifetime. Thus proper functioning over an extended time period requires proper servicing (e.g., changing oil in an engine) on a regular basis, adequate repair or replacement of failed parts or components, proper storage when not in service, and so forth. These actions are a part of maintenance and maintainability. Definitions of these and related quantities are as follows:

Maintenance comprises any actions (other than routine servicing during operation such as fueling or minor adjustments) that alter a product or system in such a way as to keep it in an operational condition or to return it to an operational condition if it is in a failed condition.

There are two primary types of maintenance actions. These are:

Preventive maintenance. These actions generally require shutdown of an operational system and are intended to increase the length of its lifetime and/or its reliability. Actions range from relatively minor servicing requiring a short downtime, such as lubrication, testing, planned replacement of parts or components, and so forth, to major overhauls requiring a significant amount of downtime. For a survey of research in this area, see Valdez-Flores and Feldman (1989).

Corrective maintenance. This comprises actions taken to restore a failed product or system to an operational state. The actions involve repair or replacement (by either new or used items) of all failed parts and components necessary for successful operation of the item.

Corrective maintenance (CM) actions are unscheduled actions intended to restore a system from a failed state into a working state. These involve either repair or replacement of failed components. In contrast, preventive maintenance (PM) actions are scheduled actions carried out to either reduce the likelihood of a failure or prolong the life of the system. There are several different kinds of PM and CM actions as indicated below.

Classification of Corrective Maintenance

In the case of a repairable product, the behaviour of an item after a repair depends on the type of repair carried out. Various types of repair action can be defined:

- **Good-as-new repair:** Here, the failure time distribution of repaired items is identical to that of a new item, and we model successive failures using an ordinary renewal process. In real life this type of repair would seldom occur.
- **Minimal repair:** A failed item is returned to operation with the same effective age as it possessed immediately prior to failure. Failures then occur according to a nonhomogeneous Poisson process with an intensity function having the same form as the hazard rate of the distribution of the time to first failure. This type of rectification model is appropriate when item failure is caused by one of many components failing and the failed component is replaced by a new one. [See Murthy (1991) and Nakagawa and Kowada (1983).]
- **Different-from-new repair (I):** Sometimes when an item fails, not only are the failed components replaced, but also others that have deteriorated sufficiently. These major overhauls result in $F_1(x)$, the failure time distribution function of all repaired items, being different from $F(x)$, the failure time dis-

tribution function of a new item. The mean time to failure of a repaired item is assumed to be smaller than that of a new item. In this case, successive failures are modeled by a modified renewal process.

- **Different-from-new repair (II):** In some instances, the failure distribution of a repaired item depends on the number of times the item has been repaired. This situation can be modeled by assuming that the distribution function after the jth repair ($j \geq 1$) is $F_j(x)$ with the mean time to failure μ_j decreasing as j increases.

Classification of preventive maintenance

Preventive maintenance (PM) actions can be divided into the following categories:

- **Clock-based maintenance:** PM actions are carried out at set times. An example of this is the "Block replacement" policy. [See Blischke and Murthy (2000).]
- **Age-based maintenance:** PM actions are based on the age of the component. An example of this is the "Age replacement" policy. [See Blischke and Murthy (2000).]
- **Usage-based maintenance:** PM actions are based on usage of the product. This is appropriate for items such as tires, components of an aircraft, and so forth.
- **Condition-based maintenance:** PM actions are based on the condition of the component being maintained. This involves monitoring of one or more variables characterizing the wear process (e.g., crack growth in a mechanical component). It is often difficult to measure the variable of interest directly; in this case, some other variable may be used to obtain estimates of the variable of interest. For example, the wear of bearings can be measured by dismantling the crankcase of an engine. However, measuring the vibration, noise, or temperature of the bearing case provides information about wear because there is a strong correlation between these variables and bearing wear.
- **Opportunity-based maintenance:** This is applicable for multicomponent systems, where a maintenance action (PM or CM) for a component provides an opportunity for carrying out PM actions on one or more of the remaining components of the system.
- **Design-out maintenance:** This involves carrying out modifications through redesigning the component. As a result, the new component has better reliability characteristics.

In general, preventive maintenance is carried out at discrete time instants. In cases where the PM actions are carried out fairly frequently, they can be treated as occurring continuously over time. Many different types of model formulations have been proposed to study the effect of preventive maintenance on the degradation and failure occurrence of items and to derive optimal preventive maintenance strategies.

Concepts, modeling, and analysis of maintenance and reliability and related areas are discussed in detail in Pecht (1995). Jardine and Buzacott (1985) discuss the relationship between reliability and maintenance. An extensive review of this area in given by Pham and Wang (1996).

1.2.3 Maintainability

Many items can be repaired quite easily. Others definitely cannot. Conceptually, design issues dealing with this aspect of maintenance are the domain of the subject of maintainability. Design issues include accessibility of parts for repair, standardization of parts, modular construction, and development of diagnostic procedures and equipment. [See Hegde and Kubat (1989), Malcolm and Foreman (1984) for further details.] Technically, maintainability is defined in probabilistic terms as follows:

Maintainability is the probability that a failed system can be made operable in a specified period of time (Kapur and Lamberson, 1977).

According to BS 4778, maintainability is defined as:

> The ability of an item, under stated conditions of use, to be retained in, or restored to, a state in which it can perform its required functions, when maintenance is performed under stated conditions and using prescribed procedures and resources.

According to IEC 60300-3-4 (1996), a complete specification of maintainability performance requirements should cover the following:

- The maintainability performance to be achieved by the design of the item
- The constraints that will be placed on the use of the item which will affect maintenance
- The maintainability program requirements to be accomplished by the supplier to ensure that the delivered item has the required maintainability characteristics
- The provision of maintenance support planning

For further discussion of maintainability, maintenance, and their relationship to reliability, see Ireson and Coombs (1988) and Blanchard et al. (1995).

1.2.4 Quality

Reliability and maintainability are only two of the many dimensions of the broader concept of **quality.** There are many definitions of quality, and it has been looked at from many perspectives. [See Evans and Lindsay (1999).] Some other quality characteristics are conformance (to standards and to specifications), performance, features, aesthetics, durability, serviceability (basically the speed and competence of repair work), repairability, and availability (the probability that a product or system is operational). Many of these are related to R&M, and some of these (e.g., conformance, performance and availability) are issues that also may be addressed in a case whose primary thrust is reliability or maintenance.

1.2.5 The Role and Importance of Reliability, Maintenance, Maintainability, and Quality

There are many factors that can lead to failure of a product or contribute to the likelihood of occurrence of a failure. These include design, materials, manufacture, quality control, shipping and handling, storage, use, environment, age, the occurrence of related previous failures, the failure of an interconnected component, part, or system, quality of repair after a previous failure, and so forth. Furthermore, because of the nature of the many factors that may be involved, the time of occurrence of a failure (i.e., the lifetime of the item) is unpredictable. Reliability and maintenance deal with estimation and prediction of the probability of failure (i.e., the randomness inherent in this event), as well as many related issues such as prevention of failure, cost of prevention, optimization of policies for dealing with failure, and the effect of environment on failure rates, to name a few.

Although the time to failure of an item is random, the ultimate failure of an item, if kept in service, is certain. Every item, no matter how well designed, manufactured, and maintained, will fail sooner or later. Furthermore, building and maintaining reliable products is an increasingly difficult task because of the increased complexity of items, new technologies and materials, and new manufacturing processes. As a result, a quantitative approach to R&M has become essential in most applications.

The consequences of failure (i.e., the effect of unreliability) are many and varied, depending on the item and the stakeholder involved, but nearly all failures have an economic impact. Stakeholders include one or more of manufacturer, dealer, buyer, insurer, other third parties, and, in extreme cases, society as a whole. As an example, failure of a braking system in an automobile may result in an accident, affecting the driver and others who may be riding in the car at the time of the accident, the owner of the car and the owner's insurer, and others who may be involved in the accident and their insurers. If the failure occurs frequently and is found to be the fault of the manufacturer (because of poor design, materials, workmanship, etc.), the cost to the manufacturer can be enormous, due to lawsuits, possible product recall, costly engineering design changes, and other less tangible consequences such as loss of reputation and resulting loss of sales (thus impacting dealers, distributors, service personnel, and others as well). Society as a whole can experience the impact if all of this places the company in serious financial jeopardy, resulting in a government bailout or if new legislation is enacted as a result of the impact of these events.

Since failure cannot be prevented entirely, it is important to minimize both its probability of occurrence and the impact of failures when they do occur. This is one of the principal roles of reliability analysis and maintenance. Increasing either reliability or maintenance entails costs to the manufacturer, the buyer, or both. There is often a trade-off between these costs. Increasing reliability by improving design, materials and production early on will lead to fewer failures and may decrease maintenance costs later on. Increasing both reliability and maintenance efforts (and hence the costs of these) will almost certainly lead to fewer failures and hence to decreased future costs that would be incurred as a result of the failures. The analysis of the various trade-offs of reliability and maintenance costs and the costs of potential future failures is an important aspect of an overall cost analysis. Optimization of reliability and maintenance policies to minimize the total cost to the manufacturer or the buyer is the ultimate goal of such analyses and various aspects of this are discussed in several of the cases.

A related issue is the cost (to both manufacturer and consumer) of a warranty. This cost is also related to R&M. The relationship of warranty costs to reliability is clear: warranty costs to the manufacturer decrease as product reliability increases. The cost of warranty also depends on the warranty policy provided by the manufacturer, which considerably complicates the analysis. [See Blischke and Murthy (1994).] The relationship to maintenance is also complex. Some of the issues involved and the methodologies used are discussed in several of the cases dealing with warranty issues.

1.3 HISTORY OF RELIABILITY AND MAINTAINABILITY

The notion of a quantitative analysis of reliability is relatively recent, dating back to about the 1940s, at which time mathematical techniques, some of which were quite new, were applied to many operational and strategic problems in World War II. Prior to this period, the concept of reliability was primarily qualitative and subjective, based on intuitive notions. Actuarial methods had been used to estimate survivorship of railroad equipment and in other applications early in the twentieth century (Nelson, 1982), and extreme value theory was used to model fatigue life of materials beginning in the 1930s. These were the forerunners of the statistical and probabilistic models and techniques that form the basis of modern reliability theory.

The needs of modern technology, especially the complex systems used in the military and in space programs, led to the quantitative approach, based on mathematical modeling and analysis. In space applications, high reliability is especially essential because of the high level of complexity of the systems and the inability to make repairs of, or changes to, most systems once they are deployed in an outer space mission. This gave impetus to the rapid development of reliability theory and methodology beginning in the 1950s. As the space program evolved and the success of the quantitative approach became apparent, the analysis was applied in many non-defense/space applications as well. Important newer areas of application are biomedical devices and equipment, aviation, consumer electronics, communications, and transportation.

A more quantitative (or mathematical) and formal approach to reliability grew out of the demands of modern technology, and particularly out of the experiences in the Second World War with complex military systems [Barlow and Proschan (1965), p. 1]. Since the appearance of this classic book, the theory of reliability has grown at a very rapid rate, as can be seen by the large number of books and journal articles that have appeared on the subject. A number of these will be referred to later; still others are listed in the bibliography given in the final chapter of the book.

Barlow (1984) provides a historical perspective of mathematical reliability theory up to that time. Similar perspectives on reliability engineering in electronic equipment, on space reliability technology, on nuclear power system reliability, and on software reliability can be found in Coppola (1984), Cohen (1984), Fussel (1984), and Shooman (1984), respectively.

For more on the history of reliability, see Knight (1991) and Villemeur (1992). The 50th Anniversary Special Publication of the IEEE Transactions on Reliability (Volume 47, Number 3-SP, September 1998) contains several historical papers dealing with various aspects of reliability (e.g., reliability prediction, component qualification) and applications, including space systems reliability, reliability physics in electronics, and communications reliability.

1.4 APPLICATIONS

The study of reliability is applicable not only to products, including hardware and software, but to services as well. Services not only include maintenance activities on products, such as the types of maintenance and repair on an automobile typically obtained from a dealer or "service station," but also include pure service industries such as restaurants, hotels, and the travel, music, and entertainment industries. The quantitative approach to the study of the reliability of intangibles such as these is not very well developed. Reliability, or, more generally, quality of services, is generally approached by use of qualitative and subjective methods, and applications of this type are not included among the cases in this book. For further information, see Rosander (1985, 1989).

The focus here, then, is on products, and an attempt has been made to cover a broad range of hardware applications. In addition, several cases are devoted to software. Products of either type may be classified in many ways. One such scheme is:

- Consumer goods, both durable and nondurable (e.g., appliances, PCs, autos, small tools, food). These range from simple to complex and are characterized by a large number of both manufacturers and buyers.
- Commercial products (e.g., commercial aircraft, large computers, medical equipment, communications systems). These are typically characterized by relatively small numbers of both manufacturers and buyers.
- Industrial products (e.g., cutting tools, presses, earth moving equipment, large turbines). Here there are also relatively few suppliers and purchasers.
- Government—military, aerospace (except private ventures), and other noninfrastructure (e.g., military aircraft, military and nonmilitary radars, rockets, buildings and monuments). These are characterized by relatively few manufacturers and a single consumer.

- Infrastructure (e.g., transportation, highways, dams, power plants, ports). Again, there are typically few suppliers and very few buyers (most often, one—a government agency).

The cases span all of these areas of application. Note, however, that the categories are not mutually exclusive. For example, consumer goods are used by buyers in all of the other categories as well, and some items do not fit uniquely into the categories specified. Furthermore, the categorization is much more complex than this list might suggest because of the widespread existence and growing complexity of multinational corporations and multilayered governments.

In all of the categories, the products themselves may range from simple to quite complex. Nonconsumer products can be very complex indeed, requiring commensurately complex reliability analyses. Some of this complexity is also reflected in the cases.

1.5 LIFE-CYCLE CONCEPTS

1.5.1 Basic Concept

In the previous section we dealt with classification of products and of consumers and purchasers of the products. It is also useful in categorizing and discussing the cases to consider the different phases in the life cycle of the product and the reliability issues associated with each of these phases. In this section, we look at this facet of reliability analysis.

There are a number of approaches to the concept of a product life cycle, depending on the perspective from which it is defined and the temporal context. The concept is quite different in meaning and importance from the point of view of the manufacturer and that of the buyer. How the life cycle is defined also depends on how broad a cycle is to be defined. For example, the life cycle of a commercial aircraft may be arbitrarily set at 30 years, during which period it may owned by more than one airline. During this period, an engine may be changed out several times, so that the life cycle of the engine is the lifetime of the original plus the lifetimes of its replacements over the 30 year period. Thus length and definition of the life cycle are quite different for the manufacturer of the aircraft, the manufacturer of the engine, and each of the buyers of the aircraft.

An important element of the analysis of a product life cycle is that this analysis provides a basis for a cost analysis from a broad perspective. Life-cycle costing may be done from the point of view of the buyer or the manufacturer. In either case, this provides important information for long-term financial planning. Such an analysis includes not only the initial cost of the item, but also operation and maintenance costs, replacement costs and many other cost elements. From the point of view of the manufacturer, these costs would include design, research and development, production, warranty costs, and all of the other costs incurred in producing and selling the item, amortized over units produced and time. For the buyer, costs include initial cost, cost of all replacements, both under warranty and after expiration of warranty, and all other costs associated with use of the item over time.

In the case studies, only the manufacturer's point of view is addressed. In the next section, we look more carefully at the life-cycle concept from this perspective and give the life-cycle definition under which the case studies will be classified. For more on life-cycle concepts, see Rink and Swan (1979) and Murthy and Blischke (2000), and for more on life-cycle costs, see Blanchard (1998) and Blischke and Murthy (2000).

1.5.2 Manufacturer

From the manufacturer's point of view, there are again many approaches. For example, rather than look at a single product, one can view the life cycle from a broader perspective, with the product life cycle embedded in that of the product line, and this, in turn, embedded in the technology life cycle (Betz, 1993). We consider the narrower perspective of a single product. One approach is to take the product life cycle to be the time from initial concept of the product to its discarding at the end of its useful life. This involves several stages, as indicated in Figure 1.1.

The process begins with an idea of building a product to meet some identified customer requirements, such as performance (including reliability) and cost targets, maintenance requirements, aesthetics, and so forth. This is usually based on a study of the market and potential demand for the product being planned. The next step is to carry out a feasibility study. This involves evaluating whether it is possible to achieve the targets within specified cost limits. If this analysis indicates that the project is feasible, an initial product design is undertaken. A prototype is then developed and tested. At this stage it is not unusual to find that achieved performance levels of the prototype product are below the target values. In this case, further product development is undertaken to overcome the problem. Once this is achieved, the next step is to carry out trials to determine performance of the product in the field and to start a preproduction run. This is required because the manufacturing process must be fine-tuned and quality control procedures established to ensure that the items produced have the same performance characteristics as those of the final prototype. After this, the production and marketing efforts begin. Items are produced and sold, and production continues until the product is removed from the market because of obsolescence and/or the introduction a replacement product.

The life-cycle concept to be used in classifying and discussing the cases is a simplification of the above scheme wherein we group the stages of the life cycle into three phases: pre-manufacture (which includes product design and development), manufacture, and post-sale support. This is discussed in more detail in the last two sections of the chapter.

The life cycle and the reliability-related issues of importance in the different phases of the life cycle are displayed in Figure 1.2. A typical scenario is as follows: A feasibility study is carried out using a specified target value for product reliability. During the design stage, product reliability is assessed in terms of part and component reliabilities. Product reliability increases as the design is improved. Howev-

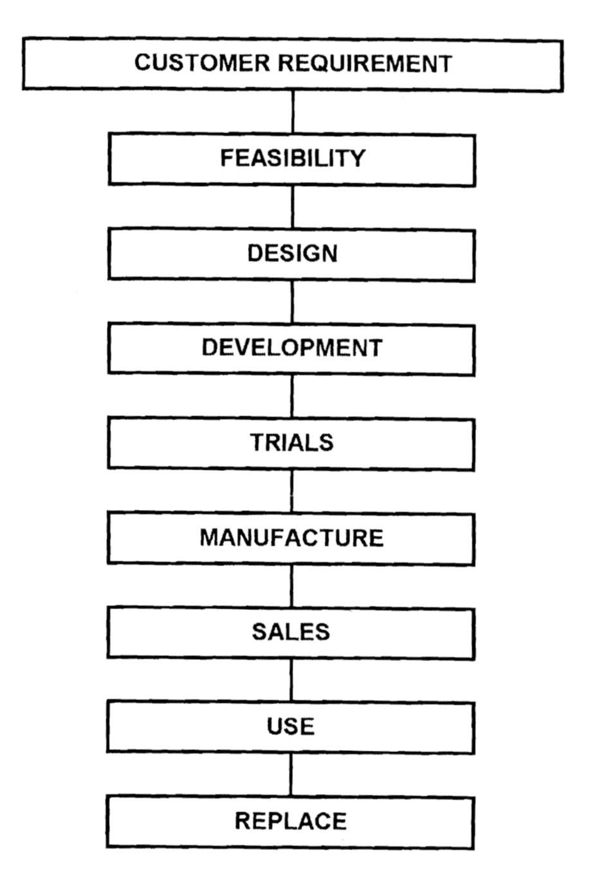

Figure 1.1 Product life cycle.

er, this improvement has an upper limit. If the target value is below this limit, then the design using available parts and components achieves the desired target value. If not, then a development program to improve the reliability through test–fix–test cycles is necessary. Here the prototype is tested until a failure occurs and the causes of the failure are analyzed. Based on this, design and/or manufacturing changes are introduced to overcome the identified failure causes. This process is continued until the reliability target is achieved.

The reliability of the items produced during the preproduction run is usually below that for the final prototype. This is caused by variations resulting from the manufacturing process. Through proper process and quality control, these variations are identified and reduced or eliminated and the reliability of items produced is increased until it reaches the target value. Once this is achieved, full-scale production commences and the items are released for sale.

It is worth noting that if the target reliability values are too high, they might not be achievable, even with further development. In this case, the manufacturer must

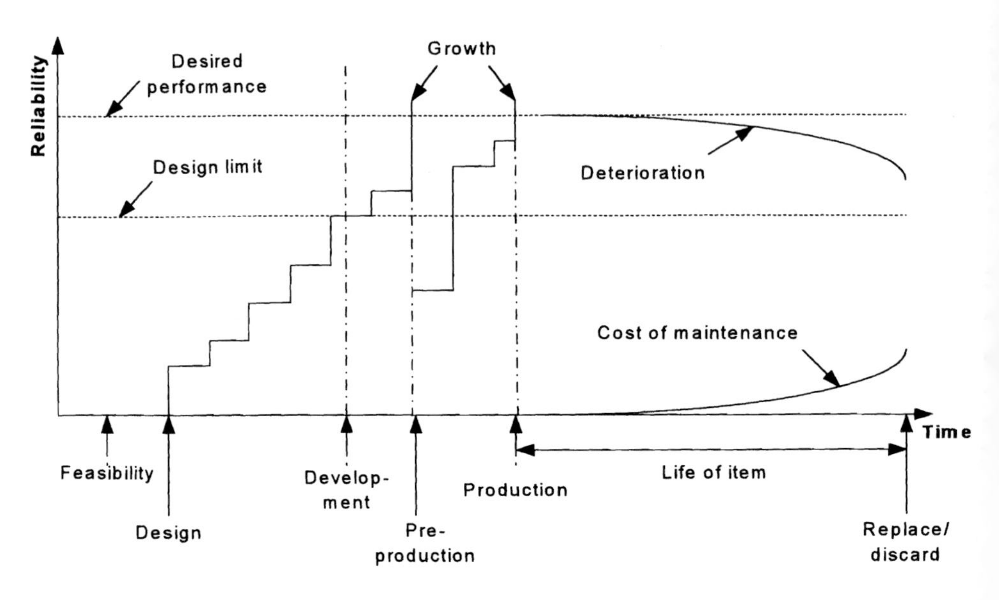

Figure 1.2 Product reliability over the life cycle (Blischke and Murthy, 2000).

revise the target value and start with a new feasibility study before proceeding with the project.

The reliability of an item in use generally deteriorates with age. This deterioration is affected by several factors, including environment, operating conditions, and maintenance. Deterioration can be slowed by increasing maintenance, but at ever-increasing cost. Eventually, deterioration becomes so rapid and/or the cost of maintenance and operation becomes so high that the item is removed from service.

Note also that the reliability of a product varies over its life cycle, and a thorough reliability analysis may be necessary at many stages.

1.6 TOOLS AND TECHNIQUES FOR THE STUDY OF RELIABILITY

In this section we briefly discuss the models, tools, and techniques that are used in the analysis and prediction of reliability. Many of the cases employ one or more of these in various stages of the product life cycle.

Since reliability is defined in probabilistic terms, the basic quantitative models are mathematical, based on the theory of probability. Implementation requires the use of data. This introduces statistical methods into the analysis. In order that the quantitative models be realistic, however, an understanding of some of the more fundamental concepts underlying item failure is necessary. As a result, many other disciplines are involved in reliability studies as well. These include engineering design, materials science, process design, stress analysis, failure analysis, test design, accelerated testing, operations analysis, and many others.

1.6.1 Systems Approach

Fundamental to the study of reliability is the systems approach. This provides a framework for integration of the different technical, commercial, and managerial aspects of the study. The systems approach involves system characterization, modeling, analysis, and optimization. In the following sections we discuss the systems approach in general, modeling aspects, and some important qualitative tools and quantitative techniques used in reliability studies.

The systems approach to solving real-world problems involves a number of steps. The successful execution of these requires concepts and techniques from many disciplines. A flow chart of the systems approach is given in Figure 1.3. Briefly, the sequence of steps in the systems approach is as follows:

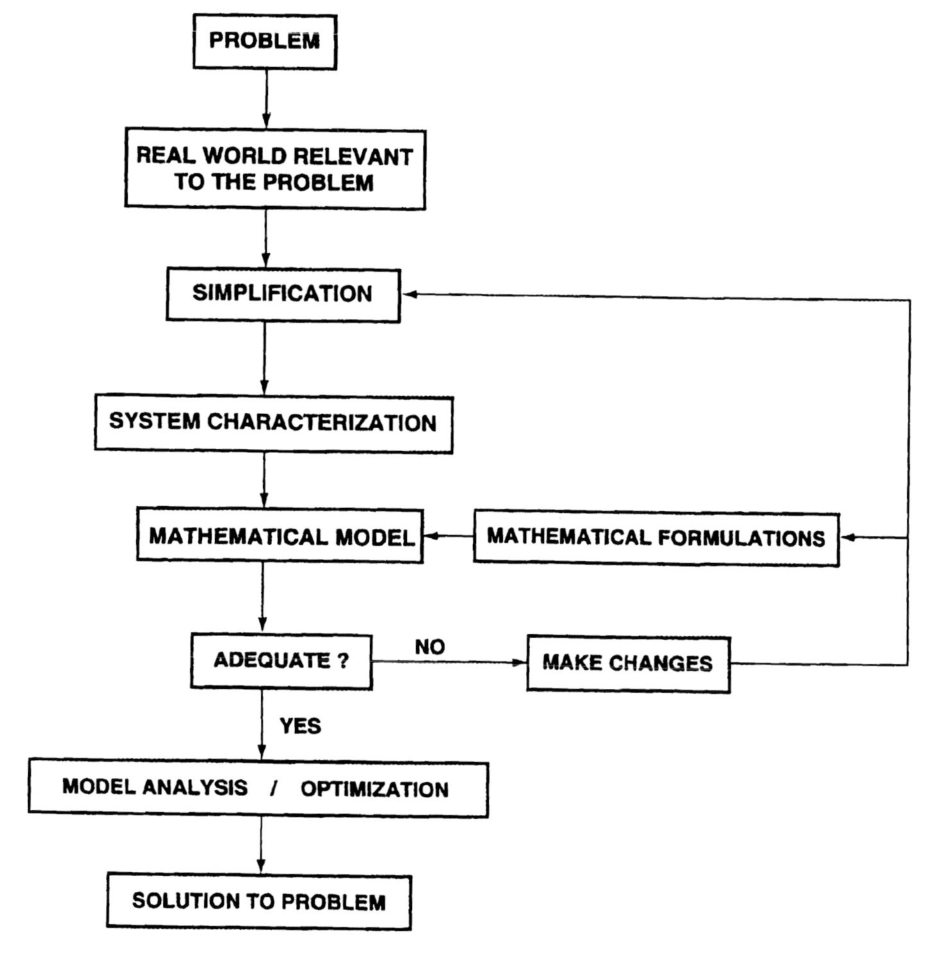

Figure 1.3 Systems approach to problem solving (Blischke and Murthy, 2000).

The process begins with a careful definition of the real-world problem that is to be addressed. In order to approach the problem analytically, it is usually necessary to make some simplifying assumptions and other simplifications that represent departures to a greater or lesser degree from the real world. Here a trade-off is made between reality and mathematical tractability. The next step, which is key to the whole process, is characterization of the system in such a way that relevant details of the problem are made apparent and appropriately modeled. Mathematical models of the system are developed and checked for adequacy. If it is felt that these do not adequately model the system, changes are made to improve the model until it is a sufficiently realistic representation of the real world and the problem being addressed to merit analysis.

Once an adequate model is obtained, mathematical techniques from probability, statistical analysis, stochastic processes, and optimization theory are needed for analysis and optimization. Reliability theory provides the concepts and tools for this purpose. In the next three sections, we look at some of these tools and techniques. The final, very important step in the process is interpretation of the results of the mathematical and statistical analyses in the context of the original real-world problem. The usefulness of the analytical solution will depend on the adequacy of the model and assumptions, validity and relevance of the data, depth and breadth of analysis done, and any of a number of other factors.

For a more detailed discussion of the systems approach in reliability studies, see Blischke and Murthy (2000).

1.6.2 Modeling

Reliability modeling involves model selection, parameter estimation, and validation. At the initial stages of reliability analysis, the modeling process begins with a qualitative model—for example, a block diagram that displays the structure of the product as a system. This exhibits the parts and components of the system and their interconnectedness, and it may be done at various levels of complexity and detail. It also provides the analyst with an indication of the types and levels of data and other information that will be required in the analysis. Block diagrams of a system may be thought of as flow charts indicating paths through the system such that the system is operational if all elements included in the path are operational. The logical structure is represented by arrangements of the blocks into series or parallel configuration, combinations of these, or, if necessary, more complex patterns. [See Barlow and Proschan (1975), Elsayed (1996), Blischke and Murthy (2000), and other reliability texts for examples of block diagrams and methods of analysis.] Blocks in a diagram may represent subsystems of a system, components, and so forth, down to the part level. For a complete analysis, the diagram must ultimately include all elements down to the most elementary part level.

The next step in the process is to assign probabilities of successful operation to the blocks in the diagram. System reliability is then calculated using the logical structural model. Evaluation of these probabilities is one of the most important aspects of reliability analysis. This is done is various ways, depending on the stage in the product life cycle and the information available. In the early stages of design, estimates of reliabilities may be very crude (e.g., only rough orders of magnitude). Even these, however, are of value, for example, in comparing various design concepts. As more information is obtained, the estimates are updated and refined. [An example of this process in a highly complex application, analysis of a rocket propulsion system, is given by Blischke (1994).]

The block diagram is also useful at the design stage in determining reliability requirements at the part and higher levels. Here we are concerned with the allocation of reliability to parts to achieve a desired system reliability. The analysis looks at the feasibility of attaining the desired objective and the trade-offs of part reliabilities that will achieve the goal.

An important technique for determining failure probabilities is the probability distribution of time to failure (also called "failure distribution" or "life distribution"). This may be obtained theoretically or estimated by use of test data. Again, distributions may be specified at the part level, component level, and so forth, up to the system level. Conceptually, at least, distributions at a given level may be obtained from those at lower levels. Other tools and techniques for reliability analysis are discussed in Section 1.6.3.

Different levels and amounts of data are available at various stages of the product life cycle. These may be used to update reliability assessments and predictions and for many related purposes. None of the cases cover the life cycle in its entirely, but many of the stages discussed previously are represented. A good introduction to reliability modeling is given by Wolstenholme (1999). For more extensive coverage of the topic, see Høyland and Rausand (1994) and Blischke and Murthy (2000).

1.6.3 Tools and Techniques for Reliability Analysis

Modeling as a tool for reliability analysis was discussed in Section 1.6.2. This is fundamental to the systems approach discussed in the previous section. Much information is required for successful implementation of the systems approach and for useful application of the models. Here we review a few of the tools and techniques used to obtain some of this information. The focus is on failure analysis—that is, analysis of the underlying causes of failure, the mechanisms of failure, and its consequences. Techniques for acquisition of reliability data and methods of analysis of reliability data are discussed in Section 1.8.

Causes of failure include overstress (e.g., fracture, buckling, deformation, de-adhesion) and wearout (e.g., corrosion, wear, diffusion, crack initiation, and propagation, radiation). Failures may be classified as mechanical (deformation, buckling, fracture, cracking, creep, and creep rupture), electrical (electrostatic discharge, dielectric breakdown, junction breakdown, and others), thermal (overheating, thermal expansion, and contraction), radiation failures (radioactivity, secondary cosmic rays), chemical (corrosion, oxidation, etc.), or combinations of two or more of these. For details of failure mechanisms as well as failure models based on these, see Blischke and Murthy (2000) and the references cited therein.

Tools used in the design stage for identifying failures and determining their consequences are as follows:

- **Failure modes and effects analysis (FMEA):** FMEA is a technique for analysis of a system in terms of its subsystems, assemblies, and so on, down to the part level, to determine failure causes. The analysis addresses issues such as how parts can conceivably fail, the mechanisms producing each failure mode, how the failures are detected, and what can be done to compensate for the failure.
- **Failure modes and effects and criticality analysis (FMECA):** This is FMEA in which criticality of each possible failure is also assessed.
- **Fault tree analysis:** A fault tree is a logic diagram that shows the relationship between a potential event affecting the system and the possible underlying causes for this event. Causes may be part or component failures, human error, environmental conditions, or combinations of these. The fault tree specifies the state (working or failed) of the system in terms of the states of its components.

See Blischke and Murthy (2000) for details and examples of these analyses.

Finally, an important issue in reliability analysis is the risk resulting from unreliable systems. Risk can be defined as the potential for realization of unwanted negative consequences of an event (for example, failure of a critical component such as a relief valve in a boiler). Risk management deals with risk assessment and risk minimization through good design, operating practices, and maintenance. Reliability modeling and analysis play an important role in managing the risks associated with unreliable systems. For example, the shuttle disaster may have been avoided if a proper failure analysis had been carried out.

1.6.4 Optimization

There are many optimization problems that may be addressed in a reliability analysis. Many of these involve trade-off studies. Most have as an objective minimizing cost or some equivalent quantity. Some examples are:

- Trade-off between increased design and analysis effort up front versus increased post-sale costs
- Reliability and maintenance cost trade-offs
- Selection of optimal warranty policies
- Selection of an optimal design that minimizes production cost (among feasible design configurations that achieve a reliability objective as discussed in the previous section)
- Selection of optimal product and process designs that achieve a stated reliability objective and minimize the time to release of a new product

Reliability optimization is based on a mathematical analysis using appropriate models and may require a significant amount of information about the system. Techniques for reliability optimization are discussed in detail in Blischke and Murthy (2000).

1.7 RELIABILITY AND MAINTENANCE DATA AND ANALYSIS

Many types of data are relevant to the estimation and prediction of reliability and maintainability. Not all are collected in many instances, and the lack of information is sometimes a serious problem in reliability analysis. At the other extreme, in some cases vast amounts of data are collected and very little of it is analyzed adequately or put into a useful form for decision makers. This problem is becoming increasingly common as computer facilities capable of storing vast data sets have evolved.

In this section we look briefly at some of the types of data that are used in R&M studies, comment on data acquisition, and give an overview of relevant data analysis techniques. Most of the cases in the book involve data collection, analysis, and interpretation.

1.7.1 Types of Data

Relevant data include experimental data, data from sampled populations, and data from other sources. Different types of data are useful at different stages of the life cycle of a product, and more are acquired as the life cycle progresses. The most important types of data in R&M studies are the following:

- **Historical data:** Test and other data on similar existing products, parts, or components
- **Vendor data:** Test data and other information on parts and components that are outsourced
- **Handbook data:** Characteristics of materials, manufactured items, and so forth, obtained from handbooks of engineering, electronics, physics, chemistry, and so on
- **Subjective data:** Information based on engineering or other expert judgment, subjective evaluations, and similar nonexperimental sources
- **Sampling data:** Data obtained from samples from well-defined populations (e.g., marketing data)
- **Test data:** Performance data on parts, components, and so on, from designed experiments
- **Observational data:** Data obtained from studies (called *observational studies*) that do not involve properly randomized designed experiments or random sampling
- **Production data:** Environmental data, production rates, and many others

- **Quality data:** Number of defectives, many other QC variables, and causes of failure
- **Operational data:** Field data on product performance obtained from buyers, experimentation, or third parties
- **Environmental data:** Data on environmental characteristics that may affect performance; these data are collected during production, during operations, or during experimentation
- **Warranty claims data:** Types, rates, and causes of failure of items under warranty, usually obtained from dealers or service centers
- **Cost data:** Costs of materials and labor and of every aspect of design, production, distribution, operation, and direct or indirect cost associated with the product

Implicit in the above list are a very large number of variables that may be observed. Some of the key variables are time to failure, number of failures, environmental variables, and others associated with test conditions, costs, and anything related to failure. Note that this discussion has focused on hardware. Still other types of data and variables may be relevant in analysis of software. See Xie (1991), Vouk (2000), and Blischke and Murthy (2000) for details. For more on maintainability data and testing, see Dhillon (1999). The cases cover most of the types of data listed above.

1.7.2 Data Structure

Data may be categorized as qualitative or quantitative. A more refined categorization includes four scales of measurement. In increasing degree of complexity, these are as follows:

- **Nominal:** Categorical data (e.g., types of parts, suppliers)
- **Ordinal:** Items are ranked (e.g., overall quality of a product on a scale of 1 to 7)
- **Interval:** Intervals of the same length have the same meaning, but there is not a well-defined zero value (e.g., temperature)
- **Ratio:** Intervals and ratios are meaningful, and there exists a unique zero (e.g., time to failure, number of items produced)

Different methods of analysis are appropriate for the different data scales. For example, counts are appropriate for nominal and ordinal data, differences and averages are meaningful only on the top two scales, and all standard arithmetic operations are appropriate for ratio data.

Data are also classified as discrete or continuous. Discrete data generally consist of counts: number of defective parts, number of warranty claims, and so forth. Data on time, voltage, length of a part, and so on, are inherently continuous. Correspondingly, discrete and continuous probability distributions are used to model random variability in these two types of variables.

Data are also classified as complete or incomplete. This is particularly important in reliability applications because incomplete data are frequently encountered. The distinction is as follows:

- **Complete data:** Exact values are known for each observation in the data set (e.g., time to failure is known for all items in the sample).
- **Incomplete data:** Exact values are not known for some or all of the observations (e.g., actual failure times are observed for some items in the sample; for others it is known only that failure had not occurred up to a recorded time).

Observations of the type just described are also called *censored.* Many reliability applications involve data of this type because of the nature of the test or observation process. A few examples are as follows: (1) A number of items are put on test at the same time, with testing stopped at some time prior to the failure of all items; (2) items in a lot are put into service at different times, and the failure times of the items observed at a set future time. In each case, the data consist of the lifetimes of the failed items and time in service (without failure) of the remaining items. There are a number of other types of censoring as well. All types of censoring have a significant impact on the analysis of the data. Standard procedures for complete data should not be used for censored data. A number of the cases feature censored data, and some of the analyses required the development of new statistical methods for this purpose.

Finally, we note that data may be grouped or ungrouped. Grouped data are data that have been categorized into classes (usually nonoverlapping intervals), with only class frequencies known. Handbook data and data from other secondary sources (e.g., technical journals) are sometimes of this type. Census data are always grouped. Grouping results in some loss of information and may require slight modifications to the techniques used for analysis.

1.7.3 Data Acquisition

For a proper analysis of data, it is necessary to know not only its structure, but also how the data were collected. The emphasis here will be on experimental data. The investigator ordinarily has little or no control over sources of data such as observational, historical, handbook, and vendor data (although demands for increasingly high quality products have led to manufacturer specifications on vendor data acquisition as well).

By experimental data we mean data collected by use of a well-developed data acquisition plan. In reliability studies, this may be test data on materials, parts, components, finished products, production processes, environment characteristics, or any other relevant variables, with data collected at any stage of the product life cycle. Statistical techniques for data collection include design of experiments (DOE, sometimes called test design in this context), sample survey methodology, special procedures used in life testing and accelerated testing, and so forth. DOE in reliability applications and related topics are discussed by Blischke and Murthy (2000). Good general texts on DOE are Lorenzen and Anderson (1993) and Montgomery (1997). For planning of life tests, see Meeker and Escobar (1998). A thorough treatment of accelerated testing is given by Nelson (1990). For sample survey methodology, see Schaeffer et al. (1996) and Thompson (1992). For modeling and test procedures in software applications, see Friedman and Voas (1995) and Vouk (2000).

In most applications, maintenance data are more difficult to obtain since most maintenance operations are carried out by the buyer. In cases where it is sufficiently important (e.g., expensive items such as aircraft, locomotives, mining equipment, large computer systems, and so forth), careful maintenance records may be kept. For most consumer goods, records are kept haphazardly, if at all. Rarely are designed experiments done in maintenance studies in this context, simply because it is nearly impossible to conduct randomized experiments under controlled conditions once the products have been manufactured and sold. As a result, obtaining valid and reliable maintenance data is a much more difficult task, except for the few special cases mentioned.

1.7.4 Data Analysis

Again, we restrict our attention to experimental data. It is important to note that, although observational data are not often encountered in properly conducted R&M studies, special care must be taken in analysis and interpretation of such data when they do occur. Significant statistical biases can result if proper randomization in employed in experimentation.

Analysis of experimental and other data begins with plots of the data and calculation of appropriate summary statistics, such as proportions, mean times to failure, means of other relevant variables, medians, standard deviations, and so forth. These sample values are used to estimate the corresponding population parameters or other population characteristics, calculate confidence intervals, test hypotheses concerning these characteristics, allocate reliabilities to components, and calculate reliability predictions, and they are also used for many other reliability-, maintainability-, and maintenance-related purposes. These statistical analyses are a part of *statistical inference*—that is, the process of inferring something about a population based on the information obtained from a sample.

In order to analyze data correctly, it is essential to know both (a) the structure of the data and (b) the plan for collection of data. Elements of data structure include scale of measurement (nominal, ordinal, etc.), whether the data are individual values or are grouped, and whether the data are complete or incomplete. Knowledge of data collection must include, as appropriate, experimental design or test plan, design of sample survey, type of censoring scheme used, randomization done, and any other factors relevant to the conduct of the experiment. Other factors that must be determined in order to analyze the data properly are the context of the study, experimental objectives, the assumptions made, models used, factors controlled, factors left uncontrolled that could influence the results, and any other features of the study that affect the results or their interpretation. A key objective in compiling cases such as those in this book is to illustrate these aspects of data analysis in the context of reliability studies.

Information on the basic notions and methodology of statistical inference may be found in most introductory statistics texts. Many of the sources for reliability and DOE include elements of basic statistical inference as well as a substantial amount of material on the acquisition and analysis of reliability and maintenance data. See especially Mann, Schafer, and Singpurwalla (1974), Kalbfleisch and Prentice (1980), Lawless (1982), Nelson (1982), Parmar and Machin (1995), Meeker and Escobar (1998), and Blischke and Murthy (2000).

1.8 ISSUES IN RELIABILITY AND MAINTENANCE

1.8.1 Interdisciplinary Nature of R&M

As noted previously, the solution of R&M problems requires integration of a number of interrelated technical, operational, commercial, and management factors and many disciplines. Some of the important issues encountered in R&M studies are listed in Section 1.8.2.

The list provides a small sample of the types of problems and issues dealt with in R&M studies. It is important to note that we have listed only the key disciplines involved in each problem area. In nearly all reliability and maintenance studies in the real world, many disciplines are involved and it is essential that the interdisciplinary nature of the issues be recognized in addressing R&M problems.

1.8.2 Issues in R&M Studies

There are many issues that may be of importance in R&M studies. We divide these into four interrelated categories: technical, operational, commercial, and management. Note that a study may have multiple objectives, dealing with a number of issues. The following list of issues, with the principal relevant discipline involved, is illustrative of the range of R&M studies.

Technical Issues

- Understanding of deterioration and failure [Material Science]
- Developing and managing a database of relevant R&M information [Information Systems]
- Effect of design on product reliability [Reliability Engineering]
- Modeling of part, component, and system reliability and their interrelationships [Probability]

- Effect of manufacturing on product reliability [Quality Variations and Control]
- Testing to obtain data for estimating part and component reliability [Design of Experiments]
- Estimation and prediction of reliability [Statistical Data Analysis]

Operational Issues

- Operational strategies for unreliable systems [Operations Research]
- Effective maintenance [Maintenance Management]

Commercial Issues

- Cost and pricing issues [Reliability Economics]
- Marketing implications [Warranty Analysis]

Management Issues

- Administration of reliability programs [Engineering Management]
- Impact of reliability decisions on business [Business Management]
- Risk to individuals and society resulting from product unreliability [Risk Theory]
- Effective management of risks from a business point of view [Risk Management]

1.9 CASE STUDIES: AN OVERVIEW

As noted previously, the objective of this book is to present real-life scenarios of reliability and maintenance from diverse areas of application and different perspectives. In the cases presented here, the authors discuss the context and background of the problems they are dealing with, the objectives of the study, and, as appropriate, the models used in addressing these problems, the data collected, data analysis and interpretation, and conclusions stated in the context of the study objectives. In the following sections, we discuss the format for a case study, the classification scheme used for the cases given in the book, and the organization of the remainder of the book.

1.9.1 Reporting a Case Study

The case studies include, as appropriate, a discussion of the problem being dealt with and the context, including reliability, managerial, and any other important aspects, models used and/or developed along with assumptions required for their use; data aspects, including methods of data collection, types of data, and analysis and interpretation; conclusions of the study, including engineering, management, and business implications, and implications regarding further research. For instructional purposes and for those who wish to pursue further the subject of the case, exercises are also included at the end of each chapter.

Case descriptions typically include, as appropriate to the study, the following material, variously organized:

- **Introduction:** Background/context; discussion of application; importance of the problem; overview.
- **Approach:** Objectives of the study; previous results and literature review; methodology.
- **System characterization:** Basis of models; assumptions; modeling at the component and system levels; statistical models.
- **Data and analysis:** Types of data; data collection (including DOE); description of results (summary statistics); methods of analysis; results (parameter estimates, confidence intervals, tests of relevant hypotheses, other statistical analyses).
- **Conclusions:** Interpretation of the results; conclusions of the study in the context of the stated objectives; general conclusions, topics for further research.

Conclusions may also include, as appropriate, comments to practitioners regarding methodological issues, pitfalls to avoid, shortcomings of studies of this type, sensitivity of results to assumptions, and so forth.

1.9.2 Classification of Cases

The discussion of reliability studies in the previous sections suggests three criteria of classification of cases, namely, type of application, phase of life cycle, and issues addressed. Here we briefly discuss each of these. Note that in each case, the categories listed are not mutually exclusive. A product may fall into more than one category, and a case may deal with several issues.

1.9.2.1 Application

Many areas of application are included in the case studies. These include the following industry sectors: aircraft, aerospace, automotive, computers, defense products, electronics, food processing, government acquisition, heavy industry, infrastructure, medicine, mining, software, and utilities. For purposes of classification, these have been grouped into the following product/application types:

1. Commercial
2. Consumer (durable and nondurable goods)
3. Industrial
4. Infrastructure
5. Government (other than infrastructure)

1.9.2.2 Life Cycle

We use a three-phase life-cycle classification. These are:

1. Premanufacturing
    - Product design (initial feasibility; conceptual design; detailed design)
    - Process design
    - Development (reliability modeling, prediction, estimation, and growth)
    - Testing (DOE; environmental testing, accelerated testing)
2. Manufacturing
    - Testing
    - Quality control
3. Post-Sale
    - Maintenance
    - Post-sale service (hot lines, etc.)
    - Logistics
    - Warranty

1.9.2.3 Issues Addressed

A broad range of issues are represented in the case studies. Most cases address multiple issues, typically with one primary and many secondary objectives. Together these provide a reasonably representative sample of the range of issues encountered in reliability studies.

Issues discussed include reliability demonstration, warranty cost analysis, identification of key failure modes, modeling of failures, reliability prediction, production downtime, plant modeling and maintenance, use of expert judgment, failure frequency, optimization of design, operation and maintenance, determination of residual life, component reliability, cost analysis, incomplete failure data, design strength, reliability improvement, laboratory analysis, repair policies, heavy censoring, simulation, use of field data, software reliability modeling and assessment, reliability allocation, resource constraints, reliability-centered maintenance, and many others.

1.9.3 List of Case Studies

The following is a list of cases included in the book, classified according to the reliability focus in the context of the phases of the life cycle. Sections of the book are designated A, B, and so on.

A. Design

Chapter 2. Space Interferometer Reliability-Based Design Evaluation, by Donald H. Ebbeler, Jr., Kim M. Aaron, George Fox, and W. John Walker

Chapter 3. Confidence Intervals for Hardware Reliability Predictions, by Chun Kin Chan and Michael Tortorella

Chapter 4. Allocation of Dependability Requirements in Power Plant Design, by Seppo Virtanen and Per-Erik Hagmark

B. Development and Testing

Chapter 5. The Determination of the Design Strength of Granite Used as External Cladding for Buildings, by Malcolm J. Faddy, Richard J. Wilson, and Gerry M. Winter

Chapter 6. Use of Sensitivity Analysis to Assess the Effect of Model Uncertainty in Analyzing Accelerated Life Test Data, by William Q. Meeker, Luis A. Escobar, and Stephen A. Zayac

Chapter 7. Virtual Qualification of Electronic Hardware, by Michael Osterman, Abhijit Dasgupta, and Thomas Stadterman

Chapter 8. Development of a Moisture Soak Model for Surface-Mounted Devices, by Loon Ching Tang and Soon Huat Ong

Chapter 9. Construction of Reliable Software in Resource-Constrained Environments, by Mladen A. Vouk and Anthony T. Rivers

Chapter 10. Modeling and Analysis of Software System Reliability, by Min Xie, Guan Yue Hong, and Claes Wohlin

Chapter 11. Information Fusion for Damage Prediction, by Nozer D. Singpurwalla, Yuling Cui, and Chung Wai Kong

C. Defect Prediction and Failure Analysis

Chapter 12. Use of Truncated Regression Methods to Estimate the Shelf Life of a Product from Incomplete Historical Data, by William Q. Meeker and Luis A. Escobar

Chapter 13. Determining Software Quality Using COQUALMO, by Sunita Chulani, Bert M. Steece, and Barry Boehm

Chapter 14. Use of Extreme Values in Reliability Assessment of Composite Materials, by Linda C. Wolstenholme

Chapter 15. Expert Judgment in the Uncertainty Analysis of Dike Ring Failure Frequency, by Roger M. Cooke and Karen A. Slijkhuis

D. Maintenance and Maintainability

Chapter 16. Component Reliability, Replacement, and Cost Analysis with Incomplete Failure Data, by Nicholas A. J. Hastings

Chapter 17. Maintainability and Maintenance—A Case Study on Mission Critical Aircraft and Engine Components,
by U. Dinesh Kumar and John Crocker

Chapter 18. Photocopier Reliability Modeling Using Evolutionary Algorithms,
by Michael Bulmer and John Eccleston

Chapter 19. Reliability Model for Underground Gas Pipelines,
by Roger M. Cooke, Eric Jager, and D. Lewandowski

Chapter 20. RCM Approach to Maintaining a Nuclear Power Plant,
by Gilles C. Zwingelstein

Chapter 21. Case Experience Comparing the RCM Approach to Plant Maintenance with a Modeling Approach,
by Xisheng Jia and Anthony H. Christer

E. Operations Optimization and Reengineering

Chapter 22. Mean Residual Life and Optimal Operating Conditions for Industrial Furnace Tubes,
by Elsayed A. Elsayed

Chapter 23. Optimization of Dragline Load,
by Peter G. A. Townson, D. N. Prabhakar Murthy, and Hal Gurgenci

Chapter 24. Reliability Improvement Process, A Case Study on Automotive Wheel Bearings,
by Karl D. Majeske, Mark D. Riches, and Hari P. Annadi

Chapter 25. Reliability of Oil Seal for Transaxle—A Science SQC Approach at Toyota,
by Kakuro Amasaka and Shunji Osaki

F. Warranty

Chapter 26. Warranty Data Analysis for Assessing Product Reliability,
by Peter C. Sander, Luis M. Toscano, Steven Luitjens, Valia T. Petkova, Antoine Huijben, and Aarnout C. Brombacher

Chapter 27. Reliability and Warranty Analysis of a Motorcycle Based on Claims Data,
by Bermawi P. Iskandar and Wallace R. Blischke

1.9.4 Organization by Application, Life Cycle, and Issues

Tables 1.1 through 1.3 provide further information concerning the content of the cases. In Table 1.1, cases are classified by area of application. The groups are consumer goods, commercial products, industrial products, government (other than infrastructure), and infrastructure. Cases having applications in more than one area are so indicated.

Table 1.1. Classification of Cases by Area of Application

	Application				
Chapter	Consumer	Commercial	Industrial	Government	Infrastructure
2				X	
3			X		
4			X		
5		X			
6		X			
7	X			X	
8	X		X		
9		X	X		
10		X	X		
11		X	X	X	
12	X				
13	X	X			
14		X			
15					X
16			X		
17		X			
18		X			
19					X
20			X		X
21	X	X			
22			X		
23			X		
24	X				
25	X				
26	X				
27	X				

Table 1.2. Classification of Cases by Life-Cycle Stage

	Premanufacture		Manufacture		Post-Sale	
Chapter	Design	Development	QA	Process Design	Maintenance	Other
2	**X**	X				
3	**X**					4,8
4	**X**	X				6
5	X	**X**				
6		**X**				
7		**X**				
8		X		**X**		
9	X	**X**			X	
10		**X**			X	
11	X	X		**X**		
12		**X**				
13		X				**1**
14						**1**
15					X	**2**
16					X	**4**
17	X				X	**5**
18					**X**	
19					**X**	
20	X				**X**	
21				X	**X**	
22	X	X			X	**6**
23					X	**6**
24	X	X	X			**7**,8
25	X		**X**			
26	X					3,7,8
27						2,**8**

Table 1.3. Products and Issues in Case Studies

Chapter	Product/System	Issues
2	Space optical interferometer	Reliability modeling and prediction, trade-offs, conceptual design, failure analysis
3	Electronic device	Failure analysis, design concept, manufacturing, SPC, management
4	Power plant unit	Design, dependability requirements
5	Granite building cladding	Design, modeling, reliability prediction, testing, fitting models, evaluating standards
6	Spring	Modeling, reliability assessment, sensitivity, accelerated testing
7	Electronic control module (automotive)	Simulation modeling, testing, reliability prediction and assessment
8	Molded plastic carrier chip in computer module	Testing, modeling, process design
9	Large software systems	Reliability modeling, assessment and growth, testing, error sources, resource constraints, data analysis
10	Software	Modeling, reliability prediction, testing, use of field data
11	Mine (explosive devices)	Reliability assessment and prediction, expert judgment, damage assessment
12	Food additive	MTTF as function of environment, special data problems
13	Software	Modeling, defect density prediction, reliability prediction
14	Complex equipment	Defect detection, modeling, estimation of defect density
15	Dikes	Historical data, expert judgment, failure analysis, reliability modeling and prediction, safety
16	Axel bushings in mine loader	Reliability assessment, optimal replacement policy
17	Aircraft	Design for reliability, reliability modeling, failure analysis, maintainability, maintenance, data collection, optimization
18	Photocopier	Reliability modeling, failure distribution, use of service data
19	Gas pipeline	Reliability modeling and prediction, inspection and repair policy, corrosion, use of field data, expert judgment
20	Nuclear power plant	Maintenance, reliability centered maintenance, failure analysis, safety
21	Chemical fertilizer	Modeling, reliability centered maintenance, production
22	Industrial furnace tubes	Optimal operating conditions, preventive maintenance, residual life

(continued)

Table 1.3. Products and Issues in Case Studies *Continued*

Chapter	Product/System	Issues
23	Dragline bucket (open-pit mine)	Availability, maintenance, optimization of operations
24	Automobile wheel bearings	FMEA, reliability improvement, testing, use of field and warranty claims data, redesign
25	Automobile oil seal	Failure analysis, design concept, manufacturing, SPC, management
26	DVD player	Failure analysis, reliability improvement, post-sale service, warranty, data collection, feedback to engineering design, use of field data
27	Motorcycle	Failure analysis, reliability modeling, warranty cost analysis

In Section 1.9.3, cases were classified according to the primary thrust of the study. Additional objectives of each study are given in Table 1.2. Groupings are Premanufacture, Manufacture, and Post-Sale, with subgroups in each case as listed in the previous section. Primary thrust is indicated by a bold font. For "Other" under Post-Sale, the subgroups are

1. Defect prediction
2. Failure analysis
3. Information feedback
4. Logistics
5. Maintainability
6. Optimization
7. Reengineering
8. Warranty

Table 1.3 provides a listing of cases by product or system analyzed as well as issues addressed. As noted, all cases address multiple issues. A few of these are indicated for each case study.

REFERENCES

Barlow, R. E. (1984). Theory of reliability: A historical perspective, *IEEE Transactions on Reliability* **33:**16–20.

Barlow, R. E., and Proschan, F (1965). *Mathematical Theory of Reliability,* Wiley, New York

Barlow, R. E., and Proschan, F. (1975). *Statistical Theory of Reliability and Life Testing,* Holt, Rinehart and Winston, New York.

Betz, F. (1993), *Strategic Technology Management,* McGraw-Hill, New York.

Blanchard, B. S. (1998), *Logistics Engineering and Management,* Prentice-Hall, Upper Saddle River, NJ.

Blanchard, B. S., Verma, D. and Peterson, E. L. (1995). *Maintainability,* Wiley, New York.

Blischke, W. R. (1994). Bayesian formulation of the best of liquid and solid reliability methodology, *Journal of Spacecraft and Rockets* **31**:297–303.

Blischke, W. R., and Murthy, D. N. P. (1994). *Warranty Cost Analysis,* Dekker, New York.

Blischke, W. R., and Murthy, D. N. P. (2000). *Reliability: Modeling, Prediction and Optimization,* Wiley, New York.

BS 4778. *Glossary of Terms Used in Quality Assurance, Including Reliability and Maintainability Terms,* British Standards Institution, London.

Cohen, H. (1984). Space reliability technology: A historical perspective, *IEEE Transactions on Reliability* **33**:36–40.

Coppola, A. (1984). Reliability engineering of electronic equipment: A historical perspective, *IEEE Transactions on Reliability* **33**:29–35.

Dhillon, B. S. (1999). *Engineering Maintainability: How to Design for Reliability and Easy Maintenance,* Gulf Publishing Co., Houston.

Elsayed, E. A. (1996). *Reliability Engineering,* Addison-Wesley, Longman, New York.

Evans, J. R., and Lindsay, W. M. (1999). *The Management and Control of Quality,* 4th edition, South-Western, Cincinnati.

Friedman, M. A., and Voas, J. M. (1995). *Software Assessment: Reliability, Safety, Testability,* Wiley, New York.

Fussell, J. B. (1984). Nuclear power system reliability: A historical perspective, *IEEE Transactions on Reliability* **33**:41–47.

Hegde, G. G., and Kubat, P (1989). Diagnostic design: A product support strategy, *European Journal of Operational Research* **38**:35–43.

Høyland, A., and Rausand, M. (1994). *System Reliability Theory,* Wiley, New York.

IEC 60300–3–4 (1996). *Dependability Management, Part 3: Application Guide, Section 4: Guide to the Specification of Dependability Requirements,* International Electrotechnical Commission, Geneva

Ireson, W. G., and Coombs, C. F., Jr. (1988). *Handbook of Reliability Engineering and Management,* McGraw-Hill, New York.

Jardine, A. K. S., and Buzacott, J. A. (1985). Equipment reliability and maintenance, *European Journal of Operational Research* **19**:285–296.

Kalbfleisch, J. D., and Prentice, R. L. (1980). *The Statistical Analysis of Failure Time Data,* Wiley, New York.

Kapur, K. C., and Lamberson, L. R. (1977). *Reliability in Engineering Design,* Wiley, New York.

Knight, C. R. (1991). Four decades of reliability progress, *Proceedings Annual Reliability and Maintainability Symposium,* 156–159.

Lawless, J. F. (1982). *Statistical Models and Methods for Lifetime Data,* Wiley, New York.

Lloyd, D. K., and Lipow, M. (1962). *Reliability: Management, Methods, and Mathematics,* Prentice-Hall, Englewood Cliffs, NJ.

Lorenzen, T. J., and Anderson, V. L. (1993). *Design of Experiments: A No-Name Approach,* Dekker, New York.

Malcolm, J. G., and Foreman. G. L. (1984). The need: Improved diagnostic—rather than improved R, *Proceedings of the Annual Reliability and Maintainability Symposium,* 315–322.

Mann, N. R., Schafer, R. E., and Singpurwalla, N. D. (1974). *Models for Statistical Analysis of Reliability and Life Data,* Wiley, New York.

Meeker, W. Q., and Escobar, L. A. (1998). *Statistical Methods for Reliability Data,* Wiley, New York.

Montgomery, D. C. (1997). *Design and Analysis of Experiments,* Wiley, New York.

Murthy, D. N. P. (1991). A note on minimal repair, *IEEE Transactions on Reliability* **40:**245–246.

Murthy, D. N. P., and Blischke, W. R. (2000). Strategic warranty management, *IEEE Transactions on Engineering Management* **47:**40–54.

Nakagawa, T., and Kowada, M. (1983). Analysis of a system with minimal repair and its application to a replacement policy, *European Journal of Operational Research* **12:**176–182.

Nelson, W. (1982). *Applied Life Data Analysis,* Wiley, New York.

Nelson, W. (1990). *Accelerated Testing,* Wiley, New York.

Parmar, M. K. B., and Machin, D. (1995). *Survival Analysis: A Practical Approach,* Wiley, New York.

Pecht, M. (Ed..) (1995). *Product Reliability, Maintainability, and Supportability Handbook,* CRC Press, Boca Raton, FL.

Pham, H., and Wang, H. (1996). Imperfect maintenance, *European Journal of Operational Research* **94:**425–438.

Rink, D. R., and Swan, J. E. (1979). Product life cycle research: A literature review, *J. Business Review* **78:**219–242.

Rosander, A. C. (1985). *Application of Quality Control in Service Industries,* Dekker, New York.

Rosander, A. C. (1989). *The Quest for Quality in Services,* Quality Press, Milwaukee, Wisconsin.

Schaeffer, R. L., Mendelhall, W., and Ott, R. L. (1996). *Elementary Survey Sampling,* 5th edition, Duxbury, New York.

Shooman, M. L. (1984). Software reliability: A historical perspective, *IEEE Transactions on Reliability* **33:**48–55.

Thompson, S. K. (1992). *Sampling,* Wiley, New York.

Valdez-Flores, C., and Feldman, R. H. (1989). Stochastically deteriorating single-unit systems, *Naval Research Logistics* **36:**419–446.

Villemeur, A. (1992). *Reliability, Maintainability and Safety Assessment, Vol. 1: Methods and Techniques,* Wiley, New York.

Vouk, M. A. (2000). Introduction to Software Reliability Engineering, *Tutorial Notes, 2000 Annual Reliability and Maintainability Symposium.*

Wolstenholme, L. C. (1999). *Reliability Modeling: A Statistical Approach,* Chapman & Hall, New York.

Xie, M. (1991). *Software Reliability Modeling,* World Scientific Press, Singapore.

PART A

Cases with Emphasis on Product Design

Chapter 2, "Space Interferometer Reliability-Based Design Evaluation"

—Deals with the use of modeling and simulation in the reliability analysis of the top-level design of a complex one-of-a-kind spaceflight system, an optical interferometer, scheduled for launch in 2009.

Chapter 3, "Confidence Intervals for Hardware Reliability Predictions"

—Uses subassembly reliability data for prediction of system reliability of a telecommunications product and discusses comparison of predicted with realized field reliability.

Chapter 4, "Allocation of Dependability Requirements in Power Plant Design"

—Presents a method of translating customer requirements into dependability allocation in the design of a diesel power plant.

CHAPTER 2

Space Interferometer Reliability-Based Design Evaluation

Donald H. Ebbeler, Jr., Kim M. Aaron, George Fox, and W. John Walker

2.1 INTRODUCTION

The Space Interferometry Mission (SIM) in NASA's Origins Program is currently scheduled for launch in 2009. SIM is being developed by the Jet Propulsion Laboratory (JPL) under contract with NASA, together with two industry partners, Lockheed Martin Missiles and Space in Sunnyvale, California and TRW Inc., Space and Electronics Group in Redondo Beach, California.

The SIM public access website, located at http://sim.jpl.nasa.gov, provides extensive background information on optical interferometry and stellar astronomy. Danner and Unwin (1999) is available online at this website, posted in the Library section as The SIM Book. Appendix A of that book provides a list of technical papers and websites organized under the topics astrometric science, planet searching, optical interferometry, and interferometry technology. Most of the technical papers are either (a) review articles covering a particular subject or (b) seminal papers for these areas. The websites provide an evolving source of up-to-date information. Lawson (1999) is also available at the SIM public access website, posted in the Library section as Principles of Long Baseline Interferometry. These proceedings, which provide an introductory overview of the technique and application of optical interferometry, are organized in book form, with each of 18 chapters including a set of references pertinent to the chapter topic. Blischke and Murthy (2000) and Kapur and Lamberson (1977) are two basic reliability texts that consider the modeling of engineering systems.

SIM will be one of the first space missions to use optical interferometry. It will

Case Studies in Reliability and Maintenance, Edited by W. R. Blischke and D. N. P. Murthy.

ISBN 0-471-41373-9

have the capability of measuring the positions of celestial objects to an unprecedented accuracy of about one billionth of a degree. In order to achieve this accuracy, which represents an improvement of almost two orders of magnitude over previous astrometric measurements, a 10-meter baseline interferometer will be flown in an earth-trailing orbit. This accuracy will allow SIM to determine the distances to stars throughout the Milky Way and to probe nearby stars for earth-sized planets.

Following orbit insertion, the solar array and the high-gain antenna will be deployed, spacecraft systems will be checked out, and tracking data will be collected in order to determine the orbit precisely. For the first 6 months, checkout and calibration of the interferometer will take place. At the end of the calibration period, the SIM interferometer will perform nearly continuous science observations for a 5-year time period as the spacecraft slowly drifts away from earth at a rate of about 0.1 AU per year. One AU, or astronomical unit, is the distance between the sun and the earth.

In order to assist in selecting an optical interferometer design likely to achieve the goal of collecting 5 years of science observations, a parametric evaluation of alternative designs meeting cost, performance, and schedule constraints was undertaken. Competing designs were compared using a Monte Carlo simulation tool based on an Excel add-in, called MCTool, created at JPL by one of the authors, George Fox. In practice, SIM design engineers have quickly learned how to utilize the visualization provided by Excel to formulate models of alternative interferometer designs, simulate reliabilities, and evaluate the robustness of alternative designs with respect to reliability.

2.2 PROBLEM DESCRIPTION

When the SIM reliability evaluation was initiated in late 1999, a baseline optical interferometry system had been selected, which will be referred to in this chapter as Classic SIM. As initially conceived, the problem was to create and evaluate a top-level model of the Classic SIM optical interferometer system. One of the authors, Kim M. Aaron, defined a Classic SIM model in terms of the relationships among a small set of top-level modeling elements. Despite the complexity introduced by redundancies and operational constraints on modeling elements, it was possible to derive analytical expressions for the reliability of this baseline system under alternative operational constraints. The success of this effort led to evaluating the model using the Excel-based Monte Carlo simulation tool in order to establish a procedure for efficiently extending the reliability evaluation to consider variants and extensions of Classic SIM when system elements and constraints are added or modified.

However, in November 2000 several independent estimates of the cost of Classic SIM led NASA to direct JPL to consider alternative, less costly designs, including the leading alternatives considered in this chapter, called ParaSIM and Shared Baseline SIM. Once the reliability modeling problem had evolved to comparing the reliabilities and corresponding robustness of less costly alternative interferometer designs and their variants, it was decided that utilizing the Excel-based Monte Carlo simulation tool was the preferred approach. As it affects interferometer reliability, the essential differences between alternative interferometer designs considered are the number and geometrical configuration of their modeling elements and how redundancy is embedded in the designs. Classic SIM included a technique to null the light of a bright star in order to observe the area close to the star. Neither ParaSIM nor Shared Baseline SIM permit nulling, and the imaging capability for both is reduced compared to Classic SIM. The science acquisition rate is a little slower for Shared Baseline SIM than for Classic SIM and significantly slower for ParaSIM. The problem considered in this chapter is an analysis of the reliability of Classic SIM and a comparison of the reliabilities of ParaSIM and Shared Baseline SIM.

2.3 ALTERNATIVE OPTICAL INTERFEROMETER DESIGNS

In this section, each of the three SIM configurations is addressed in turn. For Classic SIM, closed-form reliability solutions are developed under two alternative operational constraints. The derivations under both constraints are presented in some detail. For Classic SIM, a Monte Carlo simulation evaluation of the model was also performed, using MCTool. The same modeling elements used in the Classic SIM analytical derivations were visually modeled in Excel for the Monte Carlo evaluation. The implementation of these modeling elements in Excel is described in some detail. For ParaSIM and Shared Baseline SIM, only Monte Carlo evaluations of their visual models in Excel were carried out. The modeling elements used for these two configurations are similar to the modeling elements used in Classic SIM. The connections among the various elements are somewhat different due to the physical layout of the elements and how they are used to perform astrometry. Excel visual models for the three SIM configurations are available at the Wiley website. The Excel add-in MCTool is available by e-mail from George Fox at george.fox@jpl.nasa.gov.

2.3.1 Classic SIM

A stylized optical interferometer system for the Classic SIM design is shown in Figure 2.1. In this configuration the optical interferometer system consists of a metrology kite, a set of seven siderostat bays, an optical switchyard, and four beam combiner assemblies. The fundamental modeling elements are triple-corner cubes (TCC), kite beam launchers (KBL), siderostat bay beam launchers (SBBL), residual siderostat bay elements defined to be siderostat mirrors and corner cubes together with upper switchyard mirrors, and a beam combiner assembly defined to be a beam combiner, delay lines, and lower switchyard mirrors.

Triple-Corner Cube

A triple-corner cube is an agglomeration of reflective wedges arranged to form three retroreflecting corner cubes sharing a common vertex. A single-corner cube is normally formed using three mutually perpendicular mirrors. There are several

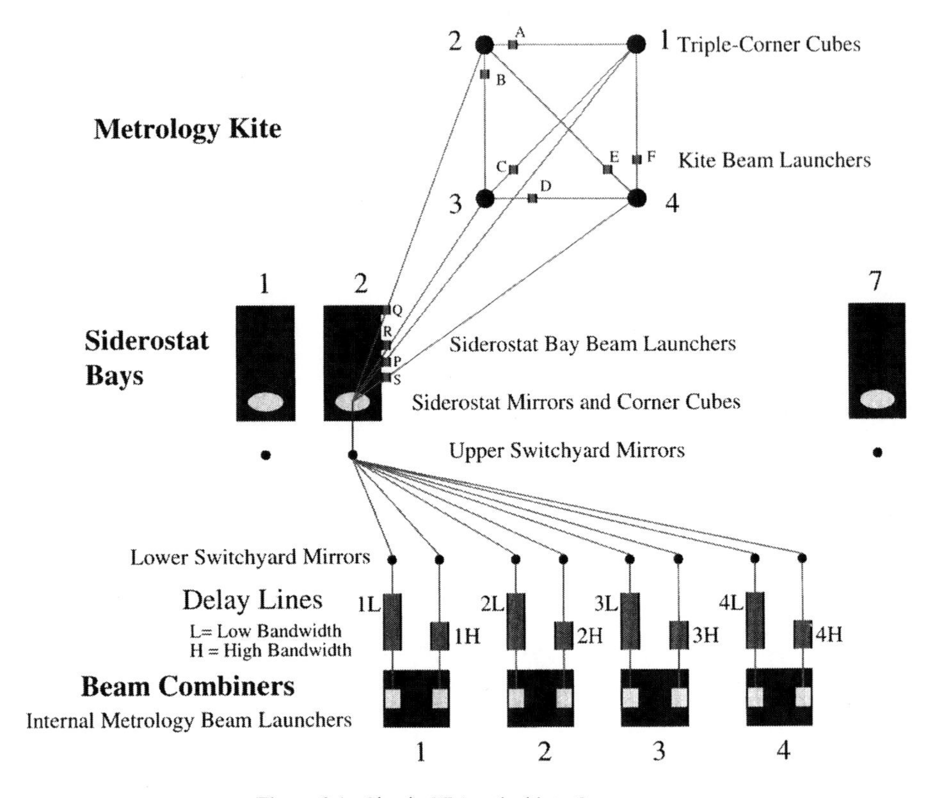

Figure 2.1. Classic SIM optical interferometer.

arrangements of reflective elements that can form a triple-corner cube. The corner cubes used in the Classic SIM configuration are formed by mounting three wedges, each with a 30° angle, on the surface of a flat mirror.

Kite Beam Launcher

A beam launcher is an assemblage of optical elements, which is fed laser light using an optical fiber. The beam launcher manipulates the light and then launches the collimated beam out toward a corner cube. The retroreflected beam then passes by or through the beam launcher to enter a second corner cube. Following retroreflection from the second corner cube, the laser light reenters the beam launcher, where it is combined interferometrically with a portion of the original laser light. By counting fringes (resolved to a small fraction of a wavelength), one can determine the relative motion of one corner cube with respect to the other corner cube with a precision of a few tens of picometers (10^{-12} m). The kite beam launchers are six beam launchers mounted on the external metrology kite and are used to measure the distances among the corners of the kite itself. These beam launchers measure the intravertex distances of the kite.

Siderostat Bay Beam Launcher
Four beam launchers, as described in the previous subsection, are mounted in each siderostat bay. These beam launchers measure the distance between the corner cube attached to the siderostat mirror and the triple-corner cubes at the four vertices of the external metrology kite.

Residual Siderostat Bay Elements
There are several elements in the optical train that manipulate the starlight after it is reflected off the surface of the siderostat mirror. Since these elements are in series, they are lumped together and assigned a combined reliability. There are also several actuators, cameras, and so on, along with thermal control of the siderostat mirror and other optical elements, which we have lumped together with residual siderostat bay elements.

Beam Combiner Assembly
A beam combiner is a group of optical elements that combines the collimated starlight from two different telescopes aimed at the same star. A half-silvered mirror (50% transmitted light, 50% reflected light) performs this beam combination function, taking the starlight from two arms of an interferometer and combining the light (there are actually two equivalent output beams), which is then focused onto detectors. The beam combiner assembly performs several other functions, such as injecting calibration signals into the optical train, providing signals used in controlling fast steering mirrors, launching metrology beams for the internal metrology system, and separating the starlight into its spectral components using prisms.

In the Classic SIM design the system will be operational if at least one kite triangle is available, at least six siderostat bays with corresponding upper switchyard mirrors are working, and at least three beam combiner assemblies are working.

2.3.1.1 Closed-Form Reliability Equation

An analytical expression for Classic SIM reliability can be derived by considering siderostat bay reliability conditioned on metrology kite states. One kite triangle must be operational for the Classic SIM optical interferometer to work. There are four kite triangles available. Each triangle comprises three TCCs and three KBLs, as shown in Figure 2.2. Each KBL is common to exactly two different triangles, so if a KBL fails there will be two triangles available. Each TCC is common to exactly three triangles, so if a TCC fails there will be only one triangle available. It is not possible to have exactly three triangles available.

Let p_T be the probability that a TCC fails. The probability of no TCC failures is $(1 - p_T)^4$, and the probability of one TCC failure is $4\,p_T(1 - p_T)^3$. Let p_K be the probability that a KBL fails. If a TCC has failed, then only one triangle is available and three specific KBLs must still be operating. The probability that three specific KBLs are working, given that a specific TCC has failed, is $(1 - p_K)^3$. If no TCCs have failed, there are three kite triangle states (one, two, or four triangles operating) under which the interferometer may be operational. The probability that all four triangles are available (all six KBLs operational) is $(1 - p_K)^6$. The probability that two

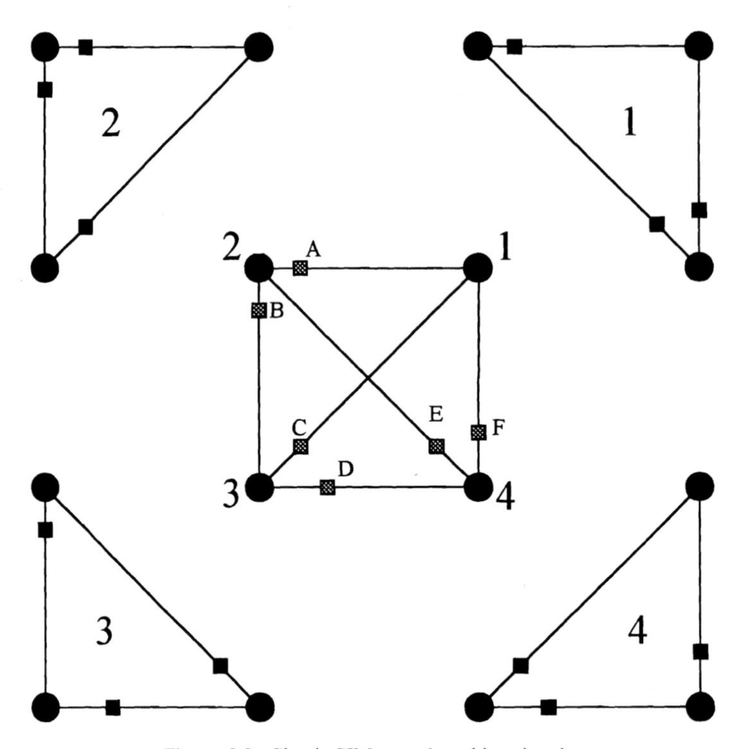

Figure 2.2. Classic SIM metrology kite triangles.

triangles are available (exactly five KBLs operational) is $6\,p_K(1-p_K)^5$. The probability that one triangle is available (three or four KBLs operational) is $4(1-p_K)^3[p_K^3 + 3\,p_K^2(1-p_K)]$. Kite reliability for the three triangle operational states is shown in Figure 2.3.

There are four SBBLs in each siderostat bay. Each SBBL is aimed at a specific kite TCC. If a particular kite triangle is being used, then the three corresponding specific SBBLs must also be working. If four kite triangles are available, then any three SBBLs must be working. If exactly two kite triangles are available, then the three SBBLs aimed at the triangle corners must be working.

There is an issue as to whether or not different siderostat bays must use the same kite triangle. The problem is that the metrology triangles all lie in the same nominal plane. The corners are randomly displaced out of plane a few microns. It is not possible to tell how far a corner is out of plane using just the kite intravertex distance measurement. If any single siderostat bay can see all four corners, then the out-of-plane displacement can be resolved. If the error multipliers are acceptable, then different siderostat bays can use different triangles. Results will be derived for both the case when any siderostat bay can use any triangle and the case when all siderostat bays must use a common triangle. The reliability will be lower if it is necessary for all siderostat bays to use a common triangle.

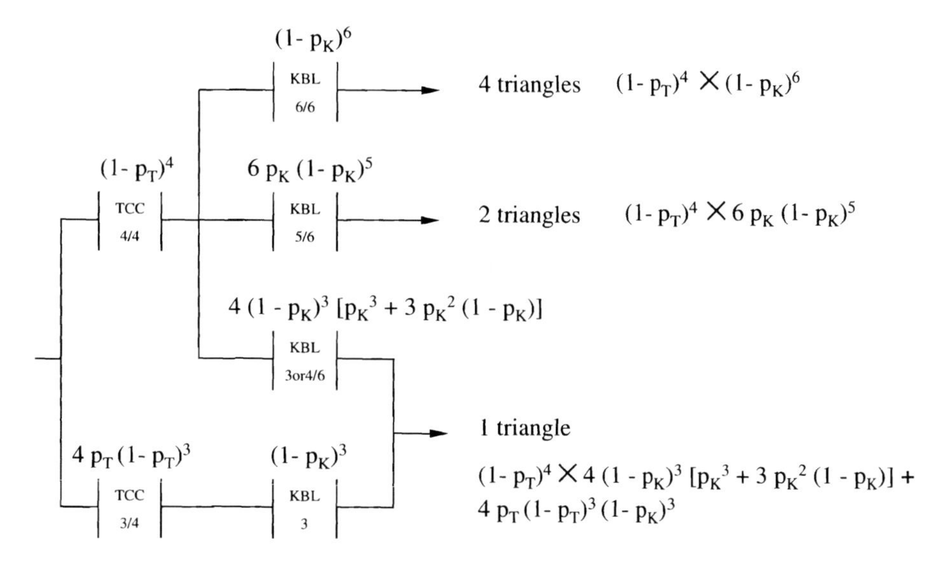

Figure 2.3. Classic SIM metrology kite reliability diagram.

Any Siderostat Bay Can Use Any Triangle
Let p_B be the probability that an SBBL fails. If all four kite triangles are working, any three SBBLs can be working. The probability that at least three are working is $(1 - p_B)^4 + 4\,p_B(1 - p_B)^3$. Let $1 - p_S$ be the probability that the rest of the siderostat bay (siderostat mirrors and corner cubes together with upper switchyard mirrors) is working. Then the probability that the siderostat bay is working, given that all four triangles are working, is $(1 - p_S)[(1 - p_B)^4 + 4\,p_B(1 - p_B)^3]$, which will be defined to be $(1 - q_4)$ for notational convenience. To form three interferometers, six operating siderostat bays are required. The probability that at least six of the siderostat bays are working, given that four triangles are available, is $(1 - q_4)^7 + 7\,q_4(1 - q_4)^6$. If two triangles are available, the corresponding probability that the siderostat bay is working is $(1 - p_S)[(1 - p_B)^4 + 2p_B(1 - p_B)^3]$, defined to be $(1 - q_2)$. The probability that at least six of the siderostat bays are working, given that exactly two triangles are available, is $(1 - q_2)^7 + 7\,q_2(1 - q_2)^6$. If only one triangle is available, the corresponding probability that the siderostat bay is working is $(1 - p_S)[(1 - p_B)^4 + p_B(1 - p_B)^3]$, defined to be $(1 - q_1)$. The probability that at least six of the siderostat bays are working, given that only one triangle is available, is $(1 - q_1)^7 + 7q_1(1 - q_1)^6$.

Let the probability of a beam combiner assembly (beam combiner, delay lines and lower switchyard mirrors) failure be p_C. At least three of the four beam combiner assemblies must be working for the system to be operational. The probability of that event is $(1 - p_C)^4 + 4p_C(1 - p_C)^3$. The resulting Classic SIM reliability structure is given in Figure 2.4. The analytical expression for Classic SIM optical interferometer reliability when any siderostat bay can use any available triangle is then given by

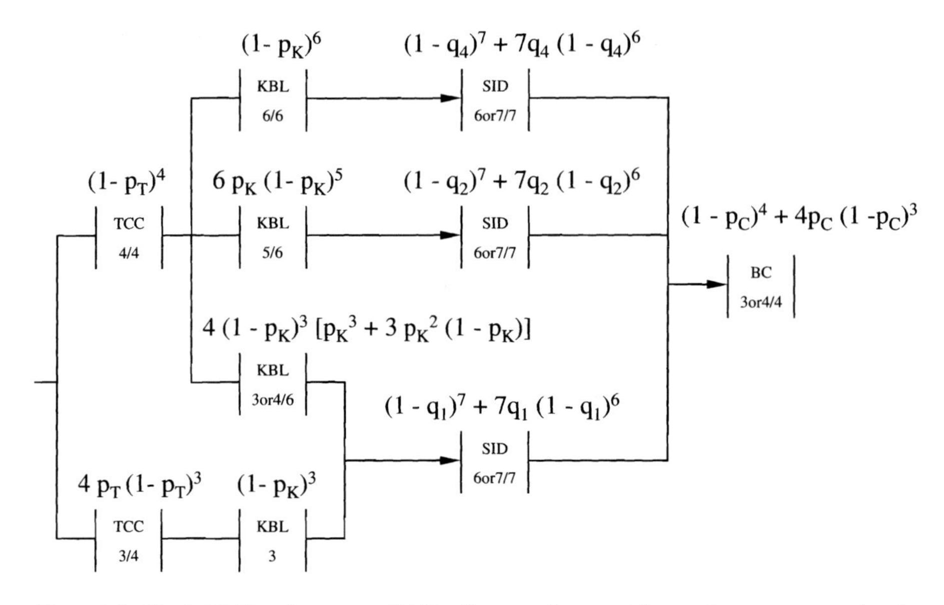

Figure 2.4. Classic SIM interferometer reliability diagram when any siderostat bay can use any triangle.

$$\begin{aligned}\{&(1-p_T)^4(1-p_K)^6[(1-q_4)^7+7q_4(1-q_4)^6]+\\ &6(1-p_T)^4p_K(1-p_K)^5[(1-q_2)^7+7q_2(1-q_2)^6]+\\ &4(1-p_T)^4(1-p_K)^3\,[p_K^3+3p_K^2(1-p_K)][(1-q_1)^7+7q_1(1-q_1)^6]+\\ &4p_T(1-p_T)^3(1-p_K)^3[(1-q_1)^7+7q_1(1-q_1)^6]\}[(1-p_C)^4+4p_C(1-p_C)^3]\end{aligned} \tag{2.1}$$

All Siderostat Bays Must Use the Same Triangle

When one triangle is available, the result is identical to the previous case with one triangle available. However, for the cases of two and four triangles available, a different approach is now necessary in order to derive the closed form expression for reliability. We will introduce a consistent notation for all three cases under the constraint that all siderostat bays must use the same triangle.

Let $(1-s_1)$ be the probability that a given siderostat bay can see a specific single triangle. This is the same as the probability $(1-q_1)$ without the constraint, $(1-p_S)[(1-p_B)^4+p_B(1-p_B)^3]$. The probability that at least six siderostat bays can see this specific triangle, in this notation, is $(1-s_1)^7+7s_1(1-s_1)^6$.

To consider the other two cases, when either two or four triangles are available, we need a general expression for the probability that at least six siderostat bays can all see at least one of the available triangles. Let E_i be the event that at least six siderostat bays can all see the ith triangle. The probability of the union of $\{E_i\}$ as a function of their intersections is given by $4P(E_1)-6P(E_1\cap E_2)+4P(E_1\cap E_2\cap E_3)-P(E_1\cap E_2\cap E_3\cap E_4)$.

When two triangles are available, $P(E_1) = (1/2)[(1 - s_1)^7 + 7s_1(1 - s_1)^6]$. A siderostat bay can see two triangles only if all four SBBLs are working. Let $(1 - s_2)$ be the probability that a given siderostat bay can see two specific triangles, $(1 - s_2) = (1 - p_S)(1 - p_B)^4$. $P(E_1 \cap E_2) = (1/6)\{(1 - s_2)^7 + 7s_2(1 - s_2)^6 + 21(1 - s_2)^5 2[(1 - p_S)p_B(1 - p_B)^3]^2\}$. $P(E_1 \cap E_2 \cap E_3)$ and $P(E_1 \cap E_2 \cap E_3 \cap E_4)$ are zero. Then the probability that at least six siderostat bays can all see at least one of the available triangles, when two triangles are available, is $2[(1 - s_1)^7 + 7s_1(1 - s_1)^6] - \{(1 - s_2)^7 + 7s_2(1 - s_2)^6 + 42(1 - s_2)^5[(1 - p_S)p_B(1 - p_B)^3]^2\}$. When four triangles are available, $P(E_1) = (1 - s_1)^7 + 7s_1(1 - s_1)^6$. A siderostat bay can see all four triangles only if all four SBBLs are working. Let $(1 - s_4) = (1 - s_2)$ be the probability that a given siderostat bay can see all four triangles, $(1 - s_4) = (1 - p_S)(1 - p_B)^4$. $P(E_1 \cap E_2) = (1 - s_4)^7 + 7s_4(1 - s_4)^6 + 21(1 - s_4)^5 2[(1 - p_S)p_B(1 - p_B)^3]^2\}$. $P(E_1 \cap E_2 \cap E_3) = P(E_1 \cap E_2 \cap E_3 \cap E_4) = (1 - s_4)^7 + 7s_4(1 - s_4)^6$. Then the probability that at least six siderostat bays can all see at least one of the triangles, when all four triangles are available, is $4[(1 - s_1)^7 + 7s_1(1 - s_1)^6] - 3\{(1 - s_4)^7 + 7s_4(1 - s_4)^6 + 84(1 - s_4)^5[(1 - p_S)p_B(1 - p_B)^3]^2\}$. The analytical expression for Classic SIM optical interferometer reliability when all siderostat bays must use the same triangle is then given by

$$
\begin{aligned}
&\{(1 - p_T)^4(1 - p_K)^6\{4[(1 - s_1)^7 + 7s_1(1 - s_1)^6] - 3[(1 - s_4)^7 + 7s_4(1 - s_4)^6 + \\
&84(1 - s_4)^5[(1 - p_S)p_B(1 - p_B)^3]^2]\} + 6(1 - p_T)^4 p_K(1 - p_K)^5\{2[(1 - s_1)^7 + \\
&7s_1(1 - s_1)^6] - [(1 - s_2)^7 + 7s_2(1 - s_2)^6 + 42(1 - s_2)^5[(1 - p_S)p_B(1 - p_B)^3]^2]\} \\
&\quad + 4(1 - p_T)^4(1 - p_K)^3[p_K^3 + 3p_K^2(1 - p_K)][(1 - s_1)^7 + 7s_1(1 - s_1)^6] + \\
&4p_T(1 - p_T)^3(1 - p_K)^3[(1 - s_1)^7 + 7s_1(1 - s_1)^6]\}[(1 - p_C)^4 + 4p_C(1 - p_C)^3]
\end{aligned}
\tag{2.2}
$$

2.3.1.2 *Monte Carlo Reliability Simulation Model*

An alternative reliability evaluation approach is to use the model visualization available in Excel, together with a Monte Carlo add-in tool developed by George Fox at JPL. Excel visual modeling employs any available Excel functions needed. For any version of Excel, the list of available functions and their description is available within the Excel program. Spreadsheet models that specify stochastic variables by function call to the add-in are developed in Excel. During simulation, the generated output values are collected for later statistical analysis. Users select the type and form of desired output tables and charts. All computations, tables, and charts are executed within Excel, hence eliminating the need for commercial software add-ins. One advantage of representing the interferometer model in an Excel spreadsheet is the ease with which additional SIM system elements and constraints, such as the shared triangle constraint, can be incorporated into the analysis. Visual characterization of the model in Excel greatly enhances the ability of the users to utilize their comprehension of the physical interferometer system in the modeling process.

The reliability models constructed in Excel consist of fundamental modeling elements related by the use of the Excel functions “And” to express the reliability of two elements in series and “Or” to express the reliability of two elements in parallel. The reliability of more complex element combinations can be modeled by multiple instantiations of the same element. For example, to represent the state that two out of three elements, (A, B, C), are operating, create "(A And B) Or (A And C) Or (B And C)." Spreadsheet cells are used to represent fundamental and composite modeling elements. Cells can be colored and labeled by element type. A legend listing element types identified by color is an aid to visualizing the embedded reliability model in the Excel spreadsheet. Groups of cells can be surrounded by boxes to show fault containment regions or other composite modeling element constructs.

To aid in model construction, modification, and operation, simulation input includes tables for each element type, listing the element probabilities of failure. Fundamental modeling element cells contain a simple formula to calculate the cell's binary (0, 1) operating state. For example, in the cell formula "=If(RAND()<Pf,0,1)," Pf is the probability of failure, RAND() is the Excel function that draws a random variable from a uniform distribution between zero and one on each iteration of the Monte Carlo simulation, and (0, 1) are the possible cell operating states on each iteration. When RAND() is less than Pf, the cell operating state is assigned 0, otherwise the cell operating state is assigned 1. Common cause failures among elements can be modeled by referencing the same RAND() call. Cells corresponding to composite modeling element constructs use the logical connectives and the states of other cells to calculate their states. For example, "=If(Or(A1,B1),1,0)" means that if either cells A1 or B1 are 1, then this cell is 1, otherwise 0.

The operating state of the system is labeled with an Excel name starting with rv_. This allows the Excel add-in, MCTool, to recognize the cell and collect its values during the simulation. MCTool provides a set of probability distribution functions for use in modeling as well as simulation execution and output cell statistics for use in analysis. MCTool distribution functions begin with RV_. Output statistics available are a histogram or cumulative distribution function and the minimum, maximum, mean, and standard deviation for each simulation.

Fundamental modeling elements in Classic SIM are triple corner cubes (TCC), kite beam launchers (KBL), siderostat bay beam launchers (SBBL), residual siderostat bay elements defined to be siderostat mirrors and corner cubes together with upper switchyard mirrors, and a beam combiner assembly defined to be a beam combiner, delay lines, and lower switchyard mirrors. KBLs and TCCs are combined to form kite triangles. Each siderostat bay contains four SBBLs, each pointing to a different TCC. Each set of SBBLs is in series with its residual siderostat bay elements. Beam combiner assemblies provide redundancy for the functioning of all siderostat bays. These elements, in combination with a kite triangle, comprise working interferometers. Assessment of the fully functioning metrology system is carried out under alternative constraints on the matching of kite triangles with siderostat bays: "Any siderostat bay can use any triangle" and "All siderostat bays must use the same triangle."

In the visual modeling approach used, elements are arranged in groups on the spreadsheet to represent the reliability connectivity. There are four TCCs on the metrology kite, each possible pair requiring a shared KBL. Graphically, a TCC is placed at each corner of the kite rectangle, with lines connecting each pair of TCCs to the KBL. The state of each of the four kite triangles is modeled as a column of cells which align with the representation of operating siderostat bays discussed later. The collection of SBBLs is modeled as a matrix of cells, with a column corresponding to each siderostat bay and each row corresponding to a particular TCC, that is, each SBBL is indexed by siderostat bay and the TCC to which it points. Beam combiner assemblies are modeled as a group, with an aggregate node representing the operation of the three out of four beam combiner assemblies required for the operation of the interferometers. For each siderostat bay, a residual siderostat bay elements reliability node is in series with the working combinations of the siderostat bay's SBBLs. The aggregate beam combiner assemblies node is then in series with all siderostat bays. The major composite structure is a matrix of cells, with rows corresponding to triangles and columns corresponding to siderostat bays, which allows the Monte Carlo simulation to generate the reliability of the system under the two constraints "Any siderostat bay can use any triangle" and "All siderostat bays must use the same triangle." The state of each cell in the matrix represents a working kite triangle–siderostat bay combination. Each row sum of this matrix is the number of working siderostat bays for each triangle. If any row has six working siderostat bays, then we have working interferometers under the constraint "All siderostat bays must use the same triangle." If at least one cell in a matrix column has the value 1, it defines a working siderostat bay with a connection to a triangle. If there exist six such working combinations, then the constraint "Any siderostat bay can use any triangle" has been satisfied. Using this model characterization, reliability under each constraint can be generated with a single simulation.

2.3.2 ParaSIM and Shared Baseline SIM

In order to compare reliabilities of ParaSIM and Shared Baseline SIM, the two designs are represented as alternative combinations of seven fundamental modeling elements that constitute reliability blocks. Each element is a complex assembly composed of electrical, mechanical, and optical subelements. In order to use the Monte Carlo simulation tool to simulate an end-of-mission reliability for a design, it is necessary to specify end-of-mission failure probabilities for each of the seven fundamental modeling elements. The composition of each modeling element will be described as follows in order to provide a physical basis for specifying their end-of-mission reference failure probabilities.

Siderostat

A siderostat mirror is a circular flat, 33 cm in diameter, which directs starlight into one or more telescopes. Each siderostat feeds one telescope in ParaSIM and two telescopes in Shared Baseline SIM. Precision gimbals give the mirror approximately 15° of angular freedom about two perpendicular axes embedded in the spacecraft structure. A small corner cube attached to each mirror provides a datum that allows the precise optical pathlength traversed by starlight within the interferometer to be measured. Also included in the siderostat modeling element are the mechanisms and actuators that facilitate gimbal articulation, as well as structural components and various electronic and thermal control items specific to the siderostat functions just described.

Telescope
Starlight collected by each siderostat is directed into one telescope in ParaSIM and into two telescopes in Shared Baseline SIM. The telescope compresses the 35-cm siderostat beam diameter in the ratio of 7:1 and passes the resultant 5-cm beam to a fast steering mirror, which in turn directs it to the appropriate beam combiner. The fast steering mirror is a 5-cm flat with a few arc-minutes of angular freedom about two perpendicular axes embedded in the spacecraft. The mechanisms that facilitate articulation of the fast steering mirror are included in the telescope modeling element. Also included are optical sensors that allow the measurement of internal pathlength traversed by starlight to be compensated for structural deflections, as well as various structural, electronic, and thermal control items specific to the telescope functions just described.

Beam Combiner
A beam combiner is an optical system that combines starlight from two telescopes in such a way that the resulting interference fringes can be observed by a fringe tracking camera. For the purpose of reliability analysis, several other components that support a beam combiner are lumped with it, and the resulting collection is defined as the beam combiner modeling element. One of these components is an optical delay line, which uses trolley, voice coil, and piezoelectric transducer (PZT) actuators to equalize the pathlengths traversed in the two arms of the interferometer. The delay line is used to adjust the optical path difference until the camera detects a stationary white light fringe pattern. This only happens when the pathlengths are equal. Equality of optical pathlengths in the two arms of the interferometer is one of two necessary conditions for fringe formation. The second condition is that the beams must be parallel. Parellelism is assured by focusing a portion of the starlight from the two beams onto two spots in the focal plane of an angle tracking camera. Signals derived by measuring the distance of these spots from nominal positions provide a measure of angular deviation that can be nulled by sending position commands to the fast steering mirror in the appropriate siderostat. Differences between the pathlengths traversed by starlight in the two arms of the interferometer are measured by an internal metrology beam launcher. In both designs there is one flight computer per beam combiner, and so one flight computer is also lumped into each beam combiner modeling element. Finally, the beam combiner modeling element also contains structural and thermal control items specific to the beam combiner functions just described.

External Metrology Beam Launcher
The external metrology beam launcher is an interferometric device for measuring the precise distance between two corner cubes. Both ParaSIM and Shared Baseline SIM use an “optical truss” that is constructed from a set of discrete beam launcher measurements and used to infer the precise distance between critical points in the presence of real-time deformations induced by dynamic and thermal effects. Also

included are various structural, electronic and thermal control items specific to the external metrology beam launcher functions just described.

Front-End Electronics
Both designs include input/output and power conditioning electronics common to the siderostats, telescopes, external metrology beam launchers, and pallet articulation mechanisms. This collection of items defines the front-end electronics modeling element.

Metrology Source
The metrology source modeling element includes the laser sources, frequency shifters, and frequency modulators needed to support the internal and external metrology beam launchers, as well as all the associated thermal control, structure, input/output, power supply electronics, and fiber distribution components.

Pallet Articulation Mechanism
In both designs, all of the foregoing elements are mounted on two pallets, each of which has two degrees of rotational freedom relative to the spacecraft. There are two pallets, designated 1 and 2, and each pallet has two actuators, designated X and Y. The pallet system is working if at least one of the two X actuators and at least one of the two Y actuators is working. The pallet articulation mechanism modeling element comprises all the dedicated structure, actuators, mechanisms, and drive electronics needed to actuate the pallets.

2.3.2.1 ParaSIM Reliability Simulation Model

The following discussion describes the logical relationships among the modeling elements that were used to construct a reliability model of ParaSIM. A stylized representation of the ParaSIM design is shown in Figure 2.5. In terms of the seven fundamental modeling elements described in Section 2.3.2, ParaSIM consists of six siderostats, six telescopes, three beam combiners, ten external metrology beam launchers, two metrology sources, four pallet articulation mechanisms, and six front-end electronics units.

All of this hardware is attached to one of two articulated pallets designated Pallet 1 and Pallet 2. Each pallet has two rotational degrees of freedom effected by actuators designated X and Y. For the ParaSIM system to be operational, at least one of the two X actuators and at least one of the two Y actuators must be working.

Geometrically, there is one siderostat and one telescope at each of the six points that define two identical parallelograms with a common side. A telescope can only be working if the associated siderostat is working. ParaSIM has three baselines. In terms of hardware, each baseline is composed of three modeling elements: two telescopes and one beam combiner. A failure in any one of these modeling elements will disable the corresponding baseline.

To have a working ParaSIM system, at least two of the three possible baseline pairs must be working. It must also be possible to measure the length of each baseline and any lack of parallelism between the baselines. Baseline length is defined as

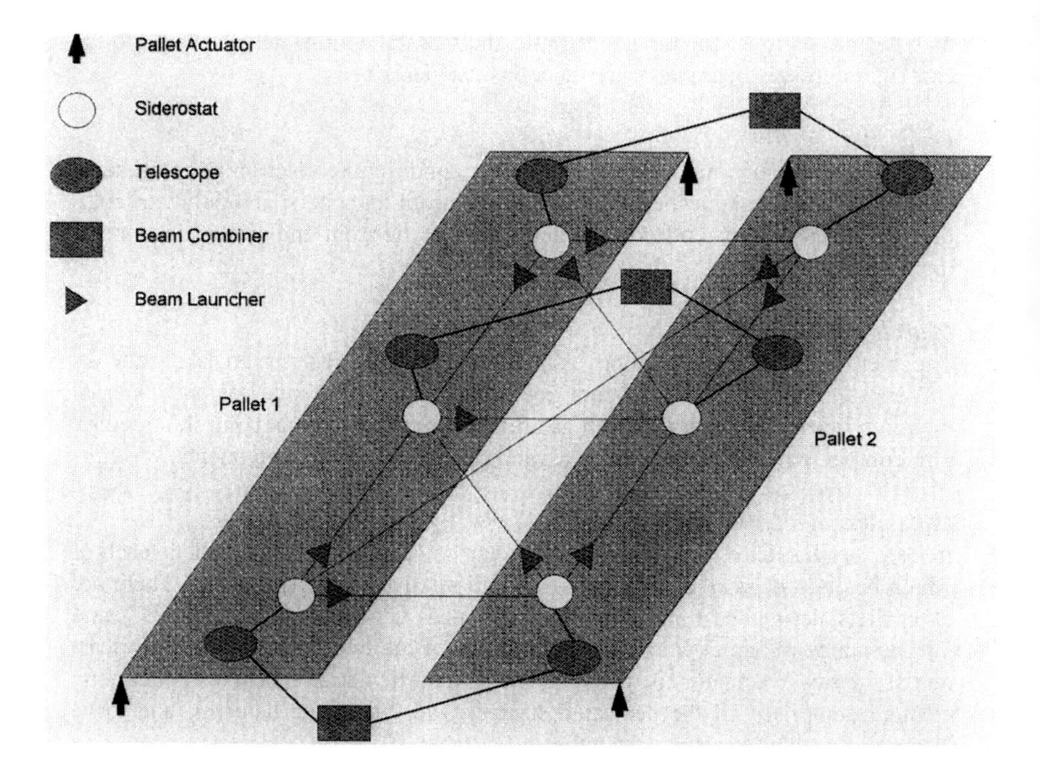

Figure 2.5. ParaSIM optical interferometer.

the distance between corner cubes embedded in the appropriate pair of siderostat mirrors. To construct the appropriate metrology truss, external metrology beam launchers are used to measure the length of the three baselines, the four parallelogram sides that connect them together, one "short diagonal" (from each of the two identical parallelograms referred to above), and one of the two "long diagonals." The minimum number of external metrology beam launchers needed to do this is ten.

The input/output and power supply electronics for the 10 external metrology beam launchers is distributed among the six front-end electronics units as follows: One of the front-end electronics units supports three beam launchers, three of them support two beam launchers, one supports one beam launcher, and one supports none. This assignment ensures that no front-end electronics failure can make it impossible to construct a viable set of external metrology beam launchers.

The definition of a viable set of external metrology beam launchers depends on which pair of baselines is operational. When two adjacent baselines are operational, only the five beam launchers in Figures 2.6(a) and 2.6(b) indicated by dark shading are essential. When the two outer baselines are operational, the set of seven working beam launchers indicated by dark shading in Figure 2.6(c) is viable. All other sets in which more than one external metrology beam launcher has failed are nonviable. Any set in which no more than one external metrology beam launcher has failed is viable.

Failure of a front-end electronics unit will result in the loss of one siderostat, one telescope, and any pallet actuator or external metrology beam launcher(s) that depend on it.

2.3.2.2 Shared Baseline SIM Reliability Simulation Model

The following discussion describes the logical relationships among the modeling elements that were used to construct a reliability model of Shared Baseline SIM. A stylized representation of the Shared Baseline SIM design is shown in Figure 2.7. In terms of the seven fundamental modeling elements described in Section 2.3.2, Shared Baseline SIM consists of four siderostats, eight telescopes, four beam combiners, 18 external metrology beam launchers, two metrology sources, four pallet articulation mechanisms, and five front-end electronics units.

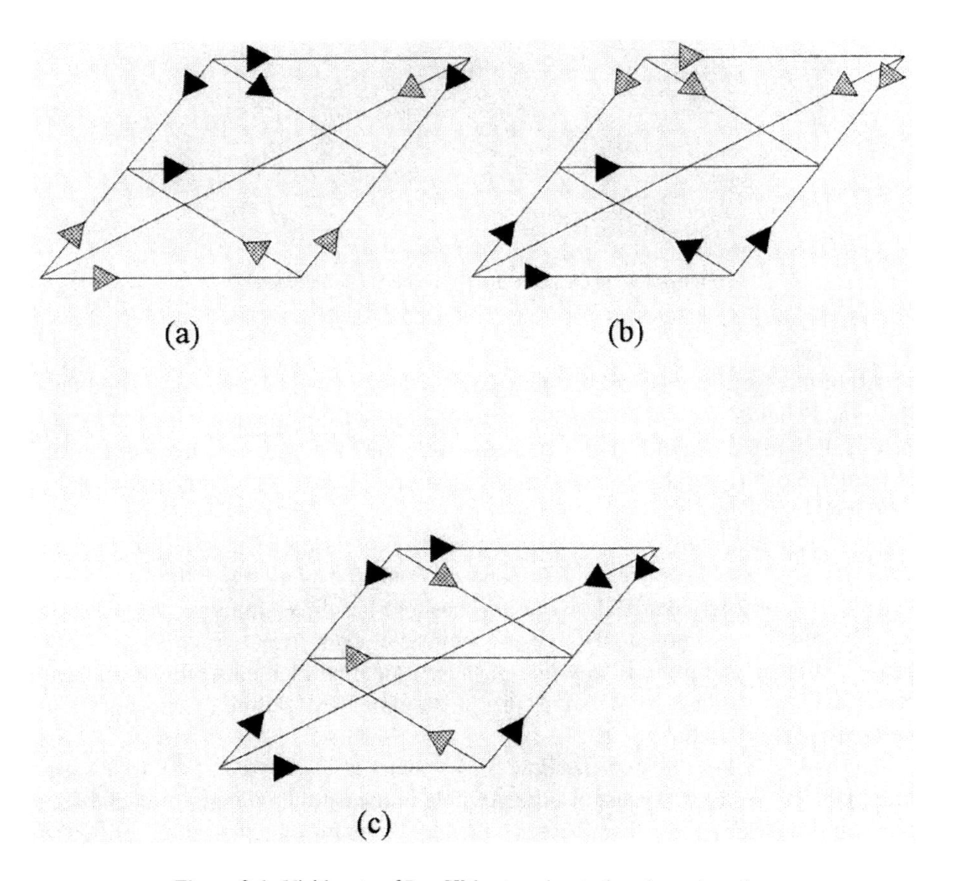

Figure 2.6. Viable sets of ParaSIM external metrology beam launchers.

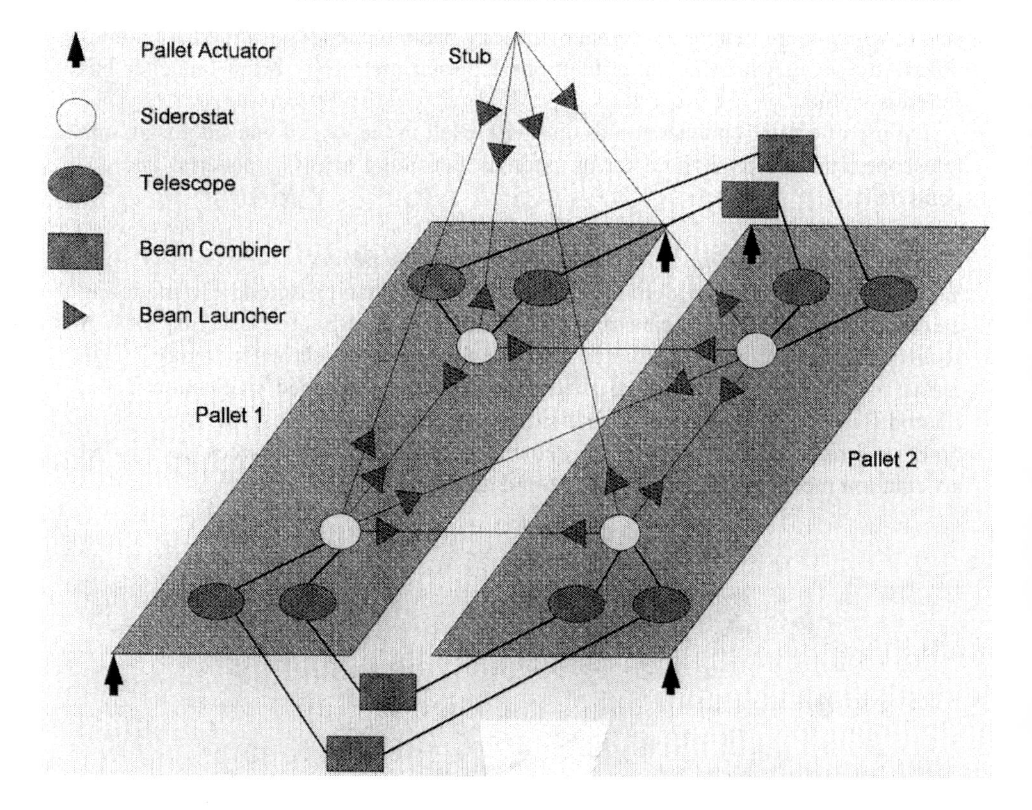

Figure 2.7. Shared Baseline SIM optical interferometer.

All of this hardware is attached to one of two articulated pallets, designated Pallet 1 and Pallet 2. Each pallet has two rotational degrees of freedom effected by actuators designated X and Y. For the Shared Baseline SIM system to be operational, at least one of the two X actuators and at least one of the two Y actuators must be working.

Geometrically, there are one siderostat and two telescopes at each of the four vertices of a base rectangle. A telescope can only be working if the associated siderostat is working. Shared baseline SIM has four baselines. In terms of hardware, each baseline is composed of three modeling elements: two telescopes and one beam combiner. A failure in any one of these modeling elements will disable the corresponding baseline. Note that in Shared Baseline SIM, failure of one siderostat will disable two baselines.

To have a working Shared Baseline SIM system, at least three of the four baselines must be working. It must also be possible to measure the length of each baseline and any lack of parallelism between the baselines. Baseline length is defined as the distance between corner cubes embedded in the appropriate pair of siderostat mirrors. External metrology beam launchers are used to measure the length of the four sides of the base rectangle, one of its diagonals, and the four slant edges of the pyramid formed by the base rectangle and a corner cube at the top of a stub boom that rises to a height of approximately 1 m above the center of the base rectangle. The minimum number of external metrology beam launchers needed to do this without redundancy is nine. If only nine were provided, failure of any single beam launcher would result in a nonviable truss, so two independent external metrology beam launchers are assigned to each edge, giving a total of 18. With this arrangement, a set of external metrology beam launchers is viable if at least one of the two launchers associated with each of the nine length measurements is operational.

The input/output and power supply electronics for the 18 external metrology beam launchers is distributed among the five front-end electronics units as follows: Three of them support four beam launchers and two of them support three beam launchers. This assignment ensures that no front-end electronics failure can make it impossible to construct a viable set of external metrology beam launchers. Failure of a front-end electronics unit will result in the loss of any and all siderostat, telescopes, pallet actuator, and external metrology beam launchers that depend on it.

2.4 EVALUATION OF ALTERNATIVE DESIGNS

2.4.1 Evaluation Approach

Each optical interferometer design is modeled as a complex system of a small number of top-level modeling elements. Each top-level element is composed of many lower-level elements whose reliability characteristics are at best not well known, particularly for the SIM application and environment. For each design, an end-of-mission reference failure probability is assigned to each top-level modeling element. These probabilities are arrived at by engineers familiar with the range of capabilities of many of the lower-level elements in other applications. They are based, in part, on (a) the best engineering judgment of the heritage and (b) the relative complexity of the top-level elements.

In our approach, the intent is to assign reference failure probabilities using the limited information available, recognizing that these reference failure probabilities are themselves uncertain and are likely to change as the design matures and more information becomes available. It is therefore extremely important to investigate the reliability robustness of any design. It is the reliability robustness exercise results which are important, not the precise value of the reference reliability. A reliability robustness investigation might validate this aspect of design quality, or it could identify opportunities for improvement.

For the Classic SIM design, the end-of-mission element reference failure probabilities are used to calculate the corresponding system end-of-mission reference reliability. A first-order sensitivity analysis is carried out in order to evaluate the reliability robustness of the design. The analysis was extended to consider the impact of reliability on science return.

To compare the ParaSIM and Shared Baseline SIM designs, the end-of-mission element reference failure probabilities are used to generate the corresponding system end-of-mission reference reliabilities. The relative reliability robustness of the two designs is evaluated from first-order sensitivity analyses.

2.4.2 Reliability Evaluations

2.4.2.1 Classic SIM

End-of-mission reference failure probabilities for the Classic SIM modeling elements are given in Table 2.1. Those failure probabilities, combined with the analytical expressions (2.1) and (2.2), can be used to derive an end-of-mission interferometer system reference reliability, assuming no failures in unmodeled Classic SIM elements. When any siderostat bay can use any triangle, that system reliability is 0.997; and when all siderostat bays must use the same triangle, it is 0.996. A Monte Carlo simulation of the Excel visual model of Classic SIM described in Section 2.3.1.2, with 10,000 replications, produced identical results.

First-order sensitivity analyses of system reliability to element failure probabilities are presented in Table 2.2. Each column corresponds to a modeling element whose failure probability is varied, holding failure probabilities for other modeling elements at their end-of-mission reference values. These reliability sensitivity

Table 2.1. Classic SIM Modeling Element End-of-Mission Reference Failure Probabilities

Rank	Modeling Element	Contents	Reference Failure Probability	Rationale
1	Triple-corner cube	Triple-corner cube	0.0005	Failure modes are meteorite impact and contamination, which have low probability of occurrence.
2	Residual siderostat bay elements	Siderostat mirrors and corner cubes, upper switchyard mirrors	0.001	Siderostat mirrors and switchyard mirrors are mechanisms with moving parts where durability is a concern; there is considerable input/output electronics in series.
3	Kite beam launcher	Kite beam launcher	0.01	Complicated, new technology, high risk of latent failure modes.
4	Siderostat bay beam launcher	Siderostat bay beam launcher	0.01	Complicated, new technology, high risk of latent failure modes.
5	Beam combiner assembly	Beam combiner, delay lines, lower switchyard mirrors	0.02	Very complex, large number of components, high risk of latent failure modes.

Table 2.2. Classic SIM Interferometer System End-of-Mission Reliability; (a) Any Siderostat Bay Can Use Any Triangle, (b) All Siderostat Bays Must Use The Same Triangle

Element Failure Probability	TCC	KBL	SBBL	RSBE	BCA
			(a)		
0	0.9968	0.9976	0.9973	0.9968	0.9990
0.01	0.9944	0.9967	0.9967	0.9941	0.9984
0.02	0.9910	0.9952	0.9952	0.9877	0.9967
0.03	0.9866	0.9931	0.9926	0.9780	0.9939
0.04	0.9812	0.9904	0.9888	0.9654	0.9900
0.05	0.9749	0.9870	0.9837	0.9500	0.9850
0.06	0.9678	0.9830	0.9767	0.9324	0.9791
0.07	0.9598	0.9782	0.9677	0.9127	0.9723
0.08	0.9510	0.9729	0.9562	0.8912	0.9646
0.09	0.9415	0.9668	0.9420	0.8682	0.9561
0.1	0.9312	0.9601	0.9249	0.8439	0.9468
			(b)		
0	0.9960	0.9967	0.9973	0.9960	0.9982
0.01	0.9936	0.9959	0.9959	0.9925	0.9976
0.02	0.9902	0.9944	0.9894	0.9853	0.9959
0.03	0.9858	0.9923	0.9754	0.9749	0.9930
0.04	0.9805	0.9896	0.9527	0.9616	0.9891
0.05	0.9742	0.9863	0.9211	0.9458	0.9842
0.06	0.9671	0.9823	0.8813	0.9277	0.9783
0.07	0.9592	0.9776	0.8343	0.9076	0.9715
0.08	0.9504	0.9722	0.7818	0.8858	0.9638
0.09	0.9409	0.9662	0.7253	0.8626	0.9553
0.1	0.9307	0.9595	0.6665	0.8381	0.9460

analyses for the Classic SIM interferometer system demonstrate a design that is very robust with respect to element failure probabilities. They do reveal the importance of targeting the element reliability of the SBBLs to achieve an acceptable system reliability in the case that all siderostat bays must use the same triangle. Even under that condition, system reliability would be unacceptable only if SBBL end-of-mission failure probability was significantly greater than the reference value of 0.01 specified in Table 2.1.

We will assume that modeling element failure rates are approximately constant for the planned 5.5-year mission and that there is a constant rate of science return after the 6-month in-orbit calibration period. Let $\{t_i\}$ be the simulated times to system failure measured in years. Compared to a perfect system, for system failure at t_i the science return is 0 for $t_i \leq 0.5$, $(t_i-0.5)/5$ for $0.5 \leq t_i \leq 5.5$, and 1 for $5.5 \leq t_i$. That can be expressed equivalently as $\text{MIN}\{\text{MAX}[0,(t_i - 0.5)/5],1\}$. The expected value of the fraction of science returned, compared to a perfect system, is then given by

$$\Sigma_{i=1,10,000} \text{MIN}\{\text{MAX}[0,(t_i - 0.5)/5],1\}/10{,}000 \tag{2.3}$$

For the modeling element end-of-mission reference failure probabilities in Table 2.1, the expected science return is 99.9% of the maximum possible end-of-mission science return when any siderostat bay can use any triangle and 99.8% of the maximum possible end-of-mission science return when all siderostat bays must use the same triangle. That is to say, with respect to the expected science return metric, the system reference reliability is nearly equivalent to a perfect system.

In general, expected science return is much less sensitive to the element failure probabilities than is the system end-of-mission reliability, since it measures contributions over the nominal collection period. To illustrate this, we will exaggerate the element failure probabilities by setting them all to 0.05. For those values, when any siderostat bay can use any triangle, the system end-of-mission reliability is only 0.788, but the expected science return is 92% of the maximum possible end-of-mission science return; when all siderostat bays must use the same triangle, the system end-of-mission reliability is only 0.726, while the expected science return is 90% of the maximum possible end-of-mission science return.

2.4.2.2 Comparison of ParaSIM and Shared Baseline SIM Reliabilities

End-of-mission reference failure probabilities for the common ParaSIM/Shared Baseline SIM modeling elements are given in Table 2.3. Monte Carlo simulations of the Excel visual models of ParaSIM and Shared Baseline SIM, described in Sections 2.3.2.1 and 2.3.2.2 for 10,000 replications, produces end-of-mission interferometer system reference reliabilities, 0.983 for ParaSIM, and 0.979 for Shared Baseline SIM.

First-order sensitivity analyses of ParaSIM and Shared Baseline SIM system reliabilities to element failure probabilities are presented in Table 2.4. Each column corresponds to a modeling element whose failure probability is varied, holding failure probabilities for other modeling elements at their end-of-mission reference values. The metrology source modeling element is excluded because it is assumed to be the same in both designs. These reliability sensitivity analyses for the ParaSIM and Shared Baseline SIM interferometer systems demonstrate designs that are very robust with respect to element failure probabilities. Only in the improbable range where front-end electronics or siderostat element failure probabilities are several orders of magnitude greater than their reference values of 0.0003 and 0.001, respectively, as specified in Table 2.3, would there be concern. Those element reliabilities should be targeted so as to guarantee that system reliability is acceptable.

A relative system reliability sensitivity analysis is given in Table 2.5. Over plausible ranges of element reliabilities there are no significant modeling element reliability discriminators, although ParaSIM tends to be slightly more reliable.

Table 2.3. ParaSIM/Shared Baseline SIM Modeling Element End-of-Mission Reference Failure Probabilities

Rank	Modeling Element	Contents	Reference Failure Probability	Rationale
1	Front-end electronics	Input/output, power conditioning and cabling for pallet articulation mechanisms, telescopes, siderostats and beam launchers	0.0003	Small parts count, no moving parts, good high reliability heritage for this kind of function
2	Pallet articulation mechanism	Actuators, mechanisms, drive electronics	0.0005	Small parts count, simple mechanism, less reliable than electronics because of moving parts
3	Siderostat	Mirrors, gimbal mechanisms, corner cubes, thermal control, structure, drive electronics	0.001	More complex mechanism than pallet articulation mechanism, but uses internal redundancy (encoders, motor windings, etc.) to limit failure probability
4	Telescope	Telescope mirror assembly, fast steering mirror, quadrature cells, thermal control, structure, drive electronics	0.003	Fewer moving parts than siderostat, but more delicate, more functions, and no internal redundancy
5	Metrology source	Laser sources, frequency shifter, frequency modulator, fiber distribution system, thermal control, structure, input/output and power supply electronics, cabling	0.005	Complicated, new technology, high risk of latent failure modes
6	Beam launcher	Optics, detectors, dither mechanisms, thermal control, structure, drive electronics	0.01	Complicated, new technology, high risk of latent failure modes, less proven heritage than metrology source
7	Beam combiner	Delay lines, fringe and angle cameras, internal metrology beam launchers, flight computers, input/output network, thermal control, structure, input/output and power supply electronics, cabling	0.03	Very complex, large number of components, high risk of latent failure modes

Table 2.4. Interferometer System End-of-Mission Reliability, (a) ParaSIM, (b) Shared Baseline SIM

Element Failure Probability	FE	PAM	SDT	TEL	BL	BC
			(a)			
0	0.987	0.989	0.988	0.991	0.995	0.997
0.01	0.982	0.987	0.980	0.983	0.990	0.994
0.02	0.972	0.988	0.972	0.973	0.979	0.992
0.03	0.958	0.987	0.965	0.964	0.964	0.990
0.04	0.944	0.985	0.951	0.950	0.948	0.986
0.05	0.937	0.984	0.942	0.937	0.929	0.983
0.06	0.924	0.980	0.928	0.922	0.902	0.976
0.07	0.900	0.975	0.905	0.913	0.886	0.970
0.08	0.882	0.976	0.890	0.893	0.865	0.965
0.09	0.863	0.971	0.872	0.869	0.834	0.956
0.1	0.838	0.970	0.854	0.860	0.815	0.954
			(b)			
0	0.988	0.988	0.992	0.989	0.988	0.993
0.01	0.951	0.988	0.951	0.982	0.988	0.993
0.02	0.914	0.986	0.919	0.970	0.984	0.990
0.03	0.869	0.984	0.880	0.953	0.977	0.987
0.04	0.836	0.984	0.842	0.933	0.975	0.983
0.05	0.800	0.982	0.819	0.919	0.967	0.974
0.06	0.777	0.980	0.765	0.896	0.957	0.971
0.07	0.734	0.976	0.742	0.863	0.945	0.962
0.08	0.693	0.976	0.706	0.853	0.930	0.953
0.09	0.672	0.971	0.680	0.817	0.917	0.947
0.1	0.646	0.969	0.656	0.789	0.905	0.936

Table 2.5. Ratio of ParaSIM to Shared Baseline SIM Interferometer System End-of-Mission Reliability

Element Failure Probability	FE	PAM	SDT	TEL	BL	BC
0	1.000	1.001	0.996	1.002	1.007	1.004
0.01	1.032	0.999	1.030	1.001	1.002	1.001
0.02	1.063	1.002	1.058	1.004	0.995	1.002
0.03	1.103	1.003	1.096	1.011	0.987	1.003
0.04	1.130	1.001	1.129	1.018	0.972	1.003
0.05	1.171	1.002	1.150	1.019	0.961	1.009
0.06	1.188	1.000	1.212	1.029	0.942	1.006
0.07	1.226	0.999	1.220	1.058	0.938	1.008
0.08	1.272	1.000	1.261	1.047	0.929	1.012
0.09	1.284	1.000	1.282	1.064	0.909	1.010
0.1	1.297	1.001	1.302	1.090	0.901	1.020

2.5 INTERPRETATIONS, CONCLUSIONS, AND EXTENSIONS

Analyses of the type presented in this chapter, which take place at the design stage and require the modeling of complex systems that are required to be reliable for a nominal operating period, can be carried out through a sequence of generic steps:

1. Define a system design in terms of top-level modeling elements, specifying their operational relationships, typically including redundancies and operational constraints. Selection of top-level modeling elements is based on knowledge of the role of physical elements in the system and how to bundle them based on the functions they support.
2. If possible, derive a closed-form analytical expression for system reliability over the nominal operating period. In any case, construct an Excel-based visual model of the design to facilitate the efficient evaluation of design variants and to support any analyses requiring time-to-failure data.
3. Specify reference failure probabilities for the end of the nominal operating period.
4. For any analyses that require time-to-failure data, it is necessary to specify the failure distributions of top-level modeling elements. At least as an approximation, it is typically assumed for nominal mission length spaceflight applications that these failure distributions are exponential. The exponential distribution parameter values are then determined from the information specified in Step 3. For alternative distributional specifications, it is likely to be necessary to introduce other relevant information in order to specify distribution parameters. The exponential assumption is suspect during extended mission operations, and failure distributions with increasing hazard functions should be considered past the nominal mission length.
5. If closed-form analytical expressions are available, system reliability at the end of the nominal operating period can be calculated. Monte Carlo simulation of the Excel-based visual model of the system also yields this system reliability as an output.
6. The reliability robustness of a given system design is considered by means of a sensitivity analysis of system reliability with respect to the reference failure probabilities specified in Step 3. At the design stage, reliability robustness is of more interest than a single computed system reliability.
7. Metrics of design quality other than reliability itself may be of great interest. For example, in this chapter it was shown how to use the Monte Carlo simulation of failure times for the Excel-based visual model of a design to compute expected science return over the nominal mission length.

The goal of this investigation was to build models for alternative SIM designs in order to ascertain whether or not the redundancy and part quality embedded in a particular configuration resulted in a design whose system reliability was robust with respect to end-of-mission element failure probabilities. The challenge was to develop an analysis procedure that could generate timely results for complex, dynamically evolving designs.

For the Classic SIM design, closed-form system reliability expressions were derived under alternative operating conditions. But the more general method, which takes advantage of an engineer's familiarity with the physical systems, uses Excel-based visual models of SIM designs in Monte Carlo simulations of the reliability robustness of the designs.

We found that Classic SIM's system reliability was extremely robust. Of interest for the ParaSIM/Shared Baseline SIM comparison is their reliability robustness and the extent to which ParaSIM system reliability exceeded Shared Baseline SIM system reliability over a range of modeling element reliabilities. Although ParaSIM is marginally more reliable, its science acquisition rate is significantly slower than that of Shared Baseline SIM.

It would be desirable to extend this analysis to compare expected science returns for the competing designs over the nominal 5.5-year mission. Also, since an operational system at the end of the nominal 5.5-year mission would most likely continue to acquire science data, it would be useful to evaluate expected science return for an extended mission. The major modeling issue to be resolved is how to credibly model element failures during an extended mission.

ACKNOWLEDGMENTS

The analysis described in this chapter was carried out at the Jet Propulsion Laboratory, California Institute of Technology, under a contract with the National Aeronautics and Space Administration.

REFERENCES

Blischke, W. R., and Murthy, D. N. P. (2000). *Reliability: Modeling, Prediction and Optimization,* Wiley, New York.

Danner, R., and Unwin, S., Executive Editors (1999). *Space Interferometry Mission: Taking the Measure of the Universe,* Jet Propulsion Laboratory, California Institute of Technology, Pasadena.

Kapur, K. C., and Lamberson, L. R. (1977). *Reliability in Engineering Design,* Wiley, New York.

Lawson, P., Editor (1999). *Principles of Long Baseline Interferometry: Course Notes from the 1999 Michelson Summer School, August 15-19, 1999,* Jet Propulsion Laboratory, California Institute of Technology, Pasadena.

EXERCISES

2.1. In the Classic SIM design, one common cause failure is a TCC micrometeoroid impact that simultaneously renders TCC 2, TCC 3, or TCC 4 and its two closest KBLs nonfunctional. Assuming no other consequences of the collision, how is the system reliability changed relative to a simple failure of a TCC?

2.2. In the Classic SIM design, an eighth siderostat bay would have provided increased redundancy and also would have significantly increased the potential science return. (a) Extend the closed form reliability solutions to include an eighth siderostat bay. (b) Extend the Excel visual simulation model to include an eighth siderostat bay.

2.3. Simply comparing alternative design reliabilities at a nominal end-of-mission cannot fully capture the relative capabilities of the designs to deliver science. Also, Shared Baseline SIM has a higher science acquisition rate than ParaSIM. Assume that, up to the nominal end-of-mission, exponential distributions are appropriate for the modeling elements and that science return begins after a 6-month in-orbit calibration period at a constant rate of return. If the optical interferometer system is still functional at the nominal end-of-mission goal, it is likely that science data collection will continue. Assume that a Weibull distribution with known shape parameter is appropriate for the modeling elements during an extended mission. Describe a procedure for simulating a science return distribution and discuss how to construct, from that distribution, alternative metrics for comparing the designs.

2.4. In this chapter an evaluation of interferometer design is based on the specification of end-of-mission modeling element reference failure probabilities, coupled with an analysis of the design's reliability robustness. Describe a procedure, based on eliciting information about the uncertain nature of these modeling element failure probabilities, for performing a probabilistic risk assessment for an interferometer design.

2.5. Describe how to incorporate end-of-mission modeling element uncertainty into a design evaluation based on expected science return.

ACRONYMS

The following acronyms are used in this chapter:

AU	Astronomical unit
BC	Beam combiner
BCA	Beam combiner assembly
BL	Beam launcher
FE	Front-end electronics
JPL	Jet Propulsion Laboratory
KBL	Kite beam launcher

MS	Metrology source
NASA	National Aeronautics and Space Administration
PAM	Pallet articulation mechanism
PZT	Piezoelectric transducer
RSBE	Residual siderostat bay elements
SBBL	Siderostat bay beam launcher
SDT	Siderostat
SID	Siderostat bay
SIM	Space Interferometry Mission
TCC	Triple-corner cube
TEL	Telescope

CHAPTER 3

Confidence Intervals for Hardware Reliability Predictions

Chun Kin Chan and Michael Tortorella

3.1 INTRODUCTION

Certainly, the routine production of reliability predictions for engineering system hardware is a major preoccupation of practicing reliability engineers. Hardware reliability prediction serves many purposes during the life of a product. At the design stage, it can be used to study the effects of thermal and electrical stresses on the design. For instance, if it is known that a device's junction temperature may be too high and excessive failures may result, design changes may be made to lower the temperature and prevent failures in the field. At the product deployment stage, hardware reliability predictions are often used to set targets for field reliability. For instance, if a product's actual field failure rate is worse than its predicted field failure rate (perhaps within some statistical tolerance), then the equipment vendor might be pressed to provide free product replacement in addition to any product warranty that may have been contractually specified.

Business and engineering goals normally require that reliability predictions be provided quickly, consistently, systematically, accurately, and with known precision. Many models (e.g., Telcordia 2001) have been developed and used that readily satisfy the first three of these requirements. The latter two, however, have provided a rich source of discussions and even controversy within the profession. Our purpose in this chapter is to describe a new, practical way of making visible the precision of hardware reliability predictions for subassemblies of electronic systems. We do not specifically address the accuracy issue in this chapter because it raises fundamental issues, beyond mathematics, whose discussion more properly belongs in forums of the reliability engineering profession.

Case Studies in Reliability and Maintenance, Edited by W. R. Blischke and D. N. P. Murthy.

ISBN 0-471-41373-9

3.2 APPROACH

Several methodologies for estimating the precision of reliability predictions have been developed over the years (Baxter, 1993; Butcher et al., 1978; El Mawaziny and Buehler, 1967; Grubbs, 1971; Kraemer, 1963; Lieberman and Ross, 1971; Madansky, 1965; Mann and Grubbs, 1972, 1974; Sarkar, 1971). All these methods contain in one disguise or other the simple idea that the variance of a sum of uncorrelated random variables is the sum of their individual variances. The challenge usually is to recover the variance of an estimator of some parameter of the component life distribution from information that is publicly available. Public information usually does not include the details of the experiments or data analyses that led to the published parameter estimates.

Our contribution has been to distill from these methods one that is easily implemented. We have done so in a computer-aided design (CAD)-linked program that provides a highly automated platform for rapid reliability prediction, including precision estimates, for hardware subassemblies. The CAD-linked program includes a database populated with component failure rate estimates, including upper 95% confidence limits, based on field tracking studies. Our approach differs from the standard approaches recommended by many device vendors in that our failure rate estimates are obtained from field operating conditions instead of only from accelerated life tests.

3.3 PROBLEM DESCRIPTION

3.3.1 General

The case we will consider is the production of a reliability prediction, with confidence interval, for a telecommunications hardware product. The information that is available to an engineer undertaking this task is a document containing failure rate estimates for categories of electronic, optical, and electro-optical components that cover the universe of such components that would be used in hardware products of this sort.[1] This publicly available information also may contain information about the quality of the failure rate estimates in some form; in our case, we will assume that a 95% upper confidence limit (UCL) for each failure rate estimate is given. Although the primary purpose of this case study is to show how to construct a confidence interval for the subassembly failure rate when the confidence intervals for the failure rate of each of its constituent components is given, we also show how the component failure rate and UCL estimates may be obtained from field reliability data. We also briefly discuss how subassembly failure rate estimates obtained from operation of the subassembly in the field may be compared with the subassembly failure rate prediction.

[1]At times, a component will be used that does not appear in the list of failure rate estimates. The engineering approaches to dealing with this issue are beyond the scope of this chapter. Suffice it to say that this provides a prime opportunity for soul-searching on the part of the constructors of the list of failure rate estimates.

3.3.2 Product Description and Reliability Characterization

The product in this study is an electronic or electro-optic subassembly, which is part of a larger system such as a digital telephone switch, Internet router, optical transport system, and so on. The reliability of a subassembly is characterized by the property that the subassembly fails if and only if one or more of its constituent components fails.

- No other cause of failure besides that of failure of a constituent component is included in the reliability model for this subassembly.
- There may be some constituent components whose failure has no effect on the system (these will be called *non-service-affecting,* abbreviated NSA). However, for purposes of forecasting the number of maintenance actions required over the useful life of the system, it is necessary to account for all components because while an NSA component failure will not be perceptible to the customer of the service, the system owner still must perform a maintenance action to restore the subassembly to full operating condition. Our case deals with this latter *maintainability* situation. Because the NSA components on a subassembly are ignored for the purpose of a reliability prediction that deals with system, or higher level of indenture subassembly, failure, it is clear how to use the methods that follow in service-focused studies also.

In the language of reliability engineering, the subassembly may be considered as a series system of its constituent components. If *all* components are included, we obtain the maintainability point of view illustrated in the case. If the customer service point of view is needed, we consider only those constituent components that are service-affecting. We further assume in this chapter that subassemblies may be replaced in the field (by spares drawn from some spares inventory) but may not be repaired in the field. Instead, they are repaired at a depot (after some delay for storage, shipping, etc.), and the repaired subassemblies are sent to the spares inventory. Therefore, for purposes of providing information to higher-level system reliability studies such as reliability modeling of the entire system or management of spares inventories, it is appropriate to describe the subassembly as a nonmaintained system.[2] That is, the subassembly's reliability is appropriately described by its life distribution or, equivalently, survivor function or failure (hazard) rate. In this chapter, it is to be understood that the term "failure rate" is interpreted as the hazard rate or force of mortality of the life distribution. In particular, we urge readers to eschew thoughts of maintained systems when they encounter this phrase in this chapter. See Ascher and Feingold (1984) for full discussion.

[2]We leave aside the question of how subsequent lifetimes of repaired subassemblies are modeled. The most frequently used models here are the renewal model and the revival, or minimal repair, model. Clearly, the process of drawing a spare at random from an inventory of subassemblies that may have been repaired some number of times yields a subassembly whose life distribution is a mixture of the distributions of successive lifetimes in whatever process is used to describe the repair. For renewal repair, this effect is nil. The minimal repair case is much more interesting.

Some reliability professionals do not accept as an adequate model for subassembly failure one that postulates that subassembly failure is caused only by constituent component failure. Some objections are easily handled, such as by treating interconnections (e.g., surface mount solder connections, through-hole solder connections, and edge-card connections for a printed circuit board) as "constituent components." However, other failure modes, such as intermittent failures or hardware–software interaction failures, may not be adequately captured by this "component failures" reliability model. In practice, a subassembly exhibiting one of these types of failures will be removed and replaced, so we may as well treat these failures as permanent, at least for the purposes of modeling the number of maintenance actions to be required of the system owner. The failure rate for such failures cannot be predicted from component failure rate information alone. Consumers of reliability prediction products need to be aware of both the values and the limitations of these models.

3.4 RELIABILITY MODELING

3.4.1 General

Given that we restrict ourselves to the subassembly reliability model described above, it is easy to see that the reliability of the subassembly (as measured by the survivor function, life distribution, or failure rate) is readily obtained from the corresponding reliability of the constituent components by standard means developed in the mathematical theory of reliability. If we assume that the lifetime random variables of each of the constituent components are mutually stochastically independent, then we obtain

$$\overline{F}(t) = \overline{F}_1(t) \cdots \overline{F}_n(t) \tag{3.1}$$

and

$$h(t) = h_1(t) + \cdots + h_n(t) \tag{3.2}$$

where F (respectively, h) represents the life distribution (respectively, hazard rate) of the subassembly, the subscripts indicate these quantities pertaining to each of the constituent components, and $\overline{F} = 1 - F$. These equations apply when the independence condition is satisfied; no particular form of the life distributions need be assumed. In practice, component life distributions are often taken to be exponential, even though this may be physically unreasonable. We believe the explanation for this has two parts, one incidental and one perhaps more convincing:

- The exponential distribution has a constant failure rate, is particularly easy to compute with, and has simple parameter estimation procedures. The force of these advantages in an environment that places a premium on simplicity and speed must not be underestimated. Even though the technology required to deal with more complicated life distributions is easily computerized, we know of no such implementation in any commercially available tool.

- Because of Drenick's theorem (Drenick, 1960), the distribution of the time to first failure of a subassembly is asymptotically exponential as the number of components on the subassembly gets large (and a condition that amounts to "no small number of failure rates dominates the group" is satisfied). This is true irrespective of the original time-to-failure distribution of the components. The reasoning then goes: Why not simplify matters and assume that the component's time-to-failure distribution is exponential to begin with? Then the time to first failure, being the minimum of independent exponentially distributed random variables, is itself exponentially distributed with a failure rate equal to the sum of the individual failure rates. So if this is the final result we are going to get (approximately), why not assume a simpler life distribution characterization to begin with? Then there is no need for complicated data analysis procedures for parameter estimation, and a great deal of work can be avoided. In addition, the uncertainty in our knowledge of the form of the life distribution and of its parameters usually is so large that it easily swamps any error that might be made by assuming the wrong distributional form.

We will adopt the exponential distribution assumption for our case study because, in addition to its ubiquity in practice, it shows our ideas in a simple setting. Other distribution assumptions will also work, but at a cost of greater complication. The general method of procedure, then, is to sum the (estimated) failure rates of the constituent components to derive an estimate of the subassembly reliability [see equation (3.2)]. This simple statement forms the core of our approach to this problem, for it is the realization that the subassembly reliability prediction is a statistic (because it is a function of statistics, namely the estimated constituent component failure rates) that enables us to begin dealing with its dispersion. The constituent component reliability estimates may be formed from field reliability data (such as the number of replacements of that component during refurbishment of the subassembly at a depot and the accumulated operating hours for those components), from accelerated life testing data, from physical first principles whose parameters are estimated in some way from data, or from some other form of data. In all cases, our estimate of the component failure rate is uncertain, to a greater or lesser degree, simply because of the statistical estimation procedure. Sample size, time in service, environmental conditions, and data quality all bear on the dispersion of the component failure rate estimates. The key problem we solve in this case is how to combine the dispersions of the constituent failure rate estimates, expressed in our example as confidence intervals or confidence limits, to derive some information about the dispersion of the subassembly failure rate estimate. We will work with upper confidence limits (UCL), which we will assume to be tabulated and publicly available[3] for each component. Lower confidence limits (LCL) and two-sided limits are handled similarly.

We note that some reliability engineers believe that a confidence limit for the subassembly failure rate estimate (prediction) can be obtained by summing the corresponding confidence limits of the constituent component failure rate estimates. We surmise that this error may have as its genesis the fact that the variance of a sum of uncorrelated random variables is the sum of the individual variances, and some may naively believe that this idea can be carried over directly to confidence limits. It is useful to examine this error closely because it points the way ahead. The confidence limit for the estimated component failure rate is indeed a function of the variance of the component failure rate estimator. So our task can be framed as follows: Recover the variance from the UCL values that are published along with the point estimates of the component hazard rates. Having done that, we can add the variances and reconvert to UCL via the distribution (or the asymptotic distribution) of the subassembly failure rate estimate. This is essentially the basis of many of the methodologies cited above. Note that an incomplete understanding of this procedure may lead to the summation error described here.

3.4.2 Confidence Intervals for a Subassembly Hazard Rate

3.4.2.1 "Parts Count" Method

To anticipate the amount of variability that is likely to be seen in field reliability results for subassemblies, we provide an (approximate) upper confidence limit (UCL) for the subassembly failure rate estimate or prediction that is obtained by summing the estimated failure rates of its constituent components. While there are many ways to express the dispersion, uncertainty, or quality of information contained in a statistic (such as a reliability prediction of this type), we choose the UCL as a simple, useful, and informative measure.[4] We will defer until Section 3.7, Comparing Field Reliability Results with Predictions, where we discuss how to interpret the UCL for the subassembly failure rate prediction.

The procedure we use for computing the approximate 95% UCL value for the subassembly hazard rate prediction is based on a method developed by Grubbs (1971) as adapted by Baxter (1993). We sketch the procedure here. We begin by examining how the UCL for the component failure rate estimate is constructed so that we can express the variance of the component failure rate estimator in terms of the (publicly known) UCL.

Let the cumulative distribution function (cdf) for the lifetime of a component be exponential with parameter (failure rate) λ and suppose that r failures are observed

[3]We emphasize the importance of public versus private information. While the table of estimated failure rates and UCLs is open to the company's general engineering community, as for instance in the Lucent *Reliability Information Notebook* (op. cit.), the data and any other construction information that produced those estimates is typically not so available. So the engineer has no access to any more knowledge about dispersion of the component failure rate estimates than the published UCL.

[4]We would also like to be able to add "widely used," but in practice this methodology is rarely used outside of Lucent Technologies.

in T device-hours of previous life testing at operating conditions and/or field experience for all components of this type. When the subassembly reliability prediction is made assuming that all components operate at the environmental conditions at which their individual failure rate estimates were constructed,[5] reliability engineers say that the "parts count" method is being used. When one or more components on the subassembly operates at an environmental condition other than nominal, adjustments to its tabulated failure rate estimate usually must be made, and reliability engineers say that the "parts stress" method is being used. We will discuss details of the parts stress method in Section 3.4.2.2.

The maximum likelihood estimate (MLE) of λ is given by $\hat{\lambda} = r/T$, and for $0 \leq \alpha \leq 1$ a $100(1-\alpha)\%$ UCL for $\hat{\lambda}$ is

$$V = \frac{\chi^2(2r, 1-\alpha)}{2T} \tag{3.3}$$

(Lawless 1982), where $\chi^2(d, p)$ is the $100p$th percentile of a chi-squared distribution on d degrees of freedom. To implement the Grubbs procedure to find a UCL for the subassembly failure rate prediction when the only information available is $\hat{\lambda}$ and V (these values would be found in a publicly accessible table, but any underlying data or procedures used to construct these estimates are usually not available), we first need to solve for the quantity $\hat{\lambda}^2/r$. The first step to this is to recover r from the above information. Noting that $T = r/\hat{\lambda}$, we obtain $V = (\hat{\lambda}/2r)\chi^2(2r, 1-\alpha)$. For large a, we have $\chi^2(2a, \cdot) \approx N_{2a,4a}(\cdot)$ where $N_{m,v}$ stands for a normal distribution with mean m and variance v [this approximation is good to within about 5% for the 99th percentile as long as $r \geq 10$ or so (Lawless, 1982]. Also, note that

$$N^{-1}_{2a,4a}(\beta) = 2\sqrt{a}N^{-1}{}_{0,1}(\beta) + 2a \tag{3.4}$$

for all β. Let z_α be defined by $N_{0,1}(z_\alpha) = 1 - \alpha$. Using the normal approximation above, we obtain

$$V \approx \frac{\hat{\lambda}}{2r}[2z_a\sqrt{r} + 2r] \tag{3.5}$$

Replacing the $\approx$ with $=$ and rearranging, we can solve for r to obtain

$$r = \left(\frac{\hat{\lambda}z_a}{V - \hat{\lambda}}\right)^2 \tag{3.6}$$

so that, finally,

$$\frac{\hat{\lambda}^2}{r} = \left(\frac{V - \hat{\lambda}}{z_a}\right)^2 \tag{3.7}$$

[5]These are referred to as "nominal conditions."

The Grubbs procedure is now implemented as follows. Assume that the subassembly comprises n components, whose lifetimes are stochastically mutually independent, in a series (reliability) configuration. There are N_i components of type i (a "type" is a component for which a distinct failure rate estimate $\hat{\lambda}_i$ and associated UCL V_i are provided in the table of component failure rate estimates; types are not arbitrary but are groups of components whose reliability should be similar because their construction and reliability physics and chemistry are similar) and k distinct types of components. In this scheme, we have

$$n = \sum_{i=1}^{k} N_i \tag{3.8}$$

The first step in the procedure is to form the quantities $s_i = (V_i - \hat{\lambda}_i)/1.645$ for each component type i. Next, form the quantities

$$m = \sum_{i=1}^{k} N_i\hat{\lambda}_i \qquad \text{and} \qquad v = \sum_{i=1}^{k} N_i s_i^2 \tag{3.9}$$

Furthermore, define

$$\delta = \frac{2m^2}{v} \tag{3.10}$$

Then an approximate 95% UCL for the subassembly hazard rate is given by

$$\frac{v}{2m}\chi^2(\delta, 0.95) \tag{3.11}$$

The value of δ as defined above will usually not turn out to be an integer, so to determine the required 95th percentile of the $\chi^2(\delta, 0.95)$ distribution it will usually be necessary to interpolate (see Exercise 3.1).

Two-sided confidence intervals can be easily obtained using this procedure as well (Exercise 3.2).

3.4.2.2 Parts Stress Method

The procedure developed in the preceding section works only when all the components on the subassembly operate at their nominal temperatures (sometimes referred to as the reference temperature)—that is, the temperature at which the tabulated component failure rate and UCL estimates are valid.[6,7] We assume that the effects of temperature on a component's failure rate may be described by an *accel-*

[6]This does not need to be the same temperature for all components on the subassembly.

[7]Other stresses besides temperature, such as electrical stress (Telcordia, 2001), may also be accommodated in an accelerated life model. We confine our attention to temperature only for conciseness.

erated life model (Lawless, 1982), which postulates that the failure rate of a component at a given temperature is related to the failure rate of that component at another temperature by a multiplicative factor that depends only on the two temperatures (and certain physical constants). Under the Arrhenius accelerated life model, if a component of type i operates at a temperature other than the nominal temperature, then the estimated failure rate for that component is multiplied by

$$A_i = \exp\left[\frac{E_{ai}}{k}\left(\frac{1}{T_{ri}} - \frac{1}{T_{ui}}\right)\right] \tag{3.12}$$

where E_{ai} is the activation energy for the type-i component in electron-volts (eV), k is Boltzmann's constant (8.62×10^{-5} eV/K), T_{ri} is the reference temperature, and T_{ui} is the use (or operating) temperature (both temperatures must be expressed in degrees kelvin)[8]. Then the type-i component failure rate estimate at the use temperature is given by

$$\lambda_i^* = A_i \hat{\lambda}_i \tag{3.13}$$

where $\hat{\lambda}_i$ is the estimated component failure rate at the reference (nominal) temperature. The multiplicative factor A_i is sometimes referred to as the acceleration factor. The corresponding 95% UCL is given by

$$V_i^* = \frac{\chi^2(2A_i w_i, 0.95)}{\chi^2(2w_i, 0.95)} V_i \tag{3.14}$$

where V_i is the 95% UCL at the reference temperature, and $w_i = (\hat{\lambda}_i/s_i)^2$. See Exercise 3.3.

Now that the 95% UCL for the type-i component failure rate estimate at the altered stress (temperature) is known, one can proceed as before to obtain the confidence limits for the subassembly failure rate prediction (estimate). The resulting degrees of freedom will be denoted by δ^*.

3.5 SUBASSEMBLY HARDWARE RELIABILITY PREDICTION

3.5.1 Composition of the Circuit Board

The first part of our case concerns the production of a reliability prediction for a new circuit board. The components on this circuit board are listed in Table 3.1 along with the results of a thermal analysis based on the anticipated environmental conditions this circuit board will encounter in operation. The thermal analysis yields an estimate of the use temperature of each device T_{ui} on the board.

[8]To convert from degrees Centigrade to degrees Kelvin, add 273. For example, 40°C is equivalent to 313 K.

Table 3.1. Thermal Analysis of Components on a Newly Designed Board

Device Type	Location	T_{ui} (°C)
Microprocessor, 30,001–50,000 gates	IC1	80
Integrated circuit, digital CMOS, 101–500 gates	IC2.1	70
Integrated circuit, digital CMOS, 101–500 gates	IC2.2	60
Integrated circuit, digital CMOS, 101–500 gates	IC2.3	65
Integrated circuit, digital CMOS, 101–500 gates	IC2.4	63
Integrated circuit, digital CMOS, 101–500 gates	IC2.5	68
Integrated circuit, digital CMOS, 101–500 gates	IC2.6	66
Oscillator	Y1	45
Oscillator	Y2	48
Random access memory, static CMOS 2048K	IC3.1	45
Random access memory, static CMOS 2048K	IC3.2	44
Film resistor	R1-R50	40
Ceramic capacitor	C1-C50	40

3.5.2 Component Failure Rate Database Example

The engineer performing the prediction has access to a table of estimated failure rates and confidence limits for the components on the circuit board. An excerpt from such a table is shown here as Table 3.2. The database table excerpt (Table 3.2) has the following fields: device type, point estimate (MLE) and 95% UCL for the device failure rate (in FITs[9]), reference temperature (i.e., the temperature at which the failure rate and UCL estimates are valid), and relevant thermal activation energy (if any). The activation energy values obtained from SR-332 are based on reliability physics studies of failure mechanisms. Activation energies associated with different failure mechanisms are determined experimentally by observing the failure times of test devices at different temperatures. Table 3.2 is an example of the kind of table that would normally be publicly available to the engineers making subassembly reliability predictions.

3.5.3 Parts Stress Analysis of a New Circuit Board

The component failure rate database in Table 3.2 can now be used to predict the failure rate of the new circuit board using the parts stress method in Section 3.4.2.2. Using T_{ui} estimates from Table 3.1 and the E_{ai}, T_{ri} values from Table 3.2, we can compute the acceleration factor for each device type according to equation (3.12). Multiplying the resulting acceleration factor A_i by the base (reference) failure rate $\hat{\lambda}_i$ from Table 3.2 gives us the device failure rate at use temperature via equation (3.13). The results are shown in Table 3.3. Column 1 in Table 3.3 shows the device types of the components on the board. Column 2 gives the number of

[9]1 FIT is one failure in 10^9 operating hours.

Table 3.2. Excerpt from Public Table of Component Failure Rate Estimates

	Estimated Failure Rate (FITs)			
Device Type i	MLE $\hat{\lambda}_i$	95% UCL V_i	Reference Temperature T_{ri} (°C)	Activation Energy E_{ai} (eV)
Microprocessor, 30,001–50,000 gates	43	206	74	0.45
Integrated circuit, digital bipolar 101–500 gates	16	33	92	0.35
Integrated circuit, digital bipolar 51–100 gates	14	69	77	0.35
Integrated circuit, digital CMOS, 101–500 gates	13	24	61	0.45
Oscillator	43	137	42	0.4
Random access memory, static CMOS 2048K	40	67	43	0.45
Integrated circuit, digital bipolar 21–50 gates	14	29	45	0.35
Integrated circuit, digital bipolar 1–20 gates	11	18	42	0.35
Film resistor	1	2	40	0.15
Tantalum capacitor	4	21	40	0.15
Silicon diode, general purpose, < 1 A	14	69	39	0.22
Integrated circuit, digital CMOS, 30,001–50,000 gates	22	103	40	0.45
Ceramic capacitor	1	3	40	0.05

components of each device type. Column 3 shows the use temperatures from Table 3.1. Columns 4, 5, and 6 show the device failure rates (point estimates), reference temperatures, and activation energies from Table 3.2, respectively. Column 7 shows the acceleration factors computed using equation (3.12). Column 8 shows the resulting component failure rates at use temperatures. Each cell in the last column shows the component failure rate at use temperature multiplied by the number of components on the circuit board. Summing all the numbers in the last column gives us the point estimate of the failure rate of the circuit board [equation (3.9)], which is found to be 415.8 FITs. This number would typically be fed into a system-level reliability model to estimate reliability figures of merit such as service accessibility (availability), maintenance actions per 100 telephone lines per year, and so on. It is also used for spares inventory management models (Chan and Tortorella, 2001).

To estimate the 95% UCL, we proceed with equation (3.14). The worksheet is shown in Table 3.4. Columns 1, 2, 3, and 5 in Table 3.4 correspond to columns 1, 2, 4, and 7 in Table 3.3. Column 4 in Table 3.4 shows the 95% UCL values from Table 3.2. Column 6 shows the standard deviations obtained from $s_i = (V_i -$

Table 3.3. Worksheet for Predicting a Point Estimate of the Failure Rate of a Newly Designed Board

Device Type	N_i	T_{ui}	$\hat{\lambda}_i$	T_{ri}	E_{ai}	A_i	λ_i^*	$N_i\lambda_i^*$
Microprocessor, 30,001–50,000 gates	1	80	43	74	0.45	1.27	55.42	55.42
Integrated circuit, digital CMOS, 101–500 gates	1	70	13	61	0.45	1.53	20.50	20.50
Integrated circuit, digital CMOS, 101–500 gates	1	60	13	61	0.45	0.97	12.98	12.98
Integrated circuit, digital CMOS, 101–500 gates	1	65	13	61	0.45	1.22	16.37	16.37
Integrated circuit, digital CMOS, 101–500 gates	1	63	13	61	0.45	1.12	14.93	14.93
Integrated circuit, digital CMOS, 101–500 gates	1	68	13	61	0.45	1.40	18.75	18.75
Integrated circuit, digital CMOS, 101–500 gates	1	66	13	61	0.45	1.28	17.13	17.13
Oscillator	1	45	43	42	0.4	1.13	49.04	49.04
Oscillator	1	48	43	42	0.4	1.29	56.20	56.20
Random access memory, static CMOS 2048K	1	45	40	43	0.45	1.11	43.74	43.74
Random access memory, static CMOS 2048K	1	44	40	43	0.45	1.05	41.53	41.53
Film resistor	50	40	1	40	0.15	1.00	0.51	25.28
Ceramic capacitor	50	40	1	40	0.05	1.00	0.88	43.92
								$m^* = \sum_{i=1}^{13} N_i\lambda_i^* = 415.8$

$\hat{\lambda}_i$)/1.645. Column 7 is obtained from $w_i = (\hat{\lambda}_i/s_i)^2$. Column 8 shows the 95% UCL values at use temperatures obtained using equation (3.14). Column 9 shows the corresponding standard deviations using $s_i^* = (V_i^* - \lambda_i^*)/1.645$. The last column computes ν^*, which is then used to compute

$$\delta^* = \frac{2(m^*)^2}{\nu^*} \tag{3.15}$$

where m^* is obtained from Table 3.3 and ν^* is from Table 3.4. Then the 95% UCL for the newly designed board is given by

$$\frac{\nu^*}{2m^*}\chi^2(\delta^*, 0.95) \tag{3.16}$$

which is found to be 695 FITs. This value is typically used to compare failure rate predictions with observed values. This comparison is discussed in detail in Section 3.7.

Table 3.4. Worksheet for Predicting the 95% UCL on the Failure Rate of a Newly Designed Board

Device Type	N_i	$\hat{\lambda}_i$	V_i	A_i	s_i	w_i	V_i^*	s_i^*	$N_i(s_i^*)^2$
Microprocessor, 30,001–50,000 gates	1	43	206	1.27	98.95	0.19	254.24	120.86	14608.23
Integrated circuit, digital CMOS, 101–500 gates	1	13	24	1.53	6.54	4.18	33.34	7.80	60.90
Integrated circuit, digital CMOS, 101–500 gates	1	13	24	0.97	6.54	4.18	23.61	6.46	41.73
Integrated circuit, digital CMOS, 101–500 gates	1	13	24	1.22	6.54	4.18	28.07	7.11	50.60
Integrated circuit, digital CMOS, 101–500 gates	1	13	24	1.12	6.54	4.18	26.20	6.85	46.89
Integrated circuit, digital CMOS, 101–500 gates	1	13	24	1.40	6.54	4.18	31.13	7.52	56.60
Integrated circuit, digital CMOS, 101–500 gates	1	13	24	1.28	6.54	4.18	29.06	7.25	52.54
Oscillator	1	43	137	1.13	56.77	0.59	148.28	60.33	3639.54
Oscillator	1	43	137	1.29	56.77	0.59	162.27	64.48	4157.37
Random access memory, static CMOS 2048K	1	40	67	1.11	16.73	5.58	72.45	17.46	304.68
Random access memory, static CMOS 2048K	1	40	67	1.05	16.73	5.58	69.63	17.08	291.73
Film resistor	50	1	2	1.00	1.15	0.19	2.40	1.15	66.20
Ceramic capacitor	50	1	3	1.00	1.15	0.59	2.76	1.15	65.77

$$\nu^* = \sum_{i=1}^{13} N_i(s_i^*)^2 = 23442.77$$

3.6 CONSTRUCTION OF COMPONENT FAILURE RATE DATABASE

3.6.1 Introduction

This section illustrates a data collection process and the subsequent analysis used to obtain component failure rate estimates $\hat{\lambda}_i$ and associated UCLs V_i for the public table. Reliability engineering specialists usually carry out the work described here, and consumers of the product normally would not see these details. We describe these details here for completeness.

Data were collected for the components on a circuit board (subassembly) at two stages of the board's product life cycle. To facilitate the following discussion, this board is subsequently referred to as CP330 because it contains 330 components. The first stage of data collection occurred at the design stage of CP330 where thermal data were collected. A thermal analysis was performed to simulate field operating conditions, and junction temperatures of the active devices were estimated using a thermal analysis software tool.

3.6.2 Thermal Analysis

The design of CP330 posed a serious thermal problem because 330 components were packed onto a small printed circuit board. According to equation (3.12), high temperatures increase component failure rates. Every active device's junction (internal) temperature must be controlled because it is the internal device temperature that is responsible for thermal failures. At the design stage of CP330, a thermal analysis was performed to make sure that the junction temperatures of the active devices in service did not exceed the component manufacturers' limits. The thermal analysis was in essence a simulation experiment to help pinpoint hot components, which must be cooled either by adding heat sinks or by moving them to a cooler part of the circuit board. The input parameters included the ambient temperature, component power dissipation estimates, thermal impedance values from manufacturer data sheets, and the air flow (cooling) parameter. The output of the thermal analysis was a reliable design with all components running at temperatures with sufficient margins below the component manufacturers' limits. Table 3.5 shows the thermal analysis results under expected field operating conditions. Column 1 shows the device type according to the descriptions in the Telcordia document SR-332. Column 2 shows the number of components of each device type on CP330. Column 3 shows the junction temperatures[10] estimated by the thermal analysis software for the active devices. Column 4 shows the device's ambient temperature, which is defined as the temperature measured by placing a probe 0.5 in. from the device (Telcordia, 2001). A typical ambient temperature value is 40°C for equipment in a central office, where telephone switches operate in a controlled environment.[11]

3.6.3 Field Tracking Study

The second stage of data collection occurred at the field deployment stage of CP330, where field reliability data were collected. A field tracking team that included participants from the equipment manufacturer and the service providers (telephone companies who owned the equipment) was formed. For each device type, the number of component failures was tracked along with the total number of operating hours it accumulated in the field. The equipment manufacturer ensured that each component replaced during the repair process went through a failure mode analysis, and each confirmed failure was included in the data. The service providers' installed CP330 boards' serial-numbered bar codes were scanned periodically to ensure that the operating-hour records were accurate. At the end of the study (time truncated), the CP330 had logged 2.3×10^7 operating unit-hours. Table 3.6 shows the field tracking data and the component failure rates estimated from the data. Column 1 repeats the device classes shown in Table 3.2. Column 2 shows the

[10]Junction temperature can be interpreted as the silicon temperature in this study. It is the internal device temperature that is responsible for thermal failures.

[11] Typical central office environment is air-conditioned with relative humidity between 40% and 60%. The average room temperature in the equipment aisles is 25°C, and the average temperature between circuit boards is 40°C.

Table 3.5. Summary of Thermal Analysis of CP330 Components

Device Type	Quantity on CP330	Junction Temperature, T_J	Ambient Temperature, T_A
Microprocessor, 30,001–50,000 gates	1	74	56
Integrated circuit, digital bipolar, 101–500 gates	14	92	59
Integrated circuit, digital bipolar, 51–100 gates	3	77	57
Integrated circuit, digital complementary metal oxide semiconductor (CMOS), 101–500 gates	26	61	50
Oscillator	2	42	40
Random access memory, static CMOS 2048K	11	43	40
Integrated circuit, digital bipolar, 21–50 gates	16	45	41
Integrated circuit, digital bipolar, 1–20 gates	57	42	40
Film resistor	86		42
Tantalum capacitor	10		38
Silicon diode, general purpose, < 1 A	3	39	39
Integrated circuit, digital CMOS, 30,001–50,000 gates	2	40	40
Ceramic capacitor	99		41

Table 3.6. CP330 Component Failure Rates from Field Tracking Data

Device Type	Confirmed Failures	Device Hours	Failure Rate (FITs)	
			MLE	95% UCL
Microprocessor, 30,001–50,000 gates	1	2.30E+07	43	206
Integrated circuit, digital bipolar 101–500 gates	5	3.22E+08	16	33
Integrated circuit, digital bipolar 51–100 gates	1	6.90E+07	14	69
Integrated circuit, digital CMOS, 101–500 gates	8	5.98E+08	13	24
Oscillator	2	4.60E+07	43	137
Random access memory, static CMOS 2048K	10	2.53E+08	40	67
Integrated circuit, digital bipolar 21–50 gates	5	3.68E+08	14	29
Integrated circuit, digital bipolar 1–20 gates	15	1.31E+09	11	18
Film resistor	1	1.98E+09	1	2
Tantalum capacitor	1	2.30E+08	4	21
Silicon diode, general purpose, < 1 A	1	6.90E+07	14	69
Integrated circuit, digital CMOS, 30,001–50,000 gates	1	4.60E+07	22	103
Ceramic capacitor	2	2.28E+09	1	3

number of confirmed failures. Column 3 shows the number of device hours. The last two columns show the failure rate MLE and associated 95% UCL (Lawless, 1982).

Using the data from Table 3.5, Table 3.6, and the activation energies from SR-332, we can now build a database table for hardware reliability prediction. This is Table 3.2 (Section 3.5).

3.7 COMPARING FIELD RELIABILITY RESULTS WITH PREDICTIONS

Let us suppose that the lifetime of the subassembly in operation has an exponential distribution with parameter (failure rate) λ_F. If we accept the reliability prediction model for the subassembly as described above, then our claim is that the life of the subassembly has an exponential distribution with parameter λ_P. If λ_P were a fixed, known quantity, the question of whether we did a good job in predicting the reliability of the subassembly could then can be cast simply as a test of the hypothesis H_0: $\lambda_F = \lambda_P$ against some suitable alternative. In the telecommunications industry, at least, users are most interested that predictions be conservative in the sense that field reliability should be at least as good as predicted. To capture this desire, we use the alternative H_1: $\lambda_F < \lambda_P$. Reliability data on the subassembly gathered from field operation usually comprises lifetime observations.[12] From these, we may also obtain the number of failed subassemblies (call it r) and the total number of subassembly-hours of operation (call this T). The standard maximum likelihood estimate of the subassembly field failure rate is then $\hat{\lambda}_F = r/(X_1 + \cdots + X_n) = r/T$, which has a chi-squared distribution with $2r$ degrees of freedom (Lawless, 1982). Then, when we take λ_P as fixed, it is straightforward to construct this hypothesis test. See Exercise 3.9 to further explore this simple case.

Our knowledge of λ_P is uncertain, however. We estimate λ_P from the reliability prediction model [either m from equation (3.9) or m^* from the development following equation (3.13), as appropriate].[13] Then $2\hat{\lambda}_P/\nu$ has approximately a chi-squared distribution with δ degrees of freedom in the parts count method; a corresponding statement holds for the starred variables in the parts stress method. The following simple methodology allows us to make good use of the dispersion information in the prediction. Let $I_p(x)$ denote the indicator function of accepting the null hypothesis for the significance level p (for example, $p = 0.05$) when the value of the predicted failure rate is x. That is, $I_p(x)$ is 1 if we accept the null hypothesis at significance level p when the predicted failure rate is x and is 0 if we reject the null hypothesis at significance level p when the predicted failure rate is x. Then the probability that we

[12]Often, direct observation of lifetimes is difficult. We will ignore those difficulties here. For a synopsis of some of the problems that can occur, see Baxter and Tortorella (1994).

[13]At this point we act as if the estimate we obtained from the prediction procedure were the value of the parameter for the subassembly's exponential distribution. This is the epistemology of the prediction: Even though the predicted failure rate is obtained as a statistical estimate, now when we want to use it as a prediction, we act as if this parameter were given to us by nature.

accept the null hypothesis at significance level p when the prediction includes dispersion according to the methodology described in Section 3.3 is

$$\int_0^{\infty} I_p(x) dP\{\hat{\lambda}_P \le x\} \qquad (3.17)$$

Exercise 3.9 asks you to work this out for the example subassembly from Exercise 3.4. Exercise 3.10, of a more theoretical nature, develops this hypothesis test further.

The subassembly reliability prediction UCL serves as a *quality of information* measure that indicates something about how well we can claim to know the reliability of the subassembly is going to turn out in operation. If the UCL is close to the prediction, this indicates that we have a right to expect that results from the field should be close to the prediction. If the UCL is far away from the prediction, this is an indication that our knowledge about the reliability of the subassembly is limited or of poor quality, and the realized field results may differ markedly from the prediction. Obviously, given the nature of the procedure, this stems from limited or poor quality knowledge about the component reliabilities (i.e., their failure rate estimates have a lot of dispersion). You may wish to experiment with equation (3.17) for different values of ν and δ with the same m to see how making the variance of $\hat{\lambda}_P$ larger or smaller affects the acceptance probability in equation (3.17).

3.8 IMPLEMENTATION

The parts stress procedure described in Section 3.4.2.2 has been made part of the standard reliability prediction methodology at Lucent Technologies. It is documented in the Lucent Technologies Reliability Information Notebook, an internal Bell Laboratories document. A corresponding CAD-linked program was developed that provides a highly automated platform for rapid reliability prediction, including precision estimates, for electronic subassemblies. This program also includes computations for both the maintenance viewpoint and the system availability viewpoint (see Section 3.3.2).

3.9 CONCLUSIONS

We have described a new, practical way of making visible the precision of hardware reliability predictions for subassemblies of electronic systems. Our contribution has been to distill from existing methods one that can be easily implemented in the parts count context where all components in an electronic subassembly operate at nominal temperatures. Then we extended the method to the parts stress domain where components operate at temperatures different from their nominal temperatures. The new procedure has been implemented in a CAD-linked program that provides a highly automated platform for rapid reliability prediction, including precision estimates, for electronic subassemblies. The heart of the CAD-linked program is a component reliability database that is populated with component failure rate estimates obtained from thermal simulation and field tracking data as described above. In order to populate the CAD-linked database with reliable data, we have demonstrated the importance of tracking an electronic subassembly throughout its product cycle from design (thermal simulation) to deployment (field returns).

REFERENCES

Ascher, H., and Feingold, H. (1984). *Repairable Systems Reliability: Modeling, Inference, Misconceptions, and Their Causes,* Marcel Dekker, New York.

Baxter, L. A. (1993). Towards a theory of confidence intervals for system reliability. *Statistics and Probability Letters* **16**:29–38.

Baxter, L. A., and Tortorella, M. (1994). Dealing with real field reliability data: Circumventing incompleteness by modeling and iteration, *Proceedings Annual Reliability and Maintainability Symposium,* pp. 255–262.

Butcher, A. C., Lampkin, H., and Winterbottom, A. (1978). Transformations improving maximum likelihood confidence intervals for system reliability, *Technometrics* **13**:467–473.

Chan, C. K. and Tortorella, M. (2001). Sparing inventory sizing for end-to-end service availability, *Proceedings Annual Reliability and Maintainability Symposium,* pp. 98–102.

Drenick, R. F. (1960). The failure law of complex equipment, *Journal of the Society for Industrial and Applied Mathematics* **8** 680–690.

El Mawaziny, A. H., and Buehler, R. J. (1967). Confidence limits for the reliability of a series system. *Journal of the American Statistical Association* **62:**1452–1459.

Grubbs, F. E. (1971). Approximate fiducial bounds for the reliability of a series system for which each component has an exponential time-to-fail distribution, *Technometrics* **13**:865–871.

Kraemer, H. C. (1963). One-sided confidence intervals for the quality indices of a complex item, *Technometrics* **5**:400–403.

Lawless, J. F. (1982). *Statistical Models and Methods for Lifetime Data,* Wiley, New York.

Lieberman, G. J. and Ross, S. M. (1971). Confidence intervals for independent exponential series systems. *Journal of the American Statistical Association* **66**:837–840.

Madansky, A. (1965). Approximate confidence limits for the reliability of series and parallel systems, *Technometrics* **7**:495–503.

Mann, N. R., and Grubbs, F. E. (1972). Approximately optimum confidence bounds on system reliability for exponential time to failure data, *Biometrika* **59**:191–204.

Mann, N. R., and Grubbs, F. E. (1974). Approximately optimum confidence bounds for system reliability based on component test data, *Technometrics* **16**:335–347.

Sarkar, T. K. (1971). An exact lower confidence bound for the reliability of a series system where each component has an exponential time to failure distribution, *Technometrics* **13**:535–546.

Telcordia (2001). *Reliability Prediction Procedure for Electronic Equipment* (SR-332), Telcordia Technologies, Issue 1.

EXERCISES

3.1. Interpret the chi-squared distribution when the number of degrees of freedom is not an integer.

3.2. Use the procedure in Section 3.4.2.1 to show that the two-sided 95% confidence interval is given by

$$\left[\frac{\nu}{2m}\chi^2(\delta, 0.025), \frac{\nu}{2m}\chi^2(\delta, 0.975)\right].$$

3.3. Derive Equation (3.14),

$$V_i^* = \frac{\chi^2(2A_i w_i, 0.95)}{\chi^2(2w_i, 0.95)} V_i$$

for the UCL at altered temperature. [*Hint:* Note that this is in the context of an accelerated life model (Lawless, 1982).]

3.4. An electronic subassembly contains components listed in the following table. Use the database in Table 3.2 and the parts count procedure in Section 3.4.2.1 to compute the hazard (failure) rate point estimate, one-sided 95% UCL, and two-sided 95% confidence interval.

Part No.	Device Type	Quantity
MC	Microprocessor, 30,001–50,000 gates	1
PAL	Integrated circuit, digital CMOS, 101–500 gates	6
OSC	Oscillator	2
SRAM	Random access memory, static CMOS 2048K	2
RES	Film resistor	50
CAP	Ceramic capacitor	50

3.5. At the design stage of the electronic subassembly in Exercise 3.4, a thermal analysis is performed to determine the operating temperatures of the components. The results of the thermal analysis are given in the following table. Use the database in Table 3.2 and the parts stress procedure in Section 3.4.2.2 to compute the hazard (failure) rate point estimate, one-sided 95% UCL, and two-sided 95% confidence interval.

Part Number	Device Type	Reference Designation (Location on the Subassembly)	Use (Operating) Temperature (°C)
MC	Microprocessor, 30,001–50,000 gates	IC1	85
PAL	Integrated circuit, digital CMOS, 101–500 gates	IC2.1	75

Part Number	Device Type	Reference Designation (Location on the Subassembly)	Use (Operating) Temperature (°C)
PAL	Integrated circuit, digital CMOS, 101–500 gates	IC2.2	65
PAL	Integrated circuit, digital CMOS, 101–500 gates	IC2.3	60
PAL	Integrated circuit, digital CMOS, 101–500 gates	IC2.4	68
PAL	Integrated circuit, digital CMOS, 101–500 gates	IC2.5	58
PAL	Integrated circuit, digital CMOS, 101–500 gates	IC2.6	56
OSC	Oscillator	Y1	43
OSC	Oscillator	Y2	45
SRAM	Random access memory, static CMOS 2048K	IC3.1	48
SRAM	Random access memory, static CMOS 2048K	IC3.2	48
RES	Film resistor	Distributed across the subassembly in 50 locations	40
CAP	Ceramic capacitor	Distributed across the subassembly in 50 locations	40

3.6. To implement the procedures discussed in Section 3.4 requires a careful database design. The database must have one or more tables that contains the following fields: Part Number, Manufacturer, Device Type, Failure Rate Point Estimate, 95% UCL, Reference Temperature, and Activation Energy. How many tables are needed to facilitate frequent update of the reliability information? What are the fields in these tables?

3.7. We note that some reliability engineers believe that a confidence limit for the subassembly failure rate can be obtained by summing the corresponding confidence limits of the constituent component failure rates. Use this incorrect method to rework Exercise 3.2 and comment on the result.

3.8. At the deployment stage of the electronic subassembly in Exercise 3.4, 3000 such subassemblies are in the field for 2 years. A careful field tracking study shows that there are two confirmed oscillator failures during this time. Discuss how the new data can be used to update Table 3.2.

3.9. Five thousand electronic subassemblies like those in Exercise 3.4 were operated in the field for 4 years, and 38 failures were recorded. Take the predicted failure rate from Exercise 3.4 as deterministic. Can we say that these results are consistent with the prediction at the .05 significance level?

3.10. Repeat Exercise 3.9 with the dispersion information on the estimated failure rate prediction included.

CHAPTER 4

Allocation of Dependability Requirements in Power Plant Design

Seppo Virtanen and Per-Erik Hagmark

4.1 INTRODUCTION

A power plant represents a complex industrial product whose design and manufacturing are carried out by a wide subcontractor network. The study case is the Wärtsilä's diesel power plant, and the main problem is to allocate quantitative dependability requirements from the diesel power plant level down to a single diesel engine and the fuel booster unit. A basic idea is that design should not be started until one knows the specific requirements for these design entities [see IEC 300-3-4 (1996)]. In this chapter we present a method that enables an effective allocation of dependability requirements to the design entities of a product in explicit form. The general term "design entity" can stand for, say, function, system, equipment, mechanism, or any kind of part.

The case study we will present is developed as a part of the research and product development program *Competitive Reliability* [see TEKES (2001)]. The allocation of dependability is carried out based on a fault tree approach that characterizes the design entities and their causal interrelations that lead to the "TOP," an event that represents the failure of the power plant. From the requirements specified by the customer and the manufacturer, we derive failure time and repair time distributions for the TOP. These distributions are allocated to the design entities using stochastic simulation. The allocation takes into account the technical complexity and importance of design entities. The approach to allocation is shown in Figure 4.1.

The structure of the chapter is as follows: In Section 4.2, we discuss the system characterization for modeling of a diesel power plant. Section 4.3 examines modeling of requirements set for TOP from the customer and manufacturer perspectives. In Section 4.4, we allocate requirements at the first level of the power plant fault

Case Studies in Reliability and Maintenance, Edited by W. R. Blischke and D. N. P. Murthy.

ISBN 0-471-41373-9

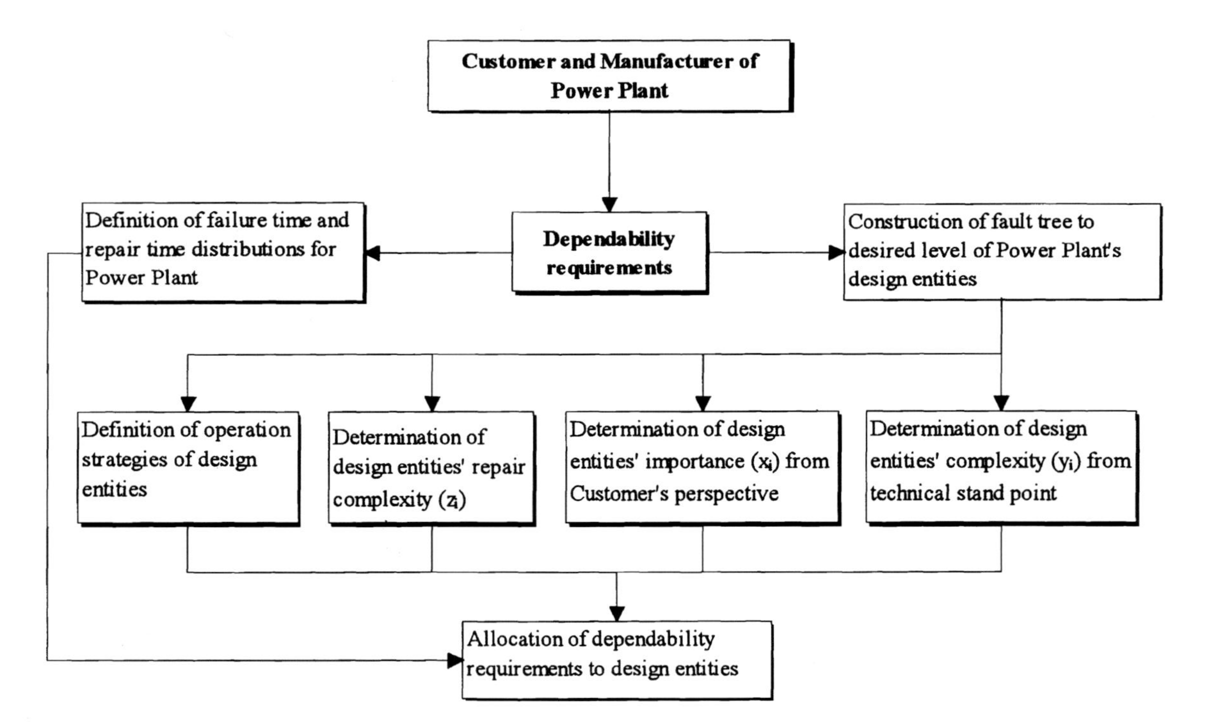

Figure 4.1. Principle of allocation of quantitative dependability requirements.

tree. Section 4.5 deals with continuous application of the allocation method to lower levels in the fault tree. Section 4.6 deals with interpretation of the results.

4.2 SYSTEM CHARACTERIZATION

Wärtsilä's modular diesel power plant projects are typically privately owned and operated. High efficiency, fuel flexibility, and proven design solutions are some of the key factors for business success. Today the life-cycle focus is stronger than ever. The best power plants are built with close cooperation between the supplier (of the plant) and the customer (owner of the plant). The customer buys five diesel engines (Wärtsilä 18V32) with the necessary auxiliary systems. The diesel power plant is a base load standard power plant for the production of power and heat. The total output of the five diesel engines is 31.5 MW of electricity and 10,000 kWh of heat recovery. The plant is controlled by an "extended" automation system (semiautomatic) that provides versatile monitoring capabilities using PLC and PC systems. Heavy fuel oil (HFO) is the main fuel, and light fuel oil (LFO) is used as standby and backup fuel.

Figure 4.2 is a fault tree representation [see Blischke and Murthy (2000)] of the diesel power plant. The plant has a hierarchical structure that consists of design entities at three levels. The state (working, failed and waiting for repair, failed and under repair, or repaired and waiting for start) of a design entity at any level is related to the state of the design entities at the lower level through logical gates and additional interrelations. A k/n gate (a design entity) has n inputs (lower-level design entities) and corresponds to the situation where the gate fails if at least k of its inputs fails simultaneously. Gates 25 (Energy Conversion) and 30 (Fuel Transfer) are of Type $k = 2$, since they fail only if (at least) two of their inputs fail simultaneously, and all other gates are of Type $k = 1$, since they fail immediately if one of the inputs fails (OR-gates). The type of the gate is indicated by [k] in the right corner of the box—except Type 1 gates, which do not have any such indicator. We start with dependability requirements at the TOP and then proceed downwards to look at the allocation for "Fuel Booster" (31) and "Eng A" (3), gate by gate.

4.3 MODELING DEPENDABILITY AND REQUIREMENTS

The treatment of the actual power plant case (Figure 4.2) begins in Section 4.3.2. First we need to specify basic definitions and the model used for dependability. Figure 4.3 shows the history of failures for an object (system or component at any level) and the repair times to correct each failure. The probability of failure is a function of the *age,* which is the cumulative running time and not calendar time. As such, repair or other stopping times will not be included in age, that is, the intervals between the failures consist entirely of running time. Repair time, which includes inspection, testing, and rectification, will be modeled independently of time to fail-

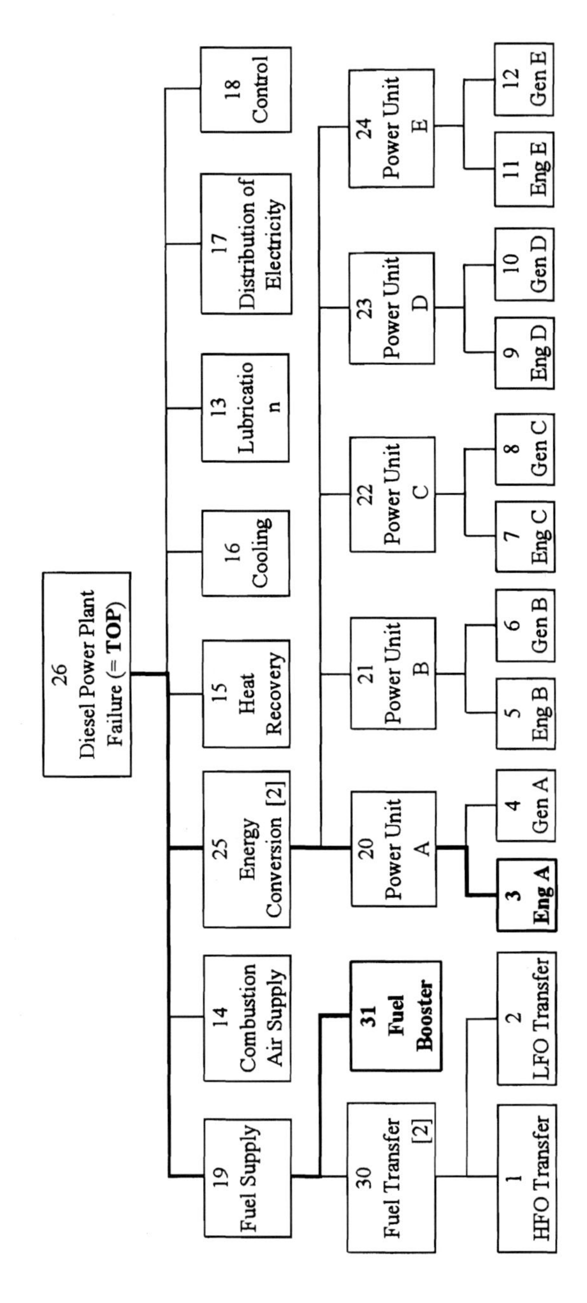

Figure 4.2. Fault tree of diesel power plant.

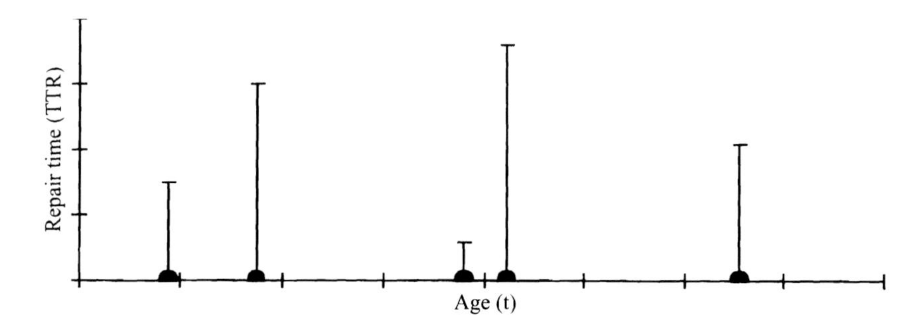

Figure 4.3. A model for dependability.

ure. No assumptions on the duration of repair will be made, and the only assumption on the qualitative type of failures is that a failure can be repaired.

4.3.1 Simulating Failures and Repair Time

The tendency of an object to fail is allowed to depend on its age. Let $\Lambda(t)$ denote a continuous function monotonically increasing with $\Lambda(0) = 0$ and approaching infinity as the age t approaches infinity. A *failure tendency* defined by $\Lambda(t)$ means that the expected number of failures during the age interval $(L, L']$ is given by the difference $\Lambda(L') - \Lambda(L)$. Thus, if we denote the probability of exactly n failures in the interval $(L, L']$ by $P(n, L, L')$, the following identities hold:

$$P(N, 0, L') = \sum_{n=0}^{N} P(N - n, 0, L) \cdot P(n, L, L') \tag{4.1}$$

$$\Lambda(L') - \Lambda(L) = \sum_{n=0}^{\infty} n \cdot P(n, L, L') \tag{4.2}$$

for $0 < L < L'$. A well-known solution of equations (4.1) and (4.2) is given by

$$P(n, L, L') = \frac{(\Lambda(L') - \Lambda(L))^n}{n!} \cdot e^{\Lambda(L) - \Lambda(L')} \tag{4.3}$$

which means that the number of failures in the interval $(L, L']$ obeys the Poisson distribution with parameter $\Lambda(L') - \Lambda(L)$. This will be our choice.

For simulation purposes, the failures must also be viewed as random points $L_1, L_2, \ldots$ over age, and they need to be modeled by a suitable stochastic point process formulation such that equation (4.3) holds. The reliability of an object of age L is the probability that the object will not fail during the subsequent age interval $(L, L + t]$:

$$R_L(t) = P(0, L, L + t) = e^{\Lambda(L) - \Lambda(L+t)} \tag{4.4}$$

If T denotes the corresponding random variable—that is, the running time to next failure—then the random variable $R_L(T)$ is uniformly distributed on the interval [0,1] [see Rubinstein (1981)]. Thus, if L_n and $L_{n+1} = L_n + T$ denote the ages at two consecutive failures, then

$$L_{n+1} = \Lambda^{-1}[\Lambda(L_n) - \ln(U)] \tag{4.5}$$

where U is the uniformly distributed random variable on the interval [0, 1]. The simulation formula (4.5) defines a point process realization of equation (4.3), a nonhomogeneous Poisson process (NHPP) with intensity function $\Lambda'(t)$ if Λ is differentiable [see Gertsbakh (2000) and Virtanen and Hagmark (1997)]. The use of the failure tendency function Λ (instead of R_L) in the simulation formula (4.5) is motivated by better numerical interpolation stability.

Repair time is a non-negative random variable modeled independently of the failure tendency by a cumulative probability function $G(x)$. For simulation, we will use the formal counterpart to the failure tendency function Λ, the "repair tendency" function $V(x) = -\ln(1 - G(x))$. Hence the simulation of (age-independent) repair time takes the form TTR $= V^{-1}(-\ln(U))$.

Let us finally discuss some simple consequences of our model. The L-parameter family of probability distributions for failure tendency defined by equation (4.4) is contained in the member $L = 0$, the probability distribution of age at first failure:

$$R_L(t) = \frac{R_0(L+t)}{R_0(L)}, \qquad R_0(t) = e^{-\Lambda(t)} \tag{4.6}$$

Formula (4.6) yields an immediate interpretation. The failure tendency of a repaired object is the same as that of a corresponding nonfailed object of the same age. That is, the failure tendency of a repaired object is statistically the same as that of the object just before it failed; in other words, repair only restores the failed object to its operational state and does not make the object better or worse than before. The model can be criticized from a practical point of view, as the failure tendency depends only on the age and not the past number of failures. (A complete model in which the failure tendency depends on past history of failures would be more complex.) However, we think our model constitutes a fair average of many cases in practice. Some exceptions can also be handled using "steering rules" in our simulation algorithm [see Hagmark (2001)].

4.3.2 Requirements for TOP

The *dependability requirements* of the customer (and/or the manufacturer) can be stated in terms of the age of the object and can involve number of failures, time between failures, reliability as a function of age, or data concerning first failure. Our model is flexible from this point of view, since it is always closely connected to failure tendency $\Lambda(t)$, which we are going to derive. Data concerning the number of failures correspond directly to $\Lambda(t)$, the failure rate or MTTF corresponds directly to

$\Lambda'(t)$ ($\equiv d\Lambda(t)/dt$), and reliability data corresponds directly to the family (4.4) and formula (4.6).

We next derive the failure tendency function and the repair time distribution for power plant TOP 26 (Figure 4.2). This requires appropriate numerical data, so that the selection of the parameters of the probability functions can be carried out in an adequate manner. Let the eight dependability requirements (agreed to by the customer and the manufacturer) for the system as a whole be as follows:

Age at the end of burn-in period	$t_a = 180$ days
Age at the end of warranty period	$t_b = 730$ days
Age at the end of useful life period	$t_d = 7300$ days
Length of age interval (for reliability)	$t_c = 30$ days
Reliability in an age interval of length t_c in period $(t_a, t_d]$: at least	Rel = 0.8
A parameter for warranty period ($s > 0$, described below)	$s = 1.2$
Mean active corrective maintenance time: at most	MACMT = 0.333 days
Maximum active corrective maintenance time: at most	$\text{ACMT}_{95} = 1$ day

Active repair time includes fault diagnosis, repair, and test. The time unit *day* is by definition 24 hours running time or active repair time.

The requirements expressed by Rel and s will now be written as inequalities for the failure tendency. First, the reliability in an interval of length t_c should be at least Rel during the period $(t_a, t_d]$. That is, using formula (4.4) we have equivalently an upper bound for the average number of failures:

$$\Lambda(L + t_c) - \Lambda(L) \leq -\ln(\text{Rel}) \qquad \text{for } t_a \leq L \leq t_d - t_c \tag{4.7}$$

Second, the expected number of failures in the warranty period $(0, t_b]$ must not be more than s times the expected number of failures in an equally long post-warranty interval $(t_b, 2t_b]$. This translates into

$$\frac{\Lambda(t_b)}{\Lambda(2 \cdot t_b) - \Lambda(t_b)} \leq s \tag{4.8}$$

The parameter s is related to the product's complexity and the level of a customer's readiness to accept more ($s > 1$) or fewer ($s < 1$) failures during the warranty period as opposed to the post-warranty period. The more complex a product is to implement technically, the more probably it can be allowed to fail during the warranty period compared to the post-warranty period. Further discussion on product complexity is given in Section 4.4.1.

4.3.3 Allowed Failure Tendency for TOP

The failure tendency functions Λ will in general be treated as dense interpolation tables, without any assumptions on the analytical form. The only exception is the

TOP, for which we use the following analytical formulation as a model for the requirements defined in Section 4.3.2:

$$\Lambda(t) = \begin{cases} \lambda \cdot (t - t_b + s \cdot t_b) + \lambda \cdot t_b \cdot (1 - s) \cdot \left(1 - \dfrac{t}{t_b}\right)^{\delta} & \text{if } t \le t_b \\ \lambda \cdot (t - t_b + s \cdot t_b) + \dfrac{\lambda_m - \lambda}{1.02} \cdot \dfrac{t_d - t_c - t_b}{\gamma} \cdot \left(\dfrac{t - t_b}{t_d - t_c - t_b}\right)^{\gamma} & \text{if } t > t_b \end{cases} \tag{4.9}$$

$$\Lambda'(t) = \begin{cases} \lambda + \lambda \cdot (s - 1) \cdot \delta \cdot \left(1 - \dfrac{t}{t_b}\right)^{\delta - 1} & \text{if } t \le t_b \\ \lambda + \dfrac{\lambda_m - \lambda}{1.02} \cdot \left(\dfrac{t - t_b}{t_d - t_c - t_b}\right)^{\gamma - 1} & \text{if } t > t_b \end{cases} \tag{4.10}$$

with γ and $\delta > 1$. The requirement parameters t_a, t_b, t_c t_d, s, and Rel are already built into the model. The failure intensity model (4.10) for TOP can take both a bathtub form (if $s > 1$) and a nonbathtub form (if $s < 1$). The constant 1.02 is an extra constant which affects the shape in the period $t > t_b$ and must be (and has been) selected before adjusting the parameters λ, δ, γ. The parameter δ affects the shape in the period $t < t_b$, and γ is concerned with the acceleration of the failure tendency in the period $t > t_b$, often assumed to take place as a result of, for example, material fatigue.

The application to our TOP begins by choosing the *design parameter* $\lambda = \Lambda'(t_b)$. Considering (4.7), we find that an absolute upper bound for λ is given by

$$\lambda_m = \frac{-\ln(\text{Rel})}{t_c} = 0.007438 \text{ per day} \tag{4.11}$$

so we can select $\lambda = 0.0065$ per day, say. This choice is based on experience from power plants of a similar type. Once λ has been chosen by the designer, the requirements (4.7) and (4.8) lead to unique values for the shape parameters, namely, $\delta = 9.224$, $\gamma = 9.619$. These shape parameter values are *optimal* in the sense that they satisfy inequality (4.8) with strict equality and satisfy inequality (4.7) with strict equality somewhere in the period $(t_a, t_d]$. [A search routine for obtaining δ and γ can be found in Hagmark (2001).] The final failure tendency of TOP is given in Figure 4.4.

Note that the model (4.9) provides some degrees of freedom. With this, the designer may still wish to adjust some other details, staying of course on the safe side of requirements. So it is possible to decrease s and/or increase Rel. The parameter δ can also be changed. Generally, the optimal δ can be increased if $s > 1$ and decreased if $s < 1$, resulting in a true inequality in (4.8). The optimal value $\delta = 9.224$ obtained above means that 93% of the failure tendency, which is planned to be eliminated during the whole warranty period $(0, t_b]$, will disappear already in the burn-in period $(0, t_a]$ as

$$\frac{\Lambda(t_a) - \lambda \cdot t_a}{\Lambda(t_b) - \lambda \cdot t_b} = 0.927 \tag{4.12}$$

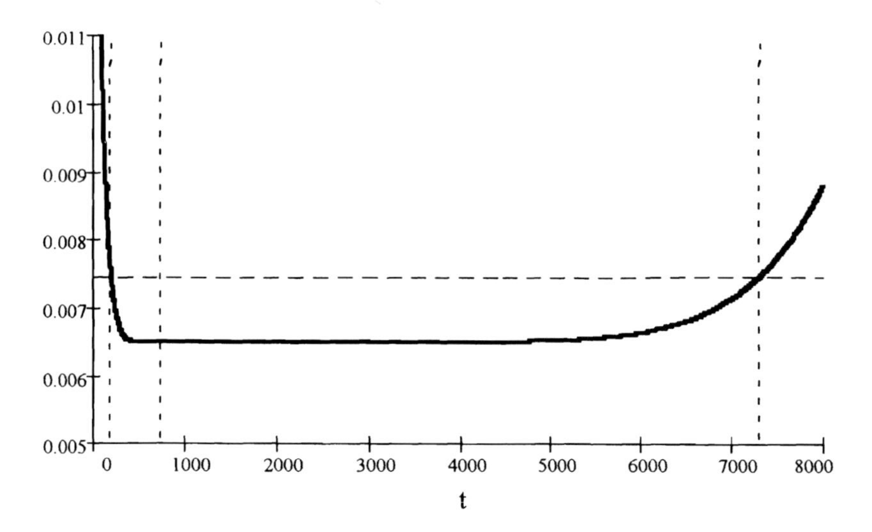

Figure 4.4. Allowed failure intensity $\Lambda'(t)$ for TOP.

The value of δ affects the percentage in equation (4.12). Since $s = 1.2$, δ can be increased. Hence, the more rapidly one plans to eliminate design, manufacturing, and installation flaws and to accumulate operation experience during the burn-in period $(0, t_a]$, the higher the value of the shape parameter δ.

4.3.4 Allowed Repair Time for TOP

The repair time distributions will in general be handled as dense interpolation tables, but for TOP we construct an analytic cumulative probability distribution that describes the TOP requirements for repair time given in Section 4.3.2, namely, MACMT = 0.333 days and ACMT_{95} = 1.0 day. The Weibull family $G(x) = 1 - \exp[-(x/\alpha)^\beta]$ is flexible enough for this purpose, since reasonable values for the Weibull parameters can be found if $1.02 \le \text{ACMT}_{95}/\text{MACMT} \le 4.67$. (The reader can verify this by elementary calculation.) The mean and 95th percentile of the Weibull distribution are

$$\text{MACMT} = \alpha \cdot \Gamma\left(\frac{1+\beta}{\beta}\right) \qquad \text{and} \qquad \text{G}(\text{ACMT}_{95}) = 0.95 \tag{4.13}$$

The corresponding values of the Weibull parameters, obtained by solving equation (4.13), are $\alpha = 0.3325$ and $\beta = 0.9964$. Thus the distribution of repair time TTR is

$$\text{G}(\text{TTR}) = 1 - e^{-\left(\frac{\text{TRR}}{0.3325}\right)^{0.9964}} \tag{4.14}$$

4.4 ALLOCATION OF REQUIREMENTS

The allocation of our power plant starts at the TOP level of the fault tree (Figure 4.2). First we define qualitatively the concepts of "importance" and "complexity" in Section 4.4.1. Then we quantify these for TOP and its input design entities (DEs), which are seen from Figure 4.2 to be the set $\{I\} = \{19, 25, 13, 14, 15, 16, 17, 18\}$ (Sections 4.4.1 and 4.4.2). This leads to the determination of *allocation coefficients* w_i, z_i for the input DEs, describing in relative terms the allowed failure tendency and the ratios between repair time of the input DEs (Section 4.4.3). The calculations are performed using authentic data for the power plant. Next the failure tendency function Λ and the cumulative probability distribution for repair time G, constructed in Sections 4.3.3 and 4.3.4 for the power plant TOP, will be transferred to corresponding functions for the input DEs. The portions are defined by the *dependability allocation principles*:

$$\Lambda_i(t) = \alpha \cdot w_i \cdot \Lambda(t), \qquad G_i(x) = G(x)^{\beta \cdot z_i} \tag{4.15}$$

Numerical experience shows that the allocation principle (4.15) is applicable for arbitrary k/n gates by appropriate choice of the *level parameters* α and β. The values of the level parameters are to be set by the designer to meet TOP requirements [equations (4.7) and (4.8)] using a "trial and error" method. The success of the choice of level parameters is immediately measured by simulation of the fault tree all the way up to TOP. This simulation/iteration process is described in Section 4.4.4.

4.4.1 Importance and Complexity

In the allocation of failure tendency, at least two aspects have to be recognized. These are "importance" (in some sense) from the customer's perspective and "complexity" (in some sense) from the technical standpoint. Heuristically speaking, the greater consequences a failure of a DE leads to, the more important this DE will be from the customer's point of view. Accordingly, the more rarely this DE is allowed to fail, that is, a high reliability should be allocated to it. And the more complex a DE is to implement technically, the more probable that it will cause the product's failure, and hence the lower the reliability that should be allocated to it.

It is often necessary to first decompose importance and complexity into several factors and to select these factors very carefully. Importance and complexity can in general be both case and DE (gate) specific. In our power plant, we need only case specificity, that is, we use the same factors for all gates. These are *importance factors*:

D = property damage
E = environmental damage
F = human damage
G = business damage

and *complexity factors*:

A = number of "parts"
B = level of human activities
C = level of "state-of-the-art"

The importance factors D, E, F, G represent possible damage types caused by failure. However, if the gate is not an OR-gate, then a general conceptual problem arises with this choice of importance factors. This question will be discussed in Section 4.5. The complexity factors A, B, C reflect the implementation of a product's technology. For example, the "state-of-the-art" factor C takes into consideration the present engineering progress in relevant fields, related to the design and development phase of the product. An aim in selecting the complexity factors has been to avoid bias and inconsistency.

Next the designer should assign relative weights (positive real numbers) to the selected importance factors according to the probabilities with which different types of damage can occur. The corresponding weights (positive real numbers) for the complexity factors again are estimates of how much complexity each factor represents. Our choice of weights for the power plant TOP 26 (Figure 4.2) will be

$$u = \begin{bmatrix} D \\ E \\ F \\ G \end{bmatrix} = \begin{bmatrix} 4 \\ 2 \\ 1 \\ 10 \end{bmatrix}, \qquad v = \begin{bmatrix} A \\ B \\ C \end{bmatrix} = \begin{bmatrix} 3 \\ 1 \\ 5 \end{bmatrix} \tag{4.16}$$

For example, we think that the power plant includes relatively much new technology and know-how (C factor) and that business damage (G factor) is the most probable consequence of a possible failure. The assessment of the weights is in practice a very intuitive process, during which experience-based simple pair or other comparisons can be very helpful. The choice of weights can, of course, be done separately for each gate to be allocated.

4.4.2 Relative Importance and Complexity for Inputs

The role of an importance or complexity factor in a DE (gate) is naturally a function of the TOP input DEs $\{I\}$. The relative importance of each of these must be assessed. The *importance matrix* X and the *complexity matrix* Y summarize the weights for TOP of the power plant. These were determined to be

$$X = \left[\begin{array}{c|c|c|c|c|c|c|c|c} \text{Importance} & 19 & 25 & 13 & 14 & 15 & 16 & 17 & 18 \\ \hline D = 4 & 1 & 10 & 1 & 2 & 2 & 1 & 3 & 2 \\ E = 2 & 10 & 5 & 5 & 5 & 2 & 1 & 1 & 1 \\ F = 1 & 1 & 5 & 1 & 1 & 2 & 1 & 10 & 1 \\ G = 10 & 3 & 10 & 2 & 2 & 1 & 1 & 10 & 10 \end{array}\right] \tag{4.17}$$

and

$$Y = \left[\begin{array}{c|c|c|c|c|c|c|c|c} \text{Complexity} & 19 & 25 & 13 & 14 & 15 & 16 & 17 & 18 \\ \hline A = 3 & 20 & 100 & 10 & 20 & 25 & 5 & 30 & 25 \\ B = 1 & 2 & 10 & 2 & 2 & 2 & 1 & 5 & 10 \\ C = 5 & 1 & 10 & 1 & 1 & 1 & 1 & 3 & 5 \end{array} \right] \tag{4.18}$$

Each row of equation (4.17) corresponds to a certain type of damage and the row values weight the input DEs against each other according to the amount of damage they cause in case of failure. For example, DE 19 causes a larger environmental damage (factor E) than other input DEs, if it fails. Likewise, each row of equation (4.18) corresponds to a complexity factor and the row values weight the input DEs against each other on this scale. For example, DE 25 (energy conversion) includes more new technology (factor C) and know-how than other input DEs.

The *importance coefficients* x_i for the inputs of TOP can now be computed. The most natural and simplest way is to define the coefficient vector as the u-weighted average of the normed horizontal rows of equation (4.17); likewise, the v-weighted average of the normed horizontal rows of equation (4.18) yields the set of *complexity coefficients* y_i, where u and v are given by equation (4.16), that is,

$$x_i = \sum_{k=1}^{4} \left[\frac{u_k}{\sum_{j=1}^{4} u_j} \cdot \frac{X_{ki}}{\sum_{i=1}^{8} X_{ki}} \right], \qquad y_i = \sum_{k=1}^{3} \left[\frac{v_k}{\sum_{j=1}^{3} v_j} \cdot \frac{Y_{ki}}{\sum_{i=1}^{8} Y_{ki}} \right], \qquad i \in \{I\} \tag{4.19}$$

The final result is

$$\left[\begin{array}{c|c|c|c|c|c|c|c|c} \text{DE} & 19 & 25 & 13 & 14 & 15 & 16 & 17 & 18 \\ \hline \text{Importance } (x) & 0.098 & 0.291 & 0.063 & 0.074 & 0.050 & 0.032 & 0.214 & 0.179 \\ \text{Complexity } (y) & 0.059 & 0.416 & 0.045 & 0.059 & 0.066 & 0.035 & 0.131 & 0.189 \end{array} \right] \tag{4.20}$$

From (4.20), we see that DE 25 (energy conversion) is the most important and also the most complex entity at the first level below TOP.

Comment. Experience from our other case studies has shown that the determination of importance proceeds very naturally even when there is very little data available on the product being designed [see Virtanen (2000)]. Furthermore, no conflicts have emerged in the determination of coefficients.

4.4.3 The Allocation Coefficients

The importance and complexity coefficients x, y defined above will now be combined to form the final *failure tendency allocation coefficients* w_i needed in equation (4.15). The aim is that the more important an entity is (x high), the less it is allowed to fail, and the more complex it is (y high), the more it is allowed to fail. A reasonable, simple way to implement these effects is the following:

$$w_i = \frac{\dfrac{y_i}{(x_i)^\varepsilon}}{\displaystyle\sum_i \frac{y_i}{(x_i)^\varepsilon}}, \qquad \varepsilon \geq 0, \qquad i \in \{I\} \tag{4.21}$$

The exponent $\varepsilon = 0$ is quite often used [see Kececioglu (1991)], and it means that consequences of failure will not be considered at all. Another choice, $\varepsilon = 1$ [see Virtanen (2000)], gives importance too strong a role from a practical point of view in some applications. As a consequence of the resulting allocation of reliability (failure tendency), one may, for example, arrive at such a strict requirement for a DE's reliability that it becomes technically and economically impossible to attain. This may be the case with a DE whose importance to the customer is greater than for other DEs. (The consequences of a failure are extensive.) A compromise $0 < \varepsilon < 1$ would ease this critical feature.

The need for a compromise exponent is indeed apparent when we compute the failure tendency coefficients for the first-level DEs of our power plant, given by the set $\{I\}$. The effect of the exponent (ε) in equation (4.21) can be seen by a comparison of the following three alternatives:

$$\begin{aligned}
w &= (0.059 \quad 0.416 \quad 0.045 \quad 0.059 \quad 0.066 \quad 0.035 \quad 0.131 \quad 0.189) \qquad \text{if } \varepsilon = 0 \\
w &= (0.073 \quad 0.300 \quad 0.069 \quad 0.084 \quad 0.115 \quad 0.074 \quad 0.110 \quad 0.173) \qquad \text{if } \varepsilon = 0.5 \\
w &= (0.079 \quad 0.188 \quad 0.093 \quad 0.105 \quad 0.175 \quad 0.140 \quad 0.081 \quad 0.139) \qquad \text{if } \varepsilon = 1
\end{aligned} \tag{4.22}$$

The energy conversion (DE 25, 2nd entry in w) is the most important entity and also technically the most complex, as seen in (4.20). The choice $\varepsilon = 0$ would completely eliminate the effect of possible damages on the allocated failure tendency, which in this case would be unreasonable. We think the failure tendency becomes too high. On the other hand, by choosing $\varepsilon = 1$ we would make the failure tendency allocated to DE 25 too low and therefore economically too demanding from the customer point of view. Hence the choice $\varepsilon = 0.5$ seems to offer a more balanced solution.

The *repair time allocation coefficients* z_i describe directly and proportionally the repair time for the DEs belonging to the set $\{I\}$. The coefficient vector has been assessed as follows: $z = (4, 10, 4, 3, 3, 3, 2, 1)$ or normalized

$$z = (0.133 \quad 0.333 \quad 0.133 \quad 0.100 \quad 0.100 \quad 0.100 \quad 0.067 \quad 0.033) \tag{4.23}$$

The greater the z-coefficient of a DE, the longer the repair time that will be allocated to it. For example, the repair time for DE 25 (energy conversion) has been estimated to be 10 times the repair time for DE 18 (control). Note that the vector z can be assessed independently of the vector w and that the repair time of a DE is not necessarily associated with how often the DE will fail.

4.4.4 Operation Strategy and the Simulation/Iteration Procedure

The level parameters α, β in the allocation principle (4.15) are still undefined. The following general instructions for the choice can be given. In case of an OR-gate, the natural first tries would be $\alpha = 1$ and $\beta \gg 1$. The value $\beta = 1$ again corresponds approximately to an AND-gate, where all input DEs are being repaired simultaneously and the repair is defined to be finished when all input DEs are finished. Thus, choosing initial values for α and β we define tentative probability distributions of the input DEs $\{I\}$. These distributions are then tested by simulation.

Our simulation procedure [see Hagmark (2001)] can consider arbitrary fault logic consisting of k/n-gates. Generality to this extent is not necessary for allocation of the actual TOP level, but it is needed later (Section 4.5) when we allocate gates deeper in the fault tree, because simulation is always done up to the (upper) TOP and not to the gate under allocation. In addition to the fault logic, there are other features that affect dependability. One of these is the *operation strategy* based on extra interrelations between TOP and DEs. We define the following three DE-specific strategies:

The DE cannot be repaired if TOP is running: $a = 1$, otherwise $a = 0$.
The DE is not running if TOP is not running: $b = 1$, otherwise $b = 0$.
TOP will not be started if the DE is in failed state: $c = 1$, otherwise $c = 0$.

Recall that our model for failure tendency (Sections 4.3 and 4.3.1) uses age as the basic variable, but the age of a DE does not increase if the DE is not running. So to have consistency with the mathematical model, the simulation must observe when a DE or TOP is not running or cannot be started although it is logically non-failed.

The operation strategies are, in principle, defined independently of the fault logic, but they will of course be applied only if the fault logic gives room. The operation strategies for the DEs belonging to the set $\{I\}$ are as follows:

DE	19	25	13	14	15	16	17	18
a	0	0	0	0	0	0	1	1
b	0	1	1	1	1	1	1	1
c	1	1	1	1	1	1	1	1

(4.24)

Since TOP happens to be an OR-gate, only the b-rules are relevant. DE 19 (with $b = 0$) continues to run during the repair of TOP. The other (nonfailed) DEs do not. The other strategies will in this case be overruled by the simple OR-logic.

The simulation generates failure instants and repair times for the input DEs, considers fault logic and operation strategies, and continuously documents the age and state of the DEs and the TOP. The simulation is repeated $N = 500$ times. From this, pseudoempirical probability distributions for the dependability of TOP are obtained. If these distributions satisfy TOP requirements, the level parameters α and β are accepted. If not, we select new values and repeat the process. After a few iterations we arrived at $\alpha = 1.0$ and $\beta = 7.3$. Figure 4.5 shows the comparison between simulated estimates and required dependability requirements. Note that the curves for the required and the simulated failure tendency are almost identical, which is typical for the first-level allocation with OR-gate.

4.4.5 Results of the TOP Allocation

In the simulation, probability functions representing the allocated requirements for the first level DEs $\{I\}$ are stored as interpolation tables. A short summary of simulation results is presented in Table 4.1. The specified requirements for the power plant are shown in the first row of the Table 4.1. When we examine the DE's mean time to failure and active corrective maintenance time, we may note that the allocation principle (4.15) guides requirements fairly evenly from the perspective of the DE's importance and complexity. The proportion of number of failures follows precisely the proportion of allocation coefficients, given in equation (4.22). Table 4.1 shows further that the repair time obtained for input DEs is systematically determined in accordance with the coefficients constructed for repair time, given in equation (4.23). Note also that the operation time for DE 19 is longer than that for the TOP. This follows from the selected operation strategy b, according to which DE 19 is running although the TOP is not.

Checking requirements
(for TOP 26)

	Simulated	Required
$R[t_a \ldots t_a+t_c]$	0.809	0.800
$R[t_d-t_c \ldots t_d]$	0.805	0.800
MACMT	0.329	0.333
$ACMT_{95}$	1.028	1.000
parameter s	1.2	1.20
availability (A)	0.9978	≥ 0.99

Note! If the results are not satisfactory, adjust the level parameters α and/or β

Number of failures / running time

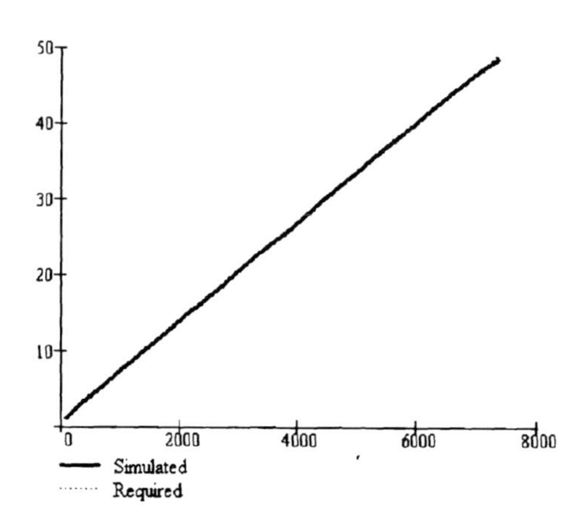

Figure 4.5. Verification of requirements (TOP 26).

Table 4.1. Dependability Requirements to the First Level of Case Fault Tree

ID No	Design Entity	Operations Time [days]	No. of Failures	Repair Time [days]	MTTF [days]	Reliability (0–180)	Reliability (0–730)	Reliability (180–210)	No. of Failures (0–180)	No. of Failures (0–730)	MACMT [days]	$ACMT_{95}$ [days]
26	Power plant	7300	48.9	16.14	149.2	0.1288	0.0033	0.8000	2.06	5.75	0.333	1.00
19	Fuel supply	7315	3.5	1.15	2102.0	0.8610	0.6570	0.9821	0.15	0.42	0.328	0.99
25	Energy conversion	7300	14.7	8.20	497.6	0.5401	0.1798	0.9349	0.62	1.72	0.553	1.29
13	Lubrication	7300	3.4	1.08	2167.5	0.8688	0.6738	0.9831	0.14	0.40	0.328	0.99
14	Combustion Air supply	7300	4.3	1.12	1704.0	0.8397	0.6169	0.9819	0.17	0.49	0.270	0.90
15	Heat recovery	7300	5.6	1.49	1312.5	0.7892	0.5156	0.9739	0.24	0.66	0.270	0.90
16	Cooling	7300	3.8	1.05	1941.5	0.8594	0.6527	0.9831	0.15	0.43	0.270	0.90
17	Distribution of electricity	7300	5.4	1.09	1356.4	0.7987	0.5318	0.9745	0.23	0.63	0.201	0.77
18	Control	7300	8.4	0.97	864.1	0.6998	0.3694	0.9614	0.36	1.00	0.116	0.55

4.5 CONTINUED ALLOCATION IN THE FAULT TREE

All first-level DEs belonging to the set $\{I\}$ of the power plant have now been assigned probability distributions for failure tendency and repair time. The next allocation to be carried out is from DE 19 (fuel supply) down to its input DEs 30 (fuel transfer) and 31 (fuel booster) (see Figure 4.2). According to the principles presented in Sections 4.4.1–4.4.3, allocation coefficients w, z are first defined for the input DEs, 30 and 31, and the initial choice of the level parameters α, β is tested by simulation. We chose the operation strategy $b = 0$ for DE 31, since it is running although the power plant is not running. Note especially that the simulation is, as always, done up to the TOP and *not* to the gate under allocation (DE 19). Consequently, the simulation results are always compared to the *original* requirements for TOP. This way ultimate flexibility will be achieved. The final values of the level parameters α, β ensure that the requirements are met more effectively than if the simulation stopped with the gate under consideration. This also implies that the distributions allocated earlier from above to DE 19 itself become unnecessary and irrelevant as soon as distributions have been allocated to its input DEs. The dependability of DE 19 is hereafter determined through its own inputs.

Allocation proceeds in the same way with DE 25 (energy conversion) to obtain corresponding probability functions for its input DEs 20, 21, 22, 23, 24 (the power units). Since DE 25 is a 2/5-gate, the operation strategies for the input DEs are explicitly given:

DE	20	21	22	23	24
a	0	0	0	0	0
b	1	1	1	1	1
c	0	0	0	0	0

(4.25)

A practical interpretation of these restrictions is as follows. Any of the power units can be repaired even if the power plant is running ($a = 0$). A power unit is (of course) not running if the power plant is not ($b = 1$), and the power plant will be started even if one of the power units is still under repair ($c = 0$). These restrictions also apply to DE 3 (diesel engine A) and DE 4 (generator A), the inputs of DE 20, which is the last DE to be allocated.

Table 4.2 presents the final dependability requirements allocated from the power plant TOP down to the fuel booster unit (DE 31) and a single diesel engine (DE 3). Specific time periods for DE 31 and DE 3 have been set so that its allowed failure tendency and repair time can be verified based on the failure information available (plant-specific, generic, and expert data). Dependability requirements specified for DE 3 have been used to guide the design process of diesel engines from the beginning to the manufacturing phase. The requirements for the booster unit play an important role in the selection of the component supplier and are stated as contract specifications.

Table 4.2. Dependability Requirements of Booster Unit and Diesel Engine

ID No	Design Entity	Operations Time [days]	No. of Failures	Repair Time [days]	MTTF [days]	Reliability (0–180)	Reliability (0–730)	Reliability (180–210)	No. of Failures (0–180)	No. of Failures (0–730)	MACMT [days]	$ACMT_{95}$ [days]
31	Booster Unit	7315	3.0	1.14	2463.1	0.8861	0.7084	0.9863	0.12	0.34	0.397	1.092
3	Diesel Engine	7242	49.2	59.84	147.3	0.1224	0.0029	0.7954	2.1	5.8	1.219	2.012

Comments. In the allocation of gate 25, a conceptual problem arises with the choice of importance factors, since this is not an OR-gate. The importance factors D, E, F are characteristically independent of the fault logic, but factor G (business damage) occurs only when the gate actually fails. It follows that the simulated occurrence of G will probably become less than its initial weight would predict. For this reason, we have done some numerical investigation, calculating the *a posteriori* value of G from the simulation data. These numerical experiments indicate that the difference in question will be minor for a $2/n$-gate. However, especially in case of a n/n-gate (AND), the difference becomes clearly significant. One solution would be to raise the initial weight of G to a level such that the simulated portion of G meets the assumption.

4.6 CONCLUSIONS

This study presents a systematic approach to the conversion of demanded quality into quantitative dependability requirements that will guide the product design process from the beginning to the manufacturing phase. Dependability requirements can be allocated to functions, systems, equipment, mechanisms, or any parts as the design work proceeds and design concepts are known. Allocation is based on the product's technical complexity and importance from the customer's perspective. The method can be applied both to tailor-made business-to-business products and to ordinary consumer goods. The following is a list of features that support the application of the developed allocation method to the product design process, as well as of some observations and limitations.

1. The reliability allocation model presented can deal with quite different types of dependability requirements for the product, leading to quite different failure tendency and intensity curves. For instance, from both the customer's and the manufacturer's perspective, there is opportunity to accept a different failure tendency during the warranty period from that after the warranty expires. In most of the reliability allocation methods, the product's failure rate is taken to be a constant [see Kececioglu (1991)].

2. Requirements can be allocated to design entities both under OR-gates (e.g., TOP 26) and k/n-gates (e.g., DE 25) and with various operation strategies. For instance, the diesel engine (DE 3) can be repaired immediately when it fails and the power plant TOP can be started even though one of its engines is still under repair. Most of the methods currently in use in reliability allocation deal with the case where the DEs are connected in series (OR-gate) [see Blischke and Murthy (2000)].

3. The allocation of repair time is determined according to the DE's repair complexity from the technical standpoint (coefficient z), not by how often a probable need for repair appears. For instance, comparisons of MACMT for energy conversion (DE 25) and fuel supply (DE 19) illustrate this feature. In common methods currently in use, allocated repair time to the DE depends on its failure tendency [see IEC 706-6 (1994)]. These methods for allocation of repair time direct one to allo-

cate longer repair time to the DE, as it will rarely fail, despite the fact that the DE is technically the easiest to repair (i.e., it will have the shortest repair time).

4. Different operation strategies, causal interrelations, and numbers of power units (diesel engines and generators) needed to achieve dependability requirements can easily be illustrated. This is important in order to avoid promising something that cannot be achieved or whose achievement will become too expensive. An appropriate operation strategy for DEs is essentially based on causal interrelation to the TOP. Our model does not handle interrelations between arbitrary DEs. Knowledge of minimal cut sets makes the strategy easier to find in a complex fault tree [see Henley and Kumamoto (1996)].

5. Requirements may only be allocated to the DEs that actively contribute to the product's operation. For example, the operation strategy where a "spare DE" starts only when the "main DE" fails and is stopped immediately when the main DE is repaired is not included in the allocation model. On the other hand, using separate pair simulation programs, we have allocated requirements for this type of gate in other case studies. The verification of simulation results is done by comparison to requirements set for the gate. When results are satisfactory, we have probability distributions to use for the "pair DE" that consists of the main and its spare DE.

6. Since customers' requirements for reliability and availability may often be contradictory, the customer must be consulted as to the priorities among the requirements. In this power plant case, the most important requirement for the customer has so far been reliability. Now assume that the customer has a requirement that the availability, A is $\geqslant 0.99$, and that achievement of this required availability

Checking requirements
(for TOP 26)

	Simulated	Required
$R[t_a \ldots t_a+t_c]$	0.382	0.800
$R[t_d-t_c \ldots t_d]$	0.407	0.800
MACMT	0.335	0.333
$ACMT_{95}$	1.04	1.00
parameter s	1.20	1.20
availability (A)	0.99007	≥ 0.9900

Note! If the results are not satisfactory, adjust the level parameters α and/or β

Number of failures / running time

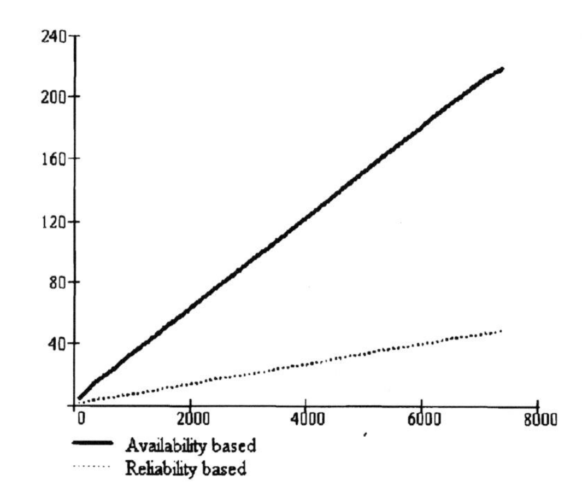

Figure 4.6. Number of failures, when required availability is more important to achieve than reliability.

is more important than that of a specified reliability. Figure 6 illustrates how big the difference between the allowed numbers of failures can be. The lower curve is based on reliability and the upper one on availability. (In this comparison, it is natural to use the same mean repair time.)

7. There are numerous reasons why a satisfactory analytical treatment of dependability is inconceivable. The fault tree is not necessarily serial (OR-gate), and there are often extra interrelations between the DEs and TOP, which concern stopping because of failure or starting after repair. Our simulation procedure considers arbitrary fault trees consisting of k/n-gates and additional operation strategies. Furthermore, when we allocate deeper in the fault tree, the dependability simulation must always be done up to the (upper) TOP and not to the gate under allocation. Only in this way can we be sure in each stage of allocation that TOP-requirements are satisfied. Finally, the probability functions for DEs cannot be modeled with analytic expressions, except for TOP. Since this would lead to information loss, we use interpolation tables with monotonic growth as the only shape limitation.

REFERENCES

Blischke, W. R., and Murthy, D. N. P (2000). *Reliability; Modeling, Prediction and Optimization,* Wiley, New York.

Gertsbakh, I. (2000). *Reliability Theory with Applications to Preventive Maintenance,* Springer-Verlag, New York.

Hagmark, P-E. (2001). *Dependability Allocation Program (Prototype),* Tampere University of Technology, Laboratory of Machine Design.

Henley, E. J., and Kumamoto, H. (1996). *Probabilistic Risk Assessment and Management for Engineers and Scientists,* 2nd edition, IEEE Press, Piscataway, NJ.

IEC 706-6 (1994). *Guide on Maintainability of Equipment. Part 6: Section 9: Statistical Methods in Maintainability Evaluation,* International Electrotechnical Commission, Geneva.

IEC 300-3-4 (1996). *Dependability Management Part 3: Application Guide, Section 4: Guide to the Specification of Dependability Requirements,* International Electrotechnical Commission, Geneva.

Kececioglu, D. (1991). *Reliability Engineering Handbook,* Vol. 2, Prentice-Hall, Upper Saddle River, NJ.

Rubinstein, R. Y. (1981). *Simulation and the Monte Carlo Method,* Wiley, New York.

TEKES (2001). *Competitive Reliability. 1996–2000,* Technology Programme Report 5/2001, National Technology Agency, Helsinki.

Virtanen, S., and Hagmark, P-E. (1997). *Reliability in Product Design—Seeking Out and Selecting Solution,* Helsinki University of Technology, Laboratory of Machine Design, Publication No. B22.

Virtanen, S. (2000). *A Method Defining Product Dependability Requirements and Specifying Input Data to Facilitate Simulation of Dependability,* Acta Polytechnica Scandinavica, Mechanical Engineering Series No. 143, Finnish Academy of Technology, Espoo.

EXERCISES

Initial Data for Exercises 4.1–4.5 Below

The fault tree to be studied:

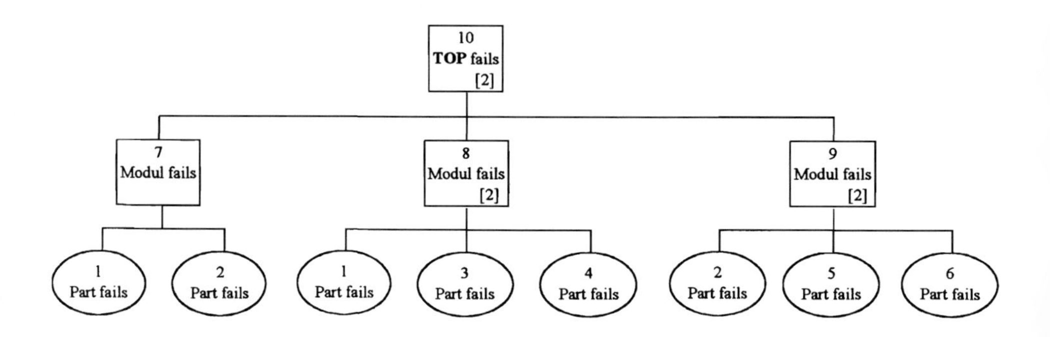

Requirements for the TOP 10 given by the customer and the manufacturer:

Age at the end of burn-in period	$t_a = 30$
Age at the end of warranty period	$t_b = 1000$
Age at the end of useful life period	$t_d = 4000$
Length of age interval (for Rel)	$t_c = 30$
Reliability in an age interval of length t_c in period $(t_a, t_d]$, at least	Rel = 0.80
A parameter for warranty period	$s = 0.8$
Mean active corrective maintenance time, at most	MACMT = 0.2
Maximum corrective maintenance time, at most	$\text{ACMT}_{95} = 0.8$

Data for importance, complexity and repair time of modules 7, 8, 9:

$X =$

Top 10	Module 7	8	9
$D = 10$	2	1	2
$E = 1$	2	1	2
$F = 5$	2	1	2
$G = 1$	1	1	1

, $Y =$

Top 10	Module 7	8	9
$A = 5$	1	2	3
$B = 2$	1	2	3
$C = 2$	1	3	3

, $z =$

Module 7	8	9
6	3	1

Allocation coefficients for parts 1, 2, 3, 4, 5, 6:

Module	7		8			9		
Part	1	2	1	3	4	2	5	6
w	0.5	0.5	0.040	0.480	0.480	0.040	0.040	0.480
z	0.5	0.5	0.2	0.4	0.4	0.2	0.4	0.4

Operation strategy:

$a = b = c = 1$ for all parts and modules 1, 2, 3, 4, 5, 6, 7, 8, 9.

4.1. Find the minimal cut sets of the fault tree above.

4.2. Construct the required failure tendency function $\Lambda(t)$ and the (Weibull) repair time distribution G(TTR) for TOP 10. Choose the design parameter $\lambda = \Lambda'(t_b) = 0.005$ and the shape parameters $\delta = \gamma = 5.0$ in formulas (4.9) and (4.10). Verify that the TOP requirements are fulfilled. Calculate mean time to first failure (MTTFF), mean time to failure (MTTF) in age period $(0, t_d]$, and the average number of failures in age periods $(0, t_b]$, $(0, t_d]$, $(t_d - t_b, t_d]$.

4.3. Determine the importance and complexity coefficients x, y, and the allocation coefficients w for modules 7, 8, 9, using the exponent $\varepsilon = 1$ in formula (4.21).

4.4. Taking into consideration allocation coefficients (w, z), the fault logic and the operation strategy, allocate failure tendency and repair time to modules 7, 8, and 9; that is, adjust α and β in formula (4.15) so that the TOP requirements remain fulfilled. Then calculate the following for *each* of the modules:

(a) Average number of failures in TOP's age periods $(0, t_b]$, $(0, t_d]$, $(t_d - t_b, t_d]$
(b) MTTF, cumulative running time and availability
(c) MACMT and ACMT_{95}

4.5. Continue allocating to the second-level parts 1, 2, 3, 4, 5, 6, so that the TOP requirements still hold. For *each* part, calculate the figures asked for in Exercise 4.4.

PART B

Cases with Emphasis on Development and Testing

Chapter 5, "The Determination of the Design Strength of Granite Used as External Cladding for Buildings"

—Is concerned with acquisition and analysis of test data for determining whether or not the flexural strength of granite cladding under extreme conditions is adequate to assure that reliability requirements are satisfied.

Chapter 6, "Use of Sensitivity Analysis to Assess the Effect of Model Uncertainty in Analyzing Accelerated Life Test Data"

—Deals with a detailed analysis of accelerated test data on fatigue life of a spring as a function of temperature, process, and displacement, with emphasis on the sensitivity of the results to model assumptions.

Chapter 7, "Virtual Qualification of Electronic Hardware"

—Describes the use of simulation of failure mechanisms to reduce the amount of physical testing required for predicting life expectancy of electronic hardware under test and field conditions.

Chapter 8, "Development of a Moisture Soak Model for Surface-Mounted Devices"

—Investigates test procedures and various models for analysis of moisture soak of integrated circuit packages soldered to a printed circuit board and its effect on product quality and reliability.

Chapter 9, "Construction of Reliable Software in Resource-Constrained Environments"

—Is concerned with efficiency of testing in software development in situations where validation and verification activities are reduced to shorten schedules, and the effect of this on software quality and reliability.

Chapter 10, "Modeling and Analysis of Software System Reliability"

—Discusses various modeling techniques for analysis of software reliability and presents a method of analysis of software failure data to estimate reliability growth during the testing phase of software development.

Chapter 11, "Information Fusion for Damage Prediction"

—Deals with the use of prior information in the form of expert judgment, in a Bayesian framework, in situations where testing is destructive and observations are costly.

CHAPTER 5

The Determination of the Design Strength of Granite Used as External Cladding for Buildings

Malcolm J. Faddy, Richard J. Wilson, and Gerry M. Winter

5.1 INTRODUCTION

Many modern buildings have prefabricated *curtain walls* which, apart from aesthetic, heat, light, and sound criteria, are designed subject to strength, deformation, weatherproofing, air infiltration, and durability criteria, which need to be specified and determined in engineering terms. Accompanying the introduction of the modular curtain wall, new materials have been developed for use as external cladding. Granite, when cut into relatively thin panels, is particularly suited to this application. Developments in the block sawing of natural stone that took place in the late 1970s have enabled granite to be cut into thin slabs relatively cheaply. Since then, granite cladding panels have become increasingly popular and are now in common use.

Granite is especially well-suited for use as an external cladding material because of its inherent advantages when compared with alternative materials. These include [see Winter (2002)] the following:

- It has a proven and unsurpassed track record. When properly designed, it is relatively free of maintenance and has significantly greater durability of finish than alternatives such as paint films and anodizing.
- It is aesthetically attractive and is available in a wide variety of colors, textures, and finishes.
- It is cost effective; when appropriately designed, initial costs are comparable with those of alternative materials.
- Building owners and the general public associate natural stone with quality.

Case Studies in Reliability and Maintenance, Edited by W. R. Blischke and D. N. P. Murthy.
ISBN 0-471-41373-9

- There is no shortage of the natural material.
- Cladding systems using thin stone panels may be designed either as a feature or to mimic the appearance of a traditionally constructed stone building.

The modular curtain wall is particularly suitable for providing the external envelope of a building where there is no external wall. However, where there is an existing external wall (lift shafts, stair wells, and so on), they are not as cost effective. In these positions, alternative cladding materials and methods are often used. If granite is used, the granite panels are generally installed as individual panels that are directly supported by the existing walls. In both this situation and on modular curtain walls, granite panels are designed as self-supporting elements and are referred to as *structural thin veneers.*

The reliability of the façade of a modern building is measured in terms of whether individual elements of the façade will fail. For granite panels, failure consists of cracks developing through the panel and at anchorage points. Such failures can result in panels falling off the façade, with potentially catastrophic consequences. In order to construct a reliable façade, it is necessary to design it so that the probability of failure of all (or at least the majority of) panels is within acceptable bounds, say less than one in a hundred. This probability depends upon the structural design criteria used.

The structural design criteria for the strength of a façade are generally determined in terms of its ability to resist the wind forces that may occur during the service life of the building. These wind forces may approach and exceed that of the maximum design load. The maximum design load is the maximum wind load such that there is a 5% chance of it being exceeded over a 50-year period. With the general acceptance of changing weather patterns, this situation is likely to be exacerbated [see SCOSS(2000)]. From this it follows that when designed in flexure, façade elements of modern buildings are one of the most highly stressed structural elements of a building. Consequently, granite panels used as structural thin veneers need to be designed as *flexural members*. The *flexural strength* of a granite is its tensile fracture strength when subjected to flexure. Granite is considered to be a brittle material so that tensile fracture is initiated by a local weakness, such as a cleavage plane or micro-crack. For a varying flexural stress, the apparent flexural strength is therefore a function of the size, the imposed loads, and the size and distribution of local weaknesses.

The thickness of the granite panels and the anchorage details are determined from their *flexural design strength.* This is the maximum strength that can be assumed for the stone for use in the design and for which the risk of failure of a single panel under the maximum design load is reduced to an acceptable level. Unfortunately, as will be seen in Section 5.4, there are flaws in the current methods used to determine the flexural design strength.

Developments in façade and natural stone technology have been relatively recent, and during this time there has been an increasing reliance on "self-regulation" by the construction industry worldwide. As a result, there are very few statutory controls that relate to the design, supply, installation, and maintenance of cladding to buildings in general and effectively none that govern or control the design, supply, installation, and maintenance of structural thin granite veneer panels. There are some exceptions, such as Singapore and New York, where statutory requirements for the periodic inspection of high-rise building cladding are in place.

The specification and design of modular curtain walls is a specialized engineering function. For historic reasons, however, it is the architect who generally specifies, details, and approves the external cladding for buildings. As a result of inappropriate design and specification, there are often significant problems with the façades of modern buildings. Apart from structural failure, curtain wall failures generally involve leakage and deficient weatherproofing. These are not usually life-threatening, although they may be potentially costly and able to cause significant inconvenience.

By contrast, in the absence of a fail-safe anchorage system, the failure of a granite panel may be catastrophic. There is, however, a cavalier attitude exhibited by many engineering and architectural practices. A lack of understanding of the natural stone industry and the behavior of natural stone veneers does not seem to be considered an impediment to specifying the selection, fabrication, and use of natural stone cladding. This attitude, when combined with the lack of statutory controls and absence of standards that set out appropriate practice, has given rise to a situation where serious deficiencies are commonplace in material selection, design, and installation of natural stone cladding panels, and especially of granite [see Winter (2002)]. Consequently, there has been an unfortunate decline in confidence in natural stone, despite the fact that it can provide reliable and relatively cheap external cladding for buildings.

The objective of this chapter is to comment on some aspects of current practices in the design processes for granite panels as a thin structural veneer and to identify appropriate alternatives that should ensure the reliability of the veneer. Of particular importance is the establishment of a rational and appropriate method for determining the design strength of granite panels. The first step is to consider the properties of granite in Section 5.2 and to establish reliability criteria in Section 5.3. Second, it is necessary to describe current practices with regard to the determination of the flexural design strength and indicate flaws in these practices; this is covered in Section 5.4. Alternative methods based on appropriate reliability criteria will be presented through a case study in Section 5.5. This will involve fitting appropriate models to the flexural strength and outlining a robust procedure for determining the flexural design strength. Finally, a general procedure will be outlined in Section 5.6.

5.2 PROPERTIES OF GRANITE

Naturally, the failure characteristics of a material depend on the structure of the material. This section describes the formation, properties, and fracture behavior of granite. First, note that *dimension stone* is any natural stone that has been selected and fabricated to specific sizes or shapes. In the dimension stone industry, the term "granite" is defined as follows [see American Society for Testing and Materials (ASTM) C 119]:

> A visibly granular, igneous rock generally ranging in colour from pink to light or dark grey and consisting mainly of quartz and feldspars, accompanied by one or more dark minerals. The texture is typically homogeneous but may be gneissic or porphyritic. Some dark granular igneous rocks, though not geologically granite, are included in this definition.

The term "granite" may be applied to any dimension stone that comprises a fused, crystalline mass, so that many may be metamorphic. The structure of such rocks may include complex microfracture patterns and sometimes comprise a mixture of two prior rock masses. If granite is considered in its least complex form, it is a material that has formed from the cooling of a molten rock mass (*magma*).

The texture of granite is described in terms of the relative sizes of the mineral grains and their relationship to each other. Since the crystallisation of individual minerals takes place between specific temperatures, the crystal size of a particular mineral is a function of the rate of cooling between these temperatures. Rapid cooling results in many centers of crystallization and small crystals, whereas slow cooling gives rise to large crystals. In the former, the inter-crystalline cracks are smaller and the cleavage planes within the feldspar crystals are not as significant. Hence, it tends to be stronger, so that it is more difficult to cut and more expensive to use. Naturally, availability and appearance also play a role in determining the cost of granite. As well, granites with larger crystals are usually chosen for use as dimension stone for aesthetic reasons. Granites with relatively uniform crystal size and mineral dispersion are generally chosen because of their (uniform) appearance.

When cooling, the rock mass loses heat from its external surface and so cools down more quickly from the outside. As the external layers of the mass cool, they contract. The outer layers are, however, supported by the hotter inner core. The reduced volumes of the external layers are held in position by the core and develop a three-dimensional crack pattern. The cracks are in the form of two patterns that radiate from the center of the core (like segments of an orange), along with a third circumferential crack pattern (like the layers of an onion). At critical distances apart, the cracks propagate throughout the rock mass to form joints. The micro-cracks within crystals are generally parallel to the joints and may be referred to as reflection cracks.

When exposed to weathering, the regularly spaced joints break down to leave large, approximately rectangular masses. The three-dimensional micro-crack pattern within the rock mass is approximately orthogonal and parallel to these joints. It is much easier to cut or split granite in a micro-crack plane than in any other direction. The micro-crack density, however, varies with direction so that the ease of cutting or splitting granite also varies with each of the micro-crack planes. For some granites this variation may be relatively small, whereas for others it is significant. The direction in which the reflection crack density is greatest, called the *rift,* is the direction in which the granite is most easily cut or split.

The panels are produced in the following manner. Large blocks are cut from the rock masses and then cut into smaller *blocks* (about 1.8 m × 1.8 m × 3 m) for transportation by truck. These are then cut into *slabs* (about 1.8 m × 3 m × 30 mm) and thence into *panels* (typically up to 1200 mm × 1200 mm × 30 mm).

Given the three-dimensional, approximately orthogonal, micro-crack pattern, if blocks are to be produced with the least effort, the blocks' sides should be parallel to the planes of the micro-cracks and therefore also parallel to the joints within the rock mass. Slabs should then be cut from the blocks in the plane of the rift. While it is industry practice for blocks to be quarried in the plane of the rift, this does not always happen. If the blocks are extracted at an angle to the rift, then the rift is not parallel to the block side. Because the direction of slabbing *must* be perpendicular to the side of a block, the slabs will then be slabbed at an angle to the rift, giving rise to a plane of weakness within the slabs and hence the panels.

The major factors that affect the strength of a granite panel are mineral composition and dispersion, crystal size, micro-cracks, and crystal orientation. The properties of these result in the granite having a compressive strength that is significantly greater than its tensile strength. Because panels are designed as "simply supported" elements, failures either occur between the anchorages as a bending failure or at the anchorages due to shear. The first of these relates to the overall strength of the panel to withstand stresses from high winds, while the second also relates to the strength of the panel at anchorage points, which is mainly a function of the strength of individual crystals.

A failure of a panel emanates from a local defect (that is, a crack) and propagates across the panel, perpendicular to the direction of the span, as a result of applied stress and the failure strength. If a granite is chosen for its uniformity, then the main source of cracks is the density of micro-cracks in the two orthogonal directions (across and along the panel) as indicated above. A secondary source of cracks is the cleavage planes within principally the feldspar crystals. The directions of these planes tend to be at random and thus may have less influence on the panel's strength. If the feldspar content is significant, then the cleavage planes may weaken the panels substantially. In what follows, the main consideration will be with panel failure rather than failure at anchorages. For further information on crystalline structure, minerals present and similar aspects, see Winter (2002).

5.3 RELIABILITY CRITERIA

The reliability of the external granite cladding of a building is assessed in terms of the risk of panels failing, and hence detaching and falling. The panels most at risk are those in "hot spots." These are locations on the surface of the building where wind forces are greatest—that is, locations subject to peak load. They can be identified through a wind pressure cladding study and are usually in local areas around corners [see Hook (1994)]. Approximately, between 100 and 200 panels may be in these places for a significantly large building upon which about 10,000 panels may be installed.

The key aspects of ensuring adequate reliability are panel thickness, anchorage design, and installation of the panels. The determination of these are highly dependent on the *flexural strength* (see Section 5.4) of each panel, so that establishing the distribution of the (flexural) strength of the panels is critical. In particular, the risk that a single panel fails can be expressed as the probability that the strength of the panel is less than the (flexural) design strength. Hence, the reliability of the granite cladding should be given in terms of the probability that a "large" proportion of the panels have strength greater than the design strength. Typically, this probability is set according to the desired risk and associated maintenance programme (the greater the risk, the more extensive and costly the maintenance). The design strength is then taken to be the maximum strength such that this probability is achieved. If the design strength is not high enough to withstand the expected wind forces, then the choice of panel size may need to be changed (greater depth, smaller length and width) in order to achieve the desired reliability. In addition (or alternatively), the choice of anchorage design and panel installation may need to be changed to provide greater support for the panels.

If it is assumed that the distribution of the strengths of panels is known, then the probability above can be obtained as follows. As the panels are individually hung, it is reasonable to assume that the panels fail independently of each other. Consequently, the number of panels with strength greater than the design strength has a binomial distribution with parameters given by the number of panels and the probability that a single panel exceeds the design strength. As the stresses on the panels at locations of peak load are greater than those for the other panels, it makes sense to take as the number of panels just the number in these locations. The remaining panels are then likely to be overengineered.

As an example, consider the situation in which there are 100 panels under consideration and it is desired that the probability that no panels fail is given by 0.99. Since the probability that no panels fail is simply the probability that a single panel doesn't fail raised to the power of 100, the design strength will be given by solving $1 - F(x_D) = 0.99^{1/100} = 0.9999$, where $F(x)$ denotes the distribution function of panel strengths and x_D denotes the design strength.

In general, if there are m panels under consideration and it is desired that at least m_0 panels have strength greater than the design strength, x_D, with probability α, then x_D is obtained by solving $\Pr(M \geq m_0) = \alpha$, where M has a binomial$(m, 1 - F(x_D))$ distribution; that is, the probabilities for M are given by the function $\Pr(M = y) = \binom{m}{y}(1 - F(x_D))^y(F(x_D))^{m-y}$. Note that the design strength is simply a lower quantile from the distribution of panel strengths.

Unfortunately, in many standards and specifications, the reliability of a complex system or a collection of items is treated as if it were the reliability of a single item. This results in a much greater probability of failure than may be desirable. For instance, in the above example, if the probability that a *single* panel does not fail is set to 0.99, then the probability that none of the hundred fail is only $0.99^{100} = 0.3660$.

In practice, the distribution of panel strengths is not known. Consequently, an appropriate model needs to be chosen to describe this distribution and then this distribution has to be fitted to data collected from sample panels, representative of the panels to be used in the particular application. The design strength is then estimated on the basis of the fitted distribution. This also needs to be taken into account in any reliability assessment.

Due to the uncertain nature of the granite (in terms of its composition and crystalline structure), there is appreciable variation both from granite to granite and from block to block across the quarry for a given granite. Consequently, there is a need for adequate testing to determine the distribution of flexural strength for the type of granite used for the given application. In addition, it is essential that the effects of the different blocks (or slabs) are estimated; hence, the sample data collected should include identifiers for each panel as to which block (or slab) it was cut from. From these, the flexural design strength can then be obtained for the mixture of panels from the blocks purchased for the current application. If the blocks (or slabs) from which each panel is obtained are recorded, then panels from blocks with greater strengths can be used in locations where the wind forces are greatest.

The procedure that will be described in Section 5.5 outlines the sampling required, the testing and calculation of the flexural strength for the sample panels, the choice of distribution and method of fitting, and the estimation of the appropriate lower quantile from the reliability criteria above. As this quantile is estimated, its uncertainty needs to be taken into account; this is done through its standard error. It should be noted that extreme quantiles are difficult to estimate with reasonable precision if only small data sets are obtained (as will be seen later). Finally, it should also be noted that this procedure is similar to the calculation of a tolerance interval, though, in this case, the standard approaches based on the assumption of either a normal distribution or an exponential distribution may not be appropriate here (as will be seen in Section 5.5). It is also possible to construct nonparametric tolerance intervals that do not require any distributional assumptions; however, these tend to be wider than those based on parametric models.

5.4 CURRENT PRACTICES

In this section, the current ASTM standards applicable to granite panels will be outlined. Also, the most common testing procedure will be described. Some of the difficulties with these will then be briefly discussed. For more information on the structural engineering concepts and theories referred to in the following, see, for example, Shanley (1967).

The meaning of terms in general use in the dimension stone industry are defined in *ASTM C 119-01 Standard Terminology Relating to Dimension Stone*. The standard includes the definition of the various types of building stone (as given above for *granite*).

The physical properties of a granite that is deemed acceptable for use as external cladding are given in *ASTM C 615-99 Standard Specification for Granite Dimension Stone*. Specifications for the supply of granite for external cladding frequently require compliance with this standard. It lists the maximum and minimum physical properties, test requirements, and test methods required for a granite to be accepted. According to ASTM C615-99, test samples are required to be taken in four conditions, wet/dry and parallel/perpendicular to the rift. Since it is industry practice to slab blocks in the plane of the rift, it is relatively difficult to obtain test specimens that are perpendicular to the rift. Consequently, it is common to obtain test samples cut in the other two crack directions; that is, they are cut perpendicular and parallel to a slab side.

At this time there are no codes of practice or standards that specifically address the determination of the flexural design strength of a thin structural veneer granite. It is common industry practice, however, to use the procedures set out in *ASTM C 99-87, Standard Test Method for Modulus of Rupture of Dimension Stone,* and *ASTM C 880-98, Standard Test Method for Flexural Strength of Dimension Stone,* for this purpose. ASTM C 99 and ASTM C 880, however, were derived in order to assess whether building stone is acceptable for external use and to provide a comparison of the strengths of different stones. Their stated purpose does not include the provision of data that might form the basis for design criteria—that is, to provide appropriate data to determine the flexural design strength.

In many procedures that determine flexural design strength, the relevance of ASTM C 99 and ASTM C 880 are confused, resulting in specifications that require testing to be undertaken in accordance with both standards. One reason for this confusion with regard to ASTM C 99 is that its test results may be misunderstood by those unfamiliar with the dimension stone industry and ASTM test requirements. This is because, to many structural engineers, the terms "modulus of rupture" and "flexural strength" are synonymous, whereas in these standards they are not. As a consequence, ASTM C 99 is applied unnecessarily. Even if this were not the case, the results of ASTM C 99 are redundant in the context of panel failures because the results of ASTM C 880 are lower than those of ASTM C 99 and are therefore always used. Consequently, ASTM C 880 has become accepted internationally and is commonly used to calculate the flexural strength of granite for the determination of the flexural design strength. Unfortunately, the purpose of ASTM C 880 is to indicate the differences in flexural strength between the various dimension stones, rather than to determine the flexural design strength of a particular stone. Only ASTM C 880 will be discussed below. For a complete discussion of both standards, see Winter (2002).

Flexural strength, as defined by ASTM C 880, is determined from a flexural test that utilizes a *test specimen* with a nominal span/depth ratio of 10:1 and two equal loads located at quarter points to fracture the stone specimen. The quarter point loading gives a constant stress field over the central half of the test specimen, and it is relatively rare for the test specimen to fail outside this region. The calculation of flexural strength gives the extreme fiber tensile stress, perpendicular to the direction of span, between the applied loads at the time of fracture, using simple bending theory [see, for example, Shanley (1967)]. Since the test specimen is loaded at its quarter points, there is a constant bending moment and zero shear between the points of load application, and so the bending stress is the principal stress.

ASTM C 880 requires that a minimum of five specimens are tested in each condition (wet/dry and parallel/perpendicular to rift). Each specimen is tested as described above. The flexural strength, f, is calculated as follows:

$$f = \frac{3WL}{4bd^2} \tag{5.1}$$

where f is the flexural strength (megapascals (MPa) or pounds per square inch (psi)), W is the failure load (newtons or pounds force (lbf)), L is the span of the test specimen (mm or in.), b is the width of the test specimen (mm or in.), and d is the thickness of the test specimen (mm or in.), with $L = 10d$ and $b \geq 1.5d$ for test purposes.

Loading has to be applied at a uniform stress rate of 4.14 MPa/minute to failure. To determine the rate of loading for each test, it is assumed that, when test specimens are tested in flexure, the stone behaves in a linear elastic manner and that the modulus of elasticity in tension and compression are the same. It also assumes that the failure plane of the specimens (that is, the fracture) originates at the extreme fiber between the two point loads and is perpendicular to the sides of the specimen. Note that, since the thickness and width of individual test specimens vary, the loading rate is different for each specimen.

For the majority of test specimens these assumptions may be acceptable. For test specimens where the failure plane on the tensile face of the specimen is significantly away from being perpendicular to the sides of the test specimen, or where the fracture plane is not between the applied loads, these assumptions are invalid. In this case, either a limit should be placed on the acceptability of data based on the location and orientation of the fracture plane or equation (5.1) should be adjusted. In the case study in Section 5.5, the equation was adjusted, resulting in lower flexural strengths for some test specimens than would have been obtained using equation (5.1).

For each of the four groups tested, the mean and standard deviation of the flexural strengths are reported.

Finally, test specimens with relatively low failure strengths should be evaluated to determine whether the failure was caused by a local defect. The effect of a local defect on panel strength and anchorage strength may differ. Particular defects apparent in test samples such as failures along well-defined cleavage planes may need to be addressed in the inspection procedure of finished slabs. Abnormal cleavage plane failures may need to be evaluated in terms of the acceptance criteria for panels.

As stated earlier, ASTM C 880 does not address the determination of the flexural design strength of a thin structural veneer granite. Flexural design strength is not a measure of either the average strength of the granite or the variability of the granite per se. It is a prediction of the strength of the weakest panel, so that for a specified level of risk the flexural design strength enables the strength of the panels with the greatest risk of failure to be determined.

Common practice for the calculation of the flexural design strength conforms to a statement by Gere (1988) at the Exterior Stone Symposium in New York Sponsored by the ASTM:

> Natural stone will act in a manner consistent with the elastic theory of materials; however, we have to compensate for its characteristics of inconsistency in strength by using safety factors when calculating stone thickness for wind load and for anchoring.

and

> Using a minimum of 5 specimens from different slabs, it is recommended by the author that the spread of these test results compared to the *average of the test results* of the five specimens be converted to safety factors as described in Table 5.1.

The flexural design strength is then calculated in the following manner. Take each group of results, combine the values for the two wet groups and similarly for the two dry groups. For each group, discard the highest and lowest values, then calculate the means and standard deviations. Again for each group, calculate the coefficient of variation and determine the safety factor from Table 5.1, then divide the mean by the safety factor (to try to account for the uncertainty of the strengths which has not been accounted for in the previous steps). The design strength is taken to be the minimum of the two results.

Unfortunately, there are a number of problems with this procedure. First, it is ad hoc and its steps seem to have been taken from other codes or standards without logical justification. Second, the strength of test specimens cut in the same direction from the panel will tend to have similar values. However, test specimens cut in perpendicular directions will tend to be consistently ordered due to the micro-crack structures. Consequently, the strength of a panel will be given by the minimum strength of the two test specimens, rather than the average. Third, because the flexural design strength is in reality an estimate of a lower quantile, using a sample of only 10 values is inadequate. To then discard the observations that give the greatest information about extremes is very undesirable. Last, the use of the coefficient of variation in this context is inappropriate because the distribution of strengths, although these are positive, may not be right skew. Also, the use of Table 5.1 transforms the continuously varying coefficient of variation into a discrete safety factor, thereby ensuring that the safety factor calculation is not robust. The level of reliabil-

Table 5.1. Safety Factors for Granite used as Structural Thin Veneer [see Gere (1988)]

Coefficient of Variation (expressed as a percentage)	Safety Factor for Wind Load	For Lateral Anchoring
Up to 10%	3	4.5
10% to 20%	4	6.0
Over 20%	6	8.0

ity achieved is not at all clear from this procedure. This is partly due to this procedure being an amalgam of methods from different sources and having no clear reliability interpretation.

Some sections of the dimension stone industry have used the following alternatives to the above procedures:

A. Obtain a 99.9% prediction interval for a single observation based on the normal distribution and assuming the sample statistics are actually population parameters. Collect the data as above, and then calculate the *mean minus three standard deviations* for each group. This still averages the values for the test specimens cut in the two directions. Also, it assumes that a normal distribution is applicable, which may not be the case, particularly with regard to the tail properties of the distribution, which is critical in calculating extreme quantiles. Both discarding the high and low values from the combined samples and, alternatively, leaving them in have been used.

B. Obtain a 99.9% prediction interval for a single observation based on the normal distribution and assuming the sample mean is the population mean: Proceed as in A except use the *mean minus 4.3 standard deviations* to allow for the fact that the standard deviation has been estimated and a t_9 distribution, rather than a standard normal distribution, should be used because the sample is small.

C. Obtain a 99.9% prediction interval for a single observation based on the normal distribution. Proceed as in B except use *mean minus 4.3 standard* deviation × (1+1/sqrt(10)) because the mean has been estimated.

D. Obtain a 99.5% tolerance interval for 99% compliance based on the normal distribution: Use *mean minus 5.61 standard deviations*.

The desired level of reliability determines the choice of confidence coefficient. Unfortunately, in all but D above, the results apply to a *single* future panel and therefore overestimate the reliability of the total cladding: The real reliability in the above is closer to 90% (provided that the assumption of normality applies). In the next section, an alternative, which can be assessed in terms of reliability goals, will be presented.

In the determination of the flexural design strength of a structural thin veneer granite cladding panel, the information available suggests that exposure to weather for the life of the cladding panel may result in some level of degradation and loss of strength. The causes of such degradation have yet to be understood, and the effects of weathering are therefore unable to be quantified.

The variables that were considered as potentially able to affect design strength were:

- Temperature variation
- Cyclic wetting and drying
- Cyclic wind loading

- Air pollution
- Freezing and thawing

At present, these factors are ignored in the current procedures and are not considered here.

5.5 CASE STUDY

In this section, one particular granite will be investigated to determine its design strength for use as a thin structural veneer. The granite chosen was *Rosa Antico,* due to the large number of specimens available. The data were collected as part of a larger study of granites used for thin structural veneers for the third author's (G. M. Winter) doctoral thesis.

Analysis of variance was used to estimate the effects of the various factors that influence the strength of the granite specimens. Following this, an analysis of the residuals was performed, and this led to the development of a model for the residual variation. The fitted model was then used in the determination of the design strength to ensure an adequate reliability of a structural thin veneer obtained using this type of granite.

5.5.1 Test Material

Rosa Antico comes from the north of Sardinia and is sourced from a number of different quarries that range in size from medium to small. The stone is considered to be fairly uniform and has a small to medium crystal size. Blocks are extracted from rock masses using black powder, then cut to size using hammer drills, feathers, and wedges.

Petrographic Description
Rosa Antico is a slightly altered, generally medium grained, mildly deformed adamellite with a hypidiomorphic inequigranular texture. The grain size may be up to 35 mm but in general does not exceed 10 mm. Plagioclase feldspar was the first felsic mineral to crystallize. Some of the larger grains are composite with twinning and zoning patterns. The pinkish alkali feldspar has a random shape and is anhedral. It is micropertihitic with a dense but generally even distribution of albite lamellae. The pinkish color probably results from the presence of finely divided flakes of hematite. Fractures through the feldspar have been filled with chlorite. The quartz is irregular in shape and has at least two generations of well-developed microfractures. Biotite, mostly euhedral, is the dominant ferromagnesian mineral. See Figure 5.1 for a photograph of a Rosa Antico sample.

Based on advice received from Italian fabricators, that test specimens, where possible, are supplied from slabs (and thus cannot be perpendicular to the rift), irrespective of what might be requested, the test program used panels that had been supplied for a particular project. The material selected for testing was therefore rep-

Figure 5.1. Rosa Antico sample.

resentative of both the material supplied for construction and the material generally used for test purposes for the establishment of the flexural strength used in a design.

Accordingly, 50 panels of polished 30-mm-thick Rosa Antico were randomly selected for testing, with each panel coming from a different slab. Specimens were cut from each panel, with two of the four test specimens cut parallel to one side of the panel and two cut perpendicular to that side. Due to various factors, some panels had fewer (down to two) and others had more specimens cut, while due to the nature of the available material, some panels had more specimens cut one way than the other. There was a total of 176 specimens. These were then tested using the procedures outlined above using ASTM C 880 with one specimen randomly chosen from each pair tested dry and the other tested wet. The testing was carried out using an *Instron 6027* test machine and a *Mohr and Federhaff* test machine. See Appendix A for the data: strengths, panel identification, condition, and cut direction.

The determination of which is the "width" and which is the "length" of each panel was made using a deflection test, where width is defined as the direction across which the panel is weaker. Thus, the strength of the widthwise specimen of a panel is almost always less than the strength of the corresponding lengthwise specimen. Since the deflection test is time-consuming and expensive, in practice, the widthwise strength should simply be taken as the minimum of the strengths of the two specimens while the lengthwise strength should be taken as the maximum. This is both practical and conservative.

5.5.2 Estimation and Comparison of Effects

The first evaluation done was to assess whether there were differences between the specimens from different panels (and hence from different slabs/blocks), whether there were differences between the specimens tested dry and those tested wet, and whether there were differences between those cut widthwise and those cut lengthwise. All three factors were treated as fixed effects, although in different circumstances it might be more appropriate to consider the slab effect as a random variable [see, for example, pp. 397–401 of Vardeman (1994) for more on such modeling]. In the context here, the slab effects are treated as *fixed* because there are samples from each location from which panels are to be sourced.

The model for the above setting is as follows:

$$X_{ijkl} = \mu + slab_i + cut_j + cond_k + \varepsilon_{ijkl} \tag{5.2}$$

where μ is the overall mean, $slab_i$ is the effect from the ith slab, cut_j is the effect from the cut direction ($j = 1$ is lengthwise and $j = 2$ is widthwise), $cond_k$ is the effect from the specimen's condition ($k = 1$ is dry and $k = 2$ is wet), ε_{ijkl} are random variables (errors/residuals) modeling the unpredictability of the strength (assumed to have zero mean, constant variance and to be uncorrelated with each other), and X_{ijkl} is the observed strength of the lth specimen to have the combination i, j, k of the factors. It is assumed that the following constraints apply:

$$\sum_i slab_i = 0, \qquad cut_1 + cut_2 = 0, \qquad cond_1 + cond_2 = 0 \tag{5.3}$$

See, for example, Vardeman (1994) for more on these types of model.

Under the assumption of a normal distribution for the random errors, the effects in the model above can be estimated and compared using a standard analysis of variance. The effects are estimated by minimizing the sum of squares of the differences of the right side and the left side of the model equation (5.2) under the constraints given by equation (5.3). The comparison of effects involves the calculation of various sums of squares for each type of factor (slab, cut direction, condition, and residual) and the determination of p values from F distributions. Differences are significant if the p value is sufficiently small. Details of this type of analysis can be found in Section 5.5.3 and Chapters 7 and 8 of Vardeman (1994).

Given the large sample size and the robustness of tests using the analysis of variance, this method was adopted to assess whether the above-mentioned differences in effects were significant, even though it was thought that the assumption of normally distributed residuals might be invalid. Initially, an interaction term between the cut direction (widthwise/lengthwise) and the condition (wet/dry) was included. With a p value of 0.136, it was not significant and thus left out of the model. There are also no engineering reasons for including such an interaction term for interactions between the slab effects and the other two factors. Table 5.2 describes the analysis without interaction terms, and Table 5.3 provides the estimates of the effects (apart from the slab effects that are shown in Figure 5.2), with their standard errors. These calculations were carried out using the statistical package MINITAB, Release 13.1.

As can be clearly seen from the p values (all are less than 0.001), there are significant differences between the slabs, between the effects of the wet and dry conditions, and the directions of cut. In particular, it can be seen that the mean strength of

Table 5.2. Analysis of Variance for Strength

Source	Degrees of Freedom	Sequential Sum of Squares	Adjusted Sum of Squares	Adjusted Mean Squares	F Ratio	P Value
Slab	49	522.001	500.580	10.216	12.03	< 0.001
Condition	1	39.126	36.005	36.005	42.41	< 0.001
Cut	1	94.629	94.629	94.629	111.45	< 0.001
Error	124	105.281	105.281	0.849		
Total	175	761.036				

Table 5.3. Estimates with Standard Errors in Parentheses

Term	Overall Mean: μ	Dry Condition: $Cond_1$	Lengthwise Cut: Cut_1
Estimate	10.864 (0.0721)	0.459 (0.0705)	0.826 (0.0782)

specimens cut widthwise is lower than for those cut lengthwise and that the mean strength of wet specimens is lower than the mean strength of dry specimens. This corresponds to what was expected.

If the circumstances had been such that the slab effects should be modeled by random variables, then the shape of the histogram of slab effects (see Figure 5.2) suggests that a non-normal model should be chosen.

5.5.3 Residual Modeling

In addition to carrying out the above tests, a residual analysis was carried out to determine if an assumption of a normal distribution for these was appropriate. As can

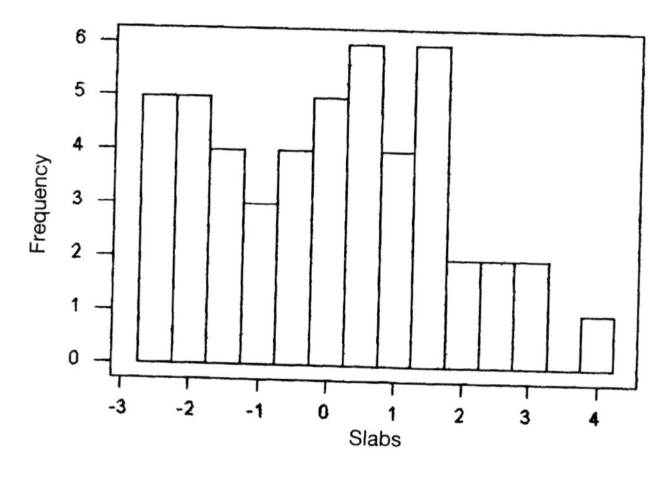

Figure 5.2. Histogram of slab effects.

be seen from the histogram and normal probability plot in Figure 5.3, there is evidence of non-normality (although not enough to compromise the results of the analysis of variance). Since the calculation of the flexural design strength is just an estimate of a lower quantile, it would be inappropriate to use a normal distribution to do this. Given the shape indicated by the histogram and normal probability plot, a skew-t distribution [see Jones (2000) for the probability density function and a full discussion of the theoretical aspects of this distribution] was chosen as being a more appropriate distribution for the residuals.

The modeling of the residual distribution was carried out as follows. Since the

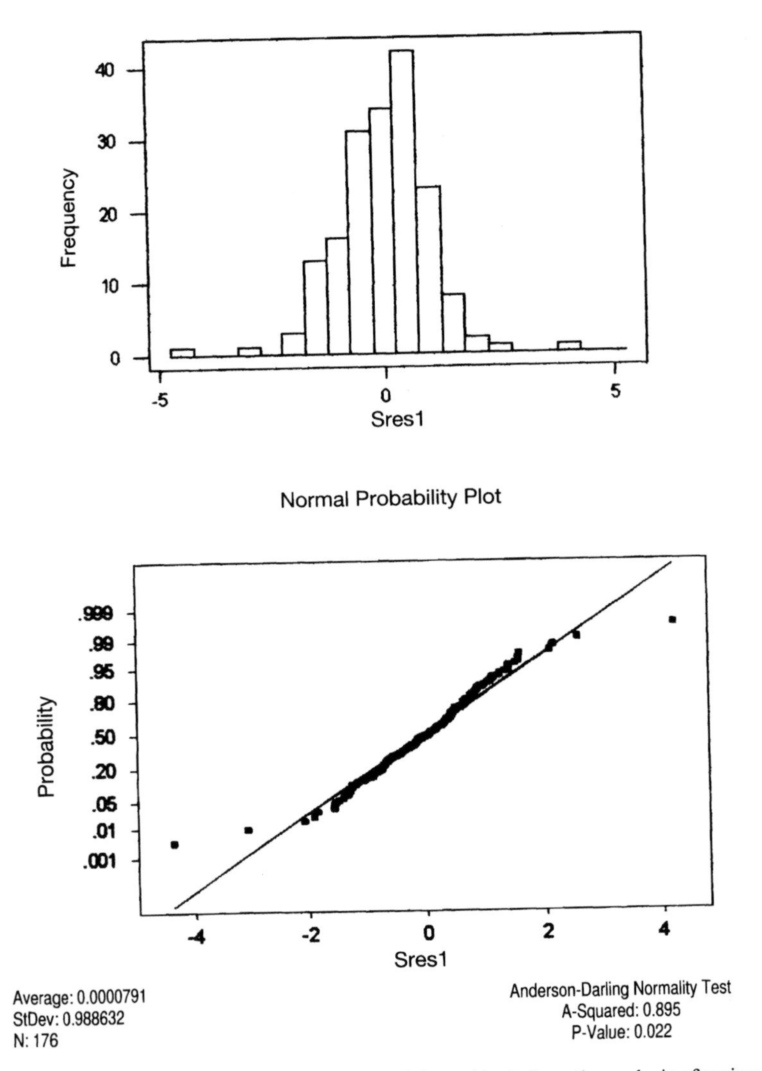

Figure 5.3. Histogram and normal probability plot of the residuals from the analysis of variance.

analysis of variance of the factor effects is robust, residuals were obtained from this analysis. Given the amount of data, it was decided to model the residual distribution within each of the four groups (wet/width, wet/length, dry/width, dry/length). Part of the motivation for doing so was that it was thought that the widthwise strengths should be left skew (as they are minima) and the lengthwise should be right skew (as they are maxima). It was thought that there should be little difference in the residual distributions from the wet and dry conditions.

Fitting a skew-*t* distribution to these residuals [by maximum likelihood—see, for example, Appendix A of Vardeman (1994)—implemented in Matlab, Version 5.3] confirmed that they were not normally distributed (twice log-likelihood ratio 17.9 on 1 d.f.: p value < 0.001) but any asymmetry was slight (twice log-likelihood ratio 0.5 on 1 d.f: p value > 0.4). However, as well as a significant difference in location due to the cut as shown by the analysis of variance with the mean strength from the widthwise cut being lower than the mean strength from the lengthwise cut, there was also a significant difference in shape of the distribution of residuals due to the cut (twice log-likelihood ratio 11.7 on 4 d.f.: p value 0.02) with those associated with the widthwise cut being left-skewed while those associated with the lengthwise cut were right skewed; Figure 5.4 shows the histograms of the two groups of residuals and their fitted skew-*t* residual distributions.

Looking at these distributions, it would appear that the lower and upper tails of the widthwise cut distribution, are similar to the upper and lower tails of the lengthwise cut distribution, respectively. Under appropriate conditions, this would be expected given that the widthwise cut is the minimum and the lengthwise cut is the maximum of the two specimens' strengths. Refitting the skew-*t* distributions with the appropriate reparameterization resulted in a comparable fit (twice log-likelihood ratio 1.4 on 4 d.f.: p value > 0.50).

There were no significant differences in shape of the residual distributions due to the conditions wet and dry (twice log-likelihood ratio 5.5 on 4 d.f.: p value > 0.24). Hence, the only significant differences due to these conditions are in respect of location. These differences were indicated earlier.

5.5.4 Design Implications

To obtain the design strength, it is necessary to consider the strength distribution for the worst case. This occurs for wet conditions across the width of the panel. Randomly mixing the wet widthwise cut strength distributions over the slab effect estimates (from the earlier analysis of variance) gave a left skew distribution of these smaller strength measurements, with some lower percentage points (with standard errors, obtained from asymptotic maximum likelihood theory, in parentheses) shown in Table 5.4.

These results show that very low percentage points are poorly estimated (as could be expected), but that the lower 0.15% point, corresponding to a reliability of 0.99 that no more than one panel in 100 will fail (see Section 5.3), is reasonably well estimated. This gives a lower 99% confidence limit for the corresponding strength of $4.25 - 2.33 \times 0.91 = 2.13$ MPa. This is then the design strength for the

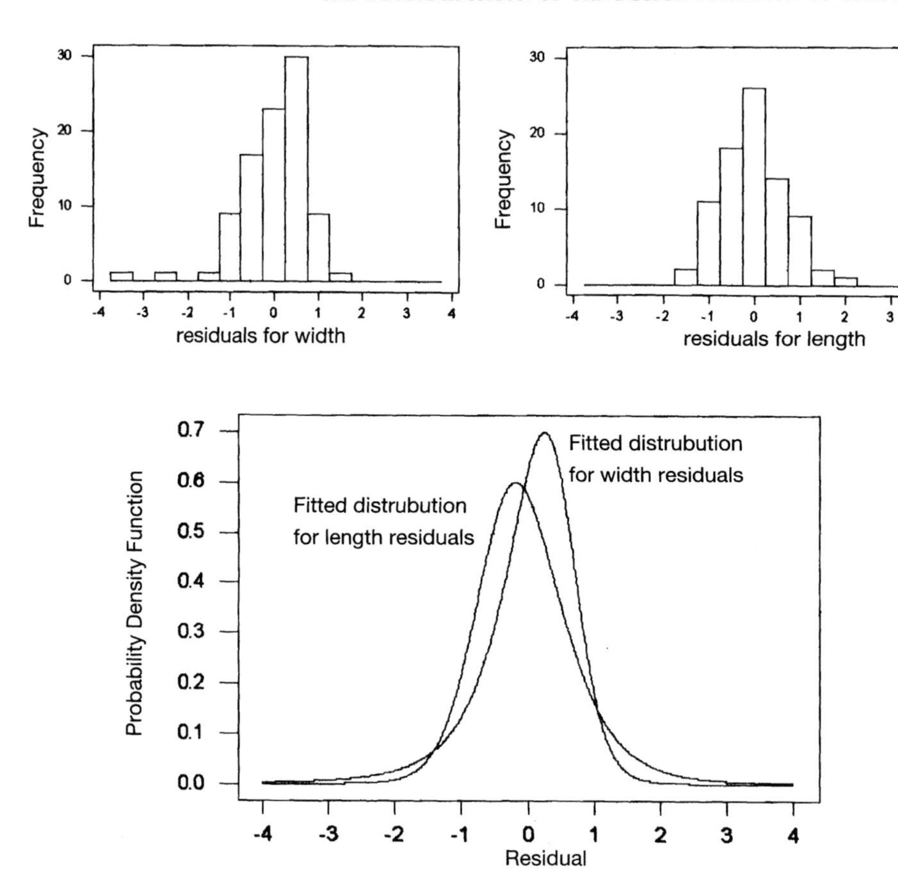

Figure 5.4. Histograms and fitted skew-*t* distributions for width and length residuals.

above design criterion. The choice of the reliability criteria above (reliability of 0.99; no more than one in 100 fails; lower 99% confidence limit) will, of course, be determined from the individual setting under consideration.

The thickness of an external structural thin veneer cladding panel is determined by its design strength. Because a value of 2.13 MPa is a relatively high value for a design strength, it may be possible to reduce the thickness of the panels from 30 mm (depending on the design load and span). However, current good practice requires panels, of the order of 1 m^2 and less, to be a minimum of 30 mm thick. Con-

Table 5.4. Estimated Percentage Points for Wet-Width Distribution, Mixed over Slab Effects with Standard Errors in Parentheses

	1%	0.5%	0.15%	0.05%
Estimated percentage points	5.79 (0.27)	5.29 (0.44)	4.25 (0.91)	3.05 (1.64)

sequently, this minimum thickness criterion would override the thickness as determined from this design strength, and thus the thickness used would remain at 30 mm.

The design strength is also used in the design of the panel anchorage system. For modular, prefabricated, façade systems the granite panels are generally either glazed in or otherwise continuously supported along their bottom and top edges, and thus the anchorage stresses are relatively low. For granite panels supported on discrete anchorages, such as hand-set installations, anchors are generally located on the lower and upper edge of the panel. All the brackets provide lateral restraint to the panels while the lower brackets also support the weight of the panels. With *kerf* brackets, which are attached to the panel by kerf slots on the lower and upper edges of the panel, the horizontal length of the slot is determined by the design strength and the design load; the lower the strength, the longer the kerf slot. Typically, a design strength of 2.13 MPa would result in each kerf slot being 100 mm for a 1-m^2 panel.

For a design strength that was much smaller than 2.13 MPa, it may be necessary to increase the thickness of the panel. Relatively small increases in panel thickness significantly increase the ability of a panel to withstand load. For example, increasing a panel thickness from 30 mm to 35 mm decreases the bending stress by over 25%. For a modular system, an increase in thickness such as this would have little or no effect on the design, manufacture, or installation of the façade modules (because they are factory assembled). Hence, increasing stone thickness would have a minimal cost impact on the installation. However, for hand-set granite installations, a lower design strength would result in an increase in installation costs (due to a change in the bracket design), especially if the panel was thicker, in which case the heavier panels would be more difficult to lift and place.

Finally, by keeping track of which panels come from which slabs (location in the quarry), stronger panels can be identified and used in locations on the building at which the greatest wind loads are expected.

5.6 CONCLUSIONS

As can be seen from the analysis above, the current approach to the determination of the flexural design strength has serious flaws. Apart from an inappropriate use of ASTM C 99 and ASTM C 880, the use of the test results from these neither reflects clear reliability goals nor provides robust estimates of the design strength. This approach has resulted in substandard design and methods in the use of granite panels as cladding. In turn, this has resulted in unnecessary failures with increased maintenance costs (and, in some cases, the replacement of a significant proportion of the panels) and potentially catastrophic consequences, leading to a drop in confidence in a reliable and reasonably maintenance-free cladding material.

A recommendation, based on the analyses presented here, is that the following procedure should be used to determine the flexural design strength for a batch of granite panels so that appropriate reliability goals can be realized.

1. Determine the number of panels in locations of peak load.
2. Determine the level of reliability desired. For example, it may be required that the design strength reflects a probability of 0.99 that no more than one panel of 100 in peak load locations will fail.
3. Choose blocks and order four test specimens from each block, two cut each way. If the number of blocks is less than 20, order additional specimens from the appropriate number of blocks to bring the total number of specimens up to 80. If the number of blocks is greater than 20, then test specimens from adjacent blocks can be skipped. A minimum of 80 test specimens should be used.
4. Specimens should be tested according to wet and dry conditions as per ASTM C 880. For specimens that do not fracture perpendicularly to the sides, the strength should be adjusted to take this into account (otherwise the strength recorded for the specimen will be higher than the actual strength). For each pair cut in the two directions from the same block/slab tested under the same conditions, label the minimum strength as "width" and the maximum as "length."
5. Carry out an analysis of variance with block/slab effect, cut direction effect, and condition effect, and then obtain residuals.
6. Fit appropriate models (for example, skew-t distributions) to the four groups of residuals.
7. Check differences between these and combine where possible, choosing the appropriate distribution.
8. From the distribution for the wet-widthwise residuals, obtain a mixture distribution over block/slab effects and estimate quantiles according to step 2 above, including standard errors.
9. Incorporate the standard error to obtain the flexural design strength as a lower 95% confidence limit.
10. If possible, ensure that the location in the quarry is available for each panel. Use the block/slab effects to determine which panels to install in peak load areas and which to install in lowest load areas on the building.

In conclusion, the above procedure will provide reliable values for the flexural design strength with clear reliability implications. In addition, it has clearly interpretable and logical steps that can be justified, unlike the procedures currently in practice.

REFERENCES

ASTM Designation: C 99-87 (Reapproved 2000). *Standard Test Method for Modulus of Rupture of Dimension Stone,* American Society for Testing and Materials.

ASTM Designation: C 119-01 (2001). *Standard Terminology Relating to Dimension Stone,* American Society for Testing and Materials.

ASTM Designation: C 615-99 (1999). *Standard Specification for Granite Dimension Stone,* American Society for Testing and Materials.

ASTM Designation: C 880-98 (1998). *Standard Test Method for Flexural Strength of Dimension Stone,* American Society for Testing and Materials.

Gere, A. S. (1988). Design Considerations for Using Stone Veneer on High-Rise Buildings, in *New Stone Technology, Design and Construction for Exterior Wall Systems,* American Society for Testing and Materials, Philadelphia, pp. 32–46.

Hook, G. (1994). Look Out Below! The Amoco Building's cladding failure (Chicago, Illinois), *Progressive Architecture* **75,** 58–61.

Jones, M.C. (2000). A skew-*t* Distribution, in *Probability and Statistical Models with Applications* (C. A. Charalambides, M. V. Koutros, and N. Balakrishnan, Editors), pp. 269–277, Chapman and Hall, London.

SCOSS—The Standing Committee on Structural Safety, (2000). Cladding: The case for improvement, *Structural Engineer,* May, U.K. **78** (9), 15–16.

Shanley, J. R. (1967). *Mechanics of Materials.* McGraw-Hill, New York.

Vardeman, S. B. (1994). *Statistics for Engineering Problem Solving,* PWS Publishing Company, Boston.

Winter, G. M. (2002). Determination of the Design Strength of Thin Structural Veneer Granite Cladding, Ph.D. thesis, The University of Queensland, Australia (in preparation).

EXERCISES

Panels made from *White Berrocal* (see Figure 5.5), a granite from Madrid, are to be installed on a building requiring about 5000 panels, with approximately 90 panels in locations of peak load. To determine the flexural design strength, test specimens were obtained from 37 slabs. For each slab, two specimens were cut parallel to one edge while two were cut perpendicular to that edge. For each pair of specimens cut in the same direction, one was tested dry and one was tested wet. For each pair of wet (dry) tested specimens from the same slab, the lower value was classified as "width" and the higher value was classified as "length." The data can be found in Appendix B.

5.1. Reliability: Find the probability of a single panel failure such that, in 90 panels, the following probability statements hold:
(a) The probability that no panels fail is 0.99.
(b) The probability that no more than one panel fails is 0.99.

5.2. Assuming there are no interactions, carry out an analysis of variance to determine residuals and estimates of all main effects.

5.3. Investigate the distribution of the residuals for each of the four groups: wet/width, wet/length, dry/width, dry/length. From this investigation, does it seem likely that the four distributions are identical?

Figure 5.5. White Berrocal sample.

5.4. Assuming that there are no differences between the distributions for the two width groups and no differences between the distributions for the two length groups, fit appropriate distributions to the residuals from the width specimens and to the residuals from the length specimens. Does it seem likely that the two fitted distributions might be reflections of each other about zero?

5.5. Assuming that the distribution of the residuals for the length group is the same as the distribution of the width group reflected about zero, re-fit such a distribution as in 5.4.

5.6. Use the fitted distribution in 5.5 to estimate the quantiles identified in 5.1. Hence, determine the flexural design strength for these panels.

5.7. Discuss what procedures should be followed in ordering the panels for this building.

APPENDIX A. ROSA ANTICO DATA

Width Specimens Tested Dry		Width Specimens Tested Wet		Length Specimens Tested Dry		Length Specimens Tested Wet	
Strength	Slab	Strength	Slab	Strength	Slab	Strength	Slab
10.58	17	9.16	17	16.58	17	11.48	17
9.50	18	8.46	18	12.55	18	11.00	18
11.14	19	8.39	19	12.70	19	11.44	19

Width Specimens Tested Dry		Width Specimens Tested Wet		Length Specimens Tested Dry		Length Specimens Tested Wet	
Strength	Slab	Strength	Slab	Strength	Slab	Strength	Slab
11.00	20	10.99	20	13.21	21	13.13	20
10.23	21	10.67	21	14.12	22	12.30	21
11.02	22	9.72	22	16.58	23	11.47	22
15.04	23	13.47	23	15.64	24	14.12	23
12.78	24	11.52	24	15.53	25	14.24	23
13.46	25	12.45	25	13.51	27	13.07	25
10.39	26	8.03	26	13.21	28	13.60	27
10.35	26	13.26	27	13.38	29	11.73	28
14.38	27	11.59	28	12.33	30	13.78	29
12.59	28	10.31	29	10.07	31	10.57	30
11.22	29	7.94	30	14.09	32	9.10	31
5.19	30	9.08	31	13.39	33	13.33	32
9.26	31	12.89	32	13.82	34	13.59	33
11.46	32	10.52	33	10.94	35	11.93	34
11.01	33	9.91	34	14.37	35	12.49	35
11.19	34	11.83	35	14.14	36	12.88	36
8.49	35	11.75	36	15.22	37	12.91	37
12.88	36	12.73	37	12.47	38	11.12	38
13.54	37	10.41	46	10.74	39	10.73	39
11.83	46	10.04	46	10.86	40	10.15	40
11.66	46	10.90	47	12.41	42	10.52	42
11.46	47	10.89	47	11.54	43	8.82	43
10.98	47	8.59	49	11.20	44	10.48	44
11.46	49	10.11	49	11.63	52	11.15	52
7.17	54	9.53	50	12.41	98	12.53	98
8.25	54	8.74	50	12.02	99	11.10	99
8.25	55	7.15	54	14.15	100	13.05	100
11.09	98	7.25	55	12.31	101	11.61	101
10.97	99	9.61	98	12.47	102	11.28	102
11.23	100	9.71	99	13.75	103	12.60	103
12.10	101	11.71	100	9.20	109	9.77	111
10.88	102	9.47	101	9.35	110	8.46	112
12.12	103	10.15	102	8.87	111	8.45	113
9.06	109	10.17	103	9.72	112	9.02	114
9.01	110	8.63	109	9.20	113	8.39	115
8.31	111	8.42	110	9.36	114	8.80	116
7.81	112	8.13	111	9.73	115	9.16	117
8.46	113	7.61	112	11.19	116	9.27	118
8.67	114	7.90	113	9.43	117		
8.27	115	8.12	114	10.61	118		
9.33	117	7.88	115				
8.36	118	8.60	116				
		8.79	117				
		8.16	118				

APPENDIX B. WHITE BERROCAL DATA

Width Specimens Tested Dry		Width Specimens Tested Wet		Length Specimens Tested Dry		Length Specimens Tested Wet	
Strength	Slab	Strength	Slab	Strength	Slab	Strength	Slab
11.95	119	11.75	119	13.82	119	12.65	119
9.25	120	8.37	120	12.84	120	11.38	120
8.16	121	7.57	121	12.64	121	11.83	121
8.95	122	7.18	122	10.54	122	11.48	122
12.02	123	11.06	123	13.55	123	11.29	123
12.14	124	10.64	124	12.39	124	10.65	124
10.02	125	8.42	125	13.10	125	10.99	125
10.38	126	9.08	126	11.68	126	11.28	126
11.79	127	10.28	127	14.86	127	12.97	127
12.32	128	11.34	128	13.59	128	12.48	128
12.79	129	11.83	129	13.67	129	12.18	129
12.60	130	11.02	130	13.91	130	11.69	130
12.70	132	11.09	132	14.00	132	12.17	132
12.11	133	10.56	133	14.93	133	10.67	133
12.82	134	9.89	134	13.42	134	11.85	134
9.63	136	7.94	136	11.70	136	9.78	136
12.00	137	11.43	137	12.38	137	11.63	137
13.50	138	9.42	138	13.52	138	9.96	138
11.71	139	10.55	139	13.11	139	11.49	139
10.75	140	9.45	140	10.90	140	11.59	140
8.96	141	8.39	141	12.18	141	11.85	141
9.46	142	7.98	142	10.36	142	9.59	142
14.56	143	12.76	143	16.48	143	13.07	143
13.97	144	12.83	144	14.34	144	13.17	144
11.75	145	10.97	145	13.92	145	12.50	145
8.76	146	9.01	146	14.91	146	13.82	146
11.78	147	10.59	147	11.91	147	10.75	147
12.51	148	12.58	148	12.91	148	12.64	148
11.57	149	13.44	149	13.99	149	14.93	149
11.77	150	8.96	150	12.43	150	11.97	150
9.26	151	9.33	151	11.95	151	10.85	151
14.61	152	13.17	152	14.82	152	13.71	152
11.99	153	11.85	153	13.38	153	12.64	153
12.07	154	11.98	154	12.91	154	12.11	154
9.43	155	8.56	155	13.77	155	13.65	155
5.78	157	7.69	157	9.08	157	13.27	157
14.64	158	10.91	158	15.09	158	12.67	158

CHAPTER 6

Use of Sensitivity Analysis to Assess the Effect of Model Uncertainty in Analyzing Accelerated Life Test Data

William Q. Meeker, Luis A. Escobar, and Stephen A. Zayac

6.1 INTRODUCTION

6.1.1 Background

This case study deals with a project in which product engineers needed information about the reliability of a spring in order to assess trade-offs and to make design decisions for a product. In general, there is a trade-off between the amount of displacement allowed in the motion of the spring and the performance of the product. More displacement leads to higher performance but shorter fatigue life. A large experiment was conducted to determine if a new processing method would improve fatigue life of the spring and to obtain a quantitative description of the displacement–life relationship.

In order to protect proprietary data and information, we generated simulated data from the fitted model for the original application, modified the scale of the data, and changed the name of one of the experimental factors. Largely, however, the nature of the application is the same as the original.

6.1.2 The Experiment

A sample of 108 springs, divided equally between the new and the old processing method, were tested until failure or 5000 kilocycles (whichever came first). Two other factors, processing temperature and stroke displacement (distance that the

Case Studies in Reliability and Maintenance, Edited by W. R. Blischke and D. N. P. Murthy.
ISBN 0-471-41373-9 © 2003 John Wiley and Sons, Inc.

spring was compressed in each cycle of the test), were varied in a 2 × 3 factorial arrangement with replication. Thus the overall experiment was a 2 × 2 × 3 factorial, and the assignment of units to levels of stress and run order were randomized. From this experiment, the engineers could develop a regression relationship to describe the effects that the experimental variables have on spring life. Stroke displacement was used as an accelerating variable. Nominal processing temperature and use conditions for these springs are 600°F and a stroke displacement of 20 mils, respectively (a mil is 1/1000 of an inch).

The goal of the experiment was to determine if the new processing method was better than the old method and to see if the spring would have a B10 life (the time by which one expects 10% of the devices to fail) equal to at least 500 megacycles (500,000 kilocycles) at use conditions. This customer specification would imply that no more than 10% of the springs would fail before the end of the technological life of the product in applications where the displacement would be 20 mils. If the specification cannot be met, then the product engineers want to know the amount of displacement that would be safe to use (the spring could still be used in the product by limiting displacement with some loss of product performance).

6.1.3 The Data

The data from the spring accelerated life test are given in Table 6.4 in Appendix B. Time is in units of kilocycles to failure. The explanatory variables are temperature (Temp) in degrees Fahrenheit, Stroke in mils, and the class variable Method, which takes the values New or Old. Springs that had not failed after 5000 kilocycles were coded as "Suspended." Note that at the condition 50 mils, 500°F, and the new processing method, there were no failures before 5000 kilocycles. All of the other conditions had at least some failures; and at 5 of the 12 conditions, all of the springs failed. At some of the conditions, one or more of the springs had not failed after 5000 kilocycles. When the number of censored observations at a condition is greater than one, it appears in the data matrix only once, with the Number column indicating the multiplicity.

Figure 6.1 is a pairs plot of the spring-accelerated life test data. The open triangles indicate right-censored observations. The plot provides a visualization of the experimental layout. Also, the plot of kilocycles versus method suggests that springs manufactured with the new method have longer lives.

6.1.4 Related Literature

Nelson (1990) comprehensively discusses useful models and statistical methods for accelerated testing. This is an important reference, and many of the ideas presented in this chapter are implicit in Nelson's extensive treatment of this subject. Chapters 18–22 of Meeker and Escobar (1998) provide some materials that complement Nel-

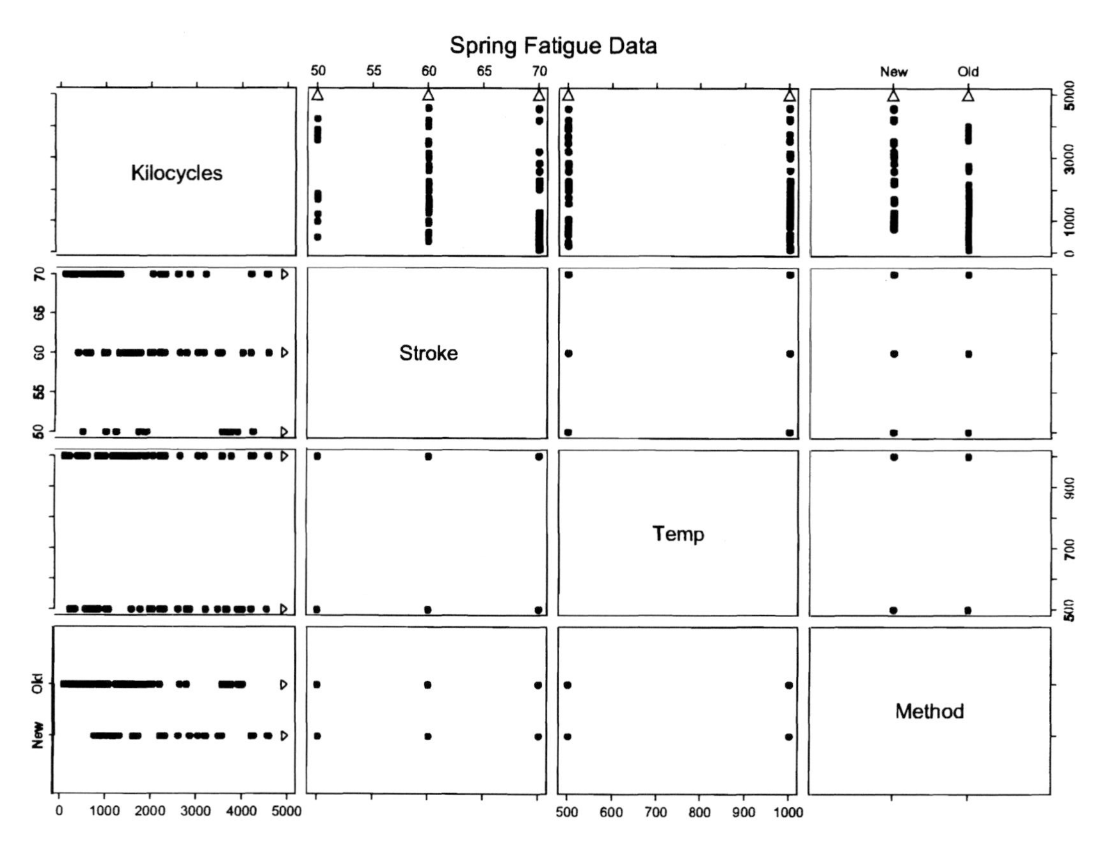

Figure 6.1. Pairs plot of the spring accelerated life test data.

son (1990). Wu and Hamada (2000) and Condra (2001) are other useful references on the use of designed experiments to improve product reliability.

6.1.5 Overview

Section 6.2 describes some initial analyses of the accelerated life test data, allowing an assessment of some of the important model assumptions. Section 6.3 illustrates the fitting of a response-surface acceleration model. Section 6.4 studies carefully the effect that stroke displacement has on spring life, including various sensitivity analyses of model assumptions. Section 6.5 makes concluding remarks and outlines some possible areas for further work.

6.1.6 Software

The analyses here were done with SPLIDA (SPlus Life Data Analysis), a collection of S-PLUS functions with a graphical interface (GUI), designed for the analysis of reliability data. The most up-to-date version of SPLIDA can be downloaded from www.public.iastate.edu/~splida. Although some of the basic analyses might be possible in advanced statistical packages like JMP, SAS, or MINITAB, the sensitivity analysis would probably require programming.

6.2 WEIBULL DISTRIBUTION AND INITIAL DATA ANALYSIS

6.2.1 Individual Analyses at Each Factor-Level Combination

Analysis of accelerated life test data usually begins by fitting, individually, one or more distributions to data from each factor-level combination (or, more precisely, at those combinations where there were failures). We will illustrate fitting models based on the Weibull distribution. Other distributions were also investigated (details are not given here), but the Weibull seemed to provide the best fit to the data.

The Weibull distribution cdf is

$$F(t) = \Pr(T \le t; \eta, \beta) = 1 - \exp\left[-\left(\frac{t}{\eta}\right)^{\beta}\right], \qquad t > 0 \tag{6.1}$$

In this parameterization, $\beta > 0$ is a shape parameter and $\eta > 0$ is a scale parameter as well as the approximate .632 quantile. The practical value of the Weibull distribution stems from its ability to describe failure distributions with many different commonly occurring shapes. See Section 4.8 of Meeker and Escobar (1998) for more information.

The Weibull distribution is a log-location-scale distribution. In particular, the logarithm of a Weibull random variable has a smallest extreme value distribution with location parameter $\mu = \log(\eta)$ and scale parameter $\sigma = 1/\beta$. In this form, the Weibull cdf is

$$F(t; \mu, \sigma) = \Phi_{\text{sev}}\left[\frac{\log(t) - \mu}{\sigma}\right], \qquad t > 0 \tag{6.2}$$

where $\Phi_{\text{sev}}(z) = 1 - \exp[-\exp(z)]$ is the cdf of the standardized ($\mu = 0$, $\sigma = 1$) smallest extreme value distribution. This parameterization is more convenient for regression modeling of simple relationships between log life and explanatory variables (log-linear models). For more discussion of log-location-scale distributions and log-linear models, see Chapters 4 and 17, respectively, in Meeker and Escobar (1998).

Maximum likelihood (ML) estimation is the standard method for parameter estimation with censored data. In large samples, ML estimators have desirable statistical properties. These methods are described in detail in Chapters 8 and 17 of Meeker and Escobar (1998) and Nelson (1990). Figure 6.2 is a Weibull probability plot showing individual ML fits for each of the 11 factor-level combinations that had failures. The straight lines on the plots are the individual ML estimates of the cdf at each factor-level combination. The differing slopes of the lines correspond to the differing Weibull shape parameter estimates. The steeper slopes correspond to using conditions that had less spread in the data. Figure 6.3 provides the same information, but without the legend and with the plot axis chosen to show the data with better resolution. The ML estimates of the Weibull parameters and corresponding standard errors are given in Table 6.1 for each condition.

6.2.2 Individual Analyses with Common Weibull Shape Parameter

Commonly used regression models for accelerated life tests assume that the scale parameter $\sigma = 1/\beta$ does not depend on the explanatory variables. Meeker and Escobar (1998) show how this simple model is implied by simple physical/chemical-based models (e.g., a one-step chemical degradation reaction). On the other hand, it is easy to find counterexamples to this model, both in physical theory and in data [e.g., Pascual and Meeker (1999)]. In any case, it is important to assess the adequacy of this model assumption. To do this, we fit a model with a separate distribution for each test condition where there were failures, but with a common Weibull shape parameter. We call this the "floating scale model." Fitting the floating scale model is similar to using the traditional one-way analysis of variance, but allows for censoring and distributions other than normal (or lognormal).

Figure 6.4 is a Weibull probability plot, similar to Figure 6.3, but with parallel lines, reflecting the common Weibull shape parameter in the floating scale model that was fitted to the data. The results of this model fit, including approximate 95% confidence intervals for the model parameters, are given in Table 6.2. The Intercept parameter estimate corresponds to the location parameter μ at the baseline condition "60Stroke; 500Temp; NewMethod," which we denote by μ_{base}. The other regression coefficients estimate $\mu_i - \mu_{\text{base}}$ where the index i corresponds to each of the other factor-level combinations (except for "50Stroke; 500Temp; NewMethod"

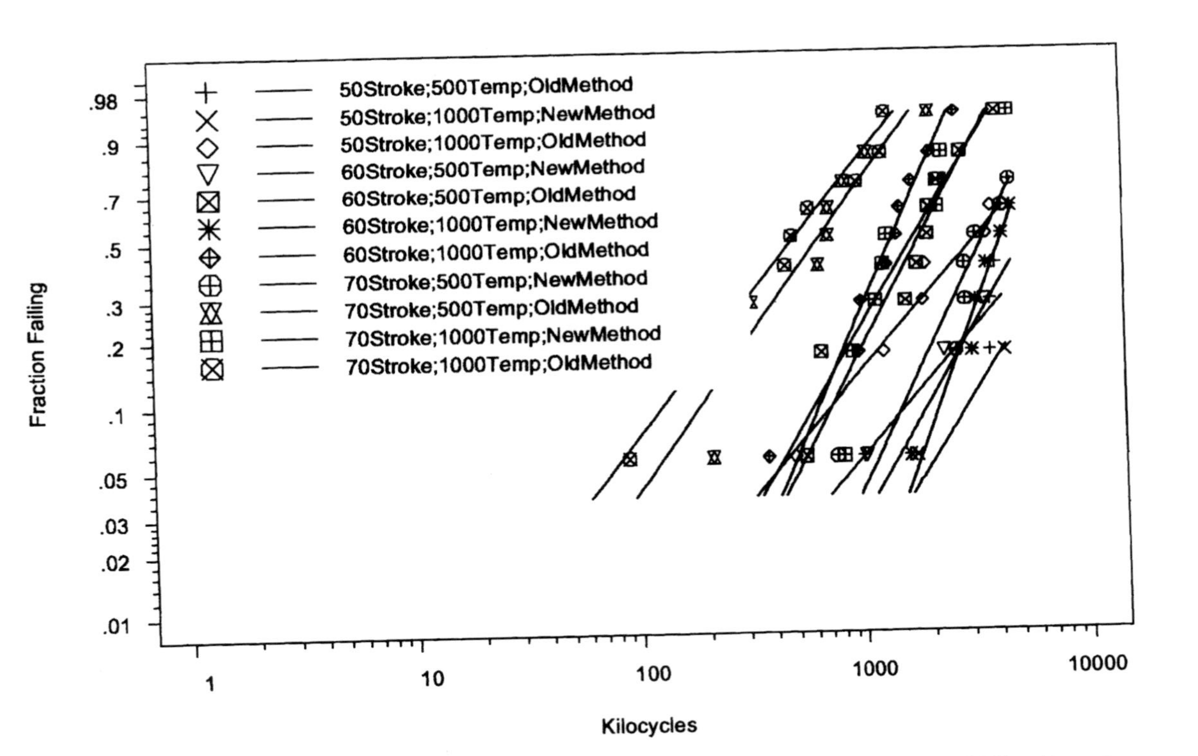

Figure 6.2. Weibull probability plot of the spring accelerated life test data with individual ML estimates of $F(t)$ and legend.

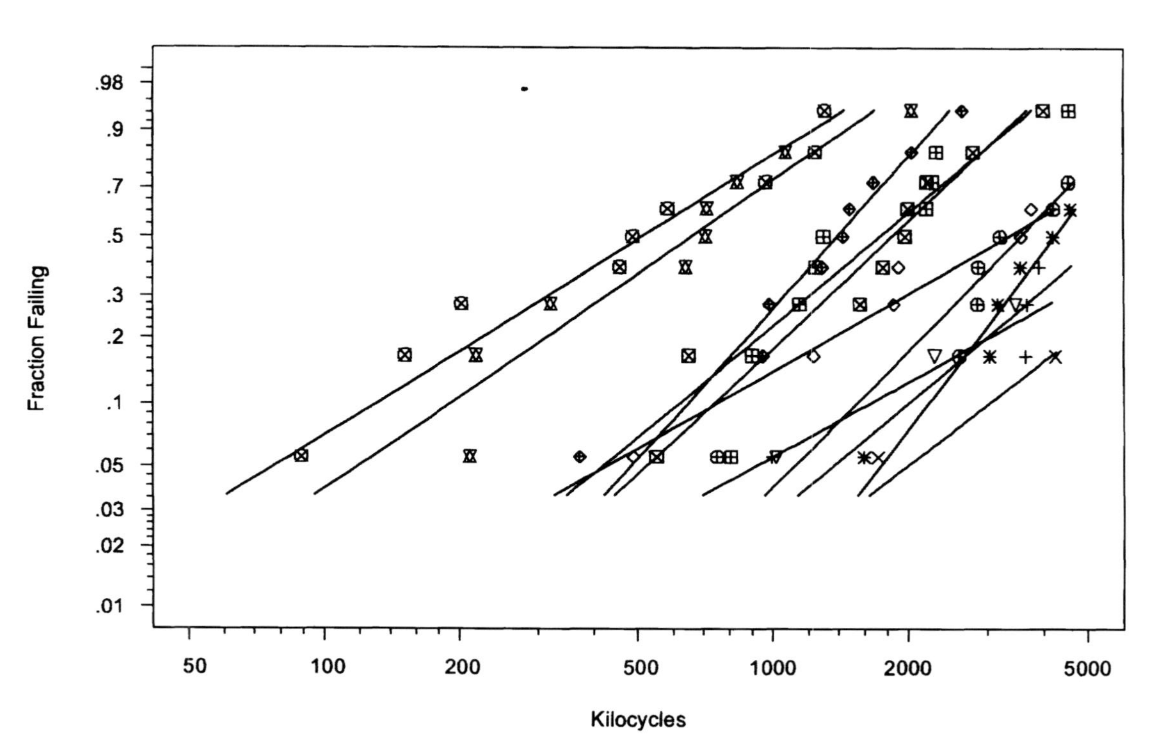

Figure 6.3. Weibull probability plot of the spring accelerated life test data with individual ML estimates of $F(t)$ using data-defined kilocycles axis and no legend.

Table 6.1. Weibull ML Estimates of the Individual Parameters (η, β) at Distinct Factor-Level Combinations for the Spring-Accelerated Life Test Data

Spring Fatigue Data
Maximum likelihood estimation results:
Response units: kilocycles

Weibull Distribution

Stroke, Temp, Method	Log likelihood	eta	se_eta	beta	se_beta
1 50Stroke;500Temp;NewMethod	NA	NA	NA	NA	NA
2 50Stroke;500Temp;OldMethod	–39.85	6639.6	2122.7	1.868	0.8845
3 50Stroke;1000Temp;NewMethod	–21.59	11104.8	7802.6	1.723	1.1772
4 50Stroke;1000Temp;OldMethod	–56.45	4430.4	1446.9	1.265	0.4536
5 60Stroke;500Temp;NewMethod	–31.17	10101.2	6480.1	1.236	0.6723
6 60Stroke;500Temp;OldMethod	–74.39	2193.1	372.5	2.067	0.5406
7 60Stroke;1000Temp;NewMethod	–55.10	4781.6	677.9	2.920	1.0591
8 60Stroke;1000Temp;OldMethod	–70.45	1606.2	228.4	2.466	0.6460
9 70Stroke;500Temp;NewMethod	–63.58	4169.5	700.5	2.250	0.7468
10 70Stroke;500Temp;OldMethod	–67.43	835.9	194.1	1.521	0.3799
11 70Stroke;1000Temp;NewMethod	–74.49	2108.8	408.3	1.831	0.4420
12 70Stroke;1000Temp;OldMethod	–66.08	665.7	169.7	1.378	0.3730

Total log likelihood = – 620.6.

where there were no failures). Table 6.2 also provides ML estimates of the common Weibull shape parameter β and $\sigma = 1/\beta$.

6.2.3 Test for Weibull Shape Parameter Homogeneity

Comparing Figures 6.3 and 6.4 show some differences among the shape parameter estimates obtained from the individual ML fits and the ML estimate obtained from the floating scale model. A formal test can be used to see if the observed differences can be explained by natural variability under the floating scale model. From Tables 6.1 and 6.2, the total log likelihood values from the corresponding models are –620.6 and –623.9, respectively. The log likelihood ratio statistic for the comparison is $Q = 2 \times [-620.6 - (-623.9)] = 6.6$. In large samples, under the null hypothesis that the Weibull shape parameter is the same in all groups, the log likelihood ratio statistic (under standard regularity conditions that are met here) has an approximate chi-square distribution with degrees of freedom equal to the difference in the number of parameters in the full and the reduced models. The difference in the number of parameters estimated in the two models (again ignoring the "50Stroke; 500Temp; NewMethod" combination where there were no failures) is $22 - 12 = 10$ degrees of freedom. The the approximate p value for the test comparing these two different models is $\Pr(\chi^2_{10} > 6.6) = 0.237$. This indicates that the differences among the slopes in Figure 6.3 can be explained by chance alone.

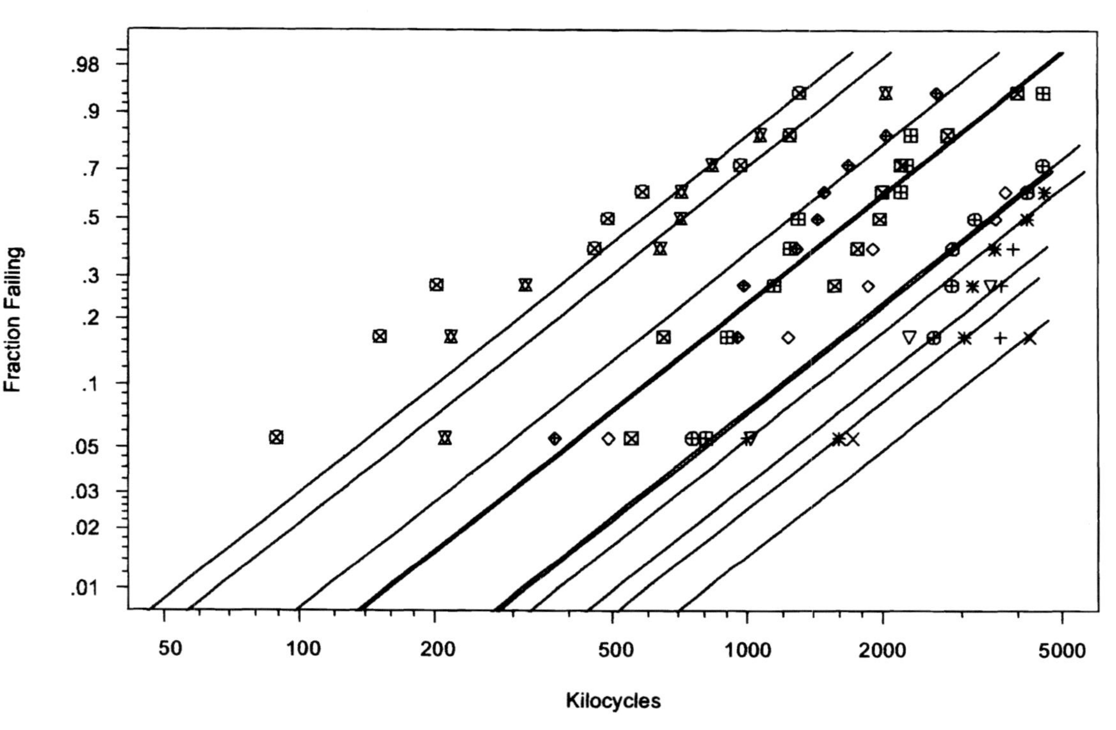

Figure 6.4. Weibull probability plot of the spring accelerated life test data with individual ML estimates of $F(t)$, assuming a common Weibull shape parameter.

Table 6.2. Weibull ML Estimate of the Floating Scale Model for the Spring Accelerated Life Test Data

Spring Fatigue Data
Maximum likelihood estimation results:
Response units: kilocycles

Weibull Distribution
Relationship(s)
1 StrokeTempMethod: class
Model formula:
Location ~ StrokeTempMethod
Log likelihood at maximum point: –623.9

			Approximate Confidence Interval	
Parameter	MLE	Standard Error	95% Lower	95% Upper
Intercept	8.9828	0.33058	8.3349	9.6307
70Stroke;500Temp;NewMethod	–0.6426	0.39322	–1.4133	0.1281
50Stroke;500Temp;OldMethod	–0.1609	0.43220	–1.0080	0.6862
60Stroke; 500Temp;OldMethod	–1.3203	0.38343	–2.0718	–0.5688
70Stroke; 500Temp;OldMethod	–2.2057	0.38629	–2.9628	–1.4486
50Stroke;1000Temp;NewMethod	0.3113	0.51700	–0.7020	1.3246
60Stroke;1000Temp;NewMethod	–0.4542	0.40150	–1.2412	0.3327
70Stroke;1000Temp;NewMethod	–1.3383	0.38483	–2.0925	–0.5840
50Stroke;1000Temp;OldMethod	–0.6151	0.40279	–1.4046	0.1743
60Stroke;1000Temp;OldMethod	–1.6525	0.38263	–2.4025	–0.9026
70Stroke;1000Temp;OldMethod	–2.4059	0.38573	–3.1619	–1.6499
sigma	0.5656	0.05429	0.4686	0.6827
weibull.beta	1.7680	0.16970	1.4649	2.1340

6.2.4 Residual Analysis

Residual analysis is used to detect possible departures from a fitted model and is an important part of any regression analysis. Censoring complicates the analysis of residuals, but the ideas and methods are basically the same as those described in standard regression textbooks. Residual analysis for censored data is described specifically in Nelson (1973), Nelson (1990), and Meeker and Escobar (1998).

Figure 6.5 is a Weibull probability plot of the residuals for the floating scale model. This plot provides a clear assessment of the Weibull distributional assumption, pooling all of the available data after they have been standardized to a common scale. Assessment of the adequacy of the distributional assumption at this stage of the modeling process has the advantage that the assessment can be done without the influence and possible bias introduced by the structure of a regression model.

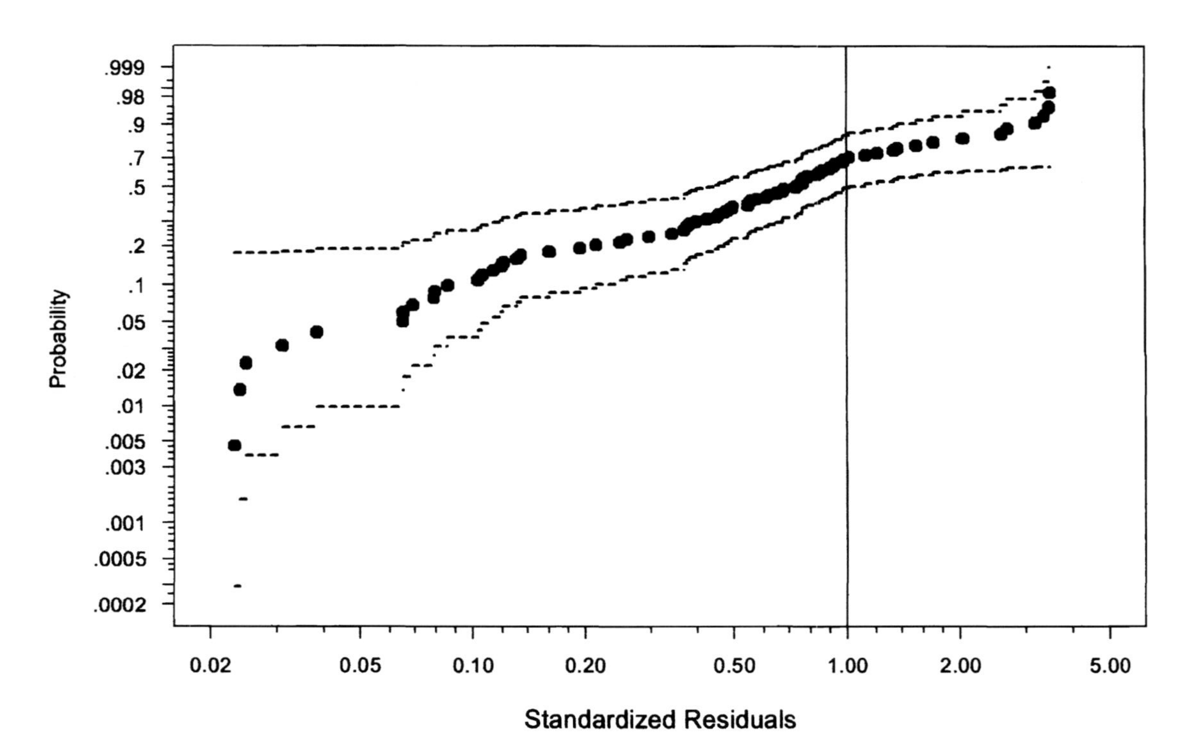

Figure 6.5. Weibull probability plot of the spring accelerated life test data residuals under the floating scale model. Residual probability plot with 95% simultaneous confidence bands; StrokeLog. TempLinear, MethodClass, Dist:Welbull.

The knee in the lower tail of the distribution in Figure 6.5 results from deviation in the two smallest residuals. This kind of deviation can be expected from chance alone as can be seen from the width of the simultaneous confidence bands in this part of the plot, as described in Section 6.3 of Meeker and Escobar (1998). The possibility of seeing such deviations can also be demonstrated by using Monte Carlo simulations like those described in Section 6.6.1 of Meeker and Escobar (1998). With repeated generation of pseudorandom Weibull samples, behavior like that seen in the lower tail of the distribution in Figure 6.5 is not uncommon. Thus it appears that the fatigue life data can be described adequately by a Weibull distribution.

A log-normal analysis (details not included here) also provided a reasonable fit to the data, but it is not as good as the Weibull fit.

6.3 RESPONSE SURFACE MODEL ANALYSIS

This section describes the construction and evaluation of a model that relates spring fatigue life to the factors in the experiment.

6.3.1 Acceleration/Response Surface Model

The response surface model suggested by the engineers was

$$\mu = \beta_0 + \beta_1 \log(\text{Stroke}) + \beta_2\text{Temp} + \beta_3 X \tag{6.3}$$

where $X = 0$ for Method = New and $X = 1$ for Method = Old (known as the "contrast treatment" method of coding dummy variables in S-PLUS). The log transformation for Stroke was chosen on the basis of previous experience and tradition. There was no previous experience relating the processing temperature to life so no transformation was used.

The results from fitting this model are given in Figure 6.6 and summarized in Table 6.3. This figure is similar to Figure 6.4 except that now the $\eta = \exp(\mu)$ values for each test condition are given by the relationship in equation (6.3). ML estimates of the regression coefficients are all statistically different from 0. This can be seen by noting that none of the confidence intervals contain zero. The response surface model allows computation of an estimate of $F(t)$ at condition "50Stroke; 500Temp; NewMethod" shown, the rightmost line on Figure 6.6, even though there were no failures at those conditions.

6.3.2 Test for Departure from the Acceleration/Response Surface Model

To assess the adequacy of the response surface model, we can use Figure 6.6 to compare the nonparametric estimates (the plotted points) with the fitted $F(t)$ lines, at each of the combinations of experimental factors. Because of the additional con-

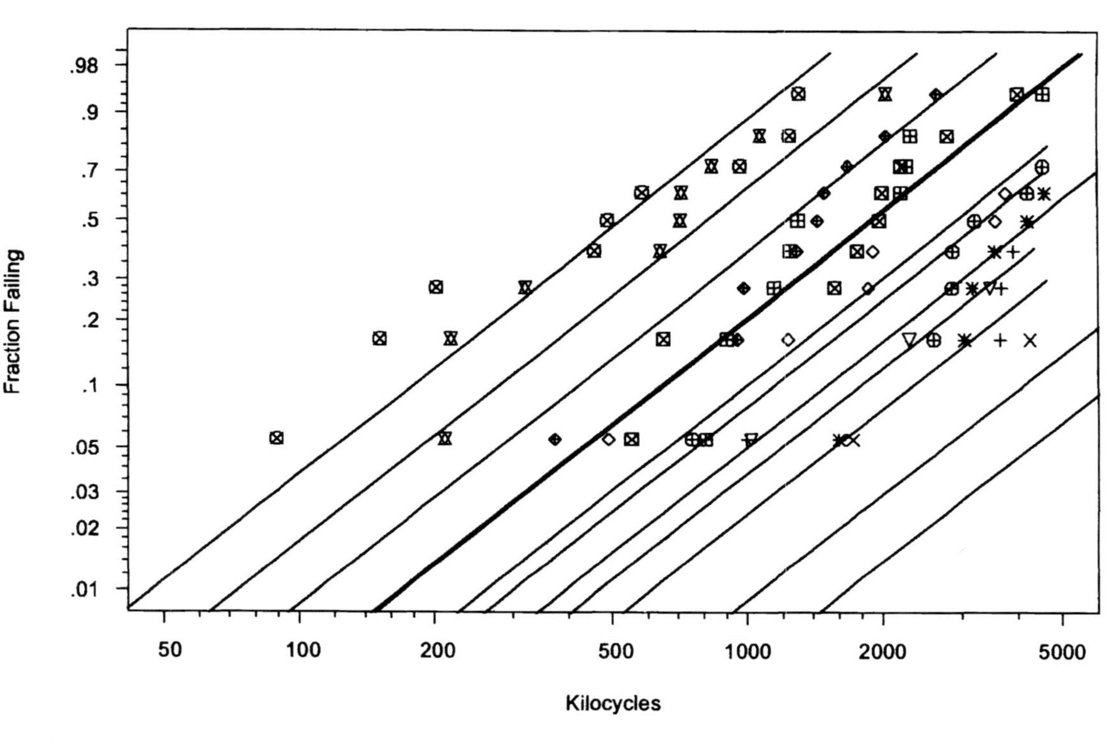

Figure 6.6. Weibull probability plot of the spring accelerated life test data with ML response surface estimates of $F(t)$.

Table 6.3. ML Estimates of the Linear Response Surface Model for the Spring Accelerated Life Test Data

Spring Fatigue Data
Maximum likelihood estimation results:
Response units: kilocycles
Weibull distribution

Relationship(s)
1 Stroke: log
2 Temp: linear
3 Method: class

Model formula:
Location ~ g(Stroke) + Temp + Method

Log likelihood at maximum point: –625.8

Parameter	MLE	Standard Error	Approximate Confidence Interval 95% Lower	Approximate Confidence Interval 95% Upper
(Intercept)	32.026951	2.4843571	27.157700	36.8962012
g(Stroke)	–5.509575	0.5872085	–6.660482	–4.3586675
Temp	–0.000883	0.0002709	–0.001414	–0.0003521
Method	–1.272388	0.1475136	–1.561510	–0.9832671
sigma	0.569491	0.0539010	0.473067	0.6855691
weibull.beta	1.755954	0.1661969	1.458642	2.1138651

straint of the relationship in equation (6.3) between the location of the distributions and the explanatory variables, there will be more deviations between the plotted points on nonparametric estimates and the fitted Weibull $F(t)$ lines in Figure 6.6 than in Figure 6.4.

To test whether such deviations are statistically important, as opposed to being explainable by the natural random variability in the data, under the relationship in (6.3), we can again do a likelihood ratio test, this time comparing the results in Tables 6.2 and 6.3. The total log likelihood values from the corresponding models are –623.9 and –625.8, respectively. The log likelihood ratio statistic for the comparison is $Q = 2 \times [-623.9 - (-625.8)] = 3.8$. The difference in the number of parameters estimated in the two models (again ignoring the "50Stroke; 500Temp; NewMethod" combination where there were no failures) is 12 – 5 = 7. The approximate p value for the test comparing these two different models is $\Pr(\chi_7^2 > 3.8) = 0.1975$, indicating that the differences between the two models can be explained by chance alone. This implies that there is not any strong evidence for lack of fit in the fitted regression model.

6.3.3 Testing for interactions

Although the relationship in equation (6.3) appears to fit the data well, it is important to make sure that there is no evidence of interaction in the data. To do such a check, we fit the following model:

$$\begin{aligned}\mu = \beta_0 + \beta_1 \log(\text{Temp}) + \beta_2\text{Stroke} + \beta_3 X \\ + \beta_4 \log(\text{Temp}) \times \text{Stroke} + \beta_5 \log(\text{Temp}) \times X + \beta_6\text{Stroke} \times X \end{aligned} \quad (6.4)$$

that adds two-factor interactions to the additive model in equation (6.3). As in equation (6.3), $X = 0$ for Method = New and $X = 1$ for Method = Old. Details of the model fit are not given here, but the total log likelihood for this model was –624.6, which is very close to the value of –623.9 for model (6.3). The log likelihood ratio statistics for the comparison is $Q = 2 \times [-624.6 - (-623.9)] = 2.4$ with a difference in the number of parameters of $8 - 5 = 3$. The corresponding p value is $\Pr(\chi^2_3 > 2.4) = 0.716$. Thus there is no evidence for interaction.

6.3.4 Residual Diagnostics

Residual plots reveal departures from a fitted model. A traditional residual plot for this purpose plots the residuals versus the fitted $\eta = \exp(\mu)$ values from the model. Such a plot for the spring fatigue life data and relationship in equation (6.3) is shown in Figure 6.7. In this example, there is one vertical line of points for each of the 12 combinations of the explanatory variables (although the fitted values for two of the combinations are so close together that it is difficult to see any separation). The triangles indicate right-censored residuals, corresponding to right-censored observations. Thus the actual residual, had there been no censoring, would have been larger (higher in Figure 6.7). Recognizing the meaning of the censored residuals, this plot does not indicate any apparent deviation from the assumed model.

6.3.5 Comparison of Old and New Springs

One of the purposes of the experiment was to compare the new and the old processing methods with respect to fatigue life. Figure 6.8 is a conditional model plot, giving estimates of the quantiles of the fatigue life distribution for the new and old methods, conditional on values of the other factors fixed at Temp = 600 and Stroke = 20, the nominal values of these variables. The densities are actually smallest extreme value densities, corresponding to Weibull distributions being plotted on a log response axis.

Figure 6.8 suggests that the new method will produce springs with a longer fatigue life. Quantifying this finding, the coefficient corresponding to MethodOld in Table 6.3 is –1.272. Under the constant Weibull shape parameter model, this quantity has the interpretation of the difference between quantiles of the distributions of

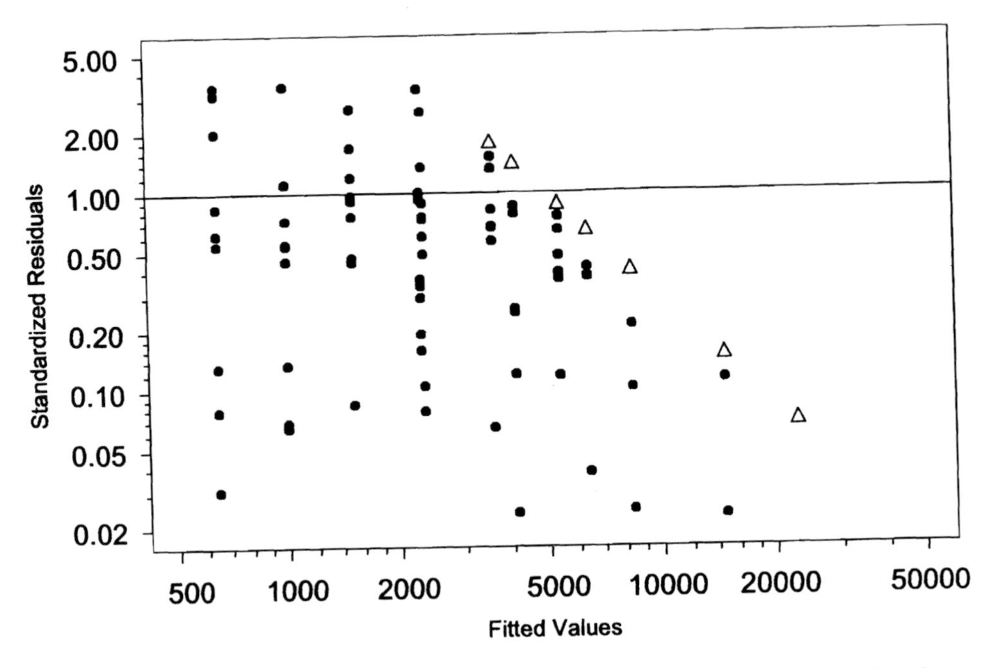

Figure 6.7. Plot of the spring-accelerated life test data residuals versus fitted values $\eta = \exp(\mu)$ from the model using equation (6.3).

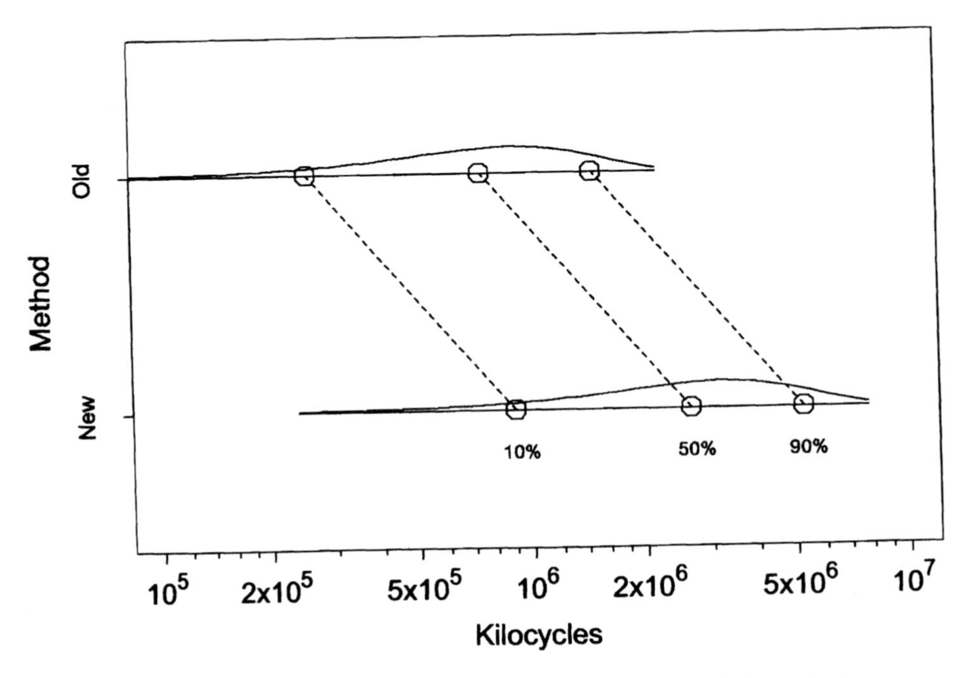

Figure 6.8. Conditional model plot showing the effect of processing method on spring life.

the old and new methods, on the log scale (e.g., $\mu_{\text{Old}} - \mu_{\text{New}}$). Alternatively, the ML estimate of any given quantile (such as B10) of the fatigue life distribution for the new method is $\exp(\mu_{\text{Old}} - \mu_{\text{New}}) = \exp[-(\mu_{\text{New}} - \mu_{\text{Old}})] = \exp(1.272) = 3.57$ times larger than that for the old method. Again taking numbers from Table 6.3, a 95% confidence interval for this improvement factor is

$$[\exp(0.9832672), \exp(1.561510)] = [2.67, 4.77]$$

Thus there is strong statistical evidence that the new processing method results in springs with a longer fatigue life distribution.

6.3.6 Estimate of B10 at Nominal Conditions

Figure 6.9 is similar to Figure 6.6, except that it shows the ML estimate for $F(t)$ for the new processing method with Temp = 600 and Stroke = 20, the nominal values of these variables. This plot provides a visualization of the distribution of primary interest and the amount of extrapolation needed to make the desired inference.

The ML estimate of the 0.10 quantile (B10) of the fatigue life for the new processing method at Temp = 600 and Stroke = 20 is 900 megacycles. A normal approximation 95% confidence interval for B10 is [237, 3,412] megacycles. The details for how to compute this estimate and the associated confidence interval are given in Section 19.2.4 of Meeker and Escobar (1998). Although the ML estimate exceeds the target value of 500 megacycles by a sizable margin, the lower confidence bound leaves some doubt as to whether the new spring will meet the reliability target if operated at 20 mils.

6.4 EFFECT OF STROKE DISPLACEMENT ON SPRING LIFE

6.4.1 Conditional Model Plot

Figure 6.10 is a conditional model plot showing the fatigue life of springs manufactured with the new method at Temp = 600, as a function of Stroke. This plot also provides a useful visualization of extrapolation to Stroke = 20 mils. Putting life on the horizontal axis is common practice in the fatigue literature when plotting cycles as a function of stress.

The confidence intervals shown in Figure 6.9 shows that there is a large amount of uncertainty in the estimate of $F(t)$ at the nominal use conditions. This is primarily due to the large amount of extrapolation. It is important to recognize, however, that the width of the confidence interval reflects only statistical uncertainty due to a limited sample size. The interval does *not* reflect possible model error. In applications (such as the present application) model error could be substantially large. We investigate potential model error in the next section.

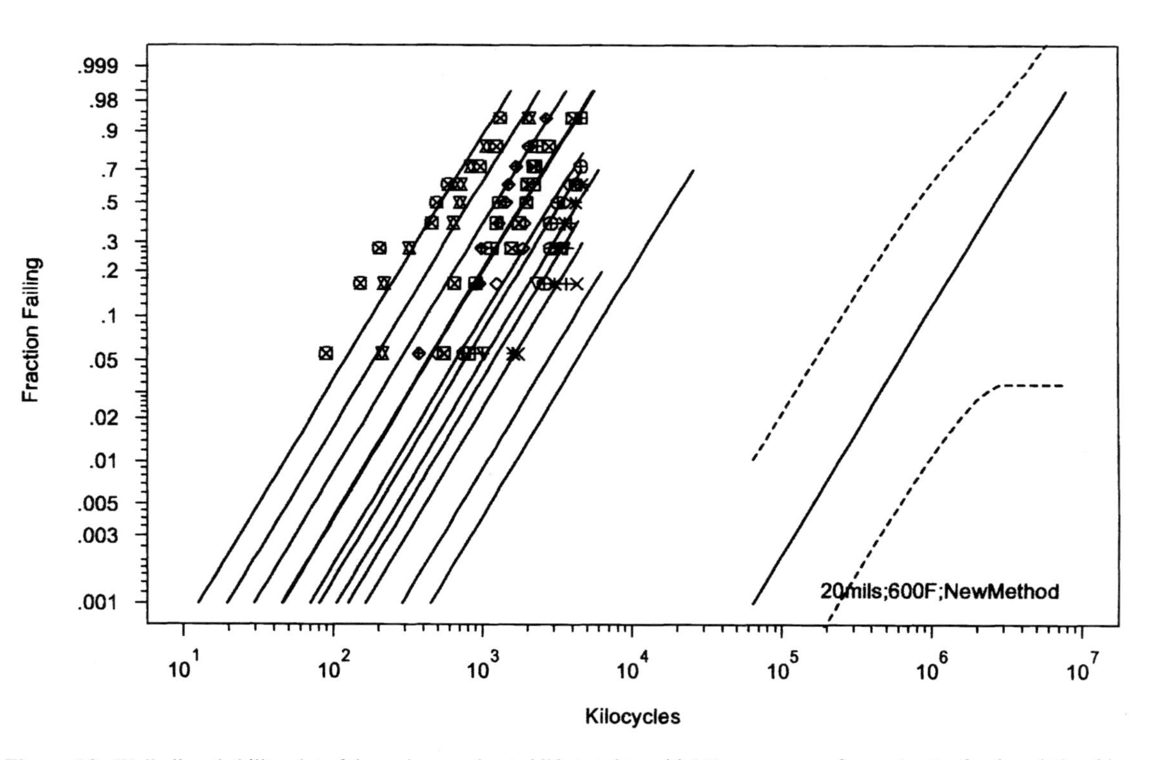

Figure 6.9. Weibull probability plot of the spring accelerated life test data with ML response surface estimates for the relationship in equation (6.3) with extrapolation to stroke displacement of 20 mils and normal approximation 95% pointwise confidence intervals.

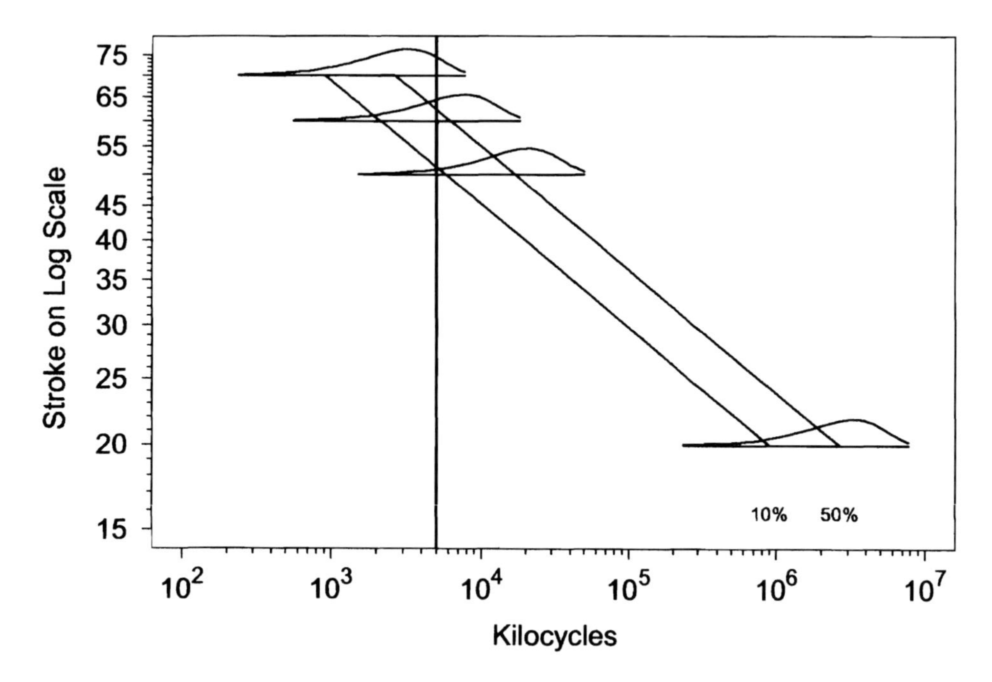

Figure 6.10. Conditional model plot showing the relationship between spring life and stroke displacement at 600°F for the new processing method.

6.4.2 Sensitivity to the Assumed Form of the Stroke–Life Relationship

The authors of the articles in Saltelli, Chan, and Scott (2000) describe different approaches to sensitivity analysis. One general, but useful, approach is to expand the formulation of the model by adding a parameter or parameters and investigating the effect of perturbing the added parameter(s), to see the effect on answers to questions of interest.

We take this approach here by extending the model in equation (6.3). We start by replacing log(Stroke) with the more general Box–Cox transformation (Box and Cox, 1964) on Stroke. In particular, we fit the model

$$\mu = \beta_0 + \beta_1 W + \beta_2 \text{Temp} + \beta_3 X \tag{6.5}$$

where

$$W = \begin{cases} \dfrac{(\text{Stroke})^{\lambda} - 1}{\lambda} & \lambda \neq 0 \\ \log(\text{Stroke}) & \lambda = 0 \end{cases} \tag{6.6}$$

and, as in equation (6.3), $X = 0$ for Method = New and $X = 1$ for Method = Old. The Box–Cox transformation (Box and Cox, 1964) was originally proposed as a simplifying transformation for a response variable. Transformation of explanatory variables, however, provides a convenient extension of our regression modeling choices. Note that because W is a continuous function of λ, equation (6.6) provides a continuum of transformations for possible evaluation and model assessment. The Box–Cox transformation parameter λ can be varied over some range of values (e.g., –1 to 2) to see the effect of different stroke–life relationships on the fitted model and inferences of interest. The results from the analyses can be displayed in a number of different ways.

For the spring fatigue life example, Figure 6.11 is a plot of the 0.10 Weibull quantile (B10) estimates versus λ between –1 and 2. Approximate confidence intervals are also given. Note that $\lambda = 0$ corresponds to the log transformation that is commonly used in fatigue life versus stress models. Also, $\lambda = 1$ corresponds to no transformation (or, more precisely, a linear transformation that affects the regression parameter values but not the underlying structure of the model). Figure 6.11 shows that fatigue life decreases by more than an order of magnitude as λ moves from 0 to 1. In particular, the ML estimate of the 0.10 quantile decreases from 900 megacycles to 84 megacycles when λ is changed from 0 to 1.

Figure 6.12 is a profile likelihood plot for the Box–Cox λ parameter, providing a visualization of what the data say about the value of this parameter. In this case the peak is at a value of λ close to 0; this is in agreement with the commonly used fa-

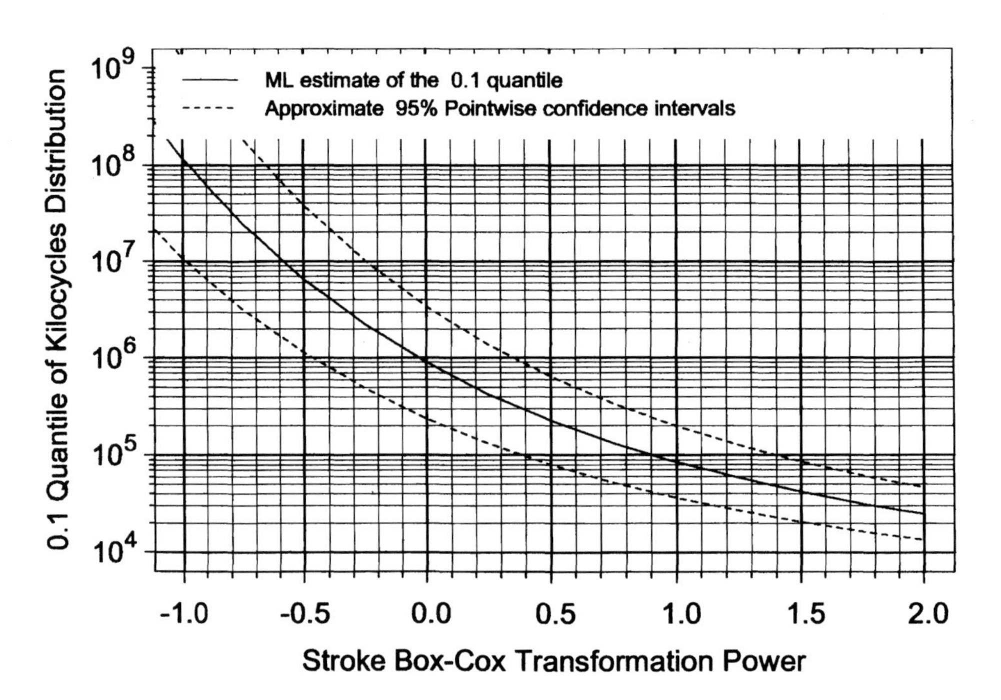

Figure 6.11. Plot of the ML estimate of the 0.10 quantile of spring life at 20 mils, 600°F, using the new method versus the Stroke displacement Box–Cox transformation parameter λ with 95% confidence limits.

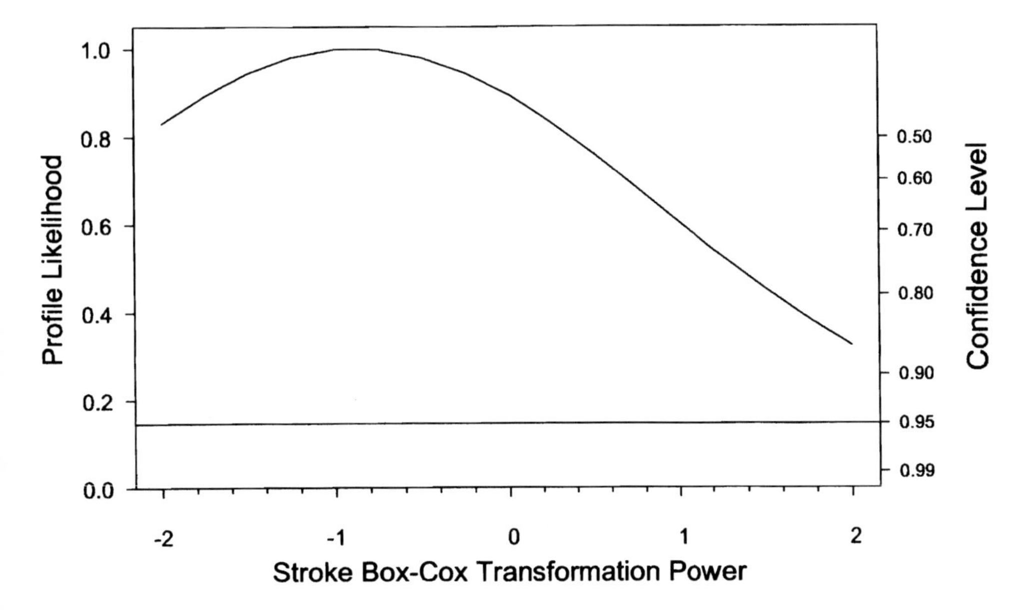

Figure 6.12. Profile likelihood plot for the Stroke Box–Cox transformation parameter λ in the spring life model.

tigue life/stress model. Values of λ close to 1 are less plausible, but cannot be ruled out, based on these data alone.

The engineers, based on experience with the same failure mode and similar materials, felt that the actual value of λ was near 0 (corresponding to the log transformation) and almost certainly less than 1. Then a conservative decision could be made by designing with an assumed value of $\lambda = 1$. Then, the somewhat optimistic evaluation in Section 6.4 would become somewhat pessimistic, relative to the 500-megacycle target.

6.4.3 Sensitivity to the Assumed Form of the Temperature–Life Relationship

The same kind of sensitivity study can be done with temperature. We omit the details here, as they are similar to the presentation in Section 6.4.2, but we present the graphical results. Figure 6.13 is a conditional model plot showing the fatigue life of springs manufactured with the new method at Stroke = 20 mils, as a function of Temp. This plot suggests that fatigue decreases somewhat with the level of this processing temperature. The nominal value of 600°F is, however, in the high-life region. The engineers were also concerned about the adequacy of the assumed relationship between fatigue life and Temp, because they had no previous experience modeling temperature. Figure 6.14 is similar to Figure 6.11, except that it is the temperature relationship that is being perturbed. The figure shows that the estimates of the fatigue life distribution are not highly sensitive to the transformation power λ. Interestingly, the

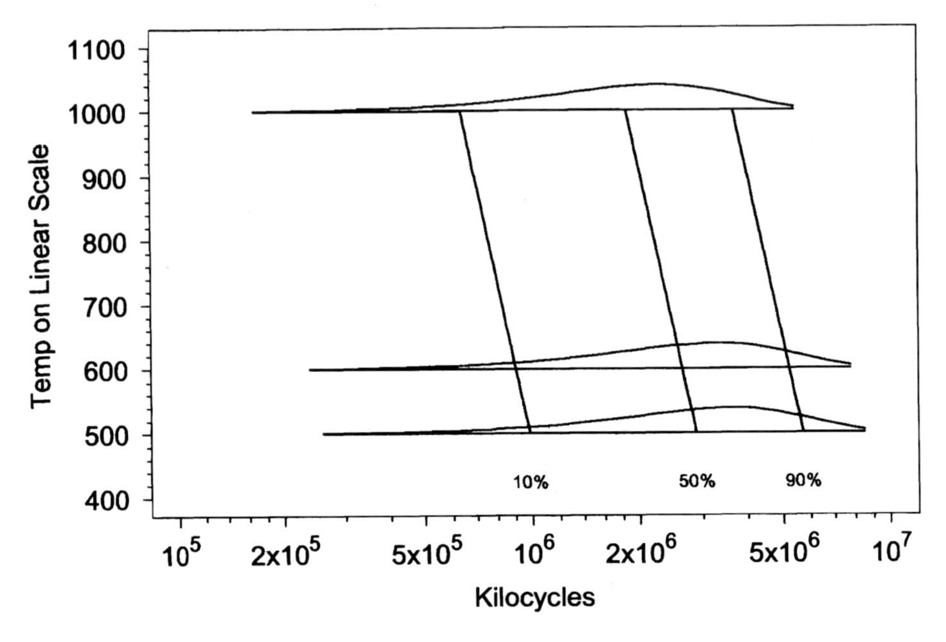

Figure 6.13. Conditional model plot showing the relationship between spring life and processing temperature for Stroke = 20 mils and the new processing method.

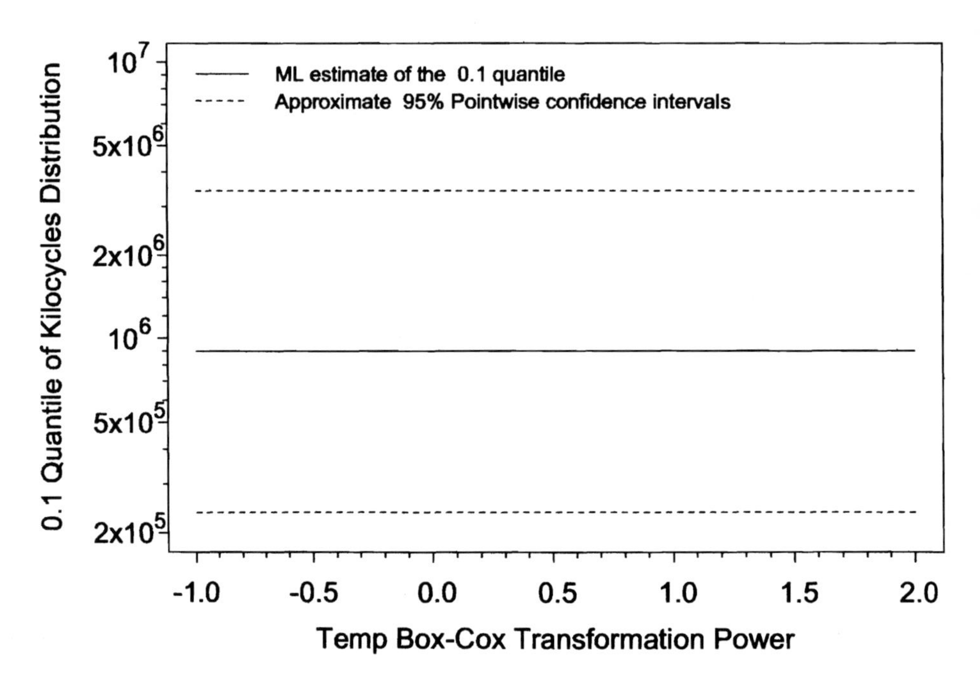

Figure 6.14. Plot of the 0.10 quantile of spring life versus the temperature Box–Cox transformation parameter λ with 95% confidence limits.

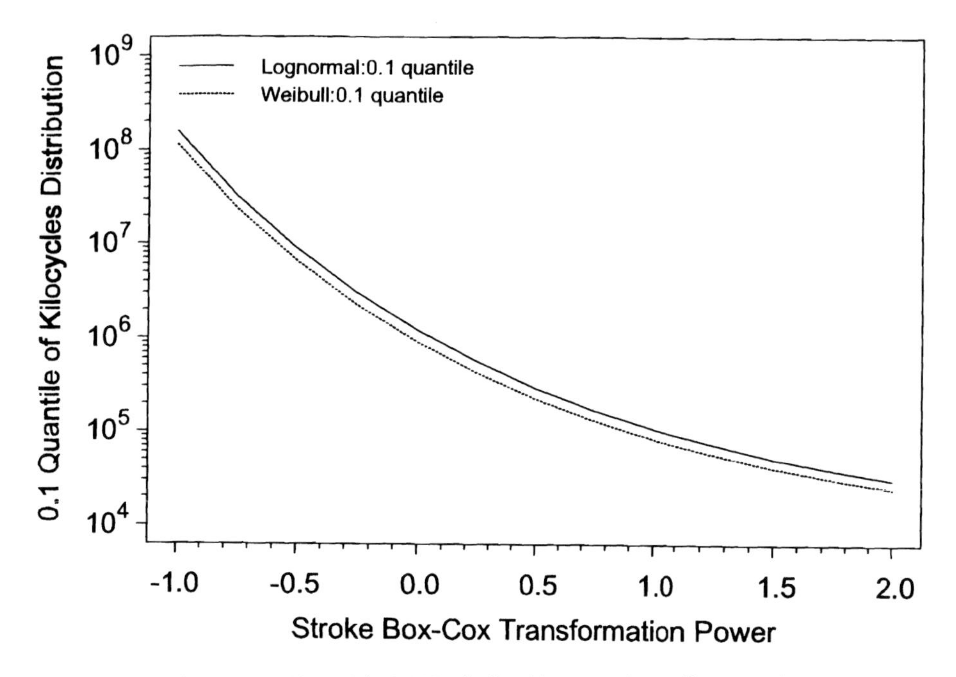

Figure 6.15. Plots of the 0.10 Weibull and log-normal quantiles versus λ.

profile likelihood for the Temp Box–Cox transformation parameter λ (not shown here) is perfectly flat. This is because the data (with only two levels of temperature) do not provide any information about the nature of the transformation.

6.4.4 Sensitivity to the Assumed Distribution

The Weibull distribution fits the data well and it is easy to demonstrate that, in the extrapolation to the lower tail of the distribution, the Weibull distribution will give predictions of life that are shorter than those given by the log-normal distribution. To confirm this we can compare directly estimates from the two different distributions. Figure 6.15 is similar to Figure 6.11, except that instead of giving confidence intervals, plots of the 0.10 Weibull and log-normal quantiles versus λ are shown. The plot shows that the log-normal distribution estimates are optimistic relative to the Weibull distribution by about a factor of 1.3 over this range of λ. The factor for the 0.01 quantile is approximately 2.

6.5 CONCLUDING REMARKS

It is often suggested that the extrapolation involved in accelerated testing requires the use of a model based on the physics or chemistry of the failure mechanism. While such a theoretical basis for extrapolation is clearly important and desirable, there are situations where important decisions need to be made with a model that is not .rmly grounded in such theory. Due to limited time and resources that preclude the timely development of such theory, alternative approaches are needed. Most commonly, empirical models are used, based on previous experience and engineering judgment. Because of the uncertainty in such models, sensitivity analyses becomes especially important.

This chapter shows how one can use sensitivity analyses to explore the effect that perturbations to the assumed model have on inferences. Even when a model based on the physics or chemistry of the failure mode is available, such sensitivity analyses are still important.

ACKNOWLEDGMENTS

We would like to thank Wayne Nelson, Wallace Blischke, and Pra Murthy for helpful comments on earlier versions of this chapter.

REFERENCES

Box, G. E. P., and Cox, D. R. (1964). An analysis of transformations (with discussion), *Journal of the Royal Statistical Society, Series B* **26:**211–252.

Condra, L. W. (2001). *Reliability Improvement with Design of Experiments,* 2nd edition, Dekker, New York.

Meeker, W. Q., and Escobar, L. A. (1998). *Statistical Methods for Reliability Data,* Wiley, New York.

Meeker, W. Q., and Escobar, L. A. (2002). *SPLIDA User's Manual.* Available from www.public.iastate.edu/~splida.

Nelson, W. (1973). Analysis of residuals from censored data, *Technometrics* **15:**697–715.

Nelson, W. (1990). *Accelerated Testing: Statistical Models, Test Plans, and Data Analyses,* Wiley, New York.

Pascual, F. G., and Meeker, W. Q. (1999). Estimating fatigue curves with the random fatigue-limit model (with discussion), *Technometrics* **41:**277–302.

Saltelli, A., Chan, K, and Scott, E. M., Editors (2000). *Sensitivity Analyses,* Wiley, New York.

Wu, C. F. J., and Hamada, M. (2000). *Experiments: Planning, Analysis, and Parameter Design Optimization,* Wiley, New York.

EXERCISES

6.1 Use SPLIDA (or other available software) to replicate the analyses in this chapter.

6.2 Compare the plots and the total log likelihood values from steps 4 and 5 in Appendix A. What does this tell you? (The difference in the number of parameters estimated in the two steps is 22 – 12 = 10. A log likelihood difference of $\chi^2_{.80,10}/2 = 6.721$ or more between the log likelihood values might be considered to be big enough to be statistically important.)

6.3 What information did the conditional model plot in Figure 6.10 provide?

6.4 Overall, which explanatory variable (Temp or Stroke) seems to have a more important effect over the range of values used in the experiment? How can you tell?

6.5 If the old processing method had to be used, what would be a safe level of stroke displacement such that one could have 95% confidence that B10 will exceed 500 megacycles (500,000 kilocycles) if the processing temperature is 600°F?

6.6 For the temperature relationship sensitivity analysis, the likelihood profile plot for the Box–Cox parameter is perfectly flat? Why?

6.7 The estimate of fatigue life at the nominal use conditions 20Stroke;600Temp; NewMethod is highly sensitive to the transformation used for the Stroke variable but not very sensitive to the transformation used for the Temp variable. Explain why.

6.8 In Section 6.4.2 it was noted that the Box–Cox transformation is a continuous function of the parameter λ. To see this, show that

$$\lim_{\lambda \to 0} \frac{x^\lambda - 1}{\lambda} = \log(x)$$

6.9 Explain why the Box–Cox transformation with $\lambda = 1$ is, in terms of inferences on the failure time distribution, the same as no transformation.

6.10 Repeat the analysis in this chapter using instead the lognormal distribution. How do the results compare within the range of the data? How do they compare outside of the range of the data?

APPENDIX A. SPLIDA COMMANDS FOR THE ANALYSES

This appendix gives explicit direction on how to use the SPLIDA (Meeker and Escobar, 2002) software to do the the analyses described in this chapter.

1. Use the data frame `NewSpring` (an example built into SPLIDA) to make the life data object `NewSpring.ld`, using Stroke, Temp, and Method as explanatory variables. Status is the censoring variable. Use `Splida ->`

Make/edit/summary/view data object -> Summary/view data object or the object browser to view NewSpring.ld.

2. First use Splida -> Multiple regression (ALT) data analysis -> Censored data pairs plot to get scatter plots for all pairs of variables. Note that censored observations are denoted in the plots by an open triangle (Δ) symbol.
3. Use Splida -> Multiple regression (ALT) data analysis -> Censored data scatter plot to make a scatter plot of the lifetimes versus Temp and lifetime versus Stroke. Use a log axes for life.
4. Use Splida -> Multiple regression (ALT) data analysis -> Probability plot and ML fit for individual conditions to obtain a probability plot analysis to allow a comparison of the failure-time distributions for the different combinations of levels of the explanatory variables. Choose the Weibull distribution for analyses.
5. Use Splida -> Multiple regression (ALT) data analysis -> Prob plot and ML fit for indiv cond: common shapes (slopes) to fit a model that has the same distribution shape at each combination of the explanatory variables, but allows the scale of the distribution to float.
6. Use Splida -> Multiple regression (ALT) data analysis -> Probability plot and fit of regression (acceleration) model to begin fitting different regression models to the data. Choose Stroke, Temp, and Method (in that order) as the explanatory variables. Specify the nominal use conditions and New method as 20;600;New under "Specify new data for evaluation" on the Basic page of the dialog box. On the Model page, specify a log relationship for Stroke and a linear relationship for Temp. Visit the Tabular output page and request tables of failure probabilities and quantiles at 20;600;New. Click Apply and examine the results.
7. Use Splida -> Regression residual analysis -> Residuals versus fitted values to obtain plots of the residuals versus the fitted values to look for evidence of lack of fit for models that you fit.
8. Use Splida -> Multiple regression (ALT) data analysis -> Conditional model plot to get a plot of the estimates of the Spring life distribution as a function of Stroke when temperature is 600°F with the New processing method. On the Plot options page, request evaluation for Stroke over the range of 15 to 90 mils.
9. Do a similar evaluation, letting temperature vary from 400°F to 1000°F when Stroke = 20 mils, for the new processing method.
10. Use Splida -> Multiple regression (ALT) data analysis -> Sensitivity analysis plot to check the sensitivity of estimates of B10 to the assumed relationships for Stroke when Stroke is 20 mils and Temp is 600°F and the processing method is "New." On the Oth-

`er inputs` page, specify the evaluation powers by entering –2, 2, .5 in the appropriate cell. Then click on "Apply."

APPENDIX B. SPRING-ACCELERATED LIFE TEST DATA

Table 6.4. The Spring-Accelerated Life Test Data

Kilocycles	Stroke	Temp	Method	Status	Weight
5000	50	500	New	Suspended	9
3464	60	500	New	Failed	1
1016	60	500	New	Failed	1
2287	60	500	New	Failed	1
5000	60	500	New	Suspended	6
2853	70	500	New	Failed	1
3199	70	500	New	Failed	1
752	70	500	New	Failed	1
2843	70	500	New	Failed	1
4196	70	500	New	Failed	1
2592	70	500	New	Failed	1
4542	70	500	New	Failed	1
5000	70	500	New	Suspended	2
997	50	500	Old	Failed	1
3904	50	500	Old	Failed	1
3674	50	500	Old	Failed	1
3644	50	500	Old	Failed	1
5000	50	500	Old	Suspended	5
2193	60	500	Old	Failed	1
2785	60	500	Old	Failed	1
4006	60	500	Old	Failed	1
1967	60	500	Old	Failed	1
1756	60	500	Old	Failed	1
650	60	500	Old	Failed	1
1995	60	500	Old	Failed	1
1563	60	500	Old	Failed	1
551	60	500	Old	Failed	1
211	70	500	Old	Failed	1
319	70	500	Old	Failed	1
712	70	500	Old	Failed	1
707	70	500	Old	Failed	1
2029	70	500	Old	Failed	1
638	70	500	Old	Failed	1
1065	70	500	Old	Failed	1
834	70	500	Old	Failed	1
218	70	500	Old	Failed	1
4241	50	1000	New	Failed	1
1715	50	1000	New	Failed	1

(continues)

Table 6.4. The Spring-Accelerated Life Test Data ***Continued***

Kilocycles	Stroke	Temp	Method	Status	Weight
5000	50	1000	New	Suspended	7
3158	60	1000	New	Failed	1
3545	60	1000	New	Failed	1
4188	60	1000	New	Failed	1
4583	60	1000	New	Failed	1
1595	60	1000	New	Failed	1
3030	60	1000	New	Failed	1
5000	60	1000	New	Suspended	3
2196	70	1000	New	Failed	1
808	70	1000	New	Failed	1
2257	70	1000	New	Failed	1
1147	70	1000	New	Failed	1
1296	70	1000	New	Failed	1
1243	70	1000	New	Failed	1
2309	70	1000	New	Failed	1
4563	70	1000	New	Failed	1
901	70	1000	New	Failed	1
489	50	1000	Old	Failed	1
3756	50	1000	Old	Failed	1
1230	50	1000	Old	Failed	1
3562	50	1000	Old	Failed	1
1898	50	1000	Old	Failed	1
1855	50	1000	Old	Failed	1
5000	50	1000	Old	Suspended	3
1670	60	1000	Old	Failed	1
1481	60	1000	Old	Failed	1
371	60	1000	Old	Failed	1
2630	60	1000	Old	Failed	1
1285	60	1000	Old	Failed	1
2031	60	1000	Old	Failed	1
951	60	1000	Old	Failed	1
1429	60	1000	Old	Failed	1
980	60	1000	Old	Failed	1
963	70	1000	Old	Failed	1
1240	70	1000	Old	Failed	1
1301	70	1000	Old	Failed	1
455	70	1000	Old	Failed	1
151	70	1000	Old	Failed	1
488	70	1000	Old	Failed	1
202	70	1000	Old	Failed	1
89	70	1000	Old	Failed	1
583	70	1000	Old	Failed	1

CHAPTER 7

Virtual Qualification of Electronic Hardware

Michael Osterman, Abhijit Dasgupta, and Thomas Stadterman

7.1 INTRODUCTION

As manufacturers of electronic products struggle to find new ways to accelerate product maturity and reduce time to market, the use of simulation techniques to assess the expected reliability of a product under field conditions has gained increased importance. In particular, in the area of electronic packaging, where the rate of growth has limited the effectiveness of traditional reliability approaches, simulation-guided testing has been shown to be an effective approach for developing and fielding reliable products.

Simulation-guided testing involves virtual qualification and physical verification. Virtual qualification is the application of simulation to determine the ability of a system to meet desired life-cycle goals. Physical verification is the process of defining and conducting physical tests on the product, or a representative product, to demonstrate that the simulation adequately modeled the anticipated use-induced failures. A schematic of the process is presented in Figure 7.1. In recent years, the application of simulation-guided testing as a method for assessing product reliability has been demonstrated on a number of electronic systems (Osterman et al., 1997, Osterman and Stadterman, 1999, Osterman and Stadterman, 2000; Cunningham et al., 2001). The overall approach to this process and its benefits will be discussed in the remainder of this section.

For this discussion, electronic hardware in the form of a circuit card assembly (CCA) will be introduced. A CCA, also referred to as a printed wiring assembly (PWA), is a common building block of electronic hardware. A CCA consists of a flat organic board typically constructed of woven glass fabric held in an epoxy matrix with metallization to form conductive paths on its surfaces. In many applications, the number of electrical connections requires the board to be constructed of

Case Studies in Reliability and Maintenance, Edited by W. R. Blischke and D. N. P. Murthy.
ISBN 0-471-41373-9

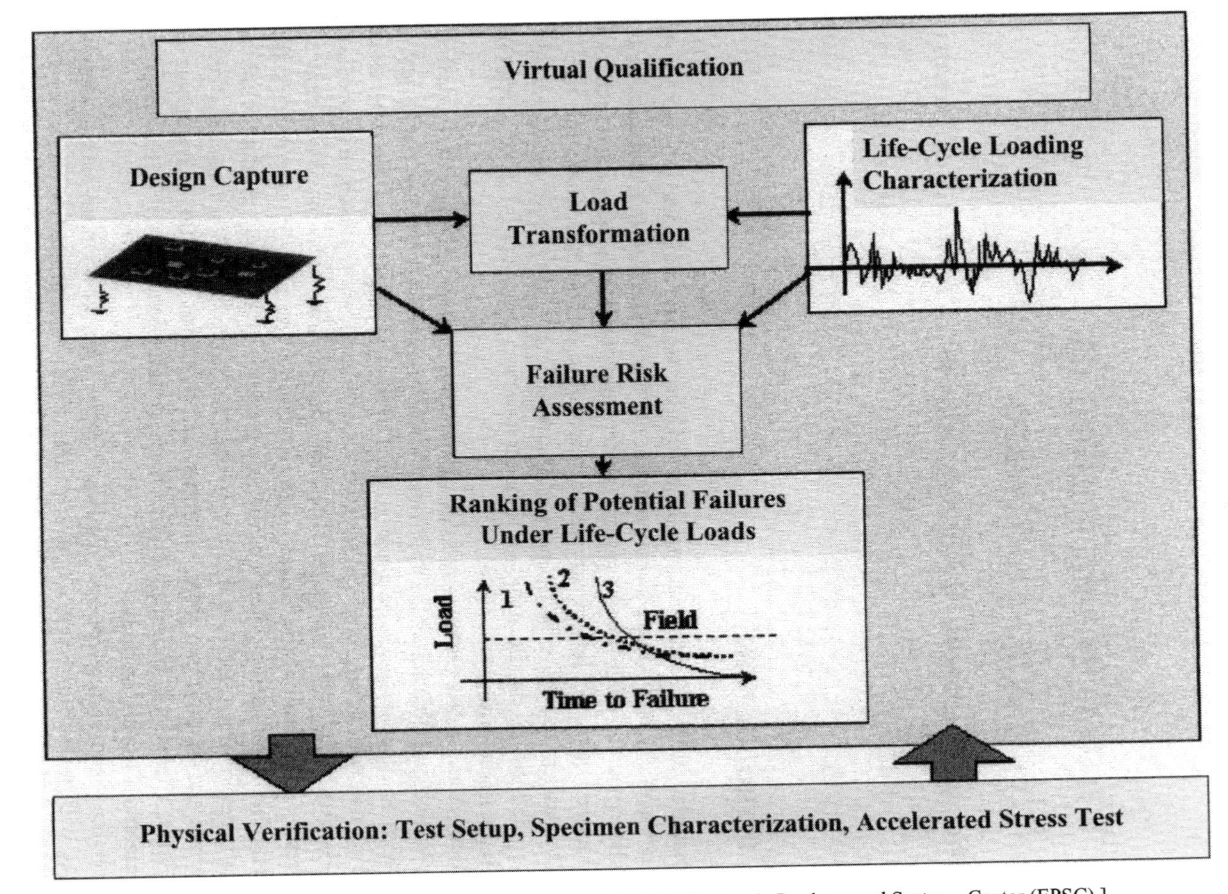

Figure 7.1. Virtual qualification process. [Courtesy of CALCE Electronic Products and Systems Center (EPSC).]

multiple conducting or signal layers laminated together. To complete the electrical functions, active and passive electronic devices [such as resistors, capacitors, inductors, crystals, transistors, integrated circuits (ICs)] and connectors are attached to the surfaces of the board.

CCAs may fail due to improper design, software error, or failure of the physical hardware. Focusing on the hardware, failure may be due to an open circuit, a short circuit, or a change or shift in operation parameters. An open circuit represents a disconnection of a conducting path, while a short circuit represents an unintended conductive path. In some cases, a short circuit can cause a runaway situation where a circuit draws a sufficient amount of current to overheat and potentially initiate melting of the conductor and burning of the surrounding material. A parameter shift can represent a change in resistance, capacitance, or inductance due to material aging, the presence of moisture, or the presence of a contaminant. The underlying physical process that causes failure is referred to as the "failure mechanism."

Failures could occur in packaged electrical devices, at package-to-board interfaces, within the board, or at the electrical connector interface for the board. Electronic device failure mechanisms include electrical overstress, stress-driven diffusive voiding (SDDV), time-dependent dielectric breakdown (TDDB), metallization corrosion, or electromigration. Failure mechanisms associated with the package include die-attach failure, die fracture, wire-bond attach failure, and case cracking, where the die is the semiconductor device that is contained within the package. Assembly-level failure mechanisms include solder-joint interconnect fatigue or fracture, plated-through hole via fatigue, conductive filament formation, and opens in connectors due to stress relaxation or growth of oxide films or contamination. Fortunately, mathematical models have been developed for many of the dominant failure mechanisms. [See Pecht (1991), Lall et al. (1995), and Tummla and Rymaszeewske (1989).]

Virtual qualification relies on the use of mathematical models, referred to as *failure models,* to estimate the life expectancy of a product. The process of reliability assessment based on virtual qualification involves the following:

- Identifying the modes by which the product can fail (e.g., electrical shorts, opens, or functional deviations that result from the activation of failure mechanisms).
- Identifying mechanisms that can produce the identified failure modes (e.g., fatigue, fracture, corrosion, voiding, wear, or shift in functional properties).
- Identifying sites where failure can potentially occur (e.g., component interconnects, board metallization, and external connections), including the geometry and material, as well as the manufacturing flaws and defects.
- Identifying failure models for the identified mechanisms and sites that can be used to estimate the time to failure.
- Identifying and evaluating appropriate stress metrics (e.g., stresses due to steady-state conditions and transient changes in voltage, current, ionic concentration, moisture content, temperature, or vibration) based on imposed op-

erational and environmental conditions as well as on the geometry, material, and potential defects and flaws.

- Evaluating time to failure and the variation of time to failure based on variations in the inputs to the failure models.
- Ranking failure sites in ascending order of time to failure and determining the minimum time to failure.
- Accepting design if analysis indicates that design meets or exceeds required time to failure with anticipated variation.

Virtual qualification provides an approach for quantifying the time to failure of a product for an assumed loading condition(s). Failure models are based on documented failure modes and mechanisms. In many cases, failure models require inputs that must be determined from the geometry, material, and applied loading conditions. Examples of loading conditions can include, but are not limited to, repetitive temperature cycles, sustained temperature exposure, repetitive dynamic mechanical load (e.g., due to vibration), sustained electrical bias, electrical transients, sustained humidity exposure, and exposure to ionic contamination. Failure models may be classified as overstress or wearout. Overstress models calculate whether failure will occur based on a single exposure to a defined loading condition. For an overstress model, the simplest formulation is comparison of an induced stress versus the strength of the material that must sustain that stress. Examples of overstress loading include voltage spikes, electrostatic shock, thermal shock, thermal runaway, and mechanical shock. Corresponding failure mechanisms include EOS, ESD, catastrophic phase transitions (solid to liquid/gas; crystalline to glassy/amorphous; etc.), and fracture. Wearout models calculate whether failure will occur due to repeated or prolonged exposure to a defined loading condition. For a wearout model, the simplest formulation is a determination of survival time due to exposure to the stress condition. Examples of wearout mechanisms are fatigue, corrosion, electromigration, TDDB, SDDV, conductive filament formation, and wear.

Another important aspect of virtual qualification is the variability of the time-to-failure results. The variability of the time to failure results based on the variability of input data can be examined with a systematic approach (i.e., Monte Carlo analysis). Furthermore, the completeness and accuracy of the virtual qualification process can be judged based on the failure mechanisms considered and the models used to quantify time to failure. With the appropriate failure models and stress simulation techniques, virtual qualification can be used to identify intrinsic design weaknesses and to conduct trade-off studies by estimating the time to failure and its distribution based on the product's environment and the operational characteristics of the product. Virtual qualification can be used for identifying and ranking likely failure sites, for determining the intrinsic (i.e., ideal) reliability of the nominal product, for developing qualification tests for the product, and for determining acceleration transforms between test and use conditions. The method should be used in combination with physical testing to verify that simulation adequately captures the product failures.

In the following sections, the virtual qualification process is demonstrated in the development of an automotive control module. In addition, physical testing of prototypes and a detailed simulation of a likely failure site are discussed.

7.2 AUTOMOTIVE MODULE CASE STUDY

Electronic hardware has become a critical part of automobiles, and the dependence on electronic systems is expected to continue to increase. As such, the reliability of electronic hardware that is responsible for operation and safety features on automobiles is a critical design concern. Furthermore, automobile manufacturers are increasing the warranty periods in an effort to maintain or expand market share. Thus, the ability to determine the reliability of electronic hardware is critical to maintaining profit and customer satisfaction.

Like many other segments of the electronics industry, automotive electronic designs have been traditionally developed based on a design–build–test–fix (DBTF) process. With tighter production schedules, DBTF can delay production if design deficiencies emerge during the testing phase. In addition, the time and cost involved in fabricating and testing multiple prototypes can be prohibitive. To reduce time and cost of development and to improve design reliability, engineers from a major automobile manufacturer investigated incorporating virtual qualification and physical verification in their development process. This section provides a discussion of that effort.

Since this effort was a demonstration program operating in the actual development process, the simulation guide testing was run in parallel with the DBTF approach. As a result, simulations were conducted on designs for which physical prototypes had been built. As a result of the virtual qualification process, the design was changed. In this text, the original design will be referred to as the alpha version, and the modified design will be referred to as the beta version. Due to time constraints, the actual order of test and simulation varied somewhat from the one presented in Figure 7.1. However, for the purpose of consistency, the discussion will follow the virtual qualification flowchart presented in Figure 7.1.

7.2.1 Design Capture

The information inputs for conducting virtual qualification are established by identifying a set of failure models for the anticipated geometries, materials, and loading conditions. Design capture entails collecting and assembling relevant information inputs from the subject of the virtual qualification process. For simplicity of illustration in this study, we will focus on failure of the electrical interconnects between packaged electronic components and the printed wiring board. In terms of interconnect failure, relevant information includes the material construction of the components and their interconnects, physical dimensions of the components, and the geometry of the interconnects. The physical dimensions and geometries are necessary to determine mechanical behavior of components under life cycle loads. Material information includes the coefficient of temperature expansion (CTE), elastic modulus, and strength and fatigue properties. In addition to the physical attributes of the components, important assembly attributes include the position of components on the printed wiring board, their power dissipation during operation, the support structure for vibration, and the expected heat removals paths.

The subject of this virtual qualification process is a body-control module that is responsible for controlling many interior functions of an automobile. The electronic module consists of a single CCA that is protected within a plastic housing (see Figure 7.2). The CCA consists of an FR4 printed circuit board (PCB) with copper circuit traces on both sides. The PCB was 152 mm wide, 144 mm long, and 1.65 mm thick. The required electrical functions are implemented through circuits created with over 400 components soldered to one side of the PCB.

The components include capacitors, resistors, field effect transistors (FETs), relays, packaged ICs, and other devices. Relevant information about the components includes physical dimensions, material construction, and power dissipation rates for the design. In this study, components of various sizes were constructed of plastic, ceramic, and metal with interconnects for attachment to the PCB. The form of the attachment is a function of the component's interconnect format. For this design, three types for interconnect formats were found. The first interconnect format, referred to as *insertion mount technology,* consists of metal leads that extend from the component. These leads are inserted through holes in the PCB and soldered to metallization on the board. The second interconnect format, termed *surface mount technology,* consists of metal leads or terminations on the component that are soldered

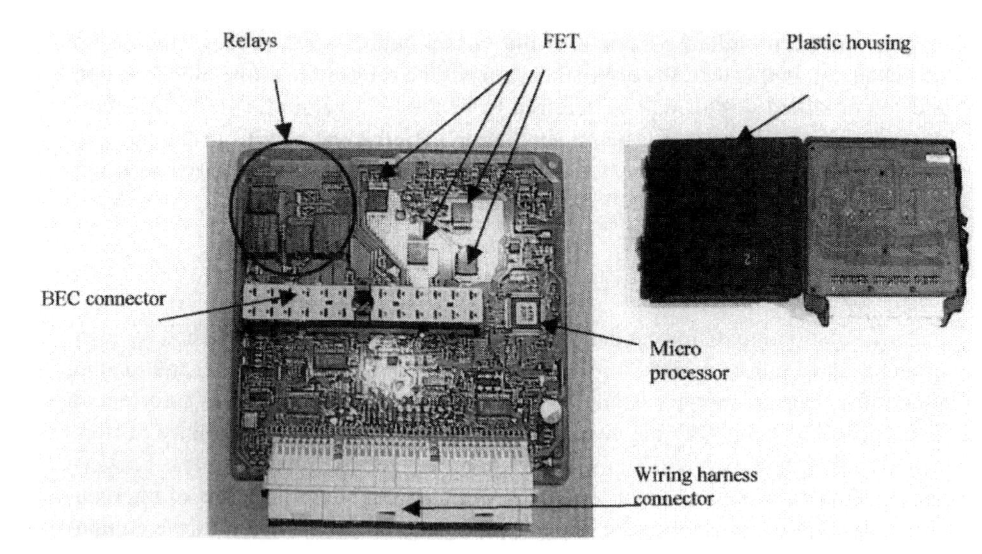

Figure 7.2. Prototype control module CCA and CCA in housing. [Courtesy of CALCE Electronic Products and Systems Center (EPSC).]

to metallization on the surface of the board. The metallization of the PCB is made to match the pattern of each component's interconnect format. Solder material for this study was a commonly used eutectic tin–lead alloy. Material properties of the components and board are summarized in Table 7.1. The plastic housing, which was intended to provide mechanical support and protection, was molded plastic (polypropylene with 20% talc). The overall dimensions of the plastic housing was 160 mm wide, 149 mm long, and 30 mm high. The complete assembly is intended to fit behind the dashboard in the passenger compartment of a vehicle.

7.2.2 Life-Cycle Load Characterization

Life-cycle load characterization is the process of defining the anticipated scenarios in which the subject of the virtual qualification will be used. For this design, the module is to be mounted behind the dashboard, and the anticipated dominant loads are assumed to come from stresses induced by vibration due the operation of the vehicle and temperature excursions due to operation and exposure to its surrounding environment. Vibration produces repetitive mechanical stress within the board, the packages, and the package-to-board interconnects that over time can lead to material fractures and electrical opens because of cyclic fatigue. Because of the use of dissimilar materials in the construction of the electronic hardware, temperature excursions also produce mechanical stresses in packages, board, and package-to-board interconnects, due to differences in thermal expansions. Temperature excursion in a product can result from operation, as well as from changes in the external ambient temperature. As with vibration, temperature excursions produce mechanical stresses that can lead to material failure. For this study, the vehicle's operational profile was assumed to be five trips per day for a total of eight hours. In examining the life-cycle loads, the focus was on quantifying the vibration and thermal loading conditions.

The severity and exposure time of the vibration environment is a function of the terrain over which the vehicle is driven. Since vibration involves the cyclic motion of a structure, it is characterized by the magnitude of cyclic displacement, velocity, and acceleration. In complex situations, vibration may be characterized in terms of frequency. A common expression for vibration is the power spectral density (PSD) curve, which plots the mean squared acceleration versus frequency. Higher PSD values indicate higher vibrational energy. Frequency is important because structures respond to vibration energy based on their geometry, material construction,

Table 7.1. Summary of Material Properties for CCA

Material	CTE (ppm/°C)	Modulus of Elasticity (GPa)
FR4	15–18	11–17
Plastic	19–22	30–60
Ceramic	7–9	120–130
Metal	5–16	120–140

and support structures. In particular, supported structures react to specific frequencies of excitation by vibrating in a particular pattern. These patterns are referred to as "characteristic mode shapes," and the frequencies at which they occur are referred to as "natural frequencies." Lower natural frequencies result in larger internal stresses within the assembly. The square of the natural frequency of a structure is proportional to its stiffness and inversely proportional to its mass. To reduce vibration-induced stress levels, it is desirable to design the structure such that its first natural frequency is higher than the frequency of the input vibrational excitation. Studies reported by Steinberg (1991) suggest that for vehicles the largest amount of input energy occurs at frequency ranges below 40 Hz. A recent study by Jih and Jung (1998) indicates that the PSD intensity falls significantly above 100 Hz with a root mean square acceleration of approximately 1.8 times the acceleration of gravity. For this study, the actual vibration profiles were considered proprietary information and a representative profile is provided in Figure 7.3. Peak vibration is assumed to occur approximately one hour per day.

In terms of temperature excursions, under operation, the electronic components will generate heat that must be removed through conduction and convection to the air and the structures that surround the plastic housing. In turn, the surrounding structures are expected to see temperature excursions due to the other heat sources such as the engine and the external environments. Engelmaier (1993) suggests that the temperature for electronics in a passenger compartment of an automobile could range from –55°C to 95°C with temperature excursion ranging between 20°C and 80°C. For this assessment, a temperature profile was extrapolated from temperature measurements monitored from within an actual vehicle over a period of 24 hours. A normalized plot of the temperature is presented in Figure 7.4.

7.2.3 Load Transformation

Load transformation is the process of quantifying relevant stresses at potential failure sites that are expected due to the life-cycle loads identified in Section 7.2.2.

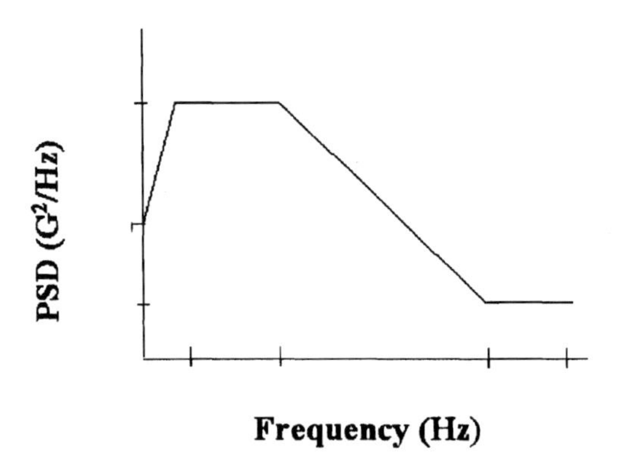

Figure 7.3. Random vibration power spectral density input curve.

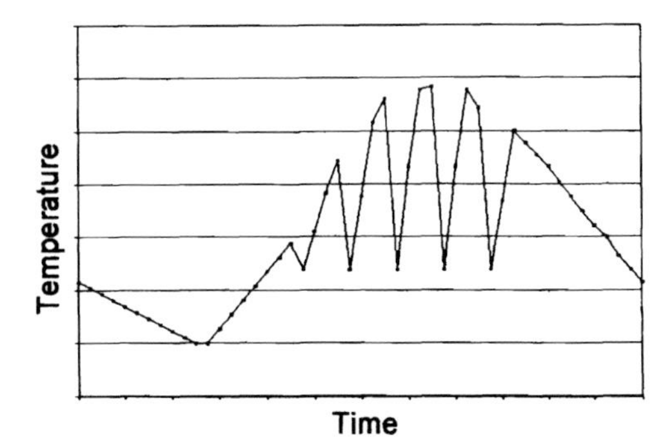

Figure 7.4. Life-cycle load (thermal). Axes are not labeled for proprietary reasons.

This is done by measurements on prototype mockups and/or by imposing the life-cycle loads on computer models of the actual product. In this study, simulation models of the initial prototype were developed by reviewing design documentation of the physical prototypes. In order to quantify the failures of the CCA, it was necessary to evaluate the response of the electronic module to anticipated life-cycle loads. As discussed in the previous section, the module is expected to see both temperature cycling and vibration loads. In the following subsections, the assessment of the operational temperatures at potential failure sites and the mechanical response of the CCA to vibration loading are discussed.

7.2.3.1 Thermal Analysis

To quantify the temperature excursion of the assembly, a thermal analysis was conducted. In conducting the thermal analysis, the thermal energy dissipated by the board and the components was assumed to be absorbed by the surrounding environment. Using component information, the position of the components on the board, and the defined board structure, a finite difference thermal analysis model was developed. The model assumed that heat was rejected to constant ambient heat sink temperature of 60°C. The total power dissipated by the board based on the individual components was determined to be 4.8 W. After multiple computational iterations to examine the effect of grid size, a maximum temperature rise of approximately 10°C above ambient was established. The temperature of the top layer of the board as calculated in the thermal analysis is depicted in Figure 7.5. The hot region of the board (the light region in Figure 7.5) was found to coincide with an area where several high-heat dissipating FETs were located. Analysis indicated that the operating case temperature of the FETs would be approximately 9°C above the board temperature. To provide a better thermal conduction path away from the high temperature region, the local conductivity was increased by adding thermal vias and increasing the copper in the lands near the FETs. This simulation activity was used to assist designers in this effort. With the modifications completed, the final simulation results were used as inputs to the failure assessment process.

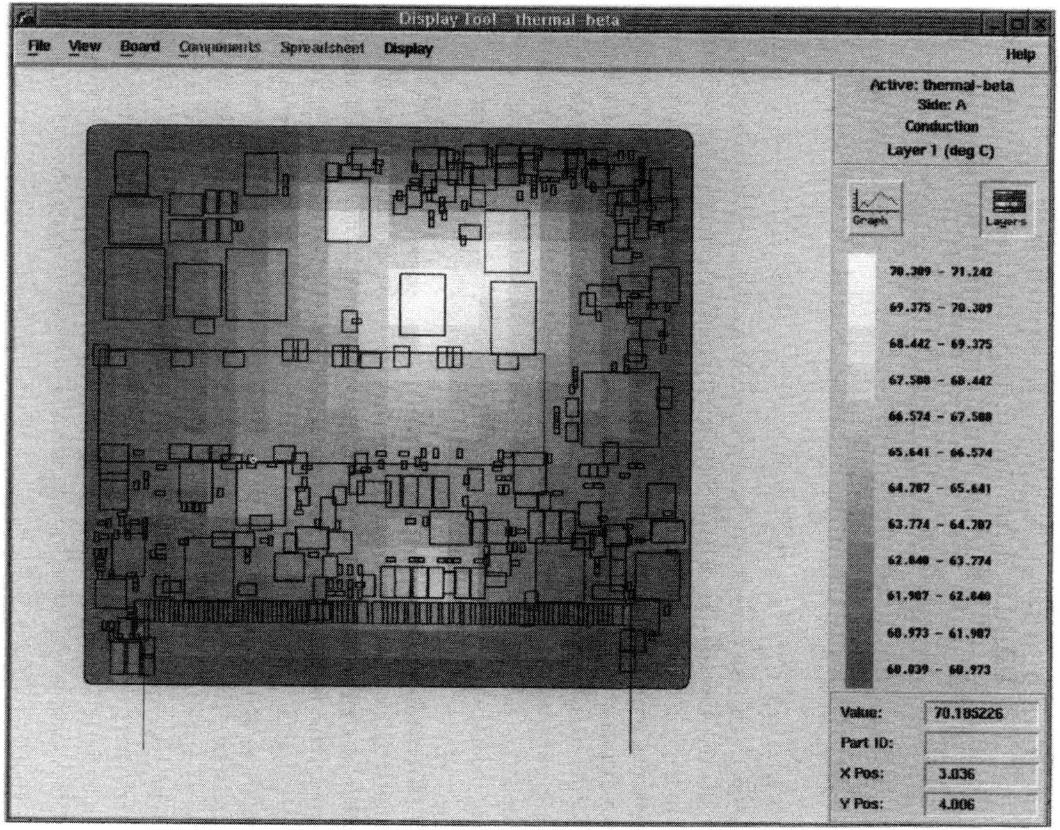

Figure 7.5. Thermal analysis boundary conditions and results.

7.2.3.2 Vibration Analysis

With regard to vibration, the plastic housing and the center connector are expected to support the CCA and act as a path for the transfer of vibrational energy. Vibration energy is expected to be transferred to the electronic module based on the response of the vehicle to road conditions and the overall transmissibility of vibration energy between the module's supports and vibration sources. After examining the housing, a finite element model of the CCA was constructed and a set of boundary conditions were applied. The model was subjected to the anticipated PSD input load. Analysis results for the alpha version of the design identified a high displacement and curvature region in an area where three large relays were to be mounted. Based on the analysis, the displacements and curvatures were found to be sufficiently large to prompt concerns with regard to the reliability of the relays and their interconnects. As a result of simulation, component substitutions were made and the support structure in the housing was modified.

After the redesign, the beta version of the CCA was assessed. The vibration analysis model for the beta version of the design is depicted in Figure 7.6. Simulation results provided a dynamic characterization of the CCA, as well as curvature and displacement results. The first fundamental mode shape is depicted in Figure 7.7. From this analysis, the calculated first natural frequency of the CCA was almost 700 Hz. Analysis of displacement based on a $0.04G^2$/Hz PSD over the 100- to

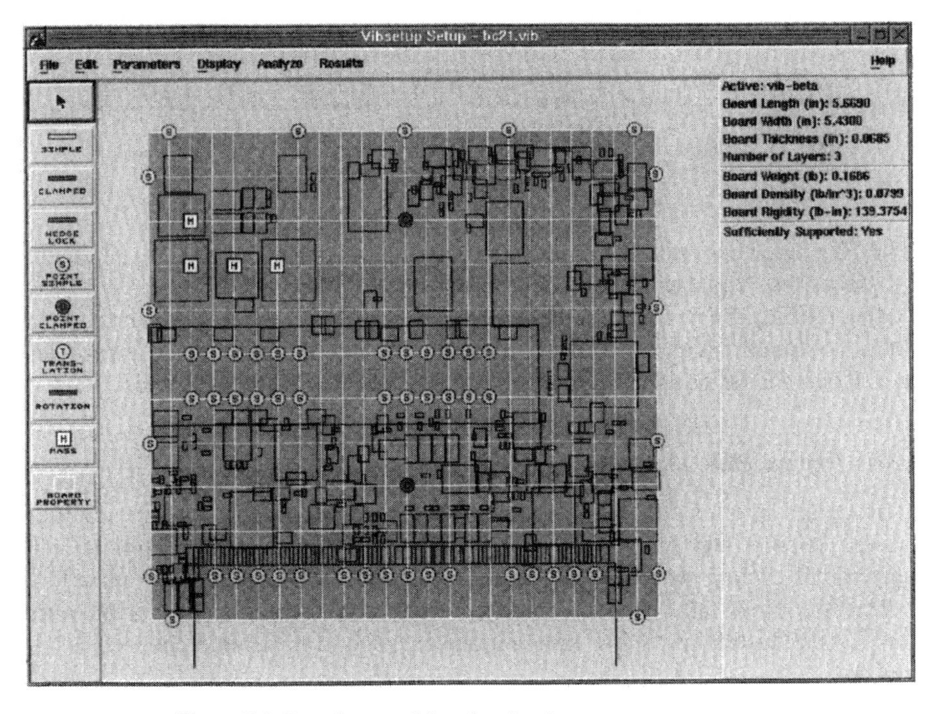

Figure 7.6. Boundary conditions for vibration response assessment.

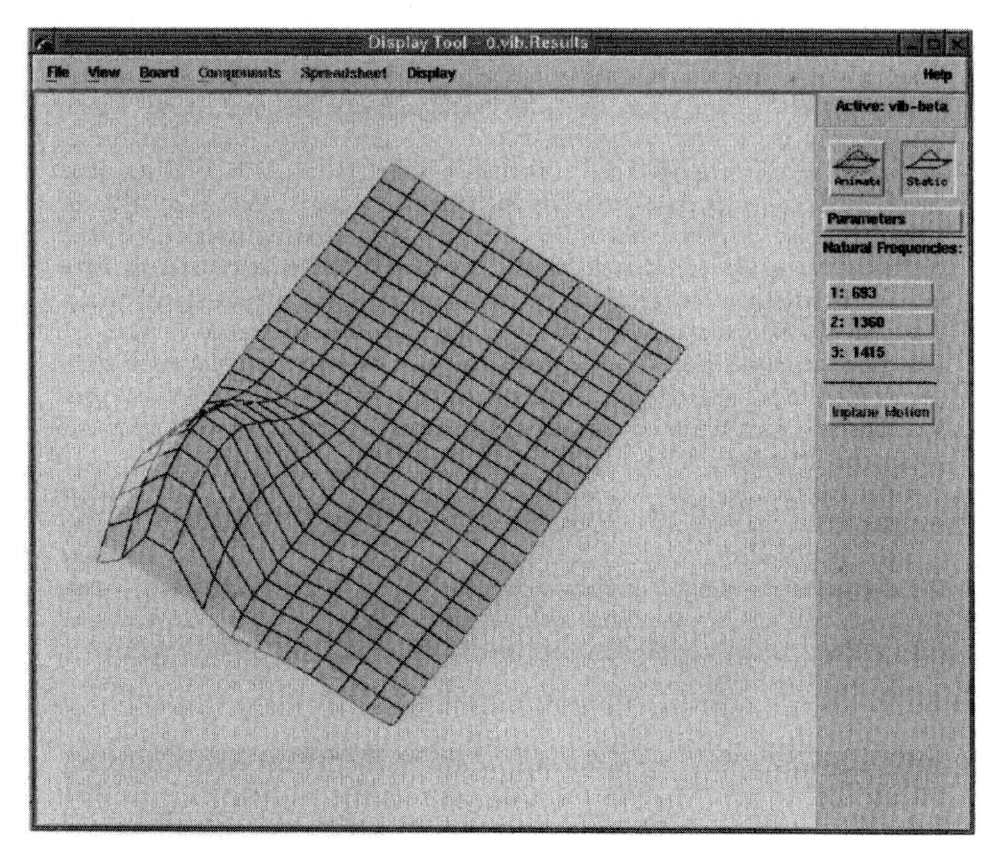

Figure 7.7. Vibration mode shapes based on boundary conditions depicted in Figure 7.6.

2000-Hz range indicated a maximum displacement of approximately 10 μm. Figure 7.8 shows the displacement of the beta version CCA based on the applied PSD loading profile. The curvature of the board under the individual components, along with the natural frequency of the assembly, and its overall displacement are provided as inputs for vibration failure models.

7.2.4 Failure Risk Assessment

As discussed in an earlier section, electronic hardware can fail from a variety of causes. Based on the anticipated loading conditions, fatigue due to vibration and temperature cycling was expected to be the most significant underlying failure mechanism. Thus, failure models for both temperature cycling and vibration were used. As an illustration, fatigue failure of the package-to-board interconnects is examined.

Fatigue failure occurs due to repeated reversals of stress that produce fissures and cracks in a material until a complete separation occurs. In general, fatigue fail-

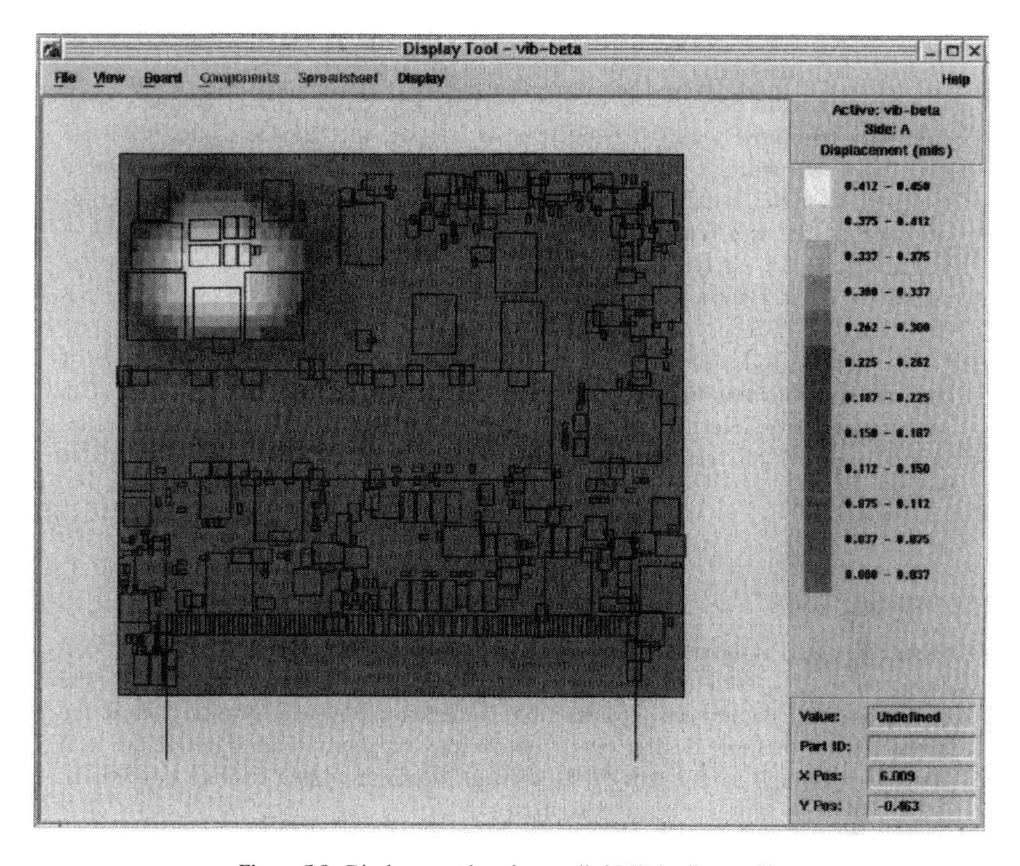

Figure 7.8. Displacement based on applied PSD loading profile.

ure models relate the number of stress reversals to failure with an appropriate stress metric. While there are a number of failure models for both vibration fatigue and temperature-cycling fatigue [Dasgupta et al. (1992), Darveaux (2000), Engelmaier (1993)], simple failure models are presented in this section to give the reader a flavor of the failure assessment process. For vibration-induced fatigue, a number of vibration cycles-to-failure (N_{vib}) may be related to the applied cyclic stress range of $\Delta\sigma$ in the form of

$$N_{\text{vib}} = C(\Delta\sigma)^b \tag{7.1}$$

where C and b are material properties. For temperature cycling, the number of temperature cycles to failure (N_{thermal}) may be related to the inelastic cyclic strain range ($\Delta\gamma$) in the form of

$$N_{\text{thermal}} = \frac{1}{2}\left(\frac{\Delta\gamma}{2\varepsilon_f}\right)^{\frac{1}{c}} \tag{7.2}$$

where ε_f and c are material properties that can vary with temperature and the rate of change of temperature .

Since vibration occurs over very short time scales and thermal fatigue occurs over very long time scales, it is difficult to compare the damage caused by these two types of loading. This damage accumulation process becomes even more complex when the load histories are nonuniform. To facilitate this process, the approach is to define a "damage index" (D) that can be used to quantify the damage per unit time for all types of loading. The cumulative damage can then be obtained by a linear superposition of the damage index at each load level, over the entire loading history.

The damage index at each load level is calculated as the ratio of the number of applied cycles to the number of survivable cycles at a particular load level. The number of survivable cycles, $N_{\text{survivable}}$, is estimated by the failure models discussed above. Thus, damage index at any time for a particular load level can be written as

$$D(t) = \frac{N_{\text{applied}}(t)}{N_{\text{survivable}}} \tag{7.3}$$

where $N_{\text{survivable}}$ is the determined by the failure model based on the particular load level and N_{applied} is determined by the frequency of the loading condition for the particular load level under consideration. If the same physical location fails in the same way by multiple loading conditions, it may be appropriate to define the total damage for a site as the sum of the damage from individual loading conditions. Mathematically, this can be expressed as

$$D_{\text{total}} = \sum_i D_i(t) \tag{7.4}$$

where each i represents a different loading condition. The time to failure, t_f, occurs when D_{total} equals 1. If D_{total} is estimated per unit time and averaged over a representative time period, then we can assume that damage will grow at a uniform rate. Then the time-to-failure is simply the reciprocal of the averaged D_{total}.

To conduct the failure assessment, the multiple load conditions established in the life-cycle load characterization process (temperature and vibration) and quantified in the load transformation process were combined in the failure risk assessment process to create a composite damage accumulation scenario that simulates the life-cycle conditions. The vibration load was assumed to have been imposed on the module for one hour every day, and a temperature cycle with a range of 60°C was assumed to be imposed five times a day. Using these data, failure models were evaluated for individual sites within the assembly, and the life expectancies of the sites were established. Results indicated that the solder interconnects of the surface mount leadless chip resistors are likely to be the locations of first interconnect failure. A graphical view of the failure results is presented in Figure 7.9. In Figure 7.9, lighter shading indicates a lower life expectancy. The simulation indicates that the

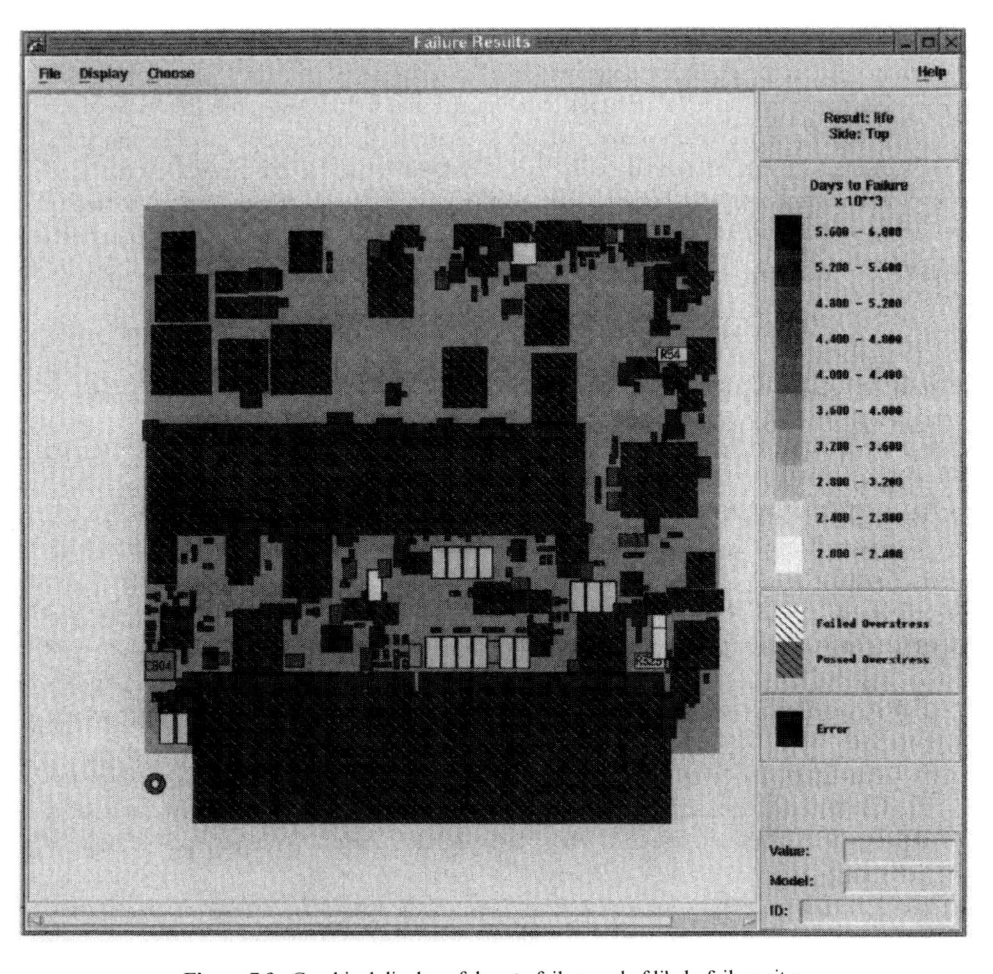

Figure 7.9. Graphical display of days to failure and of likely failure sites.

assembly should survive its life requirement under the prescribed life-cycle loading condition.

7.2.5 Physical Verification

Due to time constraints, physical testing was conducted concurrent with the virtual qualification process. As a result, the test was more of an exploration than a verification process. The testing process included the development of a fixture for the electronic module and physical testing. Both temperature cycling and vibration loads were applied to the test specimens during the physical testing. The physical tests included (a) step–stress tests to define acceptable stress levels and (b) accelerated life testing (ALT) to evaluate life expectancy by compressing the established field life through the application of elevated stress levels. For this study, eight operational electronic modules were available for testing.

To determine the level of loads at the individual failure sites, the CCAs were instrumented with sensors (strain gauges, thermal couples, and accelerometers). Placement of the individual sensors was guided by using the results of the load transformation process. For example, maximum board curvatures were expected to occur near a particular set of relays, and the hottest region of the board was expected to be near a particular set of FETs. Therefore, sensors were mounted in those locations.

In order to replicate the field environment, a special aluminum test assembly was designed to closely match the transmissibility of the mounting in the vehicle. The test assembly consisted of a mounting fixture, a cover plate, and an adapter plate to suit the bolt-hole pattern on the shaker table, as shown in Figure 7.10.

An electrical interface load simulator designed to replicate electrical load under field conditions with monitoring capability and associated wiring and connectors was used to power the test module and to monitor its operation. The simulator ran a custom software program that examined 15 critical functions of the module to detect failure. Since failures often first appear during elevated stress conditions, it was important to monitor the test module throughout the applied loading profiles. For this test, the electronic module was continuously monitored, while it was subjected to loads within the test chamber. The time and date of any failure was recorded during the testing. The simulator was programmed to check for failure once every minute.

Step–stress tests were conducted first to establish the load levels under which the module can operate, so as to avoid inadvertently failing the module by overstress during the accelerated life test. Thus, these limits determine the acceptable bounds

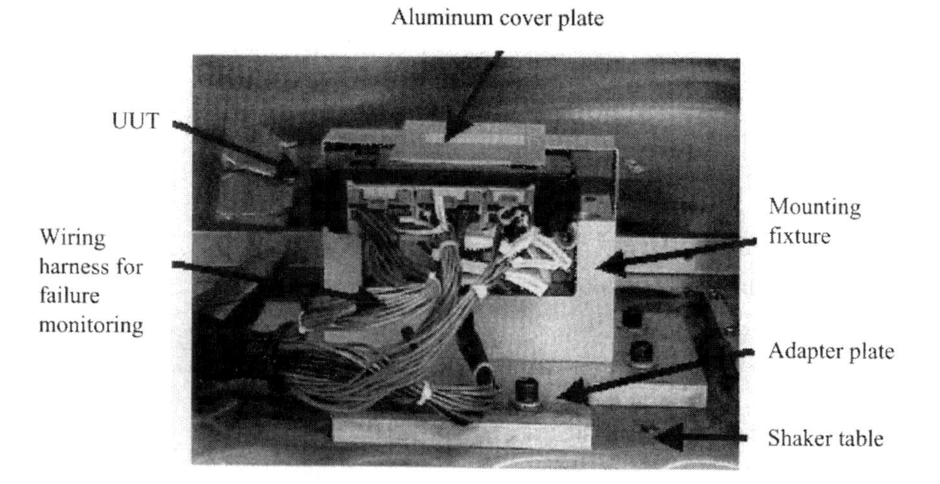

Figure 7.10. Photograph of module assembly mounted on the table of the RS shaker.

on the loads that will be used in defining the accelerated life tests. Initial load limits were established prior to conducting any physical tests by reviewing the documented limits for materials, test equipment, and measurement sensors. These limits included the following:

1. The operational temperature range for the instrumentation on the CCA was from –75°C to 175°C for the strain gauge and from –51°C to 121°C for the accelerometer.
2. The glass transition temperature of PCB was between 120°C and 140°C, and the recrystallization temperature of the plastic housing was at 120°C.
3. The maximum achievable vibration input was 60 G_{rms} due to the weight of the fixture and the available vibration shaker table.

From this information, a temperature range of –50°C to 120°C and a vibration range from 0 to 60 G_{rms} were established as the limits for the step–stress tests. With the initial limits established, three step–stress tests were conducted: (1) thermal step test to high temperature, (2) thermal step test to cold temperature, and (3) vibration step test. In the thermal step test to high temperature, the temperature was varied from 20°C to 120°C, with increments of 10°C and dwell time of 20 min at each step. For the thermal step test to cold temperature, the temperature was varied from 20°C to –50°C, with increments of –10°C and dwell time of 20 min at each step. For the vibration step test, the applied load was varied from 20 G_{rms} to 60 G_{rms}, with increments of 10 G_{rms} and dwell time of 10 min at each step. No electrical failures were observed during these tests.

Based on the step–stress test results, the accelerated life test plan was developed and consisted of two profiles:

1. *Vibration:* Loading was varied from 20 G_{rms} to 50 G_{rms} with step increments of 10 G_{rms}. Dwell time chosen to be approximately 36 min at each load level to coincide with a thermal cycle.
2. *Thermal cycling:* Loading ranged from –50°C to 120°C, with a ramp rate of 28°C/min and dwell times of 12 min at both extremes.

Due to limited availability of the load simulator, only one specimen was tested, and the test was limited to 328 thermal cycles with simultaneous vibration loading. During the test period, no electrical failures were observed. However, visual inspection revealed cracks on several 0603 leadless resistor interconnects after 328 cycles (see Figure 7.11).

To assess the impact of the accelerated test on the interconnect durability, destructive physical analysis (DPA) was performed. In this study, DPA included sectioning the test board and cross-sectioning the interconnects of selected components. Components were selected based on simulation results and the initial visual inspection. Cross sections of a 1206 resistor, a gullwing lead from a microprocessor, and the gullwing lead of a FET were conducted and the results are presented in Figure 7.12.

Figure 7.11. Visual cracks on a leadless resistor after 328 cycles.

Cross sections of some 0603 resistors revealed smaller-than-anticipated solder fillet shaped (see Figure 7.13). To better assess the influence of the solder fillet geometry, a detailed simulation of the 0603 resistor interconnect was conducted.

7.2.6 Detailed Analysis of 0603 Resistor Interconnect

Based on the test results and the examination of the actual fillet geometry of the 0603 resistors, a more detailed assessment was conducted. This assessment made

1206 Resistor (R 2807)

QFP-microprocessor (U1000)

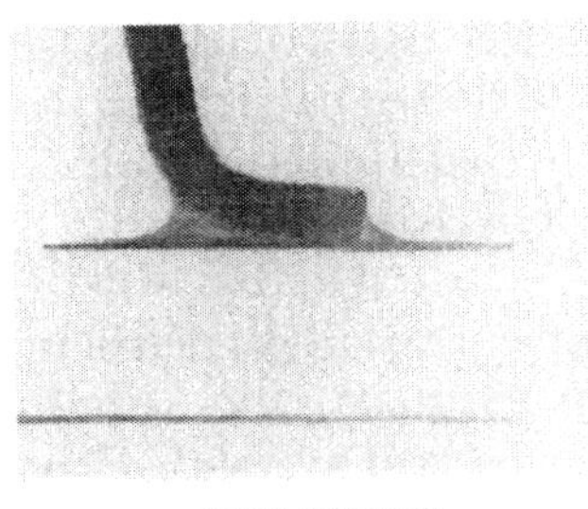

FET (U2500)

Types of Component	Ranking (1 = worst, 4 = best)
0603 LCR (R 2306)	1
1206 LCR (R 2807)	3
QFP (U1000)	2
FET (U2500)	4

Figure 7.12. Cross section of various interconnects. (Courtesy of CALCE EPSC.)

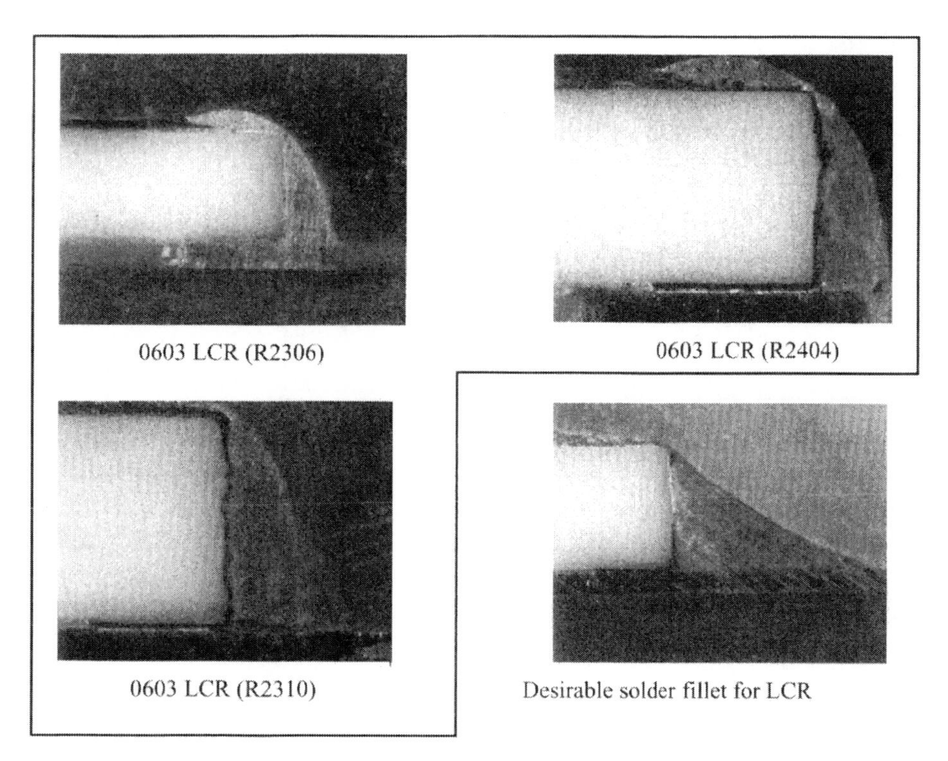

Figure 7.13. Cross-sectional views of 0603 leadless chip resistors (LCRs). (Courtesy of CALCE EPSC.)

use of general-purpose finite element software. Using the software and considering the symmetry in the loading situation, a two-dimensional model consisting of half of the resistor body, the solder joint, and board was created (see Figure 7.14). Setting the appropriate material properties and boundary conditions, a simulation analysis was conducted for the temperature cycle of the accelerated life test. The analysis provided stress and strain history within the structure and allowed for the calculation of the inelastic response of the solder.

In addition to the inelastic strain range, inelastic energy has been shown to be a good stress metric for quantifying time to failure of solder fatigue [Dasgupta et al. (1992), Solomon and Tolksdorf (1995), Darveaux (2000)0. Figure 7.15 depicts the inelastic (creep) energy calculated for the modeled solder joint. Using this information, the time-to-failure for the solder joint was estimated to be 1561 cycles. Based on the defined 36-min test cycle and assuming continuous testing, failure would be expected to occur in approximately 39 days. Using the same finite element model and the anticipated field temperature cycle, the field life was estimated to be 4816 days.

Since the objective of accelerated life testing is to demonstrate that the product can withstand field conditions for its intended useful life, one needs to relate the test

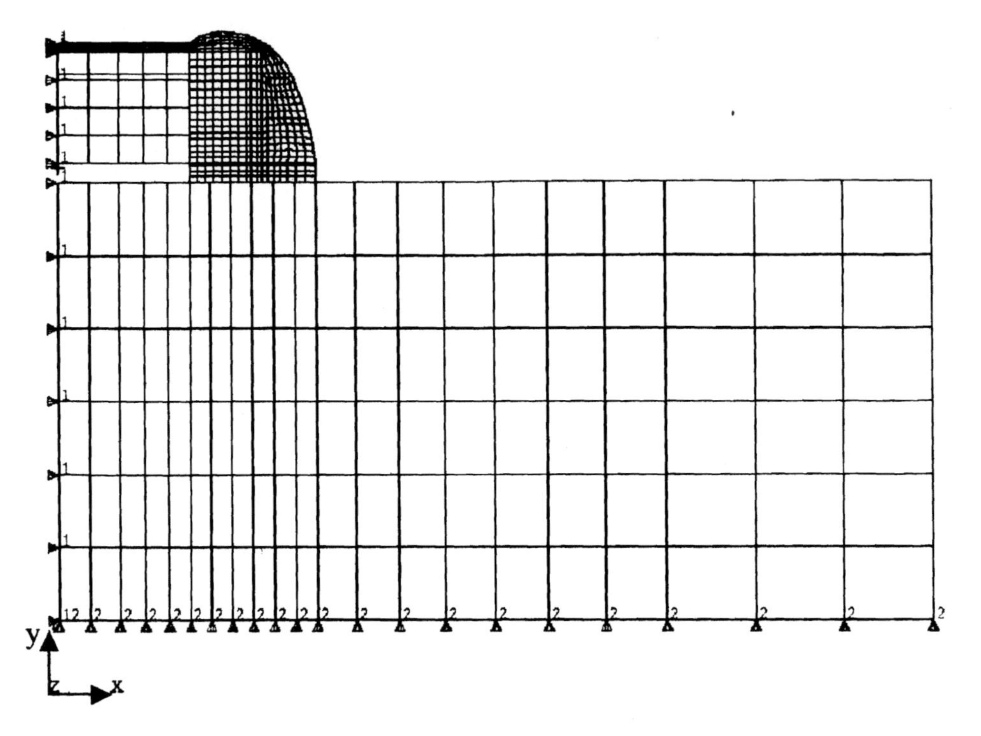

Figure 7.14. Finite-element model of leadless chip resistor (0603).

results to field life. The objective is achieved through the use of an acceleration factor (AF). An AF is defined as

$$AF = \frac{t_{\text{field}}}{t_{\text{ALT}}} \tag{7.5}$$

where t_{field} = predicted time-to-failure under life-cycle conditions (result from virtual qualification) and t_{ALT} = predicted time-to-failure under accelerated test conditions (result from virtual testing).

As previously stated, the accelerated life test was terminated after 328 thermal cycles due to time constraints in the product development schedule. Since there were no electrical failures during the accelerated life test, the test is considered as time-limited. Based on the simulation, 31 more days of continuous testing would have been required to fail the 0603 resistors. Furthermore, the simulation results indicate that the acceleration factor for the 0603 resistor under the defined test and field conditions is 123. Thus, the 8-day accelerated life test represents 2.7 years of the defined field condition.

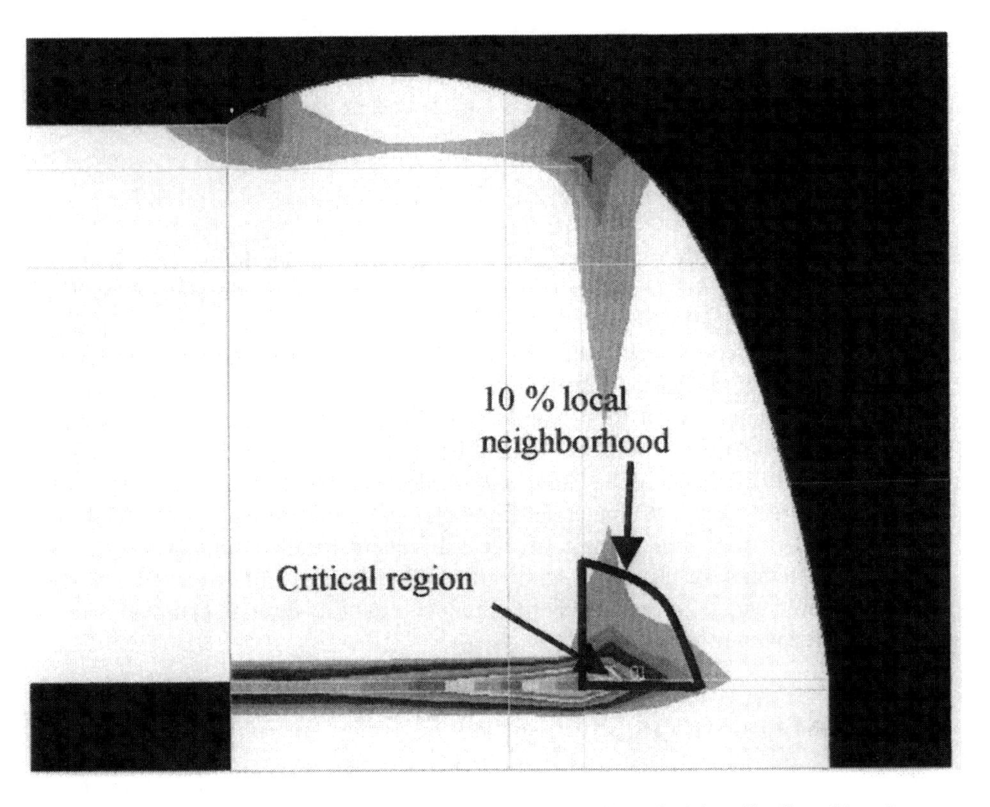

Figure 7.15. Contour plot of inelastic creep energy density (W_c) in solder joint of leadless chip resistor.

7.2.7 Discussion of Automotive Case Study

The development program discussed in this case study (virtual qualification program) was compared to an earlier development program (traditional program) for another electronic module. The earlier development program used the traditional design–build–test–fix approach. The vendor of both modules was the same, and the functionality and complexity of the case study design were significantly greater than the previous generation controller. In terms of time, the virtual qualification and simulated guided physical verification program took 109 weeks versus 130 weeks for the traditional program. The product validation testing process uncovered one hardware deficiency that was not found during the virtual qualification process. In contrast, six hardware deficiencies had to be corrected at the product verification stage in the traditional DBTF approach on the previous generation product. As discussed in this case study, the upfront analysis provided under the virtual qualification program highlighted thermal and vibration issues that were corrected between the alpha and beta versions of the design. Thus, the virtual qualification program not only allowed the design team to quantify field reliability, but also allowed it to impact reliability in the design process, accelerate design maturity, reduce overall time to market, and reduce the overall life-cycle cost of the hardware.

7.3 SUMMARY

As presented in the case study, characterizing the electronic hardware, creating life-cycle loading scenarios, assessing stress states produced by the anticipated loading scenarios, and evaluating time to failure of the hardware based on established failure models are effective methods for establishing the viability of a design as well as for improving the design. Using this process, electronic hardware design can be qualified for an application prior to building a physical prototype. In the case study, the overall virtual qualification assessment process was conducted using the calce-PWA software, a product of the University of Maryland's CALCE Electronic Products and System Center. Commercial finite element software including PATRAN and ABAQUS were used to perform the subsequent detailed simulation. As research into failures of electronic hardware continues, it can be expected that the ability to anticipate failures and quantify product reliability through computer simulations will continue to improve.

ACKNOWLEDGMENTS

The research for this work was performed at the CALCE Electronic Product and Systems Center of the University of Maryland. The Center provides a knowledge and resource base to support the development of competitive electronic components, products and systems. The Center is supported by more than 100 electronic product and systems companies from all sectors, including telecommunications, computer, avionics, automotive, and military manufacturers. The authors would also like to acknowledge James Cho for his substantial contribution in performing the analysis and conducting the experiments presented in this case study example.

REFERENCES

Cunningham, J., Valentin, R., Hillman, C., Dasgupta, A., and Osterman, M. (2001). A Demonstration of Virtual Qualification for the Design of Electronic Hardware, in *ESTECH 2001 Proceedings.*

Darveaux, R. (2000). Effect of Simulation Methodology on Solder Crack Growth Correlation, in *Proceedings of the 2000 Electronic Components and Technology Conference, IEEE,* pp. 1048–1058.

Dasgupta, A., Oyan, C., Barker, D., Pecht, M. (1992). Solder creep-fatigue analysis by an energy-partitioning approach, *ASME Trans. Electronic Packaging* **144:**152–160.

Engelmaier, W. (1993). Generic reliability figures of merit design tools for surface mount solder attachments, *IEEE Trans. CHMT* **16:**103–112.

Jih, E. and Jung, W. (1998). Vibration Fatigue of Surface Mount Solder Joints, *IEEE ITHERM98*, pp. 248–250.

Lall, P., Pecht, M., and Hakim, E. (1995). *Influence ot Temperature on Microelectronics and System Reliability: A Physics of Failure Approach,* CRC Press, New York.

Osterman, M., Stadterman, T., and Wheeler, R. (1997). CAD/E requirements and usage for reliability assessment of electronic products, *Advances in Electronic Packaging 1997,* EEP **19–1**:927–938.

Osterman, M., and Stadterman, T. (1999). Failure-Assessment Software for Circuit-Card Assemblies, *Proceedings Annual Reliability and Maintainability Symposium,* pp. 269–276.

Osterman, M., and Stadterman, T. (2000). Reliability and Performance of PWB Assemblies, Chapter 9, in *High Performance Printed Circuit Boards* (C. Harper, Editor), McGraw-Hill, New York.

Pecht, M. (1991). *Handbook of Electronic Package Design,* Dekker, New York.

Solomon, H. D., and Tolksdorf, E. D. (1995). Energy approach to the fatigue of 60/40 solder: Part I—Influence of temperature and cycle frequency, *ASME Trans. J. Electronic Packaging* **117**:130–135.

Steinberg, D. S. (1991). *Vibration Analysis for Electronic Equipment,* Wiley, New York.

Tummla, R., and Rymaszeewske, E. (1989). *Microelectronics Packaging Handbook,* Van Nostrand Reinhold, New York.

EXERCISES

7.1. How is the design-capture process tied to the failure-assessment process? What are the potential drawbacks of this relationship?

7.2. A part can survive 22,000 cycles of 20°C to 60°C cycles and 9000 cycles of –40°C to 120°C temperature cycles. If the part is subjected to the 20°C to 60°C cycle four times a day and to the –40°C to 120°C cycle two times a month, what is the life expectancy of the part in years?

7.3. Simulation indicates an acceleration factor (AF) of 30 between a test condition and the anticipated field condition for a specific product. If the product survives 100 days under the test condition, how many years should it survive in the field? What could cause it to not last the predicted amount of time?

7.4. If the number of survivable cycles is found to have a mean value of 22,000 and a standard deviation of 200, what is the six-sigma minimum time to failure for an application that has four cycles per day?

7.5. What are the pros and cons of the virtual qualification process? When is the application of the virtual qualification most appropriate?

CHAPTER 8

Development of a Moisture Soak Model for Surface-Mounted Devices

Loon Ching Tang and Soon Huat Ong

8.1 INTRODUCTION

In electronic assembly, a high-temperature operation, *solder reflow,* is used for securing surface-mounted devices (SMD) to the printed circuit board. This operation generates a new class of quality and reliability concerns such as package delamination and crack. The problem arises as moisture may enter the plastic molded compound through diffusion. This moisture inside the plastic package will turn into steam that expands rapidly when exposed to the high temperature of the VPR (vapor phase reflow), IR (infrared) soldering. Under certain conditions, the pressure from this expanding moisture can cause internal delamination, internal cracks, and other mechanical damages to the internal structure of the package. [See Kitano et al. (1988), Tay and Lin (1998), Galloway and Miles (1997), Song (1998), Song and Walberg (1998), Huang et al. (1998), Ahn and Kwon (1995), Tanaka and Nishimura (1995), and Taylor et al. (1997).] These damages can result in circuit failure, which could immediately affect the yield or could be aggravated over time, thus affecting device reliability. This has been a major quality and reliability concern for semiconductor manufacturers and users.

A proven technique in detecting various package problems is a preconditioning test as specified in JESD22-A113A (1995). The preconditioning test is a crucial test usually conducted prior to other reliability tests because this is a combination of tests to simulate the type of processes an IC package is likely to go through before being assembled. For surface-mounted devices (SMD), the procedure includes electrical DC and functional test, external visual inspection, temperature cycling, baking, and moisture soak and reflow. There are various levels of moisture soak and reflow, which are determined by the classification of the package during its qualification stage. Moisture absorption and desorption tests are seldom carried out

Case Studies in Reliability and Maintenance, Edited by W. R. Blischke and D. N. P. Murthy.
ISBN 0-471-41373-9

alone except for identifying the classification level of a plastic molded compound. This test is used for identifying the classification level of plastic SMDs so that they can be properly packed and stored. This will better protect the package and avoid any subsequent mechanical damage during reflow or any repair operation.

There are two basic issues in moisture absorption, namely, rate of absorption and the saturation level. Plastic mold compounds transport moisture primarily by diffusion. Galloway and Miles (1997) tackled the absorption and desorption of moisture in plastic mold compound through the measure of moisture diffusivity and solubility as a function of time. It was established that transportation strictly by diffusion could be modeled using the standard transient diffusion equation,

$$\frac{\partial^2 C}{\partial x^2} + \frac{\partial^2 C}{\partial y^2} + \frac{\partial^2 C}{\partial z^2} = \frac{1}{\alpha}\frac{\partial C}{\partial t}$$

where C is the concentration level, α is the diffusivity, (x, y, z) are spatial coordinates, and t is time. Using the standard separation technique results in an expression for the local concentration, which in turn gives an analytical expression for the total weight gain as a function of time. Saturation concentration, which is a function of temperature, humidity, and material, determines the maximum possible weight gain per sample volume for a particular ambient condition. This forms the basis for the functions under consideration in this case study.

Two similar international standards, EIA/JESD22-A112-A (1995) and IPC-SM–786A (1995), have been developed to classify SMD packages. Both standards define six levels of classification. Each level corresponds to the maximum floor life that a package can be exposed to in the board assembly environment prior to solder reflow without sustaining package damage during the reflow operation. Table 8.1 shows the levels defined in JESD22-A113-A for the moisture soak of the preconditioning test.

Huang et al. (1998) found that delamination and "popcorn" (a commonly adopted term to describe the failure phenomenon due to rapid expansion of water vapor trapped in plastic packages) of plastic packages were highly dependent on reflow parameters. The dominant reflow factor affecting delamination is the total heat energy applied to the package, which is represented by a combination of the total time above 100°C and the temperature throughout the soak time. The higher the temperature and the longer the dwell time over 100°C, the more likely a package is to crack if the moisture level in the package is above the critical level. The study concluded that the JEDEC-recommended vapor phase and convection reflow temperature preconditioning profiles are adequate guidelines to determine the moisture sensitivity of plastic packages, unless excessively long preheat is applied. This forms the basis for using JEDEC as the benchmark in our study.

From Table 8.1, it can be seen that there are six levels of moisture soak with three different temperature–humidity settings. Therefore, a preconditioning test of different products with different moisture sensitivity levels requires three chambers if all of the products are to be tested at the same time. Otherwise, we will have to complete the preconditioning test at one temperature–humidity setting before pro-

Table 8.1. Moisture Sensitivity Levels

Level	Floor Life	Moisture Soak
1	Unlimited at ≤ 85°C/85% RH	85°C/85% RH 168 hours
2	1 year at ≤ 30°C/60% RH	85°C/60% RH 168 hours
3	1 week at ≤ 30°C/60% RH	30°C/60% RH 192 hours
4	72 hours at ≤ 30°C/60% RH	30°C/60% RH 96 hours
5	24/48 hours at ≤ 30°C/60% RH	30°C/60% RH 48/72 hours
6	6 hours at ≤ 30°C/60% RH	30°C/60% RH 6 hours

ceeding to another test with a different setting. This will greatly increase the cycle time of the test, thus reducing productivity. Since moisture soak time takes a much longer time than the other precondition tests, it is of great value to see if the equivalent moisture soak times at harsher temperature–humidity conditions for levels 2–6 can be obtained. Given its potential economic value in terms of lower capital investment as well as shorter time to market, an investigation into the rate of moisture absorption on a selected set of devices is conducted. The aim is to develop a model for moisture gain under various conditions, from which the equivalent moisture soak time at a harsher condition can be derived. Specifically, a suitable response related to the weight gain process will be identified, and a family of response functions under preset conditions together with a physical model will be established. Analogous to the basic framework of accelerated testing models, the moisture soaking acceleration factor, and hence the equivalent moisture soak time, for the preconditioning test of a plastic package can be derived.

In the following section, we outline the experimental procedure and present a set of sample results. In Section 8.3, various choices for the moisture soak model are discussed and a regression analysis is performed to establish the most appropriate alternative. A set of acceleration factors is generated from the model. Finally, some discussions and conclusion are given in Section 8.4.

8.2 EXPERIMENTAL PROCEDURE AND RESULTS

The plastic leaded chip carrier (PLCC) package with B24 mold compound was selected as the primary test specimen for the case study. Three lots of 44-lead, 68-lead, and 84-lead PLCC were manufactured. The die and die attach pad (DAP) size were maximized so as to create a larger critical surface area and allow a higher likelihood of delamination. These are summarized in Table 8.2.

Table 8.2. Experimental Units

Lead Count	44 lead	68 lead	84 lead
Die Size	120 × 120 mils	148 × 160 mils	210 × 210 mils
DAP Size	250 × 250 mils	280 × 280 mils	300 × 300 mils

The tests were carried out in accordance with the procedures and standard set by JEDEC, EIA/JESD22-A112A (1995), and National Semiconductor specifications. The flowchart for a moisture absorption test is shown in Figure 8.1.

Votsch temperature–humidity chambers and a Mettler precision weighing scale were used for the moisture absorption and desorption tests. Before the actual commencement of the moisture test, calibrations for all the equipment were carried out. Since the weighing system is very important for this study, a gauge repeatability and reproducibility (GRR) was carried out to ensure the capability of the weighing system.

Four samples of 20 units each were prepared for the four temperature–humidity test conditions of 30°C/60% RH, 60°C/60% RH, 85°C/60% RH, and 85°C/85% RH. Typically, units are only required to undergo a 24-hour baking in a 125°C oven. However, due to the importance of accuracy in the weight of the units, all the

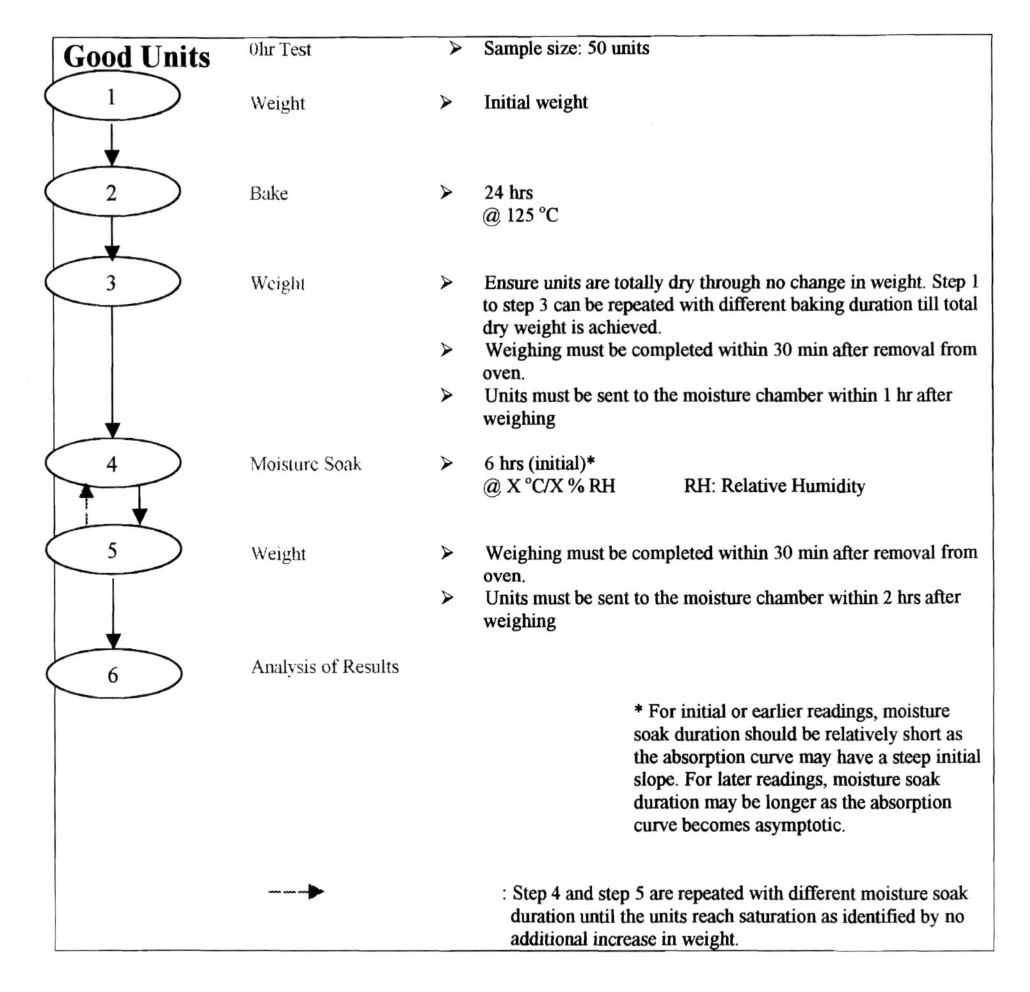

Figure 8.1. Flowchart for moisture absorption test.

Table 8.3. Typical Data for Moisture Absorption Test

84-lead PLCC Moisture Absorption Data (all weights are in grams)					
Soak Condition: 85°C /85% RH				Bake Condition: 125°C	
Duration (hr)	Weight	Weight	Average Weight	Weight Gain	% Weight Gain
0	123.6302	123.6307	123.6305	0.0000	0.0000
4	123.7218	123.7229	123.7224	0.0919	0.0743
8	123.7578	123.7578	123.7578	0.1274	0.1030
24	123.8451	123.8467	123.8459	0.2155	0.1743
48	123.9153	123.9156	123.9155	0.2850	0.2305
72	123.9500	123.9511	123.9506	0.3201	0.2589
96	123.9734	123.9740	123.9737	0.3433	0.2776
120	123.9858	123.9865	123.9862	0.3557	0.2877
144	123.9968	123.9971	123.9970	0.3665	0.2964
168	124.0040	124.0045	124.0043	0.3738	0.3024

units were overbaked to ensure that they have reached their absolute dry weight in a 125°C baking environment.

The moisture gain data for the four test conditions were recorded at regular intervals. The initial sampling interval used for the recording of the moisture gain data differs between different test conditions. The initial sampling interval is shorter for the harsher test condition to allow for an anticipated higher moisture absorption rate. All test units did not undergo moisture soaking till saturation, because this was not required. Units that underwent moisture soaking with conditions harsher than 30°C/60% RH only lasted 168 hours, because the moisture gain was already much more than the 30°C/60% RH test condition.

When test units were removed from chambers to be weighed, additional effort was made to ensure the consistency of every action, from the loading/unloading time to the steps taken in weighing the units (including the placement of the units in the weight balance). Two sets of readings were taken for all the time duration, and their average was used for subsequent analysis. Table 8.3 shows a typical data summary.

The data correspond to the standard moisture absorption response of a level 4 mold compound. For comparison of moisture absorption data between the different lead counts, the % weight gain provided a better overview of the effects of the four different test conditions on the test specimen. Figure 8.2 depicts the moisture absorption in % weight gain for all three packages with different lead counts.

8.3 THE MOISTURE SOAK MODEL

8.3.1 Choosing a Suitable Response

The logical step in constructing the moisture soak model is to fit a family of response curves to the data from which equivalent soak times under various condi-

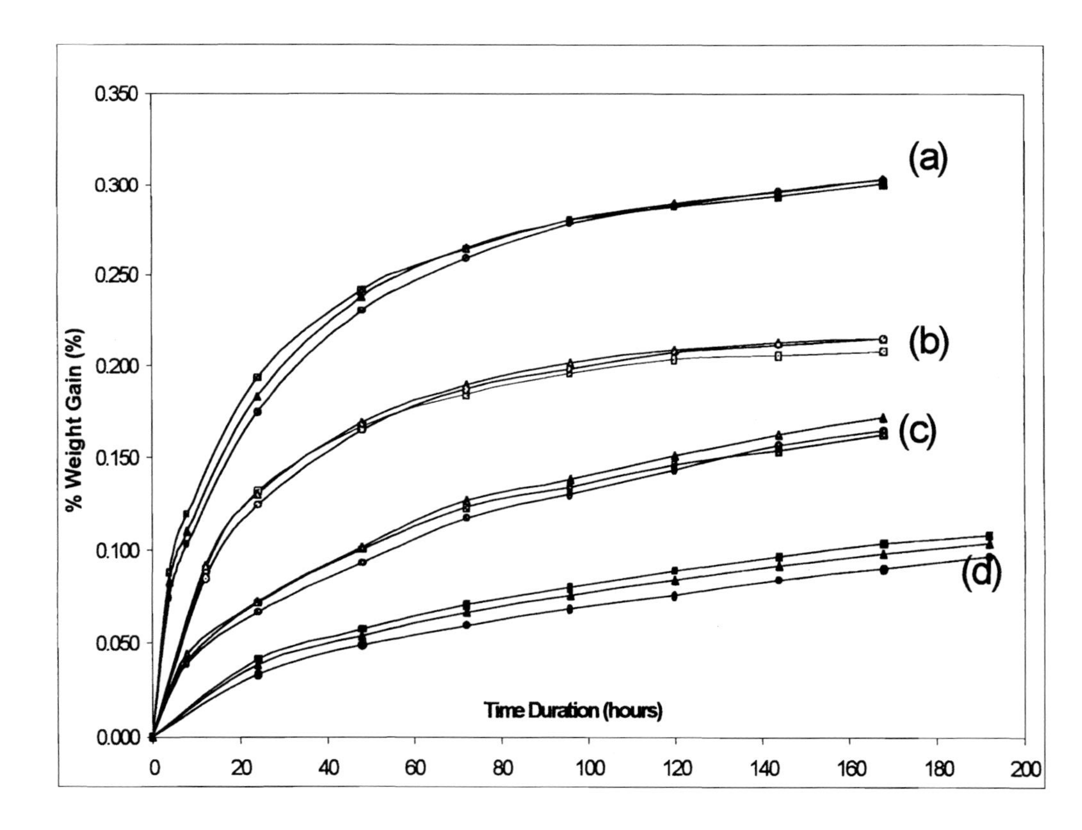

Figure 8.2. Integrated moisture absorption graph. The conditions are (a) 85/85, (b) 85/60, (c) 60/60, and (d) 30/60 for three different packaging types.

tions can be estimated. There are several possible choices for the response, namely, (a) the experimental weight gain, (b) the ratio of experimental weight gain to the maximum weight gain of the specimen, and (c) the ratio of experimental weight gain to the initial weight of the specimen. To eliminate possible dependency of the weight gain on any specific specimen, the use of a ratio—that is, (b) or (c)—is preferred. But the use of (b) requires that all specimens undergo moisture soaking until full saturation, which would not only consume too much equipment time but would also introduce additional sources of experimental errors due to prolonging the experiment. As a result, (c) is adopted and is listed as "% weight gain" together with the experimental data in Table 8.3. Moreover, normalizing the weight gain by the initial weight also allows for comparison of moisture absorption data between packages of different lead counts. This will provide a better overview of the effects of the four different test conditions on the test specimens. Figure 8.2 depicts the moisture absorption in % weight gain for all three packages with different lead counts.

8.3.2 Choosing the Family of Response Curve

Let W_t denote the weight gain at time t and let W_0 be the initial weight of a package. The reciprocal of the response chosen, W_0/W_t, is akin to a reliability function, although, in practice, weight gain will not be infinite. Nevertheless, the same method for assessing goodness-of-fit as in probability plotting can be adopted using the common choices of reliability functions. Preliminary exploration reveals that among exponential, Weibull, logistic, loglogistic, normal, and log-normal distributions (Nelson, 1990), Weibull and loglogistic distributions provide the best fit. The Weibull reliability function is given by

$$R(t) = \frac{W_0}{W_t} = \exp\left[-\left(\frac{t}{b}\right)^c\right] \Rightarrow \ln\left[\ln\left[\frac{W_t}{W_0}\right]\right] = c\ln(t) - \ln(b) \tag{8.1}$$

where b is the scale parameter and c is the shape parameter.

The loglogistic reliability function is given by

$$R(t) = \frac{W_0}{W_t} = \frac{\exp\left[-\left(\frac{\ln(t)-a}{b}\right)\right]}{1+\exp\left[-\left(\frac{\ln(t)-a}{b}\right)\right]} \Rightarrow \ln\left[\frac{W_t - W_0}{W_0}\right] = \left(\frac{\ln(t)-a}{b}\right) \tag{8.2}$$

where a is the location parameter and b is the scale parameter. The good fit to both distributions is expected because, when $W_t - W_0$ is small, we have $\ln[W_t/W_0] \approx (W_t - W_0)/W_0$.

The loglogistic and Weibull plots, resulting from equations (8.1) and (8.2) are given in Figures 8.3(a) and 8.3(b), respectively. It can be seen that both Weibull and loglogistic provide good fits, because the plots are nearly linear. The goodness-of-fit to the loglogistic distribution will be further reinforced in the combined analysis

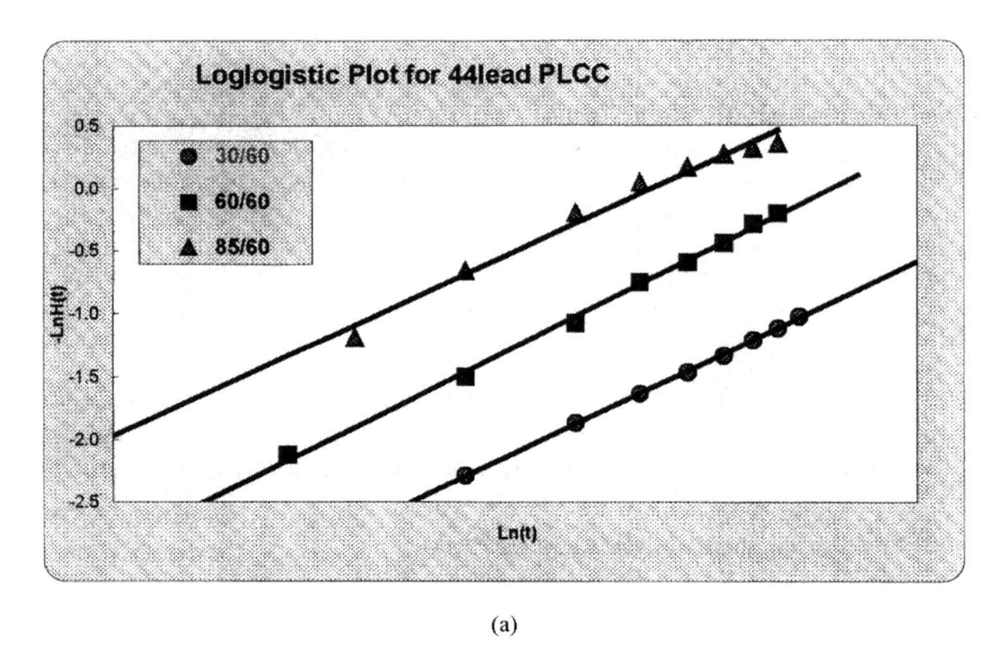

(a)

(b)

Figure 8.3. (a) Loglogistic probability plot. (b) Weibull probability plot.

presented in the following section. It is also noted that the plot for the loglogistic under different experimental conditions gives nearly parallel lines. This justifies the use of a common scale parameter, b, for different experimental conditions.

In the following, we shall use the loglogistic distribution for further analysis and discussion. However, it should be borne in mind that the same principle can be extended to other distributions.

8.3.3 Combined Analysis with the Acceleration Model

In order to estimate the equivalent moisture soak time for conditions other than those tested, we need to establish an acceleration model that best fits the moisture gain data under various relative humidity (RH) and temperature (T) combinations. From the reliability engineering literature, two models are commonly used for describing the relation between "lifetime" and the two stress variables (T and RH), namely, the Peck model and the generalized Eyring model. These models express the "nominal" product life as some function of the T and RH. The exact forms of these models are given in Nelson (1990), pp. 98–102. In the case of moisture absorption, since the time to reach saturation is akin to the time to failure, the logarithm of the "nominal" moisture soak time is equivalent to the location parameter a. In particular, for the Peck model, we have

$$L = A\left[\exp\left(\frac{E}{kT}\right)\right][RH^{-n}] \tag{8.3}$$

where L is the lifetime, A is a constant, E is the activation energy, and k is the Boltzmann constant. (These also apply to subsequent equations.) In the current context, with α_i ($i = 0, 1, 2$) as the coefficients in a linear model, we have

$$a = \alpha_0 + \alpha_1(1/kT) + \alpha_2(\ln(RH)) \tag{8.4}$$

For the generalized Eyring model, we have

$$L = \frac{A}{T}\left[\exp\left(\frac{E}{kT}\right)\right]\exp\left[RH\left(B + \frac{C}{kT}\right)\right] \tag{8.5}$$

where B and C are constant. In the current context, this becomes

$$a + (\ln(T)) = \alpha_0 + \alpha_1(1/kT) + \alpha_2(RH) + \alpha_3(RH/T) \tag{8.6}$$

A variant of the Eyring model that has been used by Intel [see Nelson (1990), p. 102] is

$$L = A\left[\exp\left(\frac{E}{kT}\right)\right][\exp(-B \cdot RH)] \tag{8.7}$$

This gives

$$a = \alpha_0 + \alpha_1(1/kT) + \alpha_2(RH) \tag{8.8}$$

In general, one could express the location parameter as a linear function of ln(T), $1/T$, RH, ln(RH), RH/T, or other similar independent variables that are variants of the above forms. This results in

$$a = f(1/T, RH, \ln(T), \ln(RH), RH/T) \tag{8.9}$$

Here, we adopt a "combined" analysis given that the loglogistic distribution function provides the best fit. From equation (8.2), we have

$$a = -b \ln\left(\frac{W_t - W_0}{W_0}\right) + \ln(t) \tag{8.10}$$

It follows that the generic form is

$$\ln\left(\frac{W_t - W_0}{W_0}\right) = \alpha_0 + \alpha_1 \ln(t) + f((1/kT), (RH), (RH/T), \ln(RH)) \tag{8.11}$$

Regression runs for equation (8.11), where $f(\cdot)$ takes the form of equations (8.4), (8.6), (8.8) are conducted. In addition, both stepwise regression and best subset regression are conducted for (8.11) to identify the best linear $f(\cdot)$. In order to investigate the effect of lead counts, a dummy variable, P, representing the type of PLCC package is also considered. The results are summarized in Table 8.4, with the coefficient of determination, R^2, and the residual root mean square, s. It can be seen that the adjusted R^2 is the best from the best subset routine and that the corresponding residual root mean square, s, is also the smallest. Moreover, the Mallow Cp statistic (Draper and Smith, 1981) from the best subset run is 5, indicating the current set of independent variables result in a residual mean square, s^2, which is approximately an unbiased estimate of the error variance.

The resulting model is

$$\ln\left(\frac{W_t - W_0}{W_0}\right) = -49.0 + 0.390 \ln(t) - 0.0302 \cdot P + 5.00\, RH/T + 6.82 \ln(T) \tag{8.12}$$

Table 8.4. Summary of Results from Combined Analysis

Model	Independent Variables	Adjusted R^2	RMS, s
Peck	$1/T$, ln(RH), ln(t)	0.975	0.08986
Generalized Eyring	ln(T), $1/T$, RH, RH/T, ln(t)	0.975	0.08988
Intel	$1/T$, RH, ln(t)	0.975	0.08986
Stepwise	ln(T), RH/T, ln(t)	0.9753	0.0894
Best subset	ln(T), RH/T, ln(t), P	0.977	0.08625

where t is time in hours, $P = (-1,0,1)$ represents three package types (i.e., 44-lead, 68-lead, and 84-lead respectively), RH is relative humidity in percentage, and T is temperature in degrees kelvin. The p values for all the above independent variables are less than 0.005.

Residual analysis is carried out with plots for residuals against fitted values, and residuals against observation orders are given in Figures 8.4(a) and 8.4(b), respectively. The latter plot is quite random, but the former shows an obvious quadratic

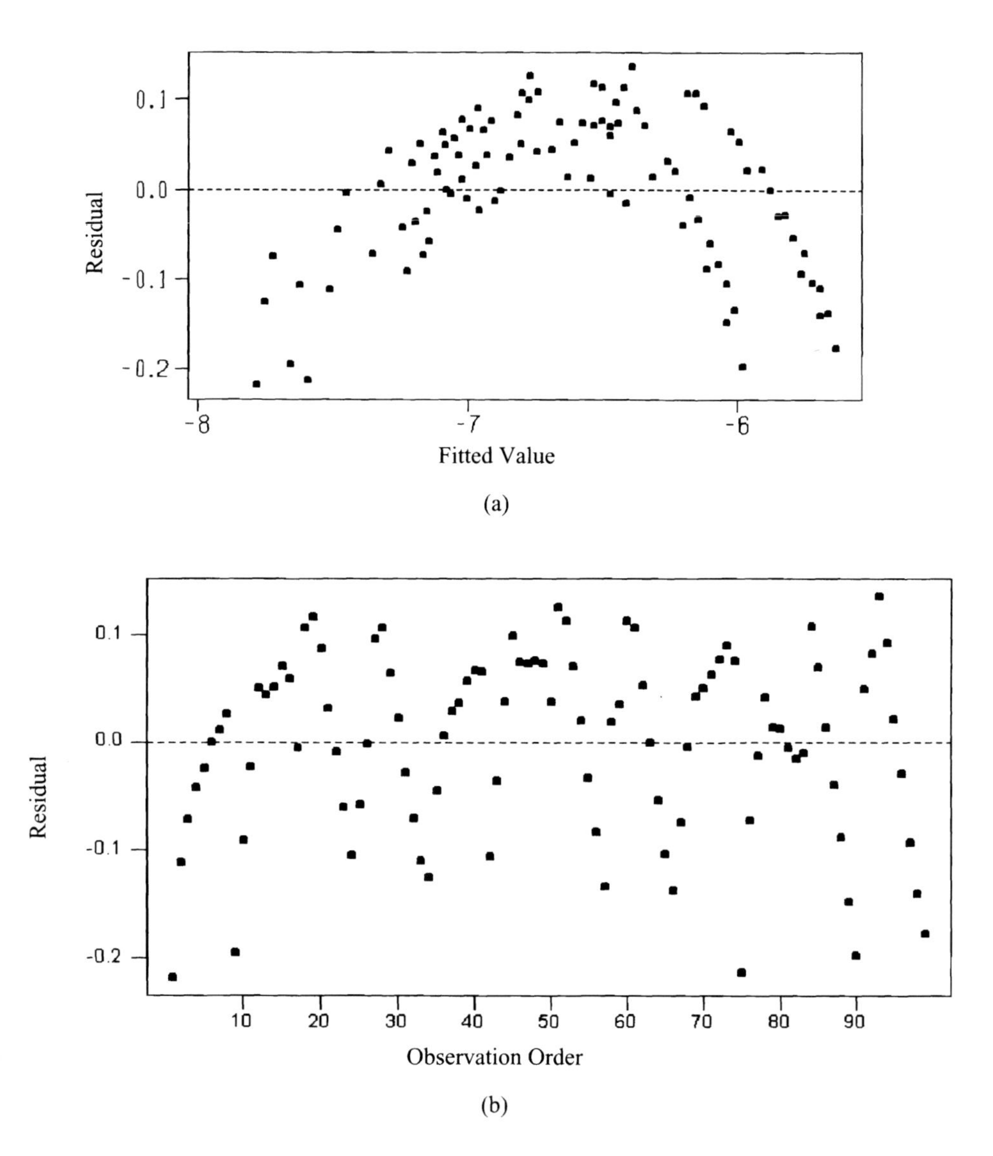

Figure 8.4. (a) Residual plots against fitted values. (b) Residual plots against observation order.

trend. Nevertheless, given the rather limited experimental conditions and a reasonably high R^2 (= 0.977), to avoid overfitting, the current model is adequate for the experimental regime (i.e., 30–85°C and 60–85% RH), under consideration.

8.3.4 Calculation of Equivalent Moisture Soak Time

Combining equations (8.10) and (8.12), we can express the location parameter of the family of loglogistic distributions as

$$a = \ln(t) - 2.564 \ln\left(\frac{W_t - W_0}{W_0}\right) = 125.64 + 0.0774 \cdot P - 12.82\, RH/T - 17.49 \cdot \ln(T) \tag{8.13}$$

From equation (8.13), the estimate for the common scale parameter, b, for the family of loglogistic distributions is 2.564 [=1/0.39 from equation (8.12); see also equations (8.10) and (8.2)]. The location parameters and their respective standard errors under various experimental conditions are given in Table 8.5. Using the same principle as in accelerated testing, the acceleration factor between two testing conditions can be computed by taking the ratio of their respective times taken in achieving the same weight gain, which, in turn, can be expressed as the exponent of the difference between the respective location parameters. For the same package type, the acceleration factor, AF, is given by

$$AF = \exp(a_1 - a_2) = \left(\frac{T_2}{T_1}\right)^{17.49} \exp\left[-12.82\left(\frac{RH_1}{T_1} - \frac{RH_2}{T_2}\right)\right] \tag{8.14}$$

From equation (8.14), the relative acceleration factor for the four testing conditions can be computed. The results are shown in Table 8.6. It should be noted that the point estimate of AF is independent of package type; however, the interval esti-

Table 8.5. Predicted Value of Location Parameters

Package	Temperature (°C)	% RH	Location Parameter, a	SE of a
1	30	60	23.1608	0.0501
1	60	60	21.7372	0.0378
1	85	60	20.6315	0.0481
1	85	85	19.7348	0.0503
0	30	60	23.0834	0.0421
0	60	60	21.6597	0.0263
0	85	60	20.554	0.0398
0	85	85	19.6573	0.0424
–1	30	60	23.0059	0.0501
–1	60	60	21.5822	0.0378
–1	85	60	20.4765	0.0481
–1	85	85	19.5799	0.0503

Table 8.6. Acceleration Factors for the Current Testing Conditions with 95% Confidence Limits

Temperature (°C)/% RH Combination	Testing Conditions (°C/% RH)		
	85/85	85/60	60/60
30/60 (levels 3,4,5,6)	30.75 (26.76,35.34)	12.54 (10.95,14.37)	4.15 (3.67,4.70)
60/60	7.41 (6.55,8.38)	3.02 (2.68,3.41)	1.00
85/60 (level 2)	2.45 (2.14,2.81)	1.00	—
85/85 (level 1)	1.00	—	—

mates are dependent on package type due to different standard errors (see the last column of Table 8.5). Given the AF, the equivalent soak time under a higher temperature/RH combination can be estimated. For example, if a level 3 package is subject to 60/60 soaking condition instead of 30/60, the soak time will be 46.3 hours instead of 192 hours as recommended by the JEDEC; a 4.15-fold reduction.

An approximate confidence interval for the acceleration factor could be obtained from the confidence interval for $a_1 - a_2$ using normal a approximation. Specifically, from Table 8.5, the standard errors for the differences between the location parameters can be computed. Then the $(1 - \gamma)$ confidence interval for AF is given by

$$(\underline{AF}, \overline{AF}) = \exp[a_1 - a_2 \pm z_{\gamma/2} \cdot SE_{a_1 - a_2}] = \exp[a_1 - a_2 \pm z_{\gamma/2} \cdot \sqrt{(SE_{a1})^2 + (SE_{a2})^2}] \tag{8.15}$$

The results for package type 1 (44-lead counts) with $\gamma = 0.05$ are given in the parentheses within their respective cell in Table 8.6.

8.4 DISCUSSION

In practice, many factors need to be considered for determining the moisture-induced crack of SMD packages. Some of the more important factors are: relative humidity and temperature of the environment; die size, relative thickness of epoxy resin in the chip and under chip pad, properties of the materials; and adhesion strength of the interfaces in the package (Lee and Earmme, 1996). In this study, we used high-pin-count PLCC packages. The relative humidity, the temperature, and the soak time were the test parameters. In the following, we compare our results with past work and the prevailing industry practice.

For comparison of weight gain data, sample data from Shook et al. (1998) for the 68-pin PLCC with the same soak condition and the time scale are extracted and shown in Table 8.7. There are almost identical considering measurement error and possible differences in materials used.

Table 8.7. Percent Weight Gain Measurements

	Percent Weight Gain Measurements (68-pin PLCC at 30°C/60% RH)			
	48 hr	72 hr	96 hr	192 hr
Shook et al. (1998)	0.047	0.057	0.063	0.085
Current work	0.046	0.056	0.064	0.087

Next we look at the acceleration factors given in Table 8.6. This is crucial as the equivalent moisture soak times are computed from these values. The result is well-aligned with the order of stress levels. The current practice for soaking level 3–5 devices at 60C/60% RH instead of the 30°C/60% RH recommended by JEDEC is to use an accelerated factor of around 4 (Song, 1998; Song and Walberg, 1998). This compares favorably with the acceleration factor of 4.15, and it also falls within the 95% confidence interval (see Table 8.6) given by our model. This, however, is not quite in agreement with Shook et al. (1998), which suggests that testing at 60C/60% RH will reduce the total required moisture soak time for levels 3–5 by a factor of 4.8 (our upper 97.5% confidence limit is 4.58) as compared to the time required at 30°C/60% RH. His result is based on the empirical evidence that the diffusion coefficient at 60°C increases the kinetics of moisture ingress by a factor of 4.8 compared to that at 30°C. He then validated the result by experimenting at 30°C/60% RH and 60C/60% RH. Our experiment, however, covers a wider range of temperature/RH combinations. Our model in equations (8.12) and (8.13) also reveals that interaction between RH and T is significant. On the other hand, it is possible that the difference in results can be due to the use of different mold compounds and some variations in the experimental process. It is widely believed that moisture absorption is dependent on the mold compound property, the amount of compound encapsulated for package, die attach epoxy, and packaging construction. For a different type of packaging, the moisture content of a package is dependent on the temperature and humidity. Nevertheless, from Figure 8.2, it is observed that the dependency on temperature is more pronounced. In particular, at 60% RH, the moisture absorption tends to be a linear function of temperature after an initial phase of rapid gain.

Next we compare the acceleration factors obtained for 85°C/85% RH and 30°C/60% RH. The acceleration factor of 30 is considered high compared to the value of 15 reported by Sun et al. (1997), who assumed a mixture of two power law models for the acceleration model without assessing the goodness of fit. In comparison with the moisture absorption graphs currently in use in the industry (see Figure 8.5), our result is closer with the industry practice. It can be seen from the projected line that a soak time of 150 hours at 30°C/60% RH will achieve the same moisture gain as a soak time of 5 hours at 85°C/85% RH. In any case, since the floor life for level 1 is unlimited, and that for level 2 is 1 year, the industry is primarily interested in evaluating the acceleration for soaking at 60°C/60% RH for replacing the recommended

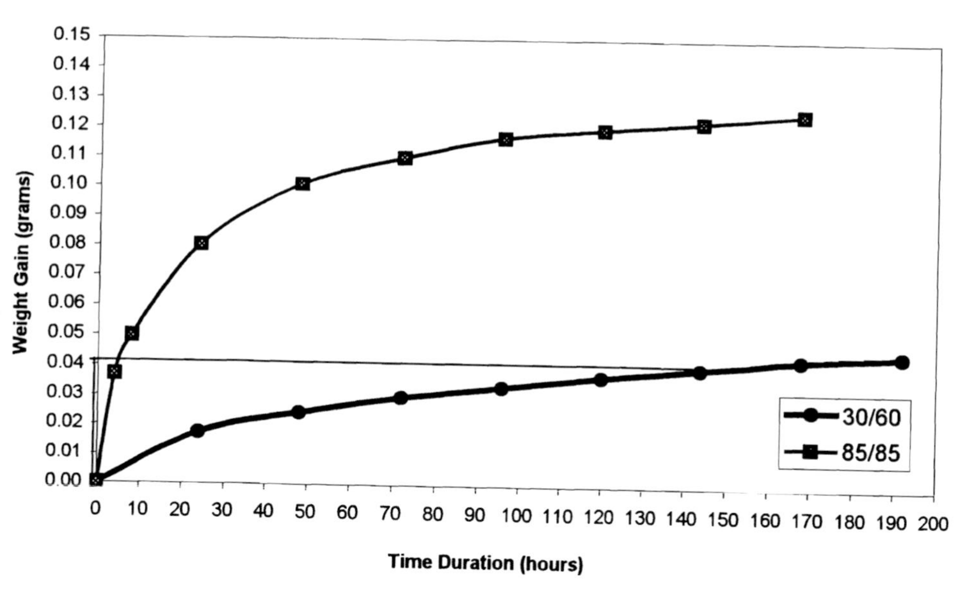

Figure 8.5. Moisture absorption graph for 30°C/60RH and 85°C/85 RH temperature–humidity setting.

30°C/60% RH for level 3 to 5. The accuracy for the acceleration factor at 85°C/85% RH is thus immaterial in practice. Moreover, it has been cautioned by Taylor et al. (1997) that certain level 3–6 packages are not designed to tolerate the high-stress soak time of 85°C/85% RH, because they fail after IR heating when the moisture content in PQFPs is higher than 60% of the equilibrium composition.

In summary, moisture soak is a front-runner for the preconditioning test to assess the moisture sensitivity level during product qualification and can also be used as an in-line process monitor. To avoid having reliability and quality issues later, it is critical that moisture sensitive devices be properly classified, identified, and packaged in dry bags until ready for PCB assembly. Several attempts have been made to derive models, through analytic or statistical means, for predicting the soak time and humidity condition for surface-mounted devices. In this case study, we use a statistical approach with physical reasoning to model the moisture absorption process so as to derive the equivalent moisture soak time for PLCC packages. In particular, it is found that the soak condition at 60°C/60% RH will reduce the soak time for level 3–5 (30°C/60% RH) by 3.7 to 4.7 times.

REFERENCES

Ahn, S. H., and Kwon, Y. S. (1995). Popcorn phenomena in a ball grid array package, *IEEE Transactions on Components, Packaging and Manufacturing Technology* **18B:**491–495.

Draper, N., and Smith H. (1981). *Applied Regression Analysis,* 2nd edition, Wiley, New York.

EIA/JESD22-A112A (1995). *Moisture-Induced Stress Sensitivity for Plastic Surface Mount Devices,* Electronic Industrial Association.

Galloway, J. E., and Miles, B. M. (1997). Moisture absorption and desorption predictions for plastic ball grid array packages, *IEEE Transactions on Components, Packaging, and Manufacturing Technology* **20:**274–279.

Huang, Y. E., Hagen, D., Dody, G., and Burnette, T. (1998). Effect of Solder Reflow Temperature Profile on Plastic Package Delamination, *IEEE/CPMT International Electronics Manufacturing Technology Symposium,* pp. 105–111.

IPC-SM–786A (1995). *Procedures for Characterizing and Handling of Moisture/Reflow Sensitive Ics,* Institute of the Interconnecting and Packaging Electronics Circuits.

JESD22-A113A (1995). *Preconditioning of Plastic Surface Mount Devices Prior to Reliability Testing,* Electronic Industrial Association.

Kitano, M., Nishimura, A., and Kawai, S. (1988). Analysis of the Packaging Cracking during Reflow Soldering Process, *IEEE International Reliability Physics Symposium,* pp. 90–95.

Lee, H., and Earmme, Y. Y. (1996). A fracture mechanics analysis of the effects of material properties and geometries of components on various types of package cracks, *IEEE Transactions on Components, Packaging, and Manufacturing Technology* **19A:**168–178.

Nelson, W. (1990). *Accelerated Testing: Statistical Models, Test Plans and Data Analysis,* Wiley, New York.

Shook, R. L., Vaccaro, B. T., and Gerlach, D. L. (1998). Method for Equivalent Acceleration of JEDEC/IPC Moisture Sensitivity Levels, *Proceedings, 36th International Reliability Physics Symposium,* pp. 214–219.

Song, J. (1998). *Moisture Sensitivity Rating of the PLCC-68L Package,* National Semiconductor Corporation.

Song, J., and Walberg, R. (1998). *A Study of Standard vs. Accelerated Soak Conditions at Levels 2A and 3,* National Semiconductor Corporation.

Sun, Y., Wong, E. C., and Ng, C. H. (1997). A Study on Accelerated Preconditioning Test, *Proceedings 1st Electronic Packaging Technology Conference,* pp. 98–101.

Tanaka, N., and Nishimura, A. (1995). Measurement of IC molding compound adhesive strength and prediction of interface delamination within package, *ASME EEP, Advances in Electronics Packaging* **10–2:**765–773.

Tay, A. A. O., and Lin, T, (1998). Moisture-Induced Interfacial Delamination Growth in Plastic Packages during Solder Reflow, *Proceedings, 48th Electronic Components and Technology Conference,* pp. 371–378.

Taylor, S. A., Chen, K., and Mahajan, R. (1997) Moisture migration and cracking in plastic quad flat packages, *Journal of Electronic Packaging* **119:**85–88.

EXERCISES

8.1. Using the data given in Table 8.3, generate the loglogistics plot that is similar to Figure 8.3a. Are the slopes under different experimental conditions approximately equal? Discuss the physical implications.

8.2. Calculate an estimate of the location parameter of the family of loglogistic distributions at 85°C and 30% RH.

8.3. Given additional resources for further experimentation, suggest other appropriate experimental conditions and justify your recommendations.

8.4. Compare and contrast various physical acceleration models given in this chapter. What are other appropriate models?

8.5. Discuss the approach adopted in this chapter in selecting the acceleration model. What are the alternative approaches in choosing an appropriate physical acceleration model?

8.6. Suppose that the Peck model was assumed. In what way will the results be different from the current analysis? Carry out the analysis to validate your answer.

8.7. From the results for the location parameter given in Table 8.5, obtain an estimate of the acceleration factor and its interval estimate for package type "0." Discuss the differences between your answers and those for type "1" packages, which are given in Table 8.6.

CHAPTER 9

Construction of Reliable Software in Resource-Constrained Environments

Mladen A. Vouk and Anthony T. Rivers

9.1 INTRODUCTION

The challenges of modern "market-driven" software development practices, such as the use of various types of incentives to influence software developers to reduce time to market and overall development cost, seem to favor a resource-constrained development approach. In a resource-constrained environment, there is a tendency to compress software specification and development processes (including verification, validation, and testing) using either mostly business model directives or some form of "short-hand" technical solution, rather than a combination of sound technical directives and a sensible business model. The business models that often guide modern "internet-based" software development efforts advocate, directly or indirectly, a lot of "corner-cutting" without explicit insistence on software process and risk management procedures, and associated process tracking tools, appropriate to these new and changing environments [e.g., extreme programming (XP), use of Web-based delivery tools (Potok and Vouk, 1999a, 1999b; Beck, 2000)]. There are also software development cultures, such as XP, specifically aimed at resource-constrained development environments (Auer and Miller, 2001; Beck 2000, 2001). Yet, almost as a rule, explicit quantitative quality and process modeling and metrics are subdued, sometimes completely avoided, in such environments. Possible excuses could range from lack of time, to lack of skills, to intrusiveness, to social reasons, and so on. As a result, such software may deliver with an excessive number of problems that are hard to quantify and even harder to relate to process improvement activity in more than an ad hoc manner.

For example, consider testing in "web-year" environments [i.e., 3-month development cycle (Dougherty, 1998)]. While having hard deadlines relatively often (say every 3 months) is an excellent idea because it may reduce the process vari-

Case Studies in Reliability and Maintenance, Edited by W. R. Blischke and D. N. P. Murthy.
ISBN 0-471-41373-9

ance, the decision to do so must mesh properly with both business and software engineering models being used or the results may not be satisfactory (Potok and Vouk, 1997, 1999a, 1999b). One frequent and undesirable effect of a business and software process mismatch is haste. Haste often manifests during both design and testing as a process analogous to a "sampling without replacement" of a finite (and sometimes very limited) number of predetermined structures, functions, and environments. The principal motive is to verify required product functions to an acceptable level, but at the same time minimize the re-execution of already tested functions, objects, and so on. This is different from "traditional" strategies that might advocate testing of product functions according to the relative frequency of their usage in the field—that is, according to their operational profile[1] (Musa, 1999; Vouk, 2000). "Traditional" testing strategies tend to allow for much more re-execution of previously tested functions/operations, and the process is closer to a "sampling with (some) replacement" of a specified set of functions. In addition, "traditional" operational profile-based testing is quite well-behaved and, when executed correctly, allows dynamic quantitative evaluation of software reliability growth based on "classical" software reliability growth metrics and models. These metrics and models can then be used to guide the process (Musa et al., 1987). When software development and testing is resource- and schedule-constrained, some traditional quality monitoring metrics and models may become unusable. For example, during nonoperational profile-based testing, failure intensity decay may be an impractical guiding and decision tool (Rivers, 1998; Rivers and Vouk, 1999).

End-product reliability can be assessed in several ways. The aspect of reliability that we are concerned with is efficient reduction of as many defects as possible in the software being shipped. In the next section we describe and discuss the notion of resource-constrained software development and testing. In Section 9.3, we present a software reliability model that formalizes resource-constrained development and testing with respect to its outcome. Empirical results, from several sources, that illustrate different constrained software development issues are presented in Section 9.4.

[1]Operational profile is a set of relative frequencies (or probabilities) of occurrence of disjoint software operations during its operational use. A detailed discussion of operational profile issues can be found in Musa et al. (1987), Musa (1993), and Musa (1999). A software-based system may have one or more operational profiles. Operational profiles are used to select verification, validation, and test cases and direct development, testing, and maintenance efforts toward the most frequently used or most risky components. Construction of an operational profile is preceded by definition of a customer profile, a user profile, a system mode profile, and a functional profile. The usual participants in this iterative process are system engineers, high-level designers, test planners, product planners, and marketing. The process starts during the requirements phase and continues until the system testing starts. Profiles are constructed by creating detailed hierarchical lists of customers, users, modes, functions and operations that the software needs to provide under each set of conditions. For each item it is necessary to estimate the probability of its occurrence (and possibly risk information) and thus provide a quantitative description of the profile. If usage is available as a rate (e.g., transactions per hour), it needs to be converted into probability. In discussing profiles, it is often helpful to use tables and graphs and annotate them with usage and criticality information.

9.2 CONSTRAINED DEVELOPMENT

Software development always goes through some basic groups of tasks related to specification, design, implementation, and testing. Of course, there are always additional activities, such as deployment and maintenance phase activities, as well as general management, and verification and validation activities that occur in parallel with the above activities. How much emphasis is put into each phase, and how the phases overlap, is a strong function of the software process adopted (or not) by the software manufacturer, and of how *resource constraints* may impact its modification and implementation (Pressman, 2001; Beck, 2000; Potok and Vouk, 1997). A typical resource constraint is development time. However, "time" (or exposure of the product to usage, in its many forms and shapes) is only one of the resources that may be in short supply when it comes to software development and testing. Other resources may include available staff, personnel qualification and skills, software and other tools employed to develop the product, business-model-driven funding and schedules (e.g., time-to-market), social issues, and so on. All these resources can be constrained in various ways, and all do impact software development schedules, quality, functionality, and so on, in various ways (e.g., Boehm, 1988; Potok and Vouk, 1997).

Specification, design, implementation, and testing activities should be applied in proportions commensurate with (a) the complexity of the software, (b) level of quality and safety one needs for the product, (c) business model, and (d) level of software and process maturity of the organization and its staff. In a resource-constrained environment, some, often all, elements are economized on. Frequently, this "economization" is analogous to a process of "sampling without replacement" of a finite (and sometimes very limited) number of predetermined software system input states, data, artifacts/structures, functions, and environments (Rivers, 1998; Rivers and Vouk, 1998).

For instance, think of the software as being a jar of marbles. Take a handful of marbles out of the jar and manipulate them for whatever purpose. For example, examine them for quality or relationship to other marbles. If there is a problem, fix it. Then put the marbles into another jar marked "completed." Then, take another sample out of the first jar, and repeat the process. Eventually, the jar marked "completed" will contain "better" marbles and the whole collection (or "heap") of marbles will be "better." However, any imperfections missed on the one-time inspection of each handful still remain. That is sampling *without replacement.* Now start again with the same (unfixed) set of marbles in the first jar. Take a handful of marbles out of the first jar and examine them. Fix their quality or whatever else is necessary, perhaps even mark them as good, and then put them back into the same jar out of which they were taken. Then, on the subsequent sampling of that jar (even if you stir the marbles), there is a chance that some of the marbles you pull out are the same ones you have handled before. You now manipulate them again, fix any problems, and put all of them back into the same jar. You have sampled the marbles in the jar *with replacement.* Obviously, sampling with replacement implies a cost overhead. You are handling again marbles you have already examined, so it takes longer to "cover" all the marbles in the jar. If you are impatient, and you do not cover all the marbles, you may miss something. However, second handling of a marble may actually reveal additional flaws and may improve the overall quality of the marble "heap" you have. So, a feedback loop or sampling with replacement may have its advantages, but it does add to the overall effort.

To illustrate the impact of resource constrained software development, it is interesting to explore the implications of three different software development approaches (Rivers, 1998; Rivers and Vouk, 1999):

1. A methodology that has virtually no resource constraints (including schedule constraints), that uses operational profiles to decide on resource distribution, and that allows a lot of feedback and cross-checking during development—essentially a methodology based on development practices akin to sampling with replacement.
2. A well-executed methodology designed to operate in a particular resource-constrained environment, based on a combination of operational profiling and effective "sampling without replacement" (Rivers, 1998, Rivers and Vouk, 1999).
3. An ad hoc, perhaps poorly executed, methodology that tries to reduce resource usage (e.g., the number of test cases with respect to the operational profile) in a manner that, in the marble example, inspects only some of the marbles out the first jar, perhaps using very large handfuls that are given more cursory inspection.

Consider Figure 9.1. It illustrates fault elimination by the three methods in a hypothetical 100-state system that has 20 faulty states. The actual computations are discussed in Section 9.3.3 in this chapter. The vertical axis indicates the fraction of the latent faults that are not "caught" by the selected fault-avoidance or fault-elimination approach, and the horizontal axis indicates expenditure of the resources. These resources are often constrained by a predetermined maximum. The resource usage is expressed as a fraction of the maximum allowed. The actual resource could be anything relevant to the project and impacting fault avoidance and elimination (e.g., time, personnel, a combination of the two, etc.). Details of the calculations suitable in the testing phase are discussed in Rivers (1998) and in Rivers and Vouk (1998). The associated constrained testing model is based on the hypergeometric probability mass function and the concept of static and dynamic software structure and construct coverage over the exposure/usage period. This usage period refers to the exposure software receives as part of static and dynamic verification and validation, execution-based testing, and field usage.

In an "ideal" situation, the fault-avoidance and fault-elimination processes are capable of avoiding or detecting all 20 faults, one or more at a time, and defect removal is instantaneous and perfect. An "ideal" approach based on sampling without replacement requires less resources to reach a desired level of "defects remaining" (vertical axis) than do methods that sample with replacement—that is recover already tested states. However, traditional operational profile testing is quite well-behaved; and when executed correctly, it may turn out test cases that are more effi-

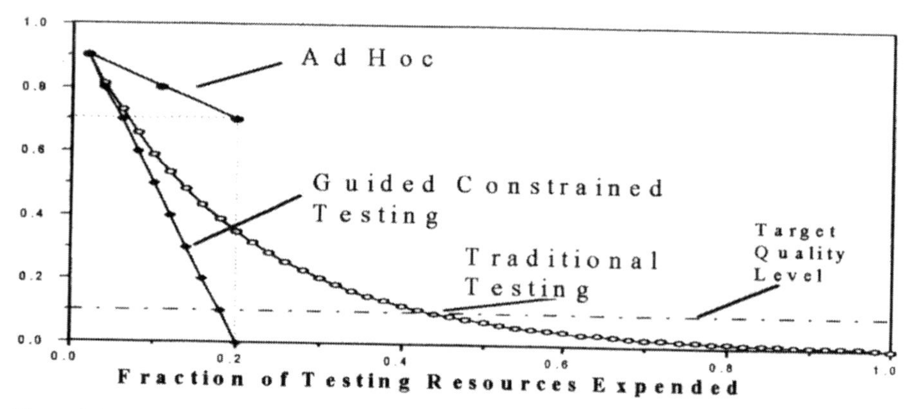

Figure 9.1. Fraction of shipped defects for "ideal" testing based on sampling with and without replacement.

cient and more comprehensive than those constructed using some nonideal coverage-based strategy. In that case, testing based on sampling without replacement will yield poorer results.

In Figure 9.1, "Ad Hoc" illustrates a failed attempt to cut the resources to about 20% of the resources that might be needed for a complete operational-profile-based test of the product. In the ad hoc method, state coverage was inadequate, and although the testing is completed within the required resource constraints, it only detects a small fraction of the latent faults (defects). In the field, this product will be a constant emitter of problems, and its maintenance will probably cost many times the resources "saved" during the testing phases.

A way to develop "trimmed" test suites is to start with test cases based on the operational profile and trim the test suite in a manner that preserves coverage of important parameters and coverage measures. One such approach is discussed by Musa (1999). Another one is to use a pairwise test case generation strategy (Cohen et al., 1996; Lei and Tai, 1998; Tai and Lei, 2002). Of course, there are many other approaches, and many of the associated issues are still research topics. The software reliability engineering (SRE) model and metrics discussed in the next section allow us to decide early in the process which of the curves the process is following. This then allows us to correct both the product and the process.

From the point of view of this chapter, of most interest are (a) the SRE activities during system and field testing and (b) operation and maintenance phases. Standard activities of *system* and *field testing* phases include: (a) execution of system and field acceptance tests, (b) checkout of the installation configurations, and (c) validation of software functionality and quality, while in the *operation and maintenance* phases the essential SRE elements are (a) continuous monitoring and evaluation of software field reliability, (b) estimation of product support staffing needs, and (c) software process improvement. This should be augmented by requiring software developers and maintainers to (a) finalize and use operational profiles, (b) actively track the development, testing, and maintenance process with respect to quality, (c) use reliability growth models to monitor and validate software reliability and availability, and (d) use reliability-based test stopping criteria to control the testing process and product patching and release schedules.

Ideally, the reaction to SRE information would be quick, and the corrections, if any, would be applied already within the life-cycle phase in which the information is collected. However, in reality, introduction of an appropriate feedback loop into the software process, along with the latency of the reaction, will depend on the accuracy of the feedback models, as well as on the software engineering capabilities of the organization. For instance, it is unlikely that organizations below the third maturity level on the Software Engineering Institute (SEI) Capability Maturity Model (CMM)[2] scale (Paulk et al., 1993b) would have processes that could react to the feedback information in less than one software release cycle. Reliable latency of less than one phase is probably not realistic for organizations below CMM level 4. This needs to be taken into account when considering the level and the economics of "corner-cutting." Since the resources are not unlimited, there is always some part of the software that may go untested, or may be verified to a lesser degree. The basic, and most difficult, trick is to decide what must be covered and what can be left uncovered or partially covered.

9.3 MODEL AND METRICS

In a practice, software verification, validation, and testing (VVT) is often driven by intuition. Generation of use cases and scenarios, and of VVT cases in general, may be less than a repeatable process. To introduce some measure of progress and repeatability, structure-based methodologies advocate generation of test cases that monotonly increase the cumulative coverage of a particular set of functional or code constructs the methodology supports. While *coverage-based* VVT in many ways implies that the "good" test cases are those that increase the overall cover-

[2]The Capability Maturity Model (CMM) for software describes the principles and practices underlying software process maturity and is intended to help software organizations improve the maturity of their software processes in terms of an evolutionary path from an ad hoc reactive and chaotic process, to a mature, disciplined, and proactive software process. The CMM is organized into five maturity levels: (1) *Initial*. The software process is characterized as ad hoc, and occasionally even chaotic. Few processes are defined, and success depends on individual effort and heroics. (2) *Repeatable*. Basic project management processes are established to track cost, schedule, and functionality. The necessary process discipline is in place to repeat earlier successes on projects with similar applications. (3) *Defined*. The software process for both management and engineering activities is documented, standardized, and integrated into a standard but still reactive software process for the organization. All projects use an approved, tailored version of the organization's standard software process for developing and maintaining software. (4) *Managed*. Detailed measures of the software process and product quality are collected. Both the software process and products are quantitatively understood and proactively controlled. (5) *Optimizing*. Continuous process improvement is enabled by quantitative feedback from the process and from piloting innovative ideas and technologies (http://www.sei.cmu.edu/cmm/cmm.sum.html).

age, what really matters is whether the process prevents or finds and eliminates faults, how efficient this process is, and how repeatable is the process of generating VVT activities (test cases) of the same or better quality (efficiency). This implies that the strategy should embody certain fault-detection guarantees. Unfortunately, for many coverage-based strategies these guarantees are weak and not well understood. In practice, designers, programmers, and testers are more likely to assess their software using a set of VVT cases derived from functional specifications (as opposed to coverage derived cases). Then, they supplement these "black box" cases with cases that increase the overall coverage based on whatever construct they have tools for. In the case of testing, usually blocks, branches, and some data-flow metrics such as p-uses[3] (Frankl and Weyuker, 1988) are employed. In the case of earlier development phases, coverage may be expressed by the cross-reference maps and coverage of user-specified functions (requirements) (Pressman, 2001). In fact, one of the key questions is whether we can tell, based on the progress, how successful the process is on delivering with respect to some target quality we wish the software to have in the field (e.g., its reliability in the field), and how much more work we need to assure that goal. In the rest of this section, we first discuss the process and the assumptions that we use in developing our model, and then we present our model.

9.3.1 Coverage

Coverage $q(M, S)$ is computed for a construct (e.g., functions, lines of code, branches, p-use, etc.) through metric M and an assessment or testing strategy S by the following expression

$$q(M, S) = \frac{\text{Number of executed constructs for } M \text{ under } S}{\text{Total number of constructs for } M \text{ under } S} \tag{9.1}$$

Notice that the above expression does not account for infeasible (unexecutable) constructs under a given testing strategy S. Hence, in general, there may exist an upper limit on the value of $q(M, S)$, $q_{max} \leq 1$. Classical examples of flow metrics are lines of code, (executable) paths, and all definition-use tuples (Beizer, 1990). But, the logic applies equally to coverage of planned test cases, coverage of planned execution time, and the like. We define the later examples as plan-flow metrics, because the object is to cover a "plan."

Assumptions. The resource constrained model of VVT discussed here is based on the work reported in Vouk (1992), Rivers and Vouk (1995), Rivers (1998), Rivers and Vouk (1998), and Rivers and Vouk (1999). We assume that:

[3]P-use (predicate data use) is use of a variable in a predicate. A predicate is a logical expression that evaluates to TRUE or FALSE, normally to direct the execution path in the software program (code) (Frankl and Weyuker, 1988).

1. Resource-constrained VVT is approximately equivalent to sampling without replacement. To increase (fulfill) coverage according to metric M, we generate new VVT cases/activities using strategy S so as to cover as many remaining uncovered constructs as possible. The "reuse" of constructs through new cases that exercise at least one new construct (i.e., increase M by at least 1) is effectively "ignored" by the metric M.
2. VVT strategy S generates a set of (VVT or test) cases $T = \{T_i\}$, where each case, T_i, exercises one or more software specification or design functions, or in some other way provides coverage of the product.
3. The order of application/execution of the cases is ignored (and is assumed random) unless otherwise dictated by the testing strategy S.
4. For each set T generated under S and monitored through M, there is a minimal coverage that results in a recorded change in M through execution of a (typical) case T_i from T. There may also be a maximum feasible coverage that can be achieved, $1 \geq q_{\max} \geq q(M, S) \geq q_{\min} \geq 0$. In general, execution of a complete entry to exit path in a program (or its specifications) will require at least one VVT case, and usually will exercise more than one M construct.
5. Faults/defects are distributed randomly throughout software elements (a software element could be a specification, design, or implementation document or code).
6. Detected faults are repaired immediately and perfectly; that is, when a defect is discovered, it is removed from the program before the next case is executed without introducing new faults.
7. The rate of fault detection with respect to coverage is proportional to the effective number (or density) of residual faults detectable by metric M under strategy S and is also a function of already achieved coverage.

Assumptions 5 and 6 may seem unreasonable and overly restrictive in practice. However, they are very common in construction of software reliability models since they permit development of an idealized model and solution which can then be modified to either theoretically or empirically account for practical departures from these assumptions (Shooman, 1983; Musa, 1987; Tohma et al., 1989; Lyu, 1996; Karcich, 1997). Except in extreme cases, the practical impact of the assumptions 5 and 6 is relatively limited and can be managed successfully when properly recognized, as we do in our complete model.

9.3.2 Testing Process

The efficiency of the fault detection process can be degraded or enhanced for a number of reasons. For example, the mapping between defects and defect sensitive constructs is not necessarily one to one, so there may be more constructs that will detect a particular defect, or the defect concentration may be higher in some parts of M-space—that is, space defined by metric M. An opposite effect takes place when, although a construct may be defect sensitive in principle, coverage of the construct may not always result in the detection of the defect. To account for this, and for any other factors that modify problem detection and repair properties of the constructs, we introduce the *testing efficiency* function g_i.

This function represents the "visibility" or "cross section" of the remaining defects to testing step i. Sometimes we will find more and sometimes less than the expected number of defects, and sometimes a defect will remain unrepaired or only partially repaired. g_i attempts to account for that in a dynamic fashion by providing a semiempirical factor describing this variability. Similar correction factors appear in different forms in many coverage-based models. For instance, Tohma et al. (1989) used the terms *sensitivity factor, ease of test,* and *skill* in developing their hypergeometric distribution model. Hou et al. (1994) used the term *learning factor,* while Piwowarski et al. (1993) used the term *error detection rate constant* to account for the problem detection variability in their coverage-based model. Malaiya et al. (1992) analyzed the fault exposure ratio (Musa et al., 1987) to account for the fluctuation of per-fault detectability while developing their exponential and logarithmic coverage-based models.

Let the total number of defects in the program P be N. Let the total number of constructs that need to be covered for that program be K. Let execution of T_i increase the number of covered constructs by $h_i > 0$. Let u_i be the number of constructs that still remain uncovered before T_i is executed. Then,

$$u_i = K - \sum_{j=1}^{i-1} h_j \tag{9.2}$$

Let D_i be the subset of d_i faults remaining in the code before T_i is executed. Let execution of T_i detect (and result in the removal of) x_i defects. Let there be at least one construct, in u_i, the coverage of which guarantees detection of each of the remaining d_i defects. That is, all items in D_i are detectable by unexecuted test cases generated under S.

The hypergeometric probability mass function (Walpole and Myers, 1978) describes the process of sampling without replacement, the process of interest in a resource constrained environment. Under our assumptions, the probability that test case T_i detects x_i defects is at least

$$p(x_i|u_i, b_i, h_i) = \frac{\binom{b_i}{x_i} \cdot \binom{u_i - b_i}{h_i - x_i}}{\binom{u_i}{h_i}} \tag{9.3}$$

where $b_i = g_i \cdot d_i$. Note that g_i is usually a positive real number, that is, it is not limited to the range of all positive integers, as equation (9.3) requires. Hence when using equation (9.3), one has to round off computed b_i to the nearest integer subject to system-level consistency constraints. In practice, there is likely to be considerable variability of g_i within a testing phase. g_i's variability stems from a variety of sources, such as the structure of the software, the software development practices employed, and, of course, the human element.

In equation (9.3), the numerator is the number of ways h_i constructs with x_i defectives can be selected. The first term in the numerator is the number of ways in which x_i defect-sensitive constructs can be selected out of b_i that can detect defects (and are still uncovered). The second term in the numerator is the number of ways in which $h_i - x_i$ non-defect-sensitive constructs can be selected out of $u_i - b_i$ that are not defect-sensitive (and are still uncovered). The denominator is the number of ways in which h_i constructs can be selected out of u_i that are still uncovered. The expected value $\bar{x}_i$ of x_i is

$$\bar{x}_i = (g_i \cdot d_i)\left(\frac{h_i}{u_i}\right) \tag{9.4}$$

(Walpole and Myers, 1978). Let the total amount of cumulative coverage up to and including $T_i - 1$ be

$$q_{i-1} = \sum_{j=1}^{i-1} h_j \tag{9.5}$$

and let the total number of cumulative defects detected up to and including $T_i - 1$ be

$$E_{i-1} = \sum_{j=1}^{i-1} x_j \tag{9.6}$$

Since $d_i = (N - E_{i-1})$, equation (9.4) becomes

$$\bar{x}_i = \bar{E}_i - \bar{E}_{i-1} = \Delta\bar{E}_i = g_i(N - \bar{E}_{i-1})\left(\frac{q_i - q_{i-1}}{K - q_{i-1}}\right) = g_i(N - \bar{E}_{i-1})\left(\frac{\Delta q_i}{K - q_{i-1}}\right) \tag{9.7}$$

where $\bar{E}_i$ denotes the expected value function. In the small step limit, this difference equation can be transformed into a differential equation by assuming that $\Delta\bar{E}_i \rightarrow d\bar{E}_i$ and $\Delta q_i \rightarrow dq_i$, that is,

$$d\bar{E}_i = g_i(N - \bar{E}_i)\left(\frac{dq_i}{K - q_i}\right) \tag{9.8}$$

Integration yields the following model:

$$E_i = N - (N - E_{\min})e^{-\int_{q_{\min}}^{q_i} \frac{g_i}{K - \varsigma} d\varsigma} \tag{9.9}$$

where $q_{\min}$ represents the minimum meaningful coverage achieved by the strategy. For example, one test case may, on the average, cover 50–60% of statements (Jacoby and Masuzawa, 1992). Function g_i may have strong influence on the shape of the failure intensity function of the model. This function encapsulates the information about the change in the ability of the test/debug process to detect faults as coverage increases.

Because the probability of "trapping" an error may increase with coverage, stopping before full coverage is achieved makes sense only if the residual fault count has been reduced below a target value. This threshold coverage value will tend to vary even between functionally equivalent implementations. The model shows that metric saturation can occur, but that additional testing may still detect errors. This is in good agreement with the experimental observations, as well as conclusions of other researchers working in the field [e.g., Chen et al. (1995)]. For some metrics, such as the fraction of planned test cases, this effect allows introduction of the concept of "hidden" constructs, reevaluation of the "plan," and estimation of the additional "time" to target intensity in a manner similar to "classical" time-based models [e.g., Musa et al. (1987)].

Now, let $g_i = a$, where "a" is a constant. Let it represent the average value of g_i over the testing period. Then the cumulative failures model of equation (9.9) becomes

$$E_i = N - (N - E_{\min})\left(\frac{K - q_i}{K - q_{\min}}\right)^a \tag{9.10}$$

To derive a failure intensity equation λ_i based on $g_i = a$, we take the derivative of equation (9.10) with respect to q_i, which yields

$$\lambda_i = \left(\frac{a}{K - q_{\min}}\right)(N - E_{\min})\left(\frac{K - q_i}{K - q_{\min}}\right)^{(a-1)} \tag{9.11}$$

9.3.3 Sampling with and without Replacement

Let us further explore the implications of a well-executed testing methodology based on "sampling without replacement" (see Figure 9.1). We did that by developing two solutions for relationship (9.7), using a reasoning similar to that described in Tohma et al. (1989) and in Rivers (1998). These solutions compute the number of defects shipped–that is, the number of defects that leave the phase in which we performed the testing.

The first case is a solution for the expected number l_{i0} of defects that would be shipped after test step i based on an "ideal" nonoperational profile-based testing strategy is (Rivers, 1998):

$$l_{i0} = N - \overline{E}_i = N - N\left(1 - \prod_{j=1}^{i}\left(1 - \frac{g_j \cdot h_j}{u_j}\right)\right) \tag{9.12}$$

In the second case, to emulate an "ideal" strategy based on sampling with replacement, we start with a variant of equation (9.4) with K in place of u_j and find the solution l_{i1} (Rivers, 1998):

$$l_{i1} = N - \overline{E}_i = N - N\left(1 - \prod_{j=1}^{i}\left(1 - \frac{g_j \cdot h_j}{K}\right)\right) \tag{9.13}$$

In an "ideal" situation, (a) the test suite has in it the test cases that are capable of detecting all faults (if all faults are M-detectable), and (b) defect removal is instantaneous and perfect. In "best" cases, g_j is consistently 1 or larger. Plots of equations (9.12) and (9.13) are shown in Figure 9.1. Equation (9.12) is denoted as "guided constrained testing" in the figure, and equation (9.13) is denoted as "traditional testing" with g_j of 1 (constant during the whole testing phase) and with $N = 20$ and $K = 100$ in both cases. The horizontal axis was normalized with respect to 500 test cases. We see, without surprise, that "ideal" methods based on sampling without replacement require fewer test steps to reach a desired level of "defects remaining" than do methods that reuse test steps (or cases) or recover already tested constructs.

However, when there are deviations from the "ideal" situation, as is usually the case in practice, the curves could be further apart, closer together, or even reversed. When testing efficiency is less than 1, or the test suite does not contain test steps (cases) capable of uncovering all software faults, a number of defects may remain uncovered by the end of the testing phase.

The impact of different (but "constant" average) g_j values on sampling without replacement (nonoperational testing) was explored using a variant of (9.10) and (9.12) (Rivers, 1998):

$$\text{Shipped defects} = N - \overline{E}_i = N - N(1 - (K - q_i)^a) \tag{9.14}$$

If g_j is constantly less than 1, some faults remain undetected. In general, the (natural) desire to constrain the number of test cases and "sample without replacement" is driven by business decisions and resource constraints. As mentioned earlier, the difficult trick is to design a testing strategy (and coverage metric) that results in a value of g_j that is larger than 1 for most of the testing phase.

9.4 CASE STUDIES

9.4.1 Issues

We use empirical data on testing efficiency to illustrate some important issues related to resource-constrained software development and testing. One issue is that of more efficient use of resources. That is, whether it is possible to constrain resources and still improve (or just maintain adequate) verification and validation processes and end-product quality as one progresses through software development phases.

The overall idea is that the designers and testers, provided they can learn from their immediate experience in a design or testing phase, can either shorten the life cycle by making that phase more efficient, or they can improve the quality of the software more than they would without the "learning" component. This implies some form of verification and validation feedback loop in the software process. Of course, all software processes recommend such a loop, but how immediate and effective the feedback is depends, to a large extent, on the capabilities and maturity of the organization (Paulk et al., 1993a, 1993b). In fact, as mentioned earlier, reliable forward-correcting within-phase feedback requires CMM level 4 and level 5 organizational capabilities. There are not many organizations like that around (Paulk et al., 1993a, 1993b). Most software development organizations operate at level 1 (ad hoc), level 2 (repeatable process), and perhaps level 3 (metrics, phase-to-phase improvements, etc.). However, all three lowest levels are reactive, and therefore the feedback may not operate at fine enough task/phase granularity to provide cost-effective verification and validation "learning" within a particular software release.

The issue is especially interesting in the context of Extreme Programming (Auer and Miller, 2001; Beck, 2000, 2001). XP is a lightweight software process that often appeals to (small?) organizational units looking at rapid prototypes and short development cycles. Because of that, it is likely to be used under quite considerable resource constraints. XP advocates several levels of verification and validation feedback loops during the development. In fact, a good problem determination method integral to XP (Beck, 2000, 2001) is pair programming.[4] Unfortunately, one of the characteristics of XP philosophy is that it tends to avoid "burdening" the process and participants with "unnecessary" documentation, measurements, tracking, and so on. This appears to include most traditional SRE metrics and methods. Hence, to date (late 2002), almost all evidence of success or failure of XP is, from the scientific point of view, anecdotal. Yet, for XP to be successful it has to cut a lot of corners with respect to the traditional software processes. That means that fault avoidance and fault elimination need to be even more sophisticated than usual. Pair programming, a standard component of XP, is an attempt to do that. However, because of its start-up cost (two programmers vs. traditional one programmer) it does imply a considerable element of fault-avoidance and fault-elimination superefficiency. In the next subsection, we will see how traditional methods appear to fare in that domain.

Some other notable issues include situations where the absence of a decreasing trend in *failure intensity* during resource constrained system testing is *not* an indica-

[4]Pair programming (PP) is a style of programming in which two programmers work side by side at one computer, continuously collaborating on the same design, algorithm, code, or test. It is important to note that pair programming includes all phases of the development process (design, debugging, testing, etc.), not just coding. Experience shows that programmers can pair at any time during development, in particular when they are working on something that is complex. The more complex the task is, the greater the need for two brains. More recently, though, overwhelming anecdotal and qualitative evidence from industry indicates that programmers working in pairs perform substantially better than do two working alone. The rise in industrial support of the practice coincides with the rise in popularity of the extreme programming (XP) (Beck, 2000; Auer and Miller, 2001) software development methodology, but also with the rise in readily available collaborative computing systems and software (e.g., dual-head video cards). Pair programming is one of the 12 practices of the XP methodology (Williams and Kessler, 2000).

tor that the software would be a problem in the field. This is something that is contrary to the concept of reliability growth based on the "classical" reliability models. There is also the issue of situations where the *business model* of an organization overshadows the software engineering practices and appears to be the sole (and sometimes poor) guidance mechanism of software development (Potok and Vouk, 1997, 1999a, 1999b).

9.4.2 Data

We use three sources of empirical information to illustrate our discussion. Two data sets are from industrial development projects (with CMM levels in the range 1.5 to 2.5), and one is from a set of very formalized software development experiments conducted by four universities (CMM level of the experiment was in the range 1.5 to 3). While formal application and use of software analysis, design, programming, and testing methods based on the object-oriented[5] (Pressman, 2001) paradigm were not employed in any of the projects, it is not clear that the quality of the first-generation object-oriented software (with minimal reuse) is any different from that of first-generation software developed using "traditional" structured methods (Potok and Vouk, 1997). Hence, we believe that the issues discussed here still apply today in the context of many practical software development shops because the issues are related more to the ability of human programmers to grasp and deal with complex problems than to the actual technology used to produce the code.

The first dataset (DataSet1) are test-and-debug data for a PL/I database application program described in Ohba (1984), Appendix F, Table 4. The size of the software was about 1.4 million lines of code, and the failure data were collected using calendar and CPU execution time. It is used, along with DataSet2, to discuss the issue of *resource conservation,* or *enhancement,* through on-the-fly "learning" that may occur during testing of software.

The second dataset (DataSet2) comes from the NASA LaRC experiment in which 20 implementations of a sensor management system for a redundant strapped down inertial measurement unit (RSDIMU) were developed to the same specification (Kelly et al., 1988). The data discussed in this chapter are from the "certification" phases. Independent certification testing consisted of a functional certification test suite and a random certification test suite. The cumulative execution coverage of the code by the test cases was determined through postexperi-

[5]Object-oriented refers to a paradigm in which the problem domain is characterized as a set of objects that have specific attributes and behaviors. The objects are manipulated with a collection of functions (called methods, operations, or services) and communicate with one another through a messaging protocol. Objects are categorized into classes and subclasses. An object encapsulates both data and the processing that is applied to the data. This important characteristic enables classes of objects to be built and inherently leads to libraries of reusable classes and objects. Because reuse is a critically important attribute of modern software engineering, the object-oriented paradigm is attractive to many software development organizations. In addition, the software components derived using the object-oriented paradigm exhibit design characteristics (e.g., function independence, information hiding) that tend to be associated with high-quality software (Pressman, 2001).

mental code instrumentation and execution tracing using BGG (Vouk and Coyle, 1989). After the functional and random certification tests, the programs were subjected to an independent operational evaluation test that found additional defects in the code. Detailed descriptions of the detected faults and the employed testing strategies are given in Vouk et al. (1990), Eckhardt et al. (1991), and Paradkar et al. (1997).

The third dataset (DataSet3) comes from real software testing and evaluation efforts conducted during late unit and early integration testing of four consecutive releases of a very large (millions of lines of code) commercial telecommunications product. Where necessary, the plots shown here are normalized to protect proprietary information.

9.4.3 Within-Phase Testing Efficiency

We obtain an empirical estimate, $\hat{g}_i(q_i)$ (Rivers, 1998; Rivers and Vouk, 1998), of the testing efficiency function, $g_i(q_i)$, at instant i using

$$\hat{g}_i(q_i) = \frac{\Delta E_i(K - q_{i-1})}{(N - E_{i-1}) \cdot \Delta q_i} \tag{9.15}$$

where ΔE_i is the difference between the observed number of cumulative errors at steps i and $i - 1$, q_i is the observed cumulative construct coverage up to step i, and Δq_i is the difference between the number of cumulative constructs covered at steps i and $i - 1$. It is interesting to note that equation (9.15), in fact, subsumes the *learning factor* proposed by Hou et al. (1994) and derived from the sensitivity factor defined by Tohma et al. (1989), that is,

$$\frac{w_i}{N} = \frac{\Delta E_i}{N - E_{i-1}} \tag{9.16}$$

where w_i is the *sensitivity factor*. It also subsumes the metric-based failure intensity, $(\Delta E_i/\Delta q_i)$. We define estimated average testing efficiency A as

$$A = \sum_{i=1}^{n} \frac{\hat{g}_i(q_i)}{n} \tag{9.17}$$

9.4.3.1 *"Learning" (DataSet1)*

As mentioned earlier, researchers have reported the presence of so-called "learning" during some testing phases of software development. That is to say, "learning" that improved the overall efficiency of the testing on the fly [e.g., Hou et al. (1994)]. If really present, this effect may become a significant factor in software process self-improvement, and it may become even more important if a product VVT team is considered as a collaborating community. Such a community needs an efficient collaborative knowledge-sharing method built into the process. Pair programming may be such a method in the scope of two developers, a paradigm that extends that concept to the whole project VVT community, and may solve a lot of problems. We explore the single-programmer issue further.

Figure 9.2 illustrates the difference between "learning factor" and "testing efficiency" functions using the same data as Hou et al. (1994) and Ohba (1984). We assume that their testing schedule of 19 weeks was planned, and that $N = 358$. Therefore, on the horizontal axis we plot the fraction of the schedule that is completed. In Figure 9.2, we show the "learning factor" based on equation (9.16) on the left and show the corresponding "testing efficiency" computed using equation (9.15) on the right. The thing to note is that the original learning factor calculations do not keep track of remaining space to be covered, and therefore tend to underestimate the probability of finding a defect as one approaches full coverage. When coverage is taken into account, *we see that the average efficiency of the testing appears to be approximately the same over the whole testing period, that is, it would appear that no "learning" really took place.* We believe that this uniformity in the efficiency is the result of a schedule (milestone) or test suite size constraint. It is very likely that the effort used a test suite and schedule that allowed very little room for change, and therefore there was really not much room to improve on the test cases. Of course, the large variability in the testing efficiency over the testing period probably means that a number of defects were not detected. Comparison of the failure intensity and the cumulative failure fits obtained using our model and Ohba's model shows that our coverage model fits the data as well as the exponential or the S-shaped models (Rivers, 1998).

9.4.3.2 Impact of Program Structure (DataSet2)

To investigate the problem further, we turn to a suite of functionally equivalent programs developed in a multi-university fault-tolerant software experiment for NASA. Figure 9.3 illustrates the testing efficiency that was measured for the certification test suite used in the NASA–LaRC experiment. The certification test

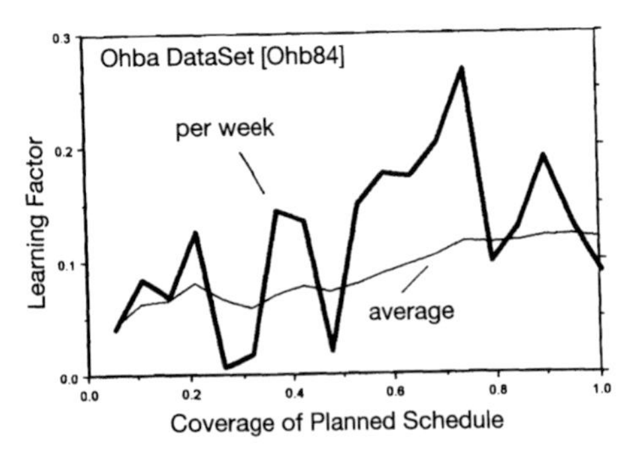

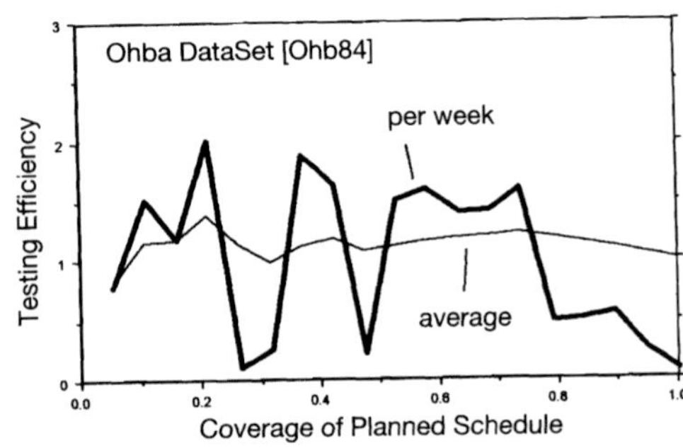

Figure 9.2. *Left:* Learning factor (Hou et al., 1994), computed using Table 4 data from Ohba (1984). *Right:* Testing efficiency, equation (1), Computed using data from Ohba (1984), Table 4.

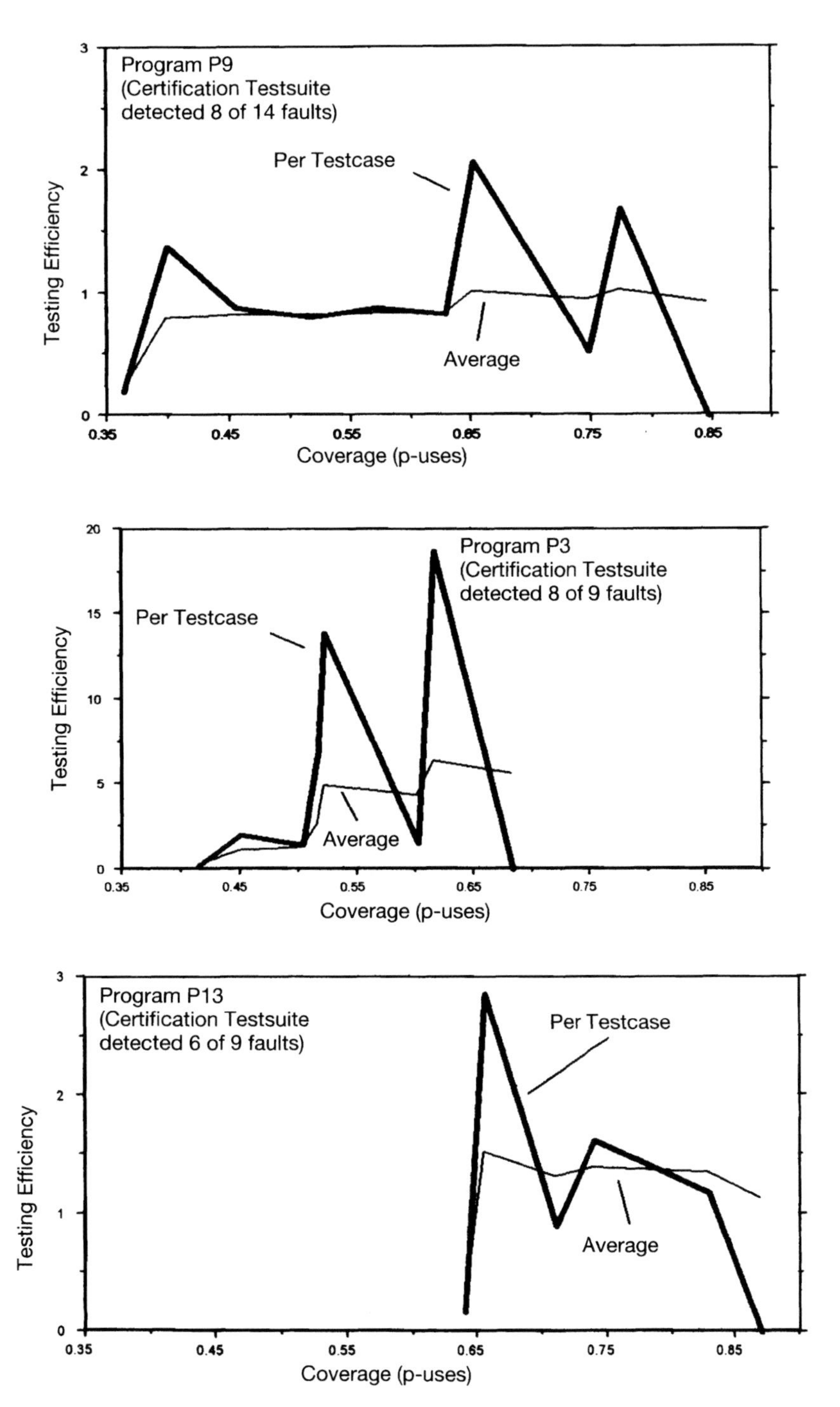

Figure 9.3. Experimental testing efficiency function g_i for program (DataSet2) P3, P9, and P13.

suite was administered after programmers finished unit testing their programs. During this phase, the programmers were allowed to fix only problems explicitly found by the certification suite. In this illustration we use p-use coverage data for three of the 20 functionally equivalent programs, and we set N to the number of faults eventually detected in these programs through extensive tests that followed the "certification" phase (Vouk et al., 1990; Eckhardt et al., 1991). In the three graphs in Figure 9.3, we plot the "instantaneous," or "per test-step," testing efficiency and its arithmetic average over the covered metric space. The first coverage point in all three graphs refers to the first test case that detected a failure, and the last coverage point in all three graphs refers to the last test case in the suite. Note that in all three cases the last efficiency is zero because no new faults were detected by the last set of test cases. *We see that the same test suite exhibits considerable diversity and variance in both its ability to cover the code and its ability to detect faults.* Similar results were obtained when statement- and branch-coverage were used (Rivers, 1998).

9.4.3.3 Discussion

From Figure 9.3 we see that average testing efficiency ranges from about 1 for P9 to about 1.5 for P13 to about 5 for P3. Since the programs are functionally equivalent and we used *exactly* the same test suite (developed on the basis of the same specifications) and data grouping in all 20 cases, the opportunity to "learn" dynamically during testing was nonexistent. *The root cause of variability is the algorithmic and implementational diversity.* That is, the observed testing efficiency range was purely due to the test suite generation approach (black-box specification-based, augmented with extreme-value and random cases, but fixed for all program tests) and the diversity in the implemented solutions stemming from the variability in the programmer microdesign decisions and skills. Yet, when we applied the "learning" models to the same data, "learning" was "reported" (Rivers, 1998; Rivers and Vouk, 1998). This is a strong indicator that

(a) a constrained testing strategy may perform quite differently in different situations,
(b) one should not assume that there is much time to improve things on-the-fly when resources are scarce, and
(c) constrained specification-based test generation needs to be supplemented with a more rigorous (and semiautomatic or automatic) test case selection approach such as predicate-based testing (Paradkar et al., 1997) or pairwise testing (Cohen et al., 1996; Lei and Tai, 1998; Tai and Lei, 2002) in an attempt to increase efficiency of the test suite as much as possible.

CMM levels for the unit development in the above projects were in the range between 1 and 2, perhaps higher in some cases. Since lightweight resource-constrained processes of XP type (Cockburn and Williams, 2000, 2001), by their nature, are unlikely to rise much beyond level 2, it is quite possible that the principal limiting factor for XP-type development is, just like in the case studies presented here, the human component—that is to say, individual analysts, designers, (pair) programmers, testers, and so on. An interesting question to ask is, "Would a process constraining and synchronizing activity like XP-embedded pair programming have enough power to sufficiently reduce the variance in the pair combination of individual programmer capabilities to provide a reliable way of predicting and achieving an acceptable cost-to-benefit ratio for large scale use of this paradigm?" Currently, there is no evidence that this is consistently the case unless the process is at least above level 3 on the CMM scale.

9.4.4 Phase-to-Phase Testing Efficiency

A somewhat less ambitious goal is to improve the resource-constrained processes from one phase to another. This is definitely achievable, as the following case study shows. Again, we use the testing efficiency metric to assess the issue. Figure 9.4 illustrates the "testing efficiency" results for releases 1 and 4 of the resource-constrained nonoperational testing phase of four consecutive releases of a large commercial software system (DataSet3). Similar results were obtained for releases 2 and 3 (Rivers, 1998). All releases of the software had their testing cycle constrained in both the time and test size domains. We obtain estimates for the parameter N by fitting the "constant testing efficiency" failure intensity model of equation (9.11) to grouped failure intensity data of each release. We used the same data grouping and parameter estimation approaches in all four cases. To obtain our parameter estimates, we used conditional least squares estimation. Logarithmic relative error was used as a weighting function. This provided a bound on the estimate corresponding to the last observed cumulative number of failures that was not lower than this observed value. The "sum of errors squared" (SSE) relationship we used is

$$SSE = \sum_{i=1}^{n} \left(\begin{array}{l} \left(\lambda_{\text{actual}} - \left(\dfrac{a}{K - q_{\min}} \right)(N - E_{\min}) \left(\dfrac{K - q_i}{K - q_{\min}} \right)^{(a-1)} \right)^2 \\ + \left(\ln(E_{\text{end}}) - \ln\left(N - (N - E_{\min}) \left(\dfrac{k - q_{\text{end}}}{k - q_{\min}} \right)^{a} \right) \right)^2 \end{array} \right) \tag{9.18}$$

where E_{end} is the number of cumulative failures we observed at juncture q_{end} within the testing phase. Note that the experimental "testing efficiency" varies considerably across the consecutive releases of this same software system despite the fact that the testing was done by the same organization. The running average is a slowly varying function of the relative number of planned test cases. The final average values of testing efficiency for each of the four releases are in Figure 9.5. It shows that the average testing efficiency increases from about 1 to about 1.9 between release 1 and release 4. We conjecture that the absolute efficiency of the test suites has also increased. Based on the above information, we feel that the average constant testing

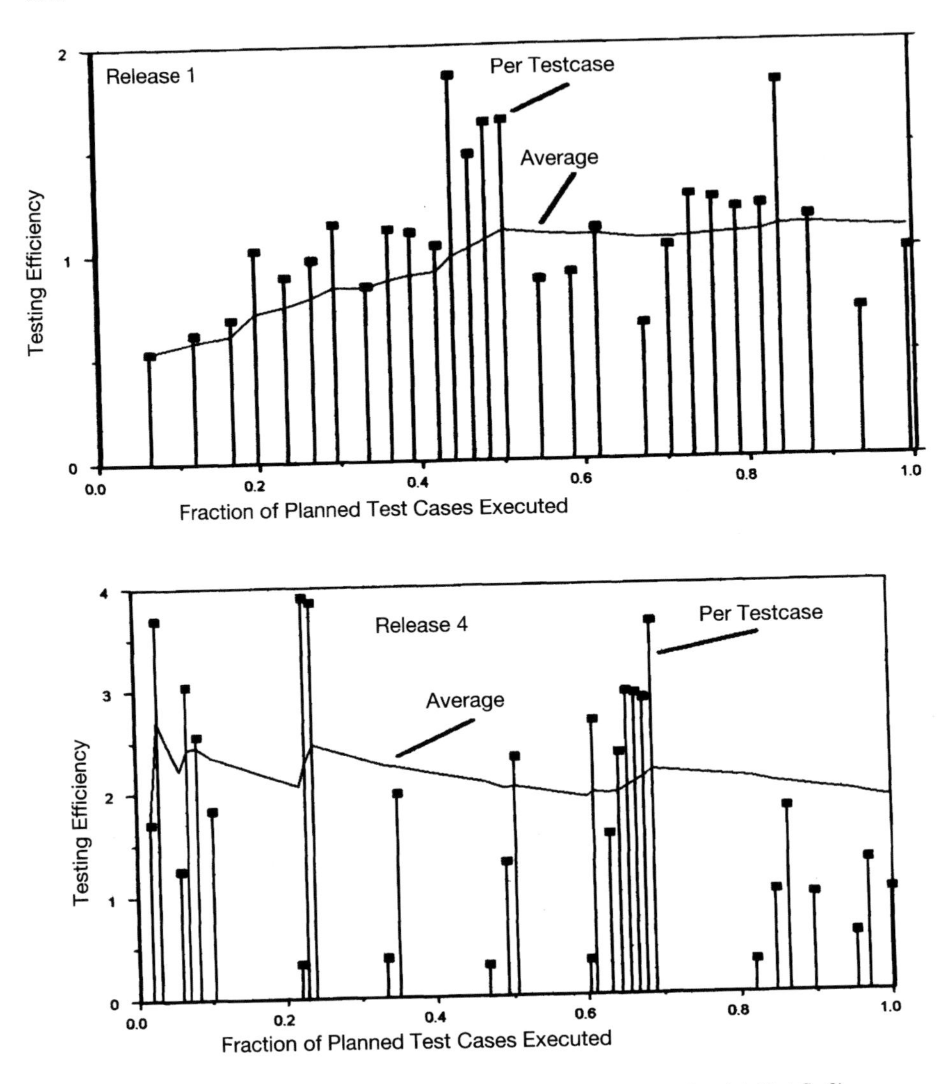

Figure 9.4. Experimental testing efficiency function for releases 1 and 4 (DataSet3).

efficiency model is probably an appropriate model for this software system and test environment, at least for the releases examined in this study.

We fit the "constant efficiency" model to empirical failure intensity and cumulative failure data for all releases. We also attempted to describe the data using a number of other software reliability models. The "constant testing efficiency" model was the only one that gave reasonable parameter estimates for all four releases (Rivers, 1998). Relative error tests show that the "constant efficiency" model fits the data quite well for all releases. Most of the time, the relative error is of the order of 10% or less.

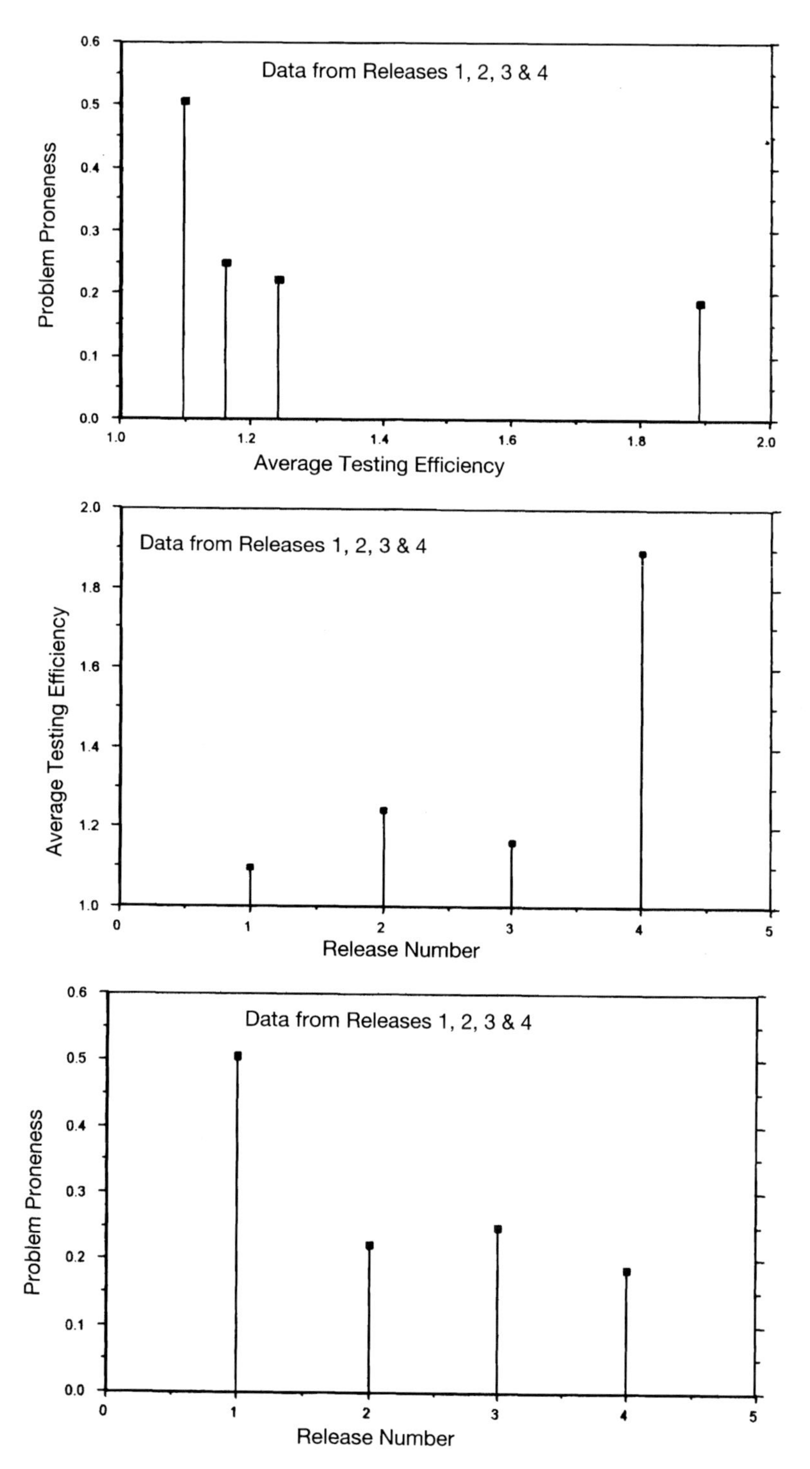

Figure 9.5. Average testing efficiency versus problem proneness, release number versus average testing efficiency, and release number versus problem proneness for releases 1–4 of DataSet 3.

9.4.5 Testing Efficiency and Field Quality

In this section we discuss the relationship we observed for our "DataSet3" between the testing efficiency and the product field quality. Our metric of field quality is "cumulative field failures" divided by "cumulative in-service time." In-service time is the sum of "clock times" due to all the licensed systems, in one site or multiple sites, that are simultaneously running a particular version of the software. To illustrate, when calculating in-service time, 1000 months of in-service time may be associated with 500 months (clock time) of two systems, or 1 month (clock time) of 1000 systems, or 1000 months (clock time) of 1 system (Jones and Vouk, 1996). Let C_j be the in-service time accumulated in Period$_j$. Then cumulative in-service time c is

$$c = \sum_{j=\text{StartDay}}^{\text{EndDay}} (C_j) \tag{9.19}$$

For example, if version X of the software is running continuously in a total of three sites on day one, then in-service time for day one is 3 days. If on day two software version X is now running in a total of seven sites, then in-service time for day two is 7 days and the cumulative in-service time for the 2 days, measured on day two, is 10 days. Similar sums can be calculated when in-service time is measured in weeks or months. Let F_j be the number of field failures observed in Period$_j$. Then the number of cumulative field failures for the same period is f, given by

$$f = \sum_{j=\text{StartCount}}^{\text{EndCount}} (F_j) \tag{9.20}$$

Our field quality metric (or field problem proneness) is z, calculated as

$$z = \frac{f}{c} \tag{9.21}$$

In the case of this study, the lower the problem proneness, the "better" the field quality.

Figure 9.5 shows the relationship between the average testing efficiency and field quality. It appears that, as the average testing efficiency increases, the field quality gets better. However, since we do not know the absolute efficiency of the test suites used, we can only conjecture based on the fact that all test suites were developed by the same organization and for the same product. It is also interesting to note that the relationship between average testing efficiency and the quality is not linear. From Figure 9.5, we see that dramatic gains were made in average testing efficiency from release 3 to release 4. We also see that after release 1, field problem proneness is reduced. This may be a promising sign that the organization's testing plan is becoming more efficient and that the company quality initiatives are producing positive results.

9.5 SUMMARY

We have presented an approach for evaluation of testing efficiency during resource-constrained nonoperational testing of software that uses "coverage" of logical constructs, or fixed schedules and test suites, as the exposure metric. We have used the model to analyze several situations where testing was conducted under constraints. Our data indicate that, in practice, testing efficiency tends to vary quite considerably even between functionally equivalent programs. However, field quality does appear to be positively correlated with the increases in the average testing efficiency. In addition to computing testing (or VTT) efficiency, there are many other metrics and ways one can use to see if one is on the right track when developing software under resource constraints. A very simple approach is to plot the difference between detected and resolved problems. If the gap between the two is too large, or is increasing, it is a definite sign that resources are constrained and the project may be in trouble.

Modern business and software development trends tend to encourage testing approaches that economize on the number of test cases to shorten schedules. They also discourage major changes to test plans. Therefore, one would expect that, on the average, the testing efficiency during such a phase is roughly constant. However, learning (improvements) can take place in consecutive testing phases, or from one software release to another of the same product. Our data seem to support both of these theories. At this point, it is not clear which metrics and testing strategies actually optimize the returns. Two of many candidate test development strategies are the pairwise testing (Cohen et al., 1996; Lei and Tai, 1998) and predicate-based testing (Paradkar et al., (1997).

ACKNOWLEDGMENTS

The authors wish to thank Marty Flournory, John Hudepohl, Dr. Wendell Jones, Dr. Garrison Kenney, and Will Snipes for helpful discussions during the course of this work.

REFERENCES

Auer, K., and Miller, R. (2001). *XP Applied,* Addison-Wesley, Reading, MA.

Beck, K. (2000). *Extreme Programming Explained: Embrace Change,* Addison-Wesley, Reading, MA.

Beck, K. (2001). Aim, Fire, *IEEE Software* **18:**87–89.

Beizer, B. (1990). *Software Testing Techniques,* Van Nostrand Reinhold, New York.

Boehm, B. W. (1988). A Spiral Model of Software Development and Enhancement, *IEEE Computer* **21**(5):61–72.

Chen, M. H., Mathur, Aditya P., and Rego, Vernon J. (1995). Effect of testing techniques on software reliability estimates obtained using a time-domain model, *IEEE Transactions on Reliability* **44(1):**97–103.

Cockburn A., and Williams, L. (2000). The Costs and Benefits of Pair Programming, in *eXtreme Programming and Flexible Processes in Software Engineering—XP2000,* Cagliari, Sardinia, Italy.

Cockburn A., and Williams, L. (2001). The Costs and Benefits of Pair Programming, in *Extreme Programming Examined,* pp. 223–248, Addison-Wesley, Boston, MA.

Cohen, D. M., Dalal, S. R., Parelius, J., and Patton, G. C. (1996). The combinatorial design approach to automatic test generation, *IEEE Software* **13**(15):83–88.

Dougherty, Dale (1998). Closing Out the Web Year, *Web Techniques* (http://www. webtechniques. com/archives/1998/12/lastpage/).

Eckhardt, D. E., Caglayan, A. K., Kelly, J. P. J., Knight, J. C., Lee, L. D., McAllister, D. F., and Vouk, M. A., (1991). An experimental evaluation of software redundancy as a strategy for improving reliability, *IEEE Trans. Software Engineering* **17**(7):692–702.

Frankl, P., and Weyuker, E. J. (1988). An applicable family of data flow testing strategies, *IEEE Transactions on Software Engineering* **11**(10):1483–1498.

Hou, R. H., Kuo, Sy-Yen, and Chang, Yi-Ping (1994). Applying Various Learning Curves to the Hypergeometric Distribution Software Reliability Growth Model, in *5th ISSRE,* pp. 196–205.

IEEE (1989). *IEEE Software Engineering Standards,* 3rd edition, IEEE, New York, 1989.

Jacoby, R., and Masuzawa, K. (1992). Test Coverage Dependent Software Reliability Estimation by the HGD Model, *Proceedings. 3rd ISSRE,* Research Triangle Park, North Carolina, pp. 193–204.

Jones, W., and Vouk, M. A. (1996). Software Reliability Field Data Analysis, in *Handbook of Software Reliability Engineering* (Lyu, M., Editor), McGraw-Hill, New York.

Karich, R. (1997). *Proceedings 8th International Symposium on Software Reliability Engineering: Case Studies,* Albuquerque, NM.

Kelly, J., Eckhardt, D., Caglayan, A., Knight, J., McAllister, D., Vouk, M. (1988). A large scale second generation experiment in multi-version software: description and early results, *Proc. FTCS* **18:**9–14.

Lei, Y., and Tai, K. C. (1998). In-Parameter-Order: A Test Generation Strategy for Pairwise Testing, in *Proceedings of the IEEE International Conference in High Assurance System Engineering.*

Lyu, M. (1996). *Handbook of Software Reliability Engineering,* McGraw-Hill, New York.

Musa, J. D., Iannino, A., and Okumoto, K. (1987). *Software Reliability: Measurement Predictions, Application,* McGraw-Hill, New York.

Musa, J. D. (1993). Operational profiles in software-reliability engineering, *IEEE Software* **10**(2):14–32.

Musa, J. D. (1999). *Software Reliability Engineering,* McGraw-Hill, New York.

Ohba, M. (1984). Software reliability analysis models, *IBM Journal of Research and Development* **28**(4):428–443.

Paradkar A., Tai, K. C., and Vouk, M. (1997). Specification-based testing using cause–effect graphs, *Annals of Software Engineering* **4:**133–158.

Paulk, M. C., Curtis, B., Chrissis, M. B., and Weber, C. V. (1993a). Capability Maturity Model for Software, Version 1.1, Technical Report CMU/SEI-93-TR-024 ESC-TR-93-

177, Software Engineering Institute, Carnegie Mellon University, Pittsburgh, Pennsylvania.

Paulk, M. C., Curtis, B., Chrissis, M. B., and Weber, C. V. (1993b). Capability Maturity Model, Version 1.1, *IEEE Software,* **10:**18–27.

Piwowarski, P., Ohba, M., and Caruso, J. (1993). Coverage Measurement Experience during Function Test, in *ICSE 93,* pp. 287–300.

Potok, T., and Vouk, M. A. (1997). The effects of the business model on the object-oriented software development productivity, *IBM Systems Journal* **36**(1):140–161.

Potok, T., and Vouk, M. A. (1999a). Productivity analysis of object-oriented software developed in a commercial environment, *Software Practice and Experience* **29**(1):833–847.

Potok, T., and Vouk, M. A. (1999b). A Model of Correlated Team Behavior in a Software Development Environment, in *Proceedings of the IEEE Symposium on Application-Specific Systems and Software Engineering Technology (ASSET'99),* Richardson, TX, pp. 280–283.

Pressman, R. S. (2001). *Software Engineering: A Practitioner's Approach,* 5th edition, McGraw-Hill, New York.

Rivers, A. T. (1998). Modeling Software Reliability During Non-Operational Testing, Ph.D. dissertation, North Carolina State University.

Rivers, A. T., and Vouk, M. A. (1995). An Experimental Evaluation of Hypergeometric Code Testing Coverage Model, in *Proceeding, Software Engineering Research Forum,* Boca Raton, FL, pp. 41–50.

Rivers, A. T., and Vouk, M. A. (1998). Resource constrained Non-Operational Testing of Software, *Proceedings ISSRE 98, 9th International Symposium on Software Reliability Engineering,* Paderborn, Germany. Received "Best Paper Award."

Rivers, A. T., and Vouk, M. A. (1999). Process Decision for Resource Constrained Software Testing, in *Proceedings ISSRE 99, 10th International Symposium on Software Reliability Engineering,* Boca Raton, FL.

Shooman, M. L. (1983). *Software Engineering,* McGraw-Hill, New York.

Tai, K. C., and Lei, Y. (2002). A test generation strategy for pairwise testing, *IEEE Transactions on Software Engineering* **28**(1):109–111.

Tohma, Yoshihiro, Jacoby, Raymond, Murata, Yukihisa, and Yamamoto, Moriki (1989). Hypergeometric Distribution Model to Estimate the Number of Residual Software Faults, in *Proceedings, COMPSAC 89,* Orlando, FL, pp. 610–617.

Vouk, M. A., and Coyle, R. E. (1989). BGG: A Testing Coverage Tool in *Proceedings, Seventh Annual Pacific Northwest Software Quality Conference,* pp. 212–233, Lawrence and Craig, Portland, OR.

Vouk, M. A., Caglayan, A., Eckhardt, D. E., Kelly, J. P. J., Knight, J., McAllister, D. F., Walker, L. (1990). Analysis of Faults Detected in a Large-Scale Multi-Version Software Development Experiment, in *Proceedings, DASC '90,* pp. 378–385.

Vouk, M. A. (1992). Using Reliability Models During Testing with Non-Operational Profiles, in *Proceedings, 2nd Bellcore/Purdue Workshop Issues in Software Reliability Estimation,* pp. 103–111.

Vouk, M. A. (2000). Software Reliability Engineering, a tutorial presented at the *Annual Reliability and Maintainability Symposium,* http://renoir. csc. ncsu. edu/Faculty/Vouk/vouk_se.html.

Walpole, R. E., and Myers, R. H. (1978). *Probability and Statistics for Engineers and Scientists,* Macmillan, New York.

Williams, L. A. and Kessler, R. R. (2000). All I ever needed to know about pair programming I learned in kindergarten, *Communications of the ACM* **43:**108–116.

EXERCISES

9.1. Construct/program and run simulations that will verify the results shown in Figure 9.1. Make sure you very clearly state all the assumptions, relationships (equations) you use, and so on. Plot and discuss the results, and write a report that compares your results with those shown in Figure 9.1.

9.2. Collect at least 5 to 10 key references related to practical and theoretical experiences and experiments with extreme programming (alternatively for the approaches called agile methodologies and practices). Write a 20-page paper discussing the selected methodology, its philosophy in the context of limited resources, and its quantitative performance in practice [quality it offers—say in faults or defects per line of code, cost effectiveness (e.g., person-hours expended vs. the time to complete the project), product quality, field failure rates or intensities, etc.]. Use the references given in this article to start your literature search, but do not limit yourself to this list—use the Web to get the latest information.

9.3. Collect data from a real project. For each project phase, obtain data such as the number of faults, fault descriptions, when each fault was detected—its exposure time, time between problem reports, failure intensity, number of test cases, inspection data, personnel resources expended, and so on. Analyze, tabulate, and plot the testing efficiency and the cost-effectiveness information for the project and discuss it. You will probably need to do some additional reading to understand the metrics better. Use the references given in this article to start your literature search.

9.4. Search the literature for two sources of data other than the ones mentioned in this chapter and perform testing efficiency analysis on the data. One source of data is at http://www.srewg.org/. Discuss your results with respect to the results shown in this article (examples of effects to compare are "learning," process improvement within and between phases, etc.)

9.5. Obtain the pairwise testing papers and read them. Obtain the pairwise toolset from http://www.srewg.org/. Use the approach to generate a suite of test cases to evaluate a networking switch that uses the following parameters (parameter name is followed by possible parameter values, full/half refers to duplex settings, TCP/UDP are protocols, size is in bytes, speed is in megabits per second). Parameters: Port1 (half/full/off), Port2 (half/full/off), Port3 (half/full/off), each port (1, 2, 3) can have speed set to (10, 100, automatic), each port (1, 2, 3) can service the following protocols (TCP, UDP), each port (1, 2, 3) can service packets of size (64, 512, 1500). Constraints: At least one port *must* be always set to full. Dependencies: All parameters can interact with/depend on each other. Answer the following: How many test cases would you have to run to cover all combinations of the parameters? Discuss the difference, if any, between the test suite obtained using pairwise testing and the "all-combinations of parameters" approach. Are there any obvious items for which PairTest is missing? If so, what are they and why are they important?

9.6. Read about operational profiles (use the references given in this article to start your literature search). Construct the top three levels of the operational profile for a typical web-browser (limit yourself to about 10 major functions/parameters with not more than 3–4 values per parameter). Construct an operational profile-based test suite for this application. Construct a pairwise test suite for this application. Compare and discuss the two as well as any other results that stem from this study.

CHAPTER 10

Modeling and Analysis of Software System Reliability

Min Xie, Guan Yue Hong, and Claes Wohlin

10.1 INTRODUCTION

Software is playing an increasingly important role in our systems. With well-developed hardware reliability engineering methodologies and standards, hardware components of complex systems are becoming more and more reliable. In contrast, such methodologies are still lacking for software reliability analysis due to the complex and difficult nature of software systems. The lack of appropriate data and necessary information needed for the analysis are the common problems faced by software reliability practitioners. In addition, practitioners often find it difficult to decide how much data should be collected at the start of their analysis and how to choose an appropriate software reliability model and method from the existing ones for that specific system.

Here the data are assumed to be failure time data during testing and debugging. The most informative type of data is the actual time each failure occurred, but grouped data might be sufficient. It could also be useful for detailed analysis if software metrics and other information can be collected during the testing. However, we will not discuss this because their main uses are not for reliability analysis, but for quality and cost prediction.

In a well-defined software development process, data should be collected throughout the software life cycle. These data usually indicate the status and progress of planning, requirement, design, coding, testing, and maintenance stages of a software project. Among all these data, failure data collected during the system testing and debugging phase are most critical for reliability analysis. During this period of time, software failures are observed and corrective actions are taken to remove the faults causing the failures. Hence there is a reliability growth phenome-

Case Studies in Reliability and Maintenance, Edited by W. R. Blischke and D. N. P. Murthy.
ISBN 0-471-41373-9

non with the progress of testing and debugging. Appropriate reliability growth models can be used to analyze this reliability growth trend (Musa et al., 1987; Xie, 1991; Lyu, 1996). The results of the analysis are often used to decide when it is the best time to stop testing and to determine the overall reliability status of the whole system, including both hardware and software. A good analysis will benefit a software company not only in gaining market advantages by releasing the software at the best time, but also in saving lots of money and resources (Laprie and Kanoun, 1992).

However, the applications of software reliability growth models are often limited by the lack of failure data. This is because the traditional approach is based on the goodness of fit of a reliability growth model to the failure data set collected during testing. With a focus on the convergence and accuracy of parameter estimation, the traditional approaches commonly suffer from divergence at the initial stage of testing and usually need a large amount of data to make stable parameter estimates and predictions. In many cases, this means that reliability analysis could only be started at a later stage of software testing, and results of the analysis might have been too late for any changes in the decisions made.

To overcome this problem, reliability analysis should be started as early as possible; to make this possible, additional information should be used in the analysis until enough data from testing can be collected. It is noted that software systems today are often put on the market in releases. A new release is usually a modification of an earlier one and is tested in a similar way. In the current practice, the data collected from the previous releases are often discarded in the reliability analysis of the current release. Although realizing that this is wasteful of information, practitioners are uncertain of how to make use of these old data. In our studies, we found that although the two consecutive releases can be considerably different, the information from an earlier release could still provide us with important and useful information about the reliability of a later release from a process development point of view; at least, the information about the testing of an earlier release will help in understanding the testing process that leads to reliability growth in the software system. Thus, a simple approach on early reliability prediction with the use of information from previous releases is introduced in the chapter.

The organization of this chapter is as follows. In Section 10.2, we first introduce a selection of traditionally used software reliability models with an emphasis on nonhomogeneous Poisson process (NHPP) models, as they are by far the most commonly used and easily interpreted existing models. In Section 10.3, a data set from the jth release was introduced and traditional modeling techniques are used for the illustration of their analysis. The convergence and accuracy problems faced by this traditional approach are explored in Section 10.4. In Section 10.5, we also discuss a new approach that can be taken at a very early stage of software testing for reliability analysis of the current jth release, making use of the previous release information. A comparison of the two approaches is also conducted as part of the case study. A brief discussion of the conclusions of the case is given in Section 10.6.

10.2 NHPP SOFTWARE RELIABILITY GROWTH MODELS

When modeling the software failure process, different models can be used. For an extensive coverage of software reliability models, see Xie (1991) and Lyu (1996), where most existing software reliability models are grouped and reviewed. In this section, we summarize some useful models that belong to the nonhomogeneous Poisson process category. Models of this type are widely used by practitioners because of their simplicity and ease of interpretation. Note that here a software failure is generally defined as the deviation from the true or expected outcome when the software is executed.

10.2.1 Introduction to NHPP Models

Let $N(t)$ be the cumulative number of failures occurred by time t. When the counting process $\{N(t); t \geq 0\}$ can be modeled by an NHPP which is a widely used counting process model for interevent arrival (Xie, 1991), NHPP software reliability models can be used for its analysis. In general, an NHPP model is a Poisson process whose intensity function is time-dependent (Rigdon and Basu, 2000; Blischke and Murthy, 2000). This type of model is also commonly called the software reliability growth model (SRGM), as the reliability is improving or the number of failures per interval is decreasing during testing with the discovery and removal of defects found. The expected cumulative number of failures in $[0, t)$ can be represented using a so-called mean value function $\mu(t)$. With different mean value functions, we have different SRGMs. The basic assumptions of NHPP SRGMs are as follows:

1. Software failures occur at random, and they are independent of each other.
2. The cumulative number of failures occurred by time t, $N(t)$, follows a Poisson process.

Here a software failure is said to have occurred when the output from the software is different from the expected or true one. During the testing and debugging of the software, a software fault is identified after each failure, and the fault that caused that failure is removed, and hence the same failure will not occur again. In this way, the failure frequency is reduced and the software can be released when it is expected to be lower than a certain acceptable level.

Usually, a software reliability growth model is specified by its mean value function $\mu(t)$ with an aim to analyze the failure data. To determine the mean value function, a number of parameters have to be estimated using the failure data collected. The maximum likelihood estimation method is the most commonly used technique for the parameter estimation of software reliability models, using the failure data, such as the number of failures per interval data or exact failure time. To perform the maximum likelihood estimation of the model parameters, the likelihood function is used:

$$L(n_1, \ldots, n_m) = \prod_{i=1}^{m} \frac{\{\mu(t_i) - \mu(t_{i-1})\}^{n_i}}{n_i!} \exp\{-[\mu(t_i) - \mu(t_{i-1})]\} \qquad (10.1)$$

where n_i ($i = 1, \ldots, m$) is the observed number of failures in interval $[t_{i-1}, t_i)$ during each of m time intervals, with $0 \leq t_0 < t_1 < \ldots < t_m$. In general, to find the maximum likelihood estimates, we take the derivative of the log-likelihood function, equate the derivative to zero, and solve the resulting equation. Often the maximum likelihood equations are very complicated and a numerical solution will be possible only using computer programs and libraries.

10.2.2 Some Specific NHPP Models

The first NHPP software reliability model was proposed by Goel and Okumoto (1979). It has formed the basis for the models using the observed number of faults per time unit. A number of other models were proposed after that, for example, the S-shaped model, the log-power model, the Musa–Okumoto model, and so on. The Goel–Okumoto model has the following mean value function

$$\mu(t) = a(1 - e^{-bt}), \qquad a > 0, b > 0 \qquad (10.2)$$

where $a = \mu(\infty)$ is the expected total number of faults in the software to be eventually detected and b indicates the failure occurrence rate.

The Goel–Okumoto model is probably the most widely used software reliability model because of its simplicity and the easy interpretation of model parameters to software-engineering-related measurements. This model assumes that there are a finite number of faults in the software and that the testing and debugging process does not introduce more faults into the software.

The failure intensity function $\lambda(t)$ is obtained by taking the derivative of the mean value function, that is,

$$\lambda(t) = \frac{d\mu(t)}{dt} = abe^{-bt} \qquad (10.3)$$

Another useful NHPP model is the S-shaped reliability growth model proposed by Yamada and Osaki (1984), which is also called the delayed S-shaped model. This model has the following mean value function

$$\mu(t) = a(1 - (1 + bt)e^{-bt}), \qquad a > 0, b > 0 \qquad (10.4)$$

The parameter a can also be interpreted as the expected total number of faults eventually to be detected, and the parameter b represents a steady-state fault detection rate per fault. This is a finite failure model with the mean value function $\mu(t)$ showing the characteristic of an S-shaped curve rather than the exponential growth curve of the Goel–Okumoto model. This model assumes that the software fault detection process has an initial learning curve, followed by growth when testers are more familiar with the software, and then leveling off as the residual faults become more and more difficult to detect.

Xie and Zhao (1993) proposed an NHPP model called the log-power model. It has the mean value function

$$\mu(t) = a \ln^b(1 + t), \qquad a > 0, b > 0 \tag{10.5}$$

This model is a modification of the traditional Duane (1964) model. An important property is that the log-power model has a graphical interpretation. If we take the logarithmic on both sides of the mean value function, we have

$$\ln \mu(t) = \ln a + b \ln \ln(1 + t) \tag{10.6}$$

If the cumulative number of failures is plotted versus the running time, the plot should tend to be on a straight line on a log–log scale. This can be used to validate the model and to easily estimate the model parameters. When the plotted points cannot be fitted with a straight line, the model is probably inappropriate, and if they can be fitted with a straight line, then the slope and intercept on vertical axis can be used as estimates of b and $\ln a$, respectively.

Note that the log-power model, as well as the Duane model for repairable systems, allows $\mu(t)$ to approach infinity. These models are infinite failure models with the assumption that the expected total number of failures to be detected is infinite. This is valid in the situation of imperfect debugging where new faults are introduced in the process of removing the detected faults.

Musa and Okumoto (1984) proposed an NHPP model called the logarithmic Poisson model. It also has an unbounded mean value function

$$\mu(t) = a \ln(1 + bt), \qquad a > 0, b > 0 \tag{10.7}$$

This model was developed based on the fact that faults that contribute more to the failure frequency are found earlier, and it often provides good results in modeling software failure data. However, unlike the log-power model, there is no graphical method for parameter estimation and model validation.

The models introduced in this section are just a glimpse of the many SRGMs that have appeared in the literature. For other models and additional discussion, see Xie (1991). Understanding the software reliability models and their selection and validation is essential for successful analysis (Musa et al., 1987; Friedman and Voas, 1995; Lyu, 1996).

10.2.3 Model Selection and Validation

Model selection and validation is an important issue in reliability modeling analysis. However, no single model has emerged to date that is superior to the others and can be recommended to software reliability practitioners in all situations. To successfully apply software reliability modeling techniques, we need to select the model that is the most appropriate for the data set we need to analyze. Goodness-of-fit tests can normally be performed to test the selected model. Some tests are reviewed in Gaudoin (1997), and related discussions can also be found in Park and Seoh (1994) and Yamada and Fujiwara (2001). It is also essential to do model comparisons in order to select a model that is the best fit for the data set we want to analyze.

Models can go wrong in many different ways. Each model has its own assumptions for its validity. To validate a model, we can compare a model prediction with the actual observation that is available later. An appropriate model should yield prediction results within the tolerable upper and lower limits to the users. When a chosen model is proven to be invalid, a reselection process should begin to find a more appropriate model.

Data collection is also critical for both model selection and validation. Good-quality data not only reflect the true software failure detection process but also form the foundation for successful analysis.

10.3 CASE STUDY

In this section, a case study is presented based on a set of failure data obtained during the system test of a large telecommunication project. The case study is used to illustrate how a traditional NHPP software reliability model introduced in Section 10.2 is implemented in practice.

The case study is based on the following background. A telecommunication company under study has been collecting failure data for each of its software releases. The data have been collected for documentation purpose without any further analysis to make use of the data. However, they decided, as they started developing release j ($j > 1$) of the system, to try applying software reliability modeling techniques to predict the software system reliability.

Although we have assumed that there is an earlier version of the software that had been tested and failure data recorded for that version, when no such information is available, the standard method assuming unknown parameters can be used. Graphical and statistical methods can be employed for model validation and parameter estimation. The software failure process is then described as failure process similar to any standard repairable system. Hence, in this section we will focus on the use of the earlier information and assume that the information is made available, which is the case in our case study.

10.3.1 Data Set from Release j

Table 10.1 shows the failure data set collected during the system testing of the jth release of a large and complex real time telecommunication system. There are 28 software system components with the size of each component between 270 and 1900 lines of code. The programming language used is a company internal lan-

Table 10.1. Number of Failures per Week from a Large Communication System

Week	Failures	Week	Failures	Week	Failures	Week	Failures
1	3	8	32	15	7	22	3
2	3	9	8	16	0	23	4
3	38	10	8	17	2	24	1
4	19	11	11	18	3	25	2
5	12	12	14	19	2	26	1
6	13	13	7	20	5	27	0
7	26	14	7	21	2	28	1

guage targeted at their specific hardware platform. For each of the components, a problem report is filled in after the detection of a failure during the testing, and corrective actions are taken to modify the code, so similar problems will not be repeated. Each problem report corresponds to the detection of one software failure. By the 28th week of release j, the number of problem reports collected from testing were counted and are summarized in Table 10.1.

Although it is desirable to know the performance of the system as early as possible, at this point the project manager of release j is interested in knowing how reliable this software system is, based on traditional reliability modeling prediction techniques.

10.3.2 Analysis of Reliability after the Release

To illustrate how traditional software reliability modeling techniques are used, first a model selection process is conducted performing a goodness-of-fit test on the data set of Table 10.1. Both the Goel–Okumoto model and S-shaped model are considered to be reasonable fits for the data set. For simplicity in illustration, we use the Goel–Okumoto model as an example in this case study. As mentioned, the Goel–Okumoto model is the most widely used model for analyzing software failure data because of the interpretation of the model parameters from a software engineering point of view.

Second, the parameters of the select software model should be estimated to determine its mean value function $\mu(t)$. The two parameters a and b of the Goel–Okumoto model can be estimated by the maximum likelihood estimation (MLE) method using the following nonlinear equations:

$$\begin{cases} \hat{a} = \dfrac{\sum_{i=1}^{k} n_i}{1 - e^{-\hat{b}t_k}} \\ \sum_{i=1}^{k} \left(\dfrac{n_i}{e^{-\hat{b}t_{i-1}} - e^{-\hat{b}t_i}} - \dfrac{\sum_{i=1}^{k} n_i}{1 - e^{-\hat{b}t_k}} \right) (t_i e^{-\hat{b}t_i} - t_{i-1} e^{-\hat{b}t_{i-1}}) = 0 \end{cases} \tag{10.8}$$

The second equation can be solved numerically using the data set in Table 10.1 with $k = 28$. Then the estimate of b can be inserted into the first equation to obtain an estimate of a. Solving the equations, we have

$$\hat{a} = 249.2 \qquad \text{and} \qquad \hat{b} = 0.0999 \tag{10.9}$$

The software reliability model is determined by its mean value function, estimated by

$$\hat{\mu}(t) = 249.2(1 - e^{-0.0999t}) \tag{10.10}$$

With the mean value function, the operational reliability $R(x|t)$ of the software system for a given time interval $(t, t + x)$ can be calculated. Note that the operational reliability is defined as the probability of no failure for a period of time x after release. Since the software has no aging and the debugging is not continued after release, the exponential distribution for software failure can be assumed during the operation (Yang and Xie, 2000). That is,

$$\begin{aligned} \hat{R}(x|t) &= \exp[-\hat{\lambda}(t)x] = \exp[-\hat{a}\,\hat{b}e^{-\hat{b}t}x] \\ &= \exp[-249.2 \cdot 0.0999 \cdot e^{-0.0999t}x] \end{aligned} \tag{10.11}$$

With $x = 0.2$ and $t = 28$, we get the estimated reliability at the end of the 28th week to be = 0.74.

In addition to reliability, other types of analysis can be conducted. For example, we can determine when to stop testing and release the system based on the criteria of whether certain failure rate level λ_0 has been achieved. To determine the release time τ, we find the smallest t that satisfies the following requirement:

$$\lambda(t) = abe^{-bt} \leq \lambda_0 \tag{10.12}$$

That is, $\tau = -[\ln(\lambda_0/ab)]/b$.

For example, with our estimated model, $\lambda(t) = 249.2 \times 0.0999 \times e^{-0.0999t}$. Suppose that the system can be released when the failure intensity is less than 1. Then $\tau = -[\ln(1/ab)]/b = 32.2$. That is, the system can be released after five more weeks of testing.

10.4 PROBLEMS AND ALTERNATIVES

10.4.1 Some Practical Problems

In the previous section, we used a case study to show how traditional modeling techniques are used for software reliability analysis. The maximum likelihood method is commonly used to obtain the parameters of the model. However, the MLE requires a large data set for an estimate to be reasonably stable; that is, when the number of failures increases, the estimates will only fluctuate slightly around the true value. A problem with the MLE method is that there might be no solution, especially at the early stages of software testing. This problem has been noticed in Knafl and Morgan (1996). A common approach is to wait until we have a large number of failures and the estimation can then be carried out. In the software industry, the availability of a large data set is often an issue. Waiting for a large data set also means that reliability analysis could only be done at a later stage of the project development cycle when critical decisions have already been made. On the other hand, we may not know how long we should wait and, when estimates can be obtained, it is possible that after the next interval, there is again no solution to the likelihood equations.

To illustrate the problems encountered in model parameter estimation using the MLE in early stages of testing when only a small number of failures is collected, again we use the data set presented in Table 10.1. We conducted model parameter estimation from the 11th week onwards. The MLEs using the data collected at the end of each week are given in Table 10.2. It can be noted that the estimates are not stable, and even may not exist at the beginning. In fact, MLE does not exist before the 12th week. It is also not until the 24th week that the estimates start to stabilize. This is very late in the testing phase, and change of any decision-making can be difficult and costly. The release date probably has already been decided by that time.

In fact, it is common that the MLE yields no solution or unstable solutions for model parameter estimations at the early stage of software testing. This is a practical problem that prevents the use of MLE in many cases.

Another problem revealed in the previous case study is perhaps the waste of information. We noticed that the system was developed and tested in a series of releases, which is quite common in the software industry today. However, the data collected from the previous releases are ignored.

To solve these problems, the case of analyzing software in previous releases needs to be studied and a procedure for early-stage software reliability predictions based on this information is needed.

Table 10.2. ML Estimates of the Model Parameter for Release *j*

Week	$\hat{a}_j$	$\hat{b}_j$	Week	$\hat{a}_j$	$\hat{b}_j$
11	NA	NA	20	NA	NA
12	211.3	0.180	21	276.8	0.0771
13	NA	NA	22	269.5	0.0819
14	NA	NA	23	NA	NA
15	NA	NA	24	261.9	0.0878
16	269.4	0.0924	25	259.6	0.0896
17	483.5	0.0335	26	256.3	0.0923
18	NA	NA	27	250.2	0.0991
19	NA	NA	28	249.2	0.0999

Note that prior to the 12th week, ML estimates do not exist.

10.4.2 The Case with *k* Releases

In fact, nowadays, most large software systems are developed in such a way that each version is a modification of an earlier release. A complete software product thus consists of a series of releases. Traditional software reliability growth models require a large amount of failure data, which is not usually available until the system has been tested for a long time. Although such an analysis is useful for the prediction of field reliability and for the estimation of software failure cost, many software developers would be more interested in estimating the software reliability as early as possible for their planning purpose. In our studies, we found that although two consecutive releases are not the same, some information from the previous release is useful for early reliability prediction of the current release.

Here we are concerned with a number of releases of a certain software system and the underlying trend of software reliability. For the case of k releases, we use N_j ($j = 1, 2, \ldots, k$) to denote the total number of faults and use b_j ($j = 1, 2, \ldots, k$) to represent the fault detection rate for the jth release in the series of k releases. If traditional reliability modeling techniques are employed, failure information of that particular release is used. During the early stage of the software fault detection process, we cannot provide any reliability predictions before enough failure data become available to build a reliability model of the release. However, after studying the characteristics of the fault detection rate, we found that the fault detection rate b_{j-1} of the $(j-1)$th release can be used for the early prediction of the jth release, which can be given as

$$b_j(t) = b_{j-1}(t), \qquad t > 0 \tag{10.13}$$

An unknown fault detection rate of most of the existing SRGMs always makes the parameter estimation nonlinear and requires regression or other estimation techniques. On the other hand, once the fault detection rate b_j is known, determination of an SRGM becomes very straightforward. A procedure for making use of this early information for reliability prediction using the Goel–Okumoto model is illustrated in the next section.

Note that equation (10.13) is an important assumption that should be verified. The justification could be subjective, by studying the actual testing process. For example, a total change of test personnel or testing strategy will certainly lead to different fault detection rate over the release. The verification could also be based on the data, assuming both parameters are unknown and estimated without using prior information. However, this will require a large data set. In fact, when we have a sufficiently large data set, the prior information may not be useful and the simple method of using a model and estimation approach as described in Section 10.3.2 will be sufficient.

10.4.3 A Proposed Approach

A procedure for accurate software reliability prediction by making use of the information from an early release of the software in the series of k releases is presented here. For illustration, we use the Goel–Okumoto model as an example. We know that the current jth release being developed has a previous $(j-1)$th release, and we have some reliability information about that release. We can apply the following procedures (see Figure 10.1) for the early reliability prediction for the jth release (current release). A similar method was presented in Xie et al. (1999).

First, according to the data collected in the previous $(j-1)$th release, a suitable software reliability model $\mu_{j-1}(t)$ is selected. Then we apply $\mu_{j-1}(t)$ to the failure data collected from the $(j-1)$th release and estimate the model parameters.

After the $(j-1)$th release, a new release j is developed, normally by the same programming group. The traditional approach is to analyze this data set indepen-

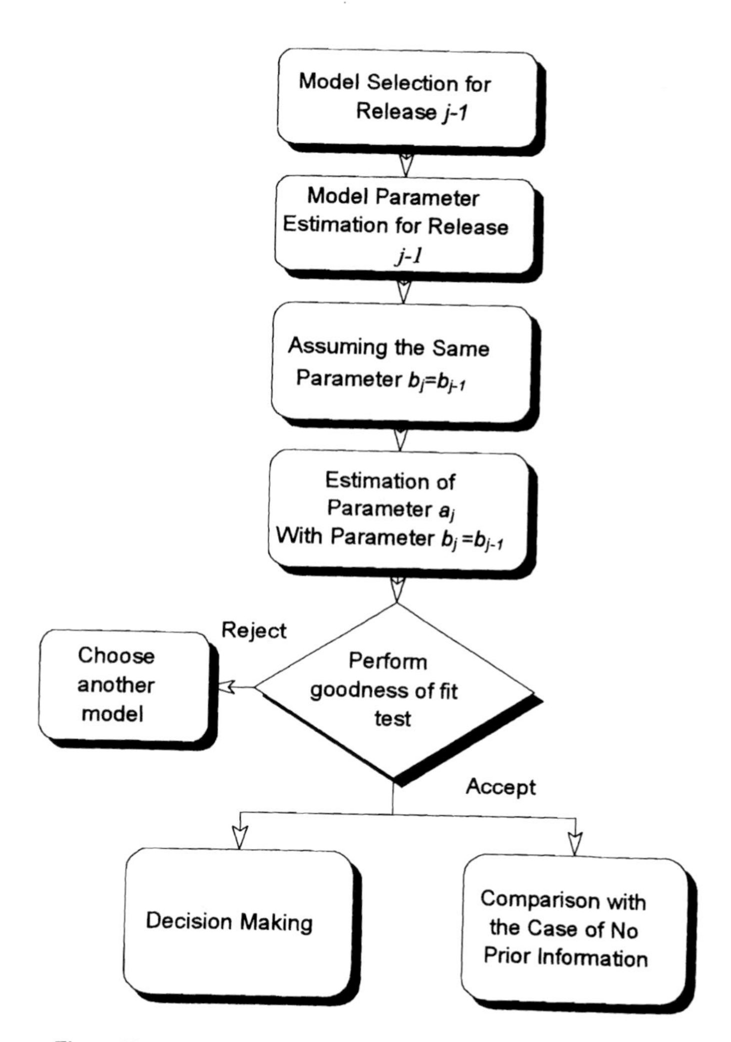

Figure 10.1. Procedures of using early information in reliability prediction.

dently to obtain the two parameters in the Goel–Okumoto model, a_j and b_j. However, we found that the fault detection rate is quite stable for the consecutive releases, assuming that the same team is performing the testing. It can be expected that the value of the fault detection rate b_j of the jth release is similar to the previous $(j-1)$th release, which is $b_j = b_{j-1}$. Parameter a_j is then estimated as a simple function of the known parameter b_j.

Figure 10.1 indicates the flow of software reliability modeling and analysis by this sequential procedure. Our focus in this chapter is on the first four boxes. The bottom four boxes will be briefly discussed here.

There are different ways to carry out goodness-of-fit tests. The methods can be different for different models. Since a NHPP model can be transformed to a homogeneous Poisson process, standard tests such as the Kolmogorov–Smirnov test, the Cramer–von Mises test, and the Anderson-Darling test can be used. If a model is rejected, another reasonable model should be selected. Usually this has to be based on experience and physical interpretation. It is important to also compare with the case when no prior information is used. This is because it is possible for earlier data to be, for example, of different type or based on different failure definition, and hence the proposed approach is no longer suitable.

For illustration of the proposed approach, the same data set used in Section 10.3 are reanalyzed in the following section, making use of additional information from the earlier releases.

10.5 CASE STUDY (CONTINUED)

In the case study introduced in Section 10.3, based on the actual failure data given in Table 10.1, the MLEs of the parameters using the data available after each week were calculated in Table 10.2. We have discussed the problems associated with this traditional reliability modeling approach in the previous section and proposed a new approach enabling reliability analysis to start as early as possible. In this section, we carry on with the case study, looking at the reliability analysis of release j from the early stage of testing based on the information from the previous release $j-1$.

10.5.1 Data Set from the Earlier Release

The system for which the failure data of release j was shown in Table 10.1 was one of the company's main products. The product went through a couple of releases and the releases under investigation are considered typical releases of this particular system. Prior to release j, the failure data of its previous release $j-1$ were also collected for documentary purpose. The number of failures detected during the testing of release $j-1$ were summarized on a weekly basis as shown in Table 10.3. The Goel–Okumoto model was chosen again for its reliability prediction. Based on Table 10.3, the MLEs of parameters a_{j-1} and b_{j-1} of release $j-1$ are obtained by solving equations (10.8). We get

Table 10.3. Number of Failures per Week from the Previous Release ($j - 1$)

Week	Failures	Week	Failures	Week	Failures	Week	Failures	Week	Failures
1	2	11	17	21	1	31	0	41	0
2	11	12	31	22	1	32	0	42	0
3	18	13	8	23	1	33	0	43	1
4	10	14	7	24	0	34	0	44	1
5	12	15	10	25	1	35	1	45	0
6	4	16	2	26	1	36	0	46	0
7	28	17	2	27	0	37	1	47	1
8	6	18	0	28	0	38	0	48	0
9	7	19	3	29	0	39	0	49	0
10	6	20	2	30	1	40	0	50	1

$$\hat{a}_{j-1} = 199.48 \quad \text{and} \quad \hat{b}_{j-1} = 0.098076 \tag{10.14}$$

In the following, this assumption will be used. Of course, when a large number of failures have occurred, it is more appropriate to use the complete failure data without assuming any prior knowledge of b. In fact, this was done in Section 10.3.2 and an estimate of b is 0.0999 in equation (10.9), which is very close to that in equations (10.14). On the other hand, the justification of making the assumption initially should be based on process- and development-related information, which is the case here; the products were developed in a similar environment and tested with the same strategy, and hence the occurrence rate per fault is expected to remain the same.

10.5.2 Earlier Prediction for Release *j*

Using the procedures discussed in Section 10.4, we can actually start reliability analysis at a very early stage of release j, making use of the information from its previous release $j - 1$. First, we estimate the two parameters a_j and b_j of the Goel–Okumoto model for release j. Note that in this case, we assume that

$$b_j = \hat{b}_{j-1} = 0.098076 \tag{10.15}$$

The estimate of parameter a_j for release j is obtained by

$$\hat{a}_j = \frac{\sum_{i=1}^{k} n_i}{1 - e^{-b_j t_k}} \tag{10.16}$$

where n_i ($i = 1, \ldots, k$) is the observed number of failures in interval $[t_{i-1}, t_i)$ as before.

Table 10.4 shows the re-estimation of parameter a_j of release j at the end of each week starting from the 11th week by using the parameter $b_j = \hat{b}_{j-1}$ from release $j - 1$.

Table 10.4. Parameter a_j Estimation Assuming $b_j = \hat{b}_{j-1}$

Week	$a_j(b_j = \hat{b}_{j-1})$	Week	$a_j(b_j = \hat{b}_{j-1})$
11	262.12	20	264.58
12	270.32	21	268.20
13	269.23	22	264.58
14	269.20	23	261.39
15	270.01	24	258.57
16	262.70	25	256.05
17	258.86	26	253.82
18	256.97	27	251.83
19	254.48	28	250.05

Second, the mean value function $\mu_j(t)$ of release j can be easily obtained using equation (10.2) and Table 10.4. Actually, with this information available from the $(j-1)$th release, the reliability prediction can be started as early as the second week of the system testing after the failure data were collected for the first week.

10.5.3 Interval Estimation

A comparison of the early reliability prediction approach with the case of traditional approach without using the previous release information is made in this section. We only need to compare the estimates of parameter a using these two approaches. However, in addition to the comparison of two point estimates, we also show the 95% confidence intervals of the values of parameter a_j using the proposed early prediction approach. In fact, for practical decision making and risk analysis, interval estimation should be more commonly used because it provides the statistical significance with the values used.

For the interval estimation, we first estimate the parameters a_j and b_j of the Goel–Okumoto model using our proposed early prediction method and can construct 95% confidence intervals for the estimated parameter a_j. To obtain confidence limits, we calculate the asymptotic variance of the MLE of the parameter a_j and we get (Xie and Hong, 2001)

$$1/I(a) = \frac{1}{E\left[-\dfrac{\partial^2 \ln L}{\partial^2 a}\right]} \tag{10.17}$$

Here, $\ln L$ is given by

$$\begin{aligned}\ln L &= \sum_{i=1}^{k} \ln\left\{\frac{[\mu(t_i) - \mu(t_{i-1})]^{n_i}}{n_i!} \exp[-(\mu(t_i) - \mu(t_{i-1}))]\right\} \\ &= \sum_{i=1}^{k} \{n_i \ln[\mu(t_i) - \mu(t_{i-1})] - [\mu(t_i) - \mu(t_{i-1})] - \ln n_i!\}\end{aligned} \tag{10.18}$$

Table 10.5. 95% Confidence Intervals for Parameter a_j Assuming $b_j = b_{j-1}$

Week	$a_j(b_j = \hat{b}_{j-1})$	a_{jU}	a_{jL}	$\hat{a}_j$ (Traditional)
12	270.32	309.07	219.29	211.32
14	269.20	306.41	231.58	N.A.
16	262.70	298.40	231.98	269.42
18	256.97	291.94	226.99	N.A.
20	254.48	289.84	222.46	N.A.
22	264.58	298.49	230.68	266.51
24	258.57	291.69	225.44	260.22
26	253.82	286.34	221.30	257.39
28	250.05	282.09	218.01	249.22

Applying this result to the Goel–Okumoto model, we get

$$I(a) = E\left[-\frac{\partial^2 \ln L}{\partial^2 a}\bigg|_{a=\hat{a}}\right] = \frac{\sum_{i=1}^{k} n_i}{\hat{a}^2} \tag{10.19}$$

For a given confidence level α, the two-sided confidence interval for parameter a is

$$a_L = \hat{a} - \frac{Z_{\alpha/2}}{\sqrt{I(\hat{a})}} \quad \text{and} \quad a_L = \hat{a} + \frac{Z_{\alpha/2}}{\sqrt{I(\hat{a})}} \tag{10.20}$$

where $Z_{\alpha/2}$ is the $[100(1 - \alpha)/2]$th standard normal percentile. Given $\alpha = 0.05$, for example, we get $Z_{\alpha/2} = Z_{0.025} = 1.96$.

The 95% confidence intervals for the prediction of parameter a_j using parameter $b_j = b_{j-1}$ are shown in Table 10.5.

From Table 10.5, we can see that the 95% confidence intervals are very wide, and this is because of the limited amount of data and the variability of the failure process.

10.6 DISCUSSION

Early software reliability prediction has attracted great interest from software managers. Most large software systems today are developed in such a way that they are a modification of an earlier release. Although two releases of software are not the same, some information should be useful for the predictions. We know that the common way of estimating parameter b of the Goel–Okumoto model usually requires programming and only numerical solutions can be obtained. This means obtaining an analytical solution for parameter b is not practical for many software managers. The proposed method of estimating the model parameter by making use of the information of a previous release solves this problem nicely. Two case studies were presented and compared in this chapter.

However, there are also some limitations. The usage of early information for reliability prediction is based on the assumption that the testing efficiency is the same, and the current software release can be analyzed using the same type of reliability models as the prior release. When these assumptions are not satisfied, the method proposed in this chapter is no longer applicable.

REFERENCES

Blischke, W. R. and Murthy, D. N. P. (2000). *Reliability; Modeling, Prediction and Optimization,* Wiley, New York.

Duane, J. T. (1964). Learning curve approach to reliability monitoring, *IEEE Transactions on Aerospace* **2:**563–566.

Friedman, M. A., and Voas, J. M. (1995). *Software Assessment: Reliability, Safety, Testability,* Wiley, New York.

Gaudoin, O. (1998). CPIT goodness-of-fit tests for the power-law process. *Communications in Statistics—A: Theory and Methods* **27:**165–180.

Goel, A. L., and Okumoto, K. (1979). Time-dependent error-detection rate model for software reliability and other performance measures, *IEEE Transactions on Reliability* **28:**206–211.

Knafl, G. J., and Morgan, J. (1996). Solving ML equations for 2-parameter Poisson-process models for ungrouped software-failure data, *IEEE Transactions on Reliability* **45:**42–53.

Laprie, J. C., and Kanoun, K. (1992). X-ware reliability and availability modelling. *IEEE Transactions on Software Engineering* **18:**130–147.

Lyu, M. R. (1996). *Handbook of Software Reliability Engineering,* McGraw-Hill, New York.

Musa, J. D., and Okumoto, K. (1984). A logarithmic Poisson execution time model for software measurement. *Proc. 7th Int. Conf. on Software Engineering,* Orlando, pp. 220–238.

Musa, J. D., Iannino, A. and Okumoto, K. (1987). *Software Reliability: Measurement, Prediction, Application,* McGraw-Hill, New York.

Park, W. J., and Seoh, M. (1994). More goodness-of-fit tests for the power-law process, *IEEE Transactions on Reliability* **R-43:**275–278.

Rigdon, S. E., and Basu, A. P. (2000). *Statistical Methods for the Reliability of Repairable Systems,* Wiley, New York.

Xie, M. (1991). *Software Reliability Modelling,* World Scientific Publishers, Singapore.

Xie, M., and Hong, G. Y. (2001). Software Reliability Modeling, Estimation and Analysis, in *Handbook of Statistics 20: Advances in Reliability* (N. Balakrishnan and C. R. Rao, Editors), Elsevier, London.

Xie, M., Hong, G. Y., and Wohlin, C. (1999). Software reliability prediction incorporating information from a similar project, *Journal of Systems and Software* **49:**43–48.

Xie, M., and Zhao, M. (1993). On some reliability growth-models with simple graphical interpretations, *Microelectronics and Reliability* **33:**149–167.

Yamada, S., and Fujiwara, T. (2001). Testing-domain dependent software reliability growth models and their comparisons of goodness-of-fit, *International Journal of Reliability, Quality and Safety Engineering* **8**:205–218.

Yamada, S., and Osaki, S. (1984). Nonhomogeneous error detection rate models for software reliability growth, in *Stochastic Models in Reliability Theory* (Osaki, S., and Hatoyama, Y., Editors), pp. 120–143, Springer-Verlag, Berlin.

Yang, B., and Xie, M. (2000). A study of operational and testing reliability in software reliability analysis, *Reliability Engineering and System Safety* **70**:323–329.

EXERCISES

10.1. It is possible to use an S-shaped NHPP model for the data set in Table 10.1. Use this model and estimate the model parameters. Discuss the pros and cons of this model.

10.2. The Duane model is widely used for repairable system reliability analysis, and software system reliability during the testing can be considered as a special case. Use the Duane model and discuss the estimation and modelling issues. In particular, explain if this model should be used.

10.3. Derive the MLE for the log-power model. Note that you can get an analytical solution for both parameters.

10.4. The Goel–Okumoto model, although commonly used, assumes that each software fault contributes the same amount of software failure intensity. Discuss this assumption and explain how it can be modified.

10.5. Software reliability model selection is an important issue. Study the failure intensity function of the Goel–Okumoto model and log-power model and give some justification for using each of them in different cases.

10.6. Use the MLE and the graphical method for the log-power model to fit the data set in Table 10.1. Compare the results.

10.7. Discuss how the proposed method of earlier estimation in Section 10.4 can be used for more complicated NHPP models, such as those with three parameters.

10.8. What is the estimated intensity function at the time of release for release j? You may use the Goel–Okumoto model with the estimated parameter in equations (10.9).

10.9. Discuss how the intensity function can be used to determine the release time. For example, assume that there is a failure rate requirement—that is, a requirement that the software cannot be released before the failure intensity reaches a certain level.

10.10. When prior information is to be used, an alternative approach is to use Bayesian analysis. Comment on the selection an of appropriate probability model, including prior distributions, in the context of the available information, and discuss the application of the Bayesian methodology in this type of study.

CHAPTER 11

Information Fusion for Damage Prediction

Nozer D. Singpurwalla, Yuling Cui, and Chung Wai Kong

11.1 INTRODUCTION

11.1.1 Problem Description

A mine is placed in very shallow waters as a deterrent against any hostile vessels that may intrude into the area. The hostiles, aware of the general location of the mine, hope to destroy it by detonating an underwater charge at some distance from the mine. An engineer (abbreviated E) wants to assess the probability that the mine will be destroyed by such an action.

The engineer can take into account many factors in determining this probability; these include the type of charge, the nature of the mine, the depth of water at the mine's location, and so on. However, in this chapter we consider only a particular type of mine and charge. Thus the engineer is interested in predicting the probability that the mine will be destroyed based on only the weight of the explosives in the charge and the distance from the mine at which the charge is detonated.

To determine this probability, data from different test conditions based on similar mines are used. At each condition, only one mine is tested. The result of each test takes on binary values: one for mine destroyed, and zero otherwise. Since the tests are destructive and costly, the amount of observed data is small. For this reason, the engineer hopes to enhance the accuracy of his prediction by incorporating information from other sources, namely a panel of experts who can provide the engineer with their inputs.

Case Studies in Reliability and Maintenance, Edited by W. R. Blischke and D. N. P. Murthy.
ISBN 0-471-41373-9

11.1.2 Outline

This chapter is presented in a way that would help practitioners solve similar problems by means of adopting the approach proposed here. It also addresses some important practical problems. These are:

1. How to systematically pool different sources of information such as binary test data and expert testimonies.
2. How to choose probability models that would accurately reflect the physical model.
3. How to apply the calculus of probability.
4. How to specify a likelihood in Bayesian analysis.

In what is to follow, Section 11.2 outlines an approach to address this damage prediction problem. Section 11.3 gives a brief overview of binary random variables and test data. Section 11.4 addresses the physical parameters used in this chapter as well as the testimonies of the experts. Sections 11.3 and 11.4 lay the foundation for Section 11.5, which outlines a detailed Bayesian approach. In Section 11.6 some camouflaged data are implemented and interpretations of the results are given. Section 11.7 concludes this case study and also provides suggestions on the variations possible to the approach presented here.

11.2 APPROACH USED

There are two schools of thought when it comes to applying statistical methods: Frequentist and Bayesian. Frequentist or classical methods do not include any scheme for the incorporation of expert testimonies, whereas Bayesian methods come with a ready-made tool: the calculus of probability. As such, we will be using a Bayesian approach in this chapter.

The Bayesian approach involves updating one's opinion on the occurrence of a *future* event by incorporating all available useful information; see, for example, Martz and Waller (1982) or Bernardo and Smith (1994). Different individuals may develop different approaches to the same problem; however, all these approaches must be based on the calculus of probability. For the problem in this chapter, the engineer may use different models to describe both the test data and the opinions of the experts. Whatever model the engineer may choose to use, it must be realistic and reflect honest belief.

In each experiment, the mine is either destroyed by the charge (deemed a success) or it is not. Therefore, the engineer may elicit from the experts their judgments on the probability of a successful experiment. However, these experts are most likely to be mine explosion scientists who can only express their probabilistic statements based on some physical parameter, say the expected kinetic energy of the shock wave. The principle of incorporating information from external sources is a powerful one, for it allows the engineer to bring in other available knowledge into his modeling and inference.

11.3 BINARY RANDOM VARIABLE AND TEST DATA

11.3.1 Bernoulli Trials and Binary Random Variables

A Bernoulli trial is a random experiment where there are only two possible outcomes. Suppose that Z is a random variable associated with a Bernoulli trial and suppose further that we define $Z = 1$ if the experiment is a success and $Z = 0$ otherwise. Since Z takes only two possible values, it is sometimes known as a binary random variable. In fact, the terms "Bernoulli random variable" and "binary random variable" can be used interchangeably. In this chapter, we will use the latter.

11.3.2 Test Data

For the scenario described in Section 11.1, let W denote the weight of the explosives in the charge, and let R denote the distance from the mine at which the charge is detonated. W and R are not variables, instead they are two pre-determined quantities that define the test condition. A binary random variable X can be used to represent the possible outcomes of the tests performed under certain values of W and R, with $X = 1$ if the mine is destroyed and $X = 0$, otherwise. Thus, if n is the number of tests conducted, then the observed test data available to the engineer can be expressed as a set of vectors $\{(W_i, R_i, x_i); i = 1, \ldots, n\}$; we denote this by d. The third entry in each vector, x_i, is the realization of X corresponding to the settings $W = W_i$ and $R = R_i$ and therefore takes values 1 or 0 as well. The observed test data are given in Table 11.1.

11.4 PHYSICAL PARAMETERS AND EXPERT TESTIMONIES

11.4.1 Modeling the Physical Parameters Involved

A key factor in determining the outcome of each experiment is Q, the magnitude of the kinetic energy of the shock wave produced by the detonation of the charge impacting on the mine. Obviously, the larger Q is, the more likely that the mine will be destroyed. Q depends on many factors, of which W and R are only two. However, it can be estimated (Cole, 1965) by $Q \approx mW^{1/3}(W^{1/3}/R)^{\gamma}$, where $m = 2410$ and $\gamma_{,} = 2.05$. The choice of these constants is due to the use of the expected shock wave energy flux density to quantify Q, as well as the choice of TNT as the explosives.

Table 11.1. Mine Explosion Test Data

i	W_i (lb.)	R_i (ft)	x_i
1	80	14	1
2	51	11.9	1
3	3.8	5	0
4	3	4	0

Therefore, Q is related to W and R via the relationship $Q = f(W, R) + \varepsilon$, where $f(W, R) = 2410W^{1/3}(W^{1/3}/R)^{2.05}$, and ε is an error term with some associated distribution. Consequently, Q has the same distributional form as that of ε.

Since Q can conceivably take any value between 0 and $+\infty$, the engineer feels that an accurate model for Q is the log-normal distribution centered around $f(W, R)$. Mathematically, this is denoted as $Q \sim \Lambda(\mu, \sigma)$, where $\mu = \ln[e^{-\sigma^2/2}f(W, R)]$ and is the mean of $\ln Q$. On the other hand, σ is the standard deviation of $\ln Q$ (Evans et al., 1993) and the engineer sets it at $\sigma = 0.2\mu$. The probability density function of Q is

$$f_Q(q) = \frac{1}{q\sigma(2\pi)^{1/2}} \times \exp\left\{\frac{-(\ln q - \mu)^2}{2\sigma^2}\right\}$$

Since $f_Q(q)$ is the prior probability density of Q in the light of W and R (i.e., W and R are known constants), a more explicit notation for $f_Q(q)$ is $f_Q\{q; (W, R)\}$ and a typical graph is shown in Figure 11.1. The use of this model captures the engineer's belief of how probable various values of Q are, given values of W and R.

11.4.2 Expert Testimonies

A number of experts, say k, are asked to give their opinions on the probability that the mine will be destroyed (success) at various magnitudes of the kinetic energy Q. Let E_i denote the ith expert, $i = 1, \ldots, k$, $P_{E_i}(X = 1|Q)$ or simply let p_i denote E_i's probability that $X = 1$ if Q were to be observed, and $p = (p_1, \ldots, p_k)$. Furthermore, since we would expect a larger Q to correspond to a larger value of $P_{E_i}(X = 1|Q)$ and vice versa, $P_{E_i}(X = 1|Q)$ should be a nondecreasing function of Q; see Figure 11.2 for some possible graphs.

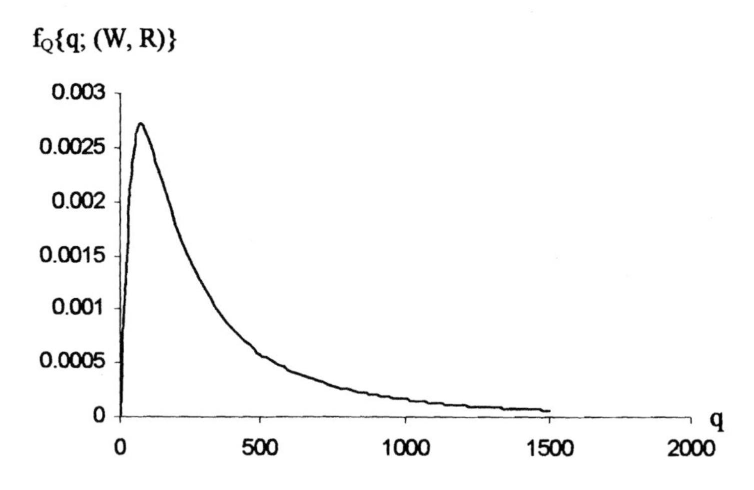

Figure 11.1. Probability density function for Q when $W = 80$ and $R = 20$.

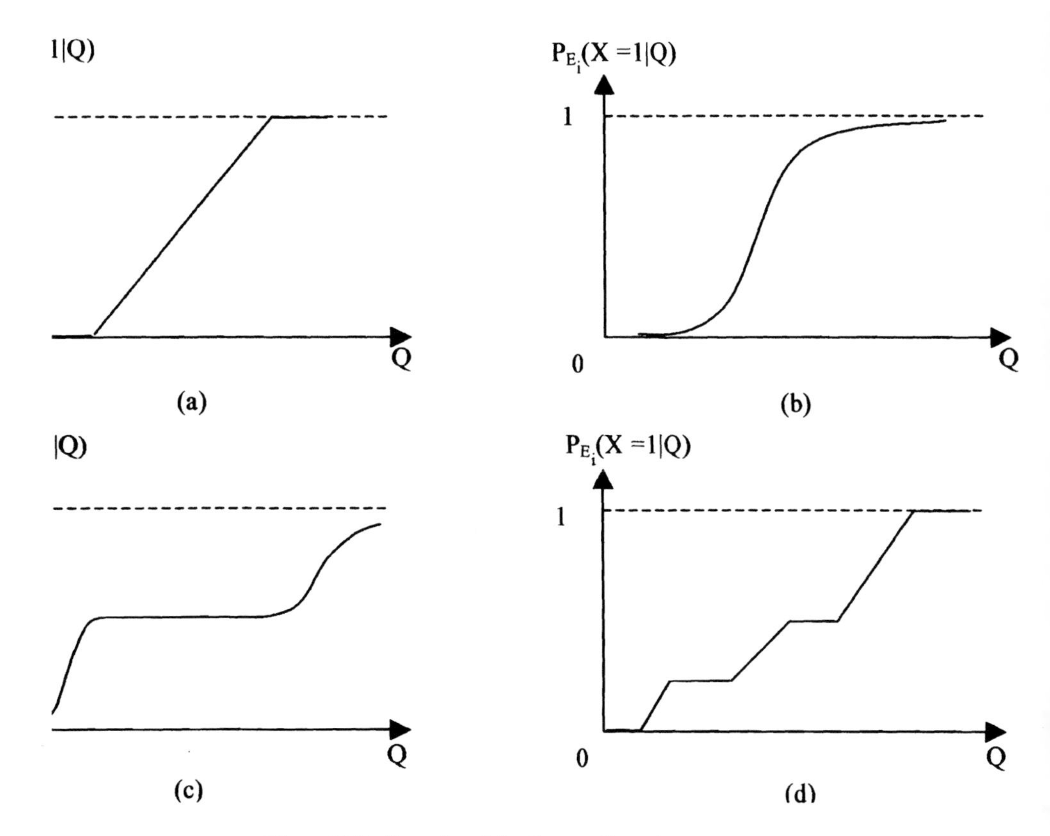

Figure 11.2. Examples of the *i*th expert's testimony.

11.4.3 Elicitation of Expert Testimonies

In this case study, the engineer consulted two experts. The engineer tells each of them that the value of Q is very likely to be bounded below and above by Q_L and Q_U, respectively. For ease of computation, the experts are only required to provide a constant value for $P_{E_i}(X=1|Q)$ for this range of values of Q.

11.5 INFORMATION FUSION: A BAYESIAN APPROACH

11.5.1 *E*'s Probability of Success Before Information Fusion

We have defined $P_{E_i}(X=1|Q)$ as the probability of success as given by the expert i. The engineer too has his own opinion, and this is based on Q as well; let this be denoted $P_E(X=1|Q)$.

The engineer knows that $P_E(X=1|Q)$ is a nondecreasing function of Q bounded between 0 and 1. To him, a sigmoidal function is a particularly appealing choice in his attempt to model $P_E(X = 1|Q)$. This is because for small (large) Q, the probability of success is close to 0 (1). As such, the engineer decides to model $P_E(X = 1|Q)$ by the Weibull distribution, where a and b are the parameters of the distribution. This is also known as the Weibitt model (see Figure 11.3).

Suppose that α and β are unknown, but that the engineer E has some prior knowledge of what they could be and describes this by $f_E(\alpha, \beta)$, the joint prior probability density of α and β. Then, by the law of total probability,

$$\begin{aligned} P_E(X = 1|Q) &= \int_\alpha \int_\beta P_E(X = 1|Q, \alpha, \beta) f_E(\alpha, \beta)\, d\alpha d\beta \\ &= \int_\alpha \int_\beta [1 - \exp(-\alpha Q^\beta)] f_E(\alpha, \beta)\, d\alpha d\beta \end{aligned} \tag{11.1}$$

From plots of $[1 - \exp(-\alpha Q^\beta)]$ versus Q, for various values of α and β, the engineer knows that the Weibull distribution has the desired sigmoidal shape for values of α and β between 5 and 10. To E, any value of α and β in this range is as good as another. A joint uniform prior distribution $f_E(\alpha, \beta) = 1/25$, for $\alpha, \beta \in [5,10]$ is thus used. With this, equation (11.1) reduces to

$$P_E(X = 1|Q) = \frac{1}{25} \int_5^{10} \int_5^{10} (1 - e^{-\alpha Q^\beta})\, d\alpha d\beta$$

This is the probability that the mine will be destroyed, in E's opinion, prior to any external input.

11.5.2 Information Fusion: Problem Formulation

The engineer's objective is to determine $P_E\{X = 1; (W^*, R^*), \mathbf{d}, \mathbf{p}\}$, where W^* and R^* denote the values of W and R at which it is desired to assess X. The use of the semicolon here is to distinguish the fact that (W^*, R^*), $\mathbf{d}$, $\mathbf{p}$ *have already been ob-*

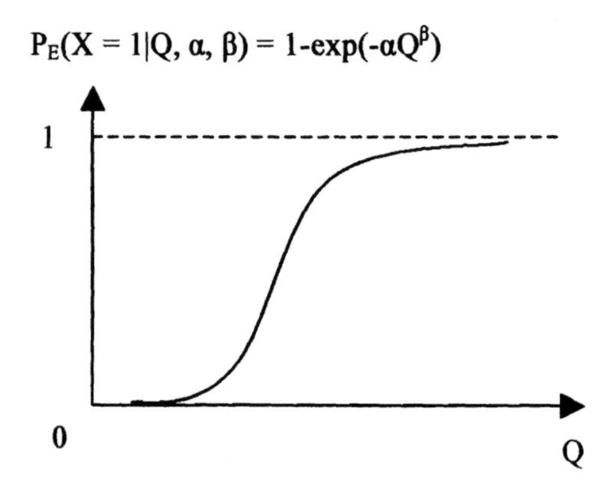

Figure 11.3. The Weibitt model for $P_E(X = 1|Q, \alpha, \beta)$.

served as compared to the use of the vertical bar to indicate the assumption that (W^*, R^*), $\mathbf{d}$, $\mathbf{p}$ were to be observed. Due to the role played by Q in obtaining the expert testimonies, it is convenient to include its consideration by applying the law of total probability on $P_E\{X = 1; (W^*, R^*), \mathbf{d}, \mathbf{p}\}$. This results in

$$P_E\{X = 1; (W^*, R^*), \mathbf{d}, \mathbf{p}\} = \int_Q P_E\{X = 1|Q; (W^*, R^*), \mathbf{d}, \mathbf{p}\} f_E\{Q; (W^*, R^*), \mathbf{d}, \mathbf{p}\}\, dQ \tag{11.2}$$

However, the distribution of Q depends only on W and R and is therefore independent of $\mathbf{d}$ and $\mathbf{p}$. Therefore, $f_E\{Q; (W^*, R^*), \mathbf{d}, p\}$ simplifies to $f_E\{Q; (W^*, R^*)\} \equiv f_Q\{q; (W^*, R^*)\}$, the probability density function of Q when $W = W^*$ and $R = R^*$.

From Section 11.4.1, we know that Q has a log-normal distribution and that the bulk of possible values of Q lies between 0 and 1500. Therefore, it is reasonable to approximate equation (11.2) by

$$P_E\{X = 1; (W^*, R^*), \mathbf{d}, \mathbf{p}\} \approx \int_0^{1500} P_E\{X = 1|Q; (W^*, R^*), \mathbf{d}, \mathbf{p}\} f_Q\{q; (W^*, R^*)\}\, dQ \tag{11.3}$$

Invoking Bayes' law on the first term of the integrand, we obtain

$$\begin{aligned} &P_E\{X = 1|Q; (W^*, R^*), \mathbf{d}, \mathbf{p}\} \\ &\quad \propto L_E\{\mathbf{p}; (X = 1|Q), (W^*, R^*), \mathbf{d}\} P_E\{X = 1|Q; (W^*, R^*), \mathbf{d}\} \end{aligned} \tag{11.4}$$

The terms in equation (11.4) can each be interpreted as follows:

- $P_E\{X = 1|Q; (W^*, R^*), \mathbf{d}\}$ is E's probability of success after taking into account the test condition as well as the observed test data.
- $L_E\{\mathbf{p}; (X = 1|Q), (W^*, R^*), \mathbf{d}\}$ is the likelihood function and represents the degree of support lent to the event $(X = 1|Q)$ by the opinions of the experts.
- $P_E\{X = 1|Q; (W^*, R^*), \mathbf{d}, \mathbf{p}\}$ is identical to the rightmost term except that $\mathbf{p}$ is also known.

As mentioned earlier in Sections 11.4.2 and 11.4.3, the experts base their judgment only on Q, which means that $\mathbf{p}$ is independent of (W^*, R^*) and $\mathbf{d}$. Therefore $L_E\{\mathbf{p}; (X = 1|Q), (W^*, R^*), \mathbf{d}\}$ should be written as $L_E\{\mathbf{p}; (X = 1|Q)\}$. Consequently, equation (11.4) becomes

$$P_E\{X = 1|Q; (W^*, R^*), \mathbf{d}, \mathbf{p}\} \propto L_E\{\mathbf{p}; (X = 1|Q)\} P_E\{X = 1|Q; (W^*, R^*), \mathbf{d}\} \tag{11.5}$$

Combining equations (11.3) and (11.5), we get

$$\begin{aligned} &P_E\{X = 1; (W^*, R^*), \mathbf{d}, \mathbf{p}\} \\ &\quad \propto \int_0^{1500} L_E\{\mathbf{p}; (X = 1|Q)\} P_E\{X = 1|Q; (W^*, R^*), \mathbf{d}\} f_Q\{q; (W^*, R^*)\}\, dQ \end{aligned} \tag{11.6}$$

where $L_E\{\mathbf{p}; (X = 1|Q)\}$ is the contribution by the expert opinions. $P_E\{X = 1|Q; (W^*, R^*), \mathbf{d}\}$, on the other hand, brings the observed data into consideration. The specification of these two unknown terms will be further articulated in Sections 11.5.3 and 11.5.4, respectively.

11.5.3 Specification of the Likelihood Based on Expert Testimonies

In this section, we derive the likelihood term of equation (11.6), $L_E\{\mathbf{p}; (X = 1|Q)\}$, a scale of comparative support lent by $\mathbf{p}$ to the future event $(X = 1|Q)$. By convention, the likelihood is often derived from the model of the prior. However, from a subjectivist point of view, it can also take any functional form or number (Singpurwalla, 2002). In this problem, the latter approach will be used.

The approach used here is based on a strategy proposed by Lindley (1988, p. 33) using the *log-odds* ratio y_i, where $y_i = \log(p_i/(1 - p_i))$. With log-odds, finding $L_E\{\mathbf{p}; (X = 1|Q)\}$ is equivalent to finding $L_E\{\mathbf{y}; (X = 1|Q)\}$.

Consider the case of just one expert. Following Lindley (1988), we suppose that y_i is unknown. Then a model for this likelihood is given by $(y_i|X = 1; \mu_1, \sigma^2) \sim N(\mu_1, \sigma^2)$; likewise, $(y_i|X = 0; \mu_0, \sigma^2) \sim N(\mu_0, \sigma^2)$, where μ_0, μ_1, and σ^2 are constants to be determined, and $N(\mu_0, \sigma^2)$ represents the normal distribution with mean μ_0 and variance σ^2. An intuitively appealing way of obtaining the values of μ_0, μ_1, and σ^2 is by considering the logarithm of the ratio of the two likelihood functions. This is of the form

$$\log \frac{L_E(y_i; X = 1)}{L_E(y_i; X = 0)} = \frac{\mu_1 - \mu_0}{\sigma^2} \left(y_i - \frac{\mu_1 + \mu_0}{2} \right) \tag{11.7}$$

Looking at equation (11.7), we see that $(\mu_0 + \mu_1)/2$ acts as a bias term for correcting y_i. Should the engineer decide that the expert i has a history of overestimation (underestimation), then this bias term would be negative (positive). The corrected term is then multiplied by $(\mu_1 - \mu_0)/\sigma^2$, with $(\mu_1 - \mu_0)/\sigma^2 > (=) [<] 1$ to reflect the fact that the engineer inflates (agrees with) [deflates] the expert i's opinion, due again to a history of ultra-caution or overconfidence. Lastly, the engineer has to decide the magnitude of σ^2; this governs the range of possible values of y_i. Once the engineer has decided on the magnitude of σ^2, the bias term, and the inflation factor, he would be able to determine the values of μ_1 and μ_0. With these values of μ_1 and μ_0, $L_E\{y_i; (X = 1|Q)\}$ is obtained by finding the height at μ_1, of the normal curve centered at y_i with variance σ^2, as shown in Figure 11.4. Likewise, $L_E\{y_i; (X = 0|Q)\}$ is the height at μ_0 of the same curve.

The case of multiple experts follows very much along the same lines, except that $L_E\{y_i; (X = 1|Q)\}$ is now replaced by $L_E\{\mathbf{y}; (X = 1|Q)\}$, with $\{\mathbf{y}; (X = 1|Q)\}$ having a multivariate normal distribution with mean vector $\boldsymbol{\mu}_1$ and correlation matrix Σ. The diagonal elements of Σ are the variances of the experts' opinions y_i and its off-diagonal elements represent the covariances between the y_i, $i = 1, \ldots, k$; see Appendix for the multivariate normal distribution. Likewise, $L_E\{\mathbf{y}; (X = 0|Q)\}$ replaces the ex-

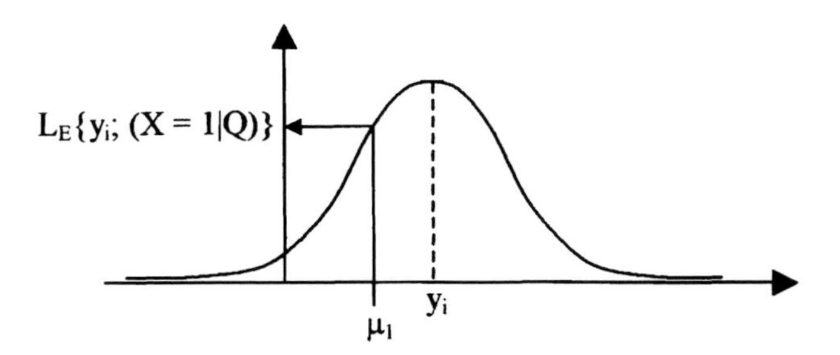

Figure 11.4. The Likelihood function of μ_1 for fixed y_i.

pression $L_E\{y_i; (X=0|Q)\}$ with $\{\mathbf{y}; (X=0|Q)\}$ having a multivariate normal distribution with mean vector $\boldsymbol{\mu}_0$ and correlation matrix Σ.

11.5.4 *E*'s Probability of Success Based on Test Data Alone

The remaining unspecified term in equation (11.6), $P_E\{X=1|Q; (W^*, R^*), \mathbf{d}\}$, is the probability of success based on test data alone. However, $P_E(X=1|Q, \alpha, \beta)$ has already been modeled by the Weibull distribution function; applying the law of total probability, we get

$$P_E\{X=1|Q; (W^*, R^*), \mathbf{d}\}$$
$$= \int_\alpha \int_\beta P_E\{X=1|Q, \alpha, \beta; (W^*, R^*), \mathbf{d}\} f_E\{\alpha, \beta; (W^*, R^*), \mathbf{d}\}\, d\alpha d\beta \quad (11.8)$$

Because the engineer is basing his opinion on the probability of success solely on Q, knowledge of Q, α, and β means that any subsequent knowledge of (W^*, R^*) and $\mathbf{d}$ will not affect his opinion on $X=1$. This means that $P_E\{X=1|Q, \alpha, \beta; (W^*, R^*), \mathbf{d}\}$ can be reduced to $P_E\{X=1|Q, \alpha, \beta\}$. Similarly, α and β, the parameters of the Weibitt model, are also independent of W^* and R^*, the parameters of an experiment that is yet to be performed. This means that $f_E\{\alpha, \beta; (W^*, R^*), \mathbf{d}\} \equiv f_E(\alpha, \beta; \mathbf{d})$. This is E's joint posterior probability density of (α, β) after observing $\mathbf{d}$. With these, Equation (11.8) becomes

$$P_E\{X=1|Q; (W^*, R^*), \mathbf{d}\} = \int_\alpha \int_\beta P_E(X=1|Q, \alpha, \beta) f_E(\alpha, \beta; \mathbf{d})\, d\alpha d\beta \quad (11.9)$$

$$= \int_5^{10} \int_5^{10} (1 - e^{-\alpha Q^\beta}) f_E(\alpha, \beta; \mathbf{d})\, d\alpha d\beta \quad (11.10)$$

To develop a model for $f_E(\alpha, \beta; \mathbf{d})$ or $f_E\{\alpha, \beta; [(W_i, R_i, x_i); i=1, \ldots, n]\}$, we first assume that the x_i's are unknown. That is, we want to evaluate $f_E\{\alpha, \beta|X_1, X_2, \ldots, X_n; [(W_i, R_i); i=1, \ldots, n]\}$. We have

$$f_E\{\alpha, \beta | X_1, X_2, \ldots, X_n; [(W_i, R_i); i = 1, \ldots, n]\}$$
$$\propto P_E\{X_1, X_2, \ldots, X_n | \alpha, \beta; [(W_i, R_i); i = 1, \ldots, n]\} f_E\{\alpha, \beta; [(W_i, R_i); i = 1, \ldots, n]\} \tag{11.11}$$

We first note that knowledge of W_i and R_i determines only the distribution of Q via the specification of the mean and variance. This knowledge does not tell us anything about individual values of Q. It is only the x_i's which play any role in updating E's opinion of the joint probability density of (α, β). Therefore, $f_E\{\alpha, \beta; [(W_i, R_i); i = 1, \ldots, n]\}$ is simply $f_E(\alpha, \beta)$. From Section 11.5.1, $f_E(\alpha, \beta) = 1/25$, a constant that can be together with the proportionality constant. If it is further assumed, given α and β and in the light of $\{(W_i, R_i); i = 1, \ldots, n\}$, that the X_i's are independent, then equation (11.11) simplifies to

$$f_E\{\alpha, \beta | X_1, X_2, \ldots, X_n; [(W_i, R_i); i = 1, \ldots, n]\}$$
$$\propto \prod_{i=1}^{n} P_E\{X_i | \alpha, \beta; [(W_i, R_i); i = 1, \ldots, n]\} \tag{11.12}$$

Since X_i is a consequence of only (W_i, R_i), (W_j, R_j)'s, $j \neq i$, have no bearing on X_i which means that $P_E\{X_i | \alpha, \beta; [(W_i, R_i); i = 1, \ldots, n]\} \equiv P_E\{X_i | \alpha, \beta; (W_i, R_i)\}$. Again, by the law of total probability and the fact that X_i is independent of (W_i, R_i) given α, β, and Q_i,

$$P_E\{X_i | \alpha, \beta; (W_i, R_i)\} = \int_{Q_i} f_E\{X_i | \alpha, \beta, Q_i = q_i; (W_i, R_i)\} P_E\{Q_i = q_i; (W_i, R_i)\}\, dQ_i$$
$$= \int_0^{1500} P_E(X_i | \alpha, \beta, Q_i = q_i) f_Q\{q_i; (W_i, R_i)\}\, dQ_i \tag{11.13}$$

Again for the simplification of computation, it is assumed that $Q_i \in [0, 1500]$.

However, X_i has actually been observed as x_i. This means that we should be evaluating $f_E\{\alpha, \beta; [(W_i, R_i, x_i); i = 1, \ldots, n]\}$ instead of $f_E\{\alpha, \beta | X_1, X_2, \ldots, X_n; [(W_i, R_i); i = 1, \ldots, n]\}$. Similarly, $P_E(X_i | \alpha, \beta, Q_i)$ should be $L_E(x_i; \alpha, \beta, Q_i)$. By specifying this likelihood based on the probability model for X_i, we have, from the Weibull distribution,

$$L_E(x_i; \alpha, \beta, Q_i) = (1 - e^{-\alpha Q_i^\beta})^{x_i} (e^{-\alpha Q_i^\beta})^{1-x_i} \tag{11.14}$$

Equation (11.14) is a result of the X_i's being binary random variables. Combining equations (11.12), (11.13), and (11.14), we get

$$f_E\{\alpha, \beta; [(W_i, R_i, x_i); i = 1, \ldots, n]\}$$
$$\propto \prod_{i=1}^{n} \left[\int_0^{1500} (1 - e^{-\alpha Q_i^\beta})^{x_i} (e^{-\alpha Q_i^\beta})^{1-x_i} f_Q\{q_i; (W_i, R_i)\}\, dQ_i \right] \tag{11.15}$$

that is,

$$f_E(\alpha, \beta; \mathbf{d}) \propto \prod_{i=1}^{n}\left[\int_0^{1500} (1 - e^{-\alpha Q_i^{\beta}})^{x_i}(e^{-\alpha Q_i^{\beta}})^{1-x_i} f_Q\{q_i; (W_i, R_i)\}\, dQ_i\right] \tag{11.16}$$

Combining equations (11.10) and (11.16), we obtain

$$P_E\{X = 1|Q; (W^*, R^*), \mathbf{d}\}$$
$$\propto \int_5^{10}\int_5^{10} (1 - e^{-\alpha Q^{\beta}})\left\{\prod_{i=1}^{n}\left[\int_0^{1500} (1 - e^{-\alpha Q_i^{\beta}})^{x_i}(e^{-\alpha Q_i^{\beta}})^{1-x_i} f_Q\{q_i; (W_i, R_i)\}\, dQ_i\right]\right\} d\alpha d\beta \tag{11.17}$$

11.5.5 *E*'s Probability of Success (Failure) After Information Fusion

Putting together equation (11.17) and $L_E\{\mathbf{y}; (X = 1|Q)\}$, as specified in Section 11.5.3, with equation (11.6), the probability of success, in *E*'s view, after incorporating expert opinions, test data, and the underlying physical information, is

$$P_E\{X = 1; (W^*, R^*), \mathbf{d}, \mathbf{p}\}$$
$$\propto \int_0^{1500} L_E\{\mathbf{y}; X = 1|Q\}\left[\int_5^{10}\int_5^{10} (1 - e^{-\alpha Q^{\beta}})\right.$$
$$\left.\left\{\prod_{i=1}^{n}\left[\int_0^{1500} (1 - e^{-\alpha Q_i^{\beta}})^{x_i}(e^{-\alpha Q_i^{\beta}})^{1-x_i} f_Q\{q_i; (W_i, R_i)\}\, dQ_i\right]\right\} d\alpha d\beta\right]$$
$$\times f_Q\{q; (W^*, R^*)\}\, dQ$$
$$= S_1, \text{ say} \tag{11.18}$$

In a similar vein, we can get *E*'s probability of failure after information fusion:

$$P_E\{X = 0; (W^*, R^*), \mathbf{d}, \mathbf{p}\}$$
$$\propto \int_0^{1500} L_E\{\mathbf{y}; X = 0|Q\}\left[\int_5^{10}\int_5^{10} e^{-\alpha Q^{\beta}}\right.$$
$$\left.\left\{\prod_{i=1}^{n}\left[\int_0^{1500} (1 - e^{-\alpha Q_i^{\beta}})^{x_i}(e^{-\alpha Q_i^{\beta}})^{1-x_i} f_Q\{q_i; (W_i, R_i)\}\, dQ_i\right]\right\} d\alpha d\beta\right]$$
$$\times f_Q\{q; (W^*, R^*)\}\, dQ$$
$$= S_0, \text{ say} \tag{11.19}$$

To obtain $P_E\{X = 1; (W^*, R^*), \mathbf{d}\}$ and $P_E\{X = 0; (W^*, R^*), \mathbf{d}\}$, we must first determine the constant of proportionality. However, since this constant is the same for both equations (11.18) and (11.19), we have

$$P_E\{X=1; (W^*, R^*), \mathbf{d}, \mathbf{p}\} = \frac{S_1}{S_1+S_0}$$
$$P_E\{X=0; (W^*, R^*), \mathbf{d}, \mathbf{p}\} = \frac{S_0}{S_1+S_0} \tag{11.20}$$

11.6 DATA ANALYSIS AND INTERPRETATION

Two experts were consulted in this case study. Based on his understanding of these two experts, the engineer assigns a correlation ratio of 0.4 to their testimonies. Their opinions are listed in Table 11.2.

In our case study, the engineer's evaluation of the expertise of the two experts result in him getting

$$\boldsymbol{\mu}_1 = \begin{pmatrix} 0.5 \\ 0.45 \end{pmatrix}, \quad \boldsymbol{\mu}_0 = \begin{pmatrix} -0.5 \\ -0.55 \end{pmatrix}, \quad \Sigma = \begin{pmatrix} 1 & 0.4 \\ 0.4 & 1 \end{pmatrix}$$

With these, the engineer is able to calculate $L_E\{\mathbf{y}; (X=1|Q)\}$ as well as $L_E\{\mathbf{y}; (X=0|Q)\}$. Putting these into equations (11.18) and (11.19), respectively, the engineer is now able to find S_1 and S_0 via numerical integration. In our case, we used the Monte Carlo method (see Appendix). Table 11.3 shows some of the camouflaged test data used, and Table 11.4 shows the predicted values of $P_E(X=1;(W, R), \mathbf{d}, \mathbf{p})$ for some combinations of (W^*, R^*).

Table 11.4 shows that when $W=3$ and $R=6$, the mine will most probably survive the test. This goes well with the available test data. Furthermore, when $W=120$ and $R=15$, then the mine is most likely to be destroyed by the charge. Again, this corresponds well with test data. However, when $W=65$ and $R=15.5$, with lit-

Table 11.2. Experts' Opinion for Various Settings

W^*	R^*	p_1	p_2
3	6	0.05	0.1
120	15	0.97	0.99
65	15.5	0.5	0.55

Table 11.3. Camouflaged Data Used

W_i	R_i	x_i
80	14	1
51	11.9	1
3.8	5	0
3	4	0

Table 11.4. Settings for Testing and Associated Probability

W^*	R^*	Probability of Outcome $X = 1$
3	6	0.0063
120	15	0.9984
65	15.5	0.5631

tle evidence from either the test data or the experts, the probability of $X = 1$ is only about 0.5631. Again, this makes sense.

11.7 CONCLUSIONS

Information fusion is an increasingly important topic and this case study shows how we can approach it. Although we are presenting a damage assessment problem here, the approach can easily be used in dose–response problems as well as in accelerated life testing. Note also that, although these methods often result in a posterior distribution that may not be integrated easily, the computational power of modern computers means that numerical methods such as the Monte Carlo method are becoming increasingly viable.

As we mentioned earlier, different practitioners may choose models that are different from those that we had chosen in this case study. Thus lies the strength of the Bayesian approach. As long as the models used are realistic, the result would be a good one.

ACKNOWLEDGMENTS

We would like to thank William W. McDonald, formerly from Naval Surface Warfare, for pointing out the problem to us. This research is supported by Grant N00014-99-1-0875 from the Office of Naval Research, and by Grant DAAD19-01-1-0502, U.S. Army Research Office.

REFERENCES

Bernardo J. M., and Smith A. F. M. (1994). *Bayesian Theory*. John Wiley & Sons, New York.

Cole R. H. (1965). *Underwater Explosions*. Dover Publications, New York.

Evans M., Hastings N., and Peacock B. (1993). *Statistical Distributions*, (2nd edition). John Wiley & Sons, New York.

Lindley D. V. (1988). The Use of Probability Statements, in *Accelerated Life Testing and Expert's Opinions in Reliability* (C. A. Clarotti and D. V. Lindley, Editors), North-Holland, New York.

Martz H. F., and Waller R. A. (1982). *Bayesian Reliability Analysis*. John Wiley & Sons, New York.

Singpurwalla N. D. (2002). Some cracks in the empire of chance: Flaws in the foundations of reliability, *International Statistical Review* (with Discussion), **7**:53–78.

EXERCISES

11.1. Suppose that you have three coins. Coin A has a 50% probability of landing heads, coin B has a 25% probability of landing heads, and coin C is two-headed. A friend picks one of the coins at random and tosses it, telling you that it landed heads.

(a) By conditioning on which coin is picked and applying the law of total probability, show that the probability of a head is 7/12.

(b) Using Bayes' law, calculate the probability that coin C was picked given a head was thrown. Repeat this calculation for coins A and B.

11.2. From your own background and experience, list and discuss at least four applications in which the binary model used in the mine study is appropriate for describing outcomes.

11.3. For each application discussed in Exercise 11.2, tell how one might obtain relevant prior information regarding the probability of each outcome.

11.4. Discuss the role of Q in the mine study during the process of information pooling.

11.5. In the data analysis of Section 11.6, an intermediate step was conducted to generate Table 11.2. What was it?

11.6. In the data analysis of Section 11.6, it was assumed that the correlation between the two experts' prior probabilities was 0.4. How would you expect the results to change if this correlation were instead taken to be 0.8? What if it were taken to be 0.1?

11.7. A Bayesian approach involves updating one's opinion on the occurrence of a future event by incorporating all available useful information. How do you think the approach used in the mine study supports this statement?

APPENDIX

The Multivariate Normal Distribution

The n-variate normal distribution is a generalization of the univariate normal distribution to n dimensions. Its density function is given by

$$f(\mathbf{x}) = [(2\pi)^{-n/2} |\Sigma|^{-1/2}]\exp\{-[(\mathbf{x} - \boldsymbol{\mu})^T \Sigma^{-1}(\mathbf{x} - \boldsymbol{\mu})]/2\}$$

where $\boldsymbol{\mu}$ is the mean vector and Σ is the variance–covariance matrix for the n variables. The exponential term is a measure of the "distance" between the input coordinate $\mathbf{x}$ and the mean vector $\boldsymbol{\mu}$, very much similar to the univariate case.

The Monte Carlo Method for Performing Numerical Integration

The Monte Carlo method uses the idea that the integral of a one-variable function is equivalent to the area bounded by the curve, the limits of integration, and the axis. This can be easily extended to the case of multiple variables. For illustrative purposes, we will examine the two-variable case.

Suppose we want to integrate $f(x_1, x_2)$ from a_1 to b_1 and from a_2 to b_2, respectively. This is equivalent to finding the volume bounded by the curve and the respectively planes. The Monte Carlo method uses the following algorithm:

Step 1: Generate random numbers y_1 and y_2 in the intervals $[a_1, b_1]$ and $[a_2, b_2]$, respectively.

Step 2: Find $f(y_1, y_2)$.

Step 3: Repeat Steps 1 and 2, say N times, and sum up each of the $f(y_1, y_2)$ calculated each time. Let S denote this sum.

Step 4: Find S/N; this gives the average 'height' of $f(y_1, y_2)$ over the desired space.

Step 5: An approximation of the volume (and hence the integral) is given by $(S/N)(b_1 - a_1)(b_2 - a_2)$.

For this method, a larger value of N will yield a better approximation.

PART C

Cases With Emphasis on Defect Prediction and Failure Analysis

Chapter 12, "Use of Truncated Regression Methods to Estimate the Shelf Life of a Product from Incomplete Historical Data"

—Investigates the relationship between time to failure (spoilage) of a perishable product and two factors, temperature and concentration of an additive to inhibit spoilage, using data in which an upper bound on observation time was known but the number of unspoiled samples was unknown.

Chapter 13, "Determining Software Quality Using COQUALMO"

—Describes a parametric quality model for use in determining trade-offs between cost, schedule, quality, and reliability of software, providing guidelines for estimation of defect density during the various phases of the software development cycle.

Chapter 14, "Use of Extreme Values in Reliability Assessment of Composite Materials"

—Analyzes defects in a sample from a batch of 1000 components manufactured in the 1990s from a laminated, fiber-reinforced composite material and intended to last at least 20 years.

Chapter 15, "Expert Judgment in the Uncertainty Analysis of Dike Ring Failure Frequency"

—Uses structured expert judgment in evaluation of the uncertainty in the prediction of failure frequencies of dike rings in The Netherlands, providing extensive discussion of the methodology and results.

CHAPTER 12

Use of Truncated Regression Methods to Estimate the Shelf Life of a Product from Incomplete Historical Data

William Q. Meeker and Luis A. Escobar

12.1 INTRODUCTION

12.1.1 Background and Motivation

An additive is used to extend the shelf life of a product (to protect the identity of this product, we will refer to it as Product A). Higher levels of the additive will extend shelf life, but will also add cost and detract from product performance. The producer would like to use the least amount of additive possible, consistent with adequate control of the risk of spoilage before use (an event that we will refer to as "failure").

Over a long period of time, data were collected to investigate the statistical relationship between Product A shelf life and concentration of additive for different levels of temperature and to monitor other product characteristics. Specimens were taken from production at chosen points in time (none of which were close together) over a period of years. If all specimens put on test had been observed until failure, classical response surface methods could have been used to estimate the effect that concentration and temperature have on Product A shelf life. If the survival times for all unfailed specimens had been recorded, then standard censored data methods, with a response-surface model, could have been used to estimate the relationship. The available data for Product A, however, had been entered into a computer database from old paper records. Only the failure times had been recorded. Notations concerning what would have been censored observations were non-numerical and

Case Studies in Reliability and Maintenance, Edited by W. R. Blischke and D. N. P. Murthy.

ISBN 0-471-41373-9

were never entered into the database. Although it was known that the upper time limit of observation had been 270 days for the product units that were monitored, there was no available information on the number of trials for which the specimens were unspoiled.

The goal of the analysis of the historical data was to estimate the relationship between life and the two explanatory variables: additive concentration and temperature. The relationship was to be used to make better decisions on how much additive should be added to the product. Although all experimentation was done at 45°C or above (to accelerate shelf life), there was interest in the distribution of shelf life down to 35°C. The higher levels of temperature were of interest to characterize the effect of high-temperature excursions that were sometimes experienced during shipping. The original Product A data are proprietary. Therefore we have used simulated data with some modification to the scaling of the parameters in order to illustrate the steps that were followed in the original analysis.

12.1.2 The Product A Shelf-Life Experimental Data

The available Product A shelf-life data are available in the SPLIDA/S-PLUS life data object ShelfLifeA.ld and also in the SPLIDA distribution as a text file. Meeker and Escobar (2002) describe the SPLIDA package. The concentration of the additive is in units of parts per million (ppm), and the storage temperature is in degrees Celsius. The time to failure (indicated by evidence of a particular observable event related to spoilage) was recorded in terms of number of days. The last two columns in the data set give the truncation information (indicating the time at which the trial would have been terminated had the product sample not failed). In this application, the right truncation time was the same for all units, but the methodology used here allows for different truncation times for different trials, as long as the truncation times are known for all observations.

Figure 12.1 shows the number of recorded shelf lives at the different levels of temperature and concentration. Figure 12.1 suggests that the allocation of samples to the particular levels of concentration and temperature over time were chosen in a somewhat haphazard manner. Recall that the actual number of samples at each condition is unknown due to the truncation. The logic for avoiding the SE (high temperature with low concentration) and NW (low temperature with high concentration) regions is that these were conditions that would not be used in practice. Experimentation below 45°C was avoided because of the large amount of time needed to observe failures.

Figure 12.2 is a plot of Product A shelf life versus temperature, conditioned on the unique values of concentration. This plot suggests the propensity for shorter shelf life at higher temperatures. Figure 12.3 is a similar plot of Product A shelf life versus concentration, conditioned on the unique values of temperature. This plot suggests the propensity for longer shelf life at higher levels of concentration. In both of these figures, at combinations of the factors implying long life, the truncation point at 270 days can be seen.

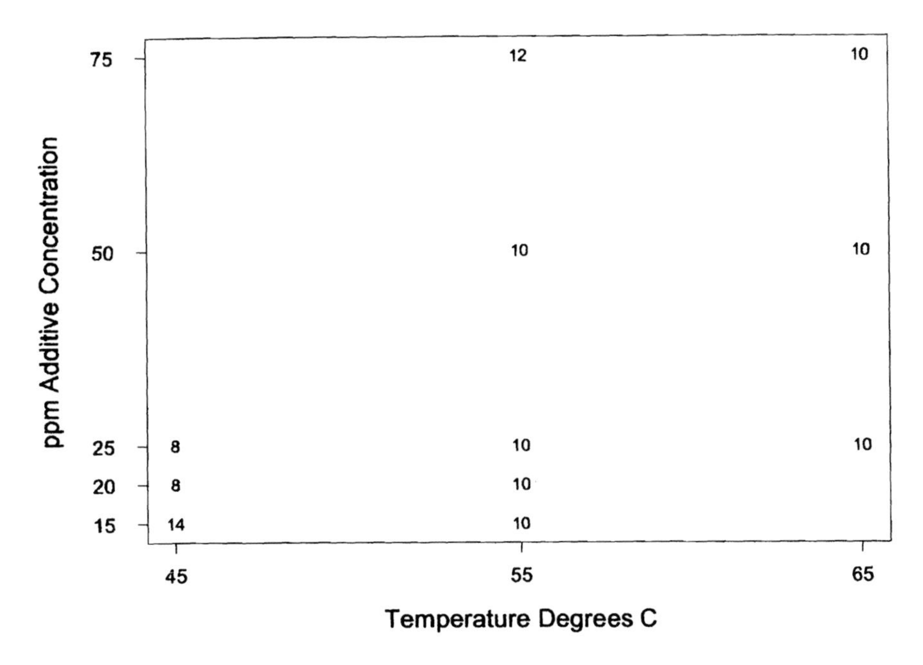

Figure 12.1. Temperature and concentration combinations used in the shelf-life study.

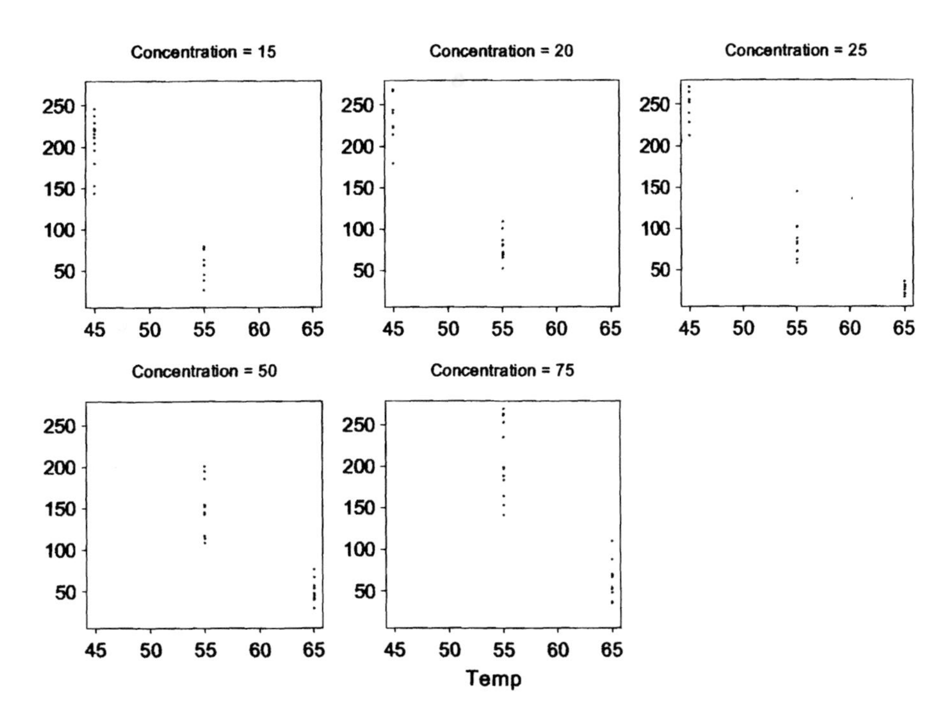

Figure 12.2. Plot of Product A shelf life versus temperature, conditioned on the unique values of concentration.

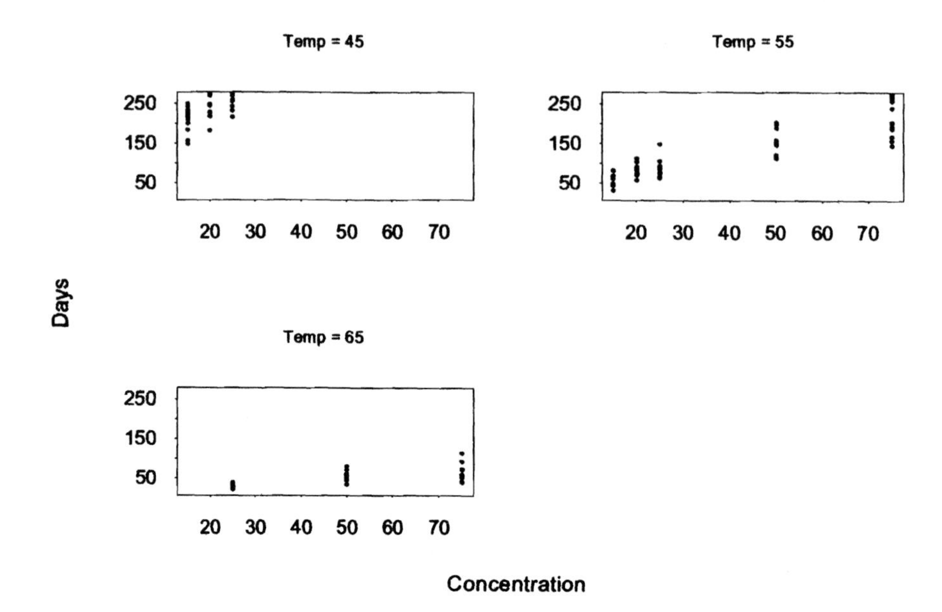

Figure 12.3. Plot of Product A shelf life versus concentration, conditioned on the unique values of temperature.

12.1.3 Related Literature

Truncated distributions and truncated data have been discussed in a number of places in the statistical literature (truncated data are data generated from a truncated distribution). The books by Cohen (1991) and Schneider (1986) describe parametric methods of estimation and inference for truncated and censored data from a single distribution. Turnbull (1976) and Nelson (1990a) provide nonparametric methods of estimation for truncated data. Kalbfleisch and Lawless (1992) describe several different reliability applications involving truncated data. Meeker (1987) shows that there is a relationship between the limited failure population (LFP) model with right censoring and right truncation (in the LFP model, there is a fraction of units in the population that will never fail, represented by a spike of probability at infinity). Kalbfleisch and Lawless (1988) illustrate the loss of efficiency of truncation relative to censoring. Escobar and Meeker (1998) present methods for computing asymptotic variances for censored and truncated experiments that would allow such comparisons in other situations. Chapter 11 of Meeker and Escobar (1998) presents a number of examples of left and right truncation, describes ML estimation, and suggests a probability plotting method to help assess distributional goodness of fit. Here we will review some of the important ideas, as they pertain to right truncation.

There are fewer references involving truncated data in regression analysis, and we were unable to find any applications in reliability data analysis. Amemiya (1973) describes economic applications and provides a detailed description of the theory behind ML estimators for regression with normally distributed residuals and *censored* data (which he refers to as "truncated"). In the last section of the paper, however, the author presents an argument that the results in the paper apply to an alternative model that is the same as the truncated data model presented here and elsewhere. Kalbfleisch and Lawless (1991) use a right-truncated regression model to analyze AIDS incubation times. Pagano et al. (1994) use truncated regression for the same application. Gross and Huber-Carol (1992) also describe regression models for truncated data.

12.1.4 Overview

Section 12.2 provides some background for estimation with truncated data, using data at a single experimental condition to make the presentation simpler. Section 12.3 describes and illustrates the use of models and methods for truncated data with explanatory variables. Section 12.4 extends the example given in Section 12.2 to give a comparison between the analysis of truncated data and censored data. Section 12.5 contains some concluding remarks and suggestions for further research.

12.2 TRUNCATED DATA BACKGROUND

12.2.1 Analysis of Truncated Data at a Single Condition

This section introduces the ideas behind the analysis of truncated data by focusing on the results of the trials with a concentration of 25 ppm and a temperature of 45°C. There were eight observations (trials with failures) at this condition. Figure 12.4 is an event plot of these data. The line at row 9 extending to 270 days indicates the unknown number of truncated observations. If we knew the number of samples that had been terminated at 270 days, the data would be right-censored instead of right-truncated. Figure 12.5 is a lognormal probability plot. The lognormal cdf and pdf are

$$F(t;\, \mu,\, \sigma) = \Phi_{\text{nor}}\left[\frac{\log(t) - \mu}{\sigma}\right] \tag{12.1}$$

$$f(t;\, \mu,\, \sigma) = \frac{1}{\sigma t}\,\phi_{\text{nor}}\left[\frac{\log(t) - \mu}{\sigma}\right], \qquad t > 0 \tag{12.2}$$

where ϕ_{nor} and Φ_{nor} are, respectively, the standardized normal pdf and cdf. The lognormal median $t_{.5} = \exp(\mu)$ is a scale parameter, and $\sigma > 0$ is the lognormal shape parameter.

If the random variable T_i is truncated when it lies above τ_i^U, then the likelihood of an observation at $t_i = \tau_i^U$ is

$$L_i(\theta) = \frac{f(t_i;\, \theta)}{F(\tau_i^U;\, \theta)} \tag{12.3}$$

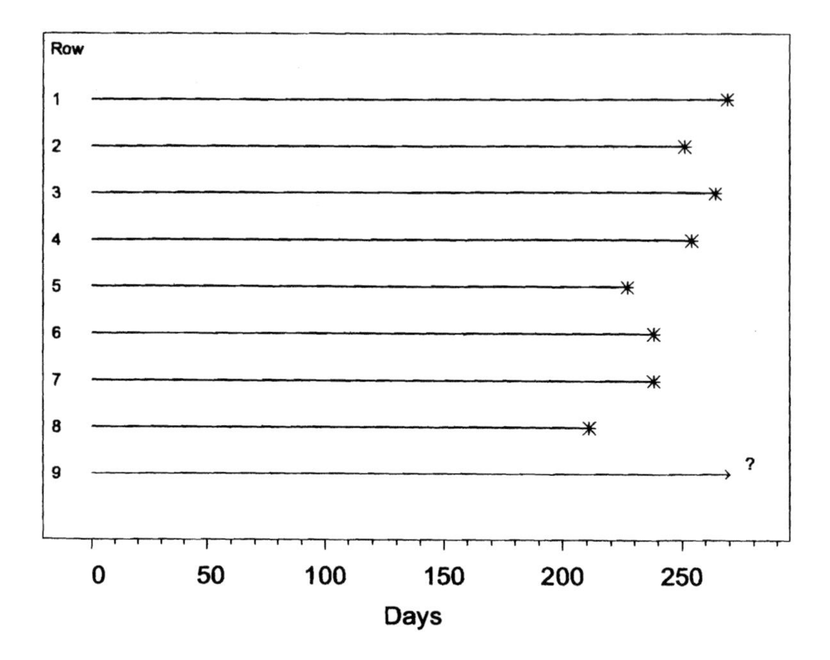

Figure 12.4. Event plot of the truncated Product A shelf life data at 25 ppm and 45°C.

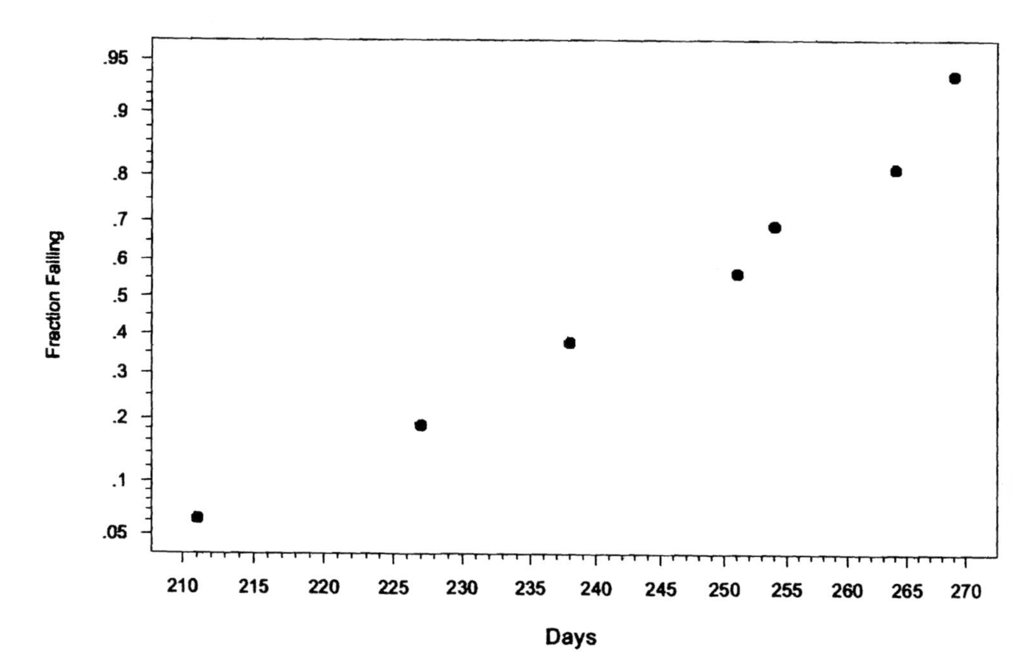

Figure 12.5. Lognormal probability plot of the truncated Product A shelf-life data at 25 ppm and 45°C.

where $\theta = (\mu, \sigma)$. The total log likelihood for a sample of n independent truncated observations is

$$\mathscr{L}(\theta) = \sum_{i}^{n} \log[L_i(\theta)] \tag{12.4}$$

The values of $\theta = (\mu, \sigma)$ that maximize (12.4) are the maximum likelihood estimates and these are denoted by $\hat{\theta} = (\hat{\mu}, \hat{\sigma})$. The ML estimation results for the truncated Product A shelf-life data at 25 ppm and 45°C are summarized in Table 12.1. The maximum of the log likelihood is at $\hat{\mu} = 5.5703$ and $\hat{\sigma} = 0.1166$. The standard errors in Table 12.1 were obtained from the observed information matrix.

Figure 12.6 is a plot of the relative likelihood function $L(\mu, \sigma)/L(\hat{\mu}, \hat{\sigma})$ for the truncated Product A shelf-life data at 25 ppm and 45°C. Although the likelihood appears to have a clear maximum, the likelihood contours indicate a ridge-like behavior running from the SW to the NE corner of this plot. This is an indication of poor identifiability of the lognormal distribution parameters. In particular, the data at this condition are not entirely capable of distinguishing between a distribution with a small value of μ with a small amount of truncation and a large value of μ with a large amount of truncation.

12.2.2 Adjustment for Probability Plotting of Truncated Data

Probability plots are useful for displaying data and parametric estimates of the cdf and for making an assessment of distributional goodness of fit. A truncated (log) location-scale distribution is not, in general, a (log) location-scale distribution. Thus special methods are needed to do a probability plot. Section 11.6 of Meeker and Escobar (1998) shows how to parametrically adjust a truncated data nonparametric estimator so that it can be used for making a probability plot. The basic idea is to use the ML estimate of an underlying parametric distribution to provide an estimate of the fraction truncated and to use this probability to adjust the nonparametric estimate, providing an estimate of the unconditional distribution that generated the truncated response. This approach can be used in situations with right truncation, left truncation, or both left and right truncation. We will describe the method as it applies to the right-truncated shelf-life data and extend the application of the

Table 12.1. Lognormal ML Estimates for the Truncated Product A Shelf-Life Data at 25 ppm and 45°C

			Approximate 95% Confidence Interval	
Parameter	ML Estimate	Standard Error	Lower	Upper
μ	5.5703	0.15216	5.27204	5.8685
σ	0.1166	0.07326	0.03406	0.3994

Log likelihood at maximum point: –33.34.

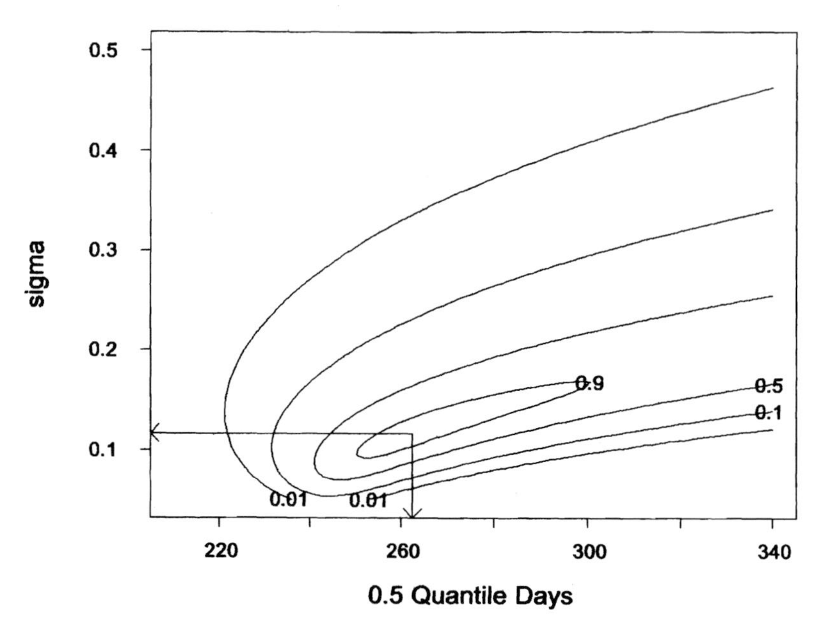

Figure 12.6. Contour plot of the lognormal likelihood function for the truncated Product A shelf-life data at 25 ppm and 45°C.

method to situations in which there are explanatory variables. With right truncation, the nonparametric estimate provides an estimate for the conditional probability given by the truncated distribution

$$F_C(t) = \Pr(T \leq t | 0 \leq T < \tau^U_{\max}) = \frac{F(t)}{F(\tau^U_{\max})}, \qquad 0 < t \leq \tau^U_{\max} \tag{12.5}$$

where $\tau^U_{\max}$ is the largest right-censoring time in the sample. Then the parametrically adjusted unconditional nonparametric (NPU) estimate of $F(t)$ is

$$\hat{F}_{NPU}(t) = \hat{F}_{NPC}(t) F(\tau^U_{\max}; \hat{\theta}), \qquad 0 < t \leq \tau^U_{\max} \tag{12.6}$$

where $\hat{F}_{NPC}$ is the nonparametric estimate of F_C. This adjusted nonparametric estimate can be plotted on probability paper in the usual way [i.e., plotting the middle probability point in the jump at each observation, as described in Section 6.4 of Meeker and Escobar (1998)].

Figure 12.7 is an adjusted probability plot for the truncated Product A shelf life data at 25 ppm and 45°C. The straight line in Figure 12.7 is the ML estimate of the untruncated lognormal distribution of the truncated Product A shelf life distribution at 25 ppm and 45°C, based on the truncated data. The points, corresponding to the adjusted nonparametric estimate, fall very much along a straight line. The agreement between the plotted points and the parametric ML estimate is very strong in

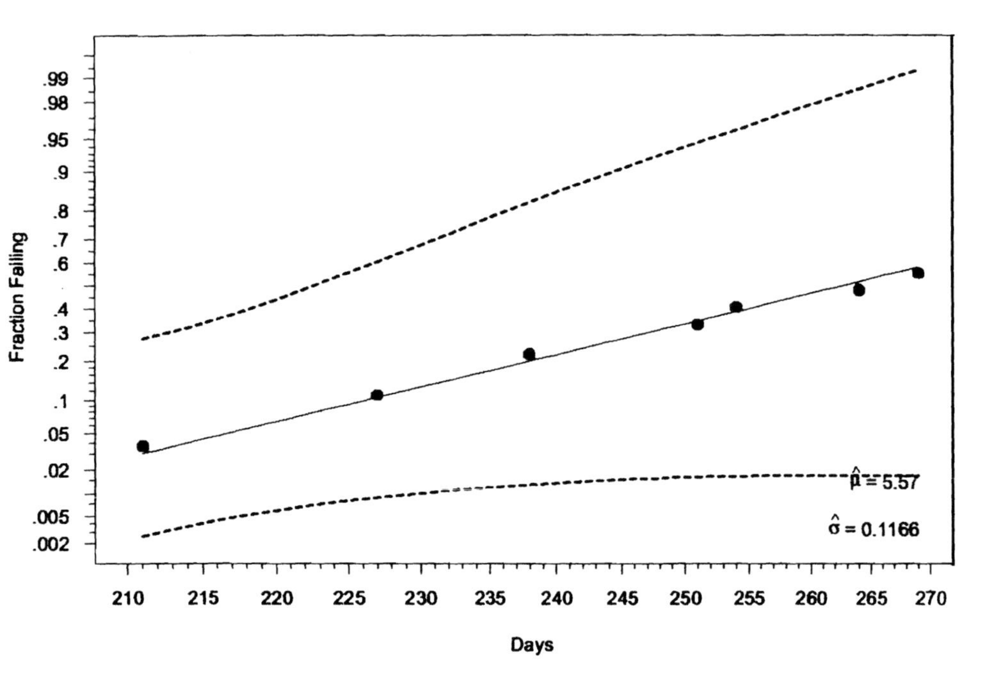

Figure 12.7. Lognormal probability plot of the truncated Product A shelf-life data at 25 ppm and 45°C.

part because of the parametric adjustment to the conditional nonparametric estimate. This indicates that there is no evidence in the data to suggest a departure from a lognormal distribution. Note, however, that in this case, there is little information in the data about the shape of the underlying distribution. A similar plot, done for the Weibull distribution, also provided an adjusted estimate of $F(t)$ that plotted almost as a straight line on the Weibull probability plot.

One conclusion from this analysis is that there is rather little information in the available data at this condition, relative to an untruncated sample or even a censored sample with the same number of reported failures. In Section 12.4, we will use these same data to do a direct comparison between truncated data and censored data to show the differences in information content.

We have presented detailed results for one combination of the explanatory variables. Similar results (with similar conclusions) were obtained for the other combinations of temperature and concentration and these will be summarized in the next section.

12.3 TRUNCATED REGRESSION MODEL FOR THE TRUNCATED PRODUCT A SHELF-LIFE DATA

This section describes models, graphical methods, and model estimation methods for the shelf life data, using all of the combinations of the explanatory variables.

The models and methods presented here are similar to those used in Nelson (1990b) and Meeker and Escobar (1998), extended to truncated data.

12.3.1 Individual Analyses at Different Experimental Conditions

Figure 12.8 is a multiple probability plot, similar to Figure 12.7, but showing the results simultaneously for all of the combinations of levels of temperature and concentration. Figure 12.9 is the same as Figure 12.8, but without the legend, making it possible to see all of the results clearly. Each of the nonparametric estimates was adjusted individually with (12.6), based on ML estimates of the lognormal $F(\tau^U_{\max}; \hat{\theta})$, using the data from the corresponding individual levels of concentration and temperature. The different slopes of the ML estimate lines in Figure 12.9 indicate some variation in the estimates of the lognormal shape parameters. Table 12.2 provides a summary of the estimation results for the individual combinations of levels.

12.3.2 Floating-Scale Model

Figure 12.10 is a multiple probability plot that is similar to Figure 12.9, but the fitted model is analogous to the standard one-way ANOVA model. In this model, there is a separate lognormal scale parameter for each combination of temperature

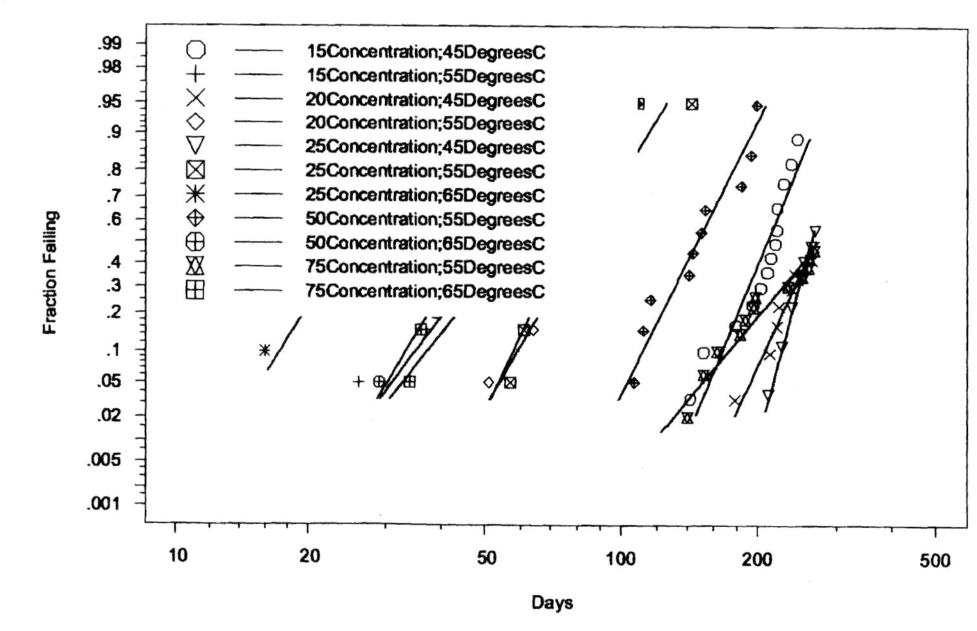

Figure 12.8. Multiple individual lognormal ML estimates and probability plots (with truncation correction) for the truncated Product A shelf-life data.

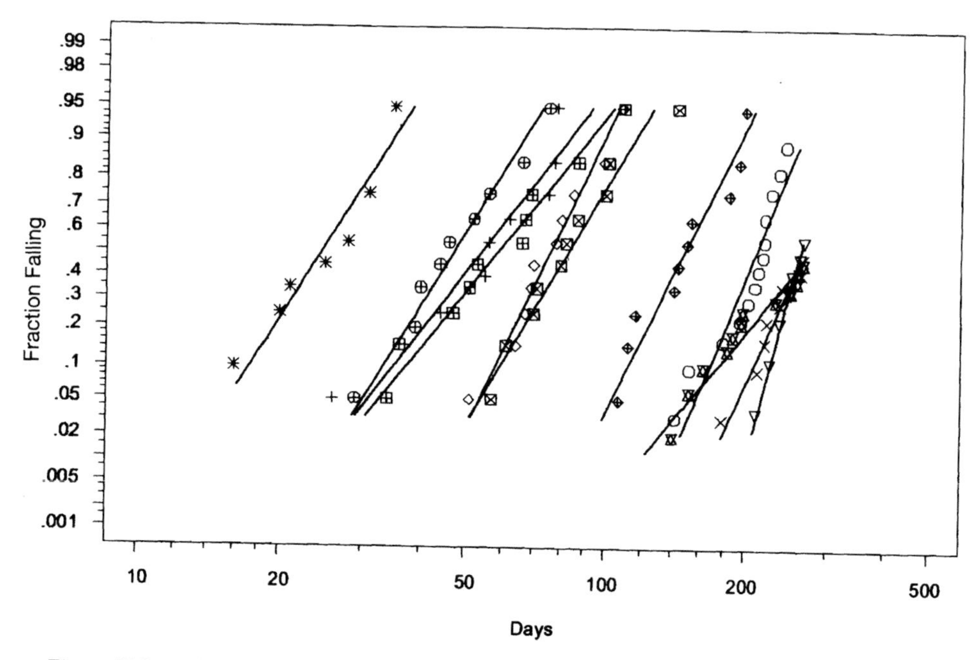

Figure 12.9. Multiple individual lognormal ML estimates and probability plots (with truncation correction) for the truncated Product A shelf-life data.

Table 12.2. ML Estimates of the Individual Lognormal Parameters (μ, σ) at Distinct Factor-Level Combinations for the Truncated Product A Shelf-Life Data

Factor Level Combination	n	Likelihood	$\hat{\mu}$	$\widehat{\text{se}}_{\hat{\mu}}$	$\hat{\sigma}$	$\widehat{\text{se}}_{\hat{\sigma}}$
15Concentration;45DegreesC	14	−67.43	5.345	0.06398	0.1775	0.04919
15Concentration;55DegreesC	10	−43.03	3.984	0.10531	0.3330	0.07447
20Concentration;45DegreesC	8	−36.58	5.591	0.32337	0.2014	0.14034
20Concentration;55DegreesC	10	−41.69	4.324	0.06553	0.2072	0.04633
25Concentration;45DegreesC	8	−33.34	5.570	0.15216	0.1166	0.07326
25Concentration;55DegreesC	10	−44.71	4.410	0.08134	0.2572	0.05753
25Concentration;65DegreesC	10	−33.16	3.199	0.08600	0.2720	0.06081
50Concentration;55DegreesC	10	−48.55	4.989	0.06864	0.2146	0.05041
50Concentration;65DegreesC	10	−39.43	3.848	0.08415	0.2661	0.05950
75Concentration;55DegreesC	12	−60.11	5.614	0.51552	0.3627	0.21500
75Concentration;65DegreesC	10	−44.28	4.065	0.11008	0.3481	0.07788

Total log likelihood = −492.3.
Response is in days.

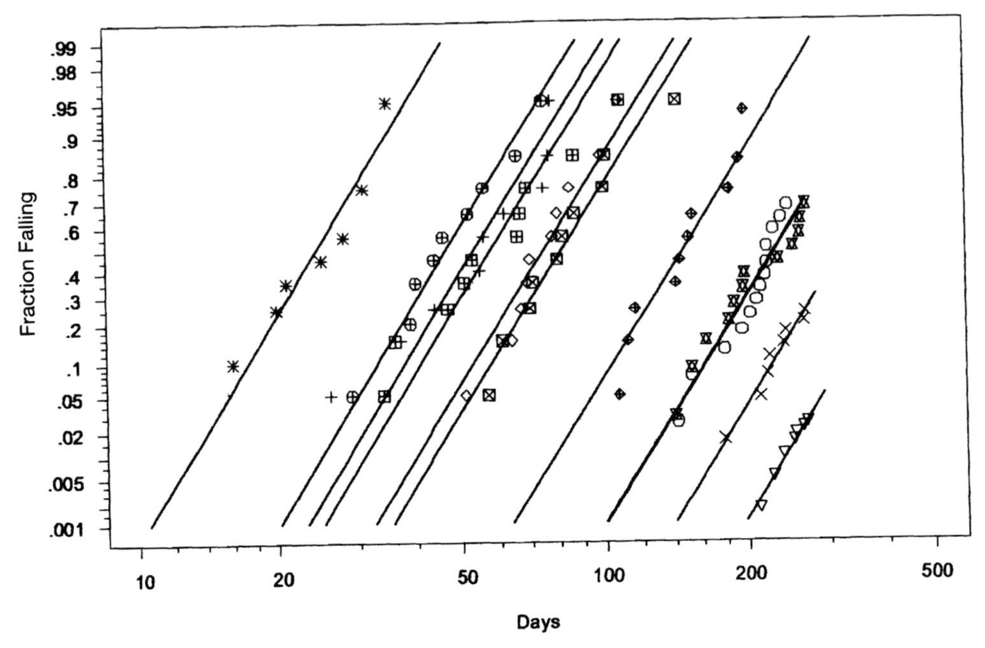

Figure 12.10. Multiple individual lognormal probability plots (with floating-scale model truncation correction) and ML estimate with common σ for the truncated Product A shelf-life data.

and concentration. In contrast with the individual analysis, however, there is a common lognormal shape parameter. We call this the "floating-scale model." The ML estimate lines are parallel because of the constraint that the lognormal shape parameters be the same.

In Figure 12.10, each of the nonparametric estimates (represented by the plotted points) was adjusted individually using (12.6), based on ML estimates of the lognormal $F(\tau^U_{\max}; \hat{\theta})$ from the constant shape parameter floating-scale model. The ML estimation results for the floating-scale model are summarized in Table 12.3. The intercept corresponds to the lognormal $\mu_{15\text{Concentration;45DegreesC}}$ parameter. The other coefficients are estimates of the deviation $\mu_i - \mu_{15\text{Concentration;45DegreesC}}$ for each of the other combinations of conditions. A comparison of the results between Figures 12.9 and 12.10 (corresponding to the individual analyses and floating-scale model, respectively) allows an assessment of whether the differences in the shape parameter estimates at different combinations of temperature and concentration can be explained by chance. The log likelihood ratio statistic for the comparison is $Q = 2\times[-492.3 - (-495.8)] = 7.0$. The difference in the number of parameters estimated in the two models is $22 - 12 = 10$. The approximate p value for the test comparing these two different models is $\Pr(\chi^2_{10} > 7) = 0.2745$, indicating that the differences in the slopes of the fitted lines relative to the the common slope of the floating-scale model can be explained by chance alone.

Table 12.3. Lognormal ML Estimate of the Floating-Scale Model for the Truncated Product A Shelf-Life Data

			Approximate 95% Confidence Interval	
Parameter	ML Estimate	Standard Error	Lower	Upper
Intercept	5.45177	0.10947	5.23721	5.6663
15Concentration;55DegreesC	−1.46793	0.13954	−1.74143	−1.1944
20Concentration;45DegreesC	0.33100	0.22543	−0.11084	0.7728
20Concentration;55DegreesC	−1.12810	0.13954	−1.40160	−0.8546
25Concentration;45DegreesC	0.67726	0.31117	0.06738	1.2871
25Concentration;55DegreesC	−1.04147	0.13954	−1.31496	−0.7680
25Concentration;65DegreesC	−2.25238	0.13954	−2.52588	−1.9789
50Concentration;55DegreesC	−0.45428	0.14070	−0.73004	−0.1785
50Concentration;65DegreesC	−1.60378	0.13954	−1.87727	−1.3303
75Concentration;55DegreesC	−0.00549	0.15241	−0.30421	0.2932
75Concentration;65DegreesC	−1.38698	0.13954	−1.66048	−1.1135
σ	0.27365	0.02256	0.23283	0.3216

Log likelihood at maximum point: −495.8.

12.3.3 Response Surface Model for Life as a Function of Concentration and Temperature

The response surface model suggested by the scientists responsible for quantifying the shelf life of Product A was

$$\mu = \beta_0 + \beta_1 \log(\text{concentration}) + \beta_2 \left(\frac{120.279}{\text{temp°C} + 273.15} \right) \tag{12.7}$$

$$\sigma = \text{constant}$$

The first and second terms in equation (12.7) correspond to a power function relationship relating life and concentration [i.e., life in days $\propto$ (concentration)$^{\beta_1}$]. The third term is based on the well-known Arrhenius rate reaction model [described in more detail in Chapter 2 of Nelson (1990b) and Chapter 18 of Meeker and Escobar (1998)], and 120.279 is the reciprocal of the gas constant in units of kJ/mole. The response surface is fit with the same lognormal distributions used in the floating scale model, except that now the scale parameter exp (μ) is controlled by equation (12.7). Table 12.4 provides a numerical summary of the lognormal ML estimation results and Figure 12.11 provides a graphical summary. The ML estimate lines are again parallel because of the constant-shape parameter restriction. In Figure 12.11, each of the nonparametric estimates (represented by the plotted points) was adjusted individually using equation (12.6), based on ML estimates of the lognormal $F(\tau^U_{\max}; \hat{\theta})$ from the response surface model in equation (12.7).

Table 12.4. ML Estimates of the Linear Response Surface Model for the Truncated Product A Shelf-Life Data

Parameter	ML Estimate	Standard Error	Approximate 95% Confidence Interval	
			Lower	Upper
β_0	−42.2604	1.97280	−46.1270	−38.3938
β_1	0.8489	0.05939	0.7324	0.9653
β_2	120.0534	5.08663	110.0838	130.0230
σ	0.2674	0.02061	0.2299	0.3110

Log likelihood at maximum point: −498.1.

The coefficient $\hat{\beta}_1 = 0.8489$ for concentration is the estimated power in a power law relationship between life and concentration. The coefficient $\hat{\beta}_2 = 120.05$ for DegreesC can be interpreted as the effective activation energy (in units of kJ/mole per °C) for the chemical reaction affecting the shelf life of Product A at a constant concentration. The value of $\hat{\sigma} = 0.2674$ implies that the variability in shelf life corresponding to a one-σ deviation in log-life, at a given temperature and concentration, is approximately 27% of the nominal response level.

Figure 12.11 can be used to assess the adequacy of the response surface model.

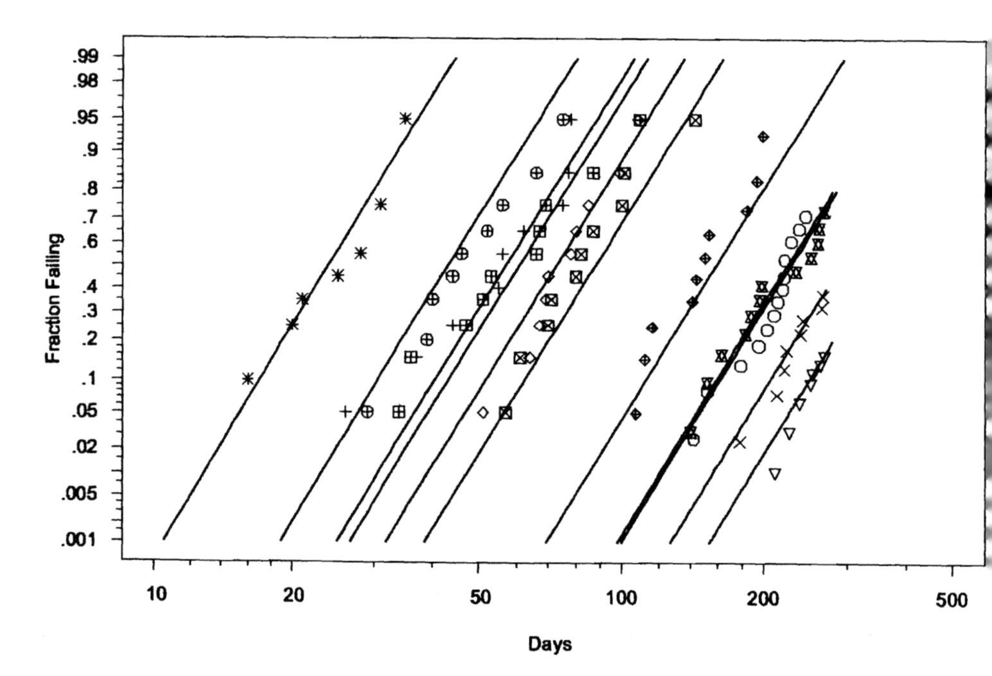

Figure 12.11. Multiple individual lognormal probability plots (with regression-model-based truncation correction) and ML estimate of the concentration–temperature response surface model for the truncated Product A shelf-life data.

The assessment is done by comparing the nonparametric estimates (the plotted points) with the fitted $F(t)$ lines, at each of the combinations of the experimental factors. The regression relationship in equation (12.7) is a further constraint in fitting a model to the data. Thus, there will be more deviations between the nonparametric estimates and the fitted $F(t)$ lines in Figure 12.11 than in Figure 12.10.

To test whether such deviations are statistically important, as opposed to being explainable by the natural variability in the data under the model in equation (12.7), we can again do a likelihood ratio test, this time comparing the results in Tables 12.3 and 12.4 (corresponding to the floating-scale model and the response surface model, respectively). The total log likelihood values from these models are –495.8 and –498.1, respectively. The log likelihood ratio statistic for the comparison is $Q = 2 \times [-495.8 - (-498.1)] = 4.6$. The difference in the number of parameters estimated in the two models is $12 - 4 = 8$. The approximate p value for the test comparing these two different models is $\Pr(\chi_8^2 > 4.6) = 0.201$, indicating that the differences in positions of the fitted lines from the response surface model, relative to the floating-scale model estimates, can be explained by chance alone.

We also fit a response surface model with a term for interaction between concentration and temperature (details are not given here), but there was no indication of interaction in the data and all inferences were similar to those provided by using the response surface model in equation (12.7).

12.3.4 Estimates of Shelf Life at Particular Storage Conditions from the Response Surface Model

Management wanted an estimate of the shelf-life distribution as a function of concentration for a nominal storage temperature of 35°C. Figure 12.12 is similar to Figure 12.11, but shows the extrapolation to the storage conditions of 25 ppm and 35°C. Figure 12.13 is a conditional model plot showing the ML estimates of the lognormal shelf life distributions for various fixed levels of concentration ranging between 15 and 75 ppm for fixed temperature at 35°C, based on the results for the response surface model summarized in Table 12.4. Table 12.5 gives the ML estimates and approximate confidence intervals for the quantiles of the shelf-life distribution at the storage conditions of 25 ppm and 35°C. Consider $t_{.01}$, the time at which a fraction .01 of the product population will fail at these conditions. The estimation results indicate that we are 95% confident that $t_{.01}$ is between 650 and 1036 days.

12.4 COMPARISON OF TRUNCATED AND CENSORED DATA ANALYSIS

Section 12.2 illustrated estimation from truncated data of Product A shelf at 25 ppm and 45°C. Here we compare the truncated data analysis with a censored data analysis, assuming that there had been a particular number of right-censored observations at 270 days. To get good agreement between the point estimates in the comparison, we used the truncated data model estimates in Section 12.2 to estimate that there

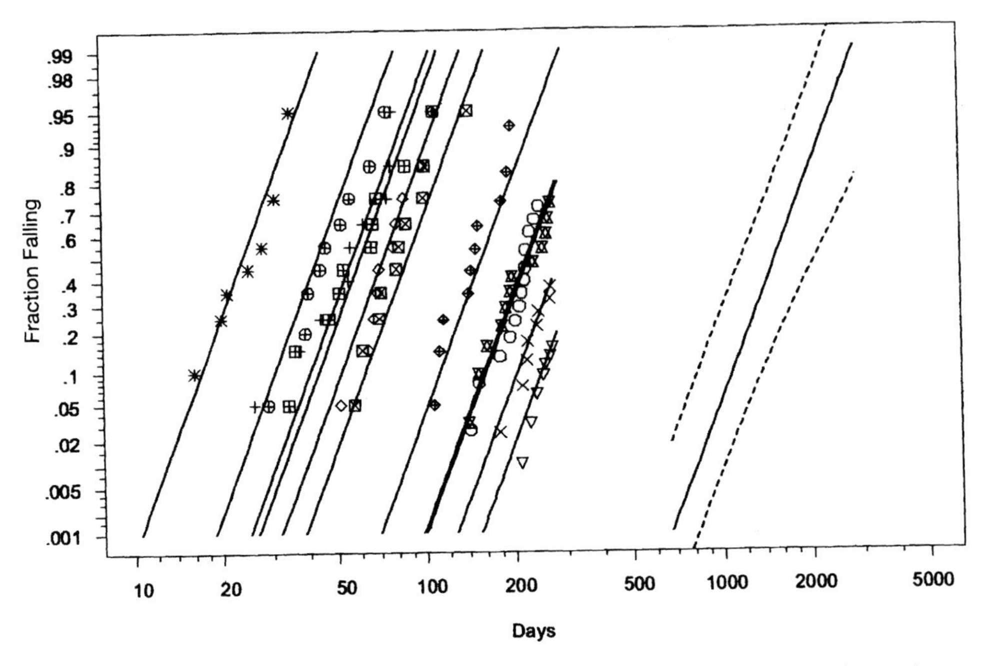

Figure 12.12. Multiple individual lognormal probability plots (with regression-model-based truncation correction) and ML estimate of the concentration–temperature response surface model for the truncated Product A shelf-life data showing extrapolation to the storage conditions of 25 ppm and 35°C.

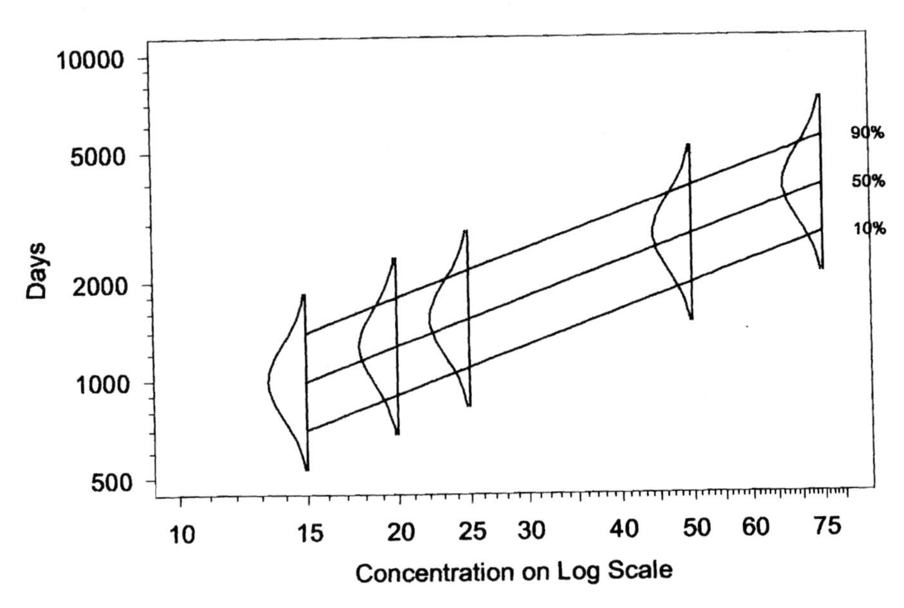

Figure 12.13. Conditional model plot of Product A shelf-life versus concentration at 35°C.

Table 12.5. ML Estimates and Approximate Confidence Intervals for the Quantiles of the Shelf-Life Distribution at the Storage Conditions of 25 ppm and 35°C

			Approximate 95% Confidence Interval	
p	$\hat{t}_p$ Days	Standard Error	Lower	Upper
0.001	668.84	79.462	529.90	844.20
0.005	767.47	91.116	608.14	968.54
0.010	820.42	97.649	649.72	1035.98
0.050	984.43	119.075	776.65	1247.80
0.100	1084.88	133.045	853.09	1379.65
0.200	1220.32	152.832	954.71	1559.84
0.300	1328.37	169.345	1034.67	1705.42
0.400	1428.23	185.145	1107.78	1841.37
0.500	1528.35	201.472	1180.36	1978.93
0.600	1635.48	219.452	1257.27	2127.46
0.700	1758.44	240.696	1344.66	2299.53
0.800	1914.12	268.467	1454.06	2519.74
0.900	2153.09	312.834	1619.52	2862.45
0.990	2847.12	451.851	2086.01	3885.94

had been six right-censored observations. Then, the actual number of trials at 25 ppm and 45°C would have been 8 + 6 = 14. We refer to these data as the "pseudo-censored" data. The main purpose of this comparison is to illustrate the important loss in precision that results from the truncated data caused by not recording the number of censored observations.

Figure 12.14 shows the ML estimation results for the pseudo-censored Product A shelf-life data at 25 ppm and 45°C. The corresponding numerical results are summarized in Table 6. Comparing Figures 12.7 and 12.14 shows the large improvement in precision that would result if the number of censored observations could be recovered. Relatedly, note the improvement in the standard errors and confidence interval widths given in Table 12.6. Some insight into the reasons for the improvements in precision can be obtained by comparing the contour plots in Figures 12.6 and 12.15. In particular, the closed likelihood contours in Figure 12.15 indicate that the shelf-life median (the 0.5 quantile) can be bounded statistically over the region of the plot, which was not possible in Figure 12.6. The ambiguity between the amount of truncation and the location of the upper tail of the shelf-life distribution has been, to a large degree, resolved by "knowing" the number of censored observations. There is, however, no good information about the fraction failing after 270 days, and the regression model fit in Section 12.3.3 appropriately quantifies the amount of uncertainty due to the lack of information about the number of censored observations. These remarks extend to the fitting of the regression model considered in Section 12.3.3.

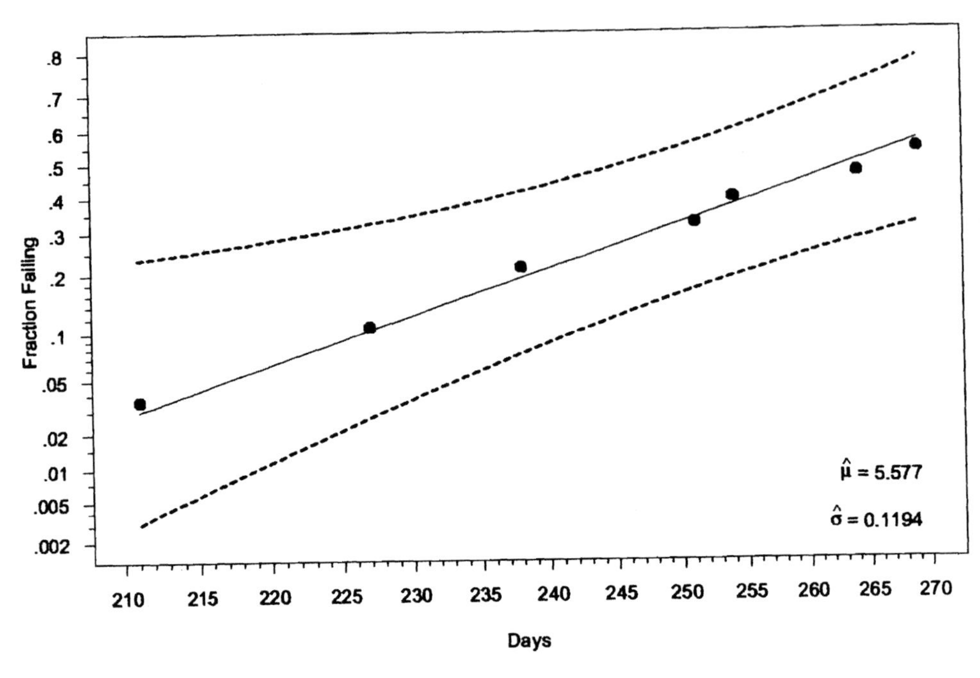

Figure 12.14. Lognormal probability plots of the pseudo-censored Product A shelf-life data at 25 ppm and 45°C.

12.5 CONCLUDING REMARKS AND EXTENSIONS

This chapter has shown how to extract useful information out of an historical set of data with a certain type of missing data. In this case, especially because there was probably a number of conditions with complete or nearly complete data (as indicated by an absence of observations near to 270 degrees at some combinations of concentration and temperature), the final answers were rather precise. This application provides a good example of how statistical methods can be adapted to deal with "messy data."

Table 12.6. Lognormal ML Estimates for the Pseudo-Censored Product A Shelf-Life Data at 25 ppm and 45°C

Parameter	ML Estimate	Standard Error	Approximate 95% Confidence Interval	
			Lower	Upper
μ	5.5766	0.03677	5.50449	5.6486
σ	0.1194	0.03282	0.06969	0.2046

Log likelihood at maximum point: –42.9.

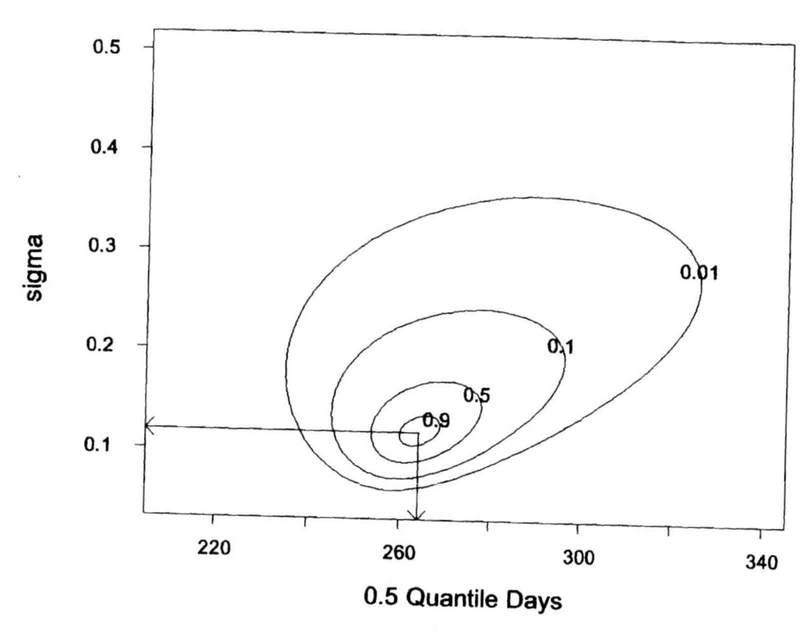

Figure 12.15. Contour plot of the lognormal likelihood function for the pseudo-censored Product A shelf-life data at 25 ppm and 45°C.

In other situations, it will be impossible to address completely important questions of interest without having to make strong assumptions or to rely on information coming from outside of the available data. For example, it may be necessary to assume that certain parameters are known. A more flexible alternative would be to obtain prior distributions for such parameters (e.g., from engineering knowledge or physical/chemical theory) and to use Bayesian methods instead. Because of recent advances in computing technology and the development of methods for Monte Carlo Markov Chain methods of inference, such methods are now practicable.

For the Product A shelf-life example, needed inferences were for storage/shipping at a given constant temperature. In other applications, however, temperature (or other environmental factors) will vary over time. In such cases, it is possible to use a model similar to the one used in this chapter to make predictions about life under a specified environmental profile. Methods for doing this are presented in Nelson (2001).

ACKNOWLEDGMENTS

We would like to thank Wally Blischke, Pra Murthy, Victor Chan, and Katherine Meeker for helpful comments on a previous version of this chapter.

REFERENCES

Amemiya, T. (1973). Regression analysis when the dependent variable is truncated normal, *Econometrica* **40:**997–1016.

Cohen, A. C. (1991). *Truncated and Censored Samples. Theory and Applications,* Marcel Dekker, New York.

Escobar, L. A., and Meeker, W. Q. (1998). The asymptotic covariance matrix for maximum likelihood estimators with models based on location-scale distributions involving censoring, truncation, and explanatory variables, *Statistica Sinica* **8:**221–237.

Gross, S. T., and Huber-Carol, C. (1992). Regression models for truncated survival data, *Scandinavian Journal of Statistics* **19:**193–213.

Kalbfleisch, J. D., and Lawless, J. F. (1988). Estimation of reliability in field-performance studies, *Technometrics* **30:**365–388 (with discussion).

Kalbfleisch, J. D., and Lawless, J. F. (1991). Regression models for right truncated data with applications to AIDS incubation times and reporting lags, *Statistica Sinica* **1:**19–32.

Kalbfleisch, J. D., and Lawless, J. F. (1992). Some useful statistical methods for truncated data, *Journal of Quality Technology* **24:**145–152.

Lawless, J. F. (1982). *Statistical Models and Methods for Lifetime Data,* Wiley, New York.

Meeker, W. Q. (1987). Limited failure population life tests: Application to integrated circuit reliability, *Technometrics* **29:**151–165.

Meeker, W. Q., and Escobar, L. A. (1998). *Statistical Methods for Reliability Data,* Wiley, New York.

Meeker, W. Q., and Escobar, L. A. (2002). *SPLIDA (S-PLUS Life Data Analysis).* www.public.iastate.edu/~splida.

Nelson, W. (1990a). Hazard plotting of left truncated life data, *Journal of Quality Technology* **22:**230–238.

Nelson, W. (1990b). *Accelerated Testing: Statistical Models, Test Plans, and Data Analyses,* Wiley, New York.

Nelson, W. (2001). Prediction of Field Reliability of Units, Each Under Differing Dynamic Stresses from Accelerated Test Data, in *Reliability* (C. R. Rao, and Balakrishnan, N., Editors), Elsevier Science, Amsterdam.

Pagano, M., Tu, X. M., De Gruttola, V., and MaWhinney, S. (1994). Regression analysis of censored and truncated data: Estimating reporting-delay distributions and AIDS incidence from surveillance data, *Biometrics* **50:**1203–1214.

Schneider, H. (1986). *Truncated and Censored Samples for Normal Populations,* Marcel Dekker, New York.

Turnbull, B. W. (1976). The empirical distribution function with arbitrary grouped, censored, and truncated data, *Journal of the Royal Statistical Society* **38:**290–295.

EXERCISES

12.1. Make scatter plots of the response versus temperature, conditioning on concentration level and vice versa. Describe how you might modify these plot to reflect the truncation.

12.2. Redo the scatter plots with different transformations on the variables. From looking at the the scatter plots, decide which scales make the relationship between life and the explanatory variables appear to be approximately linear.

12.3. Is it reasonable to assume that the shape parameter of the lognormal distribution σ does not depend on concentration? How can you tell?

12.4. Use the results of the analysis in Section 12.3 to compute the ML estimate of the .01 quantile of life at 25°C if the additive concentration is 40 ppm.

12.5. What, approximately, would be a safe level of concentration for product that needs to last at least 360 days at 25°C with the probability of failing being no more than .05?

12.6. Try fitting some alternative relationships. In particular, try the linear and square root relationships for the conservation explanatory variable. What effect does this have on the answer to Exercise 12.4?

12.7. The analyses in this chapter used the lognormal distribution. Redo the analyses with a Weibull distribution. How to the results compare? Comment both on the model diagnostics and the final answers.

12.8. Conduct the following simulation to compare the effects of censoring and truncation on estimation precision.

(a) Generate a sample from a lognormal distribution with $\eta = \exp(\mu) = 200$ hours and $\sigma = 0.5$.

(b) Find the ML estimates of the parameters and $t_{.1}$, treating any observations beyond 250 hours as right censored.

(c) Repeat the simulation 1000 times. Make appropriate plots of the sample estimates (including scatter plots to see correlation). Compute and use histograms or other graphical displays to compare the estimates from the censored and the truncated samples. Also compute the sample variances for the parameter estimates and for the estimates of $t_{.1}$.

(d) What can you conclude from this simulation experiment?

12.9. Repeat Exercise 12.8 using $t_{.9}$ in place of $t_{.1}$. What do you conclude?

12.10. Consider the results from Exercise 12.8. Provide an intuitive explanation for the reason that precision from the censored distribution is much better than that from the truncated distribution.

12.11. Derive the truncated cdf used to construct the likelihood contribution in equation (12.3).

APPENDIX. SPLIDA COMMANDS FOR THE ANALYSES

This appendix gives explicit direction on how to use the SPLIDA/S-PLUS software to do the analyses described in this chapter.

1. The data [concentration in ppm (parts per million), failure "times" in days, and truncation information (all observations were right-truncated at 270 days)] are in the SPLIDA data frame `ShelfLifeA`. Use `Splida -> Make/edit/summary/view data object -> Make life data object` to create the SPLIDA life data object `ShelfLifeA.ld`. Choose `Days` as the response, `Temp` and `Concentration` as the explanatory variables, `Truntime` as the truncation time, and `TrunType` as the Truncation ID.
2. Make a conditional scatter plot by using the SPLIDA/S-PLUS commands `ConditionalPlot("Days", "Temp", "Concentration", data = ShelfLifeA)` and `ConditionalPlot("Days", "Concentration", "Temp", data = ShelfLifeA)`.
3. To make a life data object for the data at a concentration of 25 ppm and a temperature of 45°C, use `Splida -> Multiple regression (ALT) data analysis -> Make a subset life data object`. Choose the ShelfLifeA.ld data object and highlight both explanatory variables. Then highlight `25;45` and click on "Apply."
4. To make the event plot of the data at 25 ppm and 45°C, use `Splida -> Single distribution data analysis -> Life Data event plot`. Choose the data object `ShelfLifeA.25Concentration.45DegreesC.ld` and click on "Apply."
5. To make the probability plot and fit the truncated lognormal distribution to the data at 25 ppm and 45°C, use `Splida -> Single distribution data analysis -> Probability plot with parametric ML fit`. Choose the `ShelfLifeA.25Concentration.45DegreesC.ld` data object, select the lognormal distribution, and click on "Apply."
6. To make the likelihood contour plot for the data at 25 ppm and 45°C, use `Splida -> Single distribution data analysis -> Likelihood contour/perspective/profile plots`, choose the lognormal distribution, check the contour box, and change the quantile to 0.5. Visit the `Modify plot axes` page and specify 0.05 and 0.5 for the lower and upper endpoints of the *x* axis, respectively, then click on "Apply."
7. To obtain a probability plot showing lognormal distributions fitted to the individual conditions, use `Splida -> Multiple regression (ALT) data analysis -> Probability plot and ML fit for individual conditions: common shapes (slopes)`. On the `Basic` page, choose the ShelfLife.ld life data object, the

lognormal distribution, and both Concentration and DegreesC as the explanatory variables. Click on "Apply." Visit the `Plot options` page and choose the suppress option in the legend box. Click on "Apply" again.

8. To obtain a multiple probability plot and fit of the floating-scale model, use `Splida -> Multiple regression (ALT) data analysis -> Prob plot and ML fit for indiv conditions.` On the `Basic` page, choose the lognormal distribution and both Concentration and DegreesC as the explanatory variables. Visit the `Plot options` page and choose the suppress option in the legend box. Click on "Apply."
9. The Arrhenius model will be used to fit the response model. To match the output in this chapter, it is necessary to override the default choice for the units of the Arrhenius model regression coefficient. Use `Splida -> Change SPLIDA default options (preferences),` visit the `Misc` page, and choose kJ/mole under Boltzmann/gas constant units.
10. To obtain a multiple probability plot and fit of the response surface model, use `Splida -> Multiple regression (ALT) data analysis -> Probability plot and ML fit of a regression(acceleration) model.` On the `Basic` page, choose the lognormal distribution and both Concentration and DegreesC as the explanatory variables. Enter 25;35 in the Additional levels for evaluation box. Visit the `Model` page and click on the relationships button. Choose log for Concentration and Arrhenius for DegreesC. Visit the `Plot options` page and choose the suppress option in the legend box. Visit the `Tabular output` page, check "print table" under Quantile estimates, and choose 25;35 from the "Level(s)" list. Click on "Apply."
11. To obtain a conditional model plot of the shelf life distributions as a function of concentration at 35°C, use `Splida -> Multiple regression (ALT) data analysis -> Conditional model plot.` On the `Basic` page, choose the the results object `ShelfLifeA.groupm.lognormal.CncnLog.DgrCArrh.out` and select Concentration as the variable to vary. Click on the choose button and enter 35 for the temperature. Click on "Apply."

CHAPTER 13

Determining Software Quality Using COQUALMO

Sunita Chulani, Bert M. Steece, and Barry Boehm

13.1 INTRODUCTION

Cost, schedule, and quality comprise the three most important factors in software economics modeling. Unfortunately, these three factors are highly correlated, and consequently software project managers do not have a good parametric model they can trust to determine the effects of one on the other and hence make confident decisions during the software development process. We know that beyond the "Quality is Free" point, it is difficult to increase the quality and reliability of a software product without increasing either the cost or schedule or both. Furthermore, we cannot compress the development schedule to reduce the time to market without having a negative impact on the operational reliability and quality of the software product and/or cost of development. However, software cost/schedule and quality/reliability estimation models can assist software project managers to make better decisions that impact the overall success of the software product.

Recognizing the need for a model that aids managers in making difficult trade-offs among cost/schedule/quality/reliability, we have developed COQUALMO, a quality model that extends the popular COCOMO II model (a software cost and schedule estimation model) (Boehm et al., 2000). COQUALMO helps managers determine the quality of the software product in terms of defect density that in turn is an early predictor of software reliability.

We have organized the remainder of this chapter as follows:

13.2. Software Reliability Definitions: This section introduces software reliability concepts and defines commonly used terms such as defect, fault, failure, and so on.

Case Studies in Reliability and Maintenance, Edited by W. R. Blischke and D. N. P. Murthy.
ISBN 0-471-41373-9

13.3. COQUALMO: This section provides the modeling details of COQUALMO and discusses its output which can be used as an early predictor of software reliability.
13.4. Case Study: This section provides a case study of how COQUALMO can be used in a software development project.
13.5. Conclusions: This section summarizes software reliability modeling and how practitioners can use COQUALMO to make better decisions.

13.2 SOFTWARE RELIABILITY DEFINITIONS

Because quality and reliability of software components are such predominant characteristics in software development, research on these two subjects is quite abundant. This section provides a review of the basic concepts related to software reliability modeling.

13.2.1 Basic Definitions

Reliability. The definition of software reliability is given in *IEEE's Standard Glossary of Software Engineering Terminology* as the probability that software will not cause the failure of the system for a specified time under specified conditions (IEEE, 1989). For more explanation, see Rook (1990)

Mean time to failure (MTTF). MTTF is mean time between previous failure and next failure.

Failure. A failure is a situation that arises when the system stops the service that the user wants.

Fault. The cause of a failure is a fault.

Defect. A defect is usually used as a generic term to refer to either a fault (cause) or a failure (effect). In the context of this article and COQUALMO, the defects we estimate are faults, not failures.

When software reliability modeling began in the early 1970s, most of the models focused on predicting future reliability behavior based on past failures. There were two basic concepts for observing the failures. One simply counted the number of failures during a given specific time period while the other observed the time between sequential failures.

13.3 CONSTRUCTIVE QUALITY MODEL (COQUALMO)

The COQUALMO model is based on the software defect introduction and removal model described by Boehm (1981) and is analogous to the "tank and pipe" model introduced by Jones (1975). Conceptually, the model is illustrated in Figure 13.1. Figure 13.1 shows that COQUALMO consists of both defect introduction pipes and defect removal pipes. Defects get injected into the software product through defect

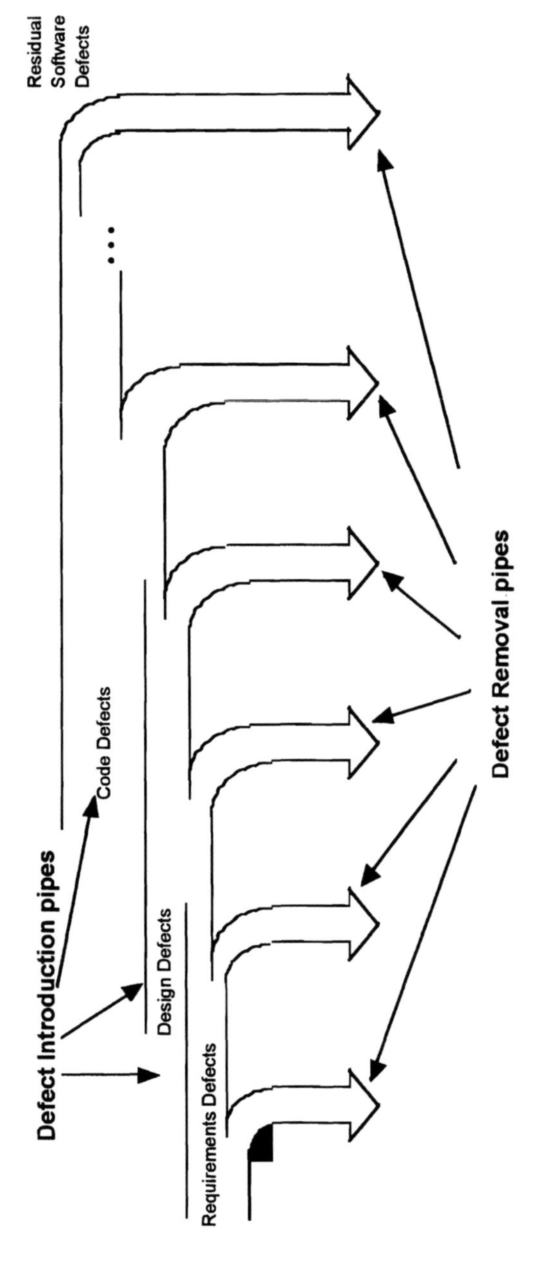

Figure 13.1. The "tank and pipe" software defect introduction and removal model.

introduction pipes and are captured by the software defect introduction model. On the other hand, defect removal techniques such as testing and inspections are captured by the software defect removal model. Sections 13.3.1 and 13.3.2 describe each of these submodels in detail.

COQUALMO, like COCOMO II, is calibrated using the Bayesian approach because changes in technologies over the years has somewhat diminished the usefulness of models based solely on historical data. The Center for Software Engineering at USC has found Bayesian models to be particularly effective in solving this problem. The Bayesian viewpoint allows the user to combine expert opinion that reflects important changes in technology while still being able to extract useful information contained in historical data.

13.3.1 The Software Defect Introduction (DI) Submodel

In COQUALMO, defects are classified by their origin as requirement defects, design defects, and coding defects. Ongoing research to enhance COQUALMO using the Orthogonal Defect Classification (Chillarege et al., 1992) scheme is underway. In its current state, the inputs to the DI (defect introduction) model are (i) size in terms of Source Lines of Code (SLOC) or Function Points and a set of DI-drivers which are the same as the COCOMO II cost drivers. As in COCOMO II (Boehm et al., 2000), these drivers can be divided into four categories as Software Platform, Product, Personnel, and Project Attributes. Figure 13.2 shows the general concept of the DI model and indicates that its output is the total number of nontrivial requirements, design and code defects. Tables 13.1 and 13.2 show the COCOMO II and COQUALMO defect introduction drivers.

The modeling equations for the DI model are

$$\text{Total number of defects introduced} = \sum_{j=1}^{3} A_j * (\text{size})^{B_j} * \prod_{i=1}^{21} (\text{DI-driver})_{ij} \quad (13.1)$$

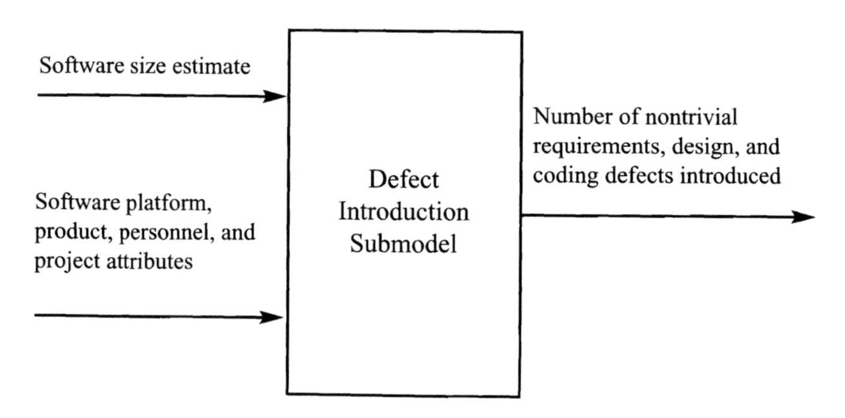

Figure 13.2. The defect introduciton submodel of COQUALMO.

Table 13.1. COQUALMO Defect Introduction Drivers

Category	Defect Introduction Driver
Platform	Required software reliability (RELY)
	Data base size (DATA)
	Required reusability (RUSE)
	Documentation match to life-cycle needs (DOCU)
	Product complexity (CPLX)
Product	Execution time constraint (TIME)
	Main storage constraint (STOR)
	Platform volatility (PVOL)
Personnel	Analyst capability (ACAP)
	Programmer capability (PCAP)[a]
	Applications experience (AEXP)
	Platform experience (PEXP)
	Language and tool experience (LTEX)
	Personnel continuity (PCON)
Project	Use of software tools (TOOL)
	Multisite development (SITE)
	Required development schedule (SCED)[a]
	Precedentedness (PREC)
	Architecture/risk resolution (RESL)
	Team cohesion (TEAM)
	Process maturity (PMAT)

[a]Example rating scales shown in Table 13.2.

where

- j identifies the three artifact types (requirements, design, and coding, for $j = 1, 2,$ and 3, respectively),
- i identifies the DI-driver category as shown in Table 13.1 ($i = 1, \ldots, 21$),
- A_j is the baseline DI rate adjustment for artifact j,
- Size is the software project measured in terms of kSLOC (thousands of source lines of code), (Park, 1992) or function points (IFPUG, 1994) and programming language,
- B_j is a factor that accounts for economies/diseconomies of scale, and
- $(\text{DI-driver})_{ij}$ is the defect introduction driver for the jth artifact and the ith factor.

B_j has been initially set to 1, as it is unclear if defect introduction rates will exhibit economies or diseconomies of scale as discussed by Banker et al. (1994) and Gulledge and Hutzler (1993). The question is, If size doubles, then will the defect introduction rate increase by more than twice the original rate? If yes, the software product indicates diseconomies of scale, implying $B_j > 1$. Or will defect introduction rate increase by a factor less than twice the original rate, indicating economies of scale, giving $B_j < 1$?

Table 13.2. Example Defect Introduction Drivers Rating Scales (from Delphi)

PCAP (Programmer Capability)	Requirements	Design	Code
VH	1.0	Fewer design defects due to easy interaction with analysts. Fewer defects introduced in fixing defects. 0.85	Fewer coding defects due to fewer detailed design reworks, conceptual misunderstandings, coding mistakes. 0.76
Nominal	Nominal level of defect introduction 1.0		
VL	1.0	More design defects due to less easy interaction with analysts. More defects introduced in fixing defects. 1.17	More coding defects due to more detailed design reworks, conceptual misunderstandings, coding mistakes. 1.32
Delphi DIR range	1.0	1.38	1.74

SCED (Required Development Schedule)	Requirements	Design	Code
VH	Fewer requirements defects due to higher likelihood of correctly interpreting specs. Fewer defects due to more thorough planning, specs, validation. 0.85	Fewer design defects due to higher likelihood of correctly interpreting specs. Fewer design defects due to fewer defects in fixes and fewer specification defects to fix. Fewer defects due to more thorough planning, specs, validation. 0.84	Fewer coding defects due to higher likelihood of correctly interpreting specs. Fewer coding defects due to requirements and design shortfalls. Fewer defects introduced in fixing defects. 0.84
Nominal	Nominal level of defect introduction 1.0		
VL	More requirements defects due to • More interface problems (more people in parallel) • More TBDs in specs, plans • Less time for validation 1.18	More design defects due to earlier TBDs, more interface problems, less time for V&V. More defects in fixes and more specification defects to fix. 1.19	More coding defects due to requirements and design shortfalls, less time for V&V. More defects introduced in fixing defects. 1.19
Delphi DIR range	1.4	1.42	1.42

For each artifact, we define a new parameter, QAF_j, such that

$$QAF_j = \prod_{i=1}^{21} (\text{DI-driver})_{ij} \tag{13.2}$$

Hence

$$\text{Total number of defects introduced} = \sum_{j=1}^{3} A_j * (\text{size})^{B_j} * QAF_j \tag{13.3}$$

Now let us discuss the impact of DI driver > 1, DI driver < 1, and DI driver $= 1$. If DI driver > 1, it has a deteriorating impact on the number of defects introduced and hence reduces the quality of the product. On the contrary, if DI driver < 1, then fewer defects are introduced, improving the quality of the product. So, for example, for a project with programmers having very high capability ratings (i.e., people with high programming ability, high efficiency and thoroughness, and very good communication and cooperation skills), only 76% of the nominal number of defects will be introduced during the coding activity. (Table 1 has PCAP—that is, programmer capability—as one of the DI drivers. The numerical values are obtained from the Delphi rounds described in Section 13.3.3; examples of PCAP and SCED are shown in Table 2). On the other hand, if the project had programmers with very low capability ratings, then 132% of the nominal number of coding defects will be introduced. This would cause the defect introduction range to be 1.32/0.76 = 1.74 for coding defects for PCAP, where the defect introduction range is defined as the ratio between the largest DI driver and the smallest DI driver. Figure 13.3 shows the Coding Defect Introduction Ranges obtained from the Delphi.

13.3.2 The Software Defect Removal (DR) Submodel

The main work of the defect removal (DR) model as defined by COQUALMO is estimating the number of defects removed using three defect removal profiles:automated analysis, peer reviews, and execution testing and tools. Figure 13.4 conceptually illustrates the DR model. The values of these three profile levels are inputs used by the DR model to estimate the number of removal defects. Each profile has six levels of increasing defect removal capability, namely, "very low," "low," "nominal," "high," "very high," and "extra high." Table 13.3 describes the three profiles and the six levels for each of these profiles.

The automated analysis profile includes code analyzers, syntax and semantics analyzers, type checkers, requirements and design consistency and traceability checkers, model checkers, formal verification and validation, and so on. The peer reviews profile covers the spectrum of all peer group review activities. The very low level is operative when no peer reviews take place, and the extra high level represents the other end of the spectrum when extensive amount of preparation with formal review roles assigned to the participants and extensive user/customer involvement. A formal change control process is incorporated with procedures for fixes. Extensive review checklists are prepared with thorough root cause analysis.

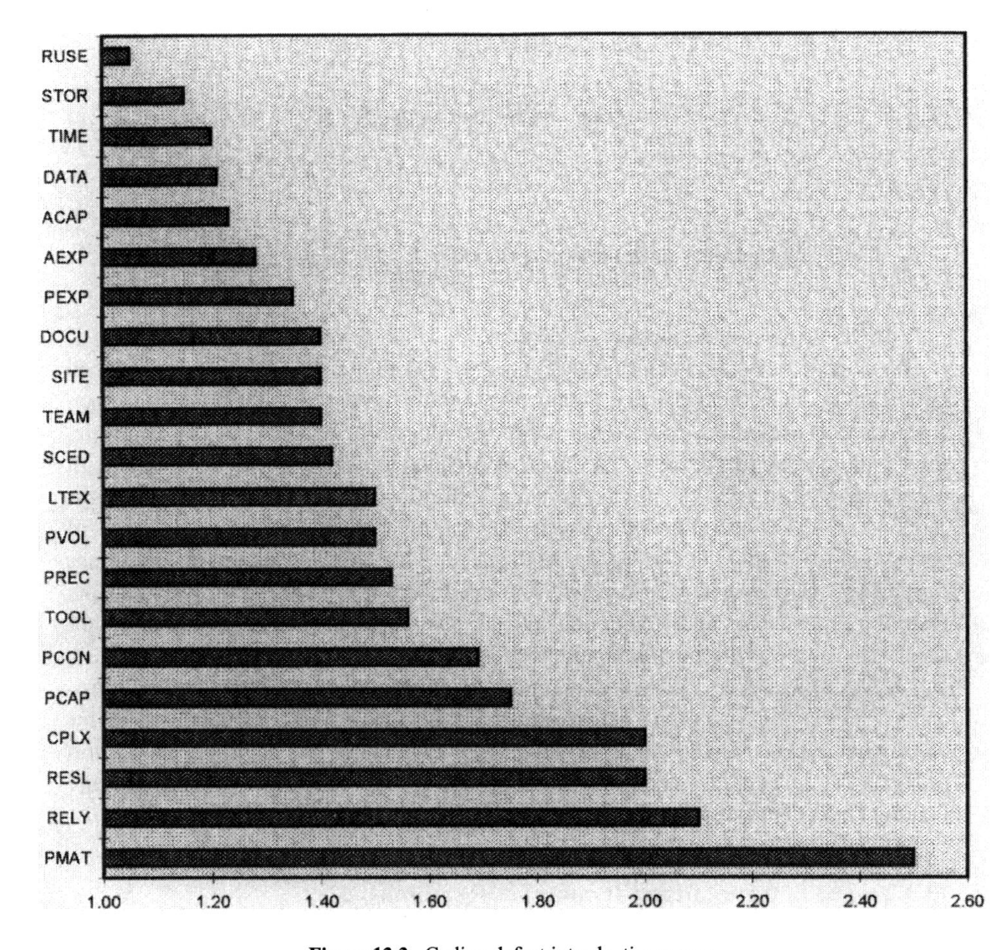

Figure 13.3. Coding defect introduction ranges.

A continuous review process improvement is also incorporated with statistical process control.

The execution testing and tools profile, as the name suggests, covers all procedures and tools used for testing with the very low level being when no testing takes place. Not much software development is done this way; that is, at least some testing is done in almost all software development efforts. The nominal level involves the use of a basic testing process with unit testing, integration testing, and system testing, with test criteria based on simple checklists and with a basic problem tracking support system in place and basic test data management. The extra high level involves the use of highly advanced tools for test oracles (i.e., inspectors, hopefully automated, that validate that test executions produce valid results) with the integration of automated analysis and test tools and distributed monitoring and analysis. Sophisticated model-based test process management is also employed at this level.

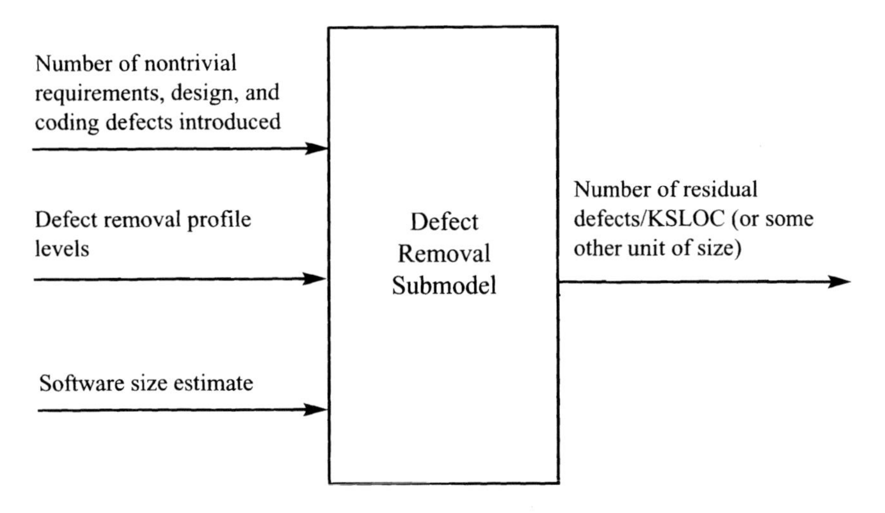

Figure 13.4. The defect removal submodel of COQUALMO.

The defect removal model for artifact *j* can be written as

$$D\,\text{Res}_{\text{Est},j} = C_j \times \text{DI}_{\text{Est},j} \times \Pi_i(1 - \text{DRF}_{ij}) \tag{13.4}$$

where

$D\,\text{Res}_{\text{Est},j}$ = estimated number of residual defects for *j*th artifact
C_j = baseline DR constant for the *j*th artifact
$\text{DI}_{\text{Est},j}$ = estimated number of defects of artifact type *j* introduced
i = 1 to 3 for each of the three DR profiles, namely, automated analysis, peer reviews, execution testing, and tools
DRF_{ij} = defect removal fraction for defect removal profile *i* and artifact type *j*

Using the nominal DIRs defined in Section 13.3.1 and the DRFs defined above, and described in detail in Boehm et al. (2000), the residual defect density is computed when each of the three profiles are at very low, low, nominal, high, very high, and extra high levels. The results are given in Table 13.4 and a detailed description can be found in Boehm et al. (2001). As an example in Table 13.4, we look at the low level for each of the three defect removal activities and for requirements of defect artifacts (shown in bold). There are 10 baseline requirements defects [as obtained from Boehm (1981) and shown in column "DI/kSLOC"]. Low levels of automated analysis, people reviews, and execution testing and tools remove 0%, 25%, and 23% requirements defects, respectively, leaving 100(.75*.77)% = 58% defects. Hence, the baseline is adjusted by 58% to give 5.8 residual defects/kSLOC as shown in column DRes/kSLOC.

Table 13.3. The Defect Removal Profiles

Rating	Automated Analysis	Peer Reviews	Execution Testing and Tools
Very low	Simple compiler syntax checking.	No peer review.	No testing.
Low	Basic compiler capabilities for static module-level code analysis, syntax, type-checking.	Ad hoc informal walkthroughs. Minimal preparation, no follow-up.	Ad hoc testing and debugging. Basic text-based debugger.
Nominal	Some compiler extensions for static module and intermodule level code analysis, syntax, type-checking. Basic requirements and design consistency, traceability checking.	Well-defined sequence of preparation, review, minimal follow-up. Informal review roles and procedures.	Basic unit test, integration test, system test process. Basic test data management, problem tracking support. Test criteria based on checklists.
High	Intermediate-level module and intermodule code syntax and semantic analysis. Simple requirements/design view consistency checking.	Formal review roles with all participants well-trained and procedures applied to all products using basic checklists, follow-up.	Well-defined test sequence tailored to organization (acceptance/alpha/beta/flight/etc.) test. Basic test coverage tools, test support system. Basic test process management.
Very high	More elaborate requirements/design view consistency checking. Basic distributed-processing and temporal analysis, model checking, symbolic execution.	Formal review roles with all participants well-trained and procedures applied to all product artifacts and changes (formal change control boards). Basic review checklists, root cause analysis. Formal follow-up. Use of historical data on inspection rate, preparation rate, fault density.	More advanced test tools, test data preparation, basic test oracle support, distributed monitoring and analysis, assertion checking. Metrics-based test process management.
Extra high	Formalized[a] specification and verification. Advanced distributed processing and temporal analysis, model checking, symbolic execution.	Formal review roles and procedures for fixes, change control. Extensive review checklists, root cause analysis. Continuous review process improvement. User/customer involvement, Statistical Process Control.	Highly advanced tools for test oracles, distributed monitoring and analysis, assertion checking Integration of automated analysis and test tools. Model-based test process management.

[a]Consistency-checkable pre-conditions and post-conditions, but not mathematical theorems.

Table 13.4. Defect Density Results from Initial Defect Removal Fraction Values

Automated Analysis	Peer Reviews	Execution Testing and Tools	Product ($1\text{-}DRF_{ij}$)	DI/kSLOC[a]	DRes/kSLOC
0.00	0.00	0.00	1.00	10	10
0.00	0.00	0.00	1.00	20	20
0.00	0.00	0.00	1.00	30	30
Total:					**60**
0.00	0.25	0.23	0.58	10	5.8
0.00	0.28	0.23	0.55	20	11
0.10	0.30	0.38	0.39	30	11.7
Total:					**28.5**
0.10	0.40	0.40	0.32	10	3.2
0.13	0.40	0.43	0.3	20	6
0.20	0.48	0.58	0.17	30	5.1
Total:					**14.3**
0.27	0.50	0.50	0.18	10	1.8
0.28	0.54	0.54	0.15	20	3
0.30	0.60	0.69	0.09	30	2.7
Total:					**7.5**
0.34	0.58	0.57	0.14	10	1.4
0.44	0.70	0.65	0.06	20	1.2
0.48	0.73	0.78	0.03	30	0.9
Total:					**3.5**
0.40	0.70	0.60	0.07	10	0.7
0.50	0.78	0.70	0.03	20	0.6
0.55	0.83	0.88	0.009	30	0.27
Total:					**1.57**

[a]1990s Baseline DIRs (defect introduction rates), excluding DI driver effects.

For example, with nominal ratings for the DI drivers and very low ratings for each of the three DR profiles, the result is a residual defect density of 60 defects/kSLOC. Similarly, with nominal ratings for the DI drivers and very high ratings for each of the three DR profiles, the result is a residual defect density of 1.57 defects/kSLOC (see Table 13.4). For further explanation of these numerical values, please refer to Boehm et al. (2000).

Thus, using the quality model described in this section, one can conclude that for a project with nominal characteristics (i.e., average ratings) for the DI drivers and the DR profiles, the residual defect density is 14.3 defects/kSLOC.

13.3.3 Calibrating COQUALMO Using the Bayesian Approach

In Sections 13.3.1 and 13.3.2, we described the defect introduction and removal models of COQUALMO. In this section, we present the calibration approach used to develop the numeric scales for the various drivers.

First let us describe the need for the use of the Bayesian approach for COQUAL-

MO. Bayesian analysis is a well-defined and rigorous process of inductive reasoning that has been used in many scientific disciplines. A distinctive feature of the Bayesian approach is that it permits the investigator to use both sample (data) and prior (expert-judgement) information in a logically consistent manner in making inferences. This is done by using Bayes' theorem to produce a "post-data" or posterior distribution for the model parameters. Using Bayes' theorem, prior (or initial) values are transformed to post-data views. This transformation can be viewed as a learning process. The posterior distribution is determined by the variations of the prior and sample information. If the variation of the prior information is smaller than the variation of the sampling information, then a higher weight is assigned to the prior information. On the other hand, if the variation of the sample information is smaller than the variation of the prior information, then a higher weight is assigned to the sample information, causing the posterior estimate to be closer to the sample information.

Using Bayes' theorem, the two sources of information can be combined as follows:

$$f(\beta|Y) = \frac{f(Y|\beta)f(\beta)}{f(Y)} \qquad (13.5)$$

where β is the vector of parameters in which we are interested and Y is the vector of sample observations from the density function $f(Y|\beta)$. In equation (13.5), $f(\beta|Y)$ is the posterior density function for β summarizing all the information about β, $f(Y|\beta)$ is the distribution of the sample information and is algebraically equivalent to the likelihood function for β, and $f(\beta)$ is the prior information summarizing the expert-judgement information about β. Equation (13.5) can be rewritten as

$$f(\beta|Y) \propto f(Y|\beta)f(\beta) \qquad (13.6)$$

In words, expression (13.6) means: *Posterior* $\propto$ *Sample* $\times$ *Prior.*

In the Bayesian analysis context, the "prior" probabilities are the simple unconditional probabilities of the sample information; while the "posterior" probabilities are the conditional probabilities given sample and prior information.

The Bayesian approach makes use of prior information that is not part of the sample data by providing an optimal combination of the two sources of information. As described in many books on Bayesian analysis (Leamer, 1978; Box and Tiao, 1973), the posterior mean, b^{**}, and variance, $\mathrm{Var}(b^{**})$, are defined as

$$b^{**} = \left[\frac{1}{s^2}X'X + H^*\right]^{-1} \times \left[\frac{1}{s^2}X'Xb + \mathrm{H}^*b^*\right] \qquad (13.7)$$

and

$$\mathrm{Var}(b^{**}) = \left[\frac{1}{s^2}X'X + H^*\right]^{-1} \qquad (13.8)$$

where X is the matrix of predictor variables, s is the sample variance, and H^* and b^* are the precision (inverse of variance) and mean of the prior information, respectively.

From equations (13.7) and (13.8), it is clear that in order to determine the Bayesian posterior mean and variance, the mean and precision of the prior information and the sampling information need to be determined. This approach alleviates the challenge of making good decisions using data that are usually scarce and incomplete and that have been successfully used in COCOMO II 2000 (Chulani, 1999).

We next describe the approach taken to determine the prior information and the sampling information, followed by the derivation of the Bayesian a posteriori model.

Prior Information

To determine the prior information for the coefficients (i.e., b^* and H^*) for COQUALMO, we conducted a Delphi exercise. Several experts from the field of software estimation were asked to independently provide their estimate of the numeric values associated with each COQUALMO driver. Roughly half of these participating experts had been lead estimation experts for large software development organizations and a few of them were originators of other proprietary software economic models. All of the participants had at least 10 years of industrial software estimation experience. Based on the credibility of the participants, we felt very comfortable using the results of the Delphi rounds as the prior information for the purposes of calibrating COQUALMO. Please refer to Boehm et al. (2000) for the Delphi results. Supporting experience is provided by Vicinanza et al. (1991) where a study showed that estimates made by experts were more accurate than model-determined estimates.

Sample Information

The sample information is derived by collecting data from the Center for Software Engineering at the University of Southern California's affiliates. These organizations represent the commercial, aerospace, and FFRDC (Federally Funded Research and Development Center) sectors of software development. The data collection activity is currently underway.

Since we do not have a reasonable sample size yet from our data collection activity, the Bayesian calibration of COQUALMO is not yet complete. We are currently using the a priori COQUALMO, and an example of its use is provided in Section 13.4.

Nevertheless, let us describe the approach of the Bayesian calibration that we will use once enough data have been collected.

Combining Prior and Sampling Information: Posterior Bayesian Update

The Bayesian paradigm will be used as a framework of formally combining prior expert judgment with the COQUALMO sample data.

Equation (13.7) reveals that if the precision of the a priori information (H^*) is larger (or the variance of the a priori information is larger) than the precision (or the variance) of the sampling information ($1/s^2X'X$), the posterior values will be closer to the a priori values. This situation can arise when the gathered data are noisy, as depicted in Figure 13.5. Figure 13.5 illustrates that the degree of belief in the prior information is higher than the degree of belief in the sample data. As a consequence, a stronger weight is assigned to the prior information, causing the posterior mean to be closer to the prior mean.

On the other hand (not illustrated), if the sampling information ($1/s^2X'X$) is more precise than the prior information (H^*), then the higher weight assigned to the sampling information causes the posterior mean to be closer to the mean of the sampling data. The resulting posterior precision will always be higher than the a priori precision or the sample data precision. Note that if the prior variance of any parameter is zero, then the parameter will be completely determined by the prior information. Although this is a restriction imposed by the Bayesian approach, it is of little concern as the situation of complete consensus very rarely arises in the software engineering domain.

13.3.4 Summary of COQUALMO

Sections 13.3.1 and 13.3.2 described the expert-determined defect introduction and defect removal submodels that compose the quality model extension to COCOMO II, namely, COQUALMO. As discussed in Section 13.3, COQUALMO is based on the tank-and-pipe model where defects are introduced through several defect source pipes described as the defect introduction model and removed through several defect elimination pipes modeled as the defect removal model. The expert-calibrated COQUALMO, when used on a project with nominal characteristics (or average ratings), predicts that approximately 14 defects per kSLOC are remaining. An independent study done on several projects verified the trends in defect rates modeled

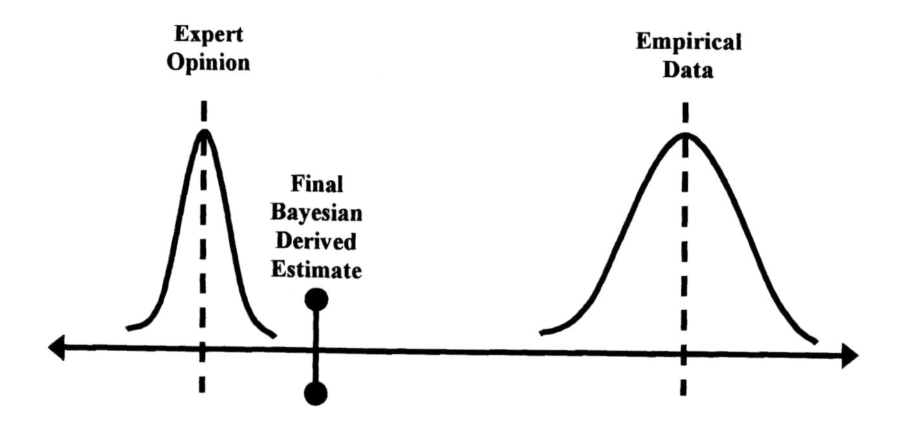

Figure 13.5. Noisy data—that is, imprecise sampling information.

by COQUALMO. When more data on actual completed projects are available, the model can be refined using the Bayesian approach used for COCOMO II calibration (Chulani, 1999). This statistical approach has been successfully used to calibrate COCOMO II.1999 to 161 projects. The Bayesian approach can be used on COQUALMO to merge expert-opinion and project data, based on the variance of the two sources of information to determine a more robust posterior model. In the meanwhile, the model described in this chapter can be used as is or can be locally calibrated to a particular organization to predict the cost, schedule, and residual defect density of the software under development (Boehm et al., 2000; Chulani, 1999). Extensive sensitivity analyses to understand the interactions between these parameters to do trade-offs, risk analysis, and return on investment can also be done.

13.4 CASE STUDY

In Section 13.3 of this chapter, we discussed the details of COQUALMO, describing the inputs and outputs of the model. In this section, we will use an example case study to illustrate how the model can be used to determine the quality of the software product in terms on defect density defined as defects/kSLOC or defects/FP.

Our case study is a large command and control software package. We considered only the application software quality although data were collected for the operating system and the system support software. The majority of the defects were analyzed and detected during the formal testing activities which were composed of validation, acceptance, and integration.

Let us assume that the model is locally calibrated to a sample of completed projects and yields the following parameters:

$$A_1 = 2.57; \quad B_1 = 1; \quad A_2 = 5; \quad B_2 = 1.2; \quad A_3 = 5.19; \quad B_3 = 1.1$$

Please note that once COQUALMO is calibrated as described in Section 13.3.3 in 2002, these values will be available.

Table 13.5 gives the detailed characteristics of our case study.

Based on the above-determined QAFs, we can now calculate the number of requirements, design, and code defects that are introduced. This is shown in Table 13.6. From Table 13.6 and the program size (110), the DIRs for requirements, design, and code defects can be computed as 1.9, 9.5, and 7.0 kSLOC, respectively.

Now, we move on to the defect removal model that uses the numbers from Table 13.6 as input. Table 13.7 has information on the three defect removal profiles. Using the model described in Section 13.3 and interpolating linearly between the rating scales for peer reviews and execution testing and tools, we can calculate the DRFs as shown in the right two columns of Table 13.7.

Table 13.8 illustrates how the defect removal model computes the estimated residual defect density using the defect removal profiles from Table 13.7. From

Table 13.5. Case Study Characteristics

Size		110 kSLOC		
Formal testing (in order of occurrence)		Validation, Acceptance, Integration		
	Rating	D-Driver$_{reqts}$	D-Driver$_{rdes}$	D-Driver$_{cod}$
RELY	H	0.84	0.83	0.83
DATA	N	1.00	1.00	1.00
RUSE	VH	1.02	1.02	1.00
DOCU	N	1.00	1.00	1.00
CPLX	N	1.00	1.00	1.00
TIME	H + 0.75	1.04	1.09	1.09
STOR	H	1.03	1.06	1.05
PVOL	N	1.00	1.00	1.00
ACAP	H	0.87	0.91	0.95
PCAP	H	1.00	0.92	0.87
PCON	H	0.91	0.89	0.88
AEXP	VH	0.81	0.82	0.88
PEXP	H	0.95	0.93	0.93
LTEX	N	1.00	1.00	1.00
TOOL	VL	1.09	1.10	1.25
SITE	H	0.94	0.94	0.95
SCED	VL	1.18	1.19	1.19
PREC	L	1.20	1.16	1.11
RESL	N	1.00	1.00	1.00
TEAM	H	0.91	0.93	0.95
PMAT	N	1.00	1.00	1.00
Quality adjustment factor (QAF) =		0.74	0.74	0.84
Actual number of requirements defects	225			
Actual number of design defects	1010			
Actual number of code defects	755			
Actual total residual defects	660			

Table 13.8, if we add all the rows in the last column we get the residual defect density for the case study project as 5.57 defects/kSLOC. Comparing this to the last row in Table 13.5, we can see that the model-determined residual defect density is very close to the actual residual defect density of 6.00 defects/kSLOC. Table 13.8 gives the estimated number of defects broken down by artifacts (i.e., number of residual requirements, design, and code defects). But the actual breakdown of re-

Table 13.6. Number of Defects Introduced for Case Study

DI_{req} (No. of requirements defects introduced)	$2.57 * (110)^{1} * 0.74 = 209$
DI_{des} (No. of design defects introduced)	$5 * (110)^{1.2} * 0.74 = 1042$
DI_{cod} (No. of code defects introduced)	$5.19 * (110)^{1.1} * 0.84 = 767$

Table 13.7. Case Study Defect Removal Profiles

Defect Removal Profile	Associated Rating	DRF	
Automated analysis	Very low; simple compiler syntax checking.	Requirements	0
		Design	0
		Code	0
Peer reviews	In between very low and low; Ad hoc informal reviews done.	Requirements	0.13
		Design	0.14
		Code	0.15
Execution testing and tools	In between high and very high; exhaustive testing done on all the artifacts but without the use of advanced test tools and metrics-based test process management.	Requirements	0.54
		Design	0.61
		Code	0.74

quirement, design and, code defects is not available to validate these numbers because the data we have are for aggregate defect counts.

The case study illustrates two important findings:

1. The trends in defect introduction rates determined from using COQUALMO on the case study are very close to the actual defect introduction rates.
2. The actual post-acceptance defect density on the case study is estimated within 7% of the actual residual defect density.

13.5 CONCLUSIONS

The theme of this chapter, software reliability, discusses, in particular, COQUALMO, a software quality estimation model which in conjunction with COCOMO II, a cost and schedule estimation model, can be used to make better management decisions relating to time to market, cost, quality, and so on.

COQUALMO is based on the tank-and-pipe model where defects are introduced through several defect source pipes described as the defect introduction model and

Table 13.8. Project A Residual Defect Density

Type of Artifact	Automated Analysis	Peer Reviews	Execution, Testing and Tools	Product (1-DRF)	DI/KSLOC	DRes/kSLOC
Requirements	0	0.13	0.54	0.40	1.9	0.80
Design	0	0.14	0.61	0.34	9.5	3.23
Code	0	0.15	0.74	0.22	7.0	1.54

removed through several defect elimination pipes modeled as the defect removal model. The output of COQUALMO is the predicted defect density of a software product during different activities of the software development life cycle. This defect density is a measure of how reliably the product will operate in the field. Obviously, if the estimated defect density is high, the assumption is that the software will not operate very reliably in the field. Conversely, if the defect density is low, we can expect the software to be quite reliable, assuming, of course, that the test sample has been representative of the system's operational profile. Management teams can use this relationship to make further trade-off analyses earlier on in the software life cycle. They may decide to add more resources to testing to improve the quality and reliability of the product, or in some other situation they may decide to ship the product earlier to be first to market.

We also discussed a case study in this chapter by taking the reader through the key steps in which COQUALMO can be used for defect density prediction. The reader should note that COQUALMO is not a fully calibrated model and is research in progress. The advantages of using COQUALMO include project planning and control, trade-off and risk analysis and budgeting.

We encourage the reader to check USC's Center for Software Engineering's website at http://sunset.usc.edu for further updates to the model.

REFERENCES

Banker, R. D., Chang, H., and Kemerer, C. F. (1994). Evidence on economies of scale in software development, *Information and Software Technology* **36:**275–282.

Boehm, B. W. (1981). *Software Engineering Economics,* Prentice-Hall, Englewood Cliffs, NJ.

Boehm, B. W., Abts, C., Brown, A. W., Chulani, S., Clark, B., Horowitz, E., Madachy, R., Reifer, D., and Steece, B. (2000). *Software Cost Estimation with COCOMO II,* Prentice-Hall, Upper Saddle River, NJ.

Box, G., and Tiao, G. (1973). *Bayesian Inference in Statistical Analysis,* Addison-Wesley, Reading, MA.

Chillarege, R., Bhandari, I. S., Chaar, J. K., Halliday, M. J., Moebus, D. S., Ray, B. K., and Wong, M. Y. (1992). Orthogonal defect classification—A concept for in-process measurements, *IEEE Transactions on Software Engineering* **18:**943–956.

Chulani, S. (1999). Bayesian Analysis of Software Cost and Quality Models, Ph.D. dissertation, University of Southern California, Los Angeles.

Gulledge, T. R., and Hutzler W. P. (1993). *Analytical Methods in Software Engineering Economics,* Springer-Verlag, Berlin.

IEEE (1989). IEEE Standard Glossary of Software Engineering Terminology, *Software Engineering Standards,* 3rd edition, IEEE, New York.

IFPUG (1994). International Function Point Users Group (IFPUG), *Function Point Counting Practices Manual,* Release 4.0.

Jones, C. J. (1975). "Programming Defect Removal," *Proceedings,* GUIDE 40, 1975.

Leamer, E. (1978). *Specification Searches, Ad Hoc Inference with Nonexperimental Data,* Wiley, New York.

Park (1992). "Software Size Measurement: A Framework for Counting Source Statements," *CMU-SEI-92-TR-20,* Software Engineering Institute, Pittsburgh, PA.

Rook, P., Editor (1990). *Software Reliability Handbook,* Elsevier Applied Science, Amsterdam.

Vicinanza S., Mukhopadhyay T., and Priebula, M. (1991). *Software Effort Estimation: An Exploratory Manual,* Release 4.0, 1994.

EXERCISES

13.1. In the case study presented in Section 13.4, if instead of simple compiler syntax checking, further investments were made to achieve Intermediate-level module and intermodule code syntax and semantic analysis and simple requirements/design view consistency checking, what would the impact be on the residual defect "checking" and "density?"

13.2. How would the quality of the project described in Section 13.4 be impacted if the SCED rating was changed from VL to VH (i.e., the schedule expansion factor was 160% of nominal)? Would the impact be different for requirements, design, and code defects? If yes, describe why?

13.3. In the case study described in Section 13.4, if you had a choice of improving just one DI driver rating to reduce the number of code defects introduced, which driver would you target?

CHAPTER 14

Use of Extreme Values in Reliability Assessment of Composite Materials

Linda C. Wolstenholme

14.1 INTRODUCTION

The subject of this investigation is a batch of components made from a composite material and in use in the transport industry. The material is a laminated, unidirectional, fiber-reinforced material, and the components are expected to have a service life of some 20 years.

There are approximately 1000 components in the batch, made during the 1990s, and for the purpose of this study they can be regarded as the population under examination. The construction and manufacturing process of more recent similar components is different and allows more ready detection of the kind of problem which occurred in the earlier batch.

These components are subject to manufacturing variation, which results in the occurrence of a feature that will be called a "defect." This can be thought of as a blemish in the material. The sizes of these defects are fixed at manufacture; they do not increase over time, but a very large defect may initiate a sequence of events leading to failure of the component. It was during some routine maintenance work, though not as a consequence of routine inspection, that one of the batch was found to have a defect large enough to raise concern that there may be similarly large defects elsewhere in the population.

The particular management problems posed here result from

(a) Detection of such defects is a costly process, involving sophisticated equipment and the dismantling of the units in which the components are installed

(b) A prohibitively high cost associated with complete replacement of the batch

Both of the above appeared to be drastic actions, given that the one component removed following discovery of the large defect may be the only such defect in the

Case Studies in Reliability and Maintenance, Edited by W. R. Blischke and D. N. P. Murthy.
ISBN 0-471-41373-9

batch; furthermore, it was a supposition that a large defect could lead to failure. However, the consequences of failure would be severe, and therefore a program of testing and analysis was required in order to establish the integrity of the batch. In the meantime, extended inspection was introduced during routine maintenance.

Here, several questions are posed and various approaches to answering these questions are considered. The principal aim is to estimate the probability that a component in service has a severe defect. The second issue for the manufacturer was to establish a test plan for a more reliable assessment of a possible reduced lifetime resulting from a large defect.

14.2 TEST DATA AND BACKGROUND KNOWLEDGE

A random sample of 93 components, almost 10% of the population, was taken out of service and subjected to ultrasonic examination. For each component, all defects of a detectable size were recorded. Eighty-seven components had one or more recordable defects, six components had none. For each component it was the largest defect that was of interest, since this would be the one to lead to failure, were such an event to occur. The component initiating this investigation will be referred to as component ZZ. It was found that none of the test sample had a defect as large as the one found in component ZZ.

A defect has a prescribed "size" criterion. A number of attributes were recorded, and from these a single meaningful measurement was constructed which describes the severity of the defect. A thorough statistical investigation of the attributes was carried out in order to be satisfied that the measure constructed suitably represented defect severity. The raw data for components with recordable defects are shown in Table 14.1. The definition of a "recordable" defect is not precise, but can be taken as a size of at least 90. For each component examined, the number of defects found is given along with the size of the most severe defect. The data are ordered according to defect size, purely for convenience. The largest defect in the sample has size 288; the largest defect found in component ZZ was 560.

Also noted is a classification of the location in which the severest defect is found. Four types of location, denoted a, b, c, and d, could be identified. These components have a complex structure. It is possible that defects are more likely in certain locations and for the probability associated with a severe defect leading to component failure to vary according to location. However, these issues will be deferred for later consideration. Figure 14.1 shows the distribution of the size of largest defect across the 87 components in Table 14.1.

14.3 MODEL FITTING AND PREDICTION

14.3.1 General Considerations

14.3.1.1 The Test Sample

It is assumed that the data refer to a randomly selected sample coming from the same population. In other words, all components have been subject to the same

Table 14.1. The Component Defect Data

Observation Number	Number of Defects	Maximum Size	Location
1	3	127	b
2	2	143	a
3	2	147	c
4	2	149	b
5	3	149	a
6	2	149	c
7	1	149	b
8	3	151	a
9	2	153	a
10	3	153	b
11	3	157	b
12	4	159	c
13	3	159	a
14	4	159	a
15	2	165	a
16	2	165	a
17	1	167	b
18	1	167	b
19	2	167	b
20	4	167	b
21	2	168	b
22	4	168	a
23	2	168	a
24	2	168	a
25	4	169	a
26	1	179	a
27	2	179	a
28	2	179	a
29	4	179	d
30	1	179	a
31	5	179	b
32	1	179	a
33	2	179	a
34	4	179	a
35	2	179	b
36	3	179	a
37	2	179	b
38	4	179	a
39	4	179	c
40	2	179	b
41	3	179	a
42	2	179	b
43	3	189	a
44	5	190	d
45	4	190	b
46	2	190	c

(continued)

Table 14.1. The Component Defect Data *Continued*

Observation Number	Number of Defects	Maximum Size	Location
47	5	190	b
48	2	192	b
49	4	192	a
50	2	192	d
51	5	192	b
52	3	192	c
53	3	192	b
54	1	194	a
55	5	194	a
56	5	198	c
57	4	200	a
58	4	200	a
59	4	200	d
60	4	204	a
61	1	204	a
62	2	204	a
63	2	204	a
64	3	204	a
65	4	208	c
66	2	208	c
67	4	208	b
68	2	208	b
69	7	208	d
70	4	208	a
71	4	208	a
72	4	208	b
73	2	208	a
74	3	209	d
75	6	213	b
76	5	213	b
77	4	213	b
78	4	213	b
79	3	219	a
80	4	222	d
81	5	226	b
82	6	226	b
83	4	226	b
84	3	230	c
85	3	248	b
86	5	277	b
87	6	288	d

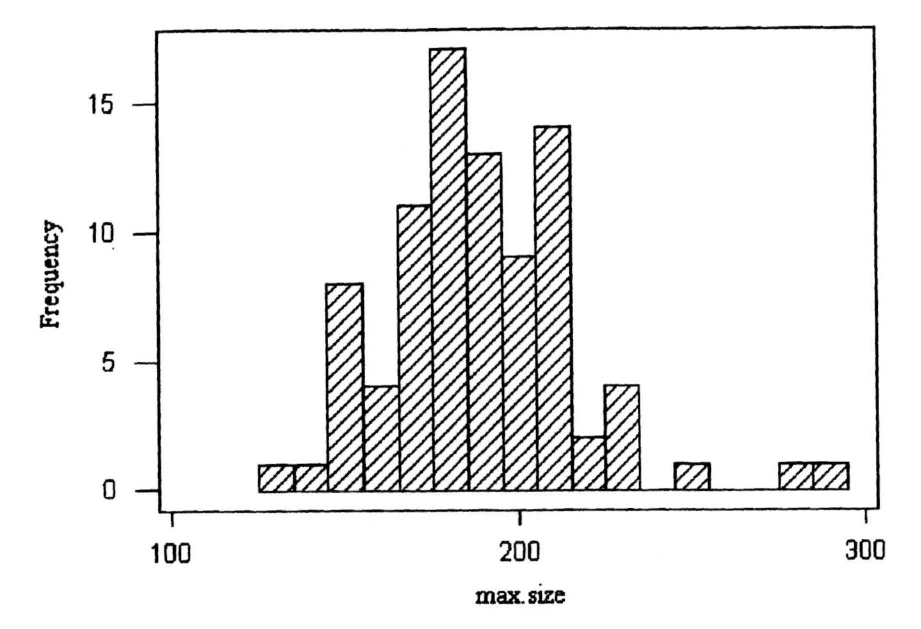

Figure 14.1. Distribution of maximum defect size.

manufacturing and in-service conditions, and all members of the population were equally likely to be selected. At this stage, it will also be assumed that the location of the severest defect is not a significant explanatory variable.

Component ZZ will be excluded from the analysis because this item of data is self-selected in the sense that it exists entirely as a result of being a cause for concern. The first objective of the analysis is to see whether the test sample would predict this event.

14.3.1.2 The Principal Question

Let there be a defect size function, X. Let m be the maximum size regarded as "not serious." Of interest is θ, the probability that a randomly selected component has a worst defect large enough to raise concerns about the integrity of the component. We therefore require

$$\theta = P(X > m)$$

14.3.1.3 Parametric versus Nonparametric Methods

The ideal is a parametric model for the random variable X that can then predict the probability of an extreme value. The advantage here is that the model can make a prediction outside the range of values observed. This does, however, depend highly on the integrity of the model. Several models may fit the bulk of the data well but vary considerably in the tails of the distributions; see, for example, pp. 56–58 of

Crowder et al. (1991) and pp. 63–65 of Wolstenholme (1999). There would also be corresponding variability in the confidence bounds put on the predicted probabilities.

If a suitable parametric model cannot be found, then there are nonparametric (distribution-free) alternatives. However, a nonparametric estimate of the distribution of X will not give a prediction outside the range of observed values. So the test data could not predict the defect in component ZZ.

A simple approach is to consider the components as severe/not severe, that is, as binary observations. This is nonparametric in principle, though parametric methods might be used in the classification of a component as severe, or potentially so.

14.3.1.4 "Missing" Observations

Throughout, allowance needs to be made for the components with no recordable defects. It may be that the worst defect is too small to observe or it might be that such components are genuinely defect-free. Given a parametric model for X which is satisfactory, it is possible to test which of these situations is the most likely by using the model to predict the occurrence of worst defects between the recordable boundary of 90 and the smallest worst defect, which in this case is 127. It is not hard to come to the conclusion that the six components with no recorded defects may well be defect-free.

We can either treat these six observations as left-censored at 90 or can adopt the formulation

$$\begin{aligned}\theta &= P(\text{observable defect present}) \times P(X > m, \text{ given observable defect present}) \\ &= \lambda\phi_m \end{aligned} \tag{14.1}$$

An estimate of λ is the proportion of the sample with observable defects and ϕ_m is estimated from a model fitted to the defect size function for the 87 components with observable defects.

14.3.1.5 Finite Population Sampling

It should also be noted that the population under consideration is finite (approximately 1000). A sample of 93 has been selected from this population and strictly speaking the "population" changes slightly at each selection. The general effect of treating the problem as though dealing with an infinite population is to make the results slightly more pessimistic than they might be. Corrections to the probability estimates to allow for the finite population will be considered.

14.3.2 Suitable Parametric Models

The choice of model for the size of the worst defect is governed by the degree of fit in statistical terms but also by plausible suitability. The "standard" distributions to consider here might be Weibull or lognormal. However, the variable of interest here is the size of the largest defect in a component, that is the maximum of several values, and this would lead naturally to consideration of extreme value distributions. This is particularly important when estimating probabilities at the upper end of the distribution because the more standard distributions may well underestimate these.

It was found that of the plausible models, the lognormal and the generalized extreme value (GEV) distributions provided the best fits to the data, though neither with a high degree of confidence.

14.3.2.1 The Lognormal Distribution

The probability density function is defined by

$$f(x) = \frac{1}{x\sigma\sqrt{2\pi}} \exp\left[-\frac{1}{2}\left(\frac{\log x - \mu}{\sigma}\right)^2\right] \tag{14.2}$$

and, following from equation (14.1), $\phi_m = \int_m^\infty f(x)dx$. When X has a lognormal distribution, $\log X$ has a normal distribution with mean μ and variance σ^2.

Figure 14.2 shows a lognormal plot of the data, excluding the six "missing" observations. In other words, we are considering here the distribution of worst defect, conditional on a defect being observable. There are fewer points on the plot than observations due to the incidence of tied values.

In general, the best method of parameter estimation is maximum likelihood (ML). Given n observations $x_1, x_2, x_3, \ldots, x_n$, assumed to come from a distribution defined by $f(x)$, the likelihood is defined by $L = \Pi_{i=1}^n f(x_i)$. In practice, the log-

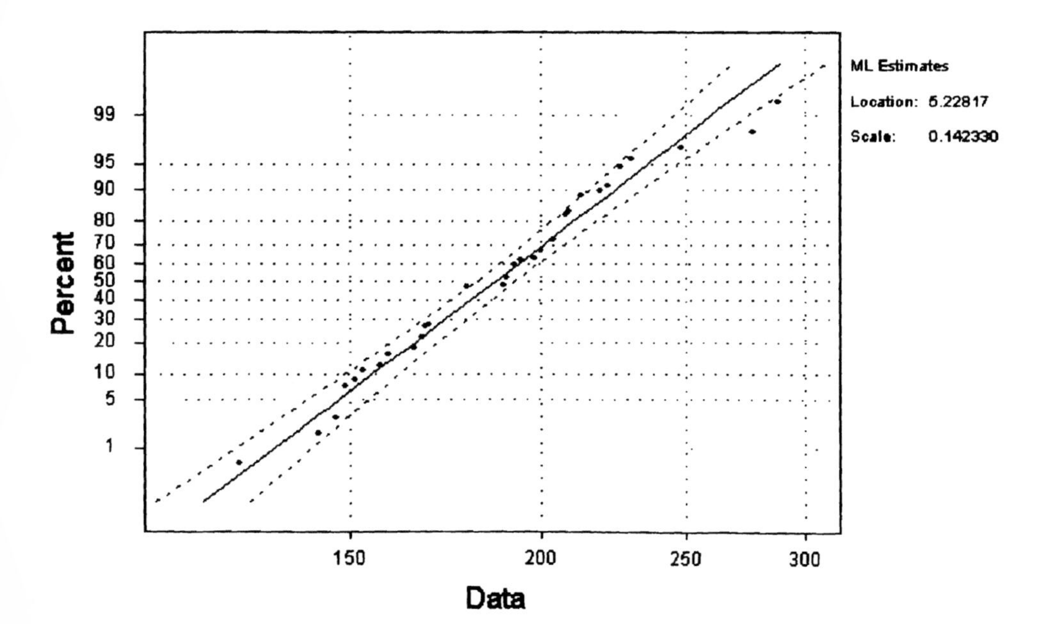

Figure 14.2. Lognormal probability plot for maximum defect size.

likelihood is easier to work with and additionally has useful properties. The log-likelihood is defined by

$$\log L = \sum_{i=1}^{n} \log f(x_i) \tag{14.3}$$

The likelihood function has a direct relationship with the probability of obtaining the given sample, and ML estimation yields parameter values that maximize this probability. These parameter estimates yield maximized values for both the likelihood and log-likelihood. A full discussion of likelihood can be found in Edwards (1992).

The ML estimates of the unknown parameters should be the values used to estimate points in the upper tail of the distribution. Fitting the lognormal distribution to the 87 worst defect observations yields $\hat{\mu} = 5.228$ and $\hat{\sigma} = 0.1423$. The dotted lines in Figure 14.2 give a 95% confidence region about the lognormal distribution line based on the ML estimates. The scatter of the points demonstrates that the fit is just acceptable. It is noted, however, that the two largest observations are higher than expected within 95% confidence.

The lognormal plot and the parameter estimation are demonstrated using standard routines in the statistical software package MINITAB. A discussion of the principles of probability plotting can be found in Wolstenholme (1999).

14.3.2.2 The Generalized Extreme Value Distribution (GEV)

In contrast to the lognormal distribution, which has two parameters, the GEV is a three-parameter distribution and therefore has increased flexibility. For the GEV distribution

$$\phi_m = 1 - \exp\left[-\left[1 + \xi\left(\frac{m - \mu}{\sigma}\right)\right]^{-1/\xi}\right] \tag{14.4}$$

where μ, σ, ξ are, respectively, "location," "scale," and "shape" parameters. It should be noted that the shape parameter determines the heaviness of the tails of the distribution; and if the shape parameter is negative, there is an upper endpoint to the density (i.e., an upper bound on the value of X). Underlying the GEV is the *extremal types theorem* [see, for example, Leadbetter et al. (1983)]. This theorem states that under suitable conditions the distribution of the maximum of a large number of independent and identically distributed random variables will be asymptotically GEV. We have here data that are certainly maxima, but it is not certain that (i) the number of defects (observable and unobservable) per component is large, (ii) the defects within a component are independent, and (iii) the defect sizes are identically distributed. There is little possibility of verifying these assumptions, but on the basis that they may be fairly close to reality, the GEV is worthy of our attention.

Using software written by Dixon (1999), the maximum likelihood estimates of the GEV parameters for the 87 observed maxima are $\hat{\mu} = 176.86$, $\hat{\sigma} = 24.32$, $\hat{\xi} = -0.1051$. Figure 14.3 demonstrates that the fit of the GEV, on a purely visual basis,

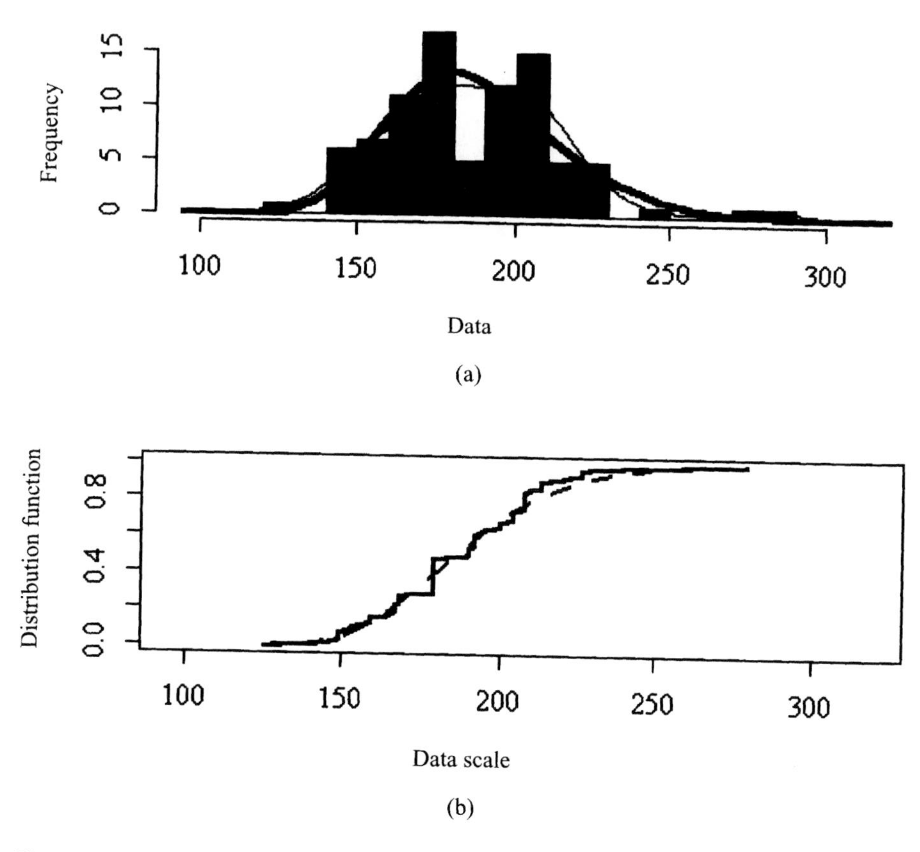

Figure 14.3. Empirical and fitted GEV distribution. (a) Histogram of severest defect data with empirical and fitted GEV probability density functions. (b) Empirical distribution function with fitted GEV distribution function.

is acceptable. Linear probability plots, referred to in Section 14.3.2.1, are not readily applicable where more than two parameters require estimation. There is also no readily available formal goodness-of-fit test for the GEV. However, it is possible to make some model comparisons.

14.3.2.3 Upper-Tail Predictions

The values of the maximized log-likelihoods may be used as a guide in comparing the fit of the models. In principle, the model yielding the largest log-likelihood would be preferred. However, since the models considered are not "nested," the comparison cannot be formalized as, say, a likelihood ratio test [see Wolstenholme (1999)]. The log-likelihood values for the log normal and for the GEV are –408.84 and –408.69, respectively, and we can say that such values do not differentiate distinctly between the models. The lognormal has the slightly higher likelihood, but the GEV gives slightly larger probabilities to high values of X and therefore provides the more cautious basis on which to estimate the probability of occurrence of severe defects.

Although both models appear to offer reasonable fit to the data, we find that neither distribution predicts with any significantly nonzero probability the occurrence of a defect as large as the one found in component ZZ. Indeed it is predicted that there is only a 1 in a 1000 chance of a largest defect exceeding 296 under the GEV model or 289 under the lognormal model. These points are very close to 288, the largest maximum in the test sample, but a very long way from the defect size of 560 found in component ZZ.

We further note now that the ML estimate of ξ is negative, and this implies that the GEV here has a finite upper bound for X; that is, the model says that higher values than this do not occur. This upper bound is estimated to be 382. Using standard errors estimated from the information matrix and a normality assumption for predictors, a 99% upper confidence limit on the X bound is 558. It can be seen that the interval estimates of the upper bound are very wide and that the result from component ZZ is at the very limit of what could be considered a result from this GEV distribution.

Given that modeling the distribution of the size of observable worst defect appears less than satisfactory, we could now adjust the fitting to include the six further "observations," left-censored at, say, 90, since there is evidence that this is the detection threshold. The log-likelihood in equation (14.3) is adjusted so that each of these left-censored observations contributes $F(x_i)$, rather than $f(x_i)$, where $F(x_i)$ is the distribution function $\int_0^{x_i} f(x)\,dx$. What happens now is that the extra weight attached to the left-hand tail of the fitted model results in probabilities in the upper tail getting smaller, and the GEV upper bound for X is lower than before. In other words, the defect in component ZZ becomes even more unusual.

We can therefore conclude that if we assume under each of the parametric analyses that component ZZ is from the same population as the sampled components, then the probability of seeing another defect of such severity is certainly less than 10^{-9}.

14.3.3 The Binary Data Approach

Alternatively, it could be concluded that the standard parametric models considered are not a sound basis for the estimation of the probability that a component randomly selected from the population has a defect as severe as that of component ZZ. Suppose there is a strong suggestion that component ZZ is different from the general population of components and that there may be a small subpopulation of components, characterized by an unusual level of manufacturing defects.

Let θ be the probability that a component has the characteristic "severe defect"; that is, the component belongs to a different population from that of components in general. We assume components to be independent and that θ is constant. For a sample of size n taken from an infinite population, the probability that y of the components are severely flawed is given by

$$P(Y = y) = \frac{n!}{(n-y)!y!} \theta^y (1-\theta)^{n-y} \tag{14.5}$$

that is, the random variable Y has a binomial distribution.

The sample proportion of severely flawed components, y/n, provides a point estimate of θ. It is, though, more important to have some bound on θ with a given level of confidence. For example, an upper value that we believe with 99% or 95% probability is not exceeded. Such a bound is provided by the solution to the equation

$$\sum_{r=0}^{y} \frac{n!}{(n-r)!r!} \theta^r (1-\theta)^{n-r} = \alpha \tag{14.6}$$

where α is usually 0.05 or 0.01, corresponding respectively to 95% and 99% confidence.

In our case study, the test samples of 93 components have been examined and, while flawed, all are considered sound from a strength point of view, so the observed y is zero. The point estimate of θ is zero, but this is uninformative. From equation (14.6), a 99% upper bound on the proportion of components in the population having severe defects is given by solving the equation

$$(1-\theta)^{93} = 0.01$$

This yields a value for θ which we will call θ_u, of 0.0483, that is, some 48 components in the batch could be of serious concern.

14.3.4 Finite Population Sampling

In any situation where the sampling from a population is without replacement, the size of the remaining population after each selection reduces by one. When the population is very large, the effect on the sampling probabilities is negligibly small and hence the assumption of constant θ in equation (14.5). Suppose, however, that the population size, N, is modest, and that there are K defective items in that population. In a sampling procedure where we take a random sample of size n from the population, without replacement, the number of defective items found follows a *hypergeometric distribution* with probabilities given by

$$P(Y = y) = \frac{\binom{K}{y}\binom{N-K}{n-y}}{\binom{N}{n}} \tag{14.7}$$

where $\binom{a}{b} = \dfrac{a!}{(a-b)!b!}$.

Table 14.2. Estimates of Proportion of Population with Severe Defect

Confidence Level	Estimated θ	Binomial θ_u	θ_u using f.p.c.f.
99%	0.0	0.0483	0.0460
95%	0.0	0.0317	0.0302

In order to make comparison with the binomial distribution, we may write $N\theta$ instead of K where θ is the proportion of defectives in the population. It can be shown that the mean of the hypergeometric is the same as for the binomial case—that is, $n\theta$—but the variance is smaller. As expected, the hypergeometric distribution tends to the binomial as N tends to infinity. A theoretical treatment of the hypergeometric distribution may be found in Stuart and Ord (1994).

We find that the variance of the hypergeometric is the variance of the binomial distribution multiplied by the factor $(N - n)/(N - 1)$. The square root of this fraction is known as the *finite population correction factor* (f.p.c.f.), and a very satisfactory adjustment to confidence bounds based on the binomial distribution to take account of finite population sampling can be obtained by multiplying confidence intervals by the square root of the f.p.c.f. This is a more tractable alternative to trying to work directly with the hypergeometric distribution. In our example, the square root of the f.p.c.f. is $[(1000 - 93)/(1000 - 1)]^{0.5} = 0.953$. Table 14.2 shows the effect on the estimated upper bounds of θ, when applying equation (14.6). The effect of using the f.p.c.f. can be seen to be small. The general wisdom is that its use becomes necessary once the sample size exceeds 10% of the population. This level is almost reached here.

14.4 STRENGTH TESTING

14.4.1 The General Approach

The analysis of the defect data does not clearly reject the possibility of components elsewhere in the population with severe defects. While engineering knowledge indicated that it was possible for a large defect to lead to failure of the component, no quantifiable evidence existed. Thus a program of testing was required in order to put boundaries on the areas of risk and hence determine a policy for component replacement and/or future monitoring strategies. Life testing would be destructive; and, given the cost of these components, minimizing the number of tests was desirable.

Tests could be constructed in the light of two factors. First, a life profile was known for components that did not have severe defects. Second, it was possible to manufacture components with severe defects deliberately introduced, in fact, damage of a nature more severe than had been seen was effected.

We will define a random variable T that represents the loading capacity of a satisfactory component. Experience showed that T was approximately normally distributed, and estimates of the mean and variance of T were given. It was considered of concern if the strength of a component fell below the lower 0.1% point of the estimated distribution of T. The latter distribution will be referred to as distribution A, and the life profile of severely flawed components will be called distribution B, which we suspect may cover generally lower strengths than distribution A.

There are a variety of outcomes to the testing of a severely flawed component, but the prime interest is the probability that a strength result looks acceptable—that is, as though from distribution A, but in fact is, by chance, an optimistic reading from distribution B. The general picture is shown in Figure 14.4, where the boundary between acceptable and unacceptable is the lower 0.1% point of distribution A and is denoted τ. The assumption is made that distribution B is also normal and has the same variance as distribution A. These assumptions were considered reasonable, purely from an engineering point of view.

Let j be the number of tests to be conducted. Determination of a suitable value for j is required in accordance with reducing the risk of failure to a very low level.

14.4.2 Statistical Inference

Consider the scenario where j results all fall in the region $T > \tau$. If distribution B is the correct strength distribution for damaged components, the probability that an individual result falls in this region is given by the area under curve B to the right of $T = \tau$. This probability will be called β. The probability that all j test results fall in this area is β^j, given that tests are statistically independent. Clearly the further the line $T = \tau$ moves to the right the smaller is β^j and the more inclined we shall be to say that the strength of these components is acceptable. However, the more $T = \tau$ is moved

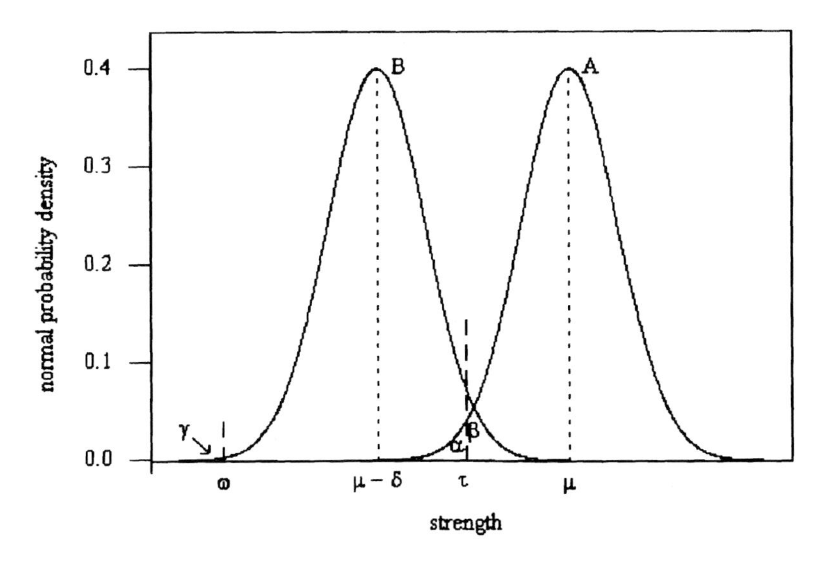

Figure 14.4. Strength distributions based on means and low probabilities of failure.

to the right, the higher the probability that results which come from distribution A will fall to the left of $T = \tau$ and cause us to reject distribution A. If the probability that a strength from distribution A falls to the left of $T = \tau$ is α, then the probability that j strains fall in this region is α^j.

The desired value of j will be determined by the size of the probabilities associated with the above outcomes. These in turn depend on α and β, which in turn are a function of the choice of τ, the shape of the strength distributions, and, critically, how far apart are these distributions. The closer B is to A, the more difficult it becomes to classify the results and a higher value of j is needed in order to achieve an acceptable level of confidence in the conclusion.

The "easy" scenario is when results are clearly unacceptable. For example, taking τ to be the 1 in a 1000 boundary—that is, $\alpha = 0.001$—three outcomes in the region $T < \tau$ would occur with probability $0.001^3 = 10^{-9}$ if the strengths belong to distribution A. It would be highly unlikely therefore that distribution A applied to these components.

More likely is at least some of the results falling in the acceptable region for distribution A but have come from distribution B. Suppose, under the hypothesis that distribution B applies for damaged components, 1 in 10 strengths would be greater than τ, so $\beta = 0.1$. If, say, nine damaged components are tested and all have strengths greater than τ, then the probability that the strengths belong to distribution B is 10^{-9}. We would be very confident that the strength of a damaged component was rather better than that represented by distribution B.

14.4.3 Parameter Interaction

It is clear that the assumptions we make about distribution B are vital to the analysis. One approach, in the absence of any other indicators, is to fix the left-hand tail of B using some kind of worst case. This might sensibly be the minimum "life" considered to be safe. We can then place distribution B so that a value below this point has very small probability. Figure 14.4 shows this point marked as ω and $\gamma = P(T < \omega)$.

We will now see how the various parameters of the problem interact with each other. Let the mean of distribution A be μ and let the mean of distribution B be $\mu - \delta$. We require

$$\beta = P(T > \tau | N(\mu - \delta, \sigma^2))$$

Since $\alpha = 0.001$, $\mu - \tau = 3.09\sigma$ and

$$P(T > \tau) = P\left(Z > \frac{\tau - (\mu - \delta)}{\sigma} = \frac{\tau - \mu + \delta}{\sigma} = \frac{\delta - 3.09\sigma}{\sigma}\right)$$

where Z is the standardized normal variable having distribution $N(0,1)$. Hence

$$\beta = P\left(Z > \frac{\delta}{\sigma} - 3.09\right) \tag{14.8}$$

Table 14.3. Values of β Corresponding to δ Expressed in Terms of σ

δ	β
4σ	0.1804
5σ	0.0281
6σ	0.0018

Note that the value of β is independent of the value of the mean, μ. We now need to express δ in terms of σ and this can be done once we have an expression for γ in terms of σ. Table 14.3 illustrates the sensitivity of β to the value chosen for δ.

We now consider the implications of the value chosen for γ. Suppose we wish to place distribution B where we can be very confident that the strength of a damaged component will not lead to failure. A corresponding value for γ might be 10^{-9}. This implies, under the normal distribution assumption, that ω is approximately 6σ away from the distribution B mean, $\mu - \sigma$. The corresponding distances for $\gamma = 10^{-8}$, 10^{-7}, and 10^{-6} are, respectively, 5.61σ, 5.20σ, and 4.75σ. These values can be found in tables such as Neave (1985).

Information from the estimated strength profiles is required in order to quantify δ, the distance between distributions A and B. In this study, the distance between the mean strength of satisfactory components and the worst-case line $T = \omega$ was expressed in terms of the estimated σ and hence δ was determined. It was found that $\mu - \omega = 10.4\sigma$ and since, for $\gamma = 10^{-9}$ we obtain $\mu - \delta - \omega = 6\sigma$, we can deduce that $\delta = 4.4\sigma$. Equation (14.8) then yields $\beta = P(Z > 1.31) = 0.0951$.

The probability that j components with life from distribution B all have a strength greater than τ is β^j. If we require β^j to be, say, 10^{-9}, then the value of j is 8.8; that is, we require nine tests and all to give a result above τ. We would then be able to say that the life of damaged components is at least as good as distribution B. Table 14.4 shows how the value of j varies with the criterion for β^j.

Clearly, a group of test results that yield values both above and below τ pose more of a problem. The probability arguments are easy to construct, but the increased uncertainty will increase the total number of tests required in order to achieve a sufficiently clear conclusion. In this case study, a suitable number of damaged components were tested and all gave a result comfortably inside distribu-

Table 14.4. Estimated Number of Tests, j (all with $T > \tau$), for $\gamma = 10^{-9}$ and $\delta = 4.4\sigma$

β^j	j
10^{-9}	8.8
10^{-8}	7.8
10^{-7}	6.9
10^{-6}	5.9

tion A, in fact near to the mean strength for satisfactory components. This test sample was also deliberately designed to have a majority of defects introduced in the component locations thought to have the highest risk of failure. In spite of this reassuring result, the manufacturer continued with an upgraded inspection program for the life of this particular batch of components.

14.5 DISCUSSION

There were two prime features of this study: how to describe the chance mechanism leading to an extreme size of defect, and how to go about assessing the integrity of components with extreme defects. The study was also highly influenced by the safety critical context of the problem and the very high cost of both testing and replacement of components.

The first fundamental question was how many components in the (fairly small) population had an extreme defect. Models could be found for the general defect population; but at the upper tail, where the principal interest lay, it was not clear that a satisfactory model could be found and hence it was not possible to place a high degree of confidence in any resulting inference. Component ZZ was so unusual that it could well be the case that it was the only such severely flawed component in the batch.

The cautious approach was to assume that component ZZ was part of a subpopulation, perhaps resulting from some period of irregularity in the manufacturing conditions. The inference then is that up to 4.6% of the population, approximately 46 components, could be similarly defective.

The response to this assessment was (a) construct a testing program to assess the performance of components with severe flaws and (b) design a monitoring scheme to ensure that where the integrity of a component is in doubt, it is removed from service.

It has been shown that, under certain assumptions about lifetime profiles, a small sample of tests on severely flawed components yields a high degree of confidence in the integrity of such components. This did, however, rely on several assumptions that might naturally be questioned. It is, for example, a matter of engineering judgment as to whether an unacceptable life profile has similar variance to an acceptable life profile, or indeed whether an unacceptable life profile would be normally distributed. Had the test results in this case been less convincing, it would have been sensible to explore how sensitive the conclusions were to the distribution assumptions.

The effect of dealing with a finite population has been explored in the binary approach to estimating the proportion of the batch having severe flaws. It can be inferred that a similar reduction in the width of confidence interval estimates may be appropriate when using a parametric lifetime model. There is no established method for doing this, and in most cases it is likely to be a minor adjustment in the light of the fact that parametric confidence intervals are in any case approximations generally.

We add a final footnote by returning to equation (14.1). If we choose to make no assumptions about the cases where no observable defects are present, equation (14.1) would provide an estimate of θ, provided that we had a sound parametric estimate of ϕ_m. Our point estimate of λ is $\hat{\lambda} = 87/93$, and via the binomial distribution the estimated variance of $\hat{\lambda}$ is $\hat{\lambda}(1 - \hat{\lambda})/93$ and hence the standard error is 0.0255. Approximations for the variance of the estimated ϕ_m and for the variance of the estimated θ can be obtained via the "delta method" [see Crowder et al. (1991) and Wolstenholme (1999)]. Hence, assuming estimators to be approximately normally distributed, confidence bounds may be placed on θ.

REFERENCES

Crowder, M. J., Kimber, A. C., Smith, R. L., and Sweeting, T. J. (1991). *Statistical Analysis of Reliability Data,* Chapman & Hall, London.

Dixon, M. (1999). *Rext Extreme Values Software,* Version 3.1, Short Course for Actuaries, City University, London.

Edwards, A. W. F. (1992). *Likelihood* (expanded edition), The John Hopkins University Press, Baltimore.

Leadbetter, M. R., Lindgren, G., and Rootzen, H. (1983). *Extremes and Related Properties of Random Sequences and Processes,* Springer-Verlag, New York.

Neave, H. R. (1985). *Elementary Statistics Tables,* George Allen & Unwin, London

Stuart, A., and Ord, J. K. (1994). *Kendall's Advanced Theory of Statistics,* Volume 1, 6th edition, Edward Arnold, London.

Wolstenholme, L. C. (1999). *Reliability Modelling, a Statistical Approach,* Chapman & Hall/CRC Press, London.

EXERCISES

14.1. The number of defects per component may be of interest. If defects occurred at random, it might be suggested that a Poisson model fitted the number per component. If the total number (observable and unobservable) were Poisson distributed, it would be the case that an observable number was also Poisson distributed. How well do the data fit this model?

14.2. By graphical means or by calculation of a correlation coefficient, assess the degree to which the number of defects per component is correlated with the maximum size of defect.

14.3. There are four distinct areas in each component where defects can occur. It may be that there are differences between the groups of maxima when segregated by location. Plot separately each group of maxima and determine whether lognormal distributions (or indeed some other model) adequately describe each group.

14.4. Under the assumption of a lognormal fit for each location group, an analysis of variance (ANOVA) could be conducted using logs of the observations. A necessary further assumption would be that the four groups of logged observations have similar variances. Conduct an F-test to see whether this latter assumption is reasonable; if it is, use one-way ANOVA to test for differences between location groups.

14.5. The Weibull distribution is an "extreme value" distribution, though most commonly applied to the smallest of a set of random variables. How might a problem involving maxima be transformed to one involving minima?

14.6. Suppose in our sample of 93 observations, one observation had occurred which was considered "severe." Use equation (14.6) with $y = 1$ to find a 95% upper bound for θ (this requires numerical iteration). Adjust the bound using the finite population correction factor.

14.7. Suppose in Section 14.4.3 that ω is given to be the point where $\gamma = 10^{-6}$ and $\mu - \omega = 8.93\sigma$. Calculate the corresponding value of δ and hence the values of j corresponding to Table 14.4.

14.8. Investigate the effect on Table 14.4 if the variance of distribution B is twice that of distribution A.

CHAPTER 15

Expert Judgment in the Uncertainty Analysis of Dike Ring Failure Frequency

Roger M. Cooke and Karen A. Slijkhuis

15.1 INTRODUCTION

The Netherlands are situated on the delta of three of Europe's biggest rivers, the Rhine, the Meuse, and the Scheldt. Large parts of the country lie lower than the water levels that may occur on the North Sea, the large rivers, and the IJsselmeer. Consequently, most of the country is protected by flood defenses.

Prior to 1953 the standard approach to the design of flood defenses was based on the highest recorded water level. In relation to this water level a safety margin of 1 meter was maintained. In 1953 a major breach occurred, killing more than 1800 people in the southwest of the country. Afterwards, an econometric analysis was undertaken by the Delta Committee in which the safety standards were based on weighing the costs of the construction of flood defenses and the possible damage caused by floods. This analysis led to a design probability of flooding of 8×10^{-6} per year. Given the technical capabilities at that time, this safety concept could not be implemented completely. In particular, the probability of a flood defense collapsing, and therefore the probability of flooding, proved to be very difficult to estimate.

For this reason a simplified concept was chosen at the time, based on design loads. The basic assumption is that every individual section of a dike has to be high enough to safely withstand a given extreme water level and the associated wave impact.

The current safety standards are laid down in the Flood Protection Act. This act foresees a change to a new safety concept based on the probability of flooding in certain dike ring areas. A dike ring area is an area that is completely surrounded by

Case Studies in Reliability and Maintenance, Edited by W. R. Blischke and D. N. P. Murthy.
ISBN 0-471-41373-9 © 2003 John Wiley and Sons, Inc.

water and therefore surrounded by flood defenses. In the past few years, models for determining the inundation probability for dike rings have been developed (Slijkhuis, 1998; van der Meer, 1997).

Because of the many new features involved in the dike ring safety concept, it was decided to perform an uncertainty analysis on the prediction of failure frequency for one illustrative dike ring, the so-called Hoeksche Waard. This involved the following steps:

1. Freezing the structural reliability model for dike section failure.
2. Performing an "in-house" quantification of uncertainty, with dependence, of all input variables.
3. Identifying those input variables whose uncertainty is anticipated to have significant impact on the uncertainty of the dike ring failure frequency.
4. Assessing the uncertainty of the selected variables with structured expert judgment.
5. Propagating the resulting uncertainties through the structural reliability model so as to obtain an uncertainty distribution on the failure frequency of the dike ring.

The motivations and goals for performing an uncertainty analysis, as opposed to a traditional reliability analysis, are set forth in the following section, which expands on uncertainty analysis. Section 15.3 discusses the structured expert judgment methodology. Section 15.4 discusses the dike ring study, and Section 15.5 presents results. A final section gathers conclusions.

15.2 UNCERTAINTY ANALYSIS

In contrast to more standard reliability analyses, the goal of this study is not to predict a failure frequency. Rather, the goal is to determine the *uncertainty* in the failure frequency for a given dike ring. The reason for shifting the focus to uncertainty in this way is because the primary source of our information in this case is expert judgment, not field data. We appeal to expert judgment because there is not sufficient data, and hence substantial uncertainty. The goal of using expert judgment, as discussed in the next section, is to obtain a rationally defensible quantification of this uncertainty.

The first step proved surprisingly difficult. The structural reliability model for a dike ring is large, involving some 300 input variables whose values are not known with certainty. Moreover, the model is under continual development, and it is very difficult to freeze a model that is known to become outdated before the uncertainty analysis is completed.

Since there are a large number of uncertain input variables in these dike ring models, it is not possible to subject all of these variables to a structured expert judgment elicitation. Instead, we must restrict the expert elicitation to those variables that are judged most important in driving the uncertainty of the dike ring failure frequency. This requires an initial "in house" quantification of uncertainties (step 2) and a selection of important variables (step 3). The techniques used in step 3 are very much in development at the moment [see, e.g., Saltelli et al. (2000)], and will not be treated here.

The present chapter focuses on step 4. The reasons for this are twofold. First, many of the questions put to the experts are of general interest and do not require lengthy exposition of the structural reliability model. These include:

- Frequencies of exceedence of extreme water levels of the North Sea
- Sea level rise
- Frequencies of exceedence of extreme discharges of the Rhine river
- Influence of climate change and human intervention in the Rhine discharge

Second, this was the major effort of the study, and the application of the structured approach led to significant insights.

15.3 EXPERT JUDGMENT METHOD

The methodology for expert judgment has been presented in Cooke (1991) and applied in many risk and reliability studies. See Goossens et al. (1998) for a recent overview. In particular, this method was used in the study of failure of gas pipelines described in this volume. This section briefly describes the expert judgment method. It is based on Frijters et al. (1999) and Cooke and Goossens (2000a,b).

The goal of applying structured expert judgment is to enhance rational consensus. Rational consensus is distinguished from "political consensus" in that it does not appeal to a "one-man one-vote" method for combining the views of several experts. Instead, views are combined via weighted averaging, where the weights are based on performance measures, and satisfy a proper scoring rule constraint (Cooke, 1991). This model for combining expert judgments bears the name "classical model" because of a strong analogy with classical hypothesis testing. We restrict our discussion to the case where experts assess their uncertainty for quantities taking values in a continuous range. There are two measures of performance: calibration and information. These are presented briefly below; for more detail see Cooke (1991).

15.3.1 Calibration

The term *calibration* was introduced by psychologists (Kahneman et al., 1982) to denote a correspondence between subjective probabilities and observed relative frequencies. This idea has fostered an extensive literature and can be operationalized in several ways. In the version considered here, the classical model treats an expert as a classical statistical hypothesis, and measures calibration as the degree to which this hypothesis is supported by observe data, in the sense of a simple significance test.

More precisely, an expert states n fixed quantiles for his/her subjective distribution for each of several uncertain quantities taking values in a continuous range. There are $n + 1$ "interquantile intervals" into which the realizations (actual values) may fall. Let

$$p = (p_1, \ldots, p_{n+1}) \tag{15.1}$$

denote the theoretical probability vector associated with these intervals. Thus, if the expert assesses the 5%, 25%, 50%, 75% and 95% quantiles for the uncertain quantities, then $n = 5$ and $p = (5\%, 20\%, 25\%, 25\%, 20\%, 5\%)$. The expert believes that there is a 20% probability that the realization falls between his/her 5% and 25% quantiles, and so on.

In an expert judgment study, experts are asked to assess their uncertainty for variables for which the realizations are known post hoc. These variables are chosen to resemble the quantities of interest, and/or to draw on the sort of expertise which is required for the assessment of the variables of interest. They are called "calibration" or "seed" variables.

Suppose we have such quantile assessments for N seed variables. Let

$$s = (s_1, \ldots, s_{n+1}) \tag{15.2}$$

denote the empirical probability vector of relative frequencies with which the realizations fall in the interquantile intervals. Thus s_2 = (number of realizations strictly above the 5% quantile and less than or equal to the 25% quantile)/N, and so on.

Under the hypothesis that the realizations may be regarded as independent samples from a multinomial distribution with probability vector p, the quantity[1]

$$2NI(s, p) = 2N \sum_{i=1}^{N} s_i \ln(s_i/p_i) \tag{15.3}$$

is asymptotically chi-square distributed with n degrees of freedom and large values are significant. Thus, if χ_n is the cumulative distribution function for a chi-square variable with n degrees of freedom, then

$$\text{CAL} = 1 - \chi_n(2NI(s, p)) \tag{15.4}$$

is the upper-tail probability, and is asymptotically equal to the probability of seeing a disagreement no larger than $I(s, p)$ on N realizations, under the hypothesis that the realizations are drawn independently from p.

We take CAL as a measure of the expert's calibration. Low values (near zero) correspond to poor calibration. This arises when the difference between s and p cannot

[1] $I(s, p)$ is called the relative Shannon information of s with respect to p. For all s, p with $p_i > 0$, $i = 1, \ldots, N$, we have $I(s, p) \geq 0$ and $I(s, p) = 0$ if and only if $s = p$ [see Kullback (1959)].

plausibly be the result of mere statistical fluctuation. For example, if $N = 10$, and we find that 8 of the realizations fall below their respective 5% quantile or above their respective 95% quantile, then we could not plausibly believe that the probability for such events was really 5%, as the expert maintains. This would correspond to an expert giving "overconfident" assessments. Similarly, if 8 of the 10 realizations fell below their 50% quantiles, then this would indicate a "median bias." In both cases, the value of CAL would be low. High values of CAL indicate good calibration.

It is well to emphasize that we are not testing or rejecting hypotheses here. Rather, we are using the standard goodness-of-fit scores to measure an expert's calibration.

15.3.2 Information

Loosely, information measures the degree to which a distribution is concentrated. This loose notion may be operationalized in many ways. For a discussion of the pros and cons of various measures, see Cooke (1991). We shall measure information as Shannon's relative information with respect to a user-selected background measure. The background measure will be taken as the uniform measure over a finite "intrinsic range." For a given uncertain quantity and a given set of expert assessments, the intrinsic range is defined as the smallest interval containing all the experts' quantiles and the realization, if available, augmented above and below by K%. The overshoot term K is chosen by default to be 10, and sensitivity to the choice of K must always be checked (see Table 15.2 below).

To implement this measure, we must associate a probability density with each expert's assessment for each uncertain quantity. When the experts have given their assessments in the form of quantiles, as above, we select that density which has minimal Shannon information with respect to the background measure and which complies with the expert's quantile assessments. When the uniform background measure is used, the minimum information density is constant between the assessed quantiles, and the mass between quantiles $i - 1$ and i is just p_i. If $f_{k,j}$ denotes the density for expert k and uncertain quantity j, then Shannon's relative information with respect to the uniform measure on the intrinsic range I_j is

$$I(f_{k,j}, U_j) = \int_{u \in I_j} f_{k,j}(u) \ln\big(f_{k,j}(u)\big)\, du + \ln(|I_j|) \tag{15.5}$$

where $|I_j|$ is the length of I_j. For each expert, an information score for all variables is obtained by summing the information scores for each variable. This corresponds to the information in the expert's joint distribution relative to the product of the background measures under the assumption that the expert's distributions are independent. Roughly speaking, with the uniform background measure, more informative distributions are gotten by choosing quantiles that are closer together whereas less informative distributions result when the quantiles are farther apart.

The calibration measure CAL is a "fast" function. With, say, 10 realizations we may typically see differences of several orders of magnitude in a set of, say 10 experts. Information, on the other hand, is a "slow" function. Differences typically lie within a factor 3. In the performance-based combination schemes discussed below, this feature means that calibration dominates strongly over information. Information serves to modulate between more or less equally well calibrated experts. The use of the calibration score in forming performance-based combinations is a distinctive feature of the classical model and implements the principle of empirical control discussed above.

15.3.3 Combination

Experts give their uncertainty assessments on query variables in the form of, say, 5%, 25%, 50%, 75%, and 95% quantiles. An important step is the combination of all experts' assessments into one combined uncertainty assessment on each query variable. The three combination schemes considered here are examples of "linear pooling"; that is, the combined distributions are weighted sums of the individual experts' distributions, with non-negative weights adding to one. Different combination schemes are distinguished by the method according to which the weights are assigned to densities. These schemes are designated "decision makers." Three decision makers are described briefly below.

Equal-Weight Decision Maker

The equal-weight decision maker results by assigning equal weight to each density. If E experts have assessed a given set of variables, the weights for each density are $1/E$; hence for variable i in this set the decision maker's density is given by

$$f_{eddm,i} = \left(\frac{1}{E}\right) \sum_{j=1 \cdots E} f_{j,i} \tag{15.6}$$

where $f_{j,i}$ is the density associated with expert j's assessment for variable i.

Global-Weight Decision Maker

The global-weight decision maker uses performance-based weights that are defined, per expert, by the product of expert's calibration score and his/her overall information score on seed variables and by an optimization routine described below [see Cooke (1991) for details]. For expert j, the same weight is used for all variables assessed. Hence, for variable i the global-weight decision-maker's density is

$$f_{gwdm,i} = \frac{\sum_{j=1 \cdots E} w_j f_{j,i}}{\sum_{j=1 \cdots E} w_j} \tag{15.7}$$

These weights satisfy a "proper scoring rule" constraint. That is, under suitable assumptions, an expert achieves his/her maximal expected weight, in the long run, by and only by stating quantiles that correspond to his/her true beliefs [see Cooke (1991)].

Item-Weight Decision Maker

As with global weights, item weights are performance-based weights that satisfy a proper scoring rule constraint, and are based on calibration and informativeness, with an optimization routine described below. Whereas global weights use an overall measure of informativeness, item weights are determined per expert and per variable in a way that is sensitive to the expert's informativeness for each variable. This enables an expert to increase or decrease his/her weight for each variable by choosing a more or less informative distribution for that variable. For the item-weight decision maker, the weights depend on the expert and on the item. Hence, the item-weight decision-maker's density for variable i is

$$f_{iwdm,i} = \frac{\sum_{j=1 \cdots E} w_{j,i} f_{j,i}}{\sum_{j=1 \cdots E} w_{j,i}} \tag{15.8}$$

15.3.4 Optimization

The proper scoring rule (Cooke, 1991) constraint entails that an expert should be unweighted if his/her calibration score falls below a certain minimum, $\alpha > 0$. The value of α is determined by optimization. That is, for each possible value of α a certain group of experts will be unweighted, namely those whose calibration score is less than α. The weights of the remaining experts will be normalized to sum to unity. For each value of α we thus define a decision maker dm_α, computed as a weighted linear combination of the experts whose calibration score exceeds α. dm_α is scored with respect to calibration and information. The weight that this dm_α would receive if he were added as a "virtual expert" is called the "virtual weight" of dm_α. The value of α for which the virtual weight of dm_α is greatest is chosen as the cutoff value for determining the unweighted expert.

15.3.5 Validation

When seed variables are available, we can use these variables to score and compare different possible combinations of the experts' distributions—or, as we shall say, different decision makers. In particular, we can measure the performance of the global- and item-weight decision makers with respect to calibration and information and can compare this to the equal-weight decision maker and to the experts themselves. This is done in the following section.

15.4 THE DIKE RING EXPERT JUDGMENT STUDY

Seventeen experts participated in this expert judgment study. They are all associated with Dutch universities or governmental institutes. The experts were acquainted with the issues, study objectives, and methods beforehand, and they were elicited individually. A typical elicitation took 3 to 4 hours. Each expert gave 5%, 25%, 50%, 75%, and 95% quantiles for 40 uncertain quantities, concerning the following:

- Frequencies per year of exceedence of extreme water levels of the North Sea
- Sea level rise
- Frequencies of exceedence of extreme discharges of the Rhine river
- Influence of climate change and human intervention in the Rhine discharge
- The significant wave height
- The significant wave period[2]
- The model term for Zwendl[3]
- The model factor for critical discharge
- The model factor for occurring discharge[4]
- The dependence between model factor for critical discharge between dike sections
- The dependence between model factor for occurring discharge between dike sections

The issue of dependence and its assessment requires more exposition than is possible here [see Cooke and Goossens (2000b)].

Model factors and model terms are used to capture an expert's uncertainty with regard to model predictions. A model factor is defined as the ratio of the realization to the model prediction. A model term is defined as the difference between the realization and the model prediction. An example of the elicitation for the model term Zwendl is given in the Appendix.

Seed Variables

The model factors and model terms afford the possibility of calibrating the experts' assessments. With some effort, realizations were found from historical and experimental records and were compared with model computations. We are interested in the relation between model predictions and realizations in extreme situations, say water levels higher than 4 meters above normal. We cannot find such realizations in the available data or in controlled experiments. However, we can find "subextreme" realizations (water levels 2.5 meters above normal). For these we can compute post

[2]Significant wave height and wave period are defined as the 67% quantile of the occurring wave height wave period distribution, respectively.

[3]Zwendl is a computer code that computes local water levels at estuary measuring stations as a function of, inter alia, North Sea storm profile and Rhine discharge profile. This model was calibrated extensively during its development. A simplified version used in the current dike ring model considers only a "standard storm profile" characterized by peak discharge and peak North Sea surcharge. The current study brought to light the fact that this simplified version was never calibrated.

[4]The critical discharge for a dike section is the maximal flux of water [liters /meter second] that a dike section can withstand without failing. The occurring discharge is the actual occurring discharge over the crown of a dike section.

hoc the model predictions. Comparing these two, we obtain realizations for the model factors and model terms. This is done in the current study and resulted in 47 seed variables. It is significant that this had *not* been done for the models in question prior to the current study. These realizations and predictions were not known to the experts (nor to the analysts) at the time of the elicitation.

Figure 15.1 gives a "range graph" for the experts' assessments of the significant wave height, and it gives the results of seven realizations.

These data were obtained from measurements of wave height distributions at the measurement station Marollegat over six months in situations where the model predictions could be calculated post hoc. We see that most experts placed their median value for this model factor at 1, indicating that the probability of over prediction was equal to the probability of under prediction. Expert 14 believed that the model for significant wave height almost certainly underpredicts the actual significant wave height. Eyeballing these data, we might say that the experts are a bit underconfident, that is, too many realizations fall within the 25% and 75% quantiles. There might be a slight tendency for the model to underpredict, but the tendency seems small relative to the experts' uncertainty.

A different picture emerges with the model term for the local water level model Zwendl shown in Figure 15.2. These data are from the measuring station Dordrecht. When a high local water level was reached, the parameters for calculating Zwendl's predicted water level were recovered post hoc. Similar pictures emerged from data from other measuring stations (Raknoord and Goidschalxoord).

All experts except numbers 1 and 12 placed their median value at zero. Only one data point is less than zero, and many fall outside the experts' 95% quantiles. This picture suggests that the model underpredicts local water levels to a degree that is significant relative to the experts' uncertainty. This result was rather surprising. Most engineers assumed that the model Zwendl had been properly calibrated, and therefore they put their median value at zero. It turned out that the original model was calibrated almost 30 years ago, according to methods and standards no longer current, and the simplified model actually in use had never been calibrated.

15.5 RESULTS

Before presenting the results, it is useful to get a picture of how well the experts agree among themselves. For a given item this can be measured as the Shannon relative information of an expert's distribution relative to the equal weight combination of all experts' distributions. Averaging these relative informations over all items, we get the data in Table 15.1.

The values for all variables range between 1.148 and 0.389, with average value over all experts of 0.62. This value should be compared to (a) the experts' own information relative to the background measure (Table 15.2, column 4) and (b) the result of performing robustness analysis on the selection of experts (see Table 15.3, column 5).

Table 15.2 gives scoring results for the experts and for various decision makers.

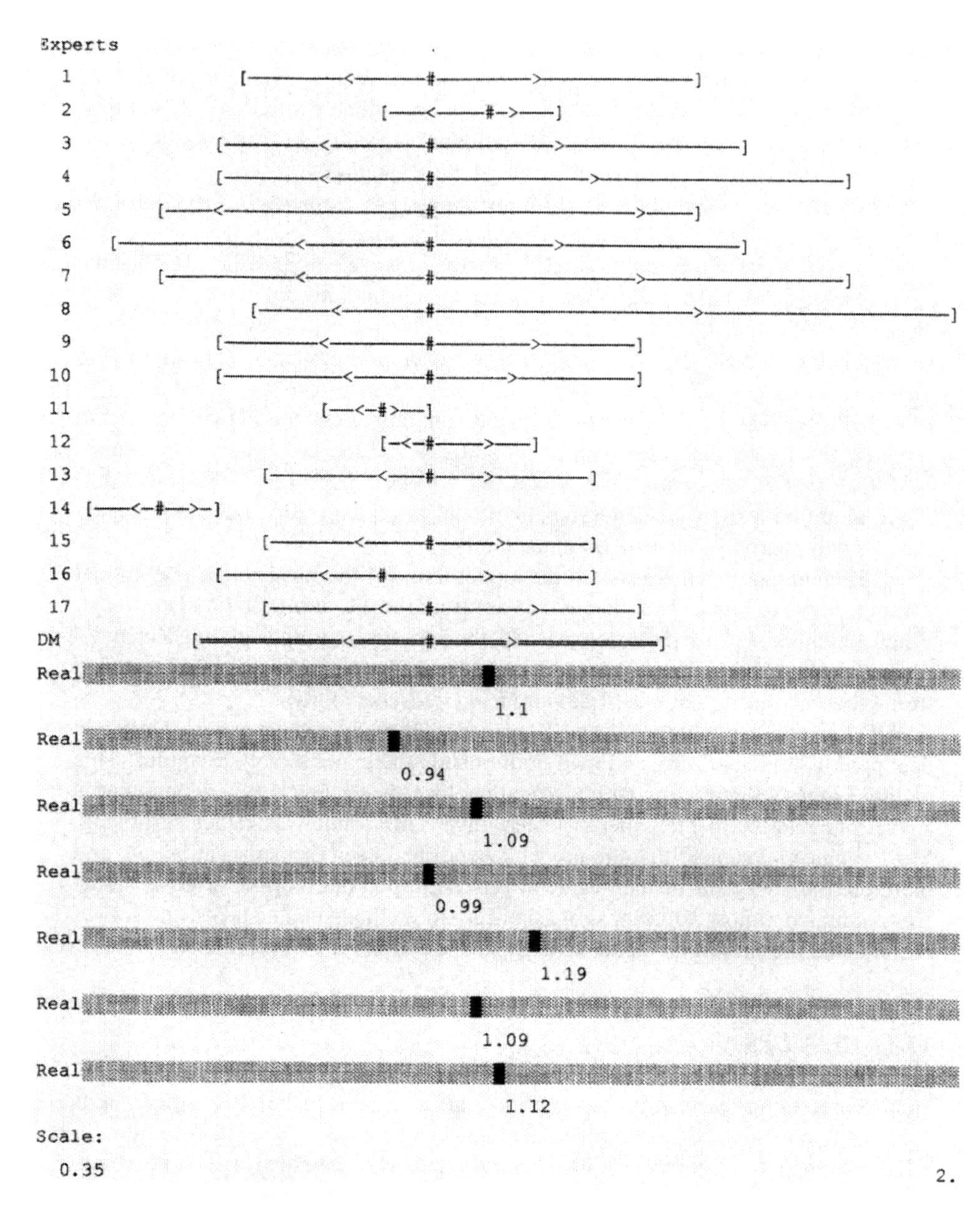

Figure 15.1. Range graph for model factor for significant wave height and seven realizations; "[" and "]" denote 5% and 95% quantiles, respectively, "<" and ">" denote 25% and 75% quantiles, respectively, "#" denotes the median, and "*dm*" is the item-weight decision maker (see Table 15.2).

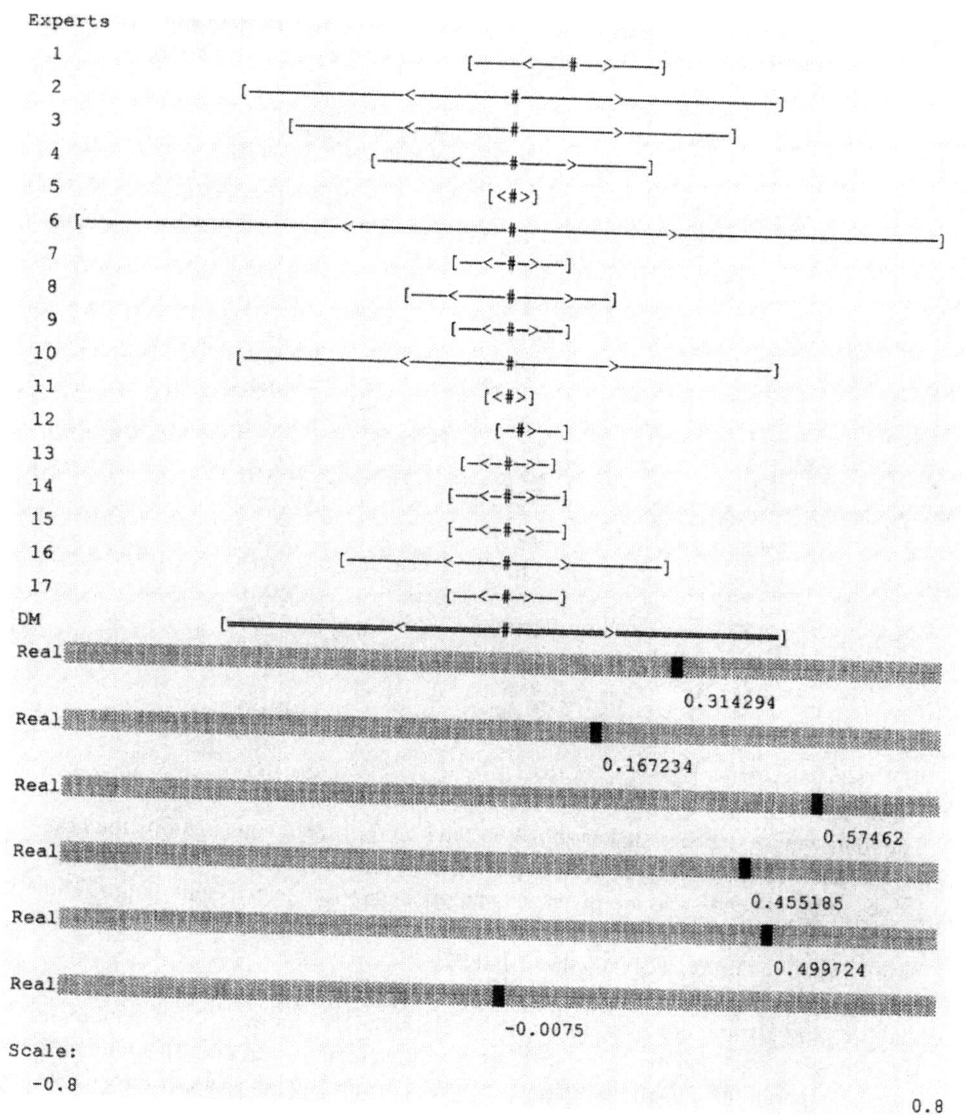

Figure 15.2. Range graph for model term for local water level model Zwendl and six realizations from the measuring station Dordrecht; "[" and "]" denote 5% and 95% quantiles, respectively, "<" and ">" denote 25% and 75% quantiles, respectively, "#" denotes the median, and "*dm*" is the item-weight decision maker (see Table 15.2).

Table 15.1. Average Relative Information of Experts with Respect to the Equal-Weight Decision Maker, for All Variables and for Seed Variables Only

	Average Relative Information with Respect to Equal-Weight *dm*	
Expert Number	All Variables	Seed Variables Only
1	0.729	0.476
2	0.578	0.574
3	0.380	0.328
4	0.687	0.677
5	0.784	0.596
6	0.428	0.368
7	0.523	0.539
8	0.478	0.444
9	0.519	0.510
10	0.427	0.318
11	0.837	0.723
12	1.039	1.093
13	0.601	0.402
14	1.148	0.999
15	0.410	0.375
16	0.489	0.272
17	0.440	0.412

The item-weight decision maker shown in the bold bordered cells exhibits the best performance (as judged by unnormalized weight[5]) and is chosen for this study. The values in the shaded areas are given as comparisons to the item-weight values. The unnormalized weights are the global weights w_j in equation (15.7). Note that column 6 is just the product of columns 2 and 5.

Column 4 of Table 15.2 should be compared to column 2 of Table 15.1. We see that the experts are much more informative relative to the uniform distribution over the intrinsic range than they are with respect to the equal weight dm. This indicates that the experts' 90% confidence bands display considerable overlap. If there were little overlap, the information scores in Table 15.1 would be closer to those in Table 15.2.

The number of seed variables, 47, is quite large; and since six or seven realizations are typically available for a single measuring station, it is doubtful whether the realizations should be considered as independent samples as required by the calibration measure. We can account for this by treating each expert's empirical probability vector s as if it were generated by a smaller number of samples. In short, we

[5]If the size of the intrinsic range is changed, then the information scores also change, hence the item-weight DM should be compared with the global-weight DM and the equal-weight DM with the same intrinsic ranges.

Table 15.2. Results of Scoring Experts and Decision Makers

Expert Number	Calibration (Effective Number of Seeds = 47)	Calibration (Effective Number of Seeds = 9)	Mean Relative Information with Respect to Background Measure: Total	Mean Relative Information with Respect to Background Measure: Seeds	Unnormalized Weight	Normalized Weight with Item-Weight *dm*
1	0.00010	0.30000	1.777	1.103	0.00011	0.00024
2	0.00010	0.30000	1.533	1.244	0.00012	0.00027
3	0.00010	0.40000	1.146	0.803	0.00008	0.00018
4	0.00010	0.10000	1.457	1.449	0.00014	0.00032
5	0.00010	0.05000	2.007	1.568	0.00016	0.00035
6	0.02500	0.80000	0.797	0.430	0.01075	0.02367
7	0.00010	0.01000	1.065	0.951	0.00010	0.00021
8	0.00010	0.40000	1.353	1.054	0.00011	0.00023
9	0.00010	0.05000	1.498	1.410	0.00014	0.00031
10	0.30000	0.95000	1.171	0.648	0.19433	0.42772
11	0.00010	0.00500	2.152	2.132	0.00021	0.00047
12	0.00010	0.00100	2.619	2.460	0.00025	0.00054
13	0.00010	0.10000	1.884	1.526	0.00015	0.00034
14	0.00010	0.00050	2.402	2.053	0.00021	0.00045
15	0.00010	0.10000	1.247	1.240	0.00012	0.00027
16	0.00100	0.60000	1.381	0.827	0.00083	0.00182
17	0.00010	0.10000	1.168	1.120	0.00011	0.00025
Item weight *dm*	0.40000	0.95000	1.039	0.616	0.24642	0.54237
Global weight *dm*	0.40000	0.90000	1.000	0.572	0.22865	0.52717
Equal weight *dm*	0.05000	0.80000	0.955	0.760	0.03798	0.15444
Item weight *K* = 50	0.40000		1.289	0.8527	0.3411	0.5314
Global weight *K* = 50	0.30000		1.470	0.9341	0.2802	0.5272
Equal weight *K* = 50	0.10000		1.218	1.014	0.1014	0.2521

replace N in equation (15.3) by a smaller number, called the *effective sample size*. An effective sample size of 9 results from considering the measurements in distinct groups (Figures 15.1 and 15.2 each represent one group) as one effective measurement. Although not really defensible, this does establish a lower bound to the effective sample size. The calibration scores in the third column may be compared with many other studies involving a comparable number of seed variables, and it emerges that these experts, as a group, are comparatively well calibrated. Consider-

Table 15.3. Robustness of the Item-Weight *dm* Against Choice of Experts

Excluded Expert Name	Relative Information to Background			Average Relative Information to Original *dm*	
	Total	Seeds Only	Calibration	Total	Seeds Only
None	1.039	0.616	0.40000	0.000	0.000
1	1.037	0.616	0.40000	0.001	0.000
2	1.034	0.616	0.40000	0.000	0.000
3	1.039	0.616	0.40000	0.000	0.000
4	0.921	0.616	0.40000	0.007	0.000
5	1.038	0.616	0.40000	0.001	0.000
6	0.968	0.512	0.30000	0.062	0.028
7	1.039	0.616	0.40000	0.000	0.000
8	1.030	0.602	0.40000	0.001	0.002
9	1.039	0.616	0.40000	0.000	0.000
10	0.951	0.625	0.02500	0.350	0.280
11	1.034	0.616	0.40000	0.001	0.000
12	1.037	0.615	0.40000	0.001	0.001
13	1.038	0.616	0.40000	0.000	0.000
14	1.037	0.613	0.40000	0.001	0.001
15	1.020	0.616	0.40000	0.001	0.000
16	1.044	0.616	0.40000	0.008	0.002
17	0.975	0.552	0.40000	0.005	0.002

ing all 47 seed variables as independent leads to the scores in the second column. Although the expert performance would be much worse in this case,[6] the global- and item-weight decision makers still reflect good calibration. To render the intuitive meaning of these scores we could say: *The hypothesis that the realizations are drawn independently from a distribution whose quantiles agree with those of the item-weight decision maker would be rejected with these 47 realizations at the 0.4 significance level (which of course is much too high for rejection in a standard hypothesis test).*

Another way to assess the decision-makers' performances is given in the last column. This is the normalized global weight after adding the decision maker to the pool of experts and scoring him as another "virtual expert."[7] For the 17 experts, the values in the last column are the global weights the experts would receive if the item-weight *dm* were added as a virtual expert. These are called "virtual weights." The item-weight *dm* has a virtual weight of 0.54237, higher than the combined

[6]We should reflect that *every* statistical hypothesis is wrong and would eventually be rejected with sufficient data. No expert will be *perfectly* calibrated, and imperfections, with sufficient data, will eventually produce a poor goodness of fit. The ratio s of scores is important when used in the performance-based weights, but absolute scores are useful for comparing performance across studies.

[7]It can be shown that adding a *dm* as a virtual expert and recomputing weights yields a new *dm*, which is identical to the initial *dm* [see Cooke (1991)].

weight of all experts. The virtual weights of the other decision makers are computed in the same way, but the corresponding weights of the 17 experts are not shown.

The last three rows of Table 15.2 show the three decision makers with 47 effective seed variables, and where the overshoot term K for the intrinsic range is changed from the default value 10 to 50. In other words, the intrinsic range for each variable is the smallest interval containing all assessed values, augmented by 50% above and below. All experts' information scores increase, as the background measure to which they are compared has become more diffuse (the increased scores are not shown). The equal weight dm's information score changes from 0.955 to 1.218, reflecting the fact that information is a slow function. Whereas the equal-weight combination scheme is not affected by changing the intrinsic range, the performance based decision makers are affected slightly, as the experts' weights change slightly. The virtual weight of the item-weight decision maker drops from 0.54237 to 0.5314.

The equal-weight decision maker is less well calibrated and less informative than the performance-based decision makers. Such a pattern emerges frequently, but not always, in such studies.

From the robustness analyses performed on this expert data, we present the results for robustness against choice of experts. Table 15.3 shows the results of removing the experts one at a time and recomputing the item weight *dm*. The second and third columns give the information scores over all items, and seed items only, for the "perturbed *dm*" resulting from the removal of the expert in column 1. The fourth column gives the perturbed *dm*'s calibration. The last two columns give the relative information of the perturbed expert with respect to the original expert. Column 5 may be compared to column 2 of Table 15.1.

The average of column 5, 0.026, should be compared with the average of column 2 in Table 15.1, 0.62. This shows that the differences between the *dm*'s caused by omitting a single expert are much smaller than the differences among the experts themselves. In this sense the robustness against choice of expert is quite acceptable. However, it will be noticed that loss of expert 10 would cause a significant degradation of calibration. Apparently this expert is very influential on the *dm*, and the good performance of this *dm* is not as robust in this respect as we would like. Of course, loss of robustness is always an issue when we optimize performance.

For selected items of interest, Table 15.4 compares the item-weight and the equal-weight *dm*'s with the in-house assessments, where available, used in step 2 of the uncertainty analysis. The boldface numbers indicate significant differences.

In general the numbers in Table 15.4 are in the same ballpark. But there are important differences. The in-house assessments tend to be more concentrated than the combined experts' assessments. For the occurring discharge at a predicted level of 100 liters per meter per second, the equal-weight *dm* has a large 95% quantile.

15.6 CONCLUSIONS

The results presented above show that the structured expert judgment approach can be a valuable tool in structural reliability uncertainty analysis. Expert judgment is

Table 15.4. Comparison of In-House, Item-Weight *dm* and Equal-Weight *dm* For Items of Interest

Frequency yearly maximum North Sea ≥ 4.5 m	**5%**	**25%**	**50%**	**75%**	**95%**
In house	0.00015	0.00029	0.00040	0.00068	0.00118
Item weights	0.00002	0.00014	0.00053	**0.00184**	**0.00299**
Equal weights	0.00001	0.00022	0.00056	**0.00162**	**0.00450**
Sea level rise in 100 years: increment to 2.5-m water level	**5%**	**25%**	**50%**	**75%**	**95%**
Item weights	–0.003	0.19	0.39	0.60	0.81
Equal weights	**–0.28**	0.22	0.36	0.58	0.99
100-year maximum North Sea level	**5%**	**25%**	**50%**	**75%**	**95%**
Item weights	3.49	3.89	4.10	4.50	5.00
Equal weights	3.12	3.80	4.24	4.66	5.50
Probability yearly maximum Rhine discharge ≥ 16,000 m³/s	**5%**	**25%**	**50%**	**75%**	**95%**
Item weights	1.10E-05	1.58E-04	6.27E-04	1.57E-03	4.11E-03
Equal weights	8.52E-06	1.59E-04	7.06E-04	1.72E-03	5.15E-03
Model term Zwendl for prediction = 4 m	**5%**	**25%**	**50%**	**75%**	**95%**
In-house	–0.25	–0.1	0.00	0.10	0.25
Item weights	**–0.91**	**–0.39**	0.00	**0.39**	**0.94**
Equal weights	**–0.61**	–0.15	0.04	0.22	**0.65**
Model factor wave height for prediction = 1 m	**5%**	**25%**	**50%**	**75%**	**95%**
In-house	0.83	0.93	1.00	1.08	1.20
Item weights	0.58	0.85	1.00	1.15	1.43
Equal weights	0.54	0.85	1.00	1.17	1.57
Model factor wave periode for prediction = 4s	**5%**	**25%**	**50%**	**75%**	**95%**
In-house	3.34	3.71	4.00	4.31	4.80
Item weights	2.77	3.40	4.49	4.60	5.33
Equal weights	2.40	3.54	4.06	4.63	**6.25**
Critical discharge for prediction = 50l/s/m	**5%**	**25%**	**50%**	**75%**	**95%**
In-house	23	36	50	69	109
Item weights	18	34	49	**118**	**207**
Equal weights	12	34	64	**115**	**185**
Occurring discharge for prediction = 100l/s/m	**5%**	**25%**	**50%**	**75%**	**95%**
In-house	46	73	100	137	218
Item weights	30	62	100	150	240
Equal weights	**23**	70	101	130	**526**

more than just subjective guessing. It can be subjected to rigorous empirical control and can contribute positively to rational consensus. The extra effort involved in defining seed variables and measuring expert performance, as opposed to simple equal weighting of experts, has paid off in this case. The experts themselves and the problem owners appreciate that this same effort contributes toward rational decision making by extending the power of quantitative analyses. Without expert judgment, any quantification of the uncertainty in dike reliability models would be impossible, and the discussion would remain stalled with safety concepts that we cannot implement.

Not only do we now have a defensible picture of the uncertainty in dike ring reliability calculations, but we can also judge where additional research effort might be expected to yield the greatest return in reducing this uncertainty and improving predictions.

The most obvious conclusion in this regard is that more effort should be devoted to calibrating the simplified version of the Zwendl model. From the engineer's viewpoint, this is the most significant result of this study. The relatively large uncertainty in the item weight *dm* for this variable throws some doubt on the reliability calculations for dike rings performed to date. In particular, there is greater probability that the local water levels are underpredicted by the deterministic models.

The greatest improvement in these results would be gained by recalibrating the model Zwendl for local high water levels. We anticipate that this would lead to higher predictions and smaller uncertainty bands. The techniques of using structured expert judgment to quantify uncertainty in combination with uncertainty propagation could be employed to many reliability models. However, the costs and effort involved are significant, and thus these techniques are most suitable in problems of high public visibility where rational consensus is important.

REFERENCES

Cooke, R. (1991). *Experts in Uncertainty,* Oxford University Press, New York.

Cooke, R. M., and Goossens, L. H. J. (2000a). Procedures guide for structured expert judgment in accident consequence modeling, *Radiation Protection Dosimetry* **90:**303–309.

Cooke R. M., and Goossens, L. H. J (2000b). *Procedures Guide for Structured Expert Judgment* European Commission, Directorate-General for Research, EUR 18820 EN, Luxembourg.

Frijters, M., Cooke, R. Slijkhuis, K., and van Noortwijk, J. (1999). *Expertmeningen Onzekerheidsanalyse, kwalitatief en kwantitatief verslag,* Ministerie van Verkeer en Waterstaat, Directoraat-Generaal Rijkswaterstaat, Bouwdienst; ONIN 1-990006.

Goossens, L. H. J., Cooke, R. M., and Kraan, B. C. P. (1998). Evaluation of weighting schemes for expert judgment studies, *Proceedings of the 4th International Conference on Probabilistic Safety Assessment and Management I,* pp. 1937–1942, Springer, New York.

Kahneman, D., Slovic, P., and Tversky, A., Editors (1982). *Judgment under Uncertainty,* Cambridge University Press, New York.

Kullback, S. (1959). *Information Theory and Statistics,* Wiley, New York.

Saltelli, A., Chan, K., and Scott, E. M. (2000). *Sensitivity Analysis,* Wiley, Chichester.

Slijkhuis, K. A. H. (1998). Beschrijvingenbundel (in Dutch) Ministerie Verkeer en Waterstat, Directoraat-Generaal Rijkswaterstaat, Bouwdienst Rijkswaterstaat, Utrecht.

Slijkhuis, K. A. H., Frijters, M. P. C., Cooke, R. M., and Vrouwenvelder, A.C.W.M. (1998). Probability of Flooding: An Uncertainty Analysis, in *Safety and Reliability* (Lyderson, Hansen and Sandtorv, Editors), pp. 1419–1425, Balkema, Rotterdam.

van der Meer, J. W. (1997). Golfoploop en golfoverslag bij dijken (wave run up and overtopping of dikes, in Dutch), Delft, Waterloopkundig Laboratorium, H2458/H3051.

EXERCISES

15.1. Suppose that a dike section has a probability of 8×10^{-6} per year of failing. Suppose a dike ring consists of 50 dike sections, and suppose that the failures of different sections are independent. What is the probability per year that the dike ring fails?

15.2. (*Continuation*) Instead of assuming independence, suppose that the event that one dike ring fails is the conjunction of two equally probable events: One event concerns only factors local to the given dike section and is independent of similar events at other dike sections; the other event concerns meteorological factors that affect all dike sections in the same way. What is the probability that the dike ring fails?

15.3. Two experts have assessed their 5%, 50%, and 95% quantiles for 10 continuous variables for which the realizations have been recovered. For the first expert, three realizations fall beneath his 5% quantile, two fall between the 5% and 50% quantile, two fall between the 50% and 95% quantile, and three fall above the 95% quantile. For the second expert, one realization falls below the 5% quantile, seven fall between then 5% and 50% quantiles, one falls between the 50% and 95% quantile, and one falls above the 95% quantile. Compute the calibration scores for these two experts.

15.4. Two experts assess their 5%, 50%, and 95% quantiles for an unknown relative frequency. For the first expert these quantiles are 0.1, 0.25, 0.4; and for the second expert these quantiles are 0.2, 0.4, 0.6. Use a uniform background measure on the interval [0, 1] and compute the Shannon relative information in each expert's assessment using the minimum information density for each expert, subject to the quantile constraints.

15.5. (*Continuation*) Compute the equal weight combination for the two experts in exercise 15.4, and compute the Shannon relative information of this combination relative to the uniform background measure on [0, 1].

15.6. (*Continuation*) Use the calibration scores from exercise 15.3 and the information scores from exercise 15.4 to compute the item-weight combination for the variable in exercise 15.4. What is the Shannon relative information for this combination relative to the uniform background measure on [0, 1]?

APPENDIX: EXAMPLE OF ELICITATION QUESTION: MODEL TERM ZWENDL

The following is a translation of the text used to elicit the distribution of the model term for Zwendl. Experts were made familiar with the notion of subjective probability distribution and quantiles. The model term is queried in four ways: First, unconditional on the actual local water level; then, conditional on a predicted local water level of 2 m, 3 m, and 4 m. This was done to see if the uncertainty in the model prediction depended on the predicted value. There was no significant pattern of dependence in the experts' responses. The unconditional assessments were used for calibrating the experts.

In the model for predicting dike ring inundation probability, Zwendl is used to compute local water levels as a function of deterministic values for the water level at the Meuse mouth and the Rhine discharge at Lobith. In calculating the failure probabilities, a model term for Zwendl will be added to the water levels computed with Zwendl. This model term is a variable whose distribution reflects the uncertainty in the Zwendl calculation in the following way:

Model term Zwendl = real local water level – local water level computed with Zwendl.

We would like to quantify the uncertainty in the output of Zwendl for the dike ring Hoeksche Waard. This uncertainty should take into account that Zwendl uses a standard water level profile for the North Sea and a single value for the Rhine discharge instead of real temporal profiles.

We would like to quantify your uncertainty for the computation of local water levels using Zwendl. We do this by asking you to state your 5%, 25%, 50%, 75%, and 95% quantiles for your subjective uncertainty distributions.

- For an arbitrary water level, what is the difference between the measured local water level and the water level computed with Zwendl (in meters):

___	___	___	___	___
5%	25%	50%	75%	95%

- Assume that Zwendl predicts a water level of 2 m above N.A.P. (the standard baseline water level) for a given location in the dike ring area of Hoeksche Waard; what is then the difference between the actually measured value of the local water level and the value computed with Zwendl (in meters):

___	___	___	___	___
5%	25%	50%	75%	95%

- Assume that Zwendl predicts a water level of 3 m above N.A.P. (the standard baseline water level) for a given location in the dike ring area of Hoeksche

Waard; what is then the difference between the actually measured value of the local water level and the value computed with Zwendl (in meters):

___	___	___	___	___
5%	25%	50%	75%	95%

- Assume that Zwendl predicts a water level of 4 m above N.A.P. (the standard baseline water level) for a given location in the dike ring area of Hoeksche Waard; what is then the difference between the actually measured value of the local water level and the value computed with Zwendl (in meters):

___	___	___	___	___
5%	25%	50%	75%	95%

PART D

Cases with Emphasis on Maintenance and Maintainability

Chapter 16, "Component Reliability, Replacement, and Cost Analysis with Incomplete Failure Data"

—Analyzes incomplete data on time to failure of axle bushes for a large ore loader used in underground metalliferous mining to develop a cost effective replacement policy, considering both age and block replacement.

Chapter 17, "Maintainability and Maintenance—A Case Study on Mission Critical Aircraft and Engine Components"

—Investigates failures of aircraft components to determine a maintenance policy that will reduce the amount of unscheduled maintenance without increasing cost in a fleet of four-engine aircraft operated by a commercial airline.

Chapter 18, "Photocopier Reliability Modeling Using Evolutionary Algorithms"

—Deals with the use of evolutionary algorithms for fitting photocopier component and system failure models, uses bootstrap resampling methods to obtain confidence intervals, and applies the results in making decisions regarding photocopier maintenance strategy.

Chapter 19, "Reliability Model for Underground Gas Pipelines"

—Models failure frequency of underground natural gas pipelines as a function of many pipe and environmental characteristics, employing historical data and expert judgment, to support inspection and maintenance decisions.

Chapter 20, "RCM Approach to Maintaining a Nuclear Power Plant"

—Presents a detailed model of a major element of a nuclear power plant, the charging pumps subsystem and the chemical and volume control system, and develops a reliability centered maintenance policy that is shown to lead to significant cost benefits.

Chapter 21, "Case Experience Comparing the RCM Approach to Plant Maintenance with a Modeling Approach"

—Compares maintenance options for a transition boiler in a high-volume, continuous-production facility that manufactures synthetic fertilizers in Northern China.

CHAPTER 16

Component Reliability, Replacement, and Cost Analysis with Incomplete Failure Data

Nicholas A. J. Hastings

16.1 INTRODUCTION

The vehicle illustrated in Figure 16.1 is an ore loader used in underground metalliferous mining. In the mine, ore is broken up by blasting, and falls to the floor of a tunnel. The loader then moves the ore to a chute or conveyor, from which it proceeds to a crusher and ultimately to the mine surface for milling and refining. The loader operates in hot, humid, and dusty conditions and is subject to a high level of vibration as it drives over a rough and debris-strewn tunnel floor. It also operates on a short cycle of loading and dumping of material and continually handles heavy loads, with a lift capacity of 7 tonnes. Under these conditions the reliability of the loaders is less than ideal, and a study was carried out with a view to reducing loader down time. The full scope of the study included the following:

- Pareto analysis to identify the most frequent causes of failure and to rank failure modes on a cost basis
- Tests to determine whether there was a trend in overall failure rate with calendar time
- Weibull analysis of various failure modes to identify burn-in, random, and wearout patterns
- Preventive replacement analysis for components subject to wearout

For the loaders as a whole, most failure modes were of a minor nature, such as electrical faults, hydraulic oil leaks, and pneumatic system air leaks. These generally showed no wearout pattern, and they were tackled by addressing issues in mainte-

Case Studies in Reliability and Maintenance, Edited by W. R. Blischke and D. N. P. Murthy.
ISBN 0-471-41373-9

Figure 16.1. Elphinstone load–haul–dump truck.

nance quality and procedures. Inspection of components for specific impending faults was carried out, but formal condition monitoring was limited to engine oil analysis. However, engine failures were not a significant problem. For some components, a distinct wearout pattern was identified, and for these the question of an optimal preventive replacement policy was examined. The main part of this chapter deals with the optimization of preventive replacement intervals for a particularly critical component, namely the oscillating axle bush. This component has been selected because it illustrates features that are relevant in comparable situations across many industries.

The data used in the study are discussed in the next section. The modeling of failures in this context is discussed in Section 16.3. In Section 16.4, we look at component replacement options and give a cost analysis of these in Section 16.5. Conclusions of the study are summarized in Section 16.6.

16.2 MAINTENANCE DATA

Maintenance activities on the mine site were recorded using a computerized maintenance management system. To set this up, the various major components of the loaders (and other equipment) had been coded using the concept of positions. A position is a location occupied by a type of component or assembly. Positions are identified by position numbers. For example, the vehicle engine as a whole was identified by position number 02ENG0101. To identify the engine position of a particular vehicle also requires the vehicle serial number. Activity types, such as repair, and replace apply to positions. Specific engines were identified by engine serial number if required.

The basic maintenance records showed, in date order, the maintenance activities carried out on the vehicles, and they indicated the position number and type of activity and the vehicle operating hours at the time of the activity. From these data, events relating to the oscillating axle bushes were extracted. Table 16.1 shows a brief extract of such data relating to a particular vehicle. This data was extracted at the time of the study, in December 1994.

16.2.1 Establishing the Reliability Data

We are interested in determining the life distribution of the axle bushes—in particular, in the failure rate pattern. From the point of view of determining a suitable maintenance strategy, it is important to know whether the bushes exhibit a wearout pattern or not. In order find this out, we need to extract, from our maintenance records, data in a suitable form for reliability analysis. Looking at the data in Table 16.1, we see from row 1 that an axle bush failure occurred in May 1991, when the loader had operated for 2662 hours. We were able to ascertain that the loader had been commissioned in July 1990 and that this was the first axle bush failure on this vehicle, so the age at failure of the bush was 2662 hours. This is shown in Table 16.2, row 2.

The next event in Table 16.1, row 2, is another bush failure at 3114 vehicle-operating hours, and this bush lasted (3114 – 2662) = 452 hours. This is shown in Table 16.2 at row 3. Then from Table 16.1, row 3 we see that a bush was replaced without failure at 5366 vehicle operating hours. This bush ran for (5366 – 3114) = 2252 hours, but did not fail. This gives rise to a "suspension" (also called "service time"), as indicated in the right-hand column of Table 16.2.

16.2.2 Failures and Suspensions

In the right-hand column of Table 16.2, "F" means that the component was replaced as a result of failure—it was a "Failure Replacement." This can include situations where a component has deteriorated to a point where it is considered prudent to replace it. "S" refers to "suspensions." Suspensions or suspended items are items for which we have successful performance data, but which have not failed nor reached

Table 16.1. Maintenance Data for an Axle Bush Position

Row	Date	Event	Vehicle Operating Hours
1	May 1991	Axle bush failure	2662
2	July 1991	Axle bush failure	3114
3	Feb. 1993	Vehicle overhaul including bush replacement	5366
4	Aug. 1993	Axle bush failure	8942

Table 16.2. Failure and Suspension Data for an Axle Bush Position

Row	Date	Event	Vehicle Operating Hours	Axle Bush Operating Hours	Failure (F) or Suspension (S)
1	July 1990	Ore loader commissioning	0		
2	May 1991	Axle bush failure	2662	2662	F
3	July 1991	Axle bush failure	3114	452	F
4	Feb. 1993	Vehicle overhaul including bush replacement	5366	2252	S
5	Aug. 1993	Axle bush failure	8942	3576	F
6	Dec. 1994	Current bush running	13396	4454	S

an unacceptable, near failure, condition. One source of suspensions can be preventive replacements carried out because the vehicle was in the shop and it was an accepted practice to replace the bush while other work was being done. This is what occurred to the axle bush referred to in row 4 of Table 16.2. It ran for 2252 hours without failure, but was then replaced even though there was nothing to indicate that it was about to fail. Other conditions where you would record "suspensions" are if the machine were taken out of service, or sold, or the whole linkage were replaced with a new (or reconditioned) one, with the old bush being removed with the old linkage. Another source of suspended items is items currently running which have not failed.

Continuing with Table 16.1, row 4, we see that a failure occurred at 8942 vehicle operating hours, corresponding to a bush life of (8942 – 5366) = 3576 hours. This gives us the entry in Table 16.2, row 5. This concludes the entries in Table 16.1. However, it is also important to include the life to date of the bush that is currently installed. To get this we need the current vehicle odometer reading (as at the date of when the data was gathered—December 1994), which was found to be 13,396 hours. Hence the current bush has been running for (13,396 – 8942) = 4454 hours and has not failed. This results in the "suspension" entry in Table 16.2, row 6.

16.2.3 Axle Bush Data for All Loaders

The data in Table 16.2 relate to a particular loader. There are six loaders in all, and there are similar types of data records for each. We made a judgment that we could amalgamate the data for the axle bushes across the loaders. The resulting data are shown in Table 16.3. In Table 16.3 the data have also been sorted by age. We see that there are quite a number of suspended items (in fact, 14 suspensions and only 11 failures). This is not unusual in practical data, since we often find many items running which have not yet failed, or we may have a preventive replacement policy in place under which items are already replaced on an age basis.

Table 16.3. Oscillating Axle Bushes Data

Item Ref: Oscillating axle bushes—position No. 050SC0102
Age Unit: OP HRS
User: RELCODE User Manual

Record Number	Age	Event Type
1	290	S
2	334	S
3	452	F
4	695	F
5	769	F
6	1668	F
7	2150	S
8	2210	S
9	2252	S
10	2467	S
11	2607	S
12	2662	F
13	3212	S
14	3260	F
15	3576	F
16	3820	S
17	3852	S
18	3984	S
19	4011	S
20	4203	S
21	4454	S
22	4636	F
23	4818	F
24	5041	F
25	5134	F

16.3 MODELING FAILURES

In this study, the RELCODE software package was used for data analysis (see references and Appendix for brief details). RELCODE fits several distribution models based on the Weibull family, by a number of different methods. The user can then choose a preferred model, or accept the model recommended by the software. We shall proceed with the case study using the distribution model recommended by the software. For the axle bush data, this is a five-parameter bi-Weibull distribution, also known as a competing risk model [see Nelson (1982), Jiang and Murthy (1995, 1997), Blischke and Murthy (2000), and Evans et al. (2000)], derived by adding two Weibull hazard functions. The particular form of bi-Weibull model used here can provide a close approximation to a bathtub curve, though it is not confined to this form, and can model any two successive or overlapping Weibull failure rate patterns.

The first of the hazard functions is a two-parameter Weibull hazard function with the equation

$$h(t) = \lambda\theta(\lambda t)^{(\theta-1)}, \qquad t \geq 0 \tag{16.1}$$

In equation (16.1), t is the component age, $h(t)$ is the hazard function at age t, λ is a scale parameter, and θ is a shape parameter. The case where θ equals 1 corresponds to a constant failure rate λ.

The second hazard function is a three-parameter Weibull hazard function, which becomes operative for $t > \gamma$. The equation is

$$h(t) = \left(\frac{\beta}{\eta}\right)\left(\frac{(t-\gamma)}{\eta}\right)^{(\beta-1)}, \qquad t \geq \gamma \tag{16.2}$$

In equation (16.2), β, η, and γ are shape, scale, and location parameters, respectively, as in the three-parameter Weibull distribution. Adding the two hazard functions gives the relevant bi-Weibull distribution, for which the hazard and reliability equations are:

Hazard

$$h(t) = \lambda\theta(\lambda t)^{(\theta-1)}, \qquad 0 < t < \gamma \tag{16.3}$$

$$h(t) = \lambda\theta(\lambda t)^{(\theta-1)} + \left(\frac{\beta}{\eta}\right)\left(\frac{(t-\gamma)}{\eta}\right)^{(\beta-1)}, \qquad t \geq \gamma \tag{16.4}$$

Reliability

$$R(t) = e^{-(\lambda t)^{\beta}}, \qquad 0 < t < \gamma \tag{16.5}$$

$$R(t) = e^{-[(\lambda t)^{\theta} + ((t-\gamma)/\eta)^{\beta}]}, \qquad t \geq \gamma \tag{16.6}$$

In equations (16.1) through (16.6), θ is not confined to values less than or equal to 1, and β is not confined to values greater than 1, although the values often conform to these ranges in practice.

The following rules apply: $\gamma \geq 0$, $\eta > 0$, $\beta > 0$, $\lambda \geq 0$, $\theta \geq 0$; if $\lambda = 0$, then $\theta = 0$. This bi-Weibull distribution reduces to an ordinary Weibull distribution if $\lambda = \theta = 0$, in which case the conventional Weibull parameters γ, η, β are used.

Figure 16.2 shows a Weibull probability scale plot of the data with a bi-Weibull curve fitted. The two-parameter Weibull plot involves fitting a single straight line to the data, but it is not unusual in practice to find that the points appear to lie on two straight lines. This is consistent with the well-known concept of multiple failure patterns, typically some combination of burn-in, random, and/or wearout failures. In the present case we have an approximately random failure

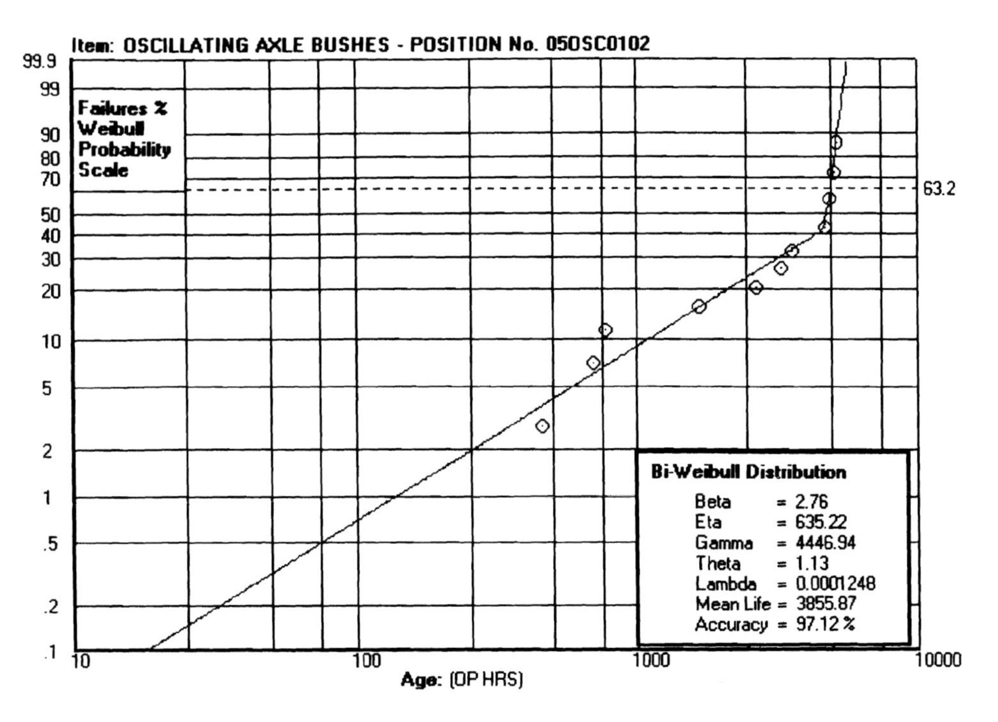

Figure 16.2. Weibull probability paper plot with bi-Weibull curve fitted.

pattern (with shape parameter 1.13) initially, followed by a wearout pattern (shape parameter 2.76).

16.3.1 Model Accuracy

The accuracy of the model was assessed by calculating the root mean square probability error. This is based on the distance of the data points from the cumulative probability of failure curve, as measured on a linear scale. This method is chosen because errors based on the probability paper scale distances can be misleading. The "accuracy" value shown in Figure 16.2 is 1 minus the root mean square probability error, multiplied by 100, so that it appears as a percentage. One hundred percent accuracy would mean that all the points were on the fitted curve. A fuller discussion of this point, along with details of relevant statistical tables, is given by Ang (1994) and Ang and Hastings (1994). The method of model selection involves fitting two-, three- and five-parameter Weibull or bi-Weibull models by linear regression, maximum likelihood, and nonlinear programming methods. The root mean square probability error is calculated in each case, and the results are judged on a scale that requires models with more parameters to achieve higher levels of accuracy than simpler models. In this case, the five-parameter bi-Weibull model emerged as the preferred model.

16.3.2 Onset of Wearout

A key factor in determining optimal replacement policies is to be able to detect the onset of wearout. In this regard, the bi-Weibull distribution, which allows for two different failure rate patterns for the one item, has an important role to play. The assumption of a single failure rate pattern, as implied by fitting a two-parameter Weibull distribution, can obscure the fact that wearout often becomes a factor late in the life of components. This can become visually apparent on a Weibull plot such as Figure 16.2, but the highly nonlinear scale of Weibull probability paper may obscure the wearout effect. Figure 16.3 shows the same data plotted on linear scales and expressed as reliability rather than cumulative probability of failure. Figure 16.3 shows a sharp drop in the reliability of the axle bushes after about 4400 operating hours.

The change in failure rate pattern is also illustrated in Figure 16.4, which shows the hazard function for the fitted bi-Weibull distribution.

Thus we conclude that the axle bushes are subject to a distinct wearout phase that comes into operation at about 4400 operating hours, indicating that a preventive replacement policy is likely to be appropriate.

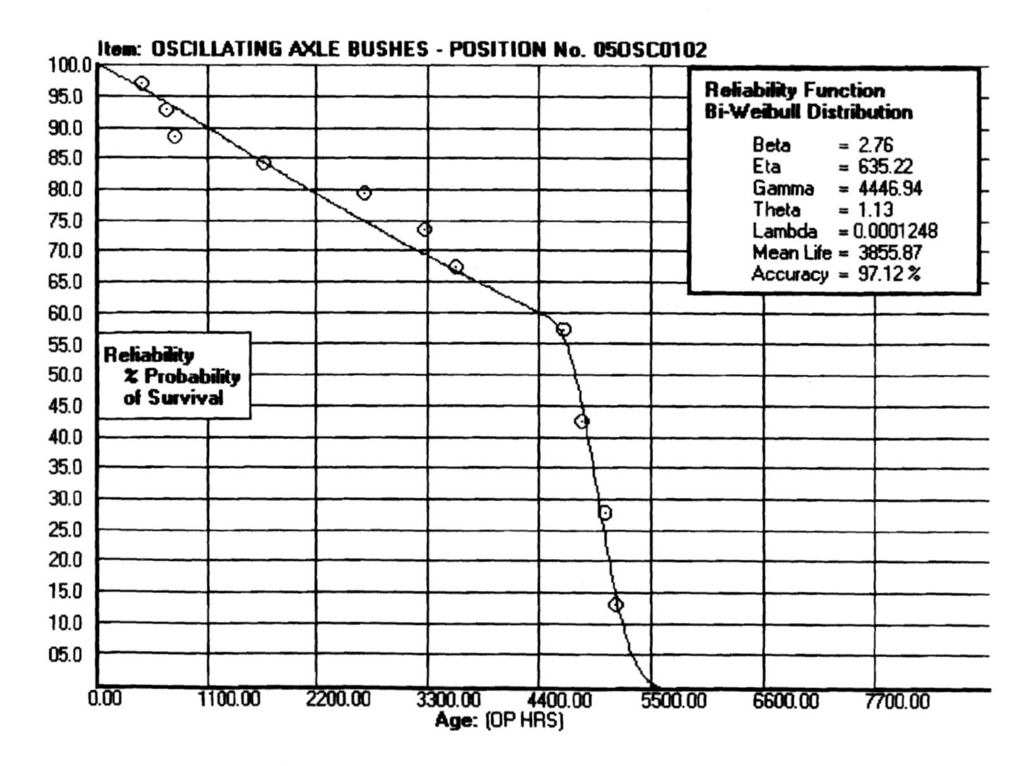

Figure 16.3. Reliability plot with bi-Weibull curve fitted.

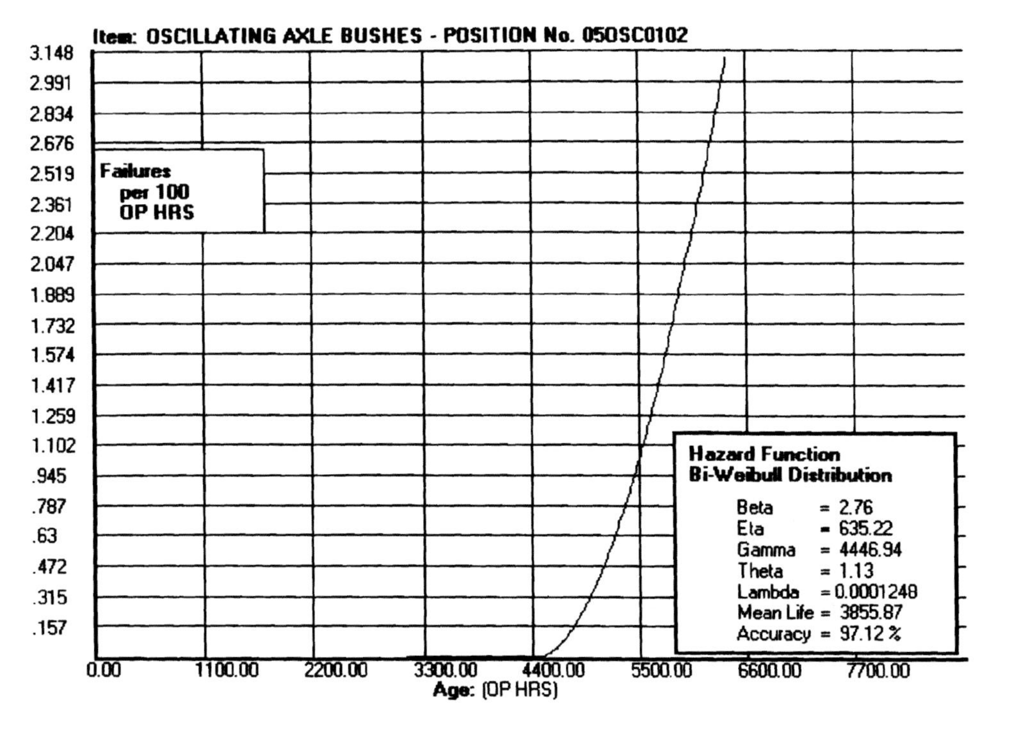

Figure 16.4. Hazard function for the fitted bi-Weibull distribution.

16.4 COMPONENT REPLACEMENT POLICY OPTIONS

16.4.1 Background

A common problem for maintenance managers is to determine a policy to adopt in regard to the replacement of components, which do, or may, fail. Background to the type of problems that are encountered and to the main methods of analysis is given by Jardine (1973). The current case of the axle bushes is typical of one of the most common forms of problem encountered in practice. The appropriate policy will depend on such factors as:

- The reliability of the component as a function of operating life, and in particular whether wearout occurs. In the present case study we have established that wearout is occurring from about 4000 operating hours onwards.
- The costs arising if we need to replace a component at an inconvenient time as the result of actual failure, or of the detection of an imminent failure condition
- The costs associated with replacement of the component before failure, at a convenient time—for example, at a routine maintenance time.

- Whether a practical method of condition monitoring exists for the component, how effective it is in predicting failure, and what options it may provide for component replacement. In the case of the axle bushes, condition monitoring was not a practical option, although operators and technicians could in some cases detect a deteriorated condition in sufficient time to prevent a failure occurring when the loader was in actual operation.

16.4.2 Failure Replacement and Preventive Replacement

In our analysis of the axle bush problem we considered two types of situations in which component replacements occur. We refer to these as *failure replacement* and *preventive replacement.* The descriptions of these are as follows:

(a) *Failure Replacement.* A failure replacement is a replacement that occurs following the failure of a component in service, or following the identification of an unfavorable condition that leads us to promptly replace the component, within a short time of the condition being detected.

(b) *Preventive Replacement.* A preventive replacement is replacement of a component that has not failed. Preventive replacements may occur because the component has reached an age deemed appropriate for preventive replacement. An example where this occurs is with "lifed" components in aircraft. Preventive replacements can also occur when items are in workshops for an annual overhaul, for example, and it seems advisable to replace components in the expectation that they will then run for another year.

16.4.3 Component Replacement Policies

Three types of component replacement policy were considered for the axle bushes. These are:

(a) *Replace Only on Failure.* Under this policy, only failure replacements are carried out and there is no preventive replacement.

(b) *Age-Based Preventive Replacement.* Under this policy, replacements occur for items that reach a certain specified age, and failure replacements occur if a component fails or reaches an imminent failure condition before the specified age. Determination of the "specified age" is the key part of the policy.

(c) *Block Replacement.* Under this policy, replacements occur for all the components under consideration at regular intervals of elapsed time. Failure replacements occur for items that fail between block replacements. The determination of the time between block replacements is the key part of the policy.

For the age-based policy it will be necessary to keep records of the ages of particular components. For the block policy and the replace-only-on-failure policy, the ages of particular components are not needed.

16.4.4 Is Preventive Replacement Worthwhile?

Preventive replacement can only be worthwhile if two conditions hold:

(a) The failure rate of the components is increasing, or will increase before another preventive replacement opportunity occurs

(b) The cost of failure replacement is greater than the cost of preventive replacement.

Preventive replacement is not appropriate if the failure rate (hazard function) is decreasing or constant. This is why we have initially carried out an analysis of the life distribution of the axle bushes and, in this case, determined that wearout is occurring.

Even if wearout occurs, the choice of policy will depend also on the cost of preventive replacement being less than the cost of failure replacement. Preventive replacement policies result in loss of useful life of the components that are removed before failure. For preventive replacement to be worthwhile, cost savings resulting from fewer failure replacements must more than compensate this loss. This can only occur if failure replacements are expensive when compared to preventive replacements. The determination of an optimal (i.e., minimum cost) policy will depend on a trade-off between these factors.

The cost of making a preventive replacement is usually less than the cost of failure replacement because we can arrange for preventive replacements to be made so as to avoid loss of production. Also, if preventive replacement is carried out as part of a routine service or overhaul, the repair cost tends to be reduced because the replacement can be done as part of the other work.

16.4.5 Cost of Lost Production

In this case study, important factors were (a) the cost of lost production that resulted from a catastrophic axle bush failure and (b) the cost of secondary mechanical damage. The extent of lost production could vary considerably, depending on just when a failure occurred. Thus, some judgment was needed to put an average figure on this cost. An advantage of computer analysis with RELCODE was that it was easy to carry out a "what if" analysis with different cost figures and see what effect this has on the replacement policy.

The loaders, of which six were in operation at the time of the study, play a critical role in the mining process, in that they move virtually all the ore. The mine works two shifts. Overtime working by the loader operators is not practicable, but maintenance workers can work outside the normal operating shifts. Loss of ore movement due to breakdown of the loaders affects the mine output. An estimate of the value of ore moved per hour when a loader is working is $15,000. However, the concept of "failure" of the axle bush covers a range of circumstances. In some cases we have a catastrophic failure, which necessitates dismantling of the loader and repair in situ. This can result in two to three hours lost production. In the worst case, there can also be secondary damage to the hydraulic system and to mechanical linkages.

On the other hand, in more than 50% of cases, deterioration of the axle bush is detected in time to move the loader to the maintenance bay. In this case, a smaller machine can still move ore, though less efficiently. Also, if a failure occurs toward the end of the working day, the maintenance crew working in the third shift can repair the loader. On balance, therefore, an assessment of the average cost C_f of failure replacement gave a figure of $5000. This amount includes the following:

- Cost of the replacement component
- Cost of labor and related overheads, including an allowance for a percentage of overtime working
- Cost of lost production in an average case

For analysis purposes we also made the following reasonable assumptions.

- The operators became aware of a failure as soon as it occurred.
- The time interval of interest (or horizon) is infinite so that one uses asymptotic results.
- A spare is available when needed.

16.4.6 Cost of Preventive Replacement

The cost C_p of preventive replacement is relatively straightforward to estimate. Preventive replacements are carried out at preplanned time, usually in conjunction with routine servicing, when the loader is out of service in any case. Thus the cost consists of:

- Cost of the replacement component
- Cost of labor and related overheads

In the present case, this was estimated at $500.

Thus the requirements for the steady-state optimality of a preventive replacement policy, namely, the presence of wearout and $C_f > C_p$, were established.

16.5 PLANNED REPLACEMENT—ANALYSIS AND COSTS

Before proceeding with analysing the replacement policy options for the axle bush, we shall look briefly at the logic underlying the various policies. First we consider the "replace-only-on-failure" strategy.

16.5.1 Replacement Only on Failure (ROOF)

The simplest form of replacement policy is to replace only on failure. That is, we carry out failure replacements only and do not do any preventive replacements. In

Table 16.4. Results for the Replace-Only-on-Failure Policy

Preventive replacement cost	\$500
Failure replacement cost	\$5000
Mean life	3855.85 hr
Cost for replacement only on failure	1.2967 \$/hr

this case, since replacements occur according to a renewal process with time between renewals given by the distribution function $F(t)$, we have from the renewal reward theorem (Ross, 1972) that the expected cost J per unit time is given by

$$J = \frac{C_f}{\mu} \tag{16.7}$$

where μ is the mean time to failure.

For the axle bush case study, we can use equation (16.7) to calculate the average cost per operating hour arising from the axle bush failures under a "replace-only-on-failure" policy. The result, which was calculated using RELCODE, is shown in Table 16.4. Incidentally, if we had not used the bi-Weibull distribution to detect the onset of wearout, we would get a much higher and quite misleading figure for the mean life. Note that the cost for the replace-only-on-failure policy is \$1.29 per operating hour. Later we shall compare this to the costs under the age-based and block replacement policies.

16.5.2 Aged-Based Preventive Replacement Policy

In an age-based preventive replacement policy, items are replaced under the following rules:

- A "preventive replacement age," denoted t_p, is set. If an item fails before age t_p, a failure replacement is made.
- If an item survives to age t_p, a preventive replacement is made at age t_p.

To implement this policy in practice, we need to record when each replacement occurs, so that the age of every component is known. In practice, the saving from preventive replacement may depend on the preventive replacement occurring at a convenient time—for example at the next routine service. At the time of such a service, any component that has reached its preventive replacement age is replaced. Items that fail before the preventive replacement age still require failure replacement.

16.5.2.1 Cost Minimization with Age-Based Preventive Replacement

The cheapest age-based preventive replacement policy is the one that has the lowest long-run cost per unit time. This cost is derived as the ratio of the average replacement cost per component and the average life per component. Here C_f is the cost of failure replacement, C_p is the cost of preventive replacement, and $F(t_p)$ is the probability of Failure Replacement, so average cost c per component is given by

$$c = C_f F(t_p) + C_p [1 - F(t_p)] \tag{16.8}$$

16.5.2.2 Truncated Mean Life

The average life per component under an age based replacement policy is also called the *truncated mean life.* It is calculated by taking into account the fact that some components (a proportion $1 - F(t_p)$) will remain in use until age t_p, but some will fail at various ages between 0 and t_p. This latter group are items that are replaced on failure. Denoting the failure probability density function by $f(t)$, the truncated mean life, say μ_{tp}, is given by

$$\mu_{tp} = t_p[1 - F(t_p)] + \int_0^{t_p} tf(t)\, dt \tag{16.9}$$

Integration by parts yields the equivalent expression

$$\mu_{tp} = \int_0^{t_p} [1 - F(t)]\, dt \tag{16.10}$$

The average cost per unit time, $J_A(t_p)$, is given by dividing equation (16.8) by (16.10) to give

$$J_A(t_p) = \frac{C_f F(t_p) + C_p[1 - F(t_p)]}{\int_0^{t_p} [1 - F(t)]\, dt} \tag{16.11}$$

The minimum cost policy is found by evaluating equation (16.11) for range values of t_p, and choosing the value that gives a minimum. The cost per unit time will vary with the preventive replacement age in the way shown in Figure 16.5, which is the graph of cost per operating hour versus preventive replacement age obtained using RELCODE The cost per unit time will have a minimum value which will occur at the optimal preventive replacement age.

The asymptotic value of C_p as t_p increases is given by equation (16.8). If the cost of failure replacement is not greater than the cost of preventive replacement, or there is no wearout, then the graph represented by Figure 16.5 will not show a minimum. In such cases, the optimal policy is to replace only on failure.

16.5.2.3 Graph, Optimal Policy, and Cost Savings

We see from Figure 16.5 that the cost per unit time falls initially quite steeply as the preventive replacement age increases. There is then a flat region around the optimal preventive replacement age, and finally the cost per unit time rises again. The lowest point on the graph is at the optimal preventive replacement age. There is an in-

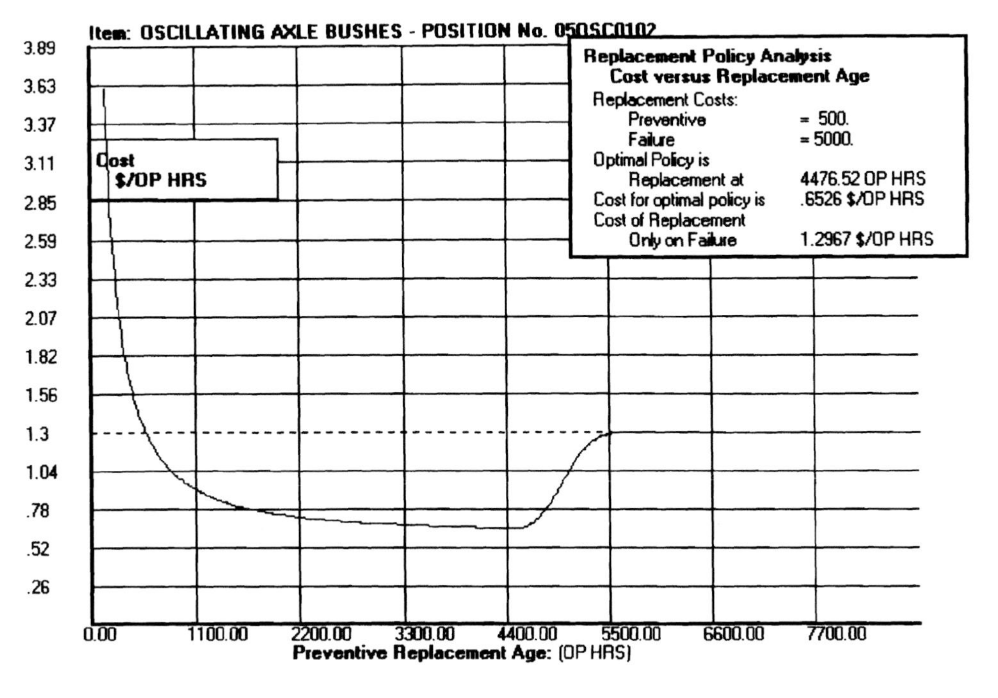

Figure 16.5. Graph of the cost per operating hour versus preventive replacement age.

terval around this value where the costs do not vary much. Thus the graph provides a sensitivity analysis for the preventive replacement age. The dashed horizontal line is drawn at the cost of a policy of replacement only on failure. This gives a visual indication of the savings from a preventive replacement policy relative to the "replace-only-on-failure" policy. The graph also shows the range of ages for which an age-based preventive replacement policy will result in lower costs than the policy of replacement only on failure. Table 16.5 summarizes the results from RELCODE. The cost of the age-based policy is a minimum when replacement occurs at 4476 operating hours, and the cost is then \$0.6526 per operating hour. This compares with the cost of \$1.2967 per operating hour for the replace only on failure policy, as given in Table 16.4. This represents a saving of nearly 50%. Given that there are six

Table 16.5. Optimal Age-Based Preventive Replacement Policy

Preventive replacement cost	\$500
Failure replacement cost	\$5000
Cheapest preventive replacement age	4476.52 hr
Cheapest cost \$/operating hour	0.6526 \$/hr
Saving of preferred policy \$/operating hour	0.6441 \$/hr
Preventive replacements as % of all replacements	60%

loaders, each with two axle bushes and each averaging 5000 hours utilization per year, the total annual cost saving is $38,644.

16.5.2.4 Specified Preventive Replacement Age

The mathematically optimal solution to the preventive replacement problem is to replace at an operating age of 4476 hours. However, in practice we would not stop the loader solely to replace a working axle bush. The cost entered for preventive replacement is based on the assumption that we carry out the axle bush replacement at a convenient time when the loader is being serviced. This is normally every 1000 hours, so this suggests that we should nominate 4000 hours as the age for preventive axle bush replacement. RELCODE will calculate the cost per operating hour for any specified preventive replacement age, so we can evaluate the cost of this policy and compare it with the optimal value. We saw from Figure 16.5 that the cost curve is quite flat in the region of the optimum, so choosing a convenient age near the minimum is a sound approach. Using 4000 hours as a specified preventive replacement age gives the result shown in Table 16.6, which is $0.6585 per operating hour. Thus the increased cost is only a fraction of a cent per operating hour compared to the optimal value given in Table 16.5. We therefore decide on 4000 hours as the preventive replacement age for the axle bushes. In practice this will mean that we replace the axle bushes at a service time, if the existing bushes have been in service for approximately 4000 hours.

16.5.2.5 Sensitivity Analysis

The graph in Figure 16.5 allows us to see at a glance how the costs vary with the choice of preventive replacement age. In fact, the graph is almost flat between about 2500 and 4500 hours, so any preventive replacement age in this range will give much the same costs. We can also obtain more precise results for any given preventive replacement age by using the "specified preventive replacement age" feature in RELCODE which will calculate the exact costs for any preferred policy. We can also easily vary the cost estimates and check the sensitivity of the policy to these changes. Further illustrations of sensitivity analysis are given in the block replacement section.

16.5.3 Block Preventive Replacement Policy

With an age-based replacement policy, we need to track the ages of individual components. This requirement can impose an extra cost. In the axle bush study,

Table 16.6. Cost Analysis for a Specified Preventive Replacement Age of 4000 Hours

Preventive replacement cost	$500
Failure replacement cost	$5000
Specified preventive replacement age	4000 hr
Cost $/operating hour for this age	0.6585

we did track the ages of the individual components, but we also investigated the costs for a block policy, for these and other components, and this option will now be discussed

In a block preventive replacement policy, components are replaced at fixed intervals of time (or operating life). Items that fail in between the block replacement times are replaced when they fail, these being failure replacements. At the time of block replacement, all items are replaced, including those that have already been subject to failure replacement since the last block replacement. We refer to the time between block replacements as the block replacement *interval.*

The block replacements are preventive replacements. The cost of a block preventive replacement may differ from that of an age-based preventive replacement. Usually a block replacement will be cheaper (per component replaced) because there are economies of scale in doing many replacements at the same time, and there is no need to track individual component ages.

In the case of the loader, a block policy involves replacing the axle bushes (there are two per vehicle) at a chosen service age of the loader, regardless of the age of the specific components in use at the time. To determine the average replacement cost per unit time, $G(x_p)$, we use an asymptotic result. The asymptotic expected cost per unit of time is given by

$$G(\tau) = \frac{C_p + C_f M(\tau)}{\tau} \tag{16.12}$$

where $M(x)$ is the renewal function associated with $F(x)$ and given by

$$M(x) = F(x) + \int_0^x M(x-t)dt \tag{16.13}$$

$M(x)$ can be expressed in closed form for only a few simple cases, but it is easily approximated or evaluated numerically [see Xie (1989), Blischke and Murthy (1994)].

For the case where no preventive replacements are made, the cost is given by equation (16.7).

16.5.3.1 Block Policy Results and Sensitivity Analysis

Analysis of block policies for the loader data has been carried out and indicates that the practical interval that gives close to minimal costs is again 4000 hours. A sensitivity analysis for this result was carried out using preventive replacement costs in the range \$400 to \$500. The results of this are shown in Table 16.7. As we would expect, Table 16.7 shows that when the preventive replacement costs are the same, the cost of the block replacement policy, at \$0.6792, is higher than the age-based policy cost of \$0.6585, Table 16.6. This is because the block policy involves some loss of useful life of second-generation components. However, the block policy is cheaper to implement, so that some reduction in the preventive replacement cost relative to the age-based case can be expected. Table 16.7 shows a sensitivity analy-

Table 16.7. Cost Sensitivity Analysis for a Block Preventive Replacement Age of 4000 Hours

Preventive replacement cost	\$400	\$450	\$500
Failure replacement cost	\$5000	\$5000	\$5000
Specified preventive replacement age	4000 hr	4000 hr	4000 hr
Cost \$/operating hour for this age	\$0.6542/hr	\$0.6667/hr	\$0.6792/hr

sis for preventive replacement costs in the range \$400 to \$500 and shows that at \$400 the block policy is worthwhile. In the case study, it was decided to continue with the age-based policy for this component, but for other less critical components a block policy was adopted.

16.5.3.2 Decision Tables for Two-Parameter Weibull Case

The oscillating axle bush component, on which the bulk of this description is based, was only one of a range of components analyzed. In over 50% of cases a two-parameter Weibull model was reasonably accurate. For this model, the optimal age-based and block preventive replacement policies are given in Tables 16.8 through 16.11 in terms of the Weibull shape parameter (β) and scale parameter (η). The tables also provide a simple sensitivity analysis to the beta value and ratio of failure to preventive replacement cost.

16.5.3.3 Decision Tables Example

The use of Tables 16.8 to 16.11 in the case study is illustrated by the following example. Reliability data (failures and suspensions) for a hydraulic seal were analyzed, and the life distribution was found to be a two-parameter Weibull with life and cost parameters shown in Table 16.12.

To get the optimal replacement policy from the tables, we calculate the ratio C_f/C_p and select an approximately corresponding value in the tables. Similarly, we approximate the β value by one shown in the tables. In this case, we get $C_f/C_p \approx 6$, $\beta \approx 2.5$.

Table 16.8. Optimal Preventive Replacement Age as a Percentage of η

Cost Ratio	β					
C_f/C_p	1.5	2	2.5	3	3.5	4
10	38	34	36	38	41	44
8	45	38	39	42	44	46
6	57	45	45	48	49	51
4	*	59	56	56	57	58
2	*	*	*	81	78	77

Note: An asterisk (*) indicates that the cost of the preventive replacement policy is more than 90% of the cost of a replace-only-on-failure policy.

Table 16.9. Cost of Optimal Age-Based Preventive Replacement Policy as a Percentage of the Cost of a Replace-Only-on-Failure Policy

Cost Ratio	β					
C_f/C_p	1.5	2	2.5	3	3.5	4
10	75	54	42	35	31	28
8	80	59	48	41	36	32
6	85	67	56	49	43	40
4	*	79	69	62	57	53
2	*	*	*	88	84	82

Note: An asterisk (*) indicates that the cost of the preventive replacement policy is more than 90% of the cost of a replace-only-on-failure policy.

Table 16.10. Optimal Preventive Block Replacement Interval as a Percentage of ETA

Cost Ratio	β					
C_f/C_p	1.5	2	2.5	3	3.5	4
10	39	33	35	37	40	43
8	47	38	38	41	43	45
6	62	45	44	45	47	49
4	*	59	54	53	54	55
2	*	*	*	*	*	*

Note: An asterisk (*) indicates that the cost of the preventive replacement policy is more than 90% of the cost of a replace-only-on-failure policy.

Table 16.11. Cost of Optimal Block Preventive Replacement Policy as a Percentage of the Cost of a Replace-Only-on-Failure Policy

Cost Ratio	β					
C_f/C_p	1.5	2	2.5	3	3.5	4
10	77	55	43	36	31	28
8	82	61	49	42	37	33
6	89	70	58	50	45	41
4	*	84	73	66	60	56
2	*	*	*	*	*	*

Note: An asterisk (*) indicates that the cost of the preventive replacement policy is more than 90% of the cost of a replace-only-on-failure policy.

Table 16.12. Distribution Parameters and Costs for the Hydraulic Seal

β	2.4
η	6412 operating hours
C_p	\$235
C_f	\$1500

From Table 16.8 we see that the optimal preventive replacement age is 45% of η, which is 2885 hours. Because there is a major service at 1000-hour intervals, replacement of the seals at 3000 hours is indicated. However, the seals are a minor component for which we do not track individual ages at present. This means that a block policy would be easier to implement, so we also look at Table 16.10, which gives the optimal block replacement age as 44% of η. The difference in timing between the two policies is only 1%, so in practice this means that we would carry out block replacement of these seals at the 3000-hour service. Before finalizing this policy, we need to consider costs. Table 16.9 shows that for the age-based policy, the cost is 56% of the replace-only-on-failure cost. Table 16.11 shows that for the block policy this increases to 58%. The cash value of this difference can be approximated by taking the replace-only-on-failure cost as C_f/η, which is \$0.23 per operating hour or \$14,036 per year for a fleet of six vehicles with two seals each, operating for 5000 hours per year. The cost difference of 2% is \$280 per year, which is not enough to justify tracking the individual seal ages. The saving of 42%, gained by adopting the block replacement policy, amounting to \$5895 per year, is worthwhile, however. A significant amount of approximation has taken place in obtaining the results for the hydraulic seal, but a more exact result can be obtained by use of the RELCODE software package described in the Appendix.

16.6 CONCLUSION

This case study has shown how, starting with practical data taken from maintenance records, we can analyze the failure pattern of a component, evaluate a number of alternative preventive replacement policies, and choose a policy that is practical and cost effective in our circumstances.

In particular we have seen how to:

- Extract data from our computerized maintenance management system and incorporate data relating to items that have run successfully without failure.
- Fit distribution models for our components including the bi-Weibull model, which helps us to identify the onset of wearout.
- Graphically display the reliability and hazard functions for our components (Figures 16.2, 16.3, and 16.4).
- Estimate costs of failure replacement and preventive replacement.
- Determine whether preventive replacement (age or block) is worthwhile.
- Graphically display the relationship between cost and preventive replacement age (Figure 16.5).
- Find the optimal (minimum cost) age-based preventive replacement policy.
- Find the optimal (minimum cost) block replacement policy.
- Find the long run average cost for any age-based replacement policy specified by the user.

- Find the long run average cost for any block replacement policy specified by the user.
- Find the long run average cost for a policy of replacement only on failure.
- Solve preventive replacement problems for components with a two-parameter Weibull life distribution.

ACKNOWLEDGMENT

This case study is based, with permission, on an extensive study of Elphinstone R1500 reliability and maintenance data carried out at Mt Isa Mine by Mr. Jeff Young. The support of MIM (Holdings) Ltd. in funding the MIM Chair in Maintenance Engineering at Queensland University of Technology is also acknowledged.

REFERENCES

Ang, J. Y. T. (1994). Model Accuracy and Goodness of Fit for the Weibull Distribution with Suspended Items, Ph.D. thesis, Monash University.

Ang, J. Y. T., and Hastings, N. A. J. (1994). Model accuracy and goodness of fit for the Weibull distribution with suspended items, *Microelectronics and Reliability* **34:**1177–1184.

Blischke, W. R., and Murthy, D. N. P. (1994). *Warranty Cost Analysis,* Marcel Dekker, New York.

Blischke, W. R., and Murthy, D. N. P. (2000). *Reliability:Modeling, Prediction, and Optimization,* Wiley, New York.

Evans, M. A., Hastings N. A. J., and Peacock, J. B. (2000). *Statistical Distributions,* 3rd edition, Wiley, New York.

Jardine A. K. S. (1973). *Maintenance, Replacement and Reliability,* Pitman, London.

Jiang, R., and Murthy, D. N. P. (1995). Reliability modelling involving two Weibull distributions, *Reliability Engineering and System Safety* **47:**187–198.

Jiang, R., and Murthy, D. N. P. (1997). Parametric study of competing risk model involving two Weibull distributions, *International Journal of Reliability, Quality, and Safety Engineering* **4:**17–34.

Nelson, W. (1982). *Applied Life Data Analysis,* Wiley, New York.

Ross, S. M. (1972). *Applied Probability Models,* Holden-Day, San Francisco.

Xie, M. (1989). On the solution of renewal-type equations, *Communications in Statistics B* **18:**281–293.

NOTATION

C_f Average cost of failure replacement

C_p Average cost of preventive replacement

$f(t)$ Failure probability density function
$F(t)$ Cumulative probability of failure
$J_A(t_p)$ Average cost per unit time for a preventive replacement policy
J Average cost per unit time for a replace-only-on-failure policy
$G(x_p)$ Average cost per unit time for a block replacement policy
$G_A(t)$ Asymptotic cost per unit time for a block replacement policy
$h(t)$ Hazard function (failure rate)
$M(t)$ Renewal function (cumulative)
$R(t)$ Reliability function
t Age of component
t_p Age of component at preventive replacement
x Time (or operating hours) since block replacement
x_p Time (or operating hours) between block replacements
β Weibull shape parameter
γ Weibull location parameter
η Weibull scale parameter
θ Bi-Weibull first-phase shape parameter
λ Bi-Weibull first-phase instantaneous failure rate
μ Mean life of component

EXERCISES

16.1. Explain what is meant by the following terms:
a. Failure replacement
b. Preventive replacement
c. Replace only on failure
d. Age-based preventive replacement
e. Block preventive replacement. What factors would favor a block policy over an age-based policy
f. Condition-based replacement
g. Cost of failure replacement
h. Cost of preventive replacement

16.2. What conditions are essential for an age-based or block preventive replacement policy to be better than a policy of replacement only on failure?

16.3. A certain type of component has been replaced only on failure. These components are installed in a number of stationary engines, with two similar components per engine. The total engine operating hours is 48,414, and the number of component failures is 9. The cost of component failure including spare part, repair, and downtime costs is estimated at \$2500 per failure. What is the average cost per component per 1000 operating hours?

16.4. In an underground mine, five similar pumps operate under a policy whereby refurbishment to good-as-new standard is carried out if a pump runs to 10,000 operating hours, or if catastrophic failure occurs before that time. Over a period, four catastrophic failures have occurred at an average age of 7500 hours, and 20 refurbishments of 10,000 hour pumps have occurred. The cost of refurbishment is estimated at \$15,000 and the extra cost associated with a catastrophic failure is \$100,000. What is the average cost per 1000 pump-hours for this policy?

16.5. Data analysis shows that a certain type of insulator used in high-voltage transmission lines has a Weibull life distribution with a shape parameter $\beta = 2.1$ and scale parameter $\eta = 16.8$ years. The cost of preventive replacement is $C_p = \$300$, and the cost of failure replacement is $C_f = \$2500$. Use Tables 16.9 and 16.10 to identify the optimal preventive replacement age and estimate the cost per insulator per year under this policy. Note that the cost of a replace-only-on-failure policy can be approximated by $J = 1.12*(C_f/\eta)$.

16.6. A block replacement policy can be applied to the insulators in Exercise 16.5, reducing the preventive replacement cost to $C_p = \$250$. If a block policy is adopted, what would then be the optimal preventive replacement age and the cost per insulator per year? Which policy is the better one, age-based or block?

APPENDIX. RELCODE SOFTWARE

In this study, the RELCODE software package was used. Reference details are as follows: RELCODE reliability, replacement, and inspection interval analysis software. Oliver Interactive Inc, www.oliver-group.com.

The data required for RELCODE are the age of the item when it failed or was suspended and whether the event was a failure or suspension. This is the same as the data given in the two right-hand columns of Table 16.3, though the data need not be sorted, and it can also be imported from Excel. For preventive replacement analysis, we also need estimates of the failure replacement cost and the preventive replacement cost. For spare parts analysis, we need to estimate the total annual component utilization of the parent equipment to which the components belong. Results similar to those obtained in this chapter can then be derived by use of the software. Tables 16.9 to 16.11 were calculated using the RELCODE software.

CHAPTER 17

Maintainability and Maintenance—A Case Study on Mission Critical Aircraft and Engine Components

U. Dinesh Kumar and John Crocker

17.1 INTRODUCTION

The way airlines, and indeed air forces, maintain and support their fleets is naturally rather sensitive information, particularly when one recognizes that this not only affects safety but has a major impact on system availability. To respect the sentiments of the airline, aircraft, and component manufactures, we have not disclosed their names. The problems and their analysis, as presented in the case study, are very similar for almost all the airlines and the aircraft manufacturers, irrespective of their size.

Before we get into the main case study, we would like to give a brief account of some of the macro-level issues surrounding the airlines and the aircraft manufacturers. One of the main driving forces with aircraft, whether civil or military, is safety. Aircraft do not float in air, unlike most ships in water. If the engines stop or the control surfaces cannot be manipulated, aircraft will generally land themselves pretty quickly under Newton's laws of gravity, usually with disastrous results.

Aircraft design tends to stretch technology to its limits. The operators want to carry higher payloads using less fuel often at higher speeds, but, of course, they do not want to lose their cargo or, indeed, their aircraft, and they want to keep their costs to a minimum. Depending on how one measures this, air travel can be shown to be the safest mode of transport. Unfortunately, it is also probably the least survivable and the most newsworthy when something does go wrong. This means that a great deal of emphasis is placed on safety and reliability.

Surprisingly, air travel has also become cheaper over the years. Between 1926 and 1993, the airfare has decreased from a dollar a mile to 13 cents a mile (Morrison and Winston, 1999). Deregulation of airlines played a major part in this

Case Studies in Reliability and Maintenance, Edited by W. R. Blischke and D. N. P. Murthy.
ISBN 0-471-41373-9

cost reduction, however, mainly at the expense of the airlines' operating profits. In the four years between 1990 and 1993, IATA (International Air Transport Agency) airlines, which has over 250 members, lost about $15.6 billion (Besse, 1999).

Maintenance accounts for about 10–15% of the operating cost. The breakdown of maintenance expenditure is shown in Figure 17.1 (Lam, 1995). Line maintenance refers to the maintenance carried out at the airports or "first line." Often this is achieved by simply replacing line replaceable units (LRU), which would then be sent to one of the lower [maintenance] echelons for subsequent recovery. Such work is referred to as "heavy maintenance" and often involves the use of sophisticated and expensive equipment. For example, on average, a typical Boeing 747 would generate a total maintenance cost of approximately $1700 per block hour. Here "block hour" refers to the total usage of the equipment including flying and taxiing time. This means that there is considerable opportunity to improve profits by removing unnecessary maintenance, provided, of course, that this does not simply move the costs to somewhere else.

17.1.1 Aircraft Maintainability

The early jet aircraft designs apparently paid little attention to maintainability. For example, the engine change on the Harrier GR3 requires the engine to be removed from the aircraft. To do this it is necessary to disconnect a variety of control systems and then remove one of the wings. This typically took around 24 hours and involved the use of heavy and bulky support equipment (Knezevic, 1999). For an aircraft that was intended to operate near the front line from rough terrain, this really was not satisfactory. But it took several years for the aircraft designers/manufacturers to realize the importance of maintainability.

The Anglo-French aircraft Jaguar was one of the first for which maintainability targets were set (Knezevic, 1999). Direct man-hour content was quoted for the first and second line tasks based on a certain number of sorties. The current generation

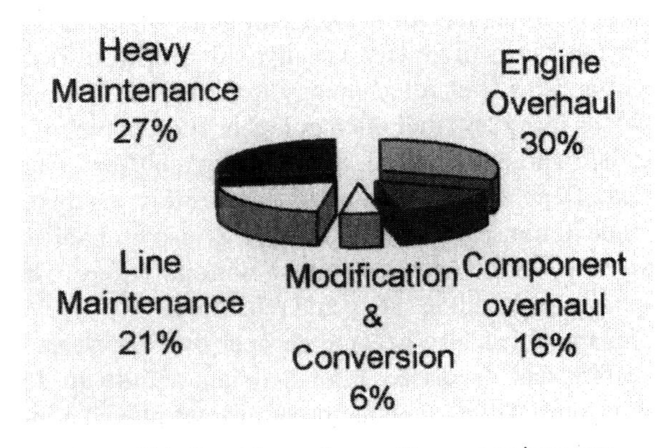

Figure 17.1. Breakdown of expenditures on maintenance.

of aircraft such as SAAB-Grippen, in the Swedish Air Force, requires around 10 minutes to turn around the aircraft. During this time, a variety of tasks, such as pre-flight inspection, refueling, reloading the gun, and mounting six air-to-air missiles, can be performed with the minimum equipment and manpower.

Maintainability studies have the following objectives:

1. To guide and direct design decisions by considering all the maintenance tasks, with a prime objective of reducing the total maintenance tasks/time
2. To predict quantitative maintainability characteristics of a system
3. To identify changes required to a system's design to meet operational requirements by reducing maintenance/inspection times

One of the common misperceptions is that maintainability is simply seen as the ability to reach a component to perform the required maintenance task (accessibility). Of course, accessibility is important. Figure 17.2 illustrates an accessibility problem in one of the older twin-engine fighter aircraft, the Gloster Javelin. Before an engine could be changed, the jet pipe had to be disconnected and removed. To remove the jet pipe, it was necessary for a technician to gain access through a hatch and then be suspended upside down to reach the clamps and pipes which had to be disconnected. The job could only be achieved by touch; the items were outside of the technician's field of view. The technician had to work his way down between the engine and the aircraft's skin, with tools in his hand. For safety reasons the mechanic has to be held by his ankles.

Maintainability should also consider factors such as visibility—that is, the ability to see a component that requires a maintenance action, testability (ability to detect system faults and fault isolation), simplicity, interchangeability, and human factors (e.g., not requiring superhuman strength to tighten or loosen nuts). Additionally, designers have to be aware of the environment in which maintainers oper-

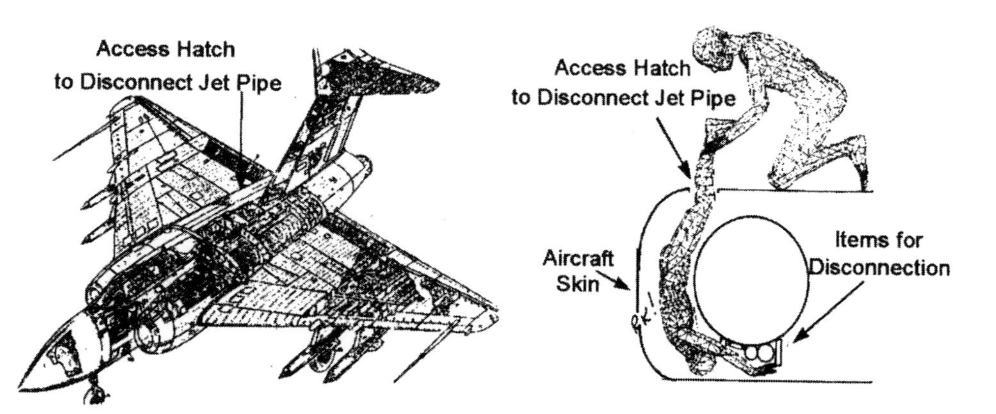

Figure 17.2. Maintainability problem in Gloster Javelin aircraft.

ate. It is much easier to maintain an item on the bench than at the airport gate, in a war zone, amongst busy morning traffic, or in any other result-oriented and schedule-driven environment.

17.1.2 Aircraft Maintenance

All aircraft must follow a maintenance program that is approved by a regulatory authority such as the FAA (Federal Aviation Administration), CAA (Civil Aviation Authority), and so on. Each airline develops its own maintenance plan based on manufacturers' recommendations and by considering its operation. Thus, two airlines may have slightly different maintenance programs for the same aircraft model used under similar operating conditions. Aircraft maintenance can be categorized as

1. Routine or scheduled maintenance
2. Non-routine or unplanned maintenance
3. Refurbishment
4. Modifications

Scheduled maintenance tasks are required at regular intervals, when life-limited components (LLC) achieve the given age, or due to airworthiness directives. The most common routine maintenance is visual inspection of the aircraft prior to a scheduled departure (known as *walk around*) by pilots and mechanics to ensure that there are no obvious problems (e.g., cracks, caps and hatches not locked shut, or even bits missing).

Routine maintenance can be classified as

1. Overnight maintenance
2. Hard time or life-limited maintenance
3. Progressive inspection

Overnight maintenance normally includes low-level maintenance checks, minor servicing, and special inspections done at the end of the working day for about one to two hours to ensure that the aeroplane is operating in accordance with the minimum equipment list. Overnight maintenance provides an opportunity to remedy passenger and crew complaints (Lam, 1995).

Hard time is the oldest primary maintenance process. Hard time requires periodic overhaul or replacement of affected systems/components and structures and is flight-, cycle-, and calendar-limited. That is, as soon as the component age reaches its hard time, it is replaced with a new component. Most of the rotating engine units are hard-timed. This is for three basic reasons: The parts are impossible to inspect in situ, their times to failure are strongly age-related, and their failure has an unacceptably high risk of catastrophic consequences.

Progressive inspections link time-related maintenance tasks with frequency of maintenance close to each other into convenient "blocks" so that the maintenance workload becomes balanced with time and maintenance can be accomplished in small "bites," making equipment more available. Different progressive inspection groups are (Kumar et al., 2000):

1. *Pre-flight.* Visual inspections carried out by the mechanic and the pilots to ensure that there are no obvious problems.
2. *A Check.* Carried out approximately every 150 flight hours, which includes selected operational checks (general inspection of the interior/exterior of the aircraft), fluid servicing, and extended visual inspection of fuselage exterior, power supply and certain operational tasks.
3. *B Check.* Occurs about every 750 flight hours and includes some preventive maintenance tasks such as engine oil spectro-analysis, oil-filter removal and checking, lubrication of parts as required, and examination of the airframe. It also incorporates A check.
4. *C Check.* Occurs every 3000 flight hours (approximately 15 months) and includes detailed inspection of airframe, engines, and accessories. In addition, components are repaired, flight controls are calibrated, and major internal mechanisms are tested. Functional and operational checks are also performed during C check. It also includes both A and B checks.
5. *D Check.* This is the most intensive form of routine maintenance and occurs about every 20,000 flight hours (6–8 years). It is an overhaul that returns the aircraft to its original condition, as far as possible. Cabin interiors, including seats, galleys, furnishings, and so on, are removed to allow careful structural inspections.

17.1.3 Case Study Approach

We have used two different approaches in the case study. A mathematical model using the Poisson process and renewal theory is developed to predict the expected number of unscheduled maintenance actions. Simulation is used to estimate the engine reliability measures such as shop visit rates and system failures.

The case study is arranged as follows. Section 17.2 is dedicated to the airline and the aircraft fleet data under consideration. Section 17.3 deals with unscheduled maintenance and the impact of no fault found on unscheduled maintenance. Section 17.4 deals with aircraft engine reliability and maintenance, in which we use the shop visit rates as a measure of engine reliability. Engine planned and unplanned maintenance is discussed in Section 17.5.

17.2 THE AIRLINE COMPANY PROFILE

The airline under consideration is a medium sized commercial airline flying to over 25 destinations within Europe. It flies over 1000 flights a week, with average flight time of around 1 hour; the maximum flight to any of their destinations is 2 hours.

On average, it carries over 3.2 million passengers per annum. Our analysis is restricted to a particular type of aircraft (we call it aircraft Alpha), and the relevant information on its characteristics and usage are as follows:

Aircraft	Alpha
Fleet size	16
Capacity	70–110 seats
Number of flights	1000 per week
Number of destinations	25
Speed	400 knots
Altitude	30,000 feet
Engine	Gas turbine
Number of engines	4
Flight time	45 minutes to 2 hours

17.2.1 The Data

Our first task is to identify the components that cause the majority of the unscheduled maintenance. The data used in the case study are collected over a period of 3 years and is a pooled data set over 16 aircraft. Each datum corresponds to the removal of a component due to some reported fault. The ages at removal of all the components are measured in terms of flying hours. The data also consist of age in terms of calendar hours and number of cycles (landings). The data do not include any planned removals or censored data. The components are subjected to predictive maintenance. The data consist of following entries:

1. Part number
2. Description
3. Aircraft
4. Position (engine)
5. Date on/off
6. Number of flying hours before the component removal
7. Number of flights before component removal
8. Original equipment manufacturer (OEM)
9. Reason for removal (initial fault)
10. Repair agency strip report finding
11. FC/FF/NFF (fault confirmed by the repair agency/fault not confirmed but found another fault/no fault found)

The important entries for this case study are entry numbers 6, 7, 9, 10, and 11. Several thousands removal reports were analyzed to identify the components that caused the majority of the unscheduled maintenance; the top 10 components that

Table 17.1. Top 10 Components for Unscheduled Maintenance

Part Name	Distribution	Parameter Estimates
TMS actuator	Weibull	$\hat{\alpha} = 778$, $\hat{\beta} = 0.7$
Attitude direction indicator	Weibull	$\hat{\alpha} = 365$, $\hat{\beta} = 0.65$
Fuel flow indicator	Weibull	$\hat{\alpha} = 1700$, $\hat{\beta} = 0.85$
Vertical ref gyro	Weibull	$\hat{\alpha} = 890$, $\hat{\beta} = 1.12$
Altitude selector	Weibull	$\hat{\alpha} = 375$, $\hat{\beta} = 0.78$
Central audio unit	Exponential	$1/\hat{\lambda} = 1020$
Navigation controller	Exponential	$1/\hat{\lambda} = 1420$
Pressure transmitter	Weibull	$\hat{\alpha} = 219$, $\hat{\beta} = 0.98$
Flight data recorder	Exponential	$1/\hat{\lambda} = 1280$
Auto-pilot computer	Exponential	$1/\hat{\lambda} = 1050$

caused the majority of the unscheduled maintenance are listed in Table 17.1 along with their time to removal distribution. The time to removal distribution is found using the software PROBCHAR™ (Knezevic, 1995).

For TMS actuator, attitude direction indicator, fuel flow indicator, vertical Ref Gyro, altitude selector, and pressure transmitter, the cumulative distribution function of time to removal random variable selected is the Weibull distribution given by

$$F(t) = 1 - \exp\left[-\left(\frac{t}{\alpha}\right)^{\beta}\right], \qquad \alpha, \beta > 0 \tag{17.1}$$

For central audio unit, navigation controller, flight data recorder, and auto pilot computer, the cumulative distribution function of time to removal random variable selected is the exponential distribution given by

$$F(t) = 1 - \exp(-\lambda t), \qquad t > 0, \lambda > 0 \tag{17.2}$$

Here α (in flying hours) and β denote the scale and shape parameters of the Weibull distribution, respectively. The notation λ is used to represent the failure rate per flying hour (thus $1/\lambda$ would give us the mean time between removals). The surprising factor here is that many of these components have shape parameter value less than 1. This corresponds to a decreasing failure rate and normally indicates a quality problem with possibly insufficient or inadequate pass-off testing. It should also be noted that the random variable under consideration is time to removal and not time to failure. Although the components are removed based on some fault reporting, one cannot say that the component that was removed actually caused the fault.

17.3 ANALYSIS OF UNSCHEDULED MAINTENANCE

Unscheduled maintenance can easily wipe out the profit of an airline. Delay and cancellation costs have become so high that no airline can afford to take any chances. The delay cost in the case of Boeing 747 is estimated to be on the order of \$1000 per minute. It is also estimated that around 28% of the delays are caused by technical problems that lead to unscheduled maintenance (Knotts, 1996).

The expected number of unscheduled maintenance actions can be derived as follows:

Exponential Time-to-Removal Distribution
For the exponential time-to-removal distribution, the expected number of unscheduled removals for a period of operation t, $UR_e(t)$, is given by the well-known expression (in this case the removals form a Poisson process with mean λt)

$$UR_e(t) = \lambda t \tag{17.3}$$

Weibull Time-to-Removal Distribution
When the time to removal is Weibull, we use the renewal function to find the expected number of removals for an operating period of t hours. The expected number of unscheduled maintenance, $UR_w(t)$, is given by (assuming that after removal, the components are replaced with components that are "*as-good-as-new*," which is seldom true)

$$UR_W(t) = \sum_{n=1}^{\infty} F^n(t) \tag{17.4}$$

where $F^n(t)$ is the n-fold convolution of the time-to-removal distribution $F(t)$ given in equation (17.1). Equation (17.4) is difficult to solve analytically. For this purpose we use the iterative method proposed by Xie (1989), by exploiting another renewal function given by

$$UR_w(t) = F(t) + \int_0^t UR_w(t-x)\, dF(x) \tag{17.5}$$

Now by setting $t = nh$, that is, the period t is divided into n equal intervals of size h, equation (17.5) can be written as

$$UR_w(nh) = \frac{1}{1-(h/2)f(0)}\left[f(nh) + \frac{h}{2}UR_w(0)\,f(nh) + h\sum_{k=1}^{n-1} UR_w[(n-k)h]\,f(kh)\right] \tag{17.6}$$

Here $f(t)$ is the probability density function of the Weibull distribution and is given by

$$f(t) = \frac{\beta}{\alpha}\left(\frac{t}{\alpha}\right)^{\beta-1} \exp\left[-\left(\frac{t}{\alpha}\right)^{\beta}\right]$$

Using equations (17.3) and (17.6), we can find the number of unscheduled maintenance tasks per year. The estimated number of unscheduled maintenance tasks per

Table 17.2. Unscheduled Maintenance Tasks Per Year

Part Name	Estimated Number of Unscheduled Maintenance Tasks per Year
TMS actuator	50
Attitude direction indicator	97
Fuel flow indicator	26
Vertical ref gyro	56
Altitude selector	110
Central audio unit	47
Navigation controller	34
Pressure transmitter	219
Flight data recorder	38
Auto-pilot computer	46

year (approximately 3000 flying hours per aircraft) for the top 10 components are listed in Table 17.2. The observed numbers of unscheduled maintenance actions over a 3-year period were close to the estimated numbers.

The total number of unscheduled maintenance per year due to just 10 components alone (of course, for a fleet of 16 aircraft) is 723. This also represents 723 potential delays/cancellations due to 10 components for a fleet of 16 aircraft. This is very high considering the fact that a modern jet aircraft has up to 4 million components (of which around half are fasteners) and over 135,000 unique components. We also noticed a large percentage of no fault found in some of the components, which is the main discussion point in the next section.

17.3.1 No Fault Found (NFF)

No fault found, NFF, is basically a reported fault for which subsequently no cause can be found. Isolating the true cause of failure of a complex system naturally demands a greater level of analytical skill, particularly where there is a fault ambiguity present. If the technical skills cannot resolve a failure to a single unit, then the probability of making errors of judgment will increase, depending on the level of ambiguity (Chorley, 1998). This problem is not unique to the Alpha airline. Data presented by Knotts (1994) quotes a Boeing figure of 40% for incorrect part removals from airframes, and British Airways estimates that NFF cost them on the order of £20 million (GBP) ($28 million) per annum. The percentage of NFF on the top 10 components is listed in Table 17.3, which was obtained from the database of the airline under investigation.

Analysis of the NFF data also yielded the following:

1. Around 19% of the unscheduled removals were NFF.
2. Four aircraft accounted for 49% of the NFF.

Table 17.3. Percentage NFF on Components

Part Name	Percentage NFF Removals
TMS actuator	26%
Attitude direction indicator	21%
Fuel flow indicator	31%
Vertical ref gyro	26%
Altitude selector	21%
Central audio unit	28%
Navigation controller	35%
Pressure transmitter	20%
Flight data recorder	18%
Auto-pilot computer	6%

3. Around 65% of the auto-pilot computers that were NFF came from a particular aircraft. All but one of these removals were carried out at a particular maintenance base.
4. Around 45% of NFF removals came from two particular bases.

In a few cases, we also noted that many of the NFF occurred within a few days of each other, for example, there were four consecutive actuator removals that occurred within 2 weeks. Similarly, four auto-pilot computers were removed from a particular aircraft within 45 days. Understanding the reasons for NFF is more complicated, because it demands a complete scrutiny of the maintenance process. Nevertheless, it is worth trying to understand the causes of NFF, because it can cost around $1000 to $3000 for a note reading "NFF" and a wait of as long as 60 days, depending on the contractor or procedure (Morgan, 1999). Chorley (1998) lists the following external factors that may cause the level of no fault found:

1. Quality and depth of training
2. Quality of technical data
3. Test equipment suitability, accuracy, and calibration
4. Design of BIT, its resolution of ambiguities and clarity
5. Intermittent faults
6. Human stress and fatigue induced error
7. Willful intent
8. The increasing impact of software

In due course, technology may provide truly intelligent products that not only detect a fault but also are capable of analyzing the actual cause and performing self-repair. In fact, this is already available in many computers. Recently, *Airbus* devised a method in which an item that is found NFF by the maintenance agency is labeled as a *rough* item and is closely monitored to unearth the cause.

By contrast, in the next section, we study the reliability, maintainability, and maintenance of aircraft engines, where the number of NFF is almost negligible.

17.4 ENGINE RELIABILITY AND MAINTENANCE POLICIES

While most of the mission critical components of the aircraft suffer from NFF, the case of an aircraft engine is entirely different. Unlike most of the subsystems in an aircraft, the propulsion system in general, but the engines in particular, suffer primarily from times to failures that can best be described by Weibull distributions with increasing hazard functions. At the same time, the number of engine removals, which result in NFF, is negligible, at least for most modern engines. This means that preventative and opportunistic maintenance are likely to have significant economic benefits by reducing the need for corrective maintenance.

Although most of the failures are age-related, there are three causes of failure that are not: foreign object damage, maintenance-induced failure, and [poor] quality control. With high by-pass, wide-chord engines, most foreign object damage is caused by bird strikes and is very seldom sufficiently serious to cause an engine failure or an in-flight shutdown. Maintenance-induced failures are often due to inadequate training or quality control within the maintenance organization; but because the cause of failure is never given as maintenance-induced failure, it is very difficult to obtain sensible data. Quality control during manufacture most commonly affects a small batch of components that may show up in a [retrospective] Weibull analysis but is, of course, impossible to predict.

With mechanical components, it is quite common for the failure of one component to cause damage to a number of other, often unrelated, components. If, for example, an HP turbine blade breaks off at its root, this piece of metal will, almost certainly, cause damage to the IP and LP turbine blades and possibly the corresponding nozzle-guide-vanes or stator blades.

Because many of the parts are replaced as a result of some form of deterioration, inspection can often be used to preempt their actual failure. Unfortunately, most of the parts in an engine cannot be easily inspected with the engine *in situ.* The opportunity to perform inspection is generally taken when the engine has been removed and disassembled for some other reason. Opportunistic maintenance therefore plays a very significant factor in the support of engines.

The *issue life* is the number of hours/cycles that a [life-limited] component still has remaining before it is due for a scheduled replacement when refitted to a module/engine. It is therefore the hard life minus the current age. The *minimum issue life* (MISL) is an age limit. If a component has exceeded this limit, then it would be reconditioned before being used as a spare for module/part exchange. Thus a low MISL would result in fewer parts being replaced before they reach their hard life than would be the case for a higher MISL.

A *soft life,* on the other hand, is not related to a hard life. A component that has exceeded its soft life would be reconditioned/replaced the next time the engine, in which it is installed, is removed for maintenance. Thus a low soft life would result in more reconditions/replacements of this type of component than a high soft life.

For the purposes of this case study, the engine in question is considered to consist of 11 modules with 22 parts spread between these modules in the way described in the engine breakdown structure shown in Figure 17.3. Corresponding information on failure rates for these 22 parts is given in Table 17.4.

The soft lives and minimum issue lives (MISL) given in Table 17.4 are for illustrative purposes only, because we will show that as these are varied, the numbers of engine removals and mean times between engine shop visits can be changed. The hard lives are for the safety critical components and, as such, are not subject to manipulation. However, their accompanying minimum issue lives are similar to the soft lives and have nothing to do with safety, and so they can be changed. All of the "lives" given are in hours, whereas the scale parameters of the Weibull distributions are given in cycles, with the average cyclic-exchange rate of 0.192 cycles per hour. Thus a value of 15,000 cycles is equivalent to 78,125 hours, on average.

The graphs in the following section were generated using a discrete-event simulation model that was developed in 1990 by CACI in Simscript II.5® under a joint Ministry of Defence/Rolls-Royce plc contract. It is called MEAROS, which is an acronym for *Modular Engine Arisings, Repair and Overhaul Simulation.* It was

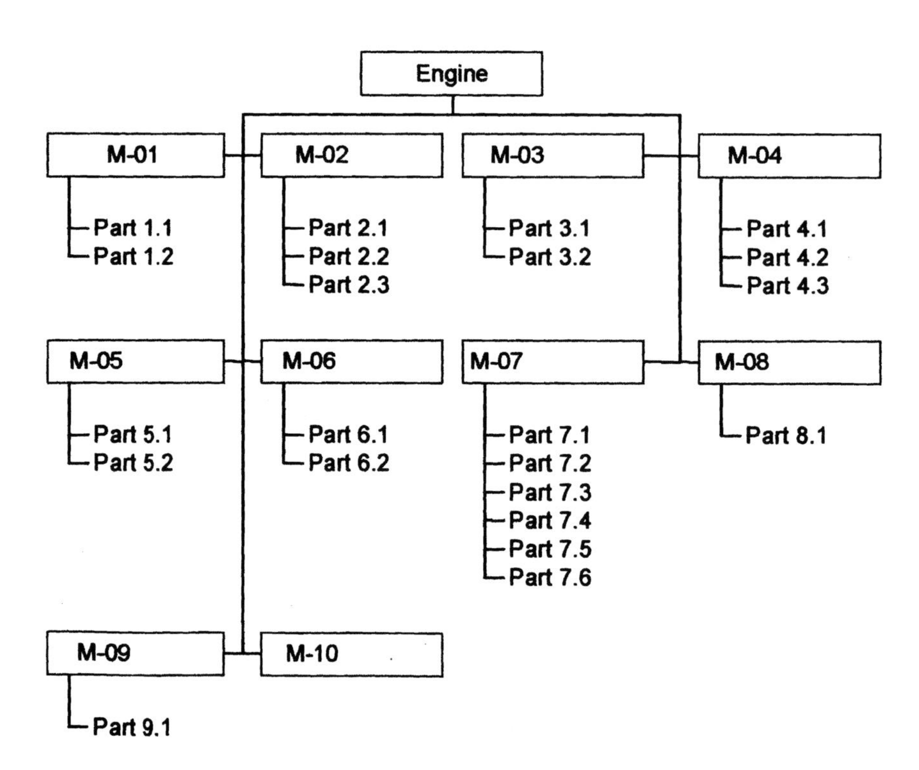

Figure 17.3. Engine breakdown structure.

Table 17.4. The Times to Failure, Soft, Hard, and Minimum Issue Lives

	Weibull Parameters					
Engine	Scale	Shape	Soft Life	Hard Life	MISL	Failure Rate per 1000 Cycles
Component						
M-01						
Part-1.1	15,000	1	15,000			0.067
Part-1.2				23,000	6,000	
M-02						
Part-2.1	16,000	5	15,000			0.068
Part-2.2				17,000	6,000	
Part-2.3				26,000	6,000	
M-03						
Part-3.1	225,000	1	15,000			0.004
Part-3.2				27,000	6,000	
M-04						
Part-4.1	14,000	2	15,000			0.081
Part-4.2				15,000	6,000	
Part-4.3				19,000	6,000	
M-05						
Part-5.1	10,000	5	15,000			0.109
Part-5.2	10,000	6	15,000			0.108
M-06						
Part-6.1			15,000			
Part-6.2	8,500	6		14,000	6,000	0.127
M-07						
Part-7.1	17,000	6	15,000			0.063
Part-7.2				27,000	6,000	
Part-7.3				27,000	6,000	
Part-7.4				18,000	6,000	
Part-7.5				25,000	6,000	
Part-7.6				16,000	6,000	
M-08						
Part-8.1	15,000	3	15,000			0.075
M-09						
Part-9.1	31,000	2	15,000			0.036
M-10	17,000	3	15,000			0.066

largely based on two earlier FORTRAN simulation models: the Royal Air Force/Ministry of Defence *AELOGSIM* model and Rolls-Royce's *ORACLE* model.

MEAROS is a multi-indenture, multi-echelon model that allows the user to define the configuration of the operating units (aircraft), the maintenance and support environments, and the rules, parameters, and assumptions that define the particular scenario being modeled. The run takes in a "snapshot" of the current status of the fleet in terms of where each of the aircraft, engines, modules, and parts are currently situated, their ages and whether they are installed, in maintenance, or in stock (as serviceable spares), and to which mark/modification standard they have been built.

Each operational aircraft is assigned to a task, for which is set a target number of flying hours to be achieved during its duration. The model uses this target along with the task duration, number of assigned aircraft, and current number of achieved flying hours, to determine the aircraft schedule. At the appropriate time, the model selects one of the available [serviceable] aircraft and flies the required mission/flight.

Before an aircraft flies for the first time, the model determines when each component in the aircraft will fail, need to be inspected, or reach its hard life. It then works out which of these events will be the first for each engine in the aircraft and hence for the aircraft itself. At the end of each mission, it checks the new aircraft age against the age at which this first event is due to happen. If the time has come, the aircraft is taken out of service and the rejected engine(s) removed. As soon as replacement engines can be found and acquired, they will be refitted and the aircraft returned to a serviceable state. The time of the next event is then determined as before.

Once an engine has been removed from an aircraft, a check is done to determine the primary cause and whether any of the other components in the engine have to be repaired, reconditioned, or replaced. This is done taking into account the likely location of where the maintenance will be done (and modified later if the "actual" location is different).

The engine is sent to the chosen site for recovery (based on capacity, capability, and the type of recovery). Resources needed to move the engine are not modeled. When it gets there, it will require resources to perform the recovery and will be put into a holding queue if these resources are not available.

Modules are removed in the sequence defined by the stripping order. Only those modules are removed which are necessary to access each of those, which are rejected (or contain rejected offspring). The depth of strip determines the time taken for recovery.

Serviceable modules may be put into stock as they are removed from the engine (for access). Whenever possible, these modules will be refitted to the same engine; but if there is a shortage and they can be used to make another engine serviceable, this will be done.

Once full sets of modules, whose modification standards satisfy a valid build combination, have been acquired, these will be used to rebuild the engine. The time to do this is also determined by the depth of strip. The rebuilt engine is then allocated to a test bed, as soon as one becomes available, for pass-off testing. If the test is successful, the serviceable engine is moved to the appropriate storage facility (supply site). Otherwise, the engine is put back in the queue for retesting.

As soon as a recovered engine, module, or part is added to stock, the model checks to see if it can be used to meet a rebuild requirement. If it is the last offspring needed, the rebuild will be started. At the end of this, the parent will be put into stock unless it is an engine, in which case it will be sent for testing first.

If a rejected component has no offspring or none of its offspring has been rejected, it will be repaired, reconditioned, or discarded. A component will be repaired only when the cause of its rejection was unplanned (primary or secondary) and it is not "beyond repair" or has not exceeded its "number of repairs" limit. A repair does not restore the component to an as-good-as-new condition, but its time-to-failure age will be modified by the "repair-effectiveness" factor (a number between 0 and 1, inclusive).

A component that is beyond repair or has been rejected due to lifing will be reconditioned (or overhauled). This is usually done at a deeper echelon and often takes longer than a repair. It also restores the component to an as-good-as-new condition by resetting the time to failure, hard life, and inspection ages to zero. It does not, however, reset the ages of any of the other components, including its offspring, if it has any.

The model therefore tracks every component in the system throughout its lifetime, both while it is in service and outside. This is important for components whose maintenance is dependent on their ages, especially when that maintenance can take many forms.

17.5 PLANNED AND UNPLANNED MAINTENANCE IN ENGINE

In the following graphs, we have normalized the results by taking as the base case the scenario with the soft life set at 10,000 hours and the minimum issue life set at 3000 hours. Each point on the plots was then divided by the corresponding base case value to give a ratio or normalized value. For example, in Figure 17.4 the time the aircraft spent waiting for spare engines in the base case was 1192 aircraft-days, compared to 44 aircraft-days in the case when the soft life was 25,000 and the MISL 7000.

In practice, the number of spares would normally be increased, so that it would be unlikely for any aircraft to actually have to wait for a spare engine. This, however, introduces a further 34 independent random variables, which for our purposes made the analysis a little too complicated. It would have been necessary to determine the "best" number of spares of each type of component to remove the aircraft waiting time. Since the numbers of spares would inevitably be different for each case, this would also affect the numbers of engine removals, module recoveries, and mean times between engine shop visits. Naturally, this would also make it difficult to determine how much of the variances could be attributed to each variable.

In Figure 17.4, we can see that the number of days an aircraft spent waiting for spare engines is dependent on both the soft life and MISL. It should be noted that all

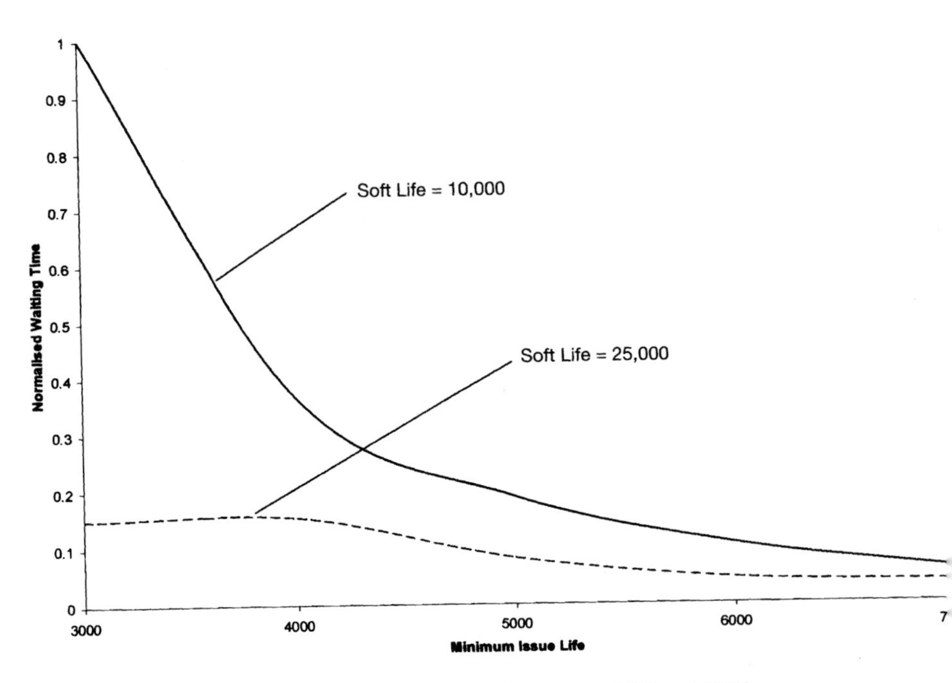

Figure 17.4. Aircraft waiting time versus soft life and MISL.

of the soft lives and MISL given in Table 17.4 were changed simultaneously. In theory, there may be some benefit from setting some higher and some lower, but to determine the "best" values for each, independently, would require a great deal of time and effort with probably little significant benefit. It would appear from the two lines that a high soft life and, rather contradictorily, a high MISL are preferable. If we now look at Figure 17.5, a similar pattern emerges. This suggests that the aircraft waiting time is more related to the number of modules that have to be recovered than to the number of engines (see Figure 17.6). Further investigations would, no doubt, indicate which particular types of module were in short supply.

The number of engine removals is the sum of both the planned and unplanned causes. By looking at Figure 17.7, we can see that the pattern in Figure 17.6 is somewhat more closely related to that of the planned than the unplanned removals. This is due to the fact that the changes in the MISL have significantly more impact on the number of planned removals than the changes in the soft life has on the unplanned removals. The actual numbers of removals are of similar magnitude; planned are twice the unplanned for the base case [3000, 10,000] but half the unplanned for the case [7000, 25,000].

One approach to measuring engine reliability is the shop visit rate (engine removal rate). The *shop visit rate* is actually rarely a "rate," but more commonly refers to the mean time to first shop visit, from first to second, or, in general, from the nth to $(n + 1)$th shop visit. In leasing contracts based on a fixed payment pe

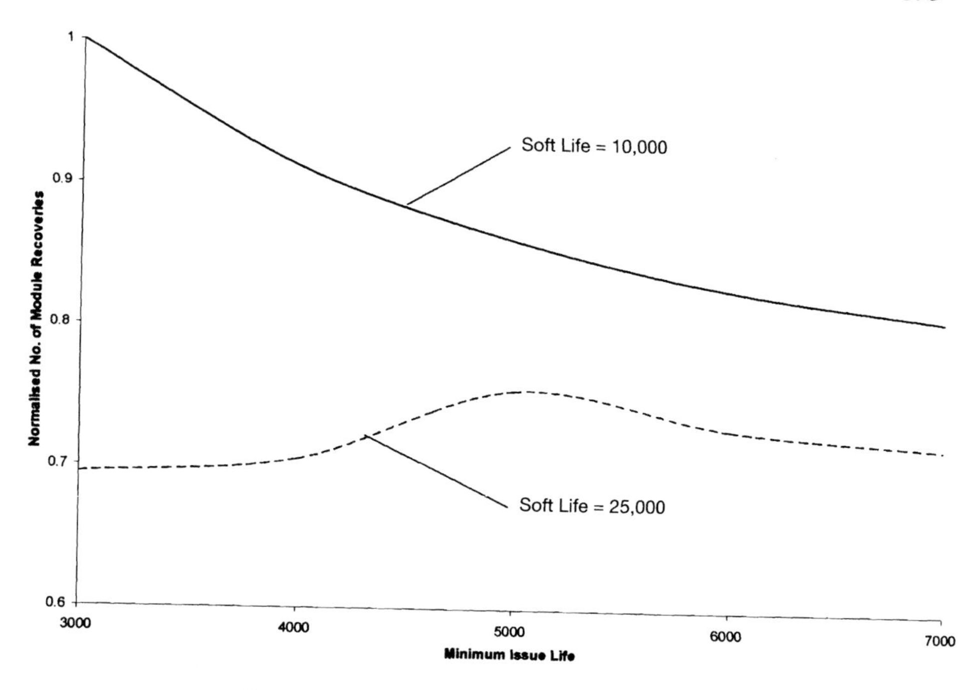

Figure 17.5. Module recoveries versus soft life and MISL.

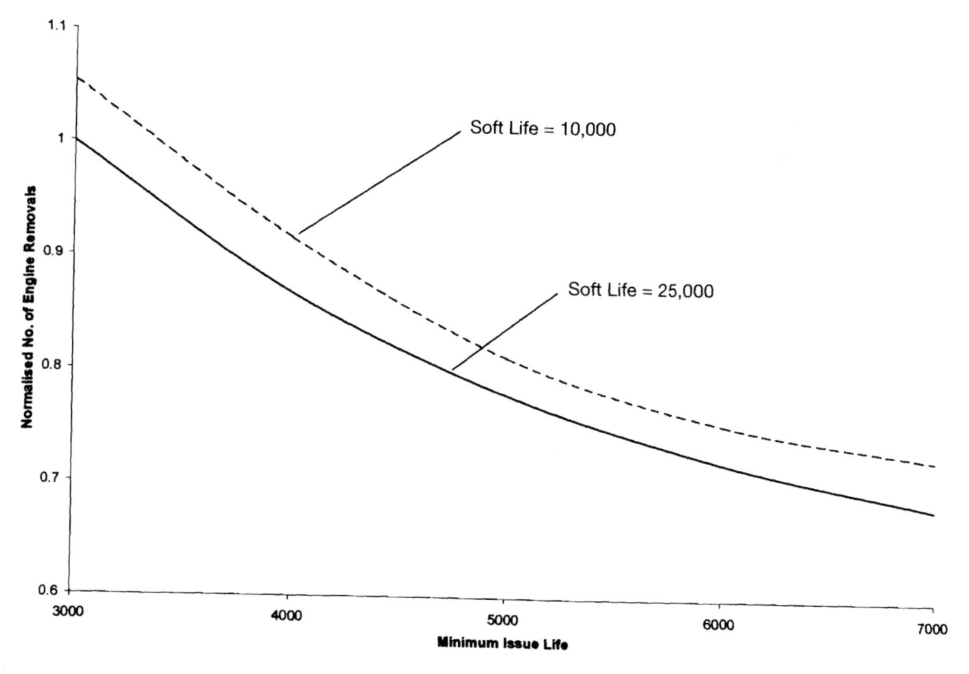

Figure 17.6. Number of engine removals versus soft life and MISL.

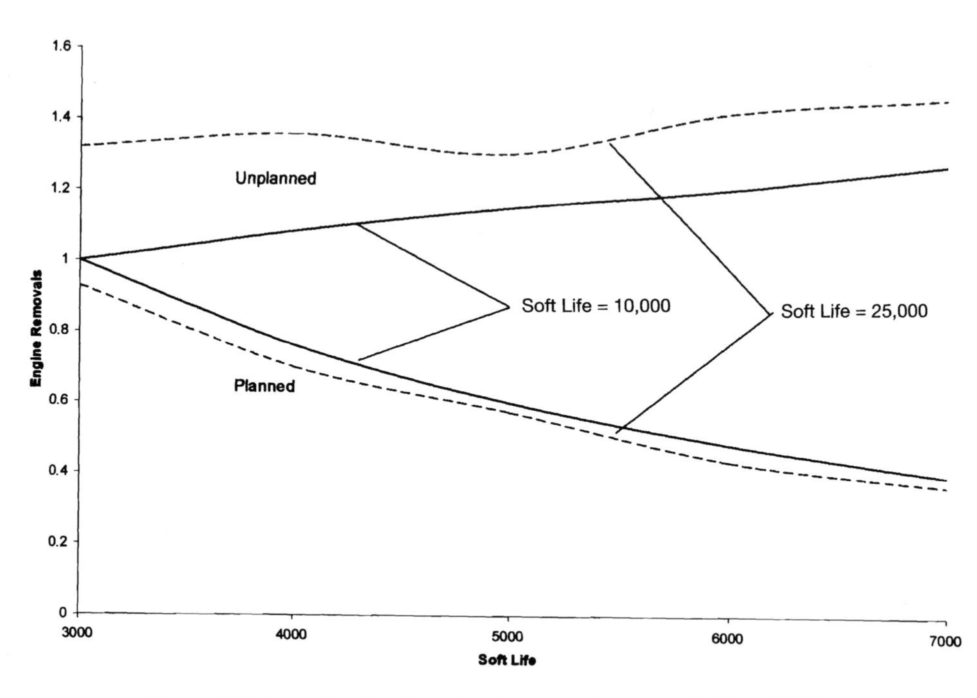

Figure 17.7. Planned and unplanned engine removals versus soft life and MISL.

hour of usage (such as Rolls-Royce's Power-by-the-Hour type of contract), the times between shop visits are critical to the success (from the lessor's viewpoint) of the contract (see Figure 17.8). Essentially, the hourly rate has to be sufficient (taking into account inflation and interest rates) to accumulate sufficient revenue to cover the cost of the shop visit.

In most cases, the time between successive shop visits tends to decrease with time, as shown in Figure 17.9. When the engine is recovered, only a subset of the components would normally be replaced. Those which have either been repaired or have been left untouched will have an age greater than zero, which means that their mean residual life will be less than that of a new component if they have a failure mode with an increasing hazard function (e.g., Weibull shape greater than one).

If the soft life on these components is very low, then they are very likely to be replaced before they fail, and hence the time to the next unplanned shop visit will tend toward that of a new engine. However, reducing the soft life will have no effect on the times to rejection due to hard lifing. This is done by changing the MISL such that a higher MISL will tend to reduce the number of planned engine rejections, and hence increase the mean time between subsequent engine shop visits. The effect is, however, dependent on the actual hard lives and whether modules are exchanged during an engine recovery. Sometimes, the modules are kept together so that the rebuilt engine has exactly the same parts as before, except, of course, when these have been replaced with new ones.

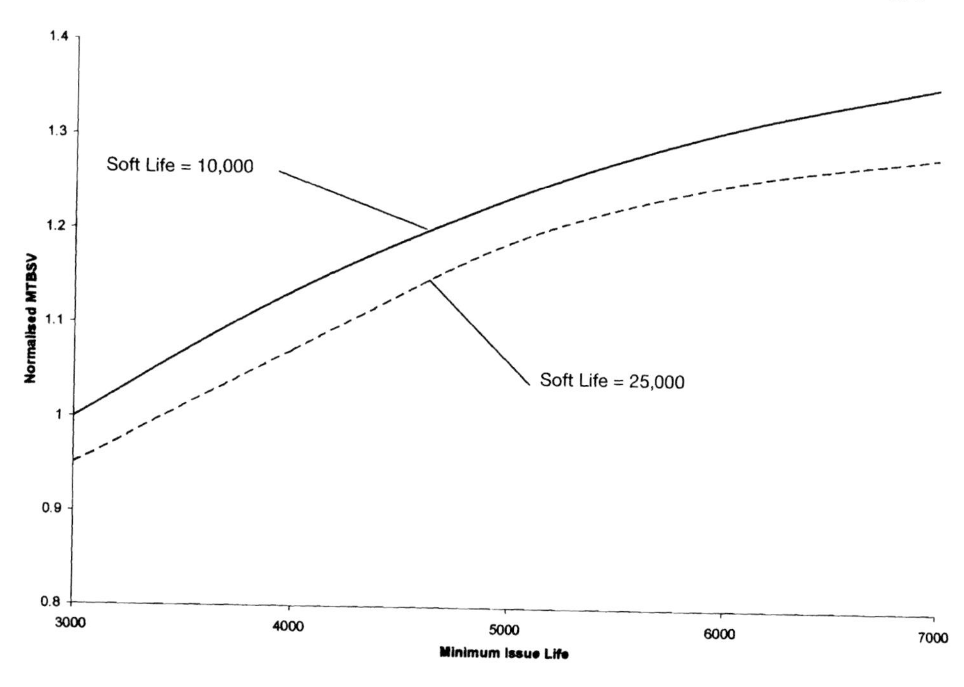

Figure 17.8. Mean time between engine shop visits versus soft life and MISL.

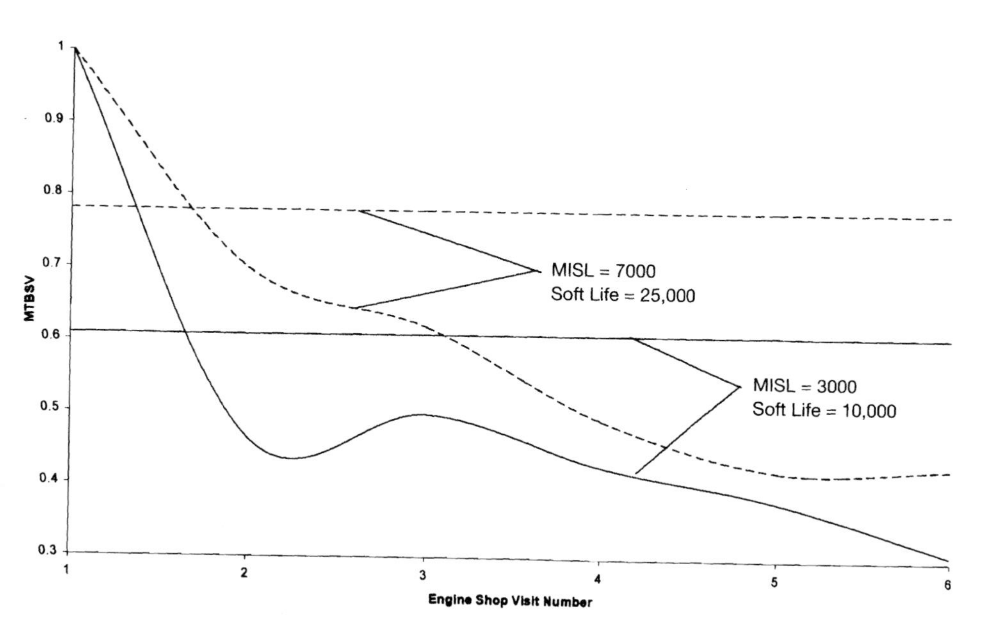

Figure 17.9. Mean times between shop visits for different soft life and MISL policies.

The curves in Figure 17.9 show how the MTBSV (mean time between shop visit) vary for each engine shop visit for the two cases: base case (MISL = 3000, soft life = 10,000) and (MISL = 7000, soft life = 25,000). It will be noted that the mean time to the first shop visit is the same for both cases (i.e., 1). This is to be expected because the lifing policies do not have an effect until after the engine has failed and is being recovered. The two horizontal lines represent the weighted average time between shop visits. In the two scenarios, all of the engines in the fleet failed at least once during the 20-year period that was simulated, but not all failed twice and, of these, not all failed a third time, and so on. Obviously, the value of this overall mean will vary with the number of [new] spare engines and the duration of the operation. The mean will approach the mean time to first shop visit as the number of spares approaches infinity (as none will need to be recovered and used again). By contrast, for a fixed, limited number of spares, the overall mean will tend to fall as the duration of the operation increases as more engines get recovered more times.

The problem for the service provider (if the service is based on a fixed rate per operational hour) is that of generating sufficient cash to fund the maintenance and support. As we have seen, different maintenance policies will almost certainly affect the frequency of maintenance but, at the same time, will also affect the [service provider's] expenditure. Replacing more components with new ones will increase the expected time to next maintenance but will make the current recovery more expensive. At the same time, the actual maintenance needed, in terms of replacing components that have been rejected following inspection, will vary from one engine recovery to the next.

The longer the engine has been in service since the last shop visit, the more components are likely to need to be replaced due to deterioration/damage. The maintenance policy, based on lifing, therefore will need to be flexible. It will be necessary to consider each engine recovery at the time of the recovery when it is known what work has to be done. The decision on what additional work should be done will have to take into account the time left for the current contract (that is, when the hourly rate can next be changed), the components that have to be replaced, and the ages and states of the remaining components. Due consideration of the current financial state will also be required along with any risks that may be associated with not meeting contractual requirements.

17.6 CONCLUSIONS

Aircraft are some of the most complex man-made systems. They are also very expensive to design, manufacture, operate, and maintain, and yet millions of people fly in them everyday. Most take them for granted, just as they would their cars or televisions.

Despite the complexity of aircraft, very few fatalities result from flying and, of these, only a small percentage are directly attributable to component failures. It could, perhaps, be argued that this is due to overprotective rules and regulations or that even this number is too high and that more should be done to reduce the risks. The current trend, certainly in the Western world, is to demand ever-greater safety and to heavily penalize anyone who does not meet these standards. At the same time, more and more people want to travel and to do so at minimum cost and minimum inconvenience.

In this case study, we analyzed two different sets of components in the first set; most of the unscheduled maintenance was due to no fault found. Currently, the percentage of NFF removals is close to 40%. Under these conditions, any planned maintenance fails to meet its objectives. Universal test equipments are being used to decrease the economic impact of NFF by the airlines and air forces. However, the universal test equipments again have their limitations, and they can only reduce the number of spares one has to stock rather than the number of NFF removals. Unless the diagnostics capabilities are improved, this problem will remain and would cost the airlines millions of dollars.

Contrary to most components of the aircraft, engine component removals seldom fall into the NFF category. Here, the number of unplanned maintenance actions depends upon the soft life and minimum issue life of the components. The selection of MISL and soft life plays a very crucial role in the amount maintenance the engine and its components requires.

ACKNOWLEDGMENTS

We would like to thank Professors Wallace Blischke and D. N. P. Murthy for their constructive comments on the previous version of this chapter, which helped us to improve the quality of the chapter considerably.

REFERENCES

Abernethy, R. B. (1983). *Weibull Analysis Handbook,* AFWAL-TR-83-2079, Wright-Paterson AFB, OH.

Anon. (2001). Keeping Down Engine Maintenance Costs, *Aircraft Economics* **53**:36–40.

Beaty, D. (1995). *The Naked Pilot: The Human Factor in Aircraft Accidents,* Airlife Publishing Ltd.

Besse, G. (1999). Policy Formulation: The IATA Position, *in Taking Stock of Air Liberalization* (M. Gaudry and R. M. Mayes, Editors), Kluwer Academic Publishers, New York.

Chorley, E. (1998). Field Data—A Life Cycle Management Tool for Sustainable Inservice Support, M.Sc. Thesis, University of Exeter, UK.

Cole, G. K. (2001). Using Proportional Hazards models, the Cumulative Hazard Function, and Artificial Intelligence Techniques to Analyse Batch Problems in Engineering Failure Distributions, dissertation submitted for the degree of MSc in Systems Operational Effectiveness, Centre for M.I.R.C.E., University of Exeter.

De Havilland, Sir Geoffrey (1961). *Sky Fever: The Autobiography of Sir Geoffrey de Havilland, CBE,* Hamish Hamilton, London.

Knezevic, J. (1999). *Systems Maintainability—Analysis, Engineering and Management,* Chapman & Hall, New York.

Knezevic, J. (1995). *Introduction to Reliability, Maintainability and Supportability,* McGraw-Hill, New York.

Knotts R. (1996). Analysis of the Impact of Reliability, Maintainability and Supportability on the Business and Economics of Civil Air Transport Aircraft Maintenance and Support, M.Phil. thesis, University of Exeter, UK.

Kumar, U. D., Crocker, J., Knezevic, J., and El-Haram, M. (2000). *Reliability, Maintenance and Logistic Support—A Life Cycle Approach,* Kluwer Academic Publishers, New York.

Lam, M. (1995). An Introduction to Airline Maintenance—in Handbook of Airline Economics, *Aviation Week Group.*

Morgan, P. (1999). No Fault Found—The Human Factors and the Way Forward, M.Sc. Thesis, University of Exeter, UK.

Morrison, S. A., and Winston, C. (1999). A Profile of the Airline Industry, in *Taking Stock of Air Liberalization* (M. Gaudry and R. M. Mayes, Editors), Kluwer Academic Publishers, New York.

Nelson, W. (1982). *Applied Life Data Analysis,* Wiley, New York

Nevell, D. (2000). Using aggregated cumulative hazard plots to visualize failure data, *Quality and Reliability Engineering International* **16:**209–219.

Xie, M. (1989). On the solution of renewal-type integral equations, *Communications in Statistics B* **18:**281–293.

EXERCISES

17.1. For the components listed in Table 17.1, estimate the number of spares required to achieve a fill rate of 95%, where the fill rate is defined as the probability that a spare will be available whenever there is a demand for one.

17.2. What should be the ideal maintenance policy for components with Weibull time to removal distribution with shape parameter less than 1, and why? Note that the random variable here is removal time rather than the failure time. The impact of reliability on removal is less significant for most of the cases where the distribution is Weibull with shape parameter less than 1.

17.3. If the airline were interested in reducing the number of unscheduled maintenance actions due to the top 10 components by 50%, how much increase in reliability would be required for each of these components?

17.4. In your opinion, what should be done whenever a component is labeled as NFF?

17.5. In your opinion, do you think that the shop visit rates reflect the true reliability of the engine? Do you think that the non-engine-caused shop visits (such as bird strike) should be used to calculate the SVR?

CHAPTER 18

Photocopier Reliability Modeling Using Evolutionary Algorithms

Michael Bulmer and John Eccleston

18.1 INTRODUCTION

Photocopiers are one of the most common and important pieces of office equipment in the modern workplace. The failure of photocopiers to operate correctly causes major inconvenience through the production of documents of poor appearance and quality, delays in production, and increased costs because of waste and loss of time. There are large variety of components in a photocopier: mechanical, electrical, and computer. These lead to the numerous modes of failure that manifest themselves in a diversity of ways including smudged or smeared copies, paper jams, damaged paper, over- or underexposure, incorrect paper feed, and incorrect alignment. The costs involve not only downtime and waste of copies but also service costs and replacement parts (as well as warranty costs).

The user does not feel the effects of failure of a photocopy alone, since the manufacture is also affected by failures. In the long to medium term, profitability and market share can be adversely affected by poor reliability and customer dissatisfaction. In the short term the manufacturer normally will be expected to cover parts and service under a warranty agreement. Typically, a service plan or contract outside warranty is offered with the sale of a photocopier. The nature of such plans is necessarily based on an understanding of the reliability of the machine, and consequently the analysis of failure data and the modeling for reliability.

In this case study we apply the evolutionary algorithm [EA; see Michalewicz (1996)] approach to photocopy failure data to determine models for the various modes of failure. The results should give insight into the reliability and quality of the photocopier and into determination of possible warranty and/or service plans. A description of the photocopier and the data regarding its service history are given in Section 18.2. Some preliminary analysis is presented in Section 18.3. A range of

Case Studies in Reliability and Maintenance, Edited by W. R. Blischke and D. N. P. Murthy.
ISBN 0-471-41373-9

potential failure models is specified in Section 18.4, and the EA methodology is then developed in Section 18.5. Section 18.6 gives results of the model fitting and introduces resampling methods to give a measure of precision for the model estimates. Summaries and recommendations in regard to the reliability and service requirements of the photocopier are given in Section 18.7. Concluding remarks regarding the use of the models are given in Section 18.8.

18.2 SYSTEM CHARACTERIZATION

The photocopier under investigation makes use of two main processes in producing copies; transferring the image onto the paper and then fixing that image. In the first of these, a rotating drum is electrically charged along a line by a charging wire. As the charged section rotates, it is exposed to the reflected image being copied. The brighter light dissipates the charge while the darker sections, the image, leave the charge unaffected. The drum surface then passes a developer where the remaining charge attracts toner. Finally, the drum surface meets the paper where an opposite charge is used to pull the toner onto the page. Drum claws then lift the paper from the drum, while a cleaning blade removes any residual toner from the surface, leaving it ready to be charged again for the next copy.

The second process involves the paper passing through a pair of rollers (the upper one of which is heated) that fuse the toner onto the page. The upper roller accumulates toner from this process. A cleaning web sits on the roller to remove this. As the web itself builds up the residual toner, it is rolled out to expose a fresh section of cleaning web. Once the web is completely finished, it must be replaced. We will model the replacement of the cleaning web in Section 18.6.

Note that the focus in this modeling is thus largely on consumables. The components to be considered are simply replaced when they are used up or worn out. We will use this to assume that the lifetimes of components can be considered to be independent times to failure. We would expect these replacements to be strongly associated with usage. The independence may not strictly hold, however, since these components function within the environment of a machine that is also aging. We will consider system failures and component failures separately.

In this case study, we propose to model the reliability of a widely used photocopier. The failure times for various components will be modeled using a range of failure distributions with particular emphasis on the Weibull distribution and its extensions. These include the standard distribution, with and without delay parameter, and combinations of Weibull distributions such as mixture models, competing risk models, multiplicative models, and sectional models. These more complex models are necessary for studying the reliability of a complex system such as a photocopier, in which different components may fail in different ways and have complicated effects on the whole.

The parameters of the models are estimated through a methodology based on evolutionary algorithms (Michalewicz, 1996). In comparison to traditional methods such as maximum likelihood and Bayesian estimation, this approach is very general and does not require initial estimates. Rather than iterating a single solution, an EA deals with a population of solutions, using analogies of biodiversity and mutation to overcome local optima. A key advantage of the EA system is the ease with which new models can be incorporated. The algorithm requires only a specification of the parameters and an expression for the reliability. This allows the user to easily explore and compare candidate models and also vary the optimality criterion used in the fitting process.

The data analyzed here were obtained from a service agent and are for a single machine. The photocopier concerned is a popular machine from a major multinational manufacturer and is targeted for medium usage in an office environment, although from these data it appears that some usage tended to be high volume at times. The service history of the machine is incomplete in that only components that were replaced twice throughout the life of the machine because of failure are included. The replacement of consumables, such as toner and developer, was not included.

The office procedure for the identification and rectification of problems with the photocopier was:

1. *Failure Mode.* An operator would notice the machine malfunctioning and contact the service provider and report the problem.
2. *Failure Cause.* The service provider would determine the reason (due to the design, manufacture, or use of the machine) that led to a failure and would then replace the component(s) as deemed necessary.

The kinds of operation and damaged copies that were considered to be failures included: smudged or smeared copies, paper damage during copying, over- or underexposure, poor paper feed, repeated paper jams, paper misalignment, and unwanted marks on copies. The most common solution to these problems was the replacement of one or more components. The dates when these replacements were made and the corresponding counter reading at the time of replacement are given in the Appendix. The data cover failures over the first 4.5 years of the photocopier.

18.3 PRELIMINARY ANALYSIS

Few items failed sufficiently often to allow an analysis and modeling at the component level. The cleaning web failed 15 times and the feed rollers 11 times; all other components failed fewer than 10 times, most much less. Accordingly, we have analyzed the data at a system level, which is by considering a system failure as when the machine is not operating properly and is serviced, and we ignore the particular components that have failed. An analysis at the component level of the cleaning web history was undertaken in addition. Modeling at the system level is needed for determining the service calls to fix failures, and modeling at the component level is needed for spare parts planning.

The time (in days) and number of copies between failures is readily obtained from the data for the system data (all the failures) and the cleaning web. Some descriptive statistics and graphs with respect to the number of failures (times and copies between failures) for the system and the cleaning web follow. There were 39 system failures (service occasions) comprising of 98 component failures.

18.3.1 System Failures

As one might expect, the number of days (time) and the number of copies between system failures, denoted n_t and n_c, respectively, were positively correlated (r = 0.753). The mean number of days between failures was 41.56, and the mean number of copies between failures was 26,647. Because of the correlation, only a time plot for copies is given. In the histograms and plots that follow, the time between failures and the number of copies between failures (the change in the counter reading on the machine denoted counter) are respectively denoted simply as day and counter.

It appears from the time plot in Figure 18.1 that the number of copies between failures decreases over time. This reflects the aging effect, so that failures occur

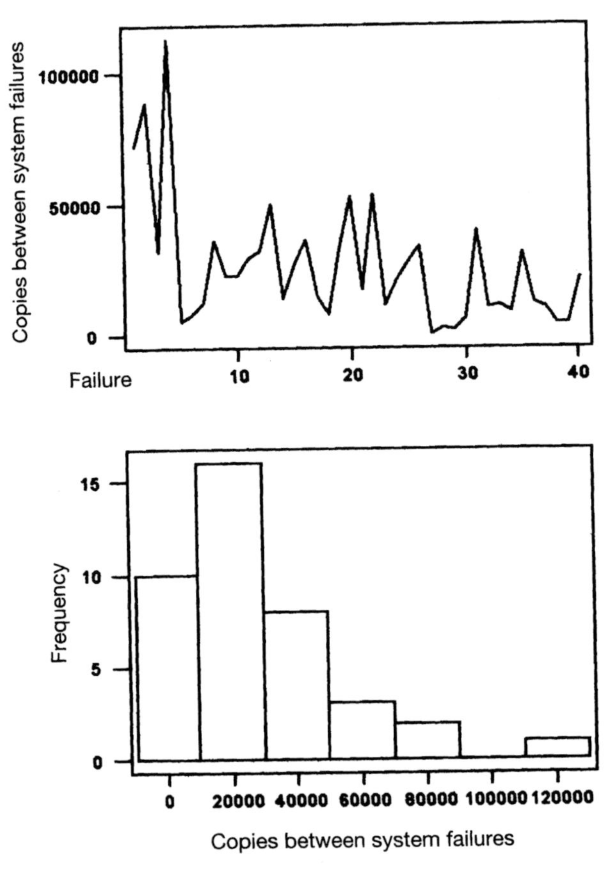

Figure 18.1. Time plot and histogram of copies between system failures.

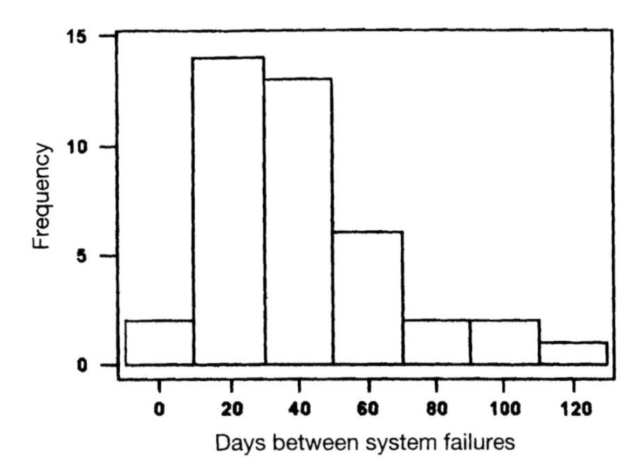

Figure 18.2. Histogram of time (days) between system failures.

more frequently as the item ages. The histogram of the number of copies between failures is highly skewed to the right. The tail corresponds to the early life of the copier, further suggesting that a mixture model may be appropriate.

Figure 18.2 shows that the histogram for day is somewhat different. This may indicate that different failure models are necessary for modeling day and counter data, as will be demonstrated later.

A further variable that may be of interest is usage, defined as the number of copies between failures divided by the number of days between failures. The mean usage between failures was 670.2 copies per day with a standard deviation of 409.8 copies per day. The last data point is a potential outlier, as can be seen in Figure 18.3, perhaps due to an incorrect recording.

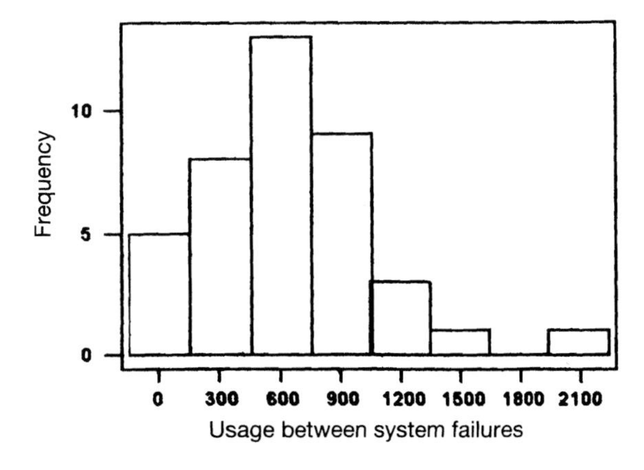

Figure 18.3. Histogram of usage between system failures.

18.3.2 Cleaning Web Failures

The component with the most observed failures was the cleaning web, and so it will be used as an example of modeling component failure. The number of days and the number of copies between failures of the cleaning web, denoted by n_{tw} and n_{cw} respectively, were strongly correlated ($r = 0.916$). The mean number of days between failures was 115.1, and the mean number of copies between failures was 72,670. These are necessarily higher than the system means because the cleaning web failures are a subset of the system failures. Figure 18.4 shows that the histograms of the counter and day data for the cleaning web are similar to those of all the data. They are skewed to the right, and the number of copies and number of days between failures of the cleaning web decreases over time.

Figure 18.5 shows a histogram of the usage, in copies per day, between failures of the cleaning web. The mean of the usage for the cleaning web was 598.8 copies per day between failures with a standard deviation of 235.8 copies per day.

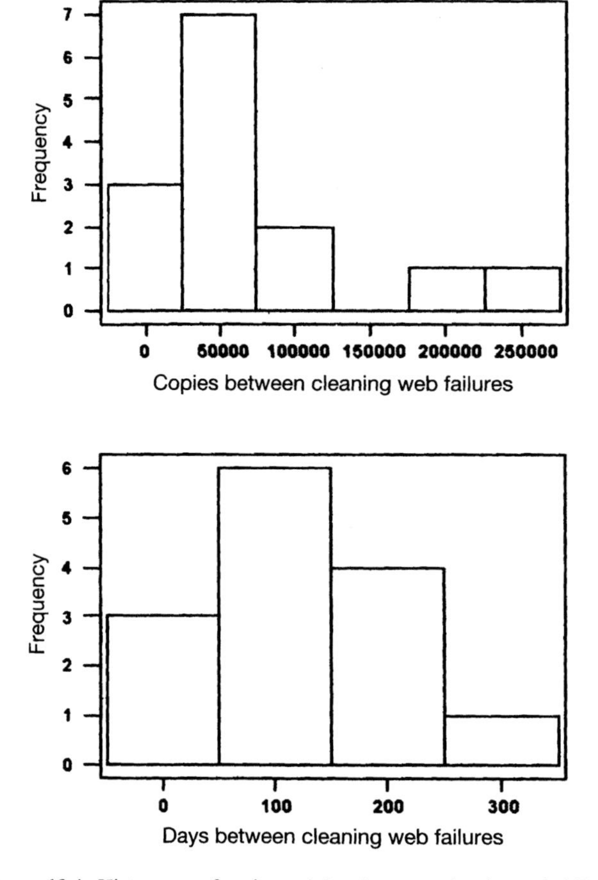

Figure 18.4. Histograms of copies and days between cleaning web failures.

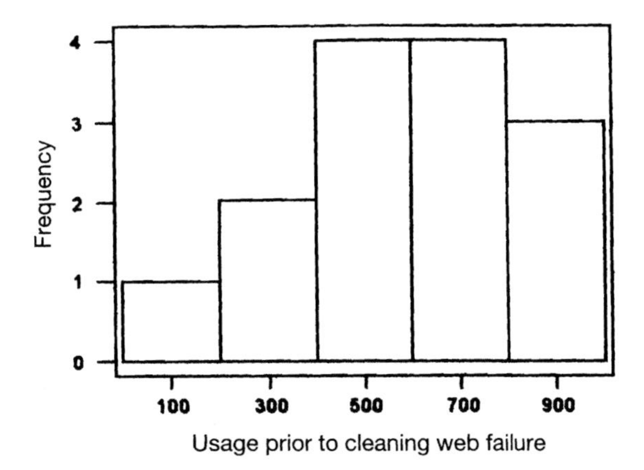

Figure 18.5. Histogram of usage prior to cleaning web failures.

The preceding descriptive statistics and graphs clearly demonstrate that as the photocopy ages, its reliability becomes questionable and eventually will be a matter of concern for the owner/user and the service provider. The distribution of copies and time between failures implies that a family of skewed distributions such as Weibull distributions are potentially useful for modeling the failure data. The usage measure, although of some descriptive use, was found not to be of value in further modeling and plays no further part in this study.

The objective then is to determine suitable models for the failure of the photocopier that could be useful in devising warranty and service contracts

18.4 WEIBULL MODELS

The family of Weibull models is one of the most widely utilized of all statistical distributions for failure and reliability modeling, and they are a vital ingredient of our approach. Brief descriptions of the models that have been explored in this case study are given below. Further discussion of the models may be found in Blischke and Murthy (2000). It is not difficult to include other models in the algorithm.

Two-Parameter Weibull

This is the simplest model considered. It has a shape parameter, β, and a scale parameter, η, with reliability function

$$R(t) = \exp(-(t/\eta)^{\beta}) \tag{18.1}$$

In this and the following models, the reliability function given is defined for $t > 0$. Similarly, the shape and scale parameters are taken to the positive; that is, $\beta > 0$ and $\eta > 0$.

Three-Parameter Weibull
This model is the same as the two-parameter Weibull but includes a delay or shift parameter, γ, with reliability function

$$R(t) = \exp\left(-\left((t-\gamma)/\eta\right)^{\beta}\right) \tag{18.2}$$

Mixture Models
An additive mixture of two two-parameter Weibull distributions gives rise to a five-parameter model with a mixture, two *shape* and two *scale* parameters denoted as p, β_1, β_2, η_1, and η_2, respectively, with reliability function

$$R(t) = p\exp(-(t/\eta_1)^{\beta_1}) + (1-p)\exp(-(t/\eta_2)^{\beta_2}) \tag{18.3}$$

where $0 \le p \le 1$. If the two failure processes are similar in nature, then a simpler model is obtained by assuming a common *shape* parameter, $\beta = \beta_1 = \beta_2$, giving a four-parameter model.

Multiplicative
A multiplicative model has a failure function that is the product of the failure functions of two Weibull distributions, and as a result a four-parameter reliability model with parameters β_1, β_2, η_1, and η_2. It models a system that fails when both of two subsystems fail, giving a reliability function

$$R(t) = 1 - \left(1 - \exp(-(t/\eta_1)^{\beta_1})\right)\left(1 - \exp(-(t/\eta_2)^{\beta_2})\right) \tag{18.4}$$

Competing Risk
A competing risk model has a reliability function that is the product of the reliability functions of two Weibull distributions with parameters β_1, β_2, η_1, and η_2. It models a system that fails when either of two subsystems fail, giving a reliability function

$$R(t) = \exp(-(t/\eta_1)^{\beta_1})\exp(-(t/\eta_2)^{\beta_2}) \tag{18.5}$$

Sectional Model
A sectional model is composed of different Weibull distributions for different time periods. The simplest sectional model is one with two Weibull distributions. This model involves six parameters, four of which are independent after considering continuity conditions. The reliability function with parameters β_1, β_2, η_1, and η_2 is

$$R(t) = \begin{cases} \exp(-(t/\eta_1)^{\beta_1}), & 0 \le t \le t_0 \\ \exp(-((t-\gamma)/\eta_2)^{\beta_2}), & t_0 < t \le \infty \end{cases} \tag{18.6}$$

where $t_0 = (\eta_1^{\beta_1}((\beta_2/\beta_1)/\eta_2)^{\beta_2})^{1/(\beta_1-\beta_2)}$ and $\gamma = (1 - \beta_2/\beta_1)t_0$.

The application of these models to a range of data sets, fitted with the EA described in the next section, is given in Bulmer and Eccleston (1998b).

18.5 EVOLUTIONARY (GENETIC) ALGORITHM

The main goal of the study is to estimate the reliability and service demands of a particular photocopier. This requires the precise estimation of the failure distributions of the system and specific components. The most commonly used methods for this task include graphical methods [for example, see Jiang and Murthy (1995)], and statistical estimation methods such as Maximum Likelihood, Bayesian, and Least Squares. The numerical techniques required for these statistical methods usually rely on reasonable initial values, and consequently the automation of these methods is problematic. Accordingly, a method of estimation and modeling which can be automated and that is not highly dependent or reliant on the choice of initial parameter values is appealing. This is particularly important for the methods we will consider, which are based on resampling.

We wish to model the data and estimate relevant parameters in some optimal way with respect to an appropriate criterion. Evolutionary algorithms (Michalewicz, 1996) have the properties of (a) ease of automation and independence from initial values and (b) flexibility that allows a wide class of possible objective functions. These concepts and ideas are described and explained more fully below.

The evolutionary algorithm described here is often referred to as a genetic algorithm for reasons that will become apparent in what follows. The basic idea of a genetic algorithm (Holland, 1975) is to iterate a population of parameter estimates, rather than work with a single point, as is done using classical methods. These solutions are "evolved" using natural selection, where fitness is measured by how close the resulting model is to the data. The idea of biodiversity helps the algorithm avoid the local optima, which may arise in the complex models considered here. An outline of the method and description of the terms we use are given below.

The genes of each individual in the population are simply its real-valued estimates for each parameter in the model. The initial population has genes that are generated randomly from a parameter space, with fitness of each individual calculated from the objective value of the corresponding model. An objective function is typically a measure of the goodness of fit of a solution to the data. A new population is created by first choosing parents from the current population at random, with probability proportional to their fitness, and then mating them to produce offspring. The solution captured by each offspring is a random affine combination of their parents' solutions. That is, the children of two parents $(x_1, x_2, \ldots, x_n)$ and $(y_1, y_2, \ldots, y_n)$ are

$$\big(cx_1 + (1-c)y_1, cx_2 + (1-c)y_2, \ldots, cx_n + (1-c)y_n\big)$$
$$\big(cy_1 + (1-c)x_1, cy_2 + (1-c)x_2, \ldots, cy_n + (1-c)x_n\big)$$

where c is a uniform random number from $[0,1]$.

This process of genetic crossover is the main evolutionary agent. However, *mutation* also plays an important role by allowing the reintroduction of lost genetic information. This work has used the "dynamic mutation" of Michalewicz and Janikow (1996), whereby each gene has a constant probability of mutating but the scale of the mutation is reduced as evolution proceeds. Given a range of values for the parameter being mutated, a deviation is chosen uniformly either above or below the current value. This deviation is then scaled by the factor

$$s = 1 - u^{\left(1 - \frac{t}{T}\right)^b} \tag{18.7}$$

where u is uniform over $[0,1]$, t is the current generation number of T generations, and b is a system parameter governing how quickly the scale of mutation is reduced. This operator is reminiscent of the Boltzmann function at the heart of simulated annealing (Van Laarhoven and Aarts, 1987), initially allowing access to the full search space and then restricting the system to local movements as the population "cools."

Floating-point methods are by no means the only approach to real-valued genetic optimization. De Jong's early experiments (De Jong, 1975) used function optimization and binary encoding to explore the behaviour of adaptive systems. A binary encoding has also been used in reliability modeling (Bulmer and Eccleston, 1998a).

An advantage of the genetic algorithm approach is that the natural selection makes no assumptions about the objective function, such as differentiability or even continuity. This means that new models and objective functions can easily be incorporated. The genetic algorithm engine requires only a specification of the parameters and an expression for the reliability function (and corresponding likelihood, if required).

Fitting Criteria

In addition to selecting a model to be fitted to the data, along with a "fitting" method, an optimality criterion used to measure the fit of a model and then determine the one of "best" fit is required. This objective function should be a measure that can be used to discriminate between models and values of parameters. In this approach the objective function can be easily altered as outlined below.

This discussion will focus on minimizing a sum of squared deviations and maximizing likelihood. The former, "least squares," method compares reliability estimates and values of the reliability function on Weibull plotting paper. The algorithm can also use the minimum sum of absolute deviations or the minimum of the maximum deviation as an objective function. The results are similar for all objective functions, and consequently the results for the latter two are not included here.

The algorithm was applied using a population of size 20, evolved over 1000 generations. This process took just under 3 seconds on a 400-MHz PowerBook laptop.

18.6 MODEL FITTING AND ANALYSIS

The genetic algorithm as outlined above was used to fit the data to the models described in the previous section. Weibull(2) denotes the two-parameter Weibull dis-

Table 18.1. Parameter Estimates for Modeling Copies Between System Failures

Model	β_1	β_2	η_1	η_2	γ	p	t_0	*lof*
Weibull (2)	1.261	—	28,219	—	—	—	—	1.050
Three-parameter Weibull	1.152	—	27,274	—	753.7	—	—	0.712
Mixture	*1.380*	*1.505*	*44,498*	*16,831*	—	*0.478*	—	*0.544*
Multiplicative	0.353	1.330	3,221,495	26,700	—	—	—	0.733
Competing risk	1.262	1.261	82,335	35,784	—	—	—	1.050
Sectional	1.259	1.970	28,251	92,588	123,987	—	219,888	1.050

tribution, and the other distributions are denoted in a clear and obvious fashion. The value of the objective function, the sum of squared deviations defined in Section 18.2, is the lack of fit between the data and the fitted model and is denoted *lof* in Tables 18.1 to 18.4.

The results are presented below in tables along with accompanying graphs. In each case, these include the graphs for the model of best fit and, for comparison, that of the standard Weibull model (with no delay parameter). The parameter estimates of each model for counter and time (days) at failure of all the data (system failure) and for the cleaning web (component failure) accompanied by the respective *lof* are given. The model of best fit is highlighted in italic type in each table.

18.6.1 Copies Between System Failures

Table 18.1 shows the parameter estimates obtained when finding least-squares fits of the models listed in Section 18.3. A mixture of two Weibull distributions was found to give the smallest *lof*, indicated by italics. The mixture indicates either that two failure mechanisms are at play or that there is a single one that undergoes a sharp change at some time. The latter is the likely explanation because more components wear and fail as the machine ages. This was seen earlier in Figure 18.1.

Figure 18.6 compares the simple Weibull model (without a delay) with the mixture obtained. As suggested by Figure 18.1, the main role of the mixture has been to better fit the failures that were separated by large numbers of copies.

Table 18.2. Parameter Estimates for Modeling Days Between System Failures

Model	β_1	β_2	η_1	η_2	γ	p	t_0	*lof*
Weibull (2)	1.710	—	46.5	—	—	—	—	2.955
Three-parameter Weibull	1.074	—	36.2	—	8.1	—	—	0.850
Mixture	7.492	2.014	14.2	54.0	—	0.177	—	0.969
Multiplicative	*5.666*	*1.464*	*11.5*	*46.0*	—	—	—	*0.551*
Competing risk	1.709	1.710	61.9	80.8	—	—	—	2.955
Sectional	14.898	1.079	11.6	36.3	8.0	—	8.6	0.851

Table 18.3. Parameter Estimates for Modeling Copies Between Cleaning Web Failures

Model	β_1	β_2	η_1	η_2	γ	p	t_0	*lof*
Weibull (2)	1.060	—	80,035	—	—	—	—	0.543
Three-parameter Weibull	1.060	—	80,035	—	0	—	—	0.543
Mixture	*0.851*	*5.230*	*79,377*	*67,923*	—	*0.674*	—	*0.092*
Multiplicative	11.013	1.060	390,693	80,090	—	—	—	0.543
Competing risk	0.630	1.266	1,046,923	96,110	—	—	—	0.502
Sectional	0.926	1.199	107,232	84,078	3,483.4	—	11,832.9	0.484

18.6.2 Days Between System Failures

The multiplicative model proved to be the best for the time (days to failure) data for all the data; see Table 18.2. This model is compared with a single Weibull in Figures 18.7 and 18.9. However, it is interesting to note that the three-parameter Weibull model is a not unreasonable fit to the data and is a simpler model than the multiplicative, as shown in Figure 18.8. A sectional model fit, also shown in Figure 18.8, is very similar to the three-parameter model.

18.6.3 Copies Between Cleaning Web Failures

The counter data for the cleaning web were well-fitted by a mixture model (Table 18.3). Note that the estimate of one of the shape parameters (β_1) is close to 1 (0.851). Thus a simple exponential model might be a more appropriate failure model in the early life of the photocopier while a two-parameter Weibull model may be more suitable later on. The histogram for the data confirms this (see Section 18.3).

18.6.4 Days Between Cleaning Web Failures

As with the days between system failures, Table 18.4 shows that the failure of the cleaning web in terms of days seems to be best fitted by a multiplicative model, and

Table 18.4. Parameter Estimates for Modeling Days Between Cleaning Web Failures

Model	β_1	β_2	η_1	η_2	γ	p	t_0	*lof*
Weibull (2)	1.481	—	129	—	—	—	—	0.752
3-parameter Weibull	0.851	—	100.3	—	23.2	—	—	0.365
Mixture	2.404	2.199	55.0	184.0		0.390	—	0.424
Multiplicative	*6.618*	*1.286*	*29.0*	*128.0*	—	—	—	*0.173*
Competing Risk	1.480	1.482	203.0	210.0	—	—	—	0.752
Sectional	3.663	0.988	56.0	103.0	20.2	—	27.7	0.362

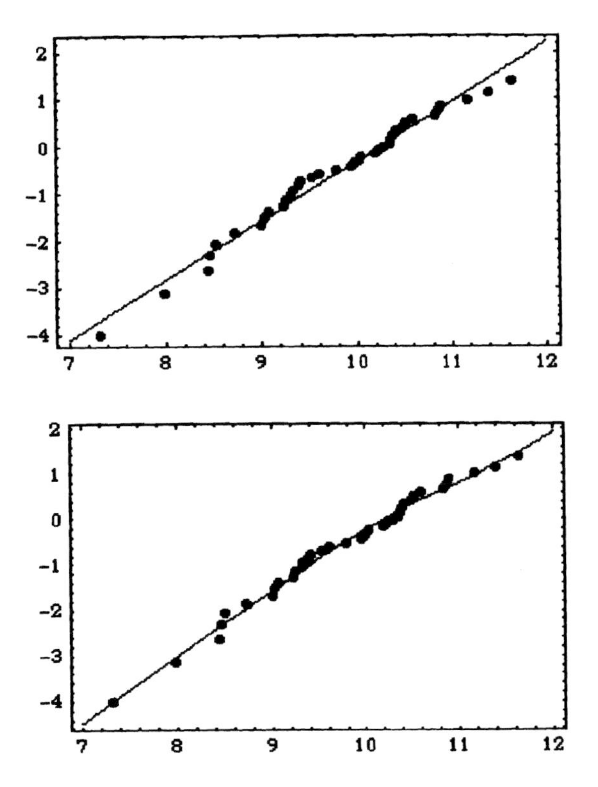

Figure 18.6. Weibull plots of copies between failures with Weibull (*top*) and mixture (*bottom*) models.

a comparison with a single Weibull is shown in Figure 18.10. Since the cleaning web is the most often failed component, the agreement with the model for days to failure for all the data is reassuring. Again the three-parameter and sectional models give comparably good fits, as shown in Figure 18.11.

18.6.5 Resampling Methods

The above results give parameter estimates that are used to summarize the observations with a given model. The choice of model can be made by selecting and applying some criterion, such as minimizing squared deviations on Weibull plotting paper. There are important issues here, in particular the possibility of overfitting the data by adding models that are unnecessarily complicated. However, even if the choice of model is clear, there remains the fundamental statistical question as to how accurately the parameter estimates reflect the true values of the failure process under investigation.

The standard approach to addressing this question is to quantify the sampling variability with a standard error. Standard errors can be calculated analytically for simple models, but this is impractical for many of the more complicated Weibull

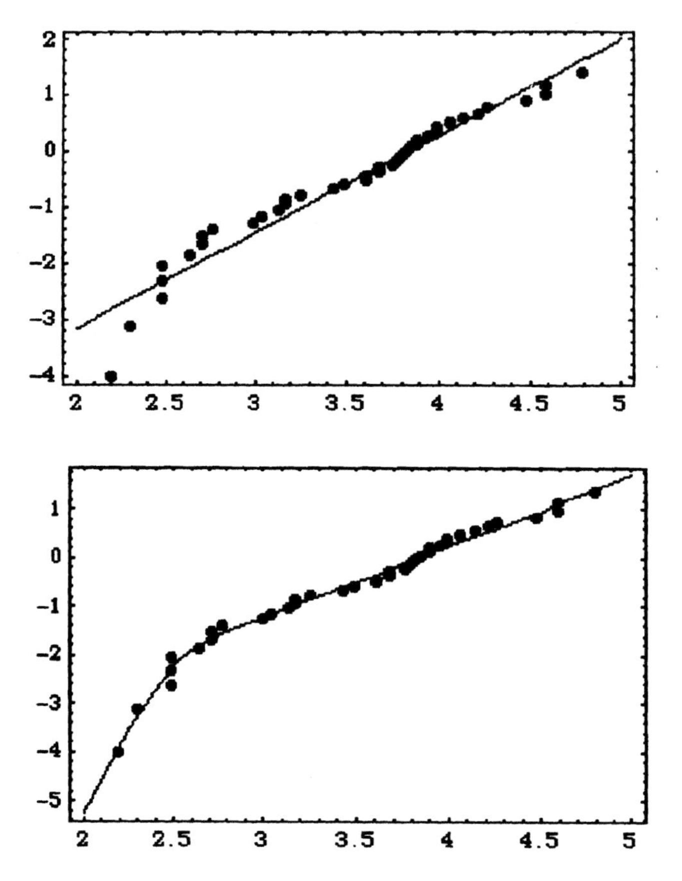

Figure 18.7. Weibull plots of days between failures with Weibull (*top*) and multiplicative (*bottom*) models.

models considered here. Instead, estimates of the standard error can be obtained through resampling techniques such as bootstrap and jackknife (Efron, 2000). The initial motivation for the EA approach was to automate the graphical approach to reliability modeling, whereby practitioners would fit asymptotes to Weibull plots in order to estimate model parameters. This would be too time-consuming to apply with resampling methods, and so we fit least squares models on Weibull plots using the EA instead.

The bootstrap method works by treating the sample as a population and taking new samples, with replacement, from this population. The model parameters are estimated for each new sample, and the observed variability in their values from sample to sample can be used to estimate the standard error. Unfortunately, for small samples, particularly those as small as the sample of cleaning web observations, the bootstrap samples can be quite unrepresentative of the original data. In particular, if there are two failure modes, then a new sample may, by chance, only

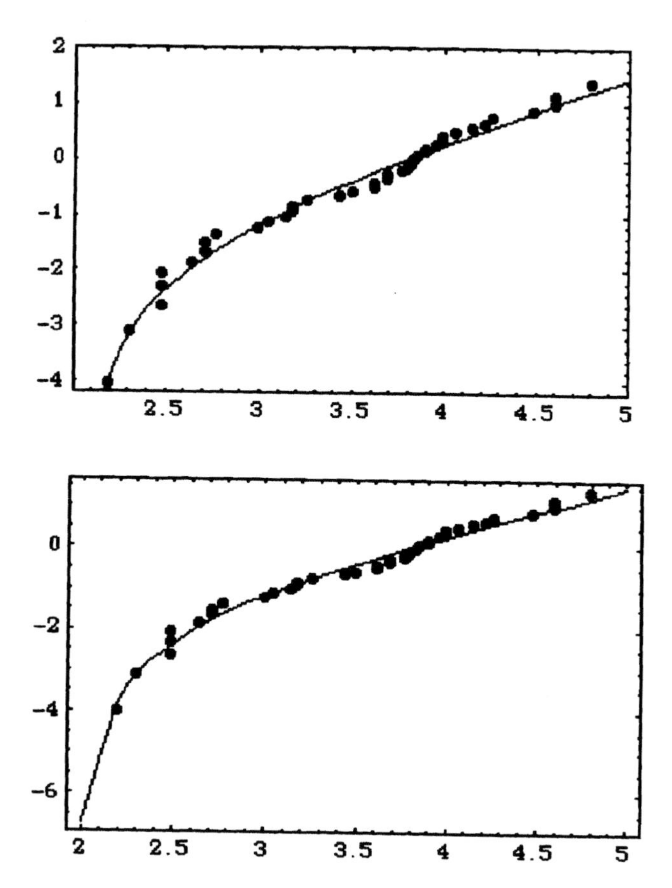

Figure 18.8. Weibull plots of days between failures with three-parameter Weibull (*top*) and Sectional (*bottom*) models.

include observations from one type of failure. The results can be highly variable and difficult to use.

An additional problem with mixtures is that if the sample departs too much from the original, then it can be difficult to determine which component is which from the parameter estimates. For example, the counter data for the cleaning web was best modeled by a Weibull mixture where $p = 0.674$. Applying the EA to a resampled set may produce a value of $p = 0.45$. It is not clear then whether this really is 0.45 or whether it should be $1 - 0.45 = 0.55$, and the two component estimates should be swapped. This is particularly difficult here since the scale parameters of the two components are also similar. Some evidence can be obtained from the shape parameters, but these have also been found to be similar in many resamples, making the bootstrap process unreliable.

An alternative is to use the jackknife technique instead. Rather than generating an entirely new sample each time, the jackknife generates n new samples of size n –

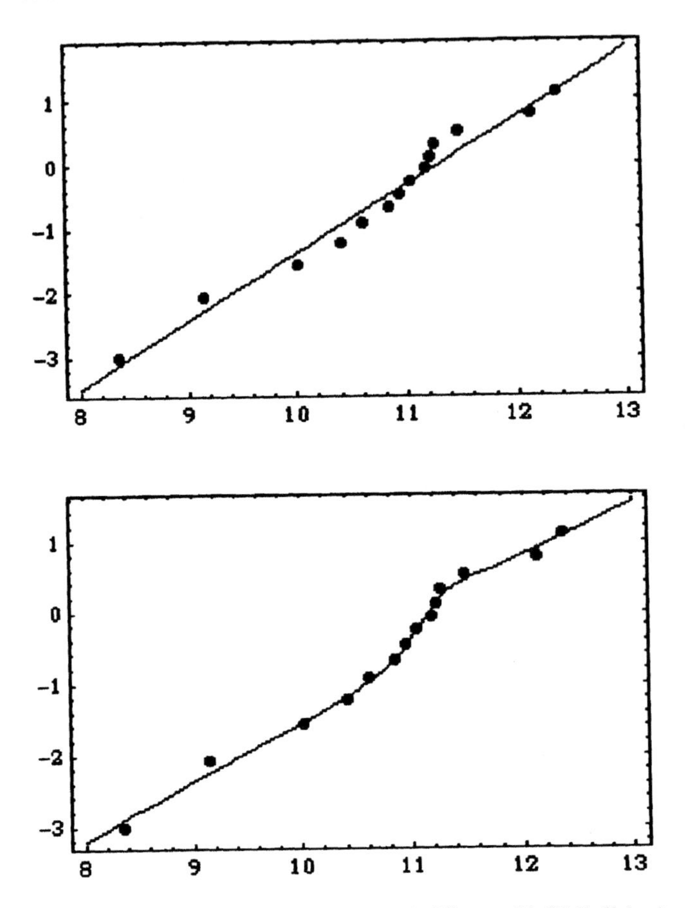

Figure 18.9. Weibull plots of copies between cleaning web failures with Weibull (*top*) and mixture (*bottom*) models.

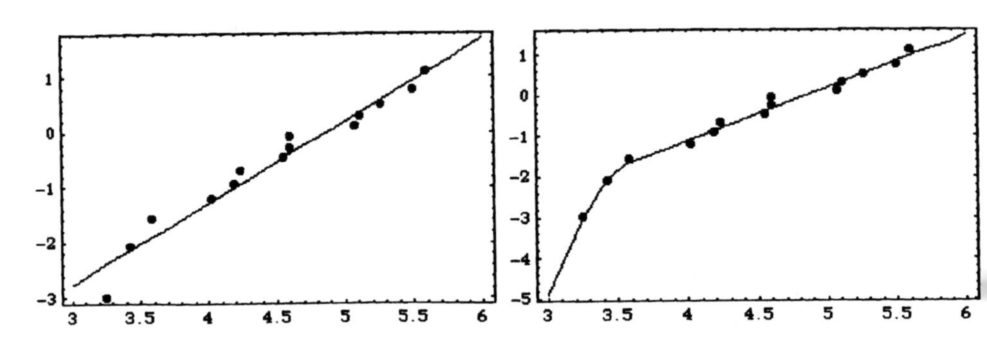

Figure 18.10. Weibull plots of days between cleaning web failures with Weibull (*left*) and multiplicative (*right*) models.

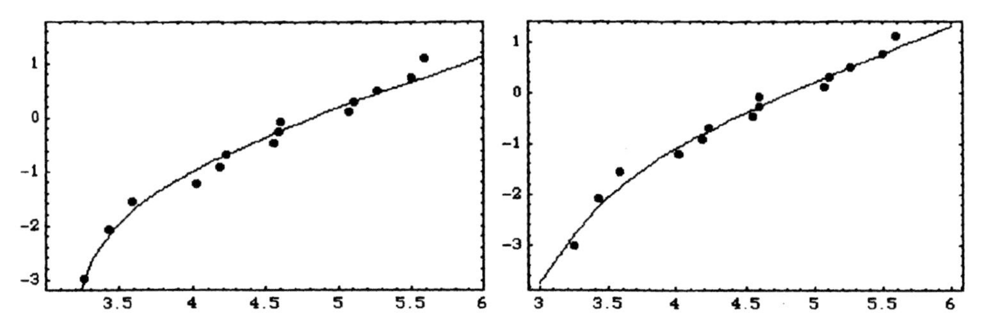

Figure 18.11. Weibull plots of days between cleaning web failures with three-parameter Weibull (*left*) and sectional (*right*) models.

1 by omitting one observation at a time from the original sample of size n. The parameter estimation process is then applied to each of the new samples, and the variability in the results is used to estimate standard error. This is an inherently more stable approach when applied to small samples.

Table 18.5 shows the results of the jackknife for the multiplicative fit to the days between system failures. It can be seen, for example, that the estimate for β_2 appears to be more accurate than the estimate for β_1. This is not surprising when compared with the Weibull plot in Figure 18.8. The first component (with $\eta_1 = 115$) captures the small number of observations with a low number of days between failures. Since there are only a few such observations, the estimates for this component are naturally less certain than for the larger failure mode. The jackknife can be used to identify this type of situation, if it is overlooked.

18.7 MODELING FAILURES OVER TIME

The underlying assumption in all of the models so far is that the times between service calls for the photocopier are independent. This is the assumption of a *renewal process,* whereby the system is like new after a failure. This is plausible for individual components, such as the cleaning web, which are replaced rather than repaired. However, at the system level it is likely that failures will become more frequent as the photocopier ages, as suggested by the time plot in Figure 18.1. The following

Table 18.5. Multiplicative Model Parameter Estimates and Jackknife Error Estimates for Days Between System Failures

	β_1	β_2	η_1	η_2
Estimate	5.666	1.464	11.5	46.0
Standard error	2.371	0.219	1.6	5.2

sections show how the expected number of failures can be estimated in these two settings.

18.7.1 Renewal Functions for the Cleaning Web

Since it is reasonable to model cleaning web replacements as a renewal process, a useful characterisation of the process is the *renewal* function, $M(t)$. The function gives the expected number of failures in $[0, t)$ and is described by an integral equation involving the failure function [see Blischke and Murthy (1994)]. Unfortunately, for all but the simplest failure functions, such as the exponential, this integral equation has no closed-form solution. Figure 18.12 shows the results of numerically solving this equation, first for the Weibull mixture model used with the copies between replacement data (see Section 18.6.3) and then for the multiplicative Weibull model used with the days between replacement data (Section 18.6.4).

Early in the life of the copier, the mixture component with small η dominates. This results in a dip in the renewal function since this component had $\beta = 5.23$. Further on, the components both tend to a linear renewal function. The same pattern holds for the model based on days between failures.

The renewal function can be used to estimate the number of components needed over certain time periods. For example, using the model based on days, in the first year it is estimated that $M(365) - M(0) = 2.83$ cleaning webs will be required for this copier, while in the second year $M(2 \cdot 365) - M(365) = 5.87 - 2.83 = 3.04$ cleaning webs will be required. Similarly, the renewal function estimates that in the first 100,000 copies 1.29 cleaning webs will be replaced, while in the second 100,000 copies 1.32 cleaning webs will be needed. Table 18.6 shows further values derived from the renewal function.

18.7.2 System Failure as a Nonstationary Point Process

Figure 18.13 summarizes the system failures by plotting the number of failures against time. The plot will be a step function, increasing by one at each time of fail-

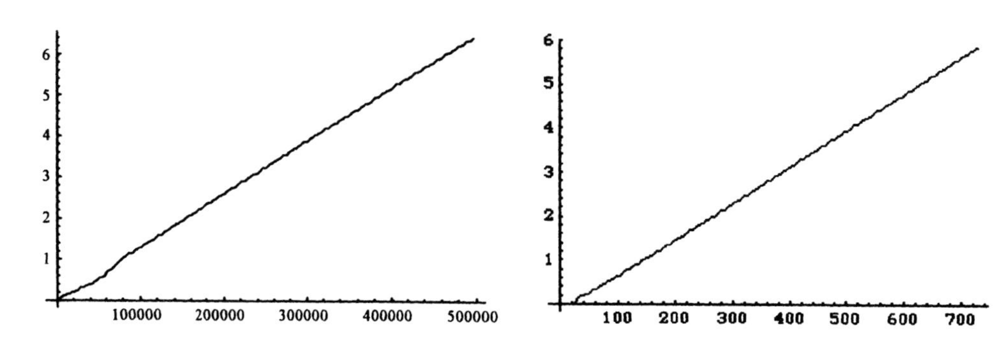

Figure 18.12. Renewal function for cleaning web replacements against number of copies (*left*, using a Weibull mixture) and against days (*right*, using a multiplicative model).

Table 18.6. Estimated Cleaning Web Replacements for 10 Years

Year:	1	2	3	4	5	6	7	8	9	10
Estimated cleaning webs:	2.83	3.04	3.04	3.04	3.04	3.04	3.04	3.04	3.04	3.04

ure. If the times between failures came independently from an exponential distribution then this plot should roughly follow a straight line. Instead it appears curved, with the steeper section corresponding to shorter times between failures, suggesting overall that failures are becoming more frequent with age.

The curve of this step function can be modeled using a power function of the form

$$\Lambda(t) = \left(\frac{t}{\beta}\right)^{\alpha} \tag{18.8}$$

The two parameters can be determined using least-squares fitting or by maximum likelihood estimation. Figure 18.13 shows the fitted power functions obtained by least squares. For time in numbers of copies, the parameter values are $\alpha = 1.64$ and $\beta = 118{,}400$. For time in days, $\alpha = 1.55$ and $\beta = 157.5$. The fitted functions can be used directly to estimate the expected number of failures over specified time intervals. For example, in the next year we would expect $\Lambda(365) - \Lambda(0) = 3.7$ failures, while in the following year we would expect $\Lambda(2 \cdot 365) - \Lambda(365) = 10.8 - 3.7 = 7.1$ failures. Table 18.7 gives a range of similar predictions. Since the estimated number of service calls is increasing each year, there will come a time where it will be cheaper to replace the copier rather than paying the servicing costs. Correspondingly, the age of the machine will be an important factor when entering into a service contract. The precise details will depend on the particular service costs.

The derivative of the power function is the estimated *intensity* function, $\lambda(t) = \Lambda'(t)$. This function gives the expected number of failures per unit time. Figure 18.14 shows the intensity functions against time in numbers of copies and in days.

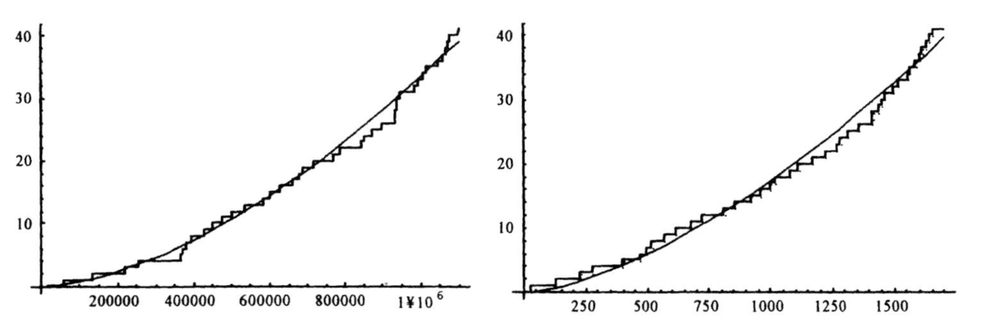

Figure 18.13. Point process for system failures (number of copies, *left*, and days, *right*) with power function fits.

Table 18.7. Estimated Service Calls for 10 Years

Year:	1	2	3	4	5	6	7	8	9	10
Estimated service calls:	3.7	7.1	9.4	11.3	13.0	14.6	16.0	17.3	18.5	19.7

Both suggest a steady increase in the rate of system failures over the life of the photocopier, though with the increase slowing relatively.

18.7.3 Predictions from Weibull Fits

In addition to the predictions given in Tables 18.6 and 18.7, the Weibull models for time between failures can be used to describe the failure process in a range of ways. For example, consider the multiplicative model for the days between cleaning web failure. This model predicts a mean time between failures of 120.1 days with a standard deviation of 90.2 days. The failure function can also be used to make further predictions relevant to inventories. For example, suppose a service provider is looking after a single copier and wants to keep inventory to a minimum without letting service times suffer. One criterion that could be used is to specify that a replacement cleaning web should be on hand 95% of the time. The failure function for the fitted model shows that chance of an interfailure time of less than 26 days is around 5%. Thus if a maintenance provider servicing this single copier can restock cleaning webs within 26 days of replacing one, the customer will only be kept waiting for a part 5% of the time. This kind of analysis is one of the main aims of modeling the failure time data; the fitted model can be used to answer a wide range of questions about likely future outcomes.

18.8 CONCLUSIONS

The EA approach has proved a useful method for fitting a range of possible distributions to observed failures. This can be used, in combination with domain-specific

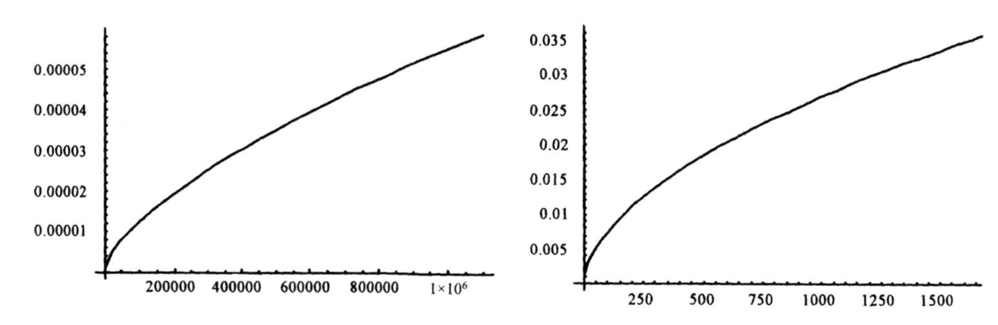

Figure 18.14. Estimated intensity functions for system failures (number of copies, *left*, and days, *right*).

judgment, to decide on a "best" model to use. It is naturally slower than traditional optimization methods, such as those based on gradient search, but its robustness is desirable for a general system for model fitting.

The resampling approach described is a simple means of augmenting parameter estimates with measures of confidence. Such resampling can be problematic for small samples or for mixtures in which one component is very rare, as discussed by Bulmer and Eccleston (1998a). However, such problems are easily visible and, on the whole, the ability to give confidence estimates greatly enriches the modeling process.

REFERENCES

Blischke, W. R., and Murthy, D. N. P. (1994). *Warranty Cost Analysis,* Marcel Dekker, New York.

Blischke, W. R., and Murthy, D. N. P. (2000). *Reliability: Modeling, Prediction and Optimization,* Wiley, New York.

Bulmer, M., and Eccleston, J. A. (1998a). *Reliability, Genetic Algorithms, and Resampling,* Centre for Statistics Research Report #88, The University of Queensland, Brisbane.

Bulmer, M. and Eccleston, J. A. (1998b). Automated reliability modelling with a genetic algorithm, *Proceedings, 6th International Applied Statistics in Industry Conference,* Melbourne, pp. 1–8.

De Jong, K. A. (1975). An Analysis of the Behavior of a Class of Genetic Adaptive Systems, Ph.D. thesis, University of Michigan, Ann Arbor.

Efron, B. (2000). The bootstrap and modern statistics, *Journal of the American Statistical Association* **95:**1293–1296.

Holland, J. H. (1975). *Adaptation in Natural and Artificial Systems,* University of Michigan Press, Ann Arbor.

Jiang, R., and Murthy, D. N. P. (1995), Modeling failure data by mixture of two Weibull distributions: A graphical approach, *IEEE Transactions on Reliability* **44:**477–488.

Michalewicz, Z. (1996). *Genetic Algorithms + Data Structures = Evolution Programs,* 3rd edition, Springer-Verlag, New York.

Michalewicz, Z., and Janikow, C. Z. (1996). GENECOP: A genetic algorithm for numerical optimisation problems with linear constraints, *Communications of the ACM* **39** (electronic supplement).

Van Laarhoven, P. J. M., and Aarts, E. H. L. (1987). *Simulated Annealing: Theory and Applications,* Reidel, Holland.

EXERCISES

18.1. *Weibull Plotting Paper.* Weibull plotting paper (WPP) plots $\ln(-\ln(R_i))$ against $\ln(t_i)$, where R_i is the estimated reliability for each time to failure t_i. Show that data following a two-parameter Weibull model will appear as a straight line on WPP. How do the intercept and slope of this line relate to the Weibull parameters?

The following exercises use Excel for model fitting and bootstrapping. You may also use your favorite statistical package for these exercises.

18.2. *Regression.* Based on Exercise 18.1, fitting a two-parameter Weibull model to failure time data can be accomplished with most statistical packages, as well as Microsoft Excel, using linear regression. The following table shows the results from Excel for fitting the days to failure of the cleaning blade.

Observation	Time to Failure	log(t)	Median Rank	log(–log(R))	
1	49	3.891820298	0.905405405	–2.308880127	Intercept
2	64	4.158883083	0.77027027	–1.343181902	–6.894355507
3	176	5.170483995	0.635135135	–0.789839834	Slope
4	227	5.424950017	0.5	–0.366512921	1.231752563
5	316	5.755742214	0.364864865	0.00819456	
6	364	5.897153868	0.22972973	0.385841654	
7	401	5.993961427	0.094594595	0.85787951	

Reproduce these results in Excel by entering the observation numbers and the (sorted) interfailure times (in days). Assuming "Observation" is in cell A1, the "log(t)" column is entered by filling down `=LN(B2)` while the "Median Rank" column uses `=1-(A2-0.3)/(7+0.4)`, with "log(–log(R))" given by `=LN(-LN(D2))`. Use the chart wizard to make an x–y scatterplot of "log(–log(R))" against "log(t)." There is an option for adding a linear regression line to the plot. Alternatively, use the functions `INTERCEPT` and `SLOPE` to obtain the parameter estimates in the worksheet.

18.3. *Bootstrap.* The procedure in Exercise 18.2 can be extended to give bootstrap estimates of the standard error of the parameter estimates. Following on from the above cells, make a column "Sample" (G) with cells given by `=CEILING(7*RAND(),1)`. That is, the column gives a bootstrap sample indicating which of the original interfailure times are to be included. The next column (H) gives the actual failure times using `=VLOOKUP(G2,A$2:B$8,2)+RAND()/1000`. The small amount of randomness is added so that we can sort them in order using Excel's `RANK` function. Column I, "Sorted," uses `=RANK(H2,H$2:H$8,1)`. The remaining columns are as in Exercise 18.2, with the median rank calculated using the ranks in column I. The intercept and slope are calculated using `INTERCEPT` and `SLOPE`.

Sample	Time to Failure	Sorted	log(t)	Median Rank	log(–log(R))	
2	64.0001	1	4.15888390	0.90540541	–2.3088801	Intercept
2	64.0002	2	4.15888672	0.77027027	–1.3431819	–7.1614363
3	176.0000	3	5.17048411	0.63513514	–0.7898398	Slope
5	316.0001	4	5.75574251	0.5	–0.3665129	1.26732663
5	316.0010	6	5.75574538	0.22972973	0.38584165	
5	316.0007	5	5.75574440	0.36486486	0.00819456	
7	401.0001	7	5.99396167	0.09459459	0.85787951	

Each time Excel recalculates the worksheet, you get a different random (bootstrap) sample and corresponding intercept and slope values, but these disappear each time. To make a record of the bootstrap estimates we can use a Data Table in Excel. For example, fill a series of 1, 2, . . . , 30 in cells G11 to G40. In cell H10 put a reference to the slope estimate (=M5 in the above example). Select cells G10 to H40 and choose **Table** from the **Data** menu. Enter the address of any blank cell for the column input (since we are not actually changing a parameter). Clicking OK will produce a table of 30 bootstrap estimates for the slope. Calculate the mean and standard deviation of these. Repeat the setup with a larger number of resamples (at least 200). Interpret the results.

18.4. *Copies Between Failures.* Repeat the analysis in Exercises 18.2 and 18.3 using the copies between failures for the cleaning blade, instead of the days between failures.

18.5. *Models.* Does the two-parameter model seem appropriate for the cleaning blade data, with failures either measured in copies or in days? Discuss any important features not captured by the straight line on WPP.

18.6. *Other Components.* The Appendix gives failure times for a range of other components replaced during maintenance of the photocopier. Repeat the study in Exercises 18.2 to 18.5 for other components.

18.7. *Renewal Assumption.* The analysis here makes a number of assumptions about the failure times. For example, it is assumed that a component is replaced because it has failed and that a new component will have the same failure distribution as a component earlier in the life of the copier. Discuss the validity of these and any related assumptions in terms of the difference between system failure and component failure for a photocopier.

APPENDIX

The data recorded from the photocopier's service history are given below. Each row describes a part that was replaced, giving the number of copies made at the time of replacement, the age of the machine in days, and the component replaced. Most services involved replacing multiple components.

Counter	Day	Component	Counter	Day	Component
60152	29	Cleaning web	220832	227	Toner filter
60152	29	Toner filter	220832	227	Cleaning blade
60152	29	Feed rollers	220832	227	Dust filter
132079	128	Cleaning web	220832	227	Drum claws
132079	128	Drum cleaning blade	252491	276	Drum cleaning blade
132079	128	Toner guide	252491	276	Cleaning blade

(continues)

Counter	Day	Component	Counter	Day	Component
252491	276	Drum	840494	1266	Feed rollers
252491	276	Toner guide	840494	1266	Ozone filter
365075	397	Cleaning web	851657	1281	Cleaning blade
365075	397	Toner filter	851657	1281	Toner guide
365075	397	Drum claws	872523	1312	Drum claws
365075	397	Ozone filter	872523	1312	Drum
370070	468	Feed rollers	900362	1356	Cleaning web
378223	492	Drum	900362	1356	Upper fuser roller
390459	516	Upper fuser roller	900362	1356	Upper roller claws
427056	563	Cleaning web	933637	1410	Feed rollers
427056	563	Upper fuser roller	933637	1410	Dust filter
449928	609	Toner filter	933637	1410	Ozone filter
449928	609	Feed rollers	933785	1412	Cleaning web
449928	609	Upper roller claws	936597	1436	Drive gear D
472320	677	Feed rollers	938100	1448	Cleaning web
472320	677	Cleaning blade	944235	1460	Dust filter
472320	677	Upper roller claws	944235	1460	Ozone filter
501550	722	Cleaning web	984244	1493	Feed rollers
501550	722	Dust filter	984244	1493	Charging wire
501550	722	Drum	994597	1514	Cleaning web
501550	722	Toner guide	994597	1514	Ozone filter
533634	810	TS block front	994597	1514	Optics PS felt
533634	810	Charging wire	1005842	1551	Upper fuser roller
583981	853	Cleaning blade	1005842	1551	Upper roller claws
597739	916	Cleaning web	1005842	1551	Lower roller
597739	916	Drum claws	1014550	1560	Feed rollers
597739	916	Drum	1014550	1560	Drive gear D
597739	916	Toner guide	1045893	1583	Cleaning web
624578	956	Charging wire	1045893	1583	Toner guide
660958	996	Lower roller	1057844	1597	Cleaning blade
675841	1016	Cleaning web	1057844	1597	Drum
675841	1016	Feed rollers	1057844	1597	Charging wire
684186	1074	Toner filter	1068124	1609	Cleaning web
684186	1074	Ozone filter	1068124	1609	Toner filter
716636	1111	Cleaning web	1068124	1609	Ozone filter
716636	1111	Dust filter	1072760	1625	Feed rollers
716636	1111	Upper roller claws	1072760	1625	Dust filter
769384	1165	Feed rollers	1072760	1625	Ozone filter
769384	1165	Upper fuser roller	1077537	1640	Cleaning web
769384	1165	Optics PS felt	1077537	1640	Optics PS gelt
787106	1217	Cleaning blade	1077537	1640	Charging wire
787106	1217	Drum claws	1099369	1650	TS block front
787106	1217	Toner guide	1099369	1650	Charging wire

CHAPTER 19

Reliability Model for Underground Gas Pipelines

Roger M. Cooke, Eric Jager, and D. Lewandowski

19.1 INTRODUCTION

Many countries invested extensively in underground gas pipelines in the 1960s and 1970s. These pipes are approaching the age at which problems of corrosion are expected to appear with increasing frequency. These countries may be facing massive investment. The Netherlands, for example, has large gas reserves and is a major exporter of natural gas. The national gas company, Gasunie, is responsible for maintaining over 11,000 km of underground gas pipelines with a current replacement value of about 15 billion US dollars. Over the last 40 years, however, several technologies have been introduced to protect pipes against corrosion and to detect and replace weak points. If, as is generally believed, these efforts have had a positive effect, then predicting these effects is important for impending decisions regarding inspection and repair. This chapter describes a recent effort to upgrade the basis for decisions regarding inspection and replacement of underground pipelines.[1]

Previous studies [see, for example, Kiefner et al. (1990)] focused on developing ranking tools that provide qualitative indicators for prioritizing inspection and maintenance activities. Such tools perform well in some situations. In The Netherlands, however, qualitative ranking tools have not yielded sufficient discrimination to support inspection and maintenance decisions. The population of gas pipelines in The Netherlands is too homogeneous. Moreover, because the status of current pipes and knowledge of effectiveness of current technologies is uncertain, it was felt that uncertainty should be taken into account when deciding which pipelines to inspect and maintain.

[1]This chapter is based on Cooke and Jager (1998), but is expanded to account for new measurement data.

Case Studies in Reliability and Maintenance, Edited by W. R. Blischke and D. N. P. Murthy.
ISBN 0-471-41373-9

We therefore desire a quantitative model of the uncertainty in the failure frequency of gas pipelines. This uncertainty is modeled as a function of observable pipeline and environmental characteristics. The following pipe and environmental characteristics were chosen to characterize a kilometer year of pipeline (Basalo, 1992; Lukezich et al., 1992; Chaker and Palmer, 1989):

Pipe Characteristics

- Pipe wall thickness
- Pipe diameter
- Ground cover
- Coating (bitumen or polyethylene)
- Age of pipe (since last inspection)

Environmental Characteristics

- Frequency of construction activity
- Frequency of drainage, pile driving, deep plowing, placing dam walls
- Percent of pipe under water table
- Percent of pipe exposed to fluctuating water table
- Percent of pipe exposed to heavy root growth
- Percent of pipe exposed to chemical contamination
- Soil type (sand, clay, peat)
- pH value of soil
- Resistivity of soil
- Presence of cathodic protection
- Number of rectifiers
- Frequency of inspection of rectifiers
- Presence of stray currents
- Number of bond sites

Although extensive failure data are available, the data are not sufficient to quantify all parameters in the model. Indeed, the data yield significant estimates only when aggregated over large populations, whereas maintenance decisions must be taken with regard to specific pipe segments. Hence, the effects of combinations of pipe and environmental characteristics on the failure frequency is uncertain and is assessed with expert judgment. The expert judgment method discussed in Chapter 15 of this volume was applied (Cooke, 1991). Fifteen experts from The Netherlands, Germany, Belgium, Denmark, The United Kingdom, Italy, France, and Canada participated in this study.

When values for the pipe and environmental characteristics are specified, the model yields an uncertainty distribution over the failure frequency per kilometer year. Thus the model provides answers to questions such as the following:

- Given a 9-inch-diameter pipe with 7-mm wall laid in sandy soil in 1960 with bitumen coating, and so on, what is the probability that the failure frequency per year due to corrosion will exceed the yearly failure frequency due to third-party interference?
- Given a 9-inch-pipe with 7-mm walls laid in 1970 in sand, with heavy root growth, chemical contamination and fluctuating water table, how is the uncertainty in failure frequency affected by the type of coating?
- Given a clay soil with pH = 4.3, resistivity 4000 [ohm-cm] and a pipe exposed to fluctuating water table, which factors or combinations of factors are associated with high values of the free corrosion rate?

In carrying out this work we had to solve three problems:

- How should the failure frequency be modeled as a function of the above physical and environmental variables, so as to use existing data to the maximal extent?
- How should existing data be supplemented with structured expert judgment?
- How can information about complex interdependencies be communicated easily to decision makers?

In spite of the fact that the uncertainties in the failure frequency of gas pipelines are large, we can nonetheless obtain clear answers to questions like those formulated above.

The next section discusses the general issue of modeling uncertainty. Sections 19.2 discusses the modeling of pipeline failures, and Sections 19.3 through 19.5 treat third-party interference, environmental damage, and corrosion. Section 19.6 presents recent data validating the model. Results are presented in Section 19.7, and Section 19.8 gathers conclusions.

19.2 MODELING PIPELINE FAILURES

The failure of gas pipelines is a complex affair depending on physical processes, pipe characteristics, inspection and maintenance policies, and actions of third parties. A great deal of historical material has been collected, and a great deal is known about relevant physical processes. However, this knowledge is not sufficient to predict failure frequencies under all relevant circumstances. This is due to lack of knowledge of physical conditions and processes and lack of data. Hence, predictions of failure frequencies are associated with significant uncertainties, and management requires a defensible and traceable assessment of these uncertainties.

Expert judgment is used to quantify uncertainty. Experts are queried about the results of measurements or experiments which are possible in principle but not in practice. Since uncertainty concerns the results of possible observations, it is essential to distinguish failure *frequency* from failure *probability*. Frequency is an observable quantity with physical dimensions taking values between zero and infinity. Probability is a mathematical notion that may be interpreted objectively or subjectively. Experts are asked to state their subjective probability distributions over frequencies and relative frequencies. Under suitable assumptions, probabilities may be transformed into frequencies and vice versa. In this model the following transformations are employed. Let N denote the number of events occurring in 1 year in a 100-km section of pipe. N is an uncertain quantity, and the uncertainty is described by a distribution over the non-negative integers. Let $\mathcal{N}$ denote the expectation of N. If we assume that the occurrence of events along the pipe follows a Poisson distribution with respect to distance, then $\mathcal{N}/100$ is the expected frequency of events in 1 km of pipe. If $\mathcal{N}/100$ is much less than 1, such that the probability of two events occurring in 1 km in 1 year is very small, then $\mathcal{N}/100$ is approximately the probability of one event occurring in 1 km in 1 year. $(1 - \mathcal{N}/100)$ is approximately the probability of no event occurring in 1 km in 1 year, and the probability of no events in the entire 100 km is approximately $(1 - \mathcal{N}/100)^{100}$.

The result becomes more accurate if we divide the 100 km into smaller pieces. Using the fact that $\lim_{x\to\infty}(1 - \mathcal{N}/x)^x = e^{-\mathcal{N}}$, we find that the probability of no event in 100 km in 1 year is $e^{-\mathcal{N}}$; the probability of no event in 1 km in 1 year is $e^{-\mathcal{N}/100}$. The probability of at least one event in 1 km is $1 - e^{-\mathcal{N}/100}$, and if $\mathcal{N}/100 \ll 1$, then this probability is approximately $\mathcal{N}/100$. To accord with more familiar usage, however, it is often convenient to suppress the distinction between small frequencies and probabilities.

19.2.1 Example of Modeling Approach

The notation in this section is similar to, but a bit simpler than, that used in the sequel.

Suppose we are interested in the frequency per kilometer year that a gas pipeline is hit (H) during third-party actions at which an overseer from Gasunie has marked the lines (O). Third-party actions are distinguished according to whether the digging is closed (CL; drilling, pile driving, deep plowing, drainage, inserting dam walls, etc) and open (OP; e.g., construction). Letting F denote frequency and P probability, we could write

$$\text{Frequency}\{\text{Hit and Oversight present per km}\cdot\text{yr}\} = F(H \cap O/kmyr)$$

$$= F(CL/kmyr)P(H \cap O|CL) + F(OP/kmyr)P(H \cap O|OP) \qquad (19.1)$$

This expression seems to give the functional dependence of $F(H \cap O)$ on $F(CL)$ and $F(OP)$, the frequencies of closed and open digs, respectively. However, equation (19.1) assumes that the conditional probabilities of hitting with oversight given closed or open digs does not depend on the frequency of closed and open digs. This may not be realistic; an area where the frequency of third-party digging is twice the population average may not experience twice as many incidents of hitting a pipe.

One may anticipate that in regions with more third-party activity, people are more aware of the risks of hitting underground pipelines and take appropriate precautions. This was indeed confirmed by the experts.

It is therefore illustrative to look at this dependence in another way. Think of $F(H \cap O)$ as a function of two continuous variables, FCL = frequency of closed digs per kilometer year, and FOP = frequency of open digs per kilometer year. Write the Taylor expansion about observed frequencies FCL_0 and FOP_0. Retaining only the linear terms, we obtain

$$F(H \cap O/kmyr) = F(FCL, FOP)$$
$$= F(FCL_0, FOP_0) + p_1(FCL - FCL_0) + p_2(FOP - FOP_0) + \cdots \quad (19.2)$$

If $P(H \cap O|CL)$ and $P(H \cap O|OP)$ do not depend on FCL and FOP, then equation (19.2) is approximately equivalent to equation (19.1). Indeed, put $p_1 = P(H \cap O|CL)$ and $p_2 = P(H \cap O|OP)$, and note that $F(FCL_0, FOP_0) = p_1 FCL_0 + p_2 FOP_0$.

The Taylor approach conveniently expresses the dependence on FCL and FOP, in a manner familiar to physical scientists and engineers. Of course it can be extended to include higher-order terms.

If we take the "zero-order term" $F(FCL_0, FOP_0)$ equal to the total number of times gas lines are hit while an overseer has marked the lines, divided by the number of kilometer years in The Netherlands, then we can estimate this term from data. FCL_0 and FOP_0 are the overall frequencies of closed and open digs. p_1 and p_2 could be estimated from data if we could estimate $F(FCL, FOP)$ for other values of FCL and FOP, but there are not enough hittings in the database to support this. As a result, these terms must be assessed with expert judgment, yielding uncertainty distributions over p_1 and p_2. Experts are queried over their subjective uncertainty regarding measurable quantities; thus they may be asked:

Taking account of the overall frequency F(FCL_0, FOP_0) *of hitting a pipeline while an overseer has marked the lines, what are the 5%, 50%, and 95% quantiles of your subjective probability distribution for the frequency of hitting a pipeline while an overseer has marked the lines if frequency of closed digs increases from* FCL_0 to FCL, *with other factors remaining the same?*

In answering this question, the expert conditionalizes his uncertainty on everything he knows, in particular the overall frequency $F(FCL_0, FOP_0)$. We configure the elicitation such that the "zero-order terms" can be determined from historical data, whenever possible.

How do we use these distributions? Of course if we are only interested in the average situation in The Netherlands, then we needn't use them at all, since this frequency is estimated from data. However, it is known that the frequency of third-party activity (with and without oversight) varies significantly from region to region. If we wish to estimate the frequency of hitting with oversight where $FCL \neq FCL_0$ and $FOP \neq FOP_0$, then we substitute these values into equation (19.2) and obtain an uncertainty distribution for $F(H \cap O)$, conditional on the 'zero-order' estimate and conditional on the values FCL, FOP. This is pure expert subjective uncertainty. If we wish, we may also include uncertainty due to sampling fluctuations in the zero-order estimate.

19.2.2 Overall Modeling Approach

The failure probability of gas pipelines is modeled as the sum of a failure probability due to third-party actions and a failure probability due to corrosion[2]:

$$\begin{aligned} &P\{\text{failure of gas pipelines/km} \cdot \text{year}\} \\ &\quad = P\{\text{direct leak due to third parties/km} \cdot \text{year}\} \\ &\qquad + P\{\text{failure due to corrosion/km} \cdot \text{year}\} \end{aligned} \tag{19.3}$$

Both terms on the right-hand side will be expressed as functions of other uncertain quantities and parameters. The parameters will be assigned specific values in specific situations, and the uncertain quantities are assigned subjective uncertainty distributions on the basis of expert assessments. This results in an uncertainty distribution over possible values of $P\{\text{failure of gas pipelines/km} \cdot \text{year}\}$, conditional on the values of the known variables.

Failure due to corrosion requires damage to the pipe coating material and (partial) failure of the cathodic and stray current protection systems. Damage to coating may come either from third parties or from the environment (Lukezich et al., 1992). The overall model may be put in the form of a fault tree as shown in Figure 19.1.

19.3 THIRD-PARTY INTERFERENCE

The model described here enables the calculation of uncertainty distributions over the probability per kilometer year of various damage categories, with or without repair, resulting from third-party interference. The underlying probability model is a so-called "marked point process." For a given 1 km section of pipe, third-party activities (within 10 m of the pipe) are represented as a Poisson process in time. Each dig-event is associated with a number of "marks"—that is, random variables that assume values in each dig-event (Figure 19.2). The random variables and their possible values are shown in Table 19.1.

For each 1-km pipe section, the following picture emerges: On the first dig the pipe was not hit, hence the damage was nonexistent ($D = n$) and no repair was carried out ($R = rn$). The second dig was an open dig with oversight; small line damage occurred, but was repaired. The third dig was closed without oversight, and the line was hit and resulted in a direct leak. By definition, repair was unable to prevent leak, hence $R = rn$.

[2]The model does not include stress corrosion cracking or hydrogen-induced cracking, because these have not manifested themselves in The Netherlands. Damage to pipelines during construction and installation is not modeled. Low probability scenarios like earthquake and flood have not been modeled, and "exotic" scenarios like sabotage, war, malfeasance and the like are neglected.

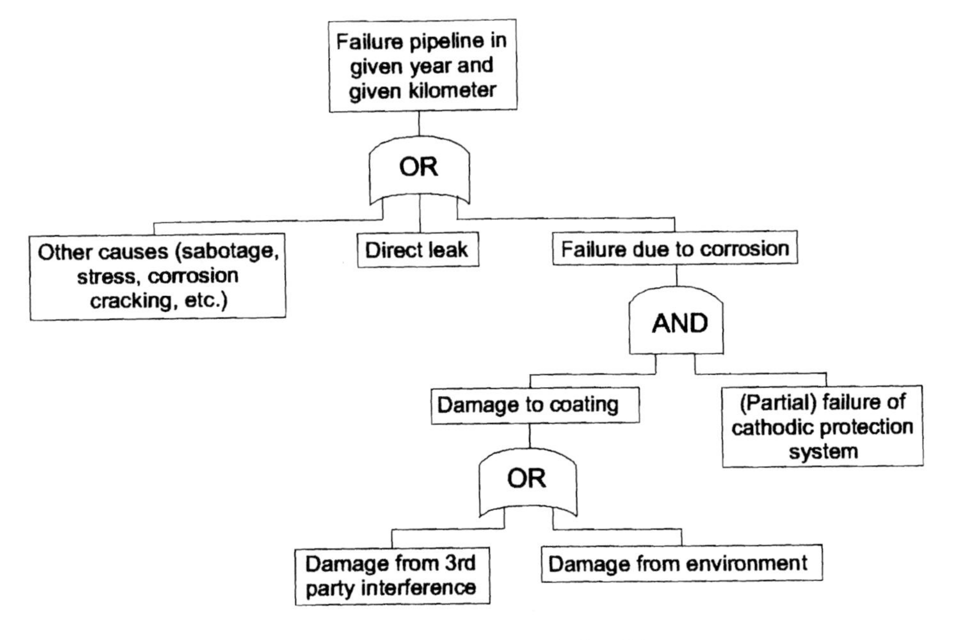

Figure 19.1. Fault tree for gas pipeline failure.

The 1-km pipe section is described by a number of parameters, which are assumed to be constant along this section:

- *t*: pipe wall thickness
- *gc*: depth of ground cover
- *f* = (*fop*, *fcl*): frequency of open and closed digs within 10 m of pipe

The values of these parameters will influence the distributions of the random variables in Table 19.1. Hence, we regard these as random variables, and their influence

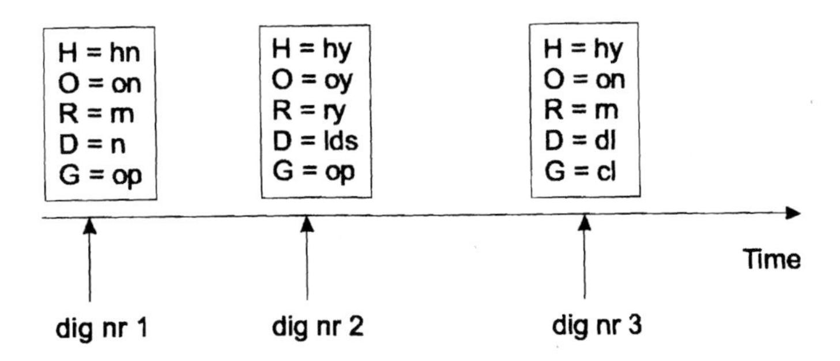

Figure 19.2. Digs as marked point process.

Table 19.1. Marks for Third-Party Digs

Variable	Meaning	Values	Interpretation
H	Pipe hit?	*hy*, *hn*	hit yes, hit no
O	Overseer notified?	*oy*, *on*	oversight yes, oversight no
R	Repair carried out?	*ry*, *rn*	repair yes, repair no
D	Damage?	*n*, *cd*3, *lds*, *ldl*, *dl*	none, coating damage, small line damage, large line damage, direct leak
G	Dig type?	*op*, *cl*	open dig, closed dig

on other random variables is described by conditionalization. In any 1-km section the values for these variables can be retrieved from Gasunie data, and the distributions of other variables can be conditionalized on these values. From a preliminary study (Geervliet, 1994) it emerged that pipe diameter and coating type did not have influence on the probability of hitting a pipe.

The damage types indicated in Table 19.1 are defined more precisely as:

- *dl*: direct leak (puncture or rupture)
- *ldl*: line damage large (at least 1 mm of pipe material removed, no leak)
- *lds*: line damage small (less than 1 mm of pipe material removed)
- *cd*3: coating damage without line damage due to third parties

Every time a gas pipeline is hit, we assume that one and only one of these damage categories is realized. Hence, $D = n$ if and only if $H = hn$. By definition, if $D = dl$, then repair prior to failure is impossible.

We wish to calculate uncertainty distributions over the probability of unrepaired damage:

$$P(D = x \cap R = rn|\mathrm{t}, gc, f);\ x \in \{cd3, lds, ldl, dl\} \tag{19.4}$$

Letting Σ_{HOG} denote summation over the possible values of H, O, G, we have

$$\begin{aligned} &P(D = x \cap R = rn|\mathrm{t}, gc, f) \\ &= \sum_{HOG} P(D = x \cap R = rn|H, O, G, t, gc, f)P(H, O, G|t, gc, f) \\ &= \sum_{OG} P(D = x \cap R = rn|H = hy, O, G, t, gc, f)P(H = hy, O, G|t, gc, f) \end{aligned} \tag{19.5}$$

since third-party damage can only occur if the pipe is hit.

For each of the four damage types, there are four conditional probabilities to consider, each conditional on four continuous-valued random variables. To keep the model tractable it is necessary to identify plausible simplifying assumptions. These are listed and discussed below. The expressions "X depends only on Y" and

"X influences only Y" mean that given Y, X is independent of every other random variable.

1. D depends only on (H, G, t).
2. R depends only on $(H, G, O, D \in \{cd3, lds, ldl\})$.
3. gc influences only H.
4. gc is independent of f.
5. G is independent of f.
6. (H, O, G) is independent of t given (gc, f).

Assumptions 1, 3, 4, and 5 speak more or less for themselves. Assumption 2 says the following: If the pipe is hit and the damage is repairable ($D \neq dl$), then the probability of repair depends only on the type of dig and the presence of oversight; it does not depend on the type of repairable damage inflicted. Assumptions 1 and 2 entail that (D, R) depends only on $(H, G, O, t, D \in \{cd3, lds, ldl\})$.

To appreciate assumption 5, suppose that the uncertainty over $f = (fop, fcl)$ is described by an uncertainty distribution, and consider the expression $P(fop, fcl|G = cl)$. Would knowing the type of dig in a given third party event tell us anything about the frequencies of open and closed digs? It might. Suppose that either all digs were open or all digs were closed, each possibility having probability 1/2 initially. Now we learn that one dig was closed; conditional on this knowledge, only closed digs are possible. Barring extreme correlations between the uncertainty over values for fop and fcl, knowing $G = cl$ can tell us very little about the values of fcl and fop. Assumption 5 says that it tells us nothing at all.

To illustrate how these assumptions simplify the calculations, we consider the event $(D = cd3 \cap R = rn)$, which we abbreviate as $(cd3 \cap rn)$. We can show[3]:

$$P(cd3 \cap rn|t, gc, f)$$
$$= \sum_{OGP} (rn|hy, O, G)P(cd3|hy, G, t)P(O, hy|G, f)P(G)P(gc|hy)/P(gc) \quad (19.8)$$

[3]Using elementary probability manipulations and assumptions 1 and 2, we obtain

$$\begin{aligned} &P(cd3 \cap rn|hy, O, G, t, gc, f)P(hy, O, G|t, gc, f) \\ &= P(cd3 \cap rn|hy, O, G, t)P(hy, O, G|t, gc, f) \\ &= P(cd3|rn, hy, O, G, t)P(rn|hy, O, G, t)P(hy, O, G|t, gc, f) \\ &= P(cd3|hy, G, t)P(rn|hy, O, G)P(hy, O, G|t, gc, f) \end{aligned} \quad (19.6)$$

Reasoning similarly with assumptions 3, 4, 5, and 6 and using Bayes' theorem, we obtain

$$\begin{aligned} &P(hy, O, G|t, gc, f) = P(hy, O, G|gc, f) \\ &= P(gc, O, G|hy, f)P(hy|f)/P(gc|f)= \\ &= P(gc|O, G, hy, f)P(O, G|hy, f)P(hy|f)/P(gc|f) \\ &= P(gc|hy)P(O, G|hy, f)P(hy|f)/P(gc) \\ &= P(gc|hy)P(O, hy|G, f)P(G|f)/P(gc) \\ &= P(gc|hy)P(O, hy|G, f)P(G)/P(gc) \end{aligned} \quad (19.7)$$

Similar expressions hold for damage types *lds* and *ldl*. For *dl*, the term $P(rn|hy, O, G)$ equals 1 because repair is not possible in this case. The terms

$$P(cd3|hy, G, t), P(G|hy), P(G), P(gc|hy) \text{ and } P(gc)$$

can be estimated from data; the other terms are assessed (with uncertainty) using expert judgment. The uncertainty in the data estimates derives from sampling fluctuations and can be added later, although this will be small relative to uncertainty from expert judgment. The term

$$P(gc|hy)/P(gc) = P(H = hy|gc)/P(H = hy)$$

is called the "depth factor"; it is estimated by dividing the proportion of hits at depth gc by the total proportion of pipe at depth gc. $P(G = cl|hy, cd3)$ is estimated as the percentage of coating damages from third parties caused by closed digs; $P(G = cl|hy)$ is the percentage of hits caused by closed digs, and $P(G)$ is the probability per kilometer year of a closed dig. This probability is estimated as the frequency of closed digs per kilometer year, if this frequency is much less than 1 (which it is).

The term $P(rn|hy, O, G)$ is assessed by experts directly when the terms $P(O, hy|G, f)$ are assessed using the Taylor approach described in Section 19.3. There are no ruptures directly caused by third-party activities in the Dutch database. To assess the probability (with uncertainty) of ruptures due to third parties, experts assess, for two different wall thickness, the percentages of direct leaks that will be ruptures. Let *RUP71* and *RUP54* denote random variables whose distributions reflect the uncertainty in these percentages for thickness 7.1 and 5.4 mm, respectively. We assume that *RUP71* and *RUP54* are comonotonic.[4] Putting

$$RUP54 = RUP71 + x(7.1 - 5.4)$$

we can solve for the the linear factor x and for some other thickness t:

$$RUPt = RUP71 + x(7.1 - t)$$

gives an assessment of the uncertainty in the probability of rupture, given direct leak, for thickness t. This produces reasonable results for t near 7.1. For $t > 10$ mm it is generally agreed that rupture from third parties is not possible (Hopkins et al., 1992).

19.4 DAMAGE DUE TO ENVIRONMENT

In dealing with damage to coating due to environmental factors per kilometer year, we revert to the frequency notation, as this frequency can be larger than 1. For both bitumen (*bit*) and polyethylene (*pe*) coatings, the probability of environmental dam-

[4]That is, their rank correlations are equal to 1.

age depends on the pipe diameter (d), on the soil type (st), and on the percentage of the pipe exposed to fluctuating water table (wt_f). Bitumen coating is also sensitive to the proportion of the 1-km length exposed to tree roots (rt) and chemical contamination (ch). The effects of these factors are captured with a first-order Taylor expansion whose linear terms $p_5, \ldots, p_{10}$ are assessed by experts.

$$F(bit) = Fo(bit) + p_5 \cdot (d - d_0) + p_6 \cdot wt_f + p_7 \cdot rt + p_8 \cdot ch + st \cdot bit \quad (19.9)$$

$$F(pe) = Fo(pe) + p_9 \cdot (d - d_o) + p_{10} \cdot wt_f + st \cdot pe \quad (19.10)$$

To determine the probability of at least one coating damage per kilometer, these frequencies are divided by 100 to determine the frequency per 10-m section. Because these frequencies will be much less than 1, and assuming that damage to different 10-m sections are independent, we have

$$P\{\text{at least 1 coating damage per km}\} = 1 - (1 - F/100) \quad (19.11)$$

On substituting equations (19.9) and (19.10) into equation (19.11), we obtain the probabilities per kilometer year of bitumen and polyethylene coating damage per kilometer year, due to environmental factors, notated $P(CDE_{bit})$ and $P(CDE_{pe})$.

19.5 FAILURE DUE TO CORROSION

19.5.1 Modeling Corrosion Induced Failures

The modeling of failure due to corrosion is more complicated than that of failure due to third parties. The probability of failure due to corrosion depends on many factors as listed in Table 19.2.

The model described here uses only pit corrosion. Given these factors, the corrosion rate is assumed constant in time (Camitz and Vinka, 1989).

For a pipeline to fail due to corrosion, two lines of defense must be breached. First the coating must be damaged, and second, depending on location, the cathodic or stray current protection system must not function as intended. Coating damage may be caused either by third-party actions or by environmental factors. These protection systems have been in place since 1970. We first elaborate the model for pipelines installed after 1970.

Table 19.2. Factors Influencing Failure Due to Corrosion

Chemical contamination of soil	Pipe diameter	Oversight
Soil type (clay, sand, peat)	Pipe inspection	Repair
Soil resistance	Stray currents	Acidity
Pipe thickness	Tree roots	Pipe age
Water table	Third party actions	

Assuming that the coating has been breached, pit corrosion will reduce the pipe wall thickness until a critical value is reached, at which point the pipe fails. This critical wall thickness—that is, the thickness at which failure occurs—is expressed as a fraction x of the original wall thickness minus the pipe material removed during the damage event. x depends on the pressure of the gas in the pipe line, and on the geometry of the pipe damage, and this relationship has been established by experiment. In this model, x is introduced as a parameter whose value depends only on the damage type; thus we distinguish x_C, x_S, and x_L for (only) coating damage, small pipe damage, and large pipe damage, respectively. Coating damage is caused either by third parties (*cd*3) or by the environment (*cde*).

A length of pipe can be inspected for corrosion, and if corrosion is found, the pipe is uncovered and repaired. Hence, after such inspection the pipe is as good as new. The effective birthday (*eb*) of a pipe section is the calendar year of the last inspection.

Given a corrosion rate (*CR*) and a damage type, we define the effective life of a pipe section as the time required for the corrosion to reduce wall thickness to the critical wall thickness. Letting t denote the original pipe wall thickness (i.e., C, S, L, $t_C = 0$, $t_S = 0.5$ mm, $t_L = 2$ mm), we obtain

$$EL(CR, i) = x_i(t - t_i)/CR \tag{19.12}$$

In this equation, CR is uncertain and x_i, t_i are parameters with uncertain indices. $EL(CR, i)$ is the time a pipe survives given corrosion rate CR after sustaining damage type i, $i \in \{C, S, L\}$.

Suppose we are interested in the event "first failure of a 1-km length of gas pipeline occurs in calendar year y." For each given value of CR, there are three years, $y_C(CR)$, $y_S(CR)$, and $y_L(CR)$, such that damage type i in year y_i, somewhere on this 1-km length of pipe, causes failure in year y. y_i is called the critical year for damage type i. The situation is pictured in Figure 19.3.

Referring to Figure 19.3, we see that failure due to damage type C is impossible; the pipe isn't old enough in year y. If small damage (S) occurs in year y_S, and not before, and if large damage has not occurred before y_L, then the pipe fails in y due to small damage in y_S. The probability of this is

$$(1 - P_S)^{Y_S - eb} P_S (1 - P_L)^{Y_L - eb} \tag{19.13}$$

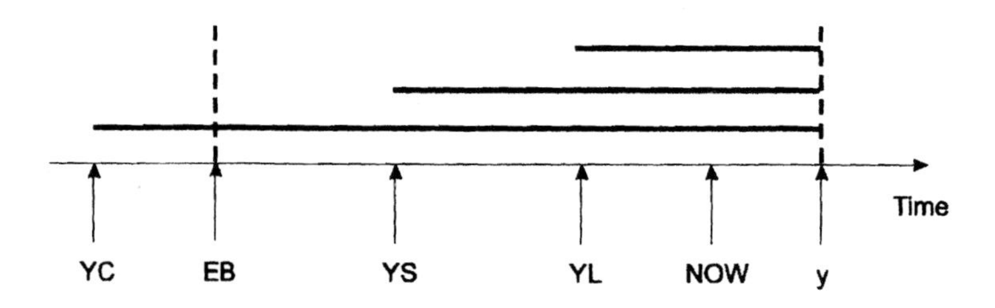

Figure 19.3. Effective lives and critical years for three damage types for fixed corrosion rate.

where we write $P_S = P(lds)$, $P_L = P(ldl)$, $P_C = P(cd3) + P(cde) - P(cd3 \cap cde)$. However, if y is "next year," then we already know that the pipeline has not already failed due to corrosion from small or large pipeline damage. Hence, we should conditionalize on the event "no small damage before y_S and no large damage before y_L." In this case, the probability of failure in year y due to small damage is simply P_S, and the probability of failure due to corrosion is, neglecting higher-order terms, $P_S + P_L$ (all of these probabilities are conditional on CR).

If y is in the future, and we conditionalize on our knowledge that no failure has occurred up to now, with $T = y - \text{now}$, $T < y_S - eb$, then the probability of failure due to corrosion between in year y is (again, conditional on CR and neglecting small terms)

$$(1 - P_S)^T (1 - P_L)^T (P_S + P_L) \tag{19.14}$$

In general, let

$$q_i(eb, y, \text{now}, CR) = MIN\{(y_i(CR) - eb), (y - 1 - \text{now})\} \tag{19.15}$$

denote the number of years between y and now in which a failure due to damage type i could have caused failure between now and year $y - 1$, conditional on CR, and let

$$1_i = \begin{cases} 1, & \text{if } y_i(CR) > eb \\ 0, & \text{otherwise} \end{cases} \tag{19.16}$$

then (sum and product are over $i \in \{C, S, L\}$);

$$P_{f|cr}(CR, eb, y, \text{now}, P_C, P_S, P_L, t, x_C, x_S, x_L) = \prod (1 - p_i)^{qi} \sum 1_i p_i \tag{19.17}$$

is approximately the probability of failure in year y due to corrosion, given CR. $P_{f|cr}$ is an uncertain quantity since the arguments written in capital letters represent uncertain quantities.

19.5.2 Pit Corrosion Rate

The free rate of pit corrosion CR_f [mm/year] is modeled to depend on the soil type (clay, sand, peat), the soil resistance (r), the acidity (pH), and the proportion of pipeline under the water table, above the water table, and fluctuating under and above the water table (wt_u, wt_a, wt_f). CR_f is the rate of corrosion that would obtain if the cathodic protection were not present. Using a zero-order corrosion rate with arguments r_0, pH_0, $wt_{u0} = wt_{f0} = 0$, $wt_{a0} = 1$, we apply the linear approximation (supported by experiment):

$$CR_f = CR_{f0} + p_{11}(r - r_0) + p_{12}(\text{pH} - \text{pH}_0) + p_{13} wt_f + p_{14} wt_u \tag{19.18}$$

The linear terms $p_{11}, \ldots, p_{14}$ are assessed with expert judgment. All of the terms in equation (19.18) depend on soil type. We distinguish three states of the cathodic protection system:

- CP_f: wholly nonfunctional, $CR = CR_f$
- CP_p: partially functional (pipe-soil potential outside prescribed range), $CR = CR_p$
- CP_{ok}: wholly functional as prescribed, $CR \sim 0$.

Before the cathodic protection system was installed in 1970, only state CP_f was available. CR_p is determined via expert judgment as a fraction of CR_f. $P(CP_i)$ is the fraction of 1-km pipe length for which cathodic protection is in state i, $i \in \{CP_f, CP_p, CP_{ok}\}$. Since the factors affecting the cathodic protection do not change from year to year, we assume that the states CP_f and CP_p affect the same portions of pipe each year.

Stray currents can induce corrosion against which cathodic protection is ineffective. In 1970 a protection system of bonds was installed to drain off strong stray currents in locations where these are known to occur. Each bond is inspected once a month; hence if a bond has failed, the stray current corrosion rate CR_{st} has been operative on the average for one-half month. In the neighborhood of a bond, the corrosion rate before 1970 due to stray currents is CR_{st}, and after 1970 it is assumed to be $CR_{st}/24$. If bs is the 17 proportion of a 1-km length of pipe in the neighborhood of a bond site and $P(SP)$ is the probability that the stray current protection system fails at one site, then $bs \cdot P(SP)$ is the probability that CR_{st} (before 1970) or $CR_{st}/24$ (after 1970) obtains, given that damage has occurred somewhere in the pipe section.

Unconditionalizing equation (19.18) on CR, we obtain the probability of failure per kilometer year due to corrosion for pipe installed after 1970:

$$P_{cor>70} = P_{f|cr}(CR_f)P(CP_f) + P_{f|cr}(CR_p)P(CP_p) + P_{f|cr}(CR_{st}/24)bsP(SP) \quad (19.19)$$

This is an uncertain quantity whose distribution is the uncertainty distribution for the failure frequency for a 1-km length of gas pipeline with specified pipe and environment parameter values.

For pipelines whose effective birthday is before 1970,

$$x_i(t - t_i) - (y - 1970) \cdot CR \quad (19.20)$$

is the thickness of pipe wall, under damage type i, exposed to corrosion at the rate obtaining before protection systems were installed. Let

$$1_{i,CR} = \begin{cases} 1, & \text{if } y - x_i(t - t_i)/CR > 1970 \\ 0, & \text{otherwise} \end{cases} \quad (19.21)$$

If $1_{i,CR} = 1$, then $y_i > 1970$; if $y_i < 1970$, then we must account for the absence of protection systems. We compute the effective life as follows:

$$
\begin{aligned}
EL(CR_f, i) &= x_i(t - t_i)/CR_f \\
EL(CR_p, i) &= 1_{i,CRp} x_i(t - t_i)/CR_p \\
&\quad + (1 - 1_{i,CRp})(x_i(t - t_i) - (y - 1970)CR_p)/CR_f \\
EL(CR_{st}, i) &= 1_{i,CR_{st}} x_i(t - t_i)24/CR_{st} \\
&\quad + (1 - 1_{i,CR_{st}})(x_i(t - t_i) - (y - 1970)CR_{st}/24)/CR_{st}
\end{aligned} \tag{19.22}
$$

$P_{cor<70}$ is obtained by using equations (19.22) instead of equation (19.12) in equation (19.17).

19.6 VALIDATION

This model was originally developed in 1996 [see Cooke and Jager (1998)]. In the last few years the Dutch gas company has launched a program of "intelligent pig runs." An intelligent pig is a device that can be sent through a large-diameter pipe to measure corrosion defects. These pig runs are quite accurate but also quite expensive. The gas company is interested in using these runs to calibrate the failure model, so that the model can be used to support the selection of pipes to be pigged in the future.

At present, data from two pig runs are available. For each run, the data consist of a list of defects, their position on the pipe, and their depth. Run A covered 66 km of a gas pipeline with bitumen coating laid in sand in 1966, with an average diameter of 12.45 in. at an average depth of 1.76 m. There were 65 incidents in which the removal of pipe material was at least 10% of the wall thickness.

Run B covered 84 km of a gas pipeline with bitumen coating laid in sand in 1965, with an average diameter of 11.45 in. at an average depth of 1.87 m. There were 92 incidents in which the removal of pipe material was at least 10% of the wall thickness. A part of the data is included in this document (Figure 19.4).

The Laplace test was applied to each data set separately to test the hypothesis that the spatial interarrival times came from an exponential distribution, against the hypothesis that the data come from a nonhomogeneous Poisson distribution. The

EVENT_NO	EVENT_NAME	CATEGORY	distance [m]	LIST_CLOCK	position [hour]	%ML	length [mm]	width [mm]	wall thickness [mm]
D - 1	Defect	External General	97.579	3:30	3:30	23	41	142	12.86
D - 2	Defect	External General	97.595	8:10	8:10	31	41	91	12.86
D - 3	Defect	External Circ Groove	97.666	7:10	7:10	18	20	145	12.86
D - 4	Defect	External Circ Groove	114.636	6:50	6:50	14	25	79	12.86
D - 57	Defect	External Pit	799.246	6:30	6:30	16	25	41	12.86

Figure 19.4. The Gasunie data.

Pipeline A data indicate a statistically significant spatial trend since 30% of observed corrosion appear in first 3 km of the pipe (the pipe is 65 km long). This spatial clustering could not be explained and is not used in the further analysis. There was no significant "spatial trend" in the Pipeline B data.

A nonparametric Kolmogorov–Smirnov test was used to test the hypothesis that the spatial interarrival intervals for events removing at least 10% of pipe wall material came from the same distribution. The hypothesis was not rejected at the 5% level. Hence, no significant difference was found between the two data sets.

The model predicts the frequency of leak due to corrosion per kilometer year. Because there have been no leaks, the model must be adapted to predict the frequency of corrosion events removing specified percentages of pipe wall material. This is done by manipulating the critical fraction (xc) of pipe wall material which must be removed in order to cause a leak due to coating damage. By setting $xc =$ 10% the "failure frequency" output by the model corresponds to the frequency of corrosion events removing at least 10% of the pipe wall material. By setting xc successively equal to 10%, 15%, 20%, . . . , 40% we obtain seven uncertainty distributions for the frequency per kilometer year of removing at least xc% of the pipe wall material. For each value of xc, we retrieve the number of events removing at least xc% of pipe wall material from the data. Dividing this number by the number of kilometer years, we obtain the empirical frequency per kilometer year of removing at least xc% of pipe wall material. We then compare these empirical frequencies with the appropriate uncertainty distribution for these frequencies from the model. The results are shown in Figures 19.5 and 19.6. We see that the model places the

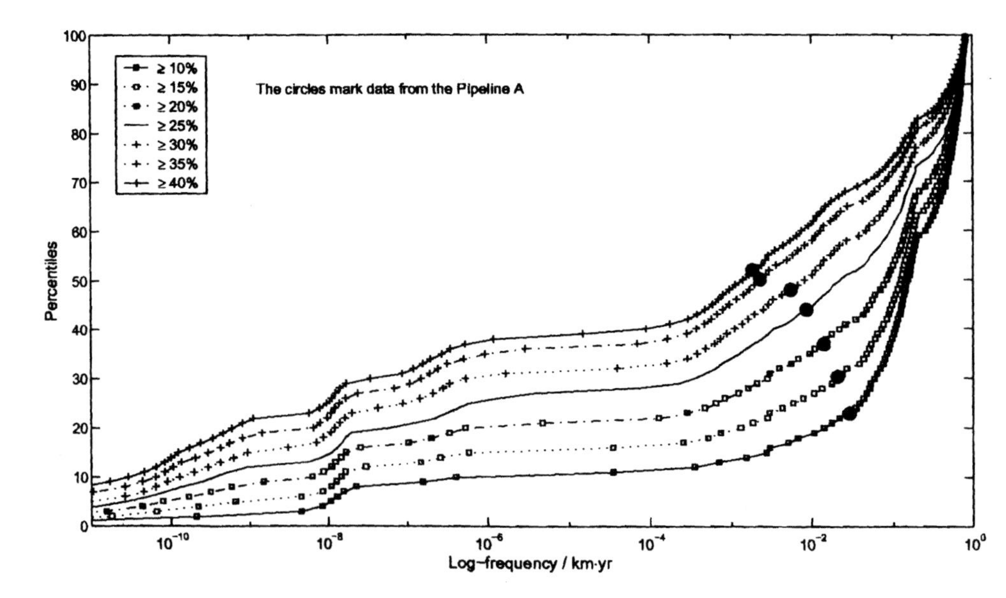

Figure 19.5. Uncertainty distributions and observed exceedence frequencies, pipe A.

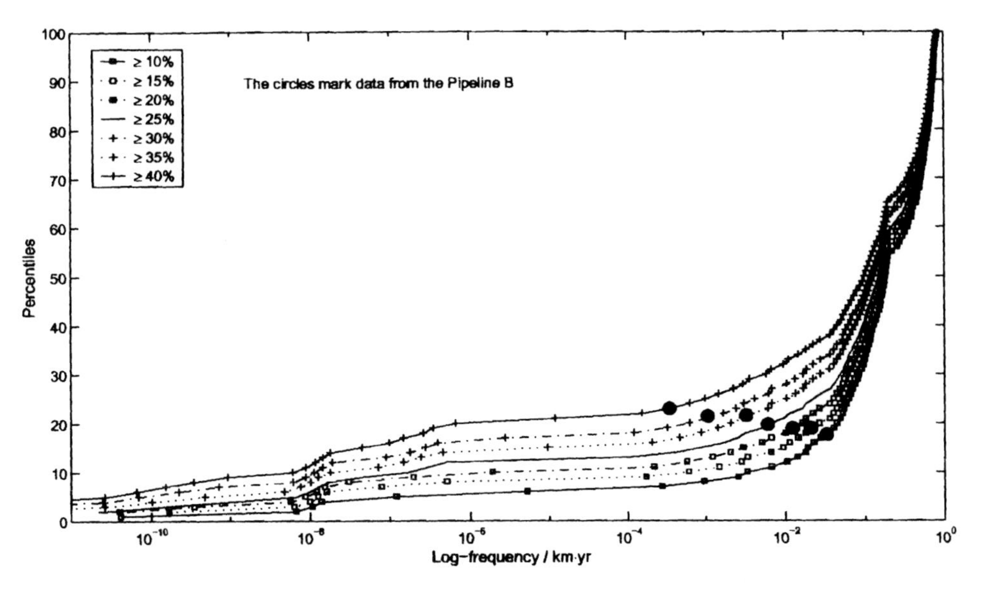

Figure 19.6. Uncertainty distributions and observed exceedence frequencies, pipe B.

observed frequencies well within the central mass of the respective uncertainty distributions.[5]

19.7 RESULTS FOR RANKING

19.7.1 Casewise Comparisons

Two types of results can be obtained with the model. First, we can perform casewise comparisons. By specifying parameter values for two or more types of kilometer-year sections of pipelines, the uncertainty distributions for the frequency of failure can be compared. Figure 19.7 compares three cases, namely,

- Bitumen-coated pipe laid in 1975 in sand
- Bitumen-coated pipe laid in 1975 in clay
- Polyethylene-coated pipe laid in 1975 in sand

Percentiles of the subjective uncertainty distribution are shown horizontally; the logarithm of the failure frequency per kilometer year is plotted vertically (the absolute values are proprietory). Other parameters are the same in all cases, and those describing the frequency of third-party intervention are chosen in accord with the

[5]The exact time at which cathodic protection was installed could not be retrieved at this writing, but is estimated to be 1970.

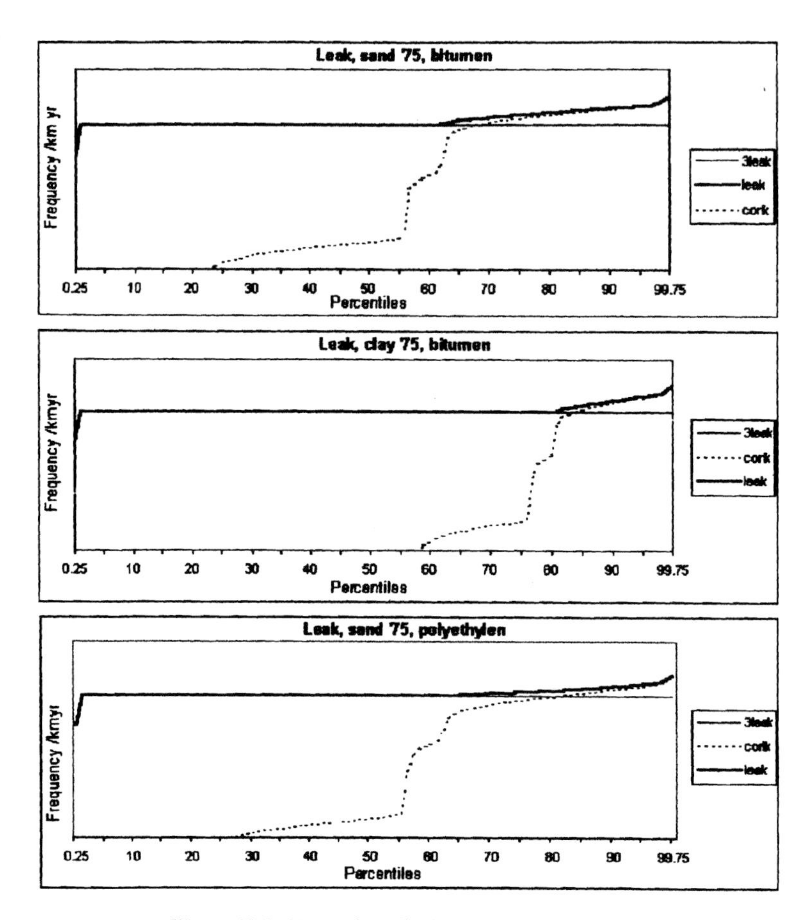

Figure 19.7. Uncertainty distributions for three cases.

generic values retrievable from the Dutch data. Each graph plots the frequency per kilometer year of failure against the percentiles of the uncertainty distribution for the failure frequency. Each graph shows three curves, a curve corresponding to failure due to corrosion (*corlk*), a curve corresponding to failure due to third-party interference (3*leak*), and a curve corresponding to the sum of these two (*leak*).

Because of the choice of third-party interference frequencies, the curves for failure due to third-party interference are constant; this means that there is no uncertainty regarding this failure frequency. The curves for failure due to corrosion are not constant. In the first graph (bitumen in sand), we see that the 66th percentile of the uncertainty distribution for failure due to corrosion corresponds to the same failure frequency as third-party intervention. In other words, there is a 0.66 probability that the frequency of failure due to corrosion will be lower than that due to third-party intervention.

In the second graph (bitumen in clay) we find a probability of 0.85 that the frequency of failure due to corrosion will be lower than that due to third-party intervention. In the third graph (polyethylene in sand), we find a probability of 0.77 that the frequency of failure due to corrosion will be lower than that due to third-party intervention. In the second graph, note that the failure frequency curve for corrosion drops off more rapidly than in the third graph. This means that very low values are more likely for bitumen-coated pipe in clay than for polyethylene-coated pipe in sand.

Comparisons of this nature can only be made on the basis of fully specified cases. We cannot conclude, for example, that "bitumen in clay is about the same as polyethylene in sand." The comparisons in Figure 19.7 depend on the values of all other environmental and pipe variables. Thus, changing the amount of root growth, the soil resistivity, the age and thickness of the pipe, or any of the other parameters might produce very different pictures. Finally, we note that the failure frequency due to corrosion is highly uncertain. Nevertheless, clear comparisons may be made by taking this uncertainty into account.

19.7.2 Importance in Specific Case

As mentioned in the introduction, we use Monte Carlo simulation to compute the uncertainty distribution of the failure frequency for given pipe and environmental characteristics. When we focus on a particular kilometer of pipe—that is, a particular set of values for all the parameters in the model—we may ask, "Which factors are important for the failure frequency in this specific case?" Since this failure frequency is uncertain, we are really asking, "Which factors are important for the uncertainty in failure frequency in this case?"

To gain insight into this type of question, a new graphic exploratory tool has been developed, termed "cobweb plots."[6] These plots enable the user to gain insight into complex relations between interdependent uncertain quantities.

We illustrate by considering the uncertainty in failure due to corrosion in a bitumen-coated pipe laying in sand for 5 years without cathodic protection.

The variable *corlk* or "leak due to corrosion" is potentially influenced by the following variables:

- crf: free corrosion rate
- crp: corrosion rate under partial functioning of cathodic protection
- crse: corrosion rate from stray currents
- ps: frequency of small unrepaired pipeline damage
- pl: frequency of large unrepaired pipeline damage
- pc3: frequency of coating damage from third parties
- pcen: frequency of coating damage from environment

The uncertainty distribution for *corlk* is built up by considering a large number of "scenarios," where each scenario is made by sampling values from all input vari-

[6]Wegman (1990) introduced a similar technique, though without conditionalization.

ables. In each scenario, unique values are assigned to all the above variables. We are interested in how the high and low values of *corlk* co-vary with high and low values of the above variables.

Cobweb plots allow the user to explore this co-variation. Suppose we plot all the values of the above variables on parallel vertical lines, with high values at the top and low values at the bottom. Each individual scenario assigns exactly one value to each variable; if we connect these values, we get a jagged line intersecting each variable line in one point. Suppose we plot jagged lines for each of 200 scenarios; the result will suggest a cobweb. It may be difficult to follow the individual lines; it is therefore convenient to "filter" or "conditionalize" on sets of lines. For example, we might conditionalize on all lines passing through high values of *corlk* and see where these lines intersect the other variables.

The first cobweb plot (Figure 19.8) shows lines for 500 scenarios. The second cobweb plot (Figure 19.9) conditionalizes on high values of *corlk*: We see that these are associated with high values of *crf* and with high values of *pcen*. *crp* and *crse* are not affected by this conditionalization; by assumption, there is no cathodic protection in this case. The third cobweb plot (Figure 19.10) conditionalizes on low values of *corlk*. We see that these are strongly associated with low values of *crf* and with *crse* but not associated with other variables. We may conclude that damage from the environment is important for high values of failure frequency due to corro-

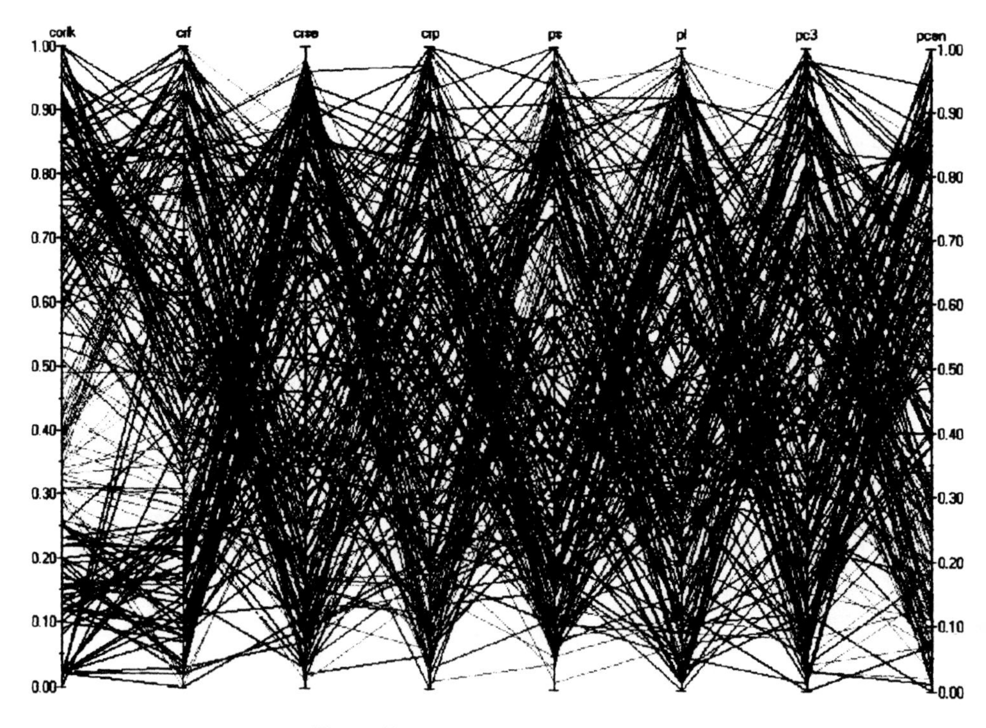

Figure 19.8. Unconditional cobweb plot.

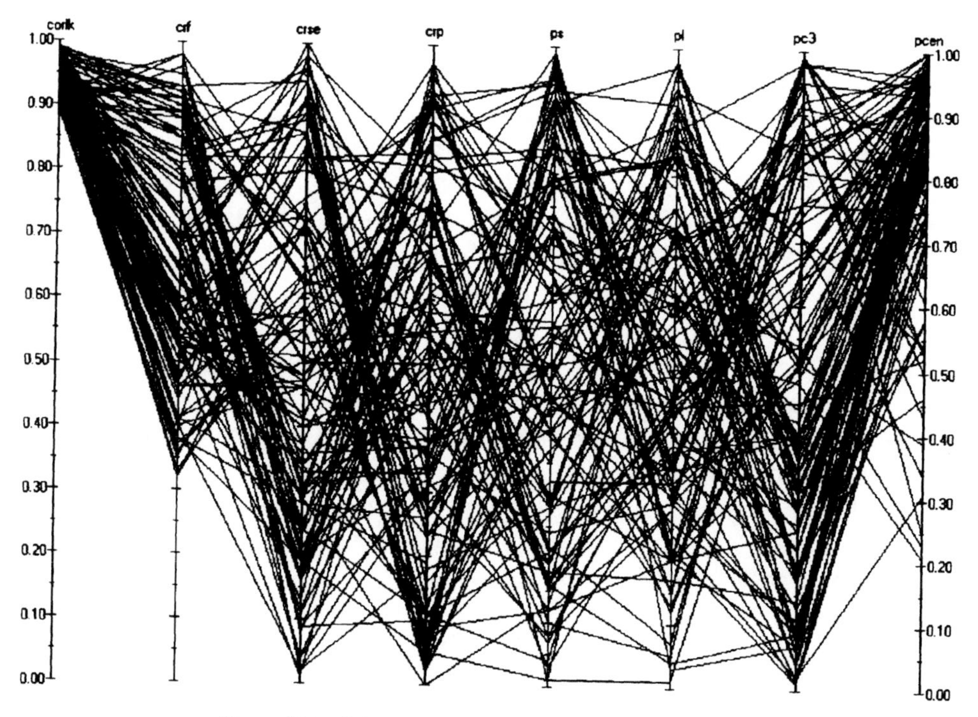

Figure 19.9. Cobweb plot conditional on high values of *corlk*.

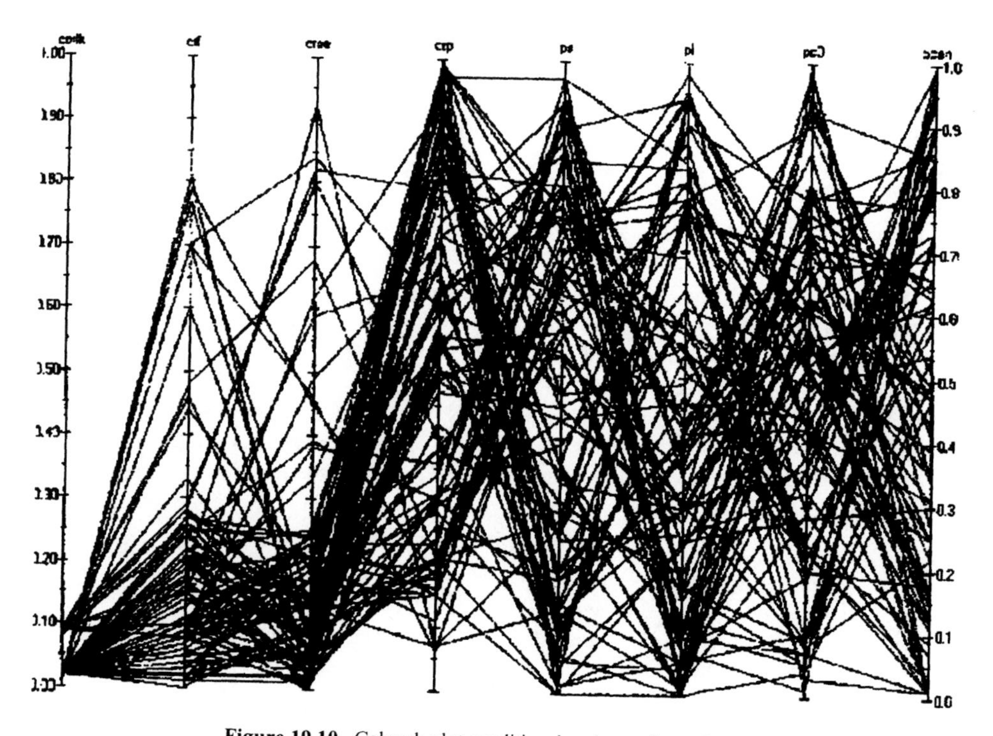

Figure 19.10. Cobweb plot conditional on low values of *corlk*.

sion, but not for low values. This sort of behavior occurs quite often; the variables that are associated with high values of some target variable are not the same as the variables associated with low values of the target variable.

19.8 CONCLUSIONS

We collect a number of conclusions.

- A ranking tool has been developed which uses failure data and structured judgment.
- The tool characterizes pipe sections according to some 20 pipe and environmental characteristics.
- The tool predicts failure frequencies per kilometer year and gives uncertainty bounds.
- These predictions allow distinctions to be made between pipe sections with different characteristics, and these distinctions are not swamped by uncertainties, despite the fact that the uncertainties are large.
- For most pipes, the risk due to corrosion is significantly less than the risk due to third-party interference.
- The depth of ground cover is of significant influence for the frequency and severity of third-party damage.

REFERENCES

Basalo, C. (1992). *Water and Gas Mains Corrosion, Degradation and Protection,* Ellis Horwood, New York.

Battelle Corporation (1990). *Methods for Prioritizing Pipeline Maintenance and Rehabilitation* by *American Gas Association.*

Bushman, J. B., and Mehalick, T. E. (1989). Statistical Analysis of Soil Characteristics to Predict Mean Time to Corrision Failure of Underground Metallic Structures, in Chaker V., and Palmer, J. D., *Effects of Soil Characteristics on Corrosion,* Report No. STP 1013, American Society for Testing and Materials, Philadelphia.

Camitz, G., and Vinka, T. G. (1989). Corrosion of Steel and Metal-Coated Steel in Swedish Soils—Effects Of Soil Parameters, in *Effects of Soil Characteristics on Corrosion* V. Chaker and J. D. Palmer (Editors), Report No. STP 1013, American Society for Testing and Materials, Philadelphia.

Cooke, R. (1991). *Experts in Uncertainty,* Oxford University Press, Oxford.

Cooke, R. M., and Jager, E. (1998). A probabilistic model for the failure frequency of underground gas pipelines, *Risk Analysis:* **18:**511–527.

Cooke, R. M. (1995). *UNICORN, Methods and Code for Uncertainty Analysis,* AEA Technology, The SRD Association, Thomson House, Risely, Warrington, Chishire, WA3 6AT, United Kingdom.

Chaker V., and Palmer J. D. (eds.) (1989). *Effects of Soil Characteristics on Corrosion,* Report No. STP 1013, American Society for Testing and Materials, Philadelphia.

Geervliet, S. (1994). *Modellering van de Faalkans van Ondergronds Transportleidingen,* in *Report for two year post graduate program, Department of Mathematics and Informatics,* performed under contract with the Netherlands Gasunie, Delft University of Technology, Delft.

Hopkins, P., Corder, I., and Corbin, P. (1992). *The Resistance of Gas Transmission Pipelines to Mechanical Damage,* European Pipeline Research Group.

Kiefner, J. F., Vieth, P. H., and Feder, P. I. (1990). *Methods for Prioritizing Pipeline Maintenance and Rehabilitation,* American Gas Association, Washington.

Lukezich, S. J., Hancock, J. R., and Yen B. C. (1992). *State-of-the-Art for the Use of Anti-Corrosion Coatings on Buried Pipelines in the Natural Gas Industry,* Topical Report 06-3699, Gas Research Institute.

Mayer, G. R., van Dyke, H. J. and Myrick, C. E. (September 21 1987). Risk analysis determines priorities among pipeline-replacement projects, *Oil and Gas Journal.*

Palmer, J. D. (1989). *Environmental Characteristics Controlling the Soil Corrosion of Ferrous Piping,* in Chaker V., and Palmer, J. D., *Effects of Soil Characteristics on Corrosion,* Report No. STP 1013, American Society for Testing and Materials, Philadelphia.

Wegman, E. J. (1990). Hyperdimensional data analysis using parallel coordinates, *Journal of the American Statistical Association* **85:**664–675.

EXERCISES

19.1. The expected frequency of failure in a pipe is 2 per kilometer year. Derive the probability of failure per kilometer year using the assumptions in Section 19.2.

19.2. Equations (19.1) and (19.2) give two expressions for the frequency of hitting a pipe with oversight. Whereas equation (19.1) simply gives a number, equation (19.2) gives this number as a function of the variables *FCL* and *FOP*. Write out the proof of equivalence sketched in the text and discuss its assumptions. In particular, suppose $P(H \cap O|CL)$ is a linear function of *FCL* and that $P(H \cap O|OP)$ is a linear function of *FOP*. Rewrite equation (19.2) using these linear functions.

19.3. Write out in detail the derivation of equation (19.6), justifying all the steps.

19.4. Write out in detail the derivation of equation (19.7), justifying all the steps.

19.5. Suppose the effective life due to coating damage is equal to the effective life due to small line damage. Derive the relation between the critical thickness for coating damage, the critical thickness for small line damage, and the pipe diameter.

19.6. Consider a cobweb plot with two independent uniform variables U, V. Draw a vertical line halfway between the lines corresponding to the two variables and consider the density of crossings on this halfway line. Show that this density is triangular (*Hint:* Show that lines crossing on the halfway line correspond to realizations (u_1, v_1), (u_2, v_2) satisfying $u_1 + v_1 = u_2 + v_2$. The density is then proportional to the length of the line $U + V = cnst$ in the unit square.)

CHAPTER 20

RCM Approach to Maintaining a Nuclear Power Plant

Gilles C. Zwingelstein

20.1 INTRODUCTION

The maintenance costs in the energy generation with nuclear power plants are a significant part of the direct costs for utilities. As an example, in France 14 billion French francs are spent each year to maintain the 55 nuclear power plants representing an electrical generation capacity of about 70,000 Mwe (megawatts per hour). To optimize the maintenance tasks, a maintenance policy based on reliability-centered maintenance (RCM) was selected to maintain about 450 systems constituting a nuclear power plant, in keeping, of course, with the required safety and reliability levels set by the regulatory body. A simplified schematic of a nuclear power plant using a pressurized water reactor is shown in Figure 20.1. Taking into account the safety concerns and the maintenance costs, most of the utilities are focusing their efforts on safety-related systems and systems important for generation and maintenance costs. The test case presented in this chapter deals with the charging pumps subsystem of chemical and volume control system (CVCS) which performs several important functions for safety and operations. The main function of the CVCS system is to inject water in the primary circuit in case of loss of coolant accident (LOCA) to prevent core meltdown. In case of main pipe rupture, the CVCS is automatically triggered and injects high-pressure water with charging pumps to maintain the water inventory in the vessel where the fuel elements are located. This subsystem includes several complex subsystems using a large variety of components such as pumps, valves, motors, sensors, and piping.

Due to the wide variety of components to be maintained, the CVCS system was selected first as a pilot project to evaluate the efficiency of a maintenance policy based on RCM.

Case Studies in Reliability and Maintenance, Edited by W. R. Blischke and D. N. P. Murthy.
ISBN 0-471-41373-9

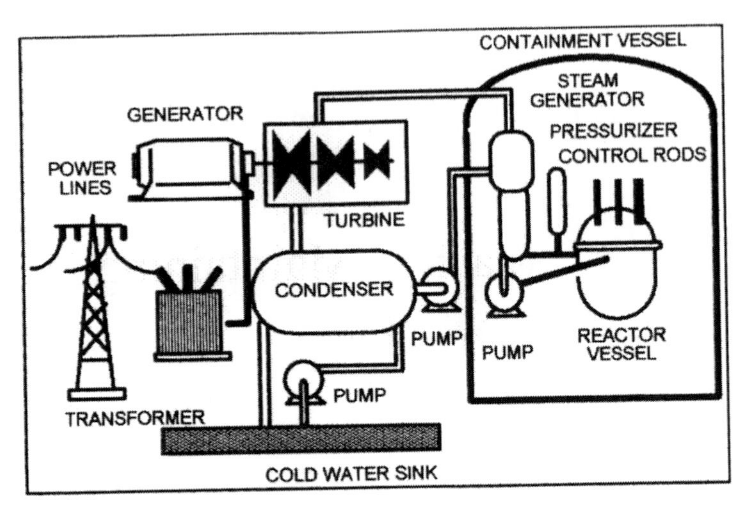

Figure 20.1. Simplified schematic of a nuclear power plant.

The outline of the remainder of the chapter is as follows. Section 20.2 is devoted to the description and modeling of the CVCS of a nuclear power plant used to demonstrate the efficiency of the RCM maintenance. An overview of RCM concepts is provided in Section 20.3, followed by a demonstration of the vital importance of field data and a description of the reliability data used and their statistical processing in Section 20.4. Section 20.5 describes the RCM process leading to the detailed maintenance program of the subsystem charging pumps of the CVCS. Finally, Section 20.6 highlights the achievements of the study and presents some figures that demonstrate the benefits of implementing a maintenance policy based on RCM. Notation used throughout the chapter is listed in the Appendix.

20.2 SYSTEM CHARACTERIZATION AND MODELING

The CVCS of a pressurized water nuclear plant controls the primary coolant inventory and water chemistry. In case of loss of coolant accident due to rupture of the main pipe, the CVCS is automatically triggered to inject high-pressure water into the primary vessel and thus to avoid core meltdown similar to what occurred during the Three Mile Island accident.

20.2.1 System Partitioning

The objectives of the system characterization and modeling are to provide a breakdown of the process into systems and subsystems, regrouping components to achieve well-identified functions at the system or subsystem levels. In the CVCS system, this task also involves the definitions of the limits of the subsystem charging pumps.

For existing processes, the partitioning method can vary according to the tools and industrial software available. The most common formal tools are the IDEFO (identification of function) method, which has replaced the SADT method since 1993), the FAST (function system analysis technique) method (Hubka, 1980), or some proprietary industry-specific methods. More details can be obtained in Feldmann (1998) and Hubka (1980).

Figure 20.2 represents the location of the CVCS system inside the nuclear power plant. For modeling and analysis, the CVCS system was then partitioned into the following eight subsystems by a group of experts:

1. Chemical and volume control tank
2. Charging pumps
3. Charging line
4. Seal primary pump injection
5. Let down (a function of the CVCS to add boric acid in the injected water)
6. Primary pump seal return
7. Residual heat removal water
8. Makeup

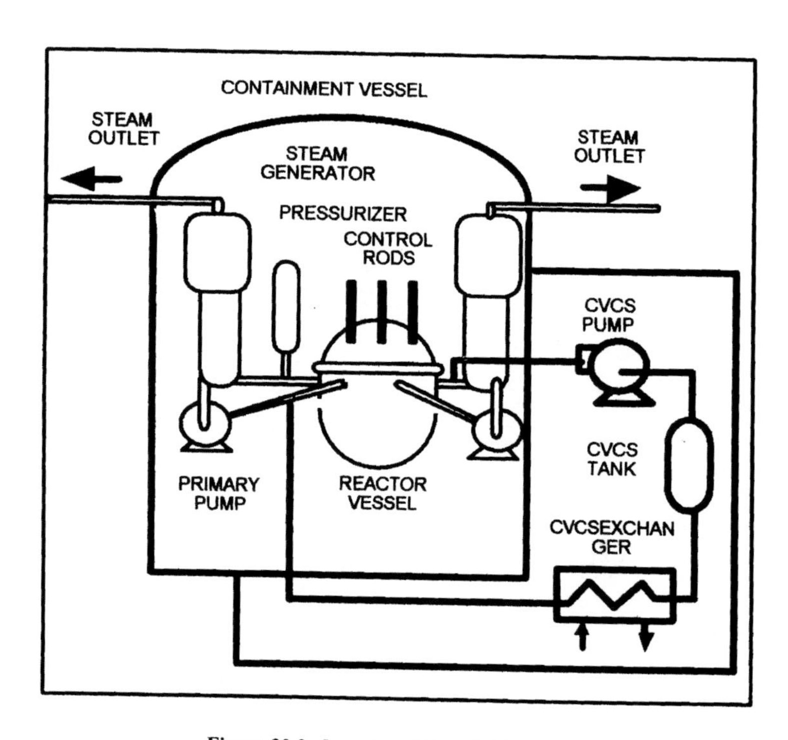

Figure 20.2. Location of the CVCS system.

Figure 20.3 represents the diagram of the charging pumps subsystem where the interfaces with other subsystems [users (e.g., systems using the outputs of the subsystem) and servers (e.g., systems providing inputs to the subsystem)] are represented by the polygonal boxes. In the schematics in Figure 20.3, RCV stands for reactor control volume and SS is the acronym for any subsystem in the CVCS partitioning.

The partial list of the users and servers (or inputs and outputs) of the CVCS pumps system is as follows:

Servers

1. SIS: safety injection system
2. RHR: residual heat removal
3. RCS: reactor control system
4. BRS: boron recycle system
5. SWTS: solid waste treatment system

Users

1. KIT: computer and data processing
2. RSP: remote shutdown panel

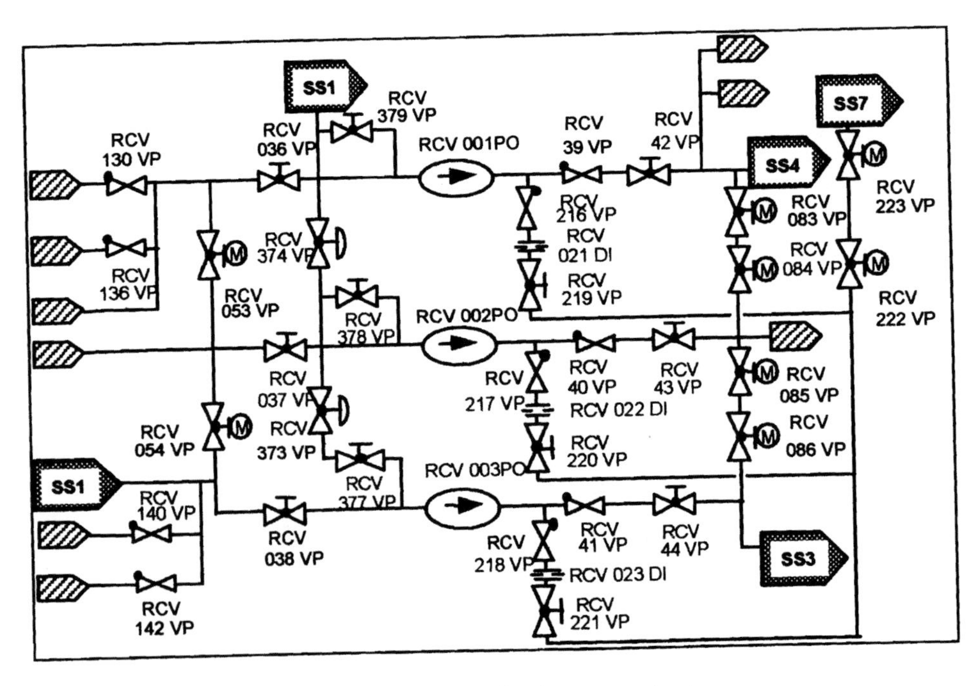

Figure 20.3. Limits of the charging pumps subsystem.

3. MCR: main control room
4. LCA: relaying 48v dc power system
5. LKA: 380v ac normal distribution system
6. LLD: 380v ac emergency supply distribution
7. NISD: nuclear island nitrogen distribution

20.2.2 System Limits and Boundaries Identification

The method for delineating the charging subsystem boundaries was made by using the schematic drawings and the piping and the instruments drawings (P&ID) of the selected system. Then the boundaries were clearly marked up on the drawings to show the boundaries and each component addressed in the RCM analysis. To assist in defining system or subsystem boundaries, the team of analysts mainly referred to the design basis document. The following guidelines were selected:

1. Air supplies are bounded by the solenoid.
2. The breaker is included in the system but not the bus.
3. The entire instrumentation loop is included in the study.
4. All the controlled components and their associated I&C must be included within the system boundaries.

20.2.3 System Function Identification and Definition

The function definition of the charging pumps subsystem describes the actions or requirements that a subsystem must accomplish, sometimes defined in terms of performance capabilities within specified limits. The functions must be clearly identified for all the process modes of operation (Blanchard and Fabrycky, 1998; Blanchard, 1998; Blanchard et al., 1995; Bloch and Geitner, 1993).

The main and auxiliary functions were determined by reviewing process system descriptions, operations procedures, and instructions, reactor coolant system (RCS) during normal operations, power changes, startup, and shutdown, including pressurizer auxiliary spray for depressurization. The CVCS also provides reactor grade water to the reactor coolant pump seals for cooling and sealing purposes. Portions of the CVCS (i.e., charging pumps system) provide an injection flow to the RCS upon receiving a safety injection signal. More precisely for the charging pumps subsystem, the following functions were taken into account for the identification of functional safety-related failures:

- Provide flow to the reactor injection system during safety injection.
- Contain the primary fluid during safety injection.
- Provide flow to the charging line or at the #1 seals of the primary pumps.
- Contain the primary fluid under accidental situation without safety injection.

20.3 AN OVERVIEW OF RCM

During the past two decades, new maintenance concepts such as TPM (total productive maintenance), RCM (reliability-centered maintenance), TQM (total quality management), and QFD (quality function deployment) have been developed and successfully implemented in many industries. An efficient preventive maintenance policy must focus on components for which the failure modes will have significant consequences on the safety, operations, quality, and maintenance costs. An essential issue is to reduce the maintenance costs in the direct operational costs of a process or large equipment. The approach used to maintain the components of the French nuclear power plant is based on RCM. This chapter describes the background, concepts, and steps necessary to implement RCM, which was developed 30 years ago in the airline industry and customized later to power plants (nuclear and fossil fuel fired). RCM is now more and more widely used in various industries. More details can be obtained in Anderson and Neri (1990), Moubray (1992), Nowlan and Heap (1978), and Zwingelstein (1996).

20.3.1 Terminology

Since this approach is to organize choices between the various maintenance types and tasks, it is important to know the meanings of terms used in RCM. Several international and national standards (ISO, IEC, IEEE, Mil-Std) carefully define maintenance-related terminology. The following definitions, which summarize the most well-accepted terms, are derived from the following references: Mil-Std-470B, Mil-Std-1390D (Navy), Mil-Std-721C, Mil-Std-781D, Mil-Std-785B, Mil-Std-1390D (Navy), Mil-Std-1390C, Mil-Std-1629A, and Mil-Std-2155.

20.3.2 Definitions

1. *Equipment, Component.* Equipment is a complete functional assembly of a given make and model (e.g., a pump, a vessel, a breaker). Component is often used to describe equipment internal assembly, subassemblies or parts (e.g., a shaft, bearing, tube).
2. *Failure.* A failure is either an interruption of functional capability of the equipment (breakdown of shutdown by an internal protection system or an equipment protection procedure) or a degradation below a defined level of performance, when such minimum is contained in the functional technical specifications for the equipment. [*Note:* When failure is used within a contract in the context of RCM, it should be defined as follows: "A failure is the presence of an unsatisfactory condition which is related to a specific situation and from the perspective of a particular observer." The particular observer should be defined.]. An equipment has failed if it is declared (or considered) inoperable by operating personnel using simple functional criteria.

3. *Failure Cause.* The physical mechanisms or reasons that produced the failure.
4. *Failure Effect.* The consequence a failure has on the operations and function of an item.
5. *Failure Mode.* The manner by which a failure is observed.
6. *Function.* Usual characteristic actions of an item.
7. *Functional Failure.* Determination of the ability of an item to perform a required function within specified limits.
8. *Maintenance.* All activities performed on equipment and control systems in order to assess, maintain, or restore their operational capabilities; maintenance is either corrective or preventive.
9. *Corrective Maintenance (CM).* All tasks performed to restore the functional capabilities of failed items—principally diagnosis and repair (rework or replace).
10. *Preventive Maintenance (PM).* All activities performed on nonfailed items to avoid or reduce the probability of failure. According to the French AFNOR standards NFX60010, it includes periodic or systematic preventive maintenance, on-condition preventive maintenance, and predictive maintenance. This includes restoration or replacement of degraded but nonfailed items.
11. *Hidden Function.* A function that is usually active and needed but is not evident to the operating crew during performance of normal duties. (A function that usually does not mandate restoration.)

PM can be divided in a variety of ways according to the needs of the program.

12. *Time-Directed PM.* Actions to restore part or all of the failure resistance (of nonfailed equipment or systems) initiated as a function of time or production, regardless of the actual condition (degradation status) of the equipment or system. It includes systematic replacement of critical components after some life limit, replacement of inexpensive components to avoid noneconomical condition assessment, and most of servicing (e.g., replacing fluids, replacing filters). This type of action has at least the advantage of a simple decision scheme: Time has or has not expired.
13. *Condition-Directed PM.* All types of restoration (of nonfailed items) are initiated as a result of condition assessment and comparison with defined acceptance criteria (potential failure). This type of PM requires additional tasks to measure the degradation level. These tasks are also considered a part of condition-directed maintenance since they are the cores of the decision process (although most of them are implemented on a regular time schedule). They include:

    In-Service Inspection (ISI)—Visual or nondestructive examination used to assess the condition of metallic structures such as massive pressure vessels or mechanical components.

In-Service Testing (IST)—Tasks that measure equipment readiness and/or performance levels in normal or emergency operating conditions (e.g., a battery endurance test or a sensor check).

Monitoring and Diagnostics (Predictive Maintenance)—Tasks that rely on variations in operations parameters as a sign or ongoing degradation, which then triggers preventive inspections or restorations to avoid failures.

14. *Overhaul.* For most complex electromechanical machinery, it is impossible to control all degradation mechanisms by using the three above methods. An overhaul tends to reconstitute reliability potential for complex machinery for a given period extending to the next one.
15. *Modifications.* This is usually associated with changes in output or plannned operations; it is also a technique to correct or prevent an intrinsic reliability problem.
16. *CM/PM Ratio.* CM/PM volume ratio is often used as an indicator to characterize the policy of a given industry.
17. *Maintenance Levels.* An important issue in defining a maintenance policy at the company level is to select the maintenance activities to be handled by its own personnel and the amount of activities, which will be performed by external contractors. Figure 20.4 describes the contents and the skills necessary for a policy involving five maintenance levels.

Level	*Tasks*	*Personnel skills*	*Means*
1	Simple adjustements	operator	Standard tools
2	Standard replacements, minor P.M tasks	Mid-qualified technicians	Portable tools
3	Breakdown diagnosis Standard replacements Minor mechanical repair	Highly qualified technicians Or engineers	Special tools
4	Overhauls Mid-range repairs	Highly specialized team	Workshop, specialized tools and equipments
5	Reconstruction or renovation Major repairs	Manufacturer Personnel	Defined by the manufacturer

Figure 20.4. Maintenance levels and associated tasks.

18. *Effective PM Task.* A PM task is *effective* if it can control a given type of equipment failure; for example, ultrasonic testing (UT) wall thickness measurement can help control pipe erosion while visual inspection cannot (holes are visible but you do not get control of the degradation).
19. *Applicable PM Tasks.* A PM task is *applicable* when it is practical; for example, shaft vibration monitoring is efficient and applicable to control pump shaft cracking failures.

An optimized PM program includes only effective and applicable tasks. The hierarchy of PM types derives from the economic incentive to control failures. Besides applicability and efficiency, the hierarchy of PM types derives from the economic incentive to control failures while taking maximum advantage of the useful life of components and reducing the number of inspections.

20.3.3 RCM Concepts

Originating in the airlines industry, RCM can be defined as a process leading to the specification of applicable and effective PM tasks that prevent failures and monitor degradation mechanisms of components that are important for safety functions, operational functions, and maintenance expenditures. The initial definition of the RCM established in the airlines industry by the MSG (maintenance steering group) is: "Reliability-centered maintenance (RCM) is disciplined method logic or methodology used to identify preventive maintenance tasks to realize the inherent reliability of equipment at least expenditures of resources."

The Electric Power Research Institute (EPRI) RCM users group, which includes USA and non-USA utilities, has proposed the following definition: "Reliability-centered maintenance (RCM) analysis is a systematic evaluation approach for developing and optimizing maintenance programs. RCM utilizes a decision logic tree to identify the maintenance requirements of equipment according to the safety and operational consequences of each failure and degradation mechanism responsible for these failures."

The International Electrotechnical Commission (IEC) working group TC 56 is proposing, for the IEC RCM standard, the following definition: "Reliability-centered maintenance (RCM) is a method for establishing a scheduled preventive maintenance program which will efficiently and effectively achieve the inherent and safety levels of equipments and structures."

Whatever the definition, the RCM methodology aims to identify components whose failures and degradations induce loss or degradation of functions performed by the most important systems of an industrial process or cause appreciable maintenance expenditures. RCM methodology involves a systematic and logical consideration of the following

1. System, subsystem, or component functions
2. Failure modes of each function

3. Importance associated with the function and its failure
4. Prioritizing process that identifies the PM tasks that cost effectively reduce failure occurrence

In contrast to the most common equipment-oriented approach to select PM tasks, RCM uses a function-oriented approach to focus maintenance efforts only on functionally critical components. A complete RCM evaluation involves a top-down functional breakdown of the process with the appropriated indenture levels to perform a thorough analysis of the propagation of the failure root causes across the various pieces of equipment of an industrial system. The ultimate objective of these analyses is to identify the most dominant failure modes and the related root causes (at each indenture level) that can be prevented or monitored by applicable, effective PM tasks.

In addition to RCM system analysis, a successful program involves RCM task implementation and an RCM "living program" that will take into account during the remaining life all the new failures, new technologies, and new regulations

The focus of RCM on recommending predictive maintenance reduces the overall chance of failure by (i) shifting the operating cycle of the equipment into the low failure-rate region and (ii) reducing maintenance-related failures with the reduced frequency of intrusive maintenance tasks.

20.3.4 RCM Benefits

The nature and results of the RCM process yield several benefits: (i) direct maintenance costs reduction by minimizing corrective maintenance and reducing preventive maintenance, (ii) process availability and efficiency improvement by reducing production losses due to component failure and achieving the maximum long-term availability factor for the process remaining life, and (iii) optimization of process resources and organization by promoting close cooperation between operations, maintenance, industrial safety, reliability, and system engineers and providing an in-depth training in system design and operations to RCM team members. Another benefit is a direct result of documenting the logical process. Recording each step of the process provides a documented basis for the selection of the specific maintenance strategies chosen using RCM. Setting up monitoring and diagnostic testing instead of fixed-frequency overhaul-type tasks is the main emphasis of RCM. Monitoring and diagnostics reduce the frequency that PMs remove equipment from service. It is important to realize that the actual effects can be measured only after several years of operation with RCM in place.

20.3.5 RCM Analysis Procedure

The requirements of the initial RCM analysis includes several major steps:

1. Process partitionning into systems
2. Defining the system boundaries
3. Defining the important functions of the system

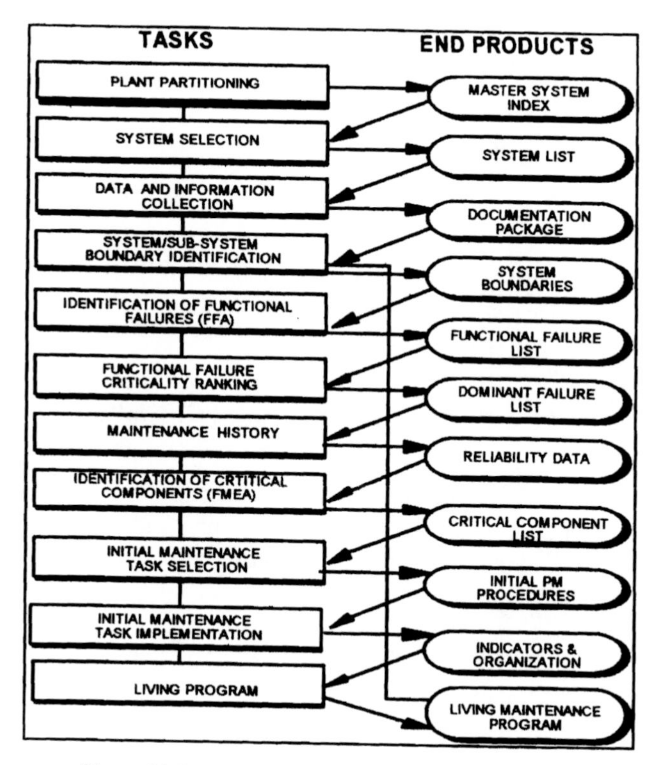

Figure 20.5. Tasks and end products of an RCM analysis.

4. Identifying the dominant system functional failure modes
5. Determining the critical components responsible for the system functional failure modes
6. Identifying applicable and effective tasks that can prevent the failures and monitor the degradation mechanisms of critical components, based on actual or potential equipment failure and equipment failure mechanisms
7. Preventive maintenance implementation
8. Implementation of maintenance indicators in the frame of the "RCM living program"

In summary, a flow chart of the overall RCM process is shown in Figure 20.5.

Independently of the "depth of the RCM analysis," and of the tools and methods that are selected to perform the steps of the initial RCM program of a given system, the quality and value of an RCM maintenance program rely essentially on the accuracy and the completeness of the definitions of the criteria associated with:

1. Function definitions
2. Selection of system failure modes

3. Selection of the critical components
4. Criticality of critical component failure mode
5. Applicability of maintenance tasks
6. Effectiveness of maintenance tasks
7. Decision logic tree for PM task selection
8. Selection of maintenance interval

The following describes the most important tools to identify the critical functional failures modes at the system levels and the tools used to identify the root causes at the component level of a critical system.

20.3.6 Selection of the Functions and Associated Critical Systems and Subsystems

System selection aims to select and prioritize process systems, which are eligible for RCM activities. The widespread use of qualitative and/or quantitative criteria taking into account maintenance costs, safety, and operational issues suggests three classes of the candidate methods:

1. Method 1: Qualitative method based on collective engineering judgement
2. Method 2: Quantitative method based on reliability and availability models
3. Method 3: Combined qualitative and quantitative criteria

20.3.7 Identification of Critical Functional Failure Modes and Their Ranking

Several analytical methods are available to study functional failures and identify the critical components:

1. Failure mode and effects analysis (FMEA)
2. Failure mode and effects and criticality analysis (FMECA)
3. Fault trees
4. Go models (tool developed by the Electric Power Research Institute, Palo Alto, CA)

For the CVCS study, FMEA, FMECA, and fault tree were used. FMEAs/FMECAs provide a well-documented analysis of component failures, causes, probability, and criticality and are largely used by most of the utilities performing RCM analysis. Dedicated to handle single failures on system components, the contents of the FMEA/FMECA are generally application-specific since no standard is available in the industry. Guidelines to design FMEAs and FMECAs can be found in Mil-Std-1629A, Hoyland and Rausand (1994), and EIC publications. In most applications, FMEA/FMECA data worksheets contain some or all of the following data:

1. System functional failures
2. System functional failure detection method
3. Component description and identification
4. Dominant failure modes
5. Failure rates
6. Mission times
7. Failure probability during a mission
8. Effects in the vicinity of the equipment
9. Functional effects on the system
10. Process effects (safety, generation, and damages)
11. Criticality of the failure
12. System functional failure detection method (evident or hidden)
13. Specific data
14. Remarks

Fault trees can identify critical components even if failure data are not available. For example, a minimal cut-set (Hoyland and Rausand, 1994) with a single failure rate event indicates that the component having this failure mode belongs to a critical component.

The definition of the criticality in the FMECA must be consistent with the criteria selected to rank the system functional failure and requires an extensive use of operations and maintenance history and personnel interviews.

Once the list of the functions of the subsystem has been completed, the likely functional failure modes are identified and documented for each function. In the present study, maintenance and operations histories were the most valuable documents to perform the likely failure mode identification. The characteristics (evident or hidden) of each failure mode were documented, and each function had to have at least one functional failure.

After a detailed analysis of the documentation associated with the CVCS system, about 80 functions were identified for the whole system. Recall that the CVCS maintains the required water inventory in the reactor. For each functional failure, one or several functional failure modes have been screened and evaluated, taking into account the feedback experience. For instance, for the functional failure #2, "Fails to contain the primary fluid during safety injection," the associated failure mode was considered to be: 2.1. Loss of primary fluid containment during safety injection.

Figure 20.6 shows a part of the FMEA table, which was used to identify the critical components inducing the failure mode 2.1. Since the 43 components listed in the second column induce a loss of the function, which is critical for safety, they will be considered as critical components. It can be noticed that this list includes valves (labeled RCV xxx VP) and pumps (labeled RCV xx PO). Note that the comments in columns 3–7 apply to all 43 lines of the table.

FMEA:SYSTEM ANALYSIS						
SYSTEM: CVCS	SUB-SYSTEM #2: CVCS					
FUNCTION FAILURE MODE	COMPONENT I.D	COMPONENT FAILURE MODE	EFFECT ON THE SYSTEM	EFFECT ON THE PLANT	SEVERITY	FAILURE EVIDENCE
2.1 LOSS OF PRIMARY FLUID CONTAINMENT DURING INJECTION	RCV 130VP RCV 136VP RCV 140VP RCV 142VP RCV 53VP RCV 54VP RCV 36VP RCV 37VP RCV 38VP RCV 377VP RCV 378VP RCV 379VP RCV 373VP RCV 374VP RCV 61 LP RCV 62 LP RCV 63 LP RCV 01 PO RCV 02 PO RCV 03 PO RCV 014 LP RCV 015 LP RCV 016 LP ↓ (43 COMPTS)	SERIOUS EXTERNAL LEAK	LOSS OF SAFETY INJECTION	PRIMARY BREAK NOT COMPENSATED	S.S SEVERE FOR SAFETY	EVIDENT

Figure 20.6. FMEA table for functional failure identification.

20.3.8 Identification of Critical Components Using FMECAs for Each Dominant Failure Mode

Since the RCM process intends to optimize maintenance tasks at the component level, FMECA were undertaken for each critical component to identify the failure modes at the part level. This was achieved as shown in Figure 20.7 for a critical motor operated valve RCV 53 VP, taking into account a standardized failure mode list:

- Fails to open
- Fails to close
- External leakage
- Internal leakage
- Incorrect valve position signal

This FMECA provided valuable information related to the failure causes and their frequency of occurrence. This is one of the more valuable contributions of the RCM to maintenance using reliability data.

In this FMCEA data sheet, column 1 indicates the function of the subset, (e.g., the motor operated valve), column 2 gives the failure mode at the functional subset level, column 3 indicates the component failure modes, and column 4 indicates all the root cause failures. In column 5, NB FAI indicates the number of failures al-

FMECA-COMPONENT ANALYSIS

CVCS SYSTEM	COMPONENT: MOTOR-OPERATED VALVES								
FUNCTIONAL SUB-SET FUNCTION	FAILURE MODE OF THE FUNCTINAL SUB SET	COMPONENT FAILURE MODE	CAUSES	NB FAI	NB DEG	NB DE Maint	SEVERI.	CRITIC	FAILURE DETECTION METHOD
1) FLUID TIGHTNESS	1) EXTERNAL LEAKAGE	1) EXTERNAL LEAKAGE	1) PACKING • LOSS OF CHARACTERITICS & WEAR OUT •LOOSENING • INCORRECT CONTACT STEM / PACKING	3	65	32			VISUAL DETECTION
			2) VENT JUNCTION • LOSS OF SEAL CHARACTERITICS • INCORRECT DESIGN •LOOSENING OF THE VENT PLUG		17	5			
			3) JUNCTION BODY-BONNET • LOSS OF GASKET CHARACTERITICS • STUD LOOSENING DESIGN		19	17			
			4) VALVE BODY (FLANGE) • EROSION • CRACK		3	1			

Figure 20.7. FMECA for failure causes of the motor-operated valve.

ready encountered; in column 6, NB DEG indicates the number of degraded states discovered both during operation and maintenance, and column 7 gives the number of degraded conditions observed only during maintenance activities (NB DE Maint).

20.3.9 RCM Limitations

Before undertaking RCM, personnel must understand the capabilities and limitations of the process. Regardless of the tools an industry uses for its maintenance optimization program, it must understand what the tool can or cannot do. Otherwise, RCM will not meet their unrealistic expectations. They could then abandon a promising program for process improvements before it has a chance to produce. RCM is a good tool for the development or optimization of a maintenance program.

RCM will not solve all of process reliability or operations and maintenance expenditure problems. RCM can address only those problems arising from inadequate, incorrect, ineffective, or redundant maintenance tasks. For incorrect performance of proper maintenance, RCM can do little to compensate for a poorly trained staff. RCM will not correct for the human element (e.g., it will not help if the wrong equipment is taken out of service for PM), nor for incorrectly installed replacement parts, nor for poor system design. It can, however, identify the need for a design change, but no amount of maintenance will overcome the built-in unreliability of poor designs.

20.3.10 Initial RCM Task Implementation

Implementation of the initial RCM PM tasks is an important issue, which ensures a successful RCM program. It includes the completion of all the new or deleted PM and monitoring activities and generally requires more effort and management involvement than that needed to undertake the RCM analysis. The activities needed during the implementation phase include the following:

1. Definition of the baseline values and action thresholds for condition-directed, predictive and monitoring tasks
2. Definition of task frequencies for new or modified tasks
3. Evaluation of the extensions of maintenance periods called for by RCM
4. Qualification of the modifications or design changes
5. Evaluation of the impacts of the changes in compliance requirements (technical specifications or environmental qualifications)
6. Packaging of elementary RCM tasks into updated maintenance procedures
7. Dedicated maintenance-personnel training to new predictive maintenance technologies
8. Interfacing RCM requirements to meet the process management information system
9. Specification and implementation of maintenance effectiveness indicators
10. Implementation of a maintenance-history database to collect RCM related data
11. Coordination of new organizational interfaces that are required to implement RCM
12. Selection of maintenance-task intervals

20.3.11 RCM Living Program

Implementation of an RCM living program is essential to verify that the initial RCM recommendations, including additional, modified, and deleted PM tasks, are effectively maintaining or even upgrading the reliability of safety-related and operational functions altogether with maintenance labor and cost reductions. Trending the maintenance effectiveness indicators will provide a rational basis to:

1. Modify and upgrade the initial RCM tasks
2. Revise the initial RCM analysis
3. Implement proven new predictive maintenance techniques
4. Consider new regulatory requirements
5. Handle unanticipated or new failures or degradation mechanisms
6. Manage the aging of components along the process life

The RCM living program is a continuous program having duration strictly related to the life cycle of the systems, structures, and components of the industrial

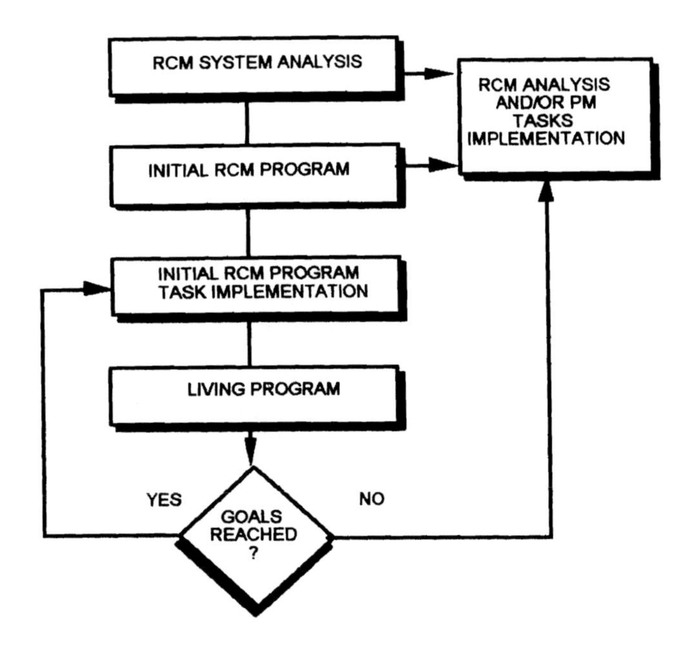

Figure 20.8. Overall RCM process diagram.

process. However, according to most utilities' experience, the consequences of the modification of a maintenance program can be significantly evaluated only after a 3- to 4-year period. A successful RCM living program also requires a dedicated organization having the responsibility of the control, review, and documentation of the decisions taken according to the feedback experience. Figure 20.8 gives the overall RCM process diagram.

20.4 FIELD DATA

20.4.1 Objectives

A history review is important in the overall RCM process because the determination of the criticality of the components requires prior knowledge of the failure modes, causes and effects, and probability of occurrence. The objectives of this task were to gather all information related to a given component from internal sources or the industry and past history of the failures recorded on each component.

20.4.2 Data and Information Collection

20.4.2.1 Data Collection

Before RCM analysis of a given system, it is necessary to gather all the information needed by the RCM team. The objectives of data and information collection are to ensure that the design and operating characteristics, requirements, and history of the system and associated components are known and considered in the RCM analysis. This data and information package allowed the RCM analyst team access to the all the information needed to determine the system functions, system boundaries, functional failure modes, component failure modes and causes, and failure and degradation rates.

The information screened for the charging pumps system includes the following:

1. Design specifications
2. Operating procedures and technical specifications
3. Maintenance and surveillance requirements
4. Drawings, descriptions, and existing studies of the system
5. All preventive and surveillance tasks performed on the system within the boundaries
6. All mechanical linked components included within a subsystem
7. The limit of the piping and valves boundaries, including the valves, if they are system isolation valves (in most of the cases, the piping is not supposed to fail)
8. Redundant components that are included in the same system or subsystem boundaries
9. Boundaries that are consistent with the process tag numbering system

20.4.2.2 Data History

For the charging pumps subsystem, the feedback maintenance experience on failure modes at subsystem level and component levels was widely used. Quantitative numbers such as MTTF, MTBF, MTTR, average service times, failure causes, and modes partitioning were helpful in evaluating failure history (Blischke and Murthy, 2000; Elsayed, 1996; Hoyland and Rausand, 1994).

To help the decision-making process, special feedback experience and statistical data processing were utilized to estimate the failure rate as a function of time and the partitioning of failure and degradation causes. Starting from the very beginning of the operations of the French nuclear program in 1978, a dedicated feedback experience data bank was implemented at each site. To ensure quality control of these data, two people were dedicated full time at each plant to screen the failure data and then to fill in the data sheets according to utility specifications. In particular, data such as status of the equipment (continuous or standby status), number of operating hours between failures, time to repair, costs to repair, losses of production, root causes, mode of failure discovery (during operations or during scheduled maintenance), classification of the technical problems (failure or degradation), and type of maintenance were recorded.

These data, through a nationwide private network, were stored in a centralized and proprietary data bank SRDF (in French: Système de Recueil des Données de Fiabilité). Then these raw data were processed by a team of experts using statistical tools according the nature of the data (Type 1 censoring, Type 2 censoring, random censoring) to derive estimates of the parameters of appropriate probability distributions (exponential, Weibull, log-logistic) and establish the confidence for the parameters. Each year, the processed data are gathered in a special report, which is used by upper management to derive key performance indicators (costs, impact on production losses, etc.). For scarce data, a Bayesian approach was implemented to update the a priori knowledge on the assumed failure laws, taking into account actual failures.

Examples associated with failures and degradations observed on an operating motor-operated valve are displayed in Figures 20.9 and 20.10. Figure 20.9 represents, on the left side, the partitioning of the causes of failures discovered during operations and maintenance over a five-year period for approximately a few hundred similar motor-operated valves and the right side shows similar information for degradations noticed during scheduled maintenance activities.

Figure 20.10 represents the behavior of the failure rate versus year, taking into account the failures discovered during operations and maintenance. The failure rate was computed as the ratio of the number of failure, N, observed each year, divided by the cumulative operating time T : $\hat{\lambda} = N/T$. To obtain the upper and the lower confidence bounds, the chi-square statistic was used [see Blischke and Murthy (2000) and Elsayed (1996)].

For missing data, some other data sources were also used to compensate for the lack of local statistics due to a reduced number of components performing similar functions in the same operating conditions and environment. Sources used included:

1. RAC
2. NPRD
3. Vendors
4. IRS (IAEA) and NRC-IE for nuclear plants
5. Industry data banks (e.g., USA, Euredata, ISPRA, Sweden (T-Book))
6. Other industries' data banks (e.g., oil, electronics (CNET, OREDA))

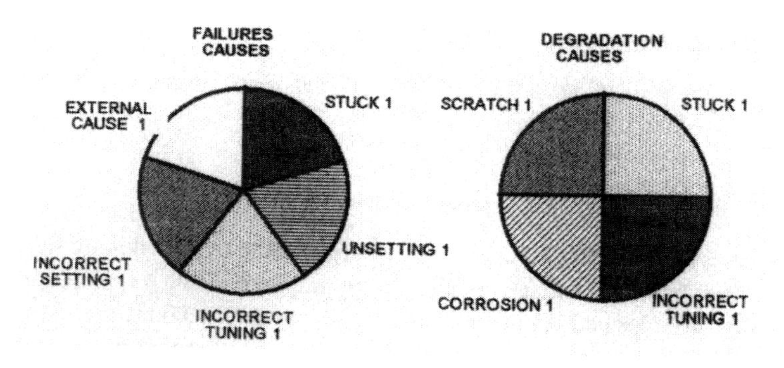

Figure 20.9. Failure and degradation causes on a motor-operated valve.

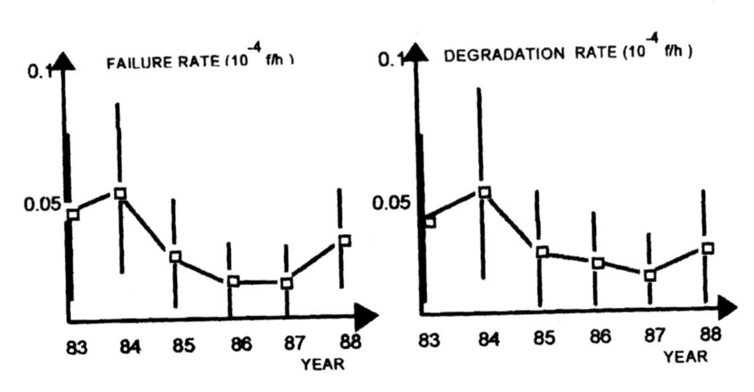

Figure 20.10. Failure and degradation rates on a motor-operated valve.

20.5 MAINTENANCE OF CVCS SYSTEM

20.5.1 RCM Maintenance Task Selection

The objectives of the RCM analysis are to identity the set of effective and applicable preventive maintenance tasks with the minimum amount of manpower and direct maintenance costs

The necessary tasks briefly described below are the following: (i) Select the critical systems, (ii) identify the failure modes of the systems and subsystems using FMEA and for each critical system and subsystem, establish a list of the dominant failure modes, (iii) perform FMECAs for each dominant failure mode to identify critical components; (iv) identify the set of applicable and efficient preventive maintenance tasks for each component failure mode; (v) implement the initial maintenance program; and (vi) monitor the performances of the RCM program using maintenance indicators during the living program. This section explains which exact RCM procedure was applied to obtain the final RCM maintenance program on a given piece of equipment of the CVCS charging pumps subsystem: a motor-operated valve.

20.5.2 Partitioning of the Process

The partitioning process used for the CVCS system was based on the function system analysis technique and also on the IDEF0 method.

20.5.3 Selection of Critical Systems and Subsystems

To select and prioritize process systems that are eligible for RCM activities, a qualitative method based on collective engineering judgment using a ranking procedure and FMEAs was applied. This was done by taking into account the maintenance history and failures of each system and subsystem. The charging pumps subsystem was declared as critical for safety with this procedure.

20.5.4 Identification of the Functions and Dominant Failure Modes of the Charging Pumps Subsystem

The objectives are now to determine the main and auxiliary functions performed by the charging pumps subsystem. The results of the functional failure analysis lead to the identification of the functions and the functional failure modes. More precisely for the charging pumps subsystem, the four functions listed in Section 20.2.2 were taken into account for the identification of functional safety-related failures. The FMEA and collective engineering judgment associated with the analyses of the past operating experience were implemented to rank the importance of the functional failures.

20.5.5 Identification of Critical Components Using FMECAs for Each Dominant Failure Mode

Since the RCM process intends to optimize maintenance tasks at the component level, FMECA were undertaken for each critical component to identify the failure modes at the part level and their root causes. This was achieved for a critical motor operated valve RCV 53 VP, taking into account the standardized failure mode list given in Section 20.3.8. As noted, the FMECA provided valuable information regarding failure causes and frequencies and was one of the most important results of the study.

20.5.6 Initial Maintenance-Task Selection

Important failure modes of the critical components must be prevented by applicable and effective maintenance tasks. The characteristics of an applicable PM task are directly related to its ability to decrease a component's failure rate. An effective maintenance task must reduce the likelihood of the failure, within cost and implementation consideration. The objectives of the initial maintenance task selection are as follows: (i) Provide a systematic approach to the identification of preventive maintenance or failure-finding tasks, promote the use of condition-directed PMs over any time-directed tasks, and define the maintenance intervals. (ii) Focus the tasks on the identified failure mechanism. (iii) Compare the RCM recommendations with the existing PM review and approve the contents of the initial maintenance RCM tasks. (iv) Define the needs for practical implementation, including maintenance organization, maintenance procedures, personnel training, and predictive maintenance tool investments.

Most of the logic tree analyses used in the industry for the selection of the maintenance tasks correspond to a customized version of the RCM decision logic tree described in the MSG-3 (Maintenance Steering Group—Version 3). The logic tree requires the analyst to answer *yes* or *no* to a series of questions dealing with (a) the evidence, the consequence, and importance of the component failure on operating safety and operational consequences and (b) specific decision logic diagrams used to guide the analyst for the selection of applicable and effective preventive maintenance tasks, failure-finding tasks, and design changes or risk failure acceptance. Since a main objective of an RCM program is to reduce the maintenance expenditures, a sequential evaluation of the applicable and effective PM tasks, classified according to their implementation cost, is performed. Following the MSG-3 recommendations, the maintenance techniques were evaluated in the following order:

1. Lubrication and servicing tasks
2. Rounds and condition monitoring
3. Condition-directed and condition-predictive tasks
4. Failure-finding tasks (functional tests)
5. Time-directed tasks
6. Replacement tasks
7. Modification of equipment design

After comparison and evaluation of the tasks, new tasks, deleted tasks, and changes, the bases for the tasks should be listed and reviewed by personnel responsible for tracking commitments. The decision to implement RCM recommendations has required a formal review and approval of all the tasks by a dedicated team involving operations, maintenance, safety, and technical support and other personnel familiar with all the details and objectives of the RCM methodology. This review included all actions necessary to implement the tasks—for example, revising operations and maintenance procedures, personnel training for RCM, and new condition-monitoring techniques and tools, specifications of dedicated operating data-collection modules, specifications of maintenance effectiveness indicators for the RCM living program, and requests for deviation from regulatory commitments. The manager responsible for implementing the RCM approved the decisions. This knowledge is essential in choosing between periodic or predictive maintenance tasks.

Once the proper PM tasks have been selected, the task-interval depends on the engineer's ability to measure and detect a reduction in the component's condition or performance before a failure occurs. The basis of the initial intervals is that the interval must be long enough to detect evidence of deterioration. These intervals were selected according to vendor recommendations or based on similar PM task reviews and the component's maintenance history. Statistical tools were implemented to estimate the nature of the failure rate [e.g., Weibull, exponential, log-normal; see Elsayed (1996), Ebeling (1997), Smith (2001)]. This knowledge is also essential to choosing between periodic or predictive maintenance tasks.

For instance, if the parameters of the Weibull law indicate that the failure rate is decreasing, we are in the infancy period and systematic preventive maintenance can be used. If the failure rate remains constant, failures occur at a constant rate and the use of preventive maintenance is questionable. If the failure rate is increasing, preventive and predictive maintenance are highly recommended. For the case of an exponential law, since a failure occurs randomly, it will be too expensive to implement a preventive maintenance strategy. If the failure PDF has a Gaussian shape, it is possible to derive a mean life of the component and to set up a safe life limit. In this case a preventive maintenance strategy is recommended. The situation will be identical for a log-logistics law.

Again, it is important to use a goodness-of-fit test to identify the nature of the reliability law of the given piece of equipment, to be sure that the maintenance policy retained is suitable according the estimated reliability law. In our case Weibull, bathtub law, log-logistic, and exponential laws were found to be the most common for the various components of he CVCS.

The initial RCM maintenance program usually leads to a set of maintenance tasks that delete, add, or modify the existing tasks. A task-by-task comparison is necessary to ensure that all the commitments and impacts of the RCM recommendations are thoroughly evaluated and reviewed and are as recommended. Conversely, if the consequence of failure has no important effect on cost, run-to-failure was the recommended strategy. This has allowed maintenance personnel to concentrate their efforts on components that are important to process operations and safety.

For the selection of maintenance tasks for the motor-operated valvc of the charging pumps subsystem, four different logic trees were selected. The first set dealt with evident and hidden failures having an impact on plant safety. The second set of logic trees was related to evident and hidden failures associated with operational and economical issues. An example of the selection task is given in Figure 20.11.

This diagram provides all the explanations required for the selection of tasks. A group of maintenance experts was created to provide the applicable and effective

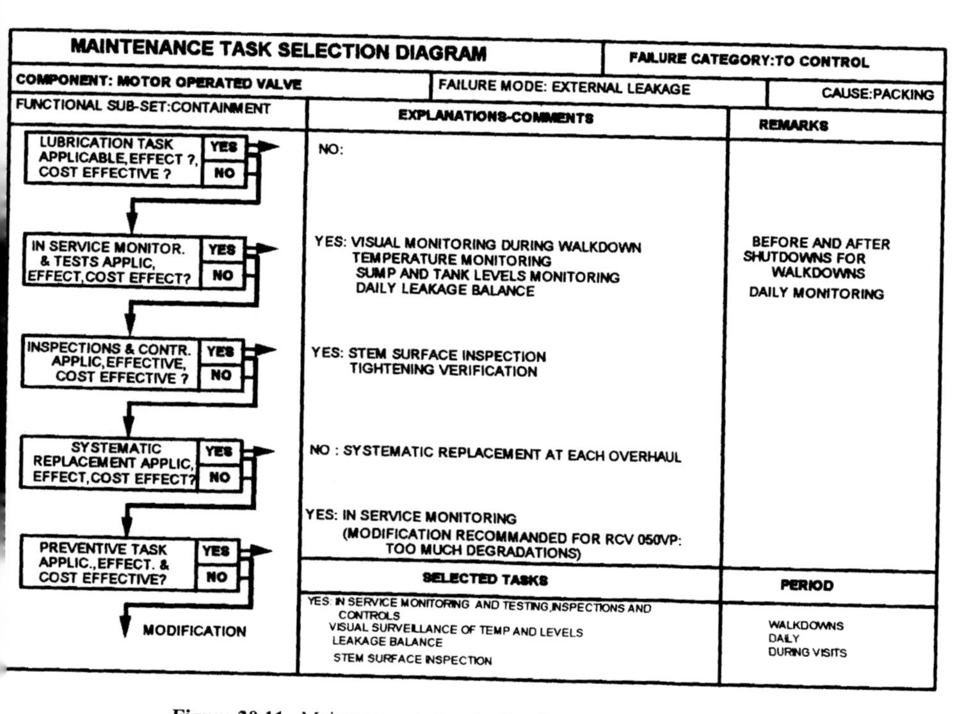

Figure 20.11. Maintenance task selection for a motor-operated valve.

tasks. For the task interval determination, the feedback experience was used together with the engineering judgment from experts in maintenance, reliability, and safety. To help the decision making process, statistical data processing was applied as described in Section 20.4 to identify the failure rate versus time and the partionning of failure and degradation causes.

20.5.7 Living RCM Program

Assessments of RCM program effectiveness must generate clearly the benefits attributable to the RCM program. While availability and capacity factors are universally recognized measures to evaluate process performance, assessment of RCM effectiveness using them directly or indirectly is difficult for several reasons: (i) Improvements in process availability and safety due to PM tasks implemented as the result of the RCM program take several years to become evident. (ii) Because so many factors directly or indirectly affect overall process safety and economy, attributing any credit to the RCM implementation requires careful analysis and quantification of the effects of the following:

1. Maintenance outage delays
2. Human errors
3. PM task implementation effectiveness
4. Organizational problems
5. Equipment or operating procedure modifications
6. New regulatory commitments
7. Unanticipated failure mechanisms

Figure 20.12 presents the life cycle of an RCM program through the life of the asset. At the very beginning, the analysis is carried out to lead to the RCM initial maintenance program, which must include the contents of the maintenance tasks to be implemented and the maintenance performance indicators. Once the RCM program has been implemented, it is necessary to evaluate the acceptability of the new maintenance scheme according to the defined maintenance indicators.

If the results are unacceptable, it is necessary to reconsider and update the first RCM analysis. If the results are satisfactory, the initial program can be kept as initially defined. However, for processes having a long life cycle, unknown failures, new operating modes, or new regulations can appear. Thus, it will be also necessary to update the initial maintenance program.

However, keeping in mind the limitations to correlate the direct effects of the implementation of an RCM program with availability, safety, and maintenance expenditures, some quantitative measures may be used with caution:

1. Maintenance labor and material costs saved
3. Change in total number of CM tasks performed

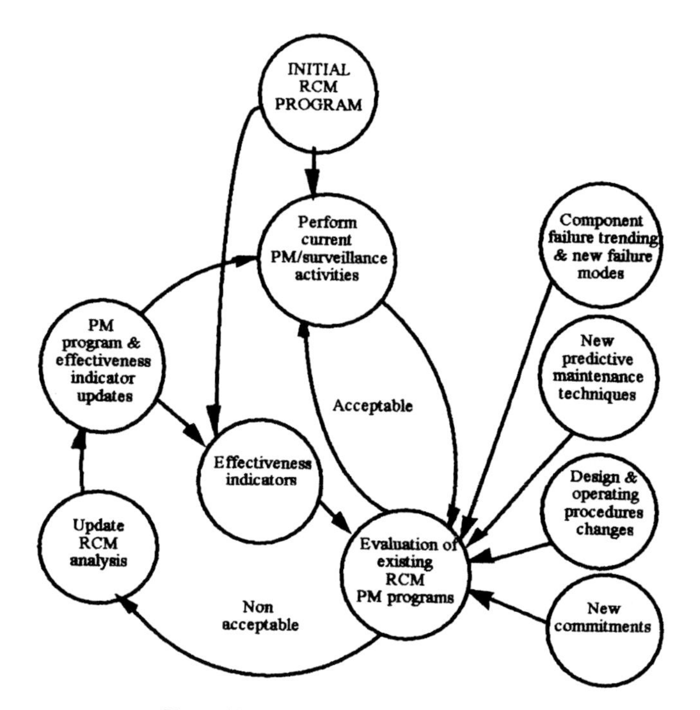

Figure 20.12. Cycle of RCM living program.

3. Number of forced outages due to component failures
4. Change in safety system unavailability

As a part of a complete RCM analysis, the strategy for noncritical components was addressed to evaluate the existing PM tasks and the consequences of deleting the tasks. If the failure rate of a noncritical component is high and if the CM is repetitive, the PM tasks or design changes should be considered.

20.6 CONCLUSIONS

After completion of the RCM studies for the complete CVCS system, the comparison between the existing tasks and the RCM recommendations is summarized in Figure 20.13.

The analysis of the tasks identified by the RCM procedure shows that the maintenance activities lead to an overall reduction of 20% in the number of components previously maintained by the classical approach.

It is worthwhile to notice that RCM improved the existing maintenance tasks in the sense of the ALARP (as low as reasonably practicable) concept. For some components, their number has been decreased (valves) but for some other components were increased (sensors).

	Total number of components	Number of critical components	Components maintained with existing PM program	Difference between the existing & RCM programs
Valves	134	55	78	- 30%
Control for valves	52	29	41	- 30%
I & C	39	24	22	+ 10%
Pumps	37	25	25	Same
Total number	262	133	166	-20 %

Figure 20.13. Comparison between previous and RCM tasks for the components of the CVCS system.

This is a situation specific to the CVCS system. The RCM program is now being applied to 80 other systems and has demonstrated that RCM can either delete or add maintenance tasks.

Implemented in 1995–1997, the first evaluations of the benefits of RCM were conducted by a engineering maintenance group of experts. As of 2001, a decrease of about 5% of the maintenance dedicated costs was reached. In parallel, a special history database was designed to collect information about failure and degradation rates. For the CVCS system, the annual savings in maintenance expenditures has been estimated to 2 million French francs (0.3 million US $). Since the RCM strategy has been utilized for about 80 systems of each nuclear power plant, the overall estimated savings per year was evaluated in year 2000 to be between 2 and 3% of the 2 billion US dollars spent each year for the maintenance of the 55 nuclear power plants

The RCM methodology applied to the charging pumps of the CVCS system provided the set of methods available to establish an RCM program. Starting from these general guidelines, each company will be able to customize its own RCM approach according to its short-term and long-term objectives, resources, budget and available maintenance history.

This structured and rational maintenance method has already demonstrated its efficiency in various industries. Based on a common-sense approach, RCM allows gaining process efficiency, improved availability, and better product quality. RCM is a very efficient tool at any stage of the system life cycle (design, construction, production, operations, and disposal) and can be used for either existing processes or new processes. It is important to recall that besides technical aspects, a successful program relies on the cooperation of all the company personnel to ensure perennially the long-term effort related to the RCM living program.

Like TPM and TQM, the culture change necessitates a "buy in" of the company personnel over a long period. If these conditions are not fulfilled, it can be worthless to undertake short-term efforts to test or to implement RCM techniques.

In conclusion, the objectives and results obtained after several years of RCM implementation using the RCM analysis described in this chapter has fully demonstrated that reliability-centered maintenance is a viable and effective way to maintain an asset. The optimization of the preventive activities tasks, keeping the same level of reliability and availability, will allow many maintenance managers to rethink their maintenance practices.

REFERENCES

Anderson, R. T., and Neri, L. I. (1990). *Reliability Centered Maintenance,* Elsevier Science Publishing, London.

Blanchard, B. S. (1998). *Logistics Engineering and Management,* 5th edition, Prentice-Hall, Upper Saddle River, NJ.

Blanchard, B. S., and Fabrycky, W. J. (1998). *Systems Engineering and Analysis,* 3rd edition, Prentice-Hall, Upper Saddle River, NJ.

Blanchard, B. S., Verma, D., and Peterson, E. L. (1995). *Maintainability,* Wiley, New York.

Blischke, W. R., and Murthy, D. N. P. (2000). *Reliability: Modeling, Prediction, and Optimization,* Wiley, New York.

Bloch, H. P., and Geitner, K. K. (1993). *An Introduction to Machinery Reliability Assessment,* Gulf Publishing, Houston, TX.

DOD Guide LCC-1, *Life Cycle Costing Procurement Guide.*

DOD Guide LCC-2, *Casebook, Life Cycle Costing Procurement Guide.*

DOD Guide LCC-3, *Life Cycle Costing For System Acquisitions.*

Ebeling, C. E. (1997). *An Introduction to Reliability Engineering,* McGraw-Hill, New York.

Elsayed, E. A. (1996). *Reliability Engineering,* Addison-Wesley-Longman, Reading, MA.

Feldmann, C. G. (1998). *The Practical Guide to Business Process Reengineering Using IDEFO,* Dorset House, New York.

Hoyland, A., and Rausand, M. (1994). *System Reliability Theory, Models and Statistical Method,* Wiley, New York.

Hubka, V. (1980). *Principles of Engineering Design,* Butterworth Scientific, London.

Mil-Std-470B, *Maintainability Program for Systems and Equipment,* DoD.

Mil-Std-1390D (Navy), *Level of Repair Analysis,* DoD.

Mil-Std-721C, *Definition of Effectiveness Terms for Reliability, Maintenability, Human Factors, and Safety,* DoD.

Mil-Std-781D, *Reliability Testing For Engineering Development, Qualification, and Production,* DoD.

Mil-Std-785B, *Reliability Program For Engineering Development, Qualification, and Production,* DoD.

Mil-Std-1390D (Navy), *Level of Repair Analysis,* DoD.

Mil-Std-1390C, *Level of Repair Analysis,* DoD.

Mil-Std-1629A, *Procedures for Performing a Failure Mode, Effects and Criticality Analysis,* DoD.

Mil-Std-2155, *Failure Reporting, Analysis, and Corrective Action System (FRACAS),* DoD.

Moubray, J. (1992). *Reliability-Centered Maintenance,* Industrial Press, New York.

Nowlan, F. S., and Heap, H. F. (1978). *Reliability-Centered Maintenance,* United Airlines.

Smith, D. J. (2001). *Reliability Maintainability and Risk,* Butterworth and Heinemann,Oxford

Zwingelstein, G. C. (1996). *La maintenance basée sur la fiabilité,* Hermes, Paris.

EXERCISES

20.1. The CVCS charging pumps subsystem can be modeled using functional analysis techniques such as IDF0 or FAST methods. Using Figure 20.3, draw the IDF0 diagrams with four levels of indentation for the following function: "Provide high-pressure flow in the nuclear reactor."

20.2. In order to identify the causes of the top event, "loss of the high-pressure injection function due to external leakage" (see Figure 20.3), plot the fault tree associated with the top event and identify the list of possible basic events. Indicate all the assumptions made to design the fault tree.

20.3. The CVCS system contains a large number of valves (motor operated valves, check valves, and manual valves). For each of this family of valves, identify the possible failure modes and, using failure data available in databases, rank the failure modes in terms of failure occurrence.

20.4. In the CVCS example for a nuclear power plant, we have defined as critical components those having a significant impact on safety, generation, and maintenance cost. Taking an example from another industry (e.g., food processing, pharmaceutical industry), define the set of criteria to identify the critical components versus the noncritical components.

20.5. Assuming that we are willing prevent external leakage on a valve with applicable and effective predictive maintenance technologies, perform an investigation on the technologies available and evaluate the costs to implement them (expressed in $/hour).

20.6. Since the CVCS system is built with many components and, in particular, many pipes made of steel, identify the list of the possible NDT (nondestructive testing) techniques applicable to detect cracks (dye penetrant, eddy current, ultrasonic, acoustic emission,) and draw a comparative table in terms of probability of crack detection.

20.7. Assuming that a component has a known safe life limit obtained by the feedback experience with a gaussian curve-like shape, what is the convenient preventive maintenance technique according to RCM principles (periodic, on-condition, or predictive maintenance)?

20.8. The Weibull law is often used to describe the reliability of mechanical systems. Assume that the reliability law is given by $R(t) = e^{-[(t-\gamma)/\eta]^{\beta}}$, with $\gamma =$ 1000 hours, $\eta = 2500$ hours for β equal respectively to 0.5, 1, and 2. Indicate the best preventive maintenance strategies to be used for the different values of β.

20.9. According to the classification of the failure consequence severity (impacts on safety, generation, or maintenance cost), what will be your recommendations for the maintenance task selection for an evident failure having an impact on safety if no aplicable or effective task is available?

20.10. In our analysis, we rank the importance of the CVCS system by taking into account the consequences of all the functional failure modes using the feedback experience and expert judgment. Carry out an analysis to propose an alternate method for ranking the importance of a system.

20.11. In our study, we have implemented a FMECA at the component level to identify the failure modes, their criticalility, and their final effects. Using the Mil-Std 1629A, redesign the FMECA table shown in Figure 20.6.

20.12. Taking into account the results of the FMECA for the motor operated valve as described in Figure 20.6, construct the fault tree associated with the top event "external leakage of the valve."

20.13. Assuming that the electric-driven pump failed with the failure mode "fails to start," provide a list of root causes and the associated applicable and effective preventive maintenance tasks.

20.14. For an asynchronous motor, list all the possible failure modes for its main function "convert an electric energy into mechanical energy" and, for each failure mode, use the logic tree to derive the RCM maintenance program assuming that the consequences of the failure modes are important for safety.

20.15. In the frame of selecting applicable and effective maintenance tasks for the prevention of cracks in a stucture, provide an inventory of all possible preventive maintenance techniques and rank them in terms of applicability, efficiency, and costs.

20.16. Assuming that a failure has a distribution function that follows a truncated gaussian distribution [i.e., $f(t) = 0$ if $t < 0$], determine the choice between periodic and predictive maintenance.

APPENDIX

We use the following notation:

CM: Corrective maintenance
PM: Preventive maintenance
IDF0: Integration definition for function modeling
IEC: International Electrotechnical Committee
IEEE: Institute of Electrical and Electronics Engineers
ISO: International Standard Organisation
Mil-Std: Military Standard (USA)
MSG-3: Maintenance steering group
MTBF: Mean time between failures
MTTR: Mean time to repair
QFD: Quality function deployment
RCM: Reliability Centered Maintenance
TPM: Total productive maintenance
TQM: Total Quality Management

CHAPTER 21

Case Experience Comparing the RCM Approach to Plant Maintenance with a Modeling Approach

Xisheng Jia and Anthony H. Christer

21.1 INTRODUCTION

This chapter reports on a study carried out in a chemical fertilizer company in 1999. The objective of the action-based research project was to introduce and implement maintenance modeling within the currently reliability-centered maintenance (RCM)-led maintenance practice observed within Northern China (Moubray, 1997). The plant concerned is Daqing Chemical Fertiliser Company, the biggest chemical fertiliser company in Northern China. Some 3000 people are employed by the company, which operates 24 hours a day, seven days a week, producing an average 1600 tons of carbamide fertilizer per day.

For most of the plant's duration, the annual maintenance and inspection practice recommended in 1976 by the U.S. company that supplied the plant has been implemented without deviation, and a high standard of record keeping and archiving of data at a level of completeness and detail previously unseen by the authors was maintained. It was this coherent and detailed archive over the previous 21 years that particularly attracted the authors. Although the records existed and were competently maintained, there was little evidence that any use or routine analysis had ever been undertaken other than the very basic. The annual maintenance inspection for the synthetic chemical fertiliser plant was first changed in 1995, when it was extended to biennial.

One of the major and important systems within the plant that had undergone a change in maintenance concept from an annual inspection since 1977 to a biennial

Case Studies in Reliability and Maintenance, Edited by W. R. Blischke and D. N. P. Murthy.
ISBN 0-471-41373-9

inspection from 1995 was the transition boiler. The function of the transition boiler is to produce and provide high-temperature (420°C) and high-pressure (105 kg/cm^2) mixed gases (N_2, H_2, and CO_2) to the synthetic fertilizer production process. The main component parts in the boiler are pipe cores, steel liners, insulating material, and gas distributors, which are shown in Figure 21.1.

There are several reasons for the 1995 change in maintenance practice for the boiler. First, an analysis of the record showed that of the 17 failures recorded in the transition boiler over the previous 21 years, 32% were attributed to the annual maintenance inspection process through faults such as misassemble, poor cleaning, damage caused by rough maintenance, and other forms of human error. These issues were to be addressed by management through training programmes and improved supervision, but in the meantime less maintenance implied less human error.

Second, the collective view and experience of the plant engineers suggested that the major accidents in the production line process over the past 21 years were mainly attributable to incorrect maintenance activity at inspection.

Third, to inspect and maintain the transition boiler required between 50 and 80 hours of otherwise productive time, which represented a substantial investment in lost production.

Finally, although the record shows there was evidence of negative aspects to the annual maintenance inspection, the plant record also shows on average that three faults were identified and "corrected" at each inspection.

In the light of this information, the view was taken that through training to improve the technical competence of operatives, and by improved supervision, the inspection interval could safely and productively be increased to 2 years. This decision was based upon the judgment and considered opinion of the engineers and management who knew the plant. No modeling was used to predict the consequences of a policy change, or otherwise provide decision support, since the availability of suitable maintenance modeling was unknown at the plant.

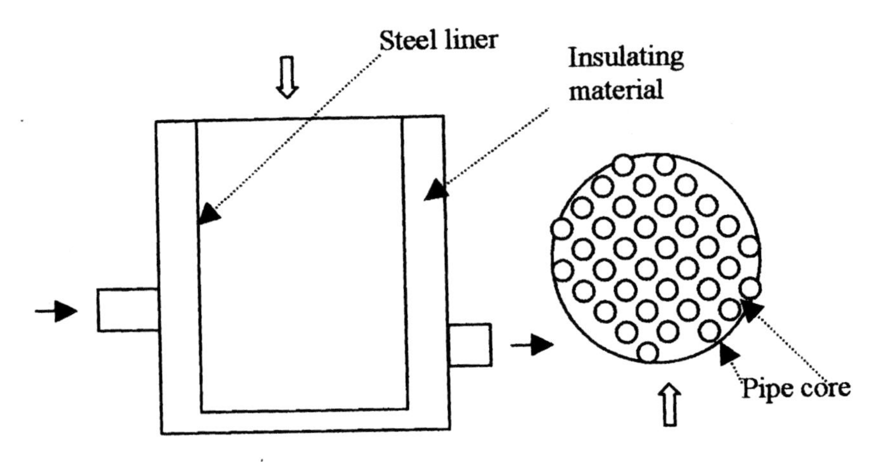

Figure 21.1. Structure of the transition boiler.

Subsequently, the plant experienced two major breakdowns in 1997 and 1999 in the boiler and a gas compressor, resulting in a considerable loss of production time and output, but also in some reported loss of confidence on the part of customers. The appropriateness of the decision to change to biennial inspections was now sharply questioned. Three options had their supporters, namely to move back to an annual inspection, to maintain the current 2-year inspection, and, for reasons we outline below, to possibly increase the period to three yearly. Having made the inspection period decision based upon subjective judgment in the spirit of the RCM process, no modeling support was immediately available to assist resolve the problem.

It was at this stage that the case study was initiated. Simply stated, the task was to advise on the best maintenance period by applying maintenance modeling techniques to understand and clarify the consequences of different periods of inspection maintenance and different levels of improvement in technical competence of operatives. This was a timely task since X. Jia was particularly interested in the use of quantitative modeling within the RCM concept.

The appropriate modeling technique for the problem was delay time modeling, which had been extensively applied and tested in case studies [see Christer and Waller (1984), Christer et al. (1995, 1997, 1998, 2000), and Desa and Christer (2000)]. A recent review by Christer (1999) summarizes the major developments, variants, and estimation techniques in DTM to date. The opportunity was taken in this study to apply both DTM and RCM techniques to the same problem as a comparative experiment.

21.2 DELAY TIME CONCEPT

The delay time concept has been well known to engineers for centuries, and the only new event is to attempt to quantify the concept and thereby use it in quantitative models of maintenance. We are interested in the relationship between the performance of equipment and maintenance intervention. To capture this relationship, the conventional reliability analysis of time to first failure, or time between failures (Barlow and Proschan, 1965), requires enrichment. Consider a repairable item of plant. It could be, say, a component, a machine, or an integrated set of machines forming a production line but viewed by management as a plant unit. The interaction between maintenance concept and equipment performance may be captured using the delay time concept.

Consider a plant failure at some time point. The likelihood is that had the plant been inspected at some point just prior to failure, it could have been seen that all was not well and a defect was present which, though the plant was still working, would ultimately lead to a failure. Such signals include excessive vibration, unusual noise, excessive heat, surface staining, smell, reduced output, increased quality variability, and so on. The first instance where the presence of a defect might reasonably be expected to be recognized by an inspection is called the initial point u of the defect, and the time h to failure from u is called the delay time of the defect, as

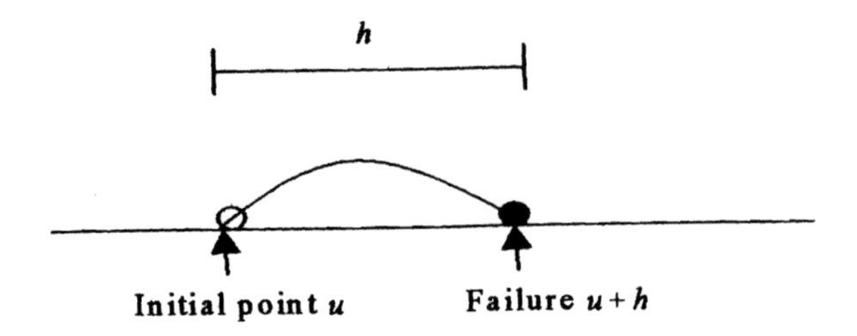

Figure 21.2. The delay time for a defect.

shown in Figure 21.2. Had an inspection taken place in $(u, u + h)$, the presence of a defect could have been noted and corrective action taken prior to a failure. Given that a defect arises, its delay time represents a window of opportunity for preventing a failure.

Suffice it to say, if the rate of arrival of defects at time u, $\lambda(u)$ say, and the probability density function of delay time h, $f(h)$, are known, then it is possible to model the desired relationship between maintenance period, maintenance quality, and the resulting consequences to cost, downtime, risk, availability, or any other measure of interest (Christer, 1999). The first stage of the modeling process is to understand and analyze any existing data and formulate the modeling assumption for the plant.

21.3 DATA RECORDS AND ANALYSIS

Data collection and discussion sessions with plant engineers was chiefly during two site visits by the authors in May and August 1999, augmented with e-mails and telephone calls as required. The considerable detail of maintenance records was in hard copy (Chinese), so it needed to be analyzed manually. Data available and studied in detail included 21 years (period 1977–1997) of objective data for the transition boiler, including:

(a) Inspection times
(b) The number and nature of faults identified at each inspection
(c) The date and nature of failures that occurred between successive inspections

In the current context, faults are defects identifiable at an inspection that can lead to failure if unattended, and they include rusting, leaks, deformation, and so on. The number of defects identified is important, but care needs to be exercised in determining this number, which is here related to the number of remedial actions required. For example, if there is a rust patch on a steel liner, removing the rust requires cleaning work and is considered to be a defect. If several rust batches are identified on the same steel liner, it may need to be replaced. However, since only one repair action is carried out, only one defect is counted. Between 1977 and 1997, 63 defects arising within the boiler and identified at inspections are recorded.

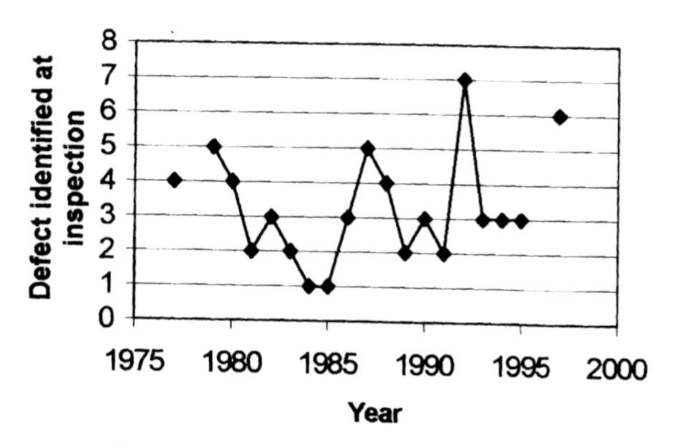

Figure 21.3. Defects identified per inspection time.

The initial data analysis of defects and failures assisted in clarifying the nature of the plant and operating practice. Figure 21.3 shows the number of defects identified at 19 inspections against the inspection years from 1977 to 1997 (there was no inspection in 1978 and 1996), and Figure 21.4 shows the cumulative number of defects identified. The linear trend in the cumulative curve, Figure 21.4, tends to support the assumption of a homogeneous process of defect arrival in the plant, though it must be remembered that the arrival pattern of defects which led to failures needs to be added before one has a view of the cumulative defect arrival pattern.

A failure occurs when the condition of the boiler is such that it can no longer perform its normal function and has to be shut down for emergency repair. Figure 21.5 presents the number of recorded failures in each year between 1977 and 1997. No trend is evident.

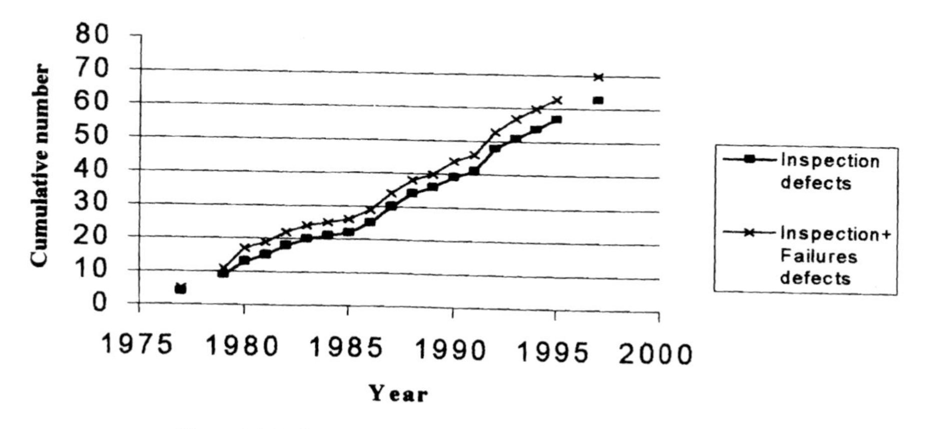

Figure 21.4. Cumulative number of defects versus operating time.

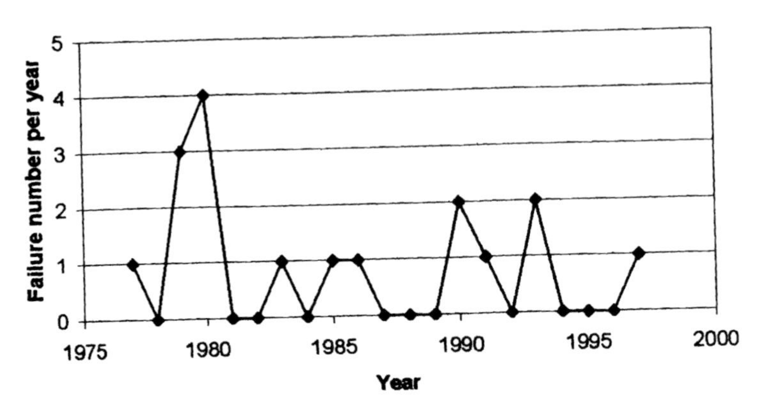

Figure 21.5. Failure numbers per calendar year.

Of the 17 failures recorded, six were failures attributable to human error associated with inspections, and four events recorded as failures were of a very minor nature, with only a few minutes of downtime required for corrective action. Both these categories of failure were excluded from the subsequent delay time analysis of failure, the first because the human error failures were not part of the failure process addressed by the delay time phenomena, and therefore need to be modeled separately, and the second because the trivial amount of downtime was because a defect had been suspected, and the plant management wanted only to confirm that corrective action could wait until the next inspection. Consequently, with four failures being part of the already recorded defect arrival process and six a separate human error process, only seven failures remain to be considered in the parameter estimation process. These seven failures each has a defect arrival point, and the cumulative defect arrival pattern for all defects—that is, those leading to inspection

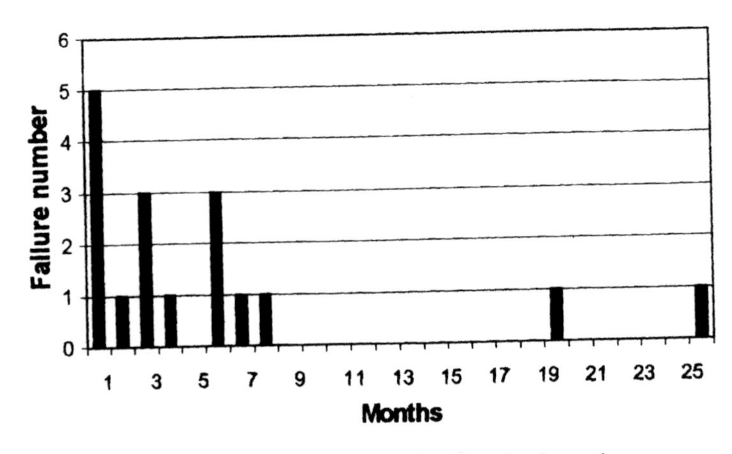

Figure 21.6. Failures per month vs time since last inspection.

repairs and failure repairs—is also shown in Figure 21.4. The near linear forms of Figure 21.4 also suggest a homogeneous arrival rate of defects to be a reasonable assumption with arrival rate $\lambda = 0.278$ per month (63 defects and 7 failures over 21 years, so $\lambda = (63 + 7)/(21 * 12) = 0.278$).

The relationship between the number of recorded failures in a month and the number of months since the last maintenance inspection is shown in Figure 21.6. Note that 29.4% of failures occurred in the first month subsequent to inspection. Further analysis revealed that all five failures were caused by defects injected at the inspection itself, which is of course contrary to the purpose of the inspection. Since 6 human error failures arise over 19 inspections, we assume that currently 0.32 failures on average are injected at each inspection.

21.4 FACTORS AFFECTING HUMAN ERROR FAILURES

Two classes of maintenance activities take place at inspections, first disassembly, inspection, and reassembly, and second, corrective actions for defects identified. Table 21.1 presents the number of defects identified at inspection activities and the total number of human error defects injected at inspection.

This table suggests that the more maintenance work undertaken (defects identified), the lower the chance for human errors. This nonintuitive result requires clarification. Influencing factors could be:

- Adequate feedback and learning: If appropriate leaning resulted from past errors, there will be less chance of a repeat occurrence. Figure 21.7 shows the cumulative number of human error failures between 1976 and 1997. The declining rate of increase supports this explanation.
- The more skillful the maintenance group, the more defects will be recognized and the lower will be the chance of human error. The record shows that the skill levels used were not uniform. Sometimes the inspection was undertaken by skilled and experienced staff, and at other times by new staff. For the repair of failures, only experienced engineers were used, resulting in fewer failures after failure repairs.

While the skill level remains variable until improved training changes the situation, because 6 human errors were identified over 19 inspections, the assumption is

Table 21.1. The Relationship Between the Number of Defects Identified at Inspection and the Number of Human Errors Made

Number of defects identified at inspection:	1	2	3	4	5	6	7
Number of inspection occasions:	2	4	6	3	2	1	1
Total number of failures due to human errors at inspection:	1	1	2	1	1	0	0

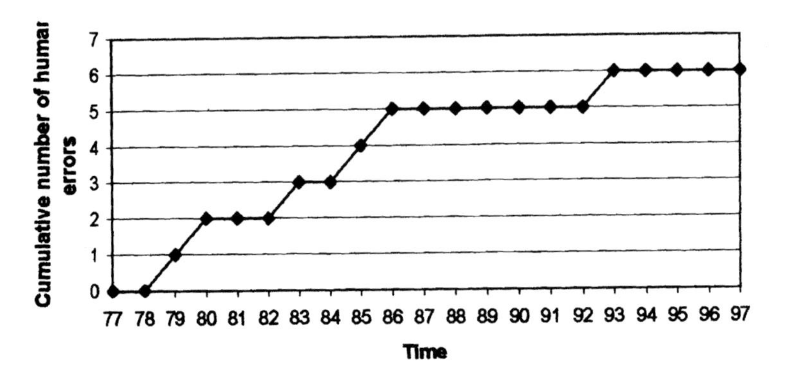

Figure 21.7. Cumulative number of human error failures.

made of a constant number of 0.32 error-based failures being injected per inspection. Training is clearly of the utmost importance.

Having explored the general problem, we now consider modeling the inspection period decision problem, and then we undertake the RCM-based approach to the same task of maintenance period choice for the transition boiler. Comparisons are then made.

21.5 MODELING ASSUMPTIONS

Based on the above data analysis, we make the following assumptions for the modeling of the transition Boiler:

(a) Defects arise within the boiler as a homogeneous poisson process (HPP), and the constant instantaneous rate of occurrence of defects is denoted by λ.

(b) Defects are independent of each other.

(c) Inspection maintenance for the boiler takes place every T months.

(d) The delay time h of a random defect is independent of its time of origin and has *pdf* $f(\bullet)$ and *cdf* $F(\bullet)$.

(e) Inspections are assumed to be imperfect with a detection rate r; that is, the defect can be identified with probability r if it is present.

(f) Repairs or maintenance undertaken during inspection maintenance periods may not be perfect in that some failures arising soon after inspection may be caused by defective maintenance or improper activities at the inspection. The expected number of failures caused by each inspection due to human errors is assumed to be $EN_H = 0.32$.

(g) Failures are identified immediately, and repairs or replacements are made at once.

(h) The objective is to identify the best inspection period T to reduce the total annual downtime or maximize availability.

21.6 DECISION MODEL

Since the major concern of the company is to reduce the joint downtime caused by failure and inspection activities, the conventional downtime measure is an appropriate criterion to model. The expected downtime per unit time measured over a long period is

$$ED(T) = \frac{d_f(EN_f(T) + EN_H) + d_p EN_p(T) + d_i}{T} \tag{21.1}$$

where

d_f is the mean down time to repair a failure

d_p is the mean time to rectify a defect identified at an inspection

d_i is the mean inspection time, including the time to disassemble, check, and reassemble the boiler

$EN_f(T)$ is the expected number of failures over T

$EN_p(T)$ is the expected number of defects identified at an inspection

EN_H is the expected number of failures caused by human error at each inspection

If the inspection is perfect—that is, $r = 1$—it has been shown (Christer et al., 1998) that

$$EN_p(T) = \lambda \int_0^T [1 - F(h)]\, dh$$

and

$$EN_f(T) = \lambda \int_0^T [1 - F(h)]\, dh$$

If the inspection is not perfect—that is, we have a detection rate $r \neq 1$—it has also been shown (Christer et al., 1998) that

$$EN_p(T) = \lim_{i \to \infty} \sum_{j=1}^{i} (1 - r)^{i-j}\, r \int_{t_{j-1}}^{t_j} \lambda\{1 - F(t_i - u)\}\, du \tag{21.2}$$

and

$$EN_f(T) = \lim_{i \to \infty} \left\{ \sum_{j=1}^{i-1} (1 - r)^{i-j} \int_{t_{j-1}}^{t_j} \lambda[F(t_i - u) - F(t_{i-1} - u)]\, du + \int_{t_{i-1}}^{t_i} \lambda F(t_i - u)\, du \right\} \tag{21.3}$$

where t_i denotes the inspection times of the observed inspections, $i = 1, 2, 3, \ldots$.

Because the downtime data are not recorded, we use the downtime estimates provided by engineers, see Table 21.2. These figures are based on the assumption that both the plant and the maintenance functions operate 24 hours a day.

Table 21.2. Subjective Estimates of Mean Downtimes

d_i	d_p	d_f
50 h	10 h	75 h

If the parameter λ and the distribution function $F(\bullet)$ are available, equations (21.2) and (21.3) complete, in the general case of $r \neq 1$, the formulation of the expected downtime model, equation (21.1).

We now address the estimation of λ and $F(\bullet)$.

21.7 ESTIMATING PARAMETERS

We have estimated the rate of arrival of defects to be $\lambda = 0.278$ per month, and that the number of human error failures injected at each inspection is currently $EN_H =$ 0.32 per maintenance inspection. It remains to estimate the quality of detection r, and the delay time distribution $F(\bullet)$ along with its parameters.

Management at the plant estimated subjectively the quality of the defect detection aspect of inspections as $r = 0.85$. This was considered to be on the high side, but acceptable to the authors as an initial value. The choice of function $F(\bullet)$ and associated parameters were made using the maximum likelihood method based upon candidate delay time distributions of negative exponential $f(h) = \beta^{-1}\exp(-h/\beta)$ and Weibull $f(h) = \alpha\beta^{-1}h^{\alpha-1}\exp(-h/\beta)^{\alpha}$. Both these distributions had proved appropriate on other occasions (Christer et al., 1995, 1997, 2000).

If T_n denotes the time of the nth inspection in operation time units of months, the data available consist of the number m_n of defects identified at the nth inspection and the k_n failures over cycle (T_{n-1}, T_n). These data exist for $n = 1, 2, \ldots, 19$ and include 63 defects identified at inspection and 7 failures. Human-error-based failures have been removed from the data and were here modeled as a separate process. Having estimates for λ and r, it remains to estimate $F(\bullet)$.

The likelihood, L, for the data collected over 21 years (19 inspections) is given by

$$L = \prod_{n=1}^{19} \{P_r(m_n \text{ defects at } T_n) \cdot P_r(k_n \text{ failures in } (T_{n-1}, T_n))\}$$

Mathematical expressions for L, which are of course dependent upon $F(\bullet)$, λ, and r, are derived as in Christer et al. (1995, 1998, 2000) and Jia (2000). For selected forms of $F(\bullet)$, the likelihood may be optimized over unknown parameters to obtain the maximum likelihood fit. The results of this process are presented in Table 21.3, where the Akaike information criteria (AIC) (Baker and Wang, 1991, 1993; Christer et al., 1995, 1997) indicates the negative exponential distribution to be a marginally better fit to the data than the Weibull. Since this distribution also satisfies the

Table 21.3. Estimated Parameters

Delay Time Distribution	Parameters		Goodness of Fit		
	$\hat{\alpha}$	$\hat{\beta}$	χ^2	$\chi_n^2(0.05)$	AIC
Weibull	0.59	371.34	2.732	7.81	47.31
Exponential	—	88.45	3.054	9.49	45.63

Data sample: 19 inspections; 63 defects, and 7 failures.
$AIC = -2 \log L + 2v$: v is the number of parameters estimated from the data.

chi-square test of acceptability with a significant level 0.05 (Jia, 2000), we now assume the delay time distribution $F(\bullet)$ to be negative exponential with parameter β = 88.45 months. Had λ and r been unknown, these too could have been estimated within the likelihood formulation, given sufficient data. It was possible here, because of the quality of record keeping, to separate out the human error aspect of inspection which caused some short-term failures after inspections. Had this separation not been possible and had human error been of sufficient magnitude, the defect arrival rate after an inspection may have needed to be modeled as a nonhomogeneous failure rate [see Christer et al. (1997)].

21.8 THE MODELS

We are now in a position to construct the maintenance model for the boiler, but first there are four cases for modeling that must be distinguished. These are:

Case 1: This assumes the current imperfect practice with $r = 0.85$ and $EN_H = 0.32$. This case models the previous performance of the boiler up to when attempts were made through training to improve the quality of the inspection and workmanship. This model still applies if the improvements are not effective.

Case 2: Here the quality of inspection is assumed to be maximally improved from $r = 0.85$ to $r = 1$, with other parameters remaining the same.

Case 3: In this case, the quality of workmanship is assumed to be maximally improved from $EN_H = 0.32$ per inspection to $EN_H = 0$, with other parameters remaining the same.

Case 4: Finally, in this case both the quality of inspection and the quality of workmanship are assumed to be improved to the perfect level of $r = 1$ and $EN_H = 0$.

With the given parameters and estimated delay time distribution $F(\bullet)$, equations (21.1), (21.2) and (21.3) can now model each of the above cases. The results are shown in Figure 21.8.

First, it is clear from Figure 21.8 that in case 1, with status quo conditions, the change from annual to two-yearly inspection was justified, and that if the level of

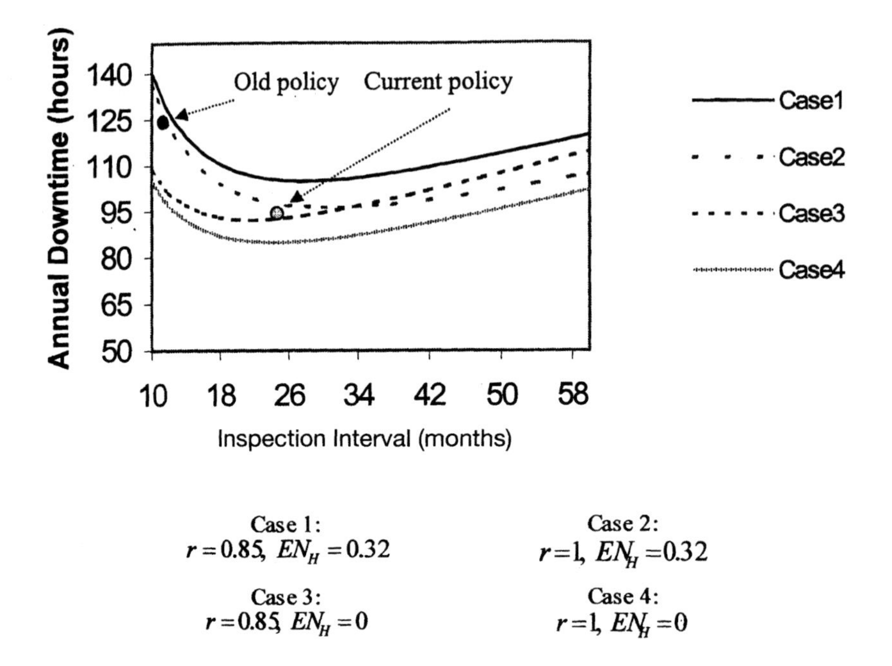

Figure 21.8. Relationship of annual downtime and inspection intervals.

inspection competence and workmanship had not improved, a move to even three-yearly inspection maintenance could be justified. However, improvement in the value of both r and EN_H are expected. We can see from Figure 21.8 that for an inspection period of up to 3 years, it is more beneficial to expend effort in lowering EN_H to 0 than in raising r to 1. In other words, it is more important here to reduce the level of defects injected through human error than to improve the quality of inspection. However, if both are improved to their maximum level, the best to be achieved is case 4, which again supports a two yearly inspection. For any value of improved workmanship and inspection quality, the performance curve will lie between case 1 and case 4. The recommendation to management is to retain the two-year inspection policy, but to improve inspection quality and technician training, with a potential to reduce downtime by some 25%.

A sensitivity analysis was undertaken for the estimated parameters assuming both Weibull and exponential distributions of delay time under different inspection situations, namely perfect inspection, imperfect inspection with an objective estimate of the detection performance, and imperfect inspection with a subjective estimate of the detection performance (Jia, 2000). The optimum biennial interval for inspection was, for this problem, not very sensitive to the choice of delay time distribution or the consequent parameter estimation process, and especially not sensitive to the perfect or imperfect inspection assumption. Also, the optimal inspection period was not sensitive, within the range considered, to the assumptions concerning the parameters r and EN_H. This robustness is welcomed and has been a common experience, provided that sensitivity analysis is conducted regarding the correct order of magnitude of parameters.

The clear benefit of this type of modeling is that for the first time, management can see the consequence of, and thereby decide upon, a change of maintenance concept. Also, for the first time management can see and decide in quantitative terms the merits and value of expenditure on training and improved supervision to maintain the quality of workmanship.

As previously mentioned, the method of maintenance modeling was unknown at the plant. For this reason we also adopt an RCM approach to explore further the difference between these approaches. That is, the authors applied the RCM approach to reexamine the boiler inspection period.

21.9 THE RCM APPROACH

Reliability-centered maintenance (RCM) can be defined as a process used to determine preventive maintenance (PM) requirements for a physical asset in its operating environment (Moubray, 1997). it originated in the U.S. aviation industry in the late 1960s and has been applied by the U.S. military services from the middle of 1970s. Because its economic benefits and cost effectiveness were believed significant, since the beginning of 1990s, RCM has experienced an unprecedented emphasis all over the world. It has been applied to many fields, such as aviation industry, military industry, energy industry, offshore oil production, and so on. RCM was initially introduced into China by military industry in the middle of the 1980s. In 1992 the first Chinese industry RCM standard, namely, GJB 1378—the methodology and requirements to develop preventive maintenance program for equipment—was developed. In this case study, the RCM process was based on GJB 1378 (1992). The initial task was completing a failure mode and effect analysis (FMEA) for the boiler. Key components in this regard were known to be the steel line, piper core, and distributor. In this particular instance, because of the extent and quality of the records, the modelers themselves were able to complete most of the FMEA, with remaining items and an overall check left to the plant engineering experts. A simplified form of the FMEA is given in Table 21.4.

For each failure cause, RCM logic requires the following five questions to be answered:

Q1: Is the failure cause evident to the operator? (Yes or No)

Q2: What is the failure consequence? (Safety, Operation, or Economic)

Q3: What tasks should be selected to prevent this failure? (Servicing, Operation monitoring, Functional check, Failure finding, Restoration, or Replacement)

Q4: What is the reason to select a specific task?

Q5: What is the suggested interval of the task?

Table 21.4. The Simplified FMEA for the Boiler

Item	Function	Failure Mmode	Failure Cause	Failure Effect
1. Steel liner	1A: Seal the inside gas and protect the outside insulator	1A1: Lose the function of protection	1A11: Rusty and deformation	Inside gas wash of the insulator to cause serious accident
			1A12: Weld cracking	Same as above
2. Pipe core	2A: Synthesize the mixture gases	2A1: Fail to synthesize the gases	2A11: Pipe deposit build up	Lower the efficiency of synthesis process
			2A12. Pipe burst	Leaking
3. Distributor	3A: Distribute the mixture gases	3A1: Cannot evenly distribute the gases	3A11: Fall down from its fixed position	Result in pipe core overheating

Table 21.5. The Answers of RCM Logic Questions

Failure cause	Q1	Q2	Q3	Q4	Q5
1A11: Rusty and deformation of steel liner	No	Operation	Functional check	Before function failure, the rust and deformation are expected to be identified at steel line.	3 years
1A12: Weld cracking of steel liner	No	Operation	Functional check	Before function failure, small weld cracking is expected to be identified by inspection.	3 years
2A11: Deposit buildup on pipe core	No	Operation	Functional check	Before function failure, the lime scales or deposit are expected to be identified at pipe core.	3 years
2A12: Burst of pipe core	No	Operation	No task	This failure is caused by human error at inspection. Must make sure there is no cleaning liquid remaining in the pipe core after cleaning	
3A11: Fall down distributor from its fixed position	No	Operation	Functional check	Before falling down, the looseness can be identified by inspection.	3 years

Table 21.6. Subjective Estimates for Boiler Components

Mean (months)	Steel Liner	Pipe Core	Distributor
Initial time	23	34	99
Delay time	30	4	19

After discussion, the experts provided answers to the above questions. These are given in Table 21.5.

Stating that the functional check interval for all the main components of the boiler should be 3 years was a surprise to the authors, since experience was mainly with one year. The plant had just moved to two-yearly PM with little accumulated operating experience, and at least one significant failure had occurred. The experts providing this opinion had not yet seen the quantitative model, Figure 21.8, but their opinion here may well have been unwittingly influenced by the authors in requesting that the engineers take part in a separate subjective assessment exercise to estimate the delay time parameters. In this exercise, two key experts gave the combined estimates of the mean initial and delay time for defects in the boiler components shown in Table 21.6.

An interpretation of these data is that a 3-year inspection would catch most of the defects arising in all components. Such an interpretation does not, of course, balance the cost and other consequences in terms of time and effect with the functional check period. For this, one again requires modeling.

In summary, the RCM analysis and expert judgment lead to the following preventive maintenance inspection policy for the boiler: Inspect every 3 years, with the inspection to include (1) checking the steel lines for rusts, deformation, and weld cracks, (2) checking the pipe core for lime scales and leaking, and (3) checking the distributors for looseness. If any defect is identified at inspection, a repair should be undertaken immediately.

This RCM analysis provides a clear decision and set of testing instructions to implement. It does not, however, provide a means of identifying the consequences or the extent of gain to be expected, or otherwise provide an economic justification for the subsequent maintenance expenditure. For this, one requires the problem to be modeled.

21.10 DISCUSSION AND CONCLUSIONS

- This case study highlights both directly and in comparison with RCM the value of modeling in maintenance in that it provides a clear picture of the expected consequences of change, along with a coherent listing of the underlying assumptions.
- No matter what assumptions are made concerning the parameters r and EN_H within the range considered, the expected downtime of the system with the current inspection interval ($T = 24$ months) is lower than that at the old in-

spection interval ($T = 12$ months). This suggests that the management made the right decision in 1995 to extend the inspection interval from 12 months to 24 months, but without modeling support, this would not be known beforehand, resulting in increased risk and stress. Nor would the superiority of a two-year interval be known without the aid of modeling.

- If the system is not changed at all—that is, all the parameters remain at the current values—the optimum inspection interval is between 28 and 40 months (see Figure 21.8). This suggests that the further extension of the inspection interval may be possible, which coincides with the expert subjective judgment (T should be 36 months) in the RCM analysis. However, considering the stability of the production system and the fact that the gain in expected downtime reduction for $T = 24$ months to $T = 36$ months is trivial, 24 months inspection interval may be more appropriate.
- The inspection quality, r, and the repair quality, EN_H, have significant effects upon the downtime of the system, and they also affect the optimum inspection interval T^*. The model shows that as r increases, T^* increases, and that as EN_H decreases (workmanship quality increases), T^* decreases. Since if inspection quality improves, the optimum inspection interval increases, and if repair quality improves, the optimum inspection interval decreases, modeling is needed to resolve the conflict of inspection and repair quality increase, such as case 4 in Figure 21.8. With the improvement in technology and management, the expert reported that there are now basically no human errors occurring in recent inspection practice. If so, and if this can be sustained, the optimum inspection interval is 24 months.
- From the viewpoint of system downtime the conclusion is that there is no reason to resume the 12-month inspection interval, and the current 2-year inspection interval should remain unchanged until new evidence shows the necessity of the further extension or reduction. Data should continue to be collected and the modeling revisited from time to time. This need not be on a regular basis as much as when management feels there may have been a change in circumstances.
- A report and conclusions has been submitted to the company, but at the time of writing it has not been formally decided if the current inspection should remain unchanged. The chief engineer in the synthetic workshop has informed us that the model prediction levels of annual downtime corresponding to the old inspection policy and to the current inspection policy basically coincide with their observation and intuition.

The modeling solution of this case study gives a clear picture of what is expected when the inspection interval is changed. It can be seen by the modeling that the downtime curve is very flat between the 24-month and the 36-month inspection intervals. This illustrates that from the viewpoint of downtime, both of management judgment and RCM analysis have made good decisions in this case. It is not necessary to change the current 24-month inspection interval. As already pointed out, this conclusion would not be known without modeling support, nor would the extent of defective inspection and repair, nor would the gains to be attained by addressing these management and training issues.

ACKNOWLEDGMENTS

The authors wish to thank Renan Jia and Guanlai Zhu, the engineers of the synthetic workshop of Daqing chemical fertilizer company, for their help in providing objective data and subjective estimations used in this chapter.

REFERENCES

Baker, R. D., and Wang, W. (1991). Determining the delay time distribution of faults in repairable machinery from failure data, *IMA Journal of Mathematics Applied in Business and Industry* **3:**259–282.

Baker, R. D., and Wang, W. (1993). Developing and testing the delay time model, *Journal of the Operational Research Society* **44:**361–374.

Barlow, R. E.. and Proschan, F. (1965). *Mathematical Theory of Reliability,* Wiley, New York.

Christer A.H. (1999). Development in delay time analysis for modelling plant maintenance, *Journal of the Operational Research Society* **50:**1120–1137.

Christer, A. H., Lee, C., and Wang, W. (2000). A data deficiency based parameter estimating problem and case study in delay time PM modelling, *International Journal of Production Economics* **67:**63–76.

Christer, A. H., and Waller, W. M. (1984). Reducing production downtime using delay time analysis, *Journal of the Operational Research Society* **35:**499–512.

Christer, A. H., Wang, W., and Baker, R. D. (1995). Modelling maintenance practice of production plant using the delay-time concept, *IMA Journal of Mathematics Applied in Business and Industry* **6:**67–83.

Christer, A. H., Wang, W., Choi, K., and Sharp, J. (1998). Modelling the preventive maintenance element of total productive maintenance: imperfect PM with incomplete repair, *IMA Journal of Mathematics Applied in Business and Industry* **9:**355–380.

Christer, A. H., Wang, W., Sharp, J., and Baker, R. D. (1997). A Stochastic Modelling Problem of High-Tech Steel Production, in *Stochastic Modelling in Innovative Manufacturing* (Christer, A. H., Osaki, S., and Thomas, L., Editors), pp. 196–214. Springer, Berlin.

GJB1378 (1992). *The Methodology and Requirements to Develop Preventive Maintenance Program for Equipment,* Chinese military industry standard, People's Republic of China.

Desa, M. I., and Christer, A. H. (2001). Modelling in the absence of data: a case study of fleet maintenance in a developing country, *Journal of the Operations Research Society* **52:**247–260.

Jia, X. (2000). OR Modelling within RCM Context, Ph.D. thesis, Salford University, UK, pp. 157–163.

Moubray, J. (1997). *Reliability Centred Maintenance,* Butterworth-Heinemann, Woburn, MA.

EXERCISES

21.1. According to the delay time concept, there are two identifiable states in the failure process, namely defective state and failure state. Give an example to illustrate these two states.

21.2. What is the definition of RCM?

21.3. What does FMEA mean?

21.4. What are the characteristics of the homogeneous Poisson process (HPP)?

21.5. In delay time modeling, suppose defects arise within a plant as a homogeneous Poisson process (HPP), and the constant instantaneous rate of occurrence of defects is $\lambda = 0.278$/month, and the delay time h of a random defect follows an exponential distribution $f(h) = \beta^{-1} \exp(-h/\beta)$, where $\beta = 88.45$ months. Inspections are assumed to be perfect, and no human errors inject defects during maintenance. If $d_f = 75$ hours, $d_p = 10$ hours, and $d_i = 50$ hours, use the downtime model (21.1) in this chapter to find the optimal inspection interval T to minimize the unit downtime of the system.

21.6. If in the above example the plant had sufficient technicians available to enable all defects identified during the inspection period to be rectified within the inspection downtime period d_{ij}, how would you intuitively expect the optimal inspection period might change?

21.7. By modifying appropriately the downtime per unit time equation, equation (21.1) of this chapter, calculate the optimal inspection interval in Exercise 21.5 when all defects identified at inspection are repaired within the inspection period d_i.

PART E

Cases with Emphasis on Operations Optimization and Reengineering

Chapter 22, "Mean Residual Life and Optimal Operating Conditions for Industrial Furnace Tubes"

—Uses test results at accelerated conditions and selected failure models to estimate the reliability and mean residual life after a given operating time, of reformer tubes in furnaces used in the petrochemical industry, to determine optimal operating conditions and preventive maintenance schedules.

Chapter 23, "Optimization of Dragline Load"

—Develops a failure model for a dragline, which is a very large and expensive piece of equipment used in coal mining, and uses the result to derive an optimal trade-off between maintenance and dragline load.

Chapter 24, "Ford's Reliability Improvement Process—A Case Study on Automotive Wheel Bearings"

—Provides a detailed presentation of the nine-step process used by Ford in reengineering for reliability improvement, with application to analysis and elimination of a noise problem in a front wheel bearing used in some Ford automobiles.

Chapter 25, "Reliability of Oil Seal for Transaxle—A Science SQC Approach at Toyota"

—Deals with a joint effort by manufacturer and supplier to improve the reliability of a transaxle unit of an automobile by analysis of the oil leakage mechanism and its relationship with process factors.

CHAPTER 22

Mean Residual Life and Optimal Operating Conditions for Industrial Furnace Tubes

Elsayed A. Elsayed

22.1 INTRODUCTION

22.1.1 Background

Furnaces used in petrochemical industry are usually subject to extreme conditions owing to the constraint of the processes involved. The tubes of such furnaces, referred to as *reformer tubes,* are arranged vertically in a reduction chamber and receive a mixture of hydrocarbons and steam at a temperature of 1710°F or above and an internal pressure of 400 psi (pounds per square inch). The gas mixture is heated in the presence of a catalyst, producing a hydrogen-rich gas at a temperature typically between 1340°F and 1460°F. The temperature in the tube walls may exceed 1832°F at the end of a process cycle when the catalyst is close to exhaustion. The hydrogen-rich gas is then used in the refinery processes. The throughput of the refinery is dependent on the amount of hydrogen gas produced that in turn depends on the operating temperature of the furnace tubes. Higher temperatures will result in higher throughput that will also result in reduction in the remaining life of the tube.

The reformer furnace tubes are generally fabricated from centrifugally cast creep-resistant austenitic steel HK grade (0.4C, 25Cr, 20Ni) or HP grade (0.4C, 25Cr, 35Ni). Long furnace tubes are made by welding together an appropriate number of short tubes to produce the desired length. The tubes are designed for a nominal life of 100,000 hours (11.4 years) based on the recommended practice of the American Petroleum Institute (API). However, the useful service life is found to range from around 30,000 to 150,000 hours depending on the actual operating conditions and the characteristics of the particular material. Other tubes are designed for similar applications with a nominal design life of 21 years. After the tubes have

Case Studies in Reliability and Maintenance, Edited by W. R. Blischke and D. N. P. Murthy.
ISBN 0-471-41373-9

been in operation for about 8 years, questions usually arise about their remaining (or residual) life and whether it will be necessary to retube the furnace. The concern is that tube ruptures could occur and result in costly repairs and operational downtime (Ibarra and Konet, 1994). Thus, it is important to accurately determine the mean residual life of the tubes so that appropriate maintenance actions are taken. Likewise, more accurate reliability estimates of the tubes are likely to result in accurate determination of the optimal inspection intervals of the tubes, thus minimizing the overall cost of the system.

The industry standards for estimating the remaining tube life are described in American Petroleum Institute Recommended Practice 530 (API, 1988). However, the accuracy of the estimate of the remaining life depends on the actual operating conditions of the tube such as temperature and pressure, humidity conditions and corrosion rate. Of course there are many uncertainties that might result in an inaccurate estimate of the remaining life. They include the actual metal temperature of the tube and the actual rupture strength of the tube metal.

The purpose of this chapter is to develop a reliability-based approach for estimating the mean residual life of furnace tubes subject to given operating conditions. The approach also determines the optimum operating conditions of the furnace tubes such that the total cost is minimized. A methodology for determining the optimum maintenance schedule for such tubes is also presented. Section 22.2 describes the failure mechanisms of the furnace tubes and methods for detecting these mechanisms. The currently used deterministic approaches for determining the mean residual life of the furnace tubes and their advantages and limitations are presented in Section 22.3. Section 22.4 presents a case study and a new reliability approach for estimating the mean residual life of the furnace tubes. The failure data at different test conditions are used in the analysis to estimate the parameters of the failure time distributions. Section 22.5 demonstrates the use of reliability approaches in determining the optimum operating conditions of the furnace tubes such that the maximum profit is attained. Optimum preventive maintenance schedules of the furnace tubes are presented in Section 22.6, and the summary is presented in Section 22.7.

22.2 FAILURE MECHANISMS AND DETECTION

22.2.1 Failure Mechanisms

Creep is the primary cause of the furnace tube damage. It usually initiates within the tube wall some two-thirds of the way through from the outer surface, making it impossible to detect by in situ metallography. This is opposite to boiler superheaters and headers where creep damage initiates at the outside surfaces, making it much easier to detect. Other tube damages occur due to carbonization, thermal shock, and recrystallization resulting from accidental overheating (May et al., 1996). These types of damage can be assessed using destructive metallographic methods. The damage is determined by sectioning of a tube and repeated careful polishing and etching to emphasize the damage and its type. Researchers have utilized the type and amount of damage to determine the remaining life of the tube, as discussed later.

22.2.2 Failure Detection

There are several methods for assessing and detecting the damage of the furnace tubes. The methods are classified as destructive and nondestructive.

22.2.2.1 Destructive Methods

When a reformer tube is removed for inspection, a small section of the tube is cut, polished, and etched. It is then examined for voids and microcracks. For example, if the voids disclosed on the tube section are arranged approximately closer than one-third in from the inner surface, then it is likely that initiation of microcracks would begin shortly. The microcracks that develop tend to propagate toward the internal surface and later to the external surface to produce leakage.

22.2.2.2 Nondestructive Methods

Infrared thermography camera can be used to document actual temperature conditions of the tube metal and can identify localized "hot spots" that could result in failure. Gong et al. (2000) propose a new method based on quartz optical fiber and digital image analysis. The optical fibers have good heat resistance and good light conductivity at high temperature. Changes in temperature and loading on the tube can be assessed by observing changes in the images of light spots sampled at different times.

Enhanced visual examination and ultrasonic testing provide data on type, location, and distribution of flaws. The data are also used to establish estimates of the maximum defect size at critical locations of the furnace tubes. Of course, other methods such as the application of magnetic particles and X-ray examinations can also be used, when appropriate, to assess creep damage in such tubes.

22.3 RESIDUAL LIFE PREDICTION: DETERMINISTIC APPROACHES

Since reformer tubes operate at high temperatures and are critical components for many industries such as nuclear power plants and chemical, petrochemical, and other process industries, it is imperative that the residual lives of these tubes be accurately estimated, especially as the tubes approach their design lives. Larson and Miller (1952) provide the first study that relates the time to rupture of the material to the operating temperature as expressed below:

$$\frac{1}{t} = Ae^{-Q/RT} \qquad \text{(for constant stress)} \qquad (22.1)$$

where t equals time to rupture, A is a constant, Q is the activation energy of the process, R is a gas constant, and T is absolute temperature. Once the constants A, Q, and R are estimated using either field data or laboratory experimentation, it becomes straightforward to estimate the time to rupture at any operating temperature. Robinson (1952) and Grant and Bucklin (1965) provide similar expressions with temperature variations. Simonen and Jaske (1985) develop a simulation model for predicting the residual life of furnace tube that is of interest to design engineers. Other qualitative and deterministic procedures for estimating the residual life of furnace tubes are summarized below.

22.3.1 Residual Life Based on Metallographic Evaluation

Researchers relate the carbide precipitation in the tubes walls to material damage and develop a semiquantitative model for prediction of the remaining life based on the tube damage as follows (Gong et al., 1999):

(a) No cavities. The primary carbides precipitate along the grain boundary. The corresponding life fraction is within 20% of the remaining life.
(b) Few cavities appear in the grain boundary of the inner side of the tube. The corresponding life fraction is about 20–40% of the remaining life.
(c) String of cavities found along the grain boundary. Few cavities form microcracks. The corresponding life fraction attains 40–60% of the remaining life.
(d) A few cavities are linked forming microcracks. Some microcracks are linked with cracks caused by carbonization in the inner surface of the tube. The consumed life fraction is about 60–75% of the life of the tubes.
(e) The microcracks link with each other, forming a macrocrack propagating toward outer surface. The crack length is about two-thirds of the thickness. The tube is postulated to fail.

Though the above approach is practical, it does not provide a precise or accurate estimate of the time to rupture or the remaining life of the tube.

22.3.2 Residual Life Based on API RP 530

The American Petroleum Institute Recommended Practice 530 (API 1988) utilizes Larson and Miller's approach (1952). Both the minimum and average Larson–Miller curves are used for the temperature range (1050–1250°F) reported for the outlet side of the tubes. The Larson–Miller curves shown in API RP 530 are extrapolated from the higher stress region of the curves. The extrapolation will decrease the accuracy in the low-stress regions. For example, at 1050°F the following time to rupture is calculated using the curves and its ranges as given in Larson and Miller (1952) and discussed in Ibarra and Konet (1994):

$$\text{tr (min curve)} = 1.06 \times 10^6 \text{ hours (life fraction} = \frac{183{,}960}{1.06 \times 10^6} = 0.18) \qquad (22.2)$$

$$\text{tr (avg curve)} = 5.69 \times 10^6 \text{ hours (life fraction} = 0.03) \qquad (22.3)$$

These results indicate that if we were certain that the tubes had operated at 1050°F for 21 years (183,960 hours), the life fraction consumed would be between 0.03 and

0.18, thus showing that the tube had suffered little creep damage in service. Clearly, if the tube had operated at higher temperatures, the life fraction would have been higher. As shown in (22.2) and (22.3), the difference in rupture life between the minimum and average Larson–Miller curve is a factor of 5. This is a significant factor when assessing the remaining life of the tube.

More importantly, Larson and Miller (1952) developed a relationship between the time and temperature at different stresses as follows:

$$T_1(20 + \log t_1) = T_2(20 + \log t_2)$$

where T_1 and T_2 are the absolute temperatures at temperature levels 1 and 2, respectively, t_1 and t_2 are the times (or lives) at these temperatures, respectively, and 20 is the material constant for ferritic steels. For example, the combinations shown in Table 22.1 should have equivalent rupture stresses (lives).

Though the API PR 530 is widely used for prediction of the remaining life, its estimate is far from accurate due to the uncertainties of the parameters.

22.3.3 Residual Life Based on Omega Method

The Materials Properties Council (MPC) of Joint Industry developed a faster, more accurate, and industry-accepted method for determining the remaining life of process equipment after long-term service. The method, referred to as the Omega Method, is based on strain rate and utilizes data generated in creep tests at temperatures and stresses as close as possible to the actual field operating conditions. The methodology has resulted in new databases for estimating the remaining life. The steps of the Omega Method (MPC, 1993) are as follows:

1. Level 1: Use Omega spreadsheets and databases to determine if the equipment falls into a "safe" region.
2. Level 2: Refine material data and stress calculations and then follow Level 1.
3. Level 3: Prioritize inspection by determining levels to achieve 1% additional strain or 20% of remaining life. If the inspection interval is too short, go to the final level.
4. Level 4: Remove sample and conduct creep rupture tests to obtain the exact data needed for Omega spreadsheet to calculate the remaining life.

Table 22.1. Equivalent Combinations of Operating and Test Conditions

Combination	Operating Conditions	Test Conditions
1	10,000 hr at 1000°F	13 hr at 1200°F
2	1000 hr at 1200°F	12 hr at 1350°F
3	1000 hr at 1350°F	17 hr at 1500°F
4	1000 hr at 300°F	1.1 hr at 400°F

22.4 RESIDUAL LIFE PREDICTION: RELIABILITY APPROACH

With exception of Dai (1995, 1996), who used the method of extrema of fuzzy functions for finding the remaining life of furnace tubes, all previous studies and theories are based on deterministic analysis. Such studies do not reflect the uncertainties in manufacturing, operating conditions, and properties of material under different environments. In this section, we utilize reliability methods to estimate the mean residual life at any given time t. We first describe the case study, and then we follow it up by demonstrating the use of reliability methods in determining the remaining life of such furnace tubes as well as the optimal operating conditions.

22.4.1 Description of the Case

A major oil company produces 100 million barrels of a sweet oil blend (SOB) per year. The production of the SOB requires hydrogen that is supplied by five hydrogen plants. The production rate is proportional to the amount of hydrogen supplied; that is, more hydrogen production results in more production of oil until the maximum capacity of the plant is reached. Therefore, it is important that hydrogen plants operate without interruption or equipment failure.

A hydrogen-producing plant operates a methane reformer furnace (MRF). Each furnace has hundreds of tubes that are filled with catalyst. Methane and steam pass through these tubes at high temperature where hydrogen is produced. The tubes are fabricated as described in previous sections, each having an internal diameter of 5.00 in., a wall thickness of 0.40 in. and a length of 45 ft. They are placed vertically in the furnace. The cost of the tubes ranges between $10M and $20M and represents a high proportion of the total cost of the furnace.

The life of the furnace tubes is dependent on its material composition, manufacturing methods, and the operating conditions of the tubes, namely, temperature and pressure. Of course, other operating conditions might have an impact on the life of the tubes, such as humidity and the presence of sulfur dioxide. However, the temperature is the most important of all, because it contributes to material creep and rupture. As mentioned earlier, increasing the hydrogen production increases the SOB production. However, increasing the hydrogen production decreases the tube's life and increases the risk of on-line tube rupture.

The cost of the furnace tubes represents a high proportion of the total cost of the furnace. Therefore, the remaining life of the tubes should be accurately estimated so that the tubes are not replaced prematurely. Moreover, the tubes should be periodically inspected for possible crack propagations.

The expected design life of the tubes when the furnace operates at the normal operating temperature of 1340°F is 100,000 hours. This design life is calculated based on the Larson–Miller design formula, which relates the properties of the tube material to the operating temperature and pressure.

Increasing the oil production requires an increase in the hydrogen production, which in turn increases the furnace burner rate. As a result, the temperature in the furnace tends to increase, which causes a significant reduction in the remaining life of the tubes. Analysis of failure data collected over 8 years of operation shows that operating a tube at 25°F above the design temperature of 1340°F results in a loss of one-half of the tube's remaining life. The engineers of the oil company are interested in estimating the reliability of the furnace, the remaining life of each set of tubes, and the operating condition that minimizes the total cost.

We intend to use a reliability approach for estimating the mean residual life of the tubes at any time t, as described in Section 22.4.2.

22.4.2 Mean Residual Life

The mean residual life function $L(t)$ is defined as

$$L(t) = E[T - t | T \geq t], \qquad t \geq 0 \tag{22.4}$$

In other words, the mean residual life function is the expected remaining life, $T - t$, given that the component, product, or a system has survived to time t. The conditional probability density function for any time $\tau \geq t$ is

$$f_{T|T\geq t}(\tau) = \frac{f(\tau)}{R(t)}, \qquad \tau \geq t \tag{22.5}$$

where $R(t) = P(T > t)$ is the reliability function. The conditional expectation of the function given in equation (22.5) is

$$E[T|T \geq t] = \int_t^{\infty} \tau f_{T|T\geq t}(\tau)\, d\tau = \int_t^{\infty} \tau \frac{f(\tau)}{R(t)}\, d\tau \tag{22.6}$$

Since the component, product, or a system has survived up to time t, the mean residual life is obtained by subtracting t from equation (22.6), thus

$$L(t) = \int_t^{\infty} (\tau - t) \frac{f(\tau)}{R(t)}\, d\tau = \int_t^{\infty} \tau \frac{f(\tau)}{R(t)}\, d\tau - t \tag{22.7}$$

or

$$L(t) = \frac{1}{R(t)} \int_t^{\infty} \tau f(\tau)\, d\tau - t \tag{22.8}$$

The objective of the case study is to determine the parameters of the failure time distributions as accelerated conditions that are then utilized to estimate the parameters of the failure time distribution at normal operating conditions. Thus, the mean residual life, as shown above, can be easily estimated at any time during the operation of the furnace. This assists in decision-making regarding replacement and maintenance of the furnace tubes.

22.4.3 Experimentation and Failure Data

Accelerated creep testing is conducted to determine the amount of deformation of test specimens as a function of time (creep test) and the measurement of the time for fracture to occur when sufficient load is present (rupture test) for materials when under constant tension loads at constant temperature. Creep testing requires a reasonably large sample size subjected to different stress levels for extended periods of time (usually in years). We now describe a typical creep test experiment for furnace tubes. Actual test data are proprietary, therefore the data presented below are generated from known failure time distributions with test durations equivalent to actual test times.

A manufacturer of reformer tubes conducted accelerated creep tests by simultaneously subjecting 50 specimens at a temperature of 1760°F and another 50 specimens at temperature of 1800°F and recorded the strain rate of the tubes. The time-to-rupture (TTR) is the time to induce 0.5% strain at the lower temperature and 0.3% strain at the higher temperature. These results are given in Tables 22.2 and 22.3, respectively.

We fit the failure data to the Weibull distributions, and the probability plots for the Weibull model are shown for the data recorded at 1800°F and 1760°F in Figures 22.1 and 22.2, respectively. Note that in the plots the data lie close to the plotted line, indicating a good fit to the Weibull distribution.

As shown in Figures 22.1 and 22.2, the Weibull model is an appropriate fit for the failure data at the two test conditions. The estimated shape parameter γ and the scale parameter for the data obtained at 1800°F are $\hat{\gamma}_1 = 1.579$ and $\hat{\theta}_1 = 9391$, while the estimated shape and scale parameters for data obtained at 1760°F are $\hat{\gamma}_2 = 1.745$ and $\hat{\theta}_2 = 19007$, respectively. The corresponding 90% confidence intervals, based on the asymptotic normality of the estimators (Elsayed, 1996), are

$$1.223 \leq \gamma_1 \leq 1.784$$

$$7{,}912 \leq \theta_1 \leq 11{,}003$$

$$1.356 \leq \gamma_2 \leq 1.784$$

$$16{,}267 \leq \theta_2 \leq 11{,}003$$

Table 22.2. Time to Rupture at 1760°F

3,515	6,989	14,298	18,846	25,016
3,578	7,771	14,401	19,152	26,256
3,968	8,969	15,652	19,503	28,523
5,225	9,057	16,196	20,108	28,659
5,392	9,172	16,265	20,193	30,901
5,694	9,408	16,484	20,306	31,393
5,768	9,747	16,553	20,370	31,992
6,197	10,231	16,793	20,610	36,498
6,339	11,351	16,932	22,668	38,821
6,871	11,374	17,209	24,971	50,753

Table 22.3. Time to Rupture at 1800°F

628	3,331	5,526	10,455	12,508
1,385	3,371	5,913	10,641	13,872
1,479	3,399	6,420	11,575	14,700
1,928	3,910	6,813	11,896	15,742
2,136	4,048	7,037	11,911	15,791
2,319	4,227	8,123	12,067	17,391
2,444	4,350	8,441	12,099	18,330
2,484	4,905	8,774	12,112	18,845
2,488	5,129	8,835	12,145	19,046
3,263	5,141	10,096	12,275	19,561

The most important assumption of using the accelerated failure time (AFT) models to estimate reliability at normal operating conditions using the accelerated conditions is that the failure time distributions at different stress levels must have approximately the same shape parameter. Since this appears to be a very reasonable assumption in our case, we use an approximate value of $\hat{\gamma} = 1.66$, the average of the two estimates. The probability density functions and reliability functions at these temperatures are shown in Figures 22.3 and 22.4, respectively.

Since temperature is the only important factor that affects the rupture of the tubes, we use the Arrhenius model (Elsayed, 1996; Nelson, 1990) to estimate the acceleration factor

$$t = ke^{c/T} \tag{22.9}$$

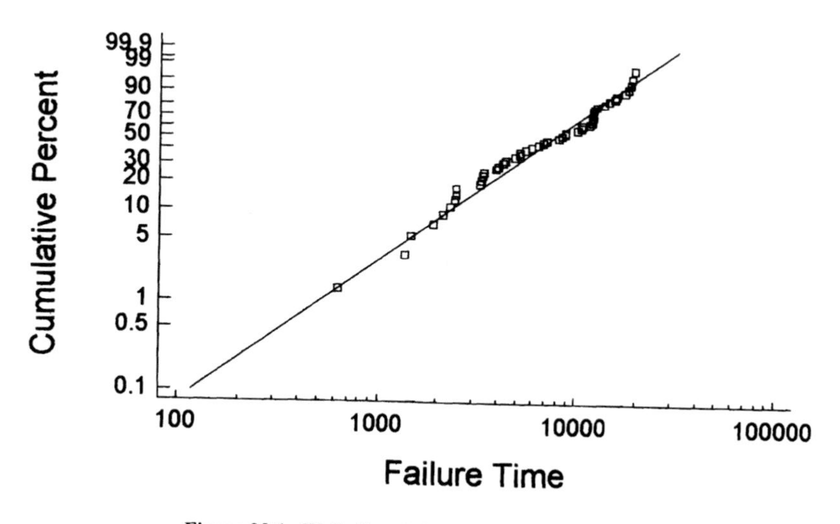

Figure 22.1. Weibull probability plot for 1800°F data.

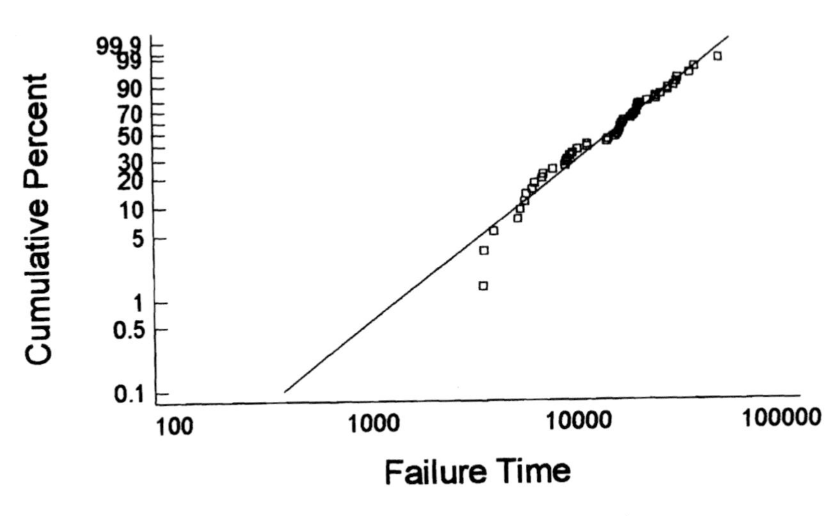

Figure 22.2. Weibull probability plot for 1760°F data.

where t is the time at which a specified portion of the population fails, k and c are constants, and T is the absolute temperature (measured in degrees Kelvin). Therefore

$$\ln t = \ln k + \frac{c}{T}$$

We determine the time at which 50% of the population fails (approximately equivalent to the median of the sample) as

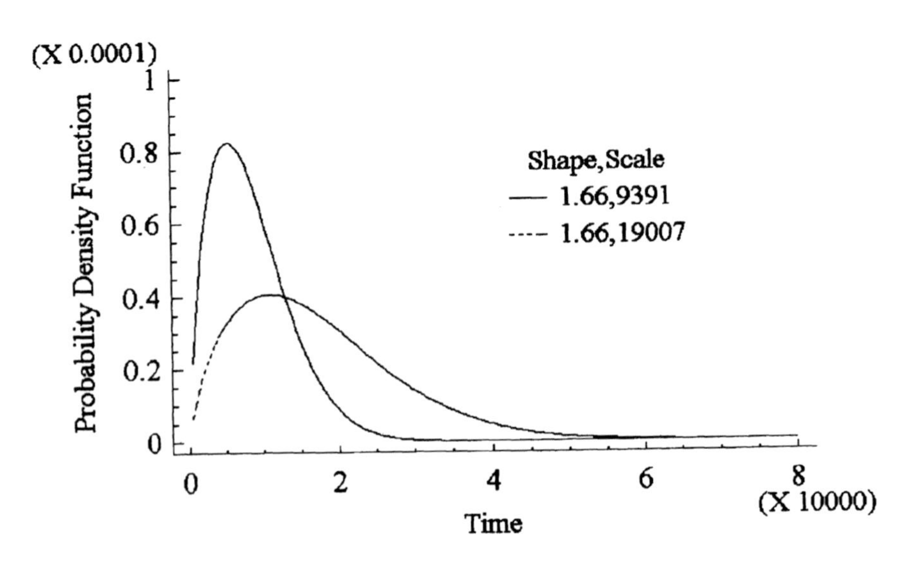

Figure 22.3. Probability density functions for the two temperatures.

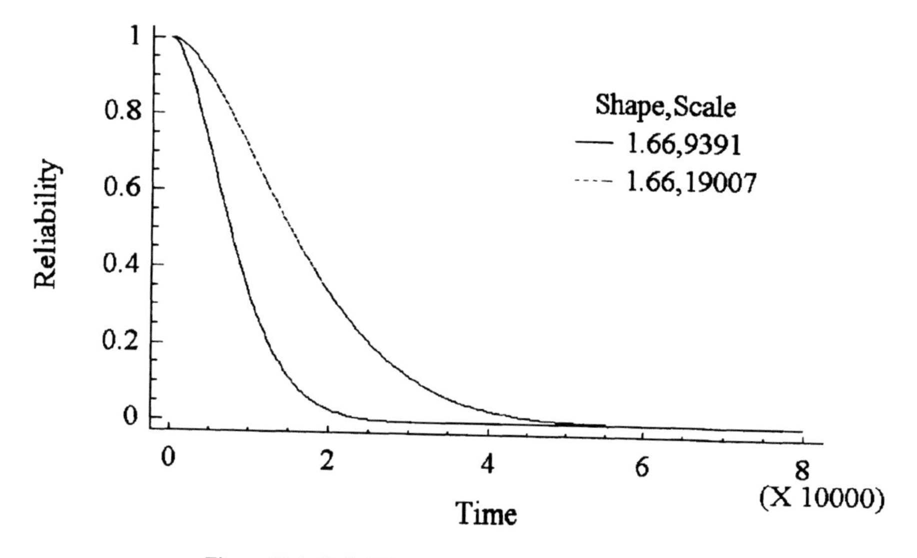

Figure 22.4. Reliability functions for the two temperatures.

$$t = \theta[-\ln(0.5)]^{1/\gamma} \tag{22.10}$$

The 50th percentiles are shown in Table 22.4. Substituting the values given in Table 22.4 in equation (22.9), we obtain

$$15{,}241 = ke^{C/1233.16}$$

$$7{,}350 = ke^{C/1255.38}$$

Solving these two equations results in $c = 50{,}803$ and $k = 1.95524 \times 10^{-14}$. Therefore, the estimated 50th percentile at given temperature T is obtained as $t_{T°C} = 1.95524 \times 10^{-14}\, e^{50803/(T+273)}$ hours, and the corresponding scale parameter can be obtained using equation (22.10). Once θ is obtained, we obtain the reliability function, shown in Figure 22.5, and the probability density function is as follows:

$$R(t) = e^{-t^{\gamma}/\theta} \tag{22.11}$$

$$f(t) = \frac{\gamma}{\theta} t^{\gamma-1}\, e^{-t^{\gamma}/\theta} \tag{22.12}$$

Table 22.4. Percentiles at Two Temperatures

Temperature:	1,760	1,800
50th Percentiles:	15,241	7,350

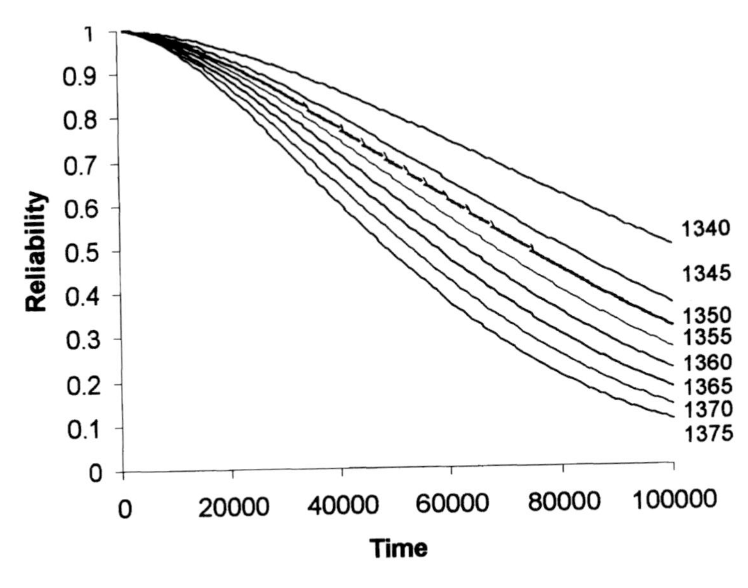

Figure 22.5. Reliability functions at different operating temperatures.

and the mean residual life at any time t is obtained using equation (22.8). Table 22.5 shows the mean residual lives (in hours) for tubes operating in temperatures ranging from 1340°F to 1375°F. The mean residual lives are shown graphically in Figure 22.6.

This approach provides more realistic estimates of the mean residual lives than those currently being used as described in Section 22.3. More importantly, the mean

Table 22.5. Mean Residual Lives at Different Operating Temperatures

Time	1340°F	1345°F	1350°F	1355°F	1360°F	1365°F	1370°F	1375°F
30,000	80,350	69,692	62,919	56,775	51,204	46,184	41,606	37,457
35,000	77,911	67,398	60,724	54,679	49,245	44,293	39,813	35,761
40,000	75,642	65,273	58,700	52,754	47,427	42,573	38,188	34,234
45,000	73,524	63,297	56,825	50,986	45,756	40,997	36,711	32,850
50,000	71,546	61,457	55,086	49,393	44,215	39,553	35,360	31,589
55,000	69,688	59,737	53,465	47,874	42,789	38,221	34,118	30,432
60,000	67,932	58,124	51,952	46,455	41,466	36,989	32,970	29,369
65,000	66,279	56,606	50,533	45,131	40,234	35,842	31,908	28,418
70,000	64,713	55,179	49,258	43,891	39,082	34,774	30,916	27,510
75,000	63,230	53,825	48,004	42,728	38,002	33,775	30,037	26,666
80,000	61,823	52,552	46,824	41,630	36,987	32,836	29,179	25,877
85,000	60,485	51,418	45,706	40,596	36,032	32,006	28,375	25,139
90,000	59,208	50,275	44,646	39,617	35,127	31,181	27,618	24,445
95,000	57,986	49,187	43,641	38,690	34,332	30,403	26,905	23,834
100,000	56,834	48,152	42,685	37,810	33,528	29,666	26,228	23,223

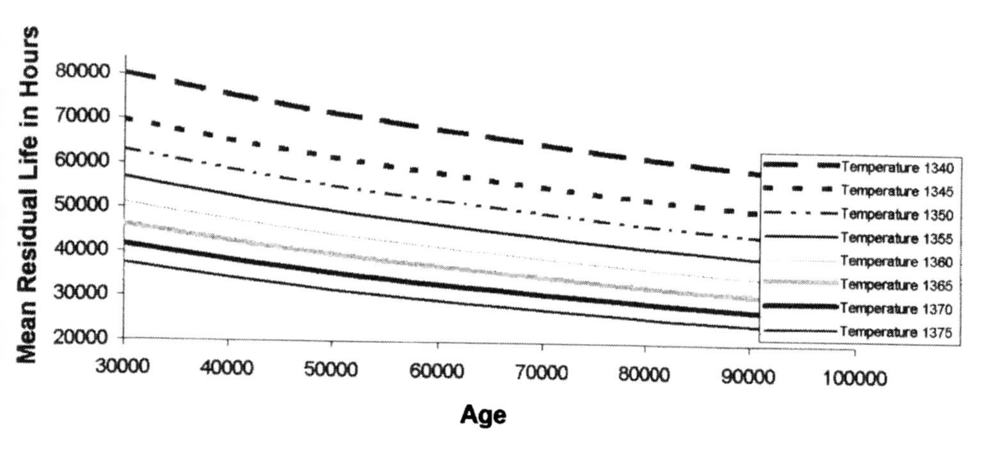

Figure 22.6. Mean residual life at different operating temperatures.

residual life is easily estimated for any operating conditions when compared with the traditional methods that do not consider the stochastic nature of the failure times. Moreover, the reliability function estimated by this approach can be used for maintenance decisions, as discussed later.

22.5 OPTIMUM OPERATING TEMPERATURE

Operating the furnace at 1340°F results in an estimated mean life of 12.72 years. The yearly production is 10^6 barrels, with a profit of $2 per barrel of refined oil. Operating at temperatures higher than 1340°F results in a 9.8% to 10% increase in oil production for every 5°F increase in operating conditions, but the expected life of the tube decreases accordingly. This will necessitate a replacement of the tubes at a cost of $25,000,000. Thus an optimal operating temperature corresponds to maximum profit obtained within the acceptable operating temperature rang of 1340°F to 1375°F. The net profit equation is given by

$$\text{Profit at temperature } T = P \times N_T \times MTTF_T - \frac{(MTTF_{1340} - MTTF_T)}{10{,}000} \times \frac{25 \times 10^6}{MTTF_{1340}} \tag{22.13}$$

where P is profit per barrel and N_T is the number of barrels produced at operating temperature T. The $MTTF_T$ is calculated as

$$MTTF_T = \theta^{1/\gamma}\Gamma\left(1 + \frac{1}{\gamma}\right) \tag{22.14}$$

The estimated scale parameters of the Weibull distributions and the calculated MTTF are utilized to calculate the profit corresponding to operating temperatures

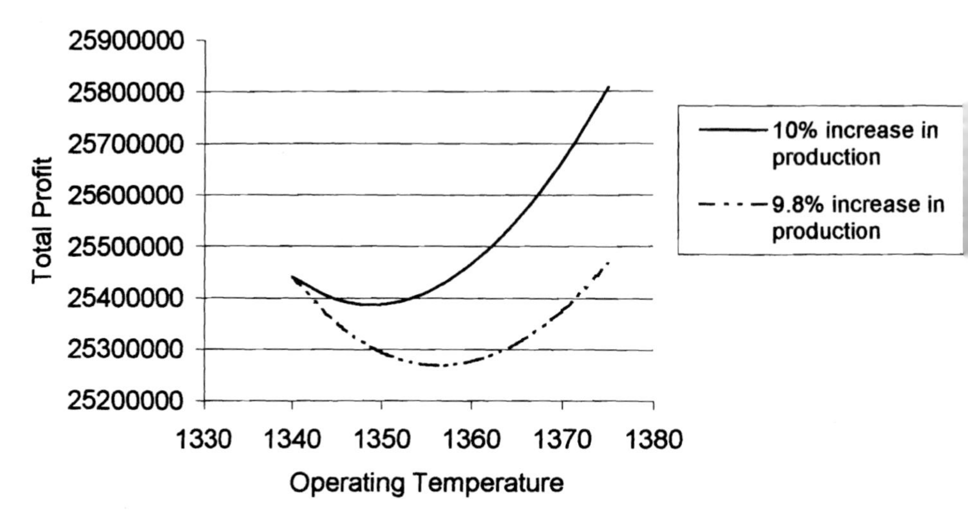

Figure 22.7. Net profit at different operating temperatures.

as shown in Figure 22.7. It is not economical to operate at 1345, 1350 and 1355 °F. At 10% increase in production, it becomes economical to operate at higher temperatures than those listed above, whereas at 9.8% increase it is more profitable to operate at 1375 °F. Indeed, the results from the profit equation indicate that it is more economical to operate at temperatures beyond 1380°F. However, the profit equation should be subjected to temperature constraints where chemical and physical conditions are imposed.

22.6 OPTIMUM PREVENTIVE MAINTENANCE

The constant interval replacement policy (CIRP) is the simplest preventive maintenance and replacement policy. Under this policy, two types of actions are performed. The first type is the preventive replacement that occurs at fixed intervals of time. Components or parts are replaced at predetermined times regardless of the age of the component or the part being replaced. The second type of action is the failure replacement upon failure. This policy is also referred to as *block replacement policy* (Elsayed, 1996).

The objective of the preventive maintenance and replacement models is to determine the parameters of the preventive maintenance policy that optimize some criterion. The most widely used criterion is the total expected replacement cost per unit time. This is accomplished by developing a total cost function per unit time as follows. Let $c(t_p)$ be the asymptotic total replacement cost per unit time as a function of t_p (Barlow et al., 1965). Then

$$c(t_p) = \frac{\text{Total expected cost in interval } (0, t_p]}{\text{Expected length of the interval}} \tag{22.15}$$

The total expected cost in the interval $(0, t_p]$ is the sum of the expected cost of failure replacements and the cost of the preventive replacement. During the interval $(0, t_p]$, one preventive replacement is performed at a cost of c_p and $M(t_p)$ failure replacements at a cost of c_f each, where $M(t_p)$ is the expected number of replacements (or renewals) during the interval $(0, t_p]$. The expected length of the interval is t_p. Equation (22.15) can be written as

$$c(t_p) = \frac{c_p + c_f M(t_p)}{t_p} \tag{22.16}$$

The expected number of failures, $M(t_p)$, during $(0, t_p]$, as t_p becomes large, may be obtained using the asymptotic form of the renewal function as

$$M(t_p) = \frac{t_p}{\mu} + \frac{\sigma^2 - \mu^2}{2\mu^2} \tag{22.17}$$

where μ and σ are the mean and standard deviation of the failure time distribution. For example, the mean and variance of the Weibull distribution are

$$\mu = \theta^{1/\gamma}\Gamma\left(1 + \frac{1}{\gamma}\right) \tag{22.18}$$

$$\sigma^2 = \theta^{2/\gamma}\left[\Gamma\left(1 + \frac{2}{\gamma}\right) - \left(\Gamma\left(1 + \frac{1}{\gamma}\right)\right)^2\right] \tag{22.19}$$

Assume that management decides to operate the furnace at 1375°F. Then the mean μ is 54,287 and the standard deviation is 38,599 and the expected number of failures is obtained using equation (22.17) as

$$M(t_p) = \frac{t_p}{54287} - 0.247722$$

Using the above expression, it is easy to determine the value of the total cost per unit time for different values of preventive and replacement costs that minimizes equation (22.16). This will correspond to the optimum preventive replacement interval.

22.7 SUMMARY

In this chapter we present a more realistic approach based on reliability engineering to determine the mean residual life of furnace tubes. The approach starts by using test results at accelerated conditions and relates them to normal operating conditions. We then obtain an expression for the reliability function and estimate the mean residual life for any operating temperature at given times. Additionally, the reliability function and mean time to failure are obtained for any operating temperature. The traditional approaches for estimating the mean residual life are limited and do not provide other reliability characteristics such as the mean time to failure and the reliability values at specified operating conditions. The methodology is also used in determining the optimal operating conditions and preventive maintenance schedule, when applicable.

REFERENCES

API (1988). *Calculations of Heater-Tube Thickness in Petroleum Refineries,* API Recommended Practice 530, 3rd edition, American Petroleum Institute, Washington, D.C.

Barlow, R. E., Proschan, F., and Hunter, L. C. (1965). *Mathematical Theory of Reliability,* Wiley, New York.

Dai, S.-H. (1995). A study on residual life prediction for pressurized furnace operated at elevated temperature by using the method of extrema of fuzzy functions, *International Journal Pressure Vessels and Piping* **63:**111–117.

Dai, S.-H. (1996). A study on the prediction of remaining life and ageing of material for pressurized tubes of industrial furnace operated at elevated temperature, *International Journal Pressure Vessels and Piping* **69:**247–252.

Elsayed, E. A. (1996). *Reliability Engineering,* Addison-Wesley, Reading, MA.

Gong J.-M., Tu, S.-T., and Ling, X. (2000). Research of welding effect on creep damage of high temperature furnace tubes, *Key Engineering Materials,* **171–174:**189–196.

Gong, J.-M., Tu, S.-T., and Yoon, K.-B. (1999). Damage assessment and maintenance strategy of hydrogen reformer furnace tubes, *Engineering Failure Analysis* **6:**143–153.

Grant, N. J., and Bucklin, A. G. (1965). *Deformation and Fracture at Elevated Temperature,* MIT Press, Boston, MA.

Ibarra, S., and Konet, R. R. (1994). Life assessment of 1¼Cr–½Mo steel catalytic performer furnace tubes using the MPC omega method," *ASME PVP,* Vol. 288, *Service Experience and Reliability Improvement: Nuclear, Fossil, and Petrochemical Plants.*

Larson, F. R., and Miller, J. (1952). A time–temperature relationship for rupture and creep stresses, *Transactions of the ASME,* **74:**765–771.

May, I. L., da Silveira, T. L., and Vianna, C. H. (1996). Criteria for the evaluation of damage and remaining life in reformer furnace tubes, *International Journal Pressure Vessels and Piping* **66:**233–241.

MPC (1993). Draft Handbook for Evaluation of Components in Creep Service, Materials Properties Council, New York.

Nelson, W. (1990). *Accelerated Testing,* Wiley, New York.

Robinson, E. L. (1952). Effect of temperature variation in the long-time rupture strength of steels, *Transactions of the ASME,* pp. 777–782.

Simonen, F. A., and Jaske, C. E. (1985). A computational model for predicting the life of tubes used in petrochemical heater service, *Journal Pressure Vessel Technology* **107:**239–246.

EXERCISES

22.1. The failure times of MOSFIT capacitors at both the accelerated stress conditions and the normal operating conditions are found to follow gamma distributions having the same shape parameter γ. The following is the probability density function of the gamma distribution

$$f(t) = \frac{t^{\gamma-1}}{\theta^{\gamma}\Gamma(\gamma)} e^{-t/\theta}$$

where θ is the scale parameter of the distribution.

(a) Develop the relationship between the hazard rates at both the accelerated stress condition and the normal operating conditions.

(b) Fifty capacitors are divided into two equal samples and subjected to two temperatures of 170°C and 210°C respectively. The failure time data are given below.

Failure times for 170°C: 4	6	8	12	12	14	14	16	22	23	28	28	33
	34	35	37	38	40	43	45	46	50	50	51	61
Failure times for 210°C: 5	5	6	9	9	10	13	16	16	18	21	21	21
	22	23	26	26	26	29	32	34	39	50	61	65

The probability density functions are

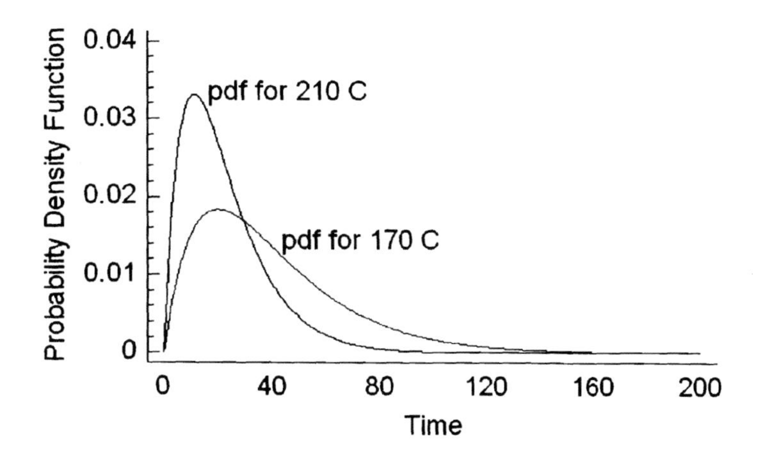

Estimate the parameters of the failure time distributions and test the hypothesis that they have the same shape parameter.

(c) Estimate the mean time to failure at normal operating temperature of 30°C.

22.2. Upon further analysis, the analyst realizes that there are five missing observations for each test temperature. They are 93, 102, 105, 110, and 156 for the 170°C sample and 52, 56, 57, 62, and 73 for the 210°C sample.
Rework Exercise 22.1 using the updated information and compare the results. Is the estimate of the mean time to failure at normal operating conditions significantly different?

22.3. Creep failure results whenever the plastic deformation in machine members accrues over a period of time under the influence of stress and temperature until the accumulated dimensional changes cause the part not to perform its function or to rupture (part failure). It is clear that the failure of the part is due to stress–time–temperature effect. Fifteen parts made of 18–18 plus stainless steel are subjected to a mechanical acceleration method of creep testing, in which the applied stress levels are significantly higher than the contemplated design stress levels, so that the limiting design strains are reached in a much shorter time than in actual service. The times (in hours) to failure at an accelerated stress of 710 MPa are (Elsayed, 1996) 30.80, 36.09, 65.68, 97.98, 130.97, 500.75, 530.22, 653.96, 889.91, 1173.76, 1317.08, 1490.44, 1669.33, 2057.95, and 2711.36

Assume that the failure times can be modeled by an exponential distribution and that the acceleration factor between the mean life at normal operating conditions and the accelerated stress condition is 20. Determine the parameter of the distribution at normal conditions and the reliability that a part will survive to 10,000 hours.
Hint: Utilize the following relationships between the accelerated conditions and normal conditions:

- Failure times:

$$t_o = A_F t_s$$

where t_o is the failure time at operating condition, t_s is the failure time at stress condition and A_F is the acceleration factor (the ratio between product life at normal conditions and life at accelerated conditions).

- Cumulative distribution functions (CDF):

$$F_o(t) = F_s\left(\frac{t}{A_F}\right)$$

- Probability density functions (pdf)

$$f_o(t) = \left(\frac{1}{A_F}\right) f_s\left(\frac{t}{A_F}\right)$$

- The failure rates:

$$h_o(t) = \left(\frac{1}{A_F}\right) h_s\left(\frac{t}{A_F}\right)$$

22.4. Derive expressions that relate the hazard rate functions, probability density functions, and the reliability functions at the accelerated stress and at the normal operating conditions when the failure time distribution at the two conditions is log-logistic in the form

$$f(t) = \frac{\lambda p(\lambda t)^{p-1}}{[1 + (\lambda t)^p]^2}$$

where λ and p are constants.

22.5. Consider a replacement policy where the time at which preventive replacement occurs depends on the age of the equipment; failure replacements are made when failures occur. Let t_p be the age of the equipment at which a preventive replacement is made, c_p the cost of a preventive replacement, and c_f the cost of a failure replacement. Assume that c_p = \$50 and c_f = \$100 and the p.d.f. of the failure time follows a Weibull distribution whose parameters are obtained using the failure times below.

319	396	506	555	666	701	762	782	818	873	905	941	977
1039	1051	1109	1144	1144	1206	1248	1257	1287	1358	1484	1504	

Determine the optimum replacement policy that minimizes the total cost per unit time.

CHAPTER 23

Optimization of Dragline Load

Peter G. A. Townson, D. N. Prabhakar Murthy, and Hal Gurgenci

23.1 INTRODUCTION

A dragline (Figure 23.1) is essentially a moving crane with a bucket at the end of a boom. It is used primarily in coal mining for removing the dirt to expose the coal. A typical dragline weighs over 4000 tonnes and costs around $100 million. The bucket volume varies around 90–120 cubic meters and the dragline is operated continuously (24 hours/day and 365 days/year) except when it is down undergoing either corrective or preventive maintenance actions. The loss of revenue for each day a dragline is out of action is estimated to be roughly $1 million. Hence, availability (measured as the expected fraction of time a dragline is in operational state) is a critical performance indicator of great significance to a mining business operating the dragline.

A performance indicator of even greater importance to a mining business is the annual output of a dragline. This is a function of the dragline (bucket) load, speed of operation, and availability. Availability depends on two factors (i) degradation of the components over time and (ii) maintenance (corrective and preventive) actions used. Degradation depends on the stresses on different components and these in turn are functions of the dragline load. As a result, availability is a function of the dragline load and the maintenance effort. Availability decreases as the dragline load increases and increases as the maintenance effort increases. This implies that the annual total output is a complex function of dragline load.

The problem of interest to the mine operators is the optimum dragline load to maximize the annual output.

Case Studies in Reliability and Maintenance, Edited by W. R. Blischke and D. N. P. Murthy.
ISBN 0-471-41373-9 © 2003 John Wiley and Sons, Inc.

Figure 23.1. Typical dragline.

23.2 APPROACH USED

There are two approaches to determining the optimal dragline load. The first approach is to experiment with the real system using different bucket sizes. This is not practical because to achieve relevant and reliable information, each bucket size experiment would have to last around a year. This would mean operating at suboptimal production rates for long periods possibly with increased breakdown maintenance and increased risk of catastrophic failures. Such a scenario would not be acceptable to the mine management. In the second approach, these problems are avoided by using a mathematical model to determine the optimal dragline load. Critical for this approach is the building of a suitable model and its validation. Once an adequate model has been built, standard analysis can be used to determine the optimal bucket size and carry out sensitivity analyses with respect to the various parameters of the model.

The approach used in this case study develops a mathematical model to determine the optimal dragline load and involves the following steps:

1. Carrying out a multilevel decomposition of the dragline to the lowest part level.
2. Examination of the maintenance records to evaluate the failure data available.
3. Appropriate system characterization to model the system in terms of a reasonable number of components based on steps 1 and 2.

4. Modeling of the components identified in step 3. This involves selecting a suitable distribution function for modeling failures and estimating the model parameters based on the data from step 2.
5. Modeling system performance using component models from step 4.
6. Modeling the effect of dragline load on component degradation and its impact on the system performance.
7. Model analysis to obtain the optimal dragline load.

The structure of the case study is as follows. In Section 23.3, we discuss the system characterization for the modeling of a dragline. Section 23.4 examines the data-related issues. Sections 23.5 and 23.6 deal with the modeling at the component and system levels. Section 23.7 deals with the modeling of the effect of the dragline load on component reliability and system performance. Section 23.8 deals with the analysis to determine optimal dragline load. Finally, we conclude with some specific recommendations and topics for further study in Section 23.9.

23.3 SYSTEM CHARACTERIZATION

23.3.1 Decomposition of Dragline

A dragline is a complex system. It can be broken down into many subsystems. Each subsystem in turn can be decomposed into assemblies, and these in turn can be broken down still further into modules and modules into parts. In other words, it can be represented as a multilevel hierarchical system with the system being at the top and parts being at the lowest level. The number of levels needed depends on the degree of detail that is appropriate and needed for the study of the problem under consideration. For the purposes of the study, we restrict our attention to subsystem level and can view the dragline as a series structure comprising of 25 components, as indicated in Table 23.1.

The series structure is appropriate because the dragline is in operational state only when all 25 components are in operational state. Whenever a component fails, the dragline becomes nonoperational and is made operational through corrective maintenance actions.

23.3.2 Component Degradation and Failures

The components degrade with age and usage. An important factor in the degradation process is the stress on the component. The stress can be electrical, mechanical, or heat, depending on the component. A component is said to fail when it is unable to perform its intended (or stated) function. The time between failures is a random variable, because the degradation of a component depends on several factors such as design, manufacture, operation, maintenance, and so on.

Table 23.1. Decomposition of Dragline

i	Component
1	Generator (hoist)
2	Generator (drag)
3	Generator (swing)
4	Generator (synchronous)
5	Motor (hoist)
6	Motor (drag)
7	Motor (swing)
8	Motor (propel)
9	Machinery (hoist)
10	Machinery (drag)
11	Machinery (swing)
12	Machinery (propel)
13	Ropes (hoist)
14	Ropes (drag)
15	Ropes (dump)
16	Ropes (suspension)
17	Ropes (fairleads)
18	Ropes (deflection sheaves)
19	Bucket and rigging
20	Frame (boom)
21	Frame (tub)
22	Frame (revolving)
23	Frame (a-frame)
24	Frame (mast)
25	Others

Note: "Others" include the following: lube system, air system, transformers, brakes, blowers, fans, cranes, winches, communication equipment, air conditioning, and safety equipment.

23.3.3 Maintenance Actions

The dragline is subjected to the following three kinds of maintenance actions.

1. *Shutdown Maintenance.* This is a major overhaul where the 25 components are inspected and all worn or damaged parts are replaced so that the components are back to "good-as-new." This requires the dragline to be out of action for 5–8 weeks and is usually carried out after the dragline has clocked a certain number of usage hours, to be referred to as T in the rest of this case study. Under the base bucket load currently used, these actions are done roughly after $T = 43{,}800$ hours of usage (or 5 years based on usage clock).
2. *Planned Maintenance.* These are minor preventive maintenance actions that are carried out regularly. Typically, this involves the dragline being down for

one shift (8 hours duration) after being in operational state for 62 shifts. In other words, it corresponds to one such action every 3 weeks.

3. *Corrective Maintenance.* These are actions to restore a failed component into an operational state. This requires repairing or replacing the failed parts of the component. Corrective maintenance is also called nonplanned maintenance since failures occur in an uncertain manner. Some corrective maintenance may occur during planned maintenance. For the purpose of this analysis, only actions carried out outside planned maintenance shifts are included under this heading.

23.3.4 Characterization of Time and Usage

We will use two different clocks to record time. The first is the "usage" clock. This clock stops whenever the dragline is in a nonoperational state. The second clock is the "real-time" clock that indicates the real age of the dragline. We assume that there is no degradation when the dragline is in a nonoperational state. As such, the degradation and the times between failures are modeled based on the usage clock.

The usage clock is reset to zero whenever a component undergoes a shutdown maintenance action because the component is rejuvenated to "good-as-new" after this action is carried out.

We use the term "operational age" to indicate the age (measured using usage clock) since the last shutdown maintenance action. In other words, it represents the time for which the item has been in operational state subsequent to the last shutdown maintenance action.

23.4 FIELD DATA

Data were collected for a dragline at an Australian coal mine. The data for the analysis were extracted from the maintenance database. In this section we discuss the data extracted from these databases for estimating the parameters of component failure distributions and for the mean time for minimal repairs.

The failure data was collected over the period (called the "data period") April 23, 1996 to July 7, 1998. The start of the period corresponded to the dragline undergoing a shutdown maintenance. In the data period there were no further shutdown maintenance actions, so every repair was either corrective or preventive maintenance. The corrective maintenance involved minimal repair (discussed in the next section). The usage time was only registered at the end of each week and not at the instant of a failure. As a result, the failure data that could be extracted from the database were the numbers of failures (and repair actions for each failure of the 25 components) over different usage time intervals.

Two tables were constructed from the data collected. The first table, the repair table, includes the number of failures (and repairs) for each component type in each

Table 23.2. Failure Data for Components 1 to 13

		1		2		3		4		5		6		7		8		9		10		11		12		13	
Clock hrs start	clock hrs end	N	T	N	T	N	T	N	T	N	T	N	T	N	T	N	T	N	T	N	T	N	T	N	T	N	T
0.00	24.00																	2	m								
24.00	110.00																	1	m								
110.00	246.00	1	m													1	m					1	m				
246.00	367.00																					1	m				
367.00	464.00									1	m	1	m							2	m			1	m		
464.00	732.00																					1	m				
732.00	854.00																	2	m	1	m	1	m	1	m		
854.00	986.00													1	m												
986.00	1,119.00																										
1,119.00	1,237.00																										
1,237.00	1,350.00	1	m	1	m									2	m			2	m			2	m			1	m
1,350.00	1,497.00																	1	m								
1,497.00	1,581.00																									1	m
1,581.00	1,710.00																	1	m								
1,710.00	1,840.00																										
1,840.00	1,943.00																										
1,943.00	2,053.00											1	m					1	m	1	m	1	m			2	m
2,053.00	2,190.00																										
2,190.00	2,318.00																									1	m
2,318.00	2,452.00											1	m							1	m					1	m
2,452.00	2,581.00			1	m									1	m			1	m								
2,581.00	2,722.00																									1	m
2,722.00	2,816.00																	2	m	2	m					1	o

usage time interval. Table 23.2 shows a limited portion of this data set for the first 13 components. For each component, there are two columns. The first indicates the number of failures in different time intervals, and the second indicates the type of maintenance action used. As indicated earlier, all failures over the data collection period involved minimal repair (denoted by m in the column). The complete data set and further discussion of the data set can be found in the Wiley site mentioned in the Preface. The second table (which is not included here but can be found in the Wiley site) is the "downtime" table. It is comprised of the downtimes and the components repaired during each of these downtimes. The construction of the tables required the sorting and combining of the maintenance data. The maintenance data were first sorted into the 25 component groups. The data for the repair table were sorted with respect to repair actions and times. The repairs of parts at lower levels than one of the 25 components in the dragline characterization chart were counted as one repair action for the component if they occurred during the same downtime period.

The "downtime" table was constructed by combining and noting all component-type repairs that occurred within the same downtime period and the downtime involved.

The usage of the machine over the data period was 14,176 hours. Also, during this period there were 1295 downtime intervals (resulting from corrective maintenance actions to rectify failures and planned preventive maintenance actions) totaling 2461 hours. This does not include downtime due to resets, calibrations, inspections, and jobs not mentioning specific repairs.

Often a repair action involved repairing two or more of the components. The time recorded was the total downtime rather than the downtime time for each component. This implied that it was not possible to estimate the mean downtime time for each component. Hence, it was assumed that the average downtime times for all components was the same. The average downtime time per component repair was calculated (by taking the ratio of the total downtime and the number of component repairs carried out) from the downtime table and is equal to 1.9 hours.

23.5 COMPONENT LEVEL: MODELING, ESTIMATION, AND ANALYSIS

We first discuss the modeling of component failures. Following this we discuss the estimation of model parameters based on the data discussed in Section 23.4. Finally, we carry out some analysis to model shutdown maintenance and to compute the component availability. These results are used in the next section to model and study performance at the system level.

We use the "black-box" approach to modeling the failure of a component. We need to differentiate between the first failure and subsequent failures. The former depends on the reliability of the component, whereas the latter depend on the reliability as well as the rectification action used.

23.5.1 Modeling First Failure

The time to first failure (X) is uncertain and is modeled by a failure distribution function

$$F(x) = P(X \leq x) \tag{23.1}$$

The time x may be measured in cycles or in operating hours, depending on the application. Obviously, the value of the probability of this failure will vary with time x and the operating environment. Various forms have been proposed for the function $F(x)$ [see Blischke and Murthy (2000)].

One distribution that has been used extensively to model the failures of electrical and mechanical components is a two-parameter Weibull distribution given by

$$F(x) = 1 - e^{-(x/\beta)^{\alpha}} \tag{23.2}$$

The distribution has two parameters, α (shape parameter) and β (scale parameter).

The failure rate function for the Weibull distribution is given by

$$r(x) = \frac{\alpha x^{(\alpha-1)}}{\beta^{\alpha}} \tag{23.3}$$

The failure rate is an increasing function if the shape parameter α is greater than 1, implying that the probability of failure in an interval $[x, x + \delta x)$ increases with x. In other words, the item failure probability increases with the operational age of the item.

[*Note:* For notational simplicity, we will use $r(x)$ instead of the expression on the right-hand side of equation (23.3) in the remainder of the case study.]

23.5.2 Modeling Subsequent Failures

The subsequent failures depend on the nature of the rectification action. In the case of a nonrepairable component, the failed component needs to be replaced by a new one, and in the case of a repairable component there can be several rectification options.

One type of rectification is the "minimal" repair, first formulated and studied by Barlow and Hunter (1961). Here a failed component is made operational by replacing the part (or parts) that caused the component failure. Since a component comprises many parts (ranging from a few to hundreds) and since a component failure is due to failure of one or a few parts, the replacement of failed parts has very little impact on the state of degradation as characterized through the failure rate. This corresponds to the failure rate after a repair being roughly the same as the failure rate just before failure. In other words, after repair of one of its parts, the system does not return to a "good-as-new" condition. The probability of a failure in a short period subsequent to a repair depends on the operational age of the component and this increases with operational age.

The number of failures, $N(t)$, over time under minimal repair policy is given by a nonstationary Poisson process with intensity function given by the failure rate [see Murthy (1991)]. This implies that the probability of experiencing n failures over the first t units of operational time is given by

$$P\{N(t) = n\} = \frac{e^{-R(t)}\{R(t)\}^n}{n!} \tag{23.4}$$

with $R(t)$ given by

$$R(t) = \int_0^t r(x)\,dx \tag{23.5}$$

and $r(x)$ given by equation (23.3). The expected number of failures over this period is given by

$$E[N(t)] = R(t) = \int_0^t r(x)\,dx \tag{23.6}$$

23.5.3 Estimation of Model Parameters

23.5.3.1 Failure Distribution Parameters

For component i, $1 \leq i \leq 25$, failures between shutdown maintenance actions are modeled by a Weibull distribution function $F_i(x)$ with parameters α_i and β_i. This is a reasonable assumption because of the nature of the failures for the components involved.

The estimates for the parameters (α_i and β_i), $1 \leq i \leq 25$, were obtained using the field data and the method of maximum likelihood or the method of least squares. The method of maximum likelihood was attempted first. In some instances it yielded an estimate for the shape parameter less than 1. This is unrealistic because it implies the component reliability improving with age. In these cases, the method of least squares was used. The Appendix gives a brief outline of the two methods. The routines to estimate α_i and β_i ($1 \leq i \leq K$) were written in Matlab. The routines required an initial guess for the parameters to be estimated. For both methods, the initial estimate for parameter α_i was 2.0 and that for parameter β_i was the mean time to failure for component i (obtained from data and given by time period/ total number of failures for component in time period) for $i = 1, \ldots, 25$. The final estimates for α_i and β_i are as shown in Table 23.3.

Proper validation of the model requires data different from that used in parameter estimation to compare model output with data. This was not possible since the sample size for each component was small, and all of the data were required for parameter estimation. Instead, a plot of the expected number of failures (based on the model) versus time for the model was prepared and visually compared with the failure data. For some of the components, the two plots were in reasonable agreement as shown in Figure 23.2 and hence one can accept with some moderate degree of confi-

Table 23.3. Estimates of Failure Distribution Parameters

i	Component	Scale	Shape	Method
1	Hoist generator	1.1076e+003	1.1904	Maximum likelihood
2	Drag generator	1.4852e + 003	1.3232	Maximum likelihood
3	Swing generator	5.3328e + 003	2.1095	Maximum likelihood
4	Synchronous motor	9.2290e + 003	1.5852	Maximum likelihood
5	Hoist motor	2.0674e + 003	1.1365	Maximum likelihood
6	Drag motor	3.8743e + 003	1.2330	Least squares
7	Swing motor	1.2466e + 003	1.1367	Least squares
8	Propel motor	6.2218e + 003	1.3211	Least squares
9	Hoist machinery	565.7741	1.1010	Least squares
10	Drag machinery	769.7037	1.1153	Least squares
11	Swing machinery	565.7506	1.1010	Least squares
12	Propel machinery	1.9862e + 003	1.5889	Maximum likelihood
13	Hoist rope	423.3406	1.0455	Maximum likelihood
14	Drag rope	302.2390	1.3213	Maximum likelihood
15	Dump rope	188.0395	1.0455	Not fitted (used hoist rope shape)
16	Suspension ropes	8.9572e + 003	3.4450	Maximum likelihood
17	Fairlead	2.8608e + 003	1.3660	Maximum likelihood
18	Deflection sheaves	3.0781e + 003	1.6706	Maximum likelihood
19	Bucket and rigging	64.6218	1.0661	Least squares
20	Boom	789.9312	1.1117	Least squares
21	Tub	3.4554e + 003	1.4647	Maximum likelihood
22	Revolving frame	372.5864	1.0938	Least squares
23	A-frame	3.0607e + 003	1.2628	Maximum likelihood
24	Mast	2.7577e + 003	1.1827	Least squares
25	Other	47.5795	1.0625	Least squares

dence that the Weibull model is a reasonable model. For the remaining, the match between the two plots was poor. These could be either due to lack of data and/or the Weibull model not being appropriate. For the remainder of the analysis, it is assumed that Weibull model is an adequate approximation for purposes of the initial analysis.

23.5.3.2 Mean Time to Repair

An estimate of the average downtime for each repair (τ_r) was obtained as discussed in Section 23.4. This value was 1.9 hours.

23.5.4 Model Analysis

23.5.4.1 Shutdown Maintenance

For a component with increasing failure rate, $r(x)$ is an increasing function of x, implying that the failure occurrences increase with operational age. As mentioned earlier, a shutdown maintenance action is used to rejuvenate the component after being in operational state for a period T. The interval between two successive shutdown

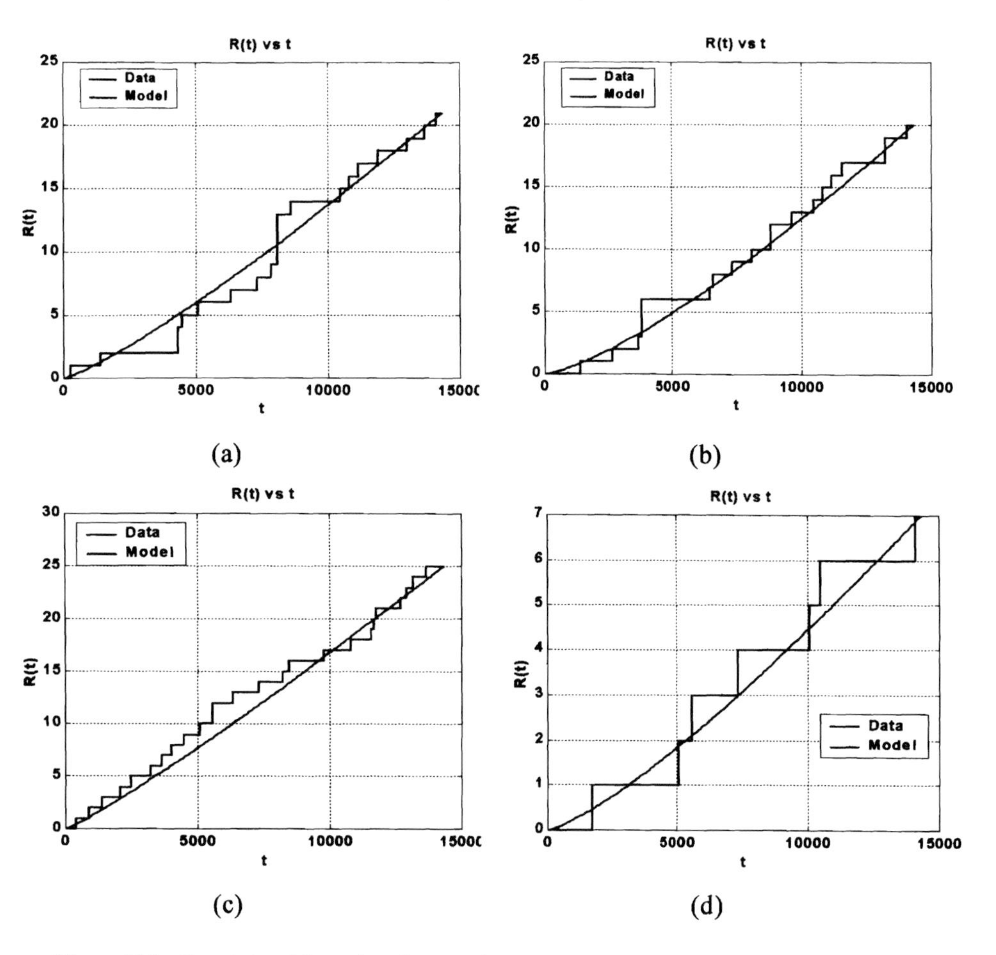

Figure 23.2. Cumulative failure plots data and fitted model. (a) Hoist generator, (b) drag generator, (c) boom, and (d) A-frame.

maintenance actions defines a cycle that repeats itself over time. Figure 23.3 shows a typical cycle where failures over the cycle are repaired minimally (through corrective maintenance) and shutdown maintenance is carried out at the end. As a result, the expected number of failures over a cycle is given by equation (23.6) with $t = T$. Note that in real time, the cycle length is longer, because it involves taking into account the time for minimal repairs over the cycle and for planned maintenance during the cycle as well as the shutdown maintenance at the end of the cycle. This is discussed in a later section.

Since shutdown maintenance action results in the component being back to new, the probability of no component failure between two shutdown maintenance actions is given by

$$S(T) = 1 - F(T) \tag{23.7}$$

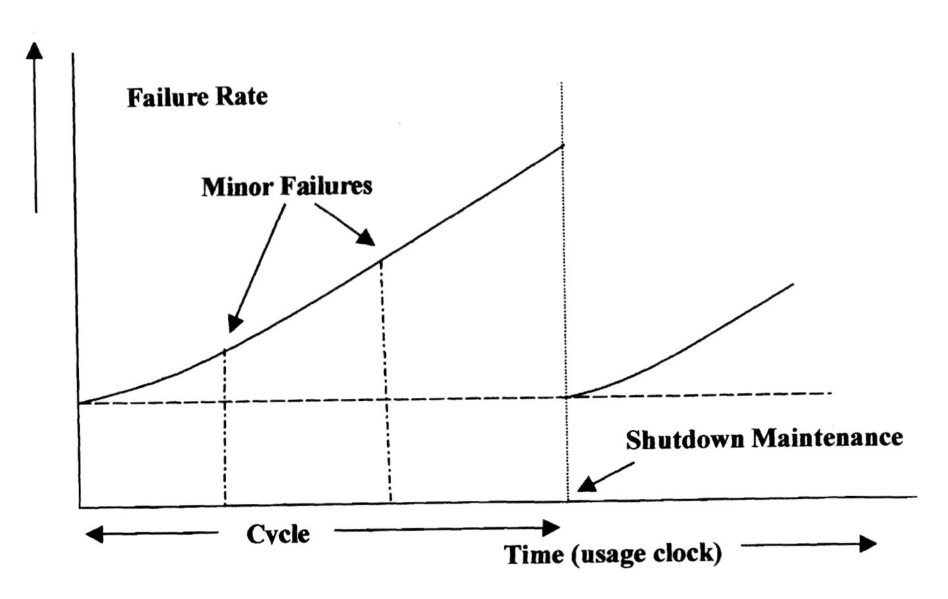

Figure 23.3. Failure rate over a cycle.

$S(T)$ is a measure of the component reliability. Note that $S(T)$ decreases as the interval T between two successive shutdown maintenance actions increases.

One chooses T so as to ensure that the reliability $S(T)$ is greater than some prespecified value δ. If $\delta = 0.95$, it implies that the probability of no failure between two successive shutdown maintenance actions is at least 0.95. This, in a sense, captures the attitude to risk. Higher values of δ correspond to a greater aversion to risk, and T decreases as δ increases.

23.5.4.2 Component Availability

The time to restore a failed component to an operational state (through minimal repair) is in general a random variable. During this time, the component is in a nonoperational state. As a result, the cycle measured using the real time clock is the sum of T (the cycle measured using the usage clock) and the times for various minimal repair actions and the time for the shutdown maintenance action. Let τ_r denote the expected time for each minimal repair and let τ_p denote the shutdown maintenance time (the time needed for carrying out the major preventive maintenance to renew the system to good-as-new) for the component. Let τ_{pm} denote the total time for planned minor preventive maintenance (time spent on planned minor preventive maintenance actions) over a cycle for the component. Then the expected cycle length (measured using real time), ECL, is given by

$$\text{ECL} = T + \tau_r \int_0^T r(x)\,dx + \tau_{pm} + \tau_p \tag{23.8}$$

The duration for which the component is in operational state is T. As a result, the availability of the component, $A(T)$, is given by the following expression:

$$A(T) = \frac{T}{\text{ECL}} = \frac{T}{T + \tau_r \int_0^T r(x)\, dx + \tau_{pm} + \tau_p} \qquad (23.9)$$

This is a measure of the fraction of time the component is in operational state over a long time horizon. Note that this is a function of T (the operational time between shutdowns), the failure rate $r(x)$ of the component, and the expected times for minimal repairs over a cycle and for preventive maintenance action at the end of the cycle during shutdown.

23.5.5 Comments

1. We focus our attention on modeling failures due to degradation with age and/or usage. Failure can also occur due to external causes—accident, operator mistakes, and so on. These can be modeled using a competing risk model [see Jiang and Murthy (1995)], and the formulations would be more complex.
2. We did not examine other distributions for modeling component failures. For some of the components, distributions other than Weibull might have been more appropriate.
3. The data for estimation were, in most cases, very limited, and hence it was not possible to carry out any rigorous statistical testing to check for the adequacy of the assumed Weibull distributions.
4. For the maximum likelihood estimates, one can obtain confidence limits based on asymptotic results. It is difficult to obtain analytical expressions for these limits, and one can compute them numerically using the estimated values. Since the data sets are small, these limits would lack credibility.
5. In some instances the preventive maintenance action can be initiated before the component has been used for a period T. This can occur if a failure occurs with the operational age very close to T or the failure is severe so that it cannot be rectified through minimal repair. These can be modeled but would require more complex mathematical formulations.

23.6 SYSTEM LEVEL MODELING AND ANALYSIS

As indicated in Section 23.2.1, the dragline is modeled as a system involving 25 components. In this section we use the component level analysis (Section 23.5) to carry out the analysis at the system level, noting that the system is a series configuration of the 25 components. We first discuss shutdown maintenance to ensure specified system reliability and the long-term system availability. Following this, we obtain an expression for the output yield.

23.6.1 Shutdown Maintenance

The time between shutdown maintenance actions is usually selected so as to ensure some specified system reliability. Since the system is in operational state only when all the components are in operational state, the failure distribution of the system is given by

$$F(x) = 1 - \prod_{i=1}^{K}(1 - F_i(x)) \tag{23.10}$$

with $K = 25$ (the number of components in the system). This follows because the system can be viewed as a series configuration. As a result, the system reliability with shutdown maintenance being done after a period T in operation is given by

$$S(T) = 1 - F(T) = \prod_{i=1}^{K}(1 - F_i(T)) \tag{23.11}$$

23.6.2 System Availability

From a reliability point of view, the system is in operational state only when all the components are in operational state and is in failed state whenever at least one of the components is in failed state. We assume that component failures are statistically independent.

We assume that the system undergoes shutdown maintenance action after being in operation for a period T. The expected number of minimal failures for the ith component over a cycle is given by equation (23.6) with $r_i(x)$ replacing $r(x)$. As a result, the expected cycle length (measured using the real time clock) is given by

$$\text{ECL} = T + \left[\sum_{i=1}^{K}\left\{\int_0^T r_i(x)\,dx\right\}\tau_r\right] + \tau_{pm} + \tau_p \tag{23.12}$$

The expected cycle length is the expected time between two major preventive maintenance actions that renew the system to back to good-as-new. The availability $A(T)$ is given by

$$A(T) = \frac{T}{T + \left[\sum_{i=1}^{K}\left\{\int_0^T r_i(x)\,dx\right\}\tau_r\right] + \tau_{pm} + \tau_p} \tag{23.13}$$

where τ_{pm} is the total time spent on preventive maintenance actions (for all 25 components) over a cycle and τ_p is the time for shutdown maintenance for the system.

Note that the availability is a function of the operation time between preventive maintenance actions (T) and the failure rate $r_i(x)$, for the K different components.

23.6.3 Output Yield

Let the average payload be V_0' and a duty cycle to move this load be of length D_0. Then over a period of length ECL, the expected number of duty cycles carried out is given by (T/D_0)—remember that T is the duration for which the component is in operational state in one cycle. The yield over a cycle length ECL is given by

$$Y = \frac{T}{D_0} V_0' \tag{23.14}$$

The average expected yield per unit time is given by

$$J(T) = \frac{Y}{\text{ECL}} = \frac{T}{\text{ECL}} \frac{V_0'}{D_0} = \frac{A(T)V_0'}{D_0} \tag{23.15}$$

As can be seen, the yield per unit time depends on the availability $A(T)$, the duty cycle D, and the bucket load V_0'.

23.6.4 Comments

1. One can use equation (23.15) to determine the optimal T. This implies that the shutdown (for major preventive maintenance) is age-based, where age is measured in terms of the usage clock. This is different from the current practice where maintenance is based on calendar or real clock time. Note that equation (23.12) links T (measured using usage clock) with ECL (measured using the real time clock). Let T^* denote the optimal T. This implies that shutdown should be planned at intervals (based on the real time clock) given by ECL obtained from (23.12) using T^*. This would need to be rounded to the nearest month, quarter, or year based on other factors (such as availability of the repair crews, repair facilities, etc.).
2. We have assumed that all the components undergo shutdown maintenance action after being in operation for a period T. We can relax this assumption so that component i undergoes a shutdown maintenance action after being in operation for a period T_i. These would need to be constrained so that $[T/T_i]$ are integers. This leads to different subcycles at the component level within each cycle at the system level.
3. The assumption that the component failures are statistically independent might not always be true. In some instances, failure of a component can induce a failure of one or more of the other components. This can be modeled using a more complex model formulation [see Murthy and Nguyen (1984) for one such model formulation].

4. The model can be modified to take into account inspections that require the dragline being in the nonoperational state. This alters the expression for ECL given by equation (23.12).

23.7 MODELING THE EFFECT OF DRAGLINE LOAD

In this section we discuss the effect of dragline load on the output yield. As indicated earlier, the stress on components increases as the dragline load increases. We first look at the effect of this on component failure distributions. Following this, we look at the effect of dragline load on the maintenance (to ensure the same degree of reliability in operation) and on the output yield. This is used in the next section to determine the optimal dragline load.

Let V_0 denote the base dragline load (i.e., the total combined load due to the bucket, rigging, and the payload for the standard load). If B is the combined weight of the empty bucket and rigging and V_0' is the payload, then

$$V_0' = V_0 - B \tag{23.16}$$

If the dragline load is altered (due to bigger bucket) from this to some other value V, it alters the stresses on the various components of the system. The stresses can be mechanical, electrical, or thermal, depending on the component under consideration. Define $v = V/V_0$. This implies that the stress increases (relative to the base dragline load level) when $v > 1$. We assume that the empty bucket and rigging load is not altered significantly so that it can be ignored; as a result, the new payload is given by

$$V' = V - B \tag{23.17}$$

In this section we first model the effect of v on the reliability of components and later on the impact on the reliability of the system.

Define $v' = V'/V_0'$. This is a measure of the increase in the payload with the increase in the dragline load. It is related to v by the following relationship:

$$m = \frac{v}{v'} = \left[\frac{1-w}{v-w}\right]v \tag{23.18}$$

where $w = B/V_0$ (the ratio of the weight of the empty bucket and rigging to the base dragline load).

23.7.1 Component Failures

Let the life of component i at base stress level corresponding to gross dragline load V_0 be X_{i0}. This is a random variable characterized by a failure distribution $F(t;\ \alpha_i, \beta_i)$, where β_i is the scale parameter. Note that we assume that the shape parameter is not affected by the changes in load. Let X_i denote the life of the component at a higher stress level corresponding to gross dragline load V. In the deterministic accelerated failure model [see Nelson (1990)] it is related to X_{i0} by the relationship

$$\frac{X_i}{X_{i0}} = \frac{1}{\psi_i(v)} \tag{23.19}$$

where $\psi_i(v)$ is a deterministic function with $d\psi_i(v)/dv > 0$ for $v \geq 1$ and $\psi_i(1) = 1$. This implies that X_i is linearly related to X_{i0} and the relationship is deterministic. As a result, the failure distribution for component i at stress level V is given by

$$F_{vi}(x;\, \alpha_i,\, \beta_{vi}) = F_i\left(x;\, \alpha_i,\, \frac{\beta_i}{\psi_i(v)}\right) \tag{23.20}$$

Note that the failure distribution of X_i is the same as that for X_{i0} except that the scale parameter has been scaled down by a factor $\psi_i(v)$ so that it is given by

$$\beta_{vi} = \frac{\beta_i}{\psi_i(v)} \tag{23.21}$$

and the shape parameter α_i is unaltered. Various forms for $\psi_i(v)$ have been proposed [see Nelson (1990)]. We consider the following form:

$$\psi_i(v) = v^{\gamma_i} \tag{23.22}$$

with $\gamma_i > 1$.

Typical values of γ for different components can be found in the literature and are as indicated below:

Component	Typical γ
Generator	7[1]
Motor	7[1]
Gearbox	9.0[2]
Bearings (roller)	3.3[3]
Boom (welds)	3.1[4]
Ropes	5.7[5,6]

[1]Guan, Z., Gurgenci, H., Austin, K., and Fry, R. "Optimisation of Design Load Levels for Dragline Buckets," ACARP #7003, Progress report (2000).

[2]AGMA Standard 2001-B88, 1988, "Fundamental rating factors and calculation methods for involute spur and helical gear teeth." American Gear Manufacturers Association, Virginia, USA.

[3]Shigley, J. E., 1986, *Mechanical Engineering Design,* McGraw-Hill, New York.

[4]British Standard BS7608, 1993, "Code of practice for fatigue design and assessment of steel structures," BSI.

[5]Chaplin, R. C,. and Potts, A. E., 1988, "*Wire Rope in Offshore Applications,*" The Marine Technology Directorate Ltd., London.

[6]Tenkate, C., and Fry, P. R., 1995, *A Summary of Component Damage Through Operation with a Heavy Bucket,* WBM Pty Ltd., Australia.

The failure rate at stress level v is related to the failure rate at the base level as follows:

$$r_{vi}(x) = \{\psi_i(v)\} r_i(x) \tag{23.23}$$

This implies that the failure rate increases as v increases so that the occurrence of failures increases.

23.7.2 Preventive Maintenance Actions

Suppose that under a base dragline load V_0, the system is subjected to shutdown maintenance actions as discussed in Section 23.4 and such actions restore the component back to "good-as-new." Let T_0 denote the operational period between shutdown maintenance actions. All failures between two shutdown maintenance actions are minimally repaired.

Under an increased dragline load V, to ensure the same reliability (probability of no failure) between two preventive maintenance actions as under a load V_0, T_0 needs to be changed to T_v. From equation (23.11), the reliability for an operational period T_0 with base bucket load V_0 is given by

$$S(T_0) = \prod_{i=1}^{K} \{1 - F_i(T_0; \alpha_i, \beta_i)\} \tag{23.24}$$

Under a dragline load V, the system reliability for an operational period T_v is given by

$$S(T_v) = \prod_{i=1}^{K} \left\{1 - F_i\left(T_v; \alpha_i, \frac{\beta_i}{\psi_i(v)}\right)\right\} \tag{23.25}$$

To ensure the same reliability,

$$\prod_{i=1}^{K} \left\{1 - F_i\left(T_v; \alpha_i, \frac{\beta_i}{\psi_i(v)}\right)\right\} = \prod_{i=1}^{K} \{1 - F_i(T_0; \alpha_i, \beta_i)\} \tag{23.26}$$

and from this one obtains T_v. Note that this is a function of T_0.

If $v > 1$, the system degradation is accelerated and hence the system reliability decreases more rapidly. This implies that shutdown maintenance actions need to be carried out more frequently (see Figure 23.4) to ensure that the system reliability is always greater than some specified value. We shall assume that the shutdown maintenance is modified according to the above relationship in the remainder of the analysis.

Since the F_i's are Weibull distributions [of the form given by equation (23.2)], (23.26) yields the following equation:

$$\sum_{i=1}^{K} \psi_i(v) \left\{\frac{T_v}{\beta_i}\right\}^{\alpha_i} = \sum_{i=1}^{K} \left\{\frac{T_0}{\beta_i}\right\}^{\alpha_i} \tag{23.27}$$

This is a nonlinear equation and needs to be solved numerically to obtain T_v.

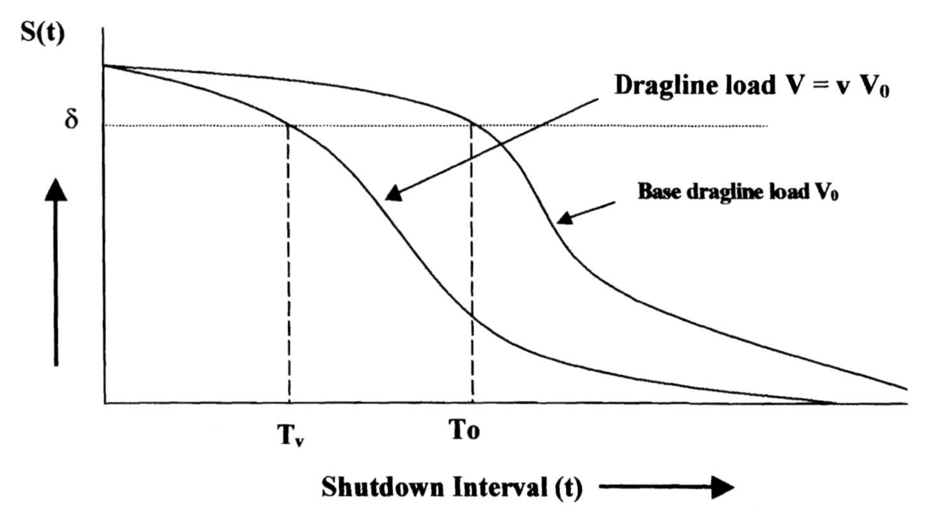

Figure 23.4. Preventive maintenance interval.

23.7.3 System Availability

The expected number of minimal repairs for component i over a cycle (the interval between two preventive maintenance actions) under dragline load V is given by

$$E[N_i(T_v)] = \int_0^{T_v} r_{vi}(x)\,dx \tag{23.28}$$

The expected cycle length is given by

$$\mathrm{ECL}(v) = T_v + \sum_{i=1}^{K}\left\{\int_0^{T_v} r_{vi}(x)\,dx\right\}\tau_{ri} + \tau_{pm} + \tau_p \tag{23.29}$$

as the expected times for minimal repairs and for shutdown maintenance are not affected by the change in the load.

The expected number of pm actions over a cycle is given by

$$N_{\mathrm{pm}} = \frac{\left[T_v + \sum_{i=1}^{K}\left\{\int_0^{T_v} r_{vi}(x)\,dx\right\}\tau_r\right]}{T_{pm} - t_{pm}} \tag{23.30}$$

where T_{pm} is the time between the starts of two successive pm actions and t_{pm} is the time for each pm action. The total time spent on pm actions over a cycle, τ_{pm}, is obtained by multiplying this by t_{pm}.

The availability is given by

$$A(T, v) = \frac{T_v}{\text{ECL}(v)} \tag{23.31}$$

with ECL(v) given by equation (23.29).

The effect of v is easily seen. As v increases, both the numerator and denominator of (23.31) decrease. The availability decreases at a faster rate if $r_i(x)$ $(1 \leq i \leq K)$ are increasing functions of x. As a result, the availability of the system decreases as v increases.

23.7.4 Output Yield

The dragline duty cycle consists of several stages. The bucket is loaded with dirt, hoisted, and swung, the dirt is dumped, and the bucket is then returned and spotted. Let $D(v)$ denote duty cycle under a gross relative bucket load v. This decreases as v increases because it takes a longer time to load the bucket and for the swing motion. One needs to determine this empirically. Let ϕ be the portion of the dragline cycle in which the bucket is empty. We can thus denote the duty cycle equation as

$$D(v) = [\phi + (1 - \phi)v]D_0 \tag{23.32}$$

This can be rewritten as

$$D(v) = [1 + a(v - 1)]D_0 \tag{23.33}$$

where $a = (1 - \phi)$; a needs to be determined empirically.

Following the approach of Section 23.4, the expected yield per unit time is given by

$$J(T, v) = \frac{T_v}{\text{ECL}(v)} \frac{V'}{D(v)} = \frac{A(T, v)v'}{D(v)} V_0' = \frac{A(t, v)}{D(v)} \frac{v}{m} V_0' = \frac{A(T, v)}{D(v)} \left[\frac{v - w}{1 - w} \right] V_0' \tag{23.34}$$

where $w = B/V_0$, $A(T, v)$ is given by equation (23.31) and $D(v)$ is given by equation (23.33).

As can be seen, the yield per unit time depends on availability $A(T, v)$, duty cycle $D(v)$, and scaled bucket load v'.

23.8 OPTIMAL DRAGLINE LOAD

The computer package Matlab was used to calculate T_v, ECL(v), $A(T, v)$, $D(v)$, and $J(T, v)$ for different values of v (ranging from 1.0 to 1.8).

23.8.1 Nominal Parameters

The base interval between shutdown maintenance, T_0, was taken as 5 years (43,800 hours) based on discussions with the maintenance managers. The length of the preventive maintenance action, τ_p, was taken as 5 weeks (840 hours). This is shutdown maintenance, and its length is assumed to be independent of the bucket weight. The other part of the preventive maintenance is the planned maintenance shifts through these 5 years. At base load operation there is a planned maintenance action every 3 weeks (504 hours) implying $T_{pm} = 504$ hrs. Each planned maintenance action varies from between 8 hours and 15 hours, and a figure of $t_{pm} = 12$ hours was chosen for the nominal value. The acceleration parameters for the 25 components are as shown in Table 23.4. These values were obtained by searching the literature on accelerated failures under increased stress. The bucket and rigging weight, B, was taken as 66 tonnes. The base dragline load (for the normal bucket) V_0 was taken as 140 tonnes, and the dragline load with a larger bucket was taken as 180 tonnes. This gives $\nu = 1.28$. The duty cycle parameter was determined as follows. A swing angle of 100

Table 23.4. Acceleration Factors γ_i for Each Component for the Nominal Case

i	Component	γ_i
1	Generator (hoist)	7.0
2	Generator (drag)	1.0
3	Generator (swing)	1.0
4	Generator (synchronous)	1.0
5	Motor (hoist)	7.0
6	Motor (drag)	1.0
7	Motor (swing)	1.0
8	Motor (propel)	1.0
9	Machinery (hoist)	9.0
10	Machinery (drag)	1.0
11	Machinery (swing)	1.0
12	Machinery (propel)	1.0
13	Ropes (hoist)	5.7
14	Ropes (drag)	1.0
15	Ropes (dump)	5.7
16	Ropes (suspension)	5.7
17	Ropes (Fairleads)	1.0
18	Ropes (deflection sheaves)	1.0
19	Bucket and rigging	1.0
20	Frame (boom)	3.1
21	Frame (tub)	3.1
22	Frame (revolving)	3.1
23	Frame (A-frame)	3.1
24	Frame (mast)	3.1
25	Others	1.0

degrees was selected. For each of the two loads the average cycle time was determined. This yielded the values $D_0 = D(1) = 60$ seconds and $D(1.28) = 70$ seconds. Using these in equation (23.35) yielded a = 0.6. In other words, the bucket is empty for roughly 40% of the duty cycle.

23.8.2 Optimal Bucket Load and Implications

T_v, ECL(v), $A(v)$, $D(v)$, and $J(T, v)$ were calculated and plotted as functions of the gross stress ratio v using the parameter values given in Table 23.2. These are shown in Figure 23.5. As can be seen from the plot of yield versus v, the maximum yield occurs at $v \approx 1.3$. This is equivalent to a dragline load of 182 tonnes or a payload of 116 (= 182 – 66) tonnes.

Note that the availability decreases as v increases, as to be expected. Initially (for $0 < v < 1.4$) the decline is roughly linear and later on it is nonlinear, indicating the effects of higher stress. The same is true of ECL and T_v.

The analysis does not take into account the cost or manpower requirements because the focus has been on yield. Changing the objective function from yield per unit time to some economic measure such as life cycle profit will affect the results. This would require various cost-related information such as sale price, spare part costs, labor costs, and so on. Some of these are highly variable (for example, sale price) and influenced by other factors such as currency fluctuations, general economy, competitor's actions, and so on.

23.8.3 Impact on Shutdown Maintenance

The corresponding interval between shutdown maintenance for the optimal load is $T_v \approx 25{,}000$ hours (approximately 2.85 years). This is approximately 60% of the current base shutdown interval of 5 years (43,800 hours). This highlights the fact that higher stresses on the components cause serious degradation of the components and to ensure the same reliability requires that shutdown actions be done more frequently.

23.8.4 Sensitivity Studies

Sensitivity studies allow one to assess the robustness of the optimal solutions to variations in the model parameters (shape and scale parameters of the component failure distributions and the mean downtime). We varied one parameter at a time while keeping the others at their nominal values. The parameters were varied by ±10% of the estimated values. The results indicated that the optimal dragline load was fairly robust to variations in the parameter values.

23.9 CONCLUSIONS AND RECOMMENDATIONS

As indicated in the Introduction, the aim of this investigation was to study the effect of dragline load on dragline availability, maintenance, and output and to determine

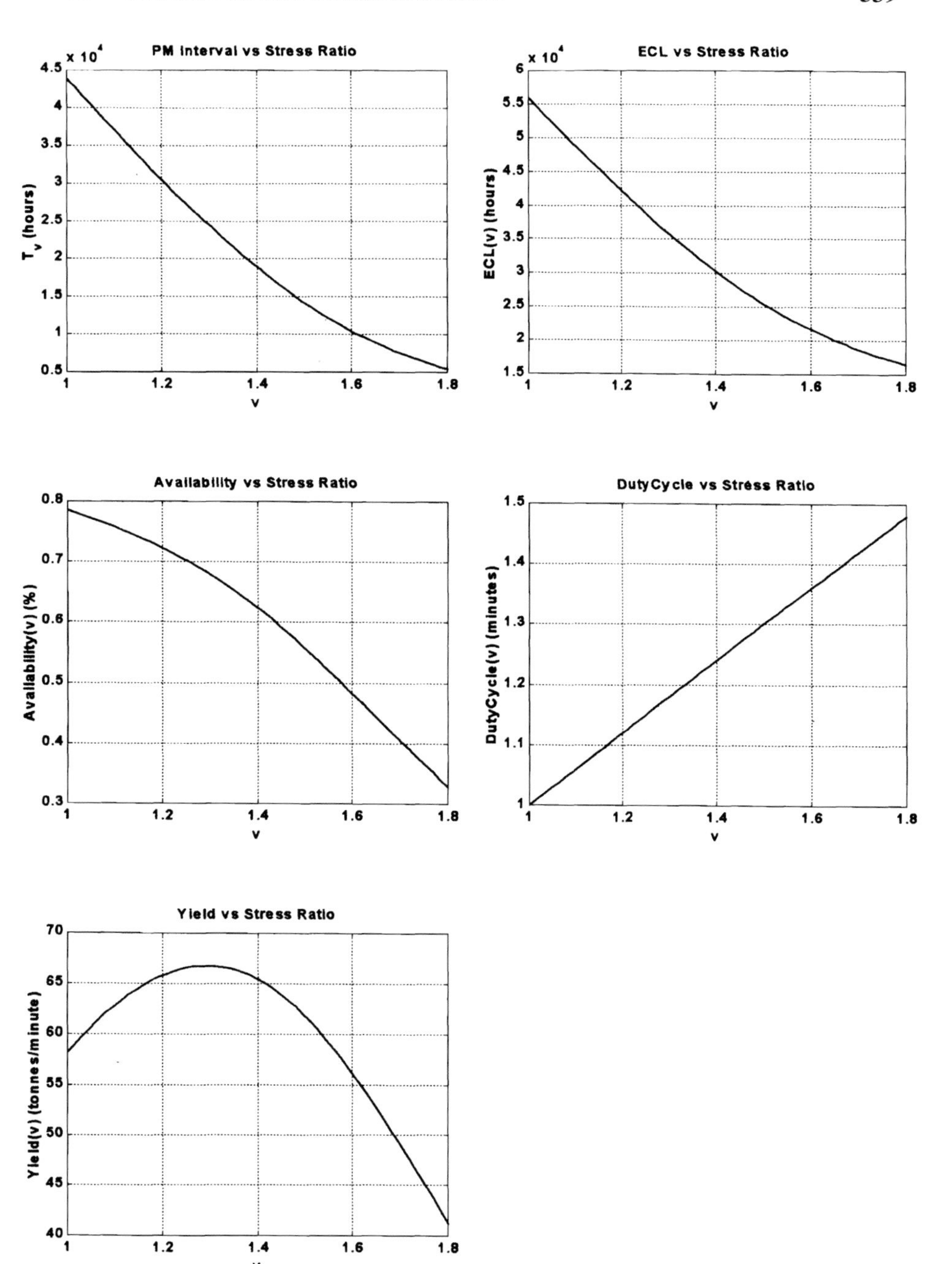

Figure 23.5. Graphs of T_v, ECL(v), $A(v)$, $D(v)$, and $J(T, v)$ for the nominal case.

the optimal dragline load to maximize the yield per unit time. This was achieved by building a mathematical model using field data obtained from different maintenance databases.

The results indicate that the maximum yield is obtained with $\nu \approx 1.3$. This corresponds to a total suspended load of 182 tonnes or a payload of 116 tonnes as opposed to the current payload of 74 tonnes.

As the analysis indicates, this increase in output is obtained at the expense of greater maintenance effort—that is, the dragline is subjected to shutdown maintenance actions at shorter intervals and more frequent inspection and minor pm action in between. The analysis indicates that to obtain the same reliability at the increased bucket size as with the standard bucket, the usage hours between shutdown maintenance actions must be reduced to 2.85 usage years (25,000 hours). This is equivalent to 4.1 calendar years (36,000 hours).

Some of the limitations of the study are as follows:

1. The validation of the Weibull model for some of the components has not been done properly. This was mainly due to lack of data. Data from other draglines can be used to obtain better estimates for the parameters and to validate the models.
2. The failure data were grouped data of failures over usage times. This implies a loss of information as ideally one should record the age at failure. However, this might not be practical.
3. The modeling of the minimal repairs was very simplistic. A more comprehensive model would include the mean repair time for each of the 25 components.

In spite of these limitations, the study reveals that the output yield can be increased by increasing the dragline load. The study also highlights that the model is reasonably robust to variations in the parameter values. This implies that one can accept the results of the study with a modest degree of confidence.

ACKNOWLEDGMENTS

The authors acknowledge the support of the Australian Coal Association Research Program (ACARP) and the Centre for Mining Technology and Equipment (CMTE) for this work.

REFERENCES

Barlow, R. E., and Hunter, J. (1961). Optimal preventive maintenance policies, *Operations Research* **8:**90–100.

Blischke, W. R., and Murthy, D. N. P. (2000). *Reliability,* Wiley, New York.

Jiang R., and Murthy, D. N. P. (1995). Reliability modeling involving two Weibull distributions, *Reliability Engineering and System Safety* **47:**187–198.

Moss, T. R., and Strutt, J. E. (1993). Data sources for reliability design and analysis, *Journal of Process Mechanical Engineering* **207**:13–19.

Murthy, D. N. P. (1991). A note on minimal repair, *IEEE Transactions on Reliability* **40**:245–246.

Murthy, D. N. P., and Nguyen, D. G. (1984). Study of multi-component systems with failure interaction, *European Journal of Operational Research* **21**:330–338.

Nelson, W. (1990). *Accelerated Testing,* Wiley, New York.

Rigdon, S. E., and Basu, A. P (2000). *Statistical Methods for the Reliability of Repairable Systems,* Wiley, New York.

EXERCISES

23.1. The model has 25 components. It can be reduced to six components by grouping all similar components so that the four generators are grouped together and similarly for the motors, machinery, and ropes. This implies more data for each component. Carry out the analysis based on this simplified model.

23.2. We assumed that the failure of each component is given by a Weibull distribution and that all failures between shutdowns are repaired minimally. As a result, the failures occur according to a point process with Weibull intensity function. For a given component there might be other distributions (such as lognormal, gamma, etc.) that model the failures better. Select a component and determine if there is some other distribution that is better than Weibull for modeling the failures.

23.3. Often there is more than one mode of failure for a component. For example, in the case of electric motors the failure could be due to brushes, windings, or bearings. In this case a competing risk model involving two or more distributions is more appropriate. Consider a twofold Weibull competing risk model given by $F(t) = 1 - (1 - F_a(t))\,(1 - F_b(t))$, where $F_a(t)$ and $F_b(t)$ are two basic Weibull distributions. Estimate the model parameters for this model for the electric motors and generators.

23.4. An alternate approach to modeling component failures between shutdowns is through a point process with intensity function $\lambda(t)$, which is a polynomial of the form $\lambda(t) = \Sigma_{j=0}^{n} a_j t^j$. Estimate the parameters $[a_j,\ 1 \le j \le n]$ using the method of maximum likelihood for $n = 1$, 2, and 3. Compare the model fits with that for the Weibull.

23.5. We have assumed that a shutdown for overhaul restores the system back to new. This might not be a valid assumption. One way of overcoming this model deficiency is as follows. After the jth shutdown the system is subjected to a major overhaul so that the system is rejuvenated but not back to new. The failure distribution of a component is given by a Weibull distribution with the same shape parameter, but the scale parameter is modified so that it is given by φ^k times the scale parameter of a new component with $\varphi < 1$. The φ can be different for different components. Carry out the analysis of this model.

23.6. In our analysis, we assumed that the change in the bucket size affected only the components (generator, motor, machinery, and ropes) associated with the hoist operation in the duty cycle. Carry out the analysis assuming that the loading affects components associated with other operations (drag, swing) of the duty cycle.

23.7. The model only deals with the volume produced with constant demand. In real life the demand fluctuates and a simple model is to treat the demand as being either average or high. In this case, depending on the demand pattern, the bucket load changes. How would the modeling and analysis need to be modified to take this into account.

23.8. In our analysis, we only needed the mean time to repair, and this was obtained by computing the sample mean from the field data. Carry out an analysis of the repair time data to determine which distribution models it best. This allows one to determine the probability that the system is down for greater than some specified value.

23.9. The data available for estimation are grouped data—that is, failures over different time intervals. If the data available were the actual failure times (based on usage clock), how would the method of maximum likelihood change for estimating the parameters?

23.10. The parameter estimates used in the case study are point estimates. If the data set is large, one can derive confidence limits based on the Cramér–Rao lower bound for the maximum likelihood estimator. For details, see Blischke and Murthy (2000). Derive the confidence limits for the parameters of the model using this approach.

23.11. For data sets, one can get confidence limits based on simulation using point estimates in the simulation model. Carry out such a study to obtain the confidence limits for the parameters.

23.12. For the non-Weibull models of Exercise 23.2, derive expressions to obtain the parameter estimates.

23.13. Estimate the parameters for the simplified model indicated in Exercise 23.1. (Note that the data need to be pooled.)

23.14. In the case study, the validation of the model was done based on visual inspection of the fit between the model and data as indicated in Figure 23.3 for some of the components. Many different statistical methods for model validation are available in the literature [see Blischke and Murthy (2000)]. Discuss how these could be used in validating the model at the component level.

23.15. Sensitivity studies are an important element of modeling. If the analysis of the model is very sensitive to changes in some of the model parameters, then these parameters need to be estimated precisely. If the data set is small, this is usually not possible. In this case, one needs to be wary of the model analysis and implications. Carry out the model analysis by varying the parameter values by ± 10% of the estimated values and study the impact on the optimal bucket load. If the variations in the optimal bucket load are of the same magnitude, then it indicates that the model is fairly robust.

23.16. We have assumed that preventive maintenance actions are carried out at set times (based on either the real or usage clock). In real life, often such actions can be carried out when an opportunity arises and is close to the planned preventive maintenance time. How would you model this? How does it affect the analysis?

23.17. We have assumed that all failures between shutdowns are repaired minimally. This might not be the case. Suppose that a failed component can be repaired minimally with probability p (close to 1) or replaced by a new one with probability $(1 - p)$. How does this affect the model analysis?

APPENDIX

We use the following notation:

K	Number of components
t_{ij}	Time (based on usage clock) of the jth failure of component i, $1 \leq i \leq K$
M_i	Number of times component i's failures are recorded over the data interval
N_{ij}	Number of component i's failures over the interval $[t_{i(j-1)}, t_{ij})$, $1 \leq j \leq M_i$ and $1 \leq i \leq K$ [Note: $t_{i0} = 0$]
TN_{ij}	Total number of component i's failures over the interval $[0, t_{ij})$ $[= \sum_{m=1}^{j} N_{im}]$
α_i	Shape parameter of Weibull distribution for component i, $1 \leq i \leq K$
β_i	Scale parameter of Weibull distribution for component i, $1 \leq i \leq K$
$r_i(x; \alpha_i, \beta_i)$	Failure rate for component i, $1 \leq i \leq K$
x	Operational age of component [Age measured using usage clock]

Method of Maximum Likelihood

For component i ($1 \leq i \leq K$) the probability of n_{ij} failures occurring over the jth PM interval $[t_{i(j-1)}, t_{ij})$ is given by

$$p_{ij}(\alpha_i, \beta_i) = P\{N_{ij} = n_{ij}\} = \frac{\left\{\int_{t_{i(j-1)}}^{t_{ij}} r_i(x;\, \alpha_i,\, \beta_i)\, dx\right\}^{n_{ij}} e^{-\int_{t_{i(j-1)}}^{t_{ij}}(x;\, \alpha_i, \beta_i)\, dx}}{n_{ij}}$$

The likelihood function is given by

$$L_i(\alpha_i, \beta_i) = \prod_{j=1}^{M_i} p_{ij}(\alpha_i, \beta_i)$$

The maximum likelihood estimators $\hat{\alpha}_i$ and $\hat{\beta}_i$ are the values of α and β which maximize $L_i(\alpha_i, \beta_i)$. This requires a computational method.

Method of Least Squares

For component i ($1 \leq i \leq K$) the estimates are obtained by minimizing

$$\Psi_i(\alpha_i, \beta_i) = \left\{\prod_{j=1}^{M_i} [TN_{ij} - \int_0^{t_{ij}} r_i(x;\, \alpha_i,\, \beta_i)\right\}^2$$

Again, a computational method is needed to obtain the estimates.

CHAPTER 24

Ford's Reliability Improvement Process—A Case Study on Automotive Wheel Bearings

Karl D. Majeske, Mark D. Riches, and Hari P. Annadi

24.1 INTRODUCTION

Engineers at Ford Motor Company continually evaluate current passenger vehicles (automobiles, light trucks, and sport utility vehicles) for design changes intended to increase quality and reliability. Historically, these changes were targeted toward preventing failures during the basic warranty period of 3 years and 36,000 miles or 5 years and 50,000 miles, with the intent of reducing warranty costs. However, in the more competitive global automobile market, vehicles must function properly for longer than the basic warranty coverage for manufacturers to retain customers. Ford now defines the useful life or design lifetime of their passenger vehicles—the point at which 95% of vehicles should still function—as 10 years or 150,000 miles.

Ford has developed a standard process to add structure and consistency to product changes intended to improve product reliability. This case study presents the reliability improvement process used by Ford that engineers have successfully implemented to improve product field performance in the time domain. This methodology is presented in general and we illustrate its application in the context of a wheel bearing used in some Ford automobiles. Figure 24.1 shows the automotive wheel bearing analyzed in this case study. The bearing assembly contains a collection of lubricated steel balls that roll in the outer bearing housing and allow the wheel to rotate. The hub flange provides a large rigid surface used to attach the wheel bearing to the wheel. The bearing cap is made of plastic and covers the portion of the anti-lock brake system (ABS) internal to the wheel bearing. The bearing cap design utilizes a friction seal to prevent contamination ingress between the cap and the outer bearing housing. The wheel bearing contains an ABS sensor that de-

Case Studies in Reliability and Maintenance, Edited by W. R. Blischke and D. N. P. Murthy.

ISBN 0-471-41373-9

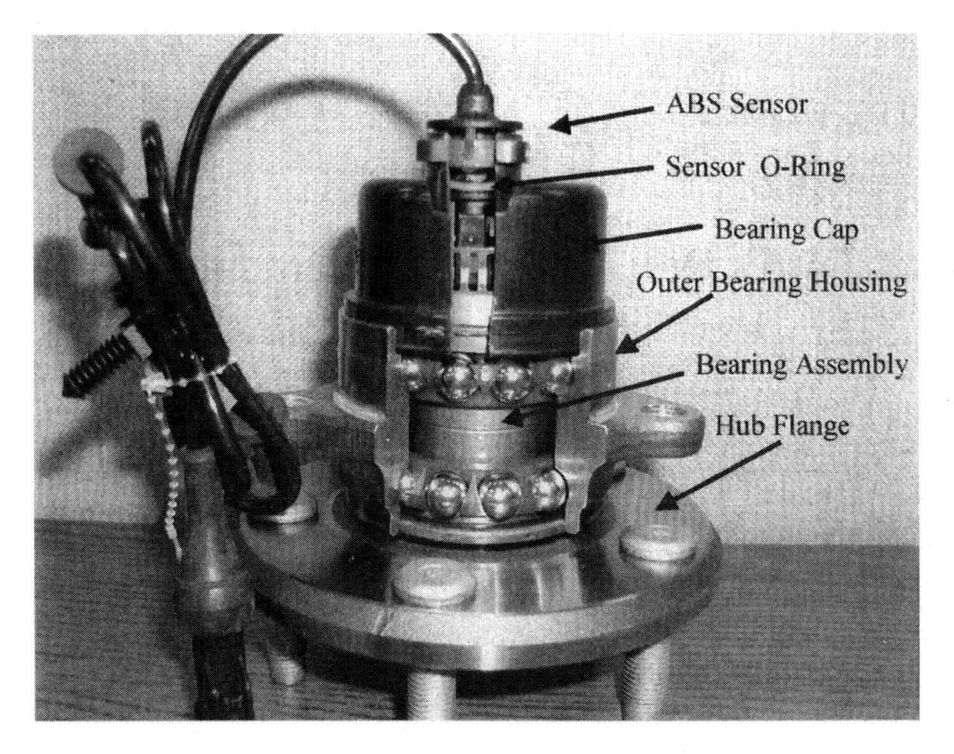

Figure 24.1. Automotive wheel bearing.

tects wheel rotation and transmits the rate of rotation to the ABS control unit. When a wheel is not rotating—the brakes are locked—the control unit will release the brakes for a very short time interval to allow the wheel to roll again that provides the driver the ability to steer the vehicle. The wheel bearing uses an O-ring to seal the ABS sensor opening and prevent contamination from entering the bearing. These wheel bearings were experiencing a condition that resulted in what customers perceived as excessive noise. This condition affected about 3% of the bearings and was purely a customer inconvenience, that is, passenger safety was never at risk. Nevertheless, the bearing was covered by warranty and they were being replaced at company expense. Therefore, warranty and reliability engineers initiated Ford's reliability improvement process to address this noisy failure mode.

The remainder of this case study has the following outline. Section 24.2 describes the nine steps of the Ford reliability improvement process. Section 24.3 develops the quantitative methods used in the methodology and provides citations for literature containing more details of these specific techniques. Section 24.4 contains an application of the methodology to an automotive wheel bearing. The application follows the outline of the Ford process and includes supporting data and analysis from the wheel bearing redesign. Section 24.5 contains conclusions and recommendations on how to effectively use the methodology.

24.2 APPROACH USED

The Ford reliability improvement process consists of nine phases, which are numbered phase 0 to phase 8 and discussed in Sections 24.2.1–24.2.9 below. This methodology has the breadth to cover most types of problem resolution, yet is flexible enough to apply to the wide variety of design issues encountered in automobile manufacturing. Each problem is unique and, to a certain extent, requires a specific solution process. The Ford methodology is not intended to be overly rigid and allows the engineer to customize the process to meet their needs. For some steps, the methodology suggests the improvement team select from multiple analytic techniques that further lends to the flexibility.

24.2.1 Identify Priority Concerns

In the first step of this phase, Ford personnel identify design issues for potential resolution. This step is somewhat open-ended, so engineers will not miss potential issues because early identification doesn't fit in the methodology. The Ford process doesn't provide quantitative criteria for problem identification; rather, it encourages justifying the issue quantitatively. In general, the identification of issues should be data driven, without any strict guidelines on what constitutes an issue. Engineers can identify a design issue using warranty data, Ford customer survey data, or third-party survey data such as *The Initial Quality Survey.*[1]

Warranty claims provide a readily available source of data on all vehicles sold to use in the problem identification phase. Warranty and reliability engineers continually receive detailed data on any vehicle repaired under warranty. Unfortunately, vehicles leave warranty coverage before they exceed their useful life, which limits the ability to identify high mileage failures. Warranty claims provide a source for identifying design issues that limit product functionality.

Another source of data is obtained through customer surveys, which Ford sends to each customer purchasing or leasing a new vehicle. Customers have the opportunity to rate items at the system level (e.g., exterior appearance, power train, seats, etc.) on a five-level satisfaction scale. The survey also contains space for open-ended or specific comments. Ford also holds inspection clinics, conducted at various locations, which allow customers to show their vehicles to Ford engineers and discuss concerns one-on-one. The data from the surveys and inspection clinics allow engineering and marketing functions to identify customer annoyance issues that may not appear in warranty claim data. External surveys provide data similar to those of the Ford customer survey, but provides comparisons to vehicles produced from a variety of manufacturers worldwide. These data allow Ford personnel to judge their product's performance relative to their competitors and can lend a different perspective to certain issues.

The second step of phase 0 is to identify a team champion. Warranty and relia-

[1] *The Initial Quality Survey*, an across manufacturer comparative study of new vehicle quality, is published annually by J. D. Power and Associates, Agoura Hills, CA.

bility engineers forward their findings to the product line engineering director (or equivalent) for review. If the director decides to initiate the redesign process, they identify a team champion who assumes responsibility for completing the project.

Phase 0 completion activities: problem identified, upper management approval for redesign project, team champion named.

24.2.2 Set-Up Team

The team champion establishes a team with the skills required to deliver a solution to the concern identified in phase 0. This team has a leader (different from the team champion) and consists of a cross-functional group (product design engineer, reliability engineer, manufacturing or process engineer, product cost analyst, etc.) with specific product knowledge. The team champion obtains local management commitment for the each member. Before being a member of the team, an employee's immediate supervisor must agree to provide the employee time and resources to work on the design issue.

Phase 1 completion activities: identify team members, obtain management commitment for each team member.

24.2.3 Detail Failure Modes and Identify Uncontrollable or Noise Variables

This phase, consisting of three steps, represents the detailed analysis of the problem or design issue. At the conclusion of this phase, the team will develop a collection of documents. The documents the team may choose to include are: block diagram, function tree, reliability and robustness checklist, and parameter diagram.

Step 1: Perform a comprehensive analysis of warranty or survey data. Failures covered under warranty have a direct impact on company profits, therefore warranty claims remain the primary source for data. The Ford process suggests using either Pareto charts or hazard plots of warranty data as supporting evidence of a problem. The team should compare actual field performance with design intent or compare projected warranty costs with budgeted values.

Step 2: Review actual field failures or customer identified annoyance parts. Any part replaced under warranty becomes the property of Ford Motor Company. To receive payment for the repair, the service provider must return the item to the design center, making these parts readily available. The purpose of these reviews is to identify failure modes of which Ford defines two types: hard failure (product stops functioning) and soft failure (product function degraded to unacceptable level). Ultimately, the team will propose design changes to eliminate the failures identified in this step. The team also creates a list of noise factors (uncontrollable variables) and control factors (controllable variables) for the product. The team then identifies the range for these variables in the field. For example, if the noise factor is outside temperature, the team needs to identify the coldest and warmed temperatures the product will experience.

Step 3: Determine the best-in-class (BIC) product. The team conducts an analysis of Ford and competitor products to determine BIC. The team compares the current Ford product to the BIC product, focusing on basic design and field performance. The team then initiates a purchase order for enough of the BIC product to do performance testing. The team will use the BIC product as a benchmark for any proposed product changes.

Phase 2 completion activities: problem statement, identify failure mode(s), list of noise factors with range of exposure, list of control factors with range of exposure, communicate failure modes to appropriate design process.

24.2.4 Analyze Failure Mechanisms

The purpose of this phase is to determine root causes of field failures by drawing on past experience and judgment. In addition, the team utilizes the 8D problem solving process (described in Section 24.3.3.1) to lend structure to their activities as they uncover various potential causes of the failure modes. The team must also demonstrate the ability to reproduce failure modes in the bench test environment. This step is critical to verifying that new designs have eliminated the failure mode. Some recommended techniques are brainstorming sessions and formal design change reviews. Once failure mechanisms have been fully explored, the team updates the failure mode and effects analysis (FMEA) documentation.

Once failure mechanisms have been identified and understood, the team estimates the associated cost of quality. While these costs can be both direct (warranty repair costs) and indirect (lost customer due to product dissatisfaction), this analysis should focus on direct quantifiable costs that could be saved through product changes. This cost information becomes an input to any future cost–benefit analysis associated with proposed product design changes. Lastly, the team investigates the BIC product for possible solutions to the failure mechanisms.

Phase 3 completion activities: failure mechanisms understood and documented, FMEA update initiated, cost of failure mechanisms quantified, BIC product evaluated for potential solutions.

24.2.5 Establish System Design Specification (SDS) and Worldwide Corporate Requirements (WCR) Compliance

The team reviews and updates design documentation to prevent future designs from containing the failure mechanism. The team reviews the SDS and WCR design guides to compare the current design with the standards and notes any differences. The team also assesses the standards to determine whether, if completely followed, they would prevent the failure mechanisms identified. Lastly, the team develops a plan for how the product should comply with the design standards.

Phase 4 completion activities: comparison of current design to design standards, plan to achieve design compliance.

24.2.6 Develop Test and Establish Performance of Current Design

Ultimately, Ford engineers design for a useful life of 10 years and 150,000 miles. However, a bench test will accelerate this usage period to a matter of a few days or weeks. The team needs to translate customer usage (time and mileage) into the metric(s) of the bench test. The team also needs to have a formal definition of hard and soft failures. Ideally, these should be quantitative or objective criteria that can be uniformly applied to many parts by many people. Finally, the team must verify that the bench test can reproduce the failure mode. After establishing a valid test, the team tests the current (failing) design to establish a baseline performance level. These data will provide a benchmark for testing of newly designed or redesigned parts. The team must verify that all the noise variables are represented in the bench test and that the bench test meets specified performance criteria.

Phase 5 completion activities: bench test that reproduces failure modes, detailed documentation of bench test, bench test performance of current design.

24.2.7 Develop Noise Factor Management Strategy

The team develops a strategy, or approach for isolating the product from the adverse affects of uncontrollable (noise) variables. This results in a product design robust to environmental variables that customers cannot control or predict (e.g., temperature, wind, or precipitation). The team documents the strategy using the noise factor management matrix.

Phase 6 completion activities: strategy for design robust to noise variables.

24.2.8 Develop New Design and Verify Using Reliability Disciplines

In this phase, the team produces the new design for the product. This design should be robust to operating conditions, no longer exhibit the failure mode currently encountered in field performance, and function for the useful life of the product. The team conducts bench tests on prototypes of the new design. These tests must show, with statistical significance, that the product will meet the usage criteria while no longer experiencing the failure mode.

Phase 7 completion activities: new product design, FMEA complete.

24.2.9 Sign-off Design and Update Lessons Learned

This phase represents the conclusion of Ford's reliability improvement process. The team completes final updates of all necessary documentation (FMEAs, system design specs, etc.) and releases the new design into production (following the usual appropriate steps which are not detailed here). The team then updates the lessons learned, which facilitates continual improvement through structured corporate memory. Lastly, upper management formally recognizes the team for completing a successful project.

24.3 QUANTITATIVE ANALYSIS METHODS

The applications of statistical methods in reliability divide into two categories: repairable and nonrepairable systems. The appropriate method depends on the characteristics of the product, service, or component under study. Nonrepairable systems reliability describes the situation when a single failure, or observation in the data, represents the end of a product or component lifetime. Repairable systems techniques apply when a product can have multiple failures or observations in the data; this implies that after repair, products return to some useful state. Thompson (1981) provides a thorough treatment of the use of distribution functions for modeling time to failure of nonrepairable systems, and he discusses stochastic process models that apply to repairable systems. The Ford methodology applies to automotive components and assemblies and treats them as nonrepairable systems.

24.3.1 Estimating Product Lifetime with Field Data

This case study uses maximum likelihood estimation to fit random variable models to field failure and bench test data. Lawless (1982) and Meeker and Escobar (1998) provide functional forms for random variables often applied to lifetime data (e.g., Weibull, exponential, and lognormal). Field failure data often results in partial or incomplete data on lifetimes, which are called *censored values*. Lifetime data can possess three types of censored data: Right-censored indicates that the product has not yet failed, thus placing a lower limit on lifetime; left-censored values arise when the product has failed but has an unknown origin and an unknown lifetime, and interval-censored data have an unknown lifetime bounded between two values. Maximum likelihood estimation (mle) allows incorporating the information contained in censored values. To use mle to estimate random variable model parameters for a sample of size n, denote product lifetimes as the random variables $T_1, T_2, \ldots, T_n$, with observed values $t_1, t_2, \ldots, t_n$. Assuming that the n component lifetimes are independent (this is analogous to assuming a random sample) and have probability density function $f(\,)$, the likelihood function for complete (uncensored) data is

$$L = \prod_{i=1}^{n} f(t_i) \tag{24.1}$$

The maximum likelihood estimates are the parameter values that maximize the likelihood function. To simplify the maximization problem, one can work with the log-likelihood function

$$\log(L) = \sum_{i=1}^{n} \log(f(t_i)) \tag{24.2}$$

Fitting random variable models to a population of automobiles using warranty data results in right-censored observations (vehicles without warranty claims have not observed a failure but have a know lifetime). To extend maximum likelihood estimation to the case of right-censored data, redefine the observed values $t_1, t_2, \ldots, t_n$ to represent either lifetimes or right censored values. In addition, define an indicator variable

$$\delta_i = \begin{cases} 1 & \text{if } t_i \text{ represents an observed lifetime} \\ 0 & \text{if } t_i \text{ represents a right-censored lifetime} \end{cases} \tag{24.3}$$

to denote right-censored observations. A right-censored observation means that the product lifetime exceeds the observed value, which represents the survivor function evaluated at the censor time

$$P(T > t) = S(t)$$

Incorporating right-censored data results in the likelihood function

$$L = \prod_{i=1}^{n} [f(t_i)]^{\delta_i}[S(t_i)]^{1-\delta_i} \tag{24.4}$$

with an associated log-likelihood function of

$$\log(L) = \sum_{i=1}^{n} [\delta_i \log(f(t_i)) + (1 - \delta_i)\log(S(t_i))] \tag{24.5}$$

24.3.2 Modeling Automobile Warranty Claims and Costs

Many manufacturers of durable goods, such as automobiles, appliances, and personal computers, provide their customers basic warranty coverage—that is, coverage included in the purchase price. Products that can experience at most one warranty claim can be modeled as nonrepairable systems using random variable-based cost models. Blischke and Murthy (1994) provide a comprehensive review of warranty cost models. Letting T represent product lifetime and letting t_c represent the warranty limit, one can estimate the probability that a given product experiences a warranty claim as the cumulative distribution function evaluated at the warranty limit

$$P(T < t_c) = F(t_c) \tag{24.6}$$

Many authors have suggested models for predicting automobile warranty claims. Kalbfleisch, Lawless, and Robinson (1991) developed a Poisson model for predicting automobile warranty claims in the time domain. They extended the model to predict cost by multiplying the expected number of claims (a function of time) by the expected cost per claim (also as a function of time). Moskowitz and Chun (1994) suggested using a bivariate Poisson model to predict claims for a two-dimensional warranty. They fit the cumulative Poisson parameter λ with various functions of time t and mileage m. Hu and Lawless (1996) suggested a technique for modeling warranty claims as truncated data that assumes that warranty claims follow a Poisson process. Sarawgi and Kurtz (1995) provided a method for predicting the number of product failures during a one-dimensional warranty coverage based on bench test data.

To use statistical methods for warranty data analysis, one must define a population for study. Forming populations of vehicles based on time of sale would serve sales, marketing, and accounting needs. However, product design, manufacturing, and assembly changes take effect on assembly or manufacturing dates. Therefore, automotive warranty and reliability engineers form populations based on the date of assembly, which allows them to compare field performance of vehicles before and after an engineering change. Due to the random nature of sales lag—the time from final assembly to customer delivery [see Robinson and McDonald (1991)]—a collection of vehicles built on the same day may sell at different times.

Many automobile manufacturers track warranty claims in terms of cumulative repairs per thousand vehicles, which we denote as $R(t)$ [see Wasserman (1992) and Robinson and McDonald (1991)]. To obtain $R(t)$ values, first define N_i as the number of vehicles with at least i months in service (MIS), which can decrease as i increases for a given set of vehicles due to sales lag. Let f_i represent the number or frequency of claims observed in month i of service by the N_i vehicles. Then calculate

$$R(t) = \sum_{i=0}^{t} \frac{f_i}{N_i} * 1000 \tag{24.7}$$

Wasserman (1992) developed a dynamic linear predictive model for $R(t)$ using data from multiple model years of a product. Singpurwalla and Wilson (1993) developed a bivariate failure model for automobile warranty data indexed by time t and mileage m. They derived the marginal failure distributions and presented a method for predicting $R(t)$ using a log–log model.

Robinson and McDonald (1991) suggested plotting $R(t)$ on log–log paper, fitting a line to the observed data, and using the fit line to make predictions. This approach is used by many automobile manufacturers who track claims using a statistic similar to $R(t)$ defined by equation (24.7). To apply the technique, the manufacturer takes a log transform of $R(t)$ and fits the simple linear regression model

$$\log(R(t)) = \beta_0 + \beta_1 \log(t) + \varepsilon \tag{24.8}$$

To predict warranty claims through some future time in service t, the manufacturer transforms the value predicted by the fit line as

$$\hat{R}(t) = \exp(\hat{\beta}_0 + \hat{\beta}_1 \log(t)) \tag{24.9}$$

24.3.3 Analytic Methods

Ford's reliability improvement process incorporates many quantitative tools—some widely used techniques and others less well known. This section presents background information and references for many of the methods used in Ford's process.

24.3.3.1 8D Problem-Solving Process

Many organizations use structured approaches to problem solving, and Ford is no different in this regard. Ledolter and Burrill (1999) and Pande, Neuman, and Cavanagh (2000) show structured approaches to problem solving. Ford uses a methodology they call the 8D Problem-Solving Process, which consists of eight steps. The following is a synopsis of the process.

D1: Establish Team

Establish a small group of people with the process and/or product knowledge, allocated time, authority, and skill in the required technical disciplines to solve the problem and implement corrective actions. The group must have a designated champion and team leader. The group begins the team building process.

D2: Describe the Problem

Describe the internal/external customer problem by identifying "what is wrong with what" and detail the problem in quantifiable terms.

D3: Develop Interim Containment Action (ICA)

Define, verify, and implement an interim containment action (ICA) to isolate effects of the problem from any internal/external customer until permanent corrective actions (PCAs) are implemented. Validate the effectiveness of the containment actions.

D4: Define and Verify Root Cause and Escape Point

Isolate and verify the root cause by testing each possible cause against the problem description and test data. Also isolate and verify the place in the process where the effect of the root cause should have been detected and contained (escape point).

D5: Choose and Verify Permanent Corrective Actions (PCAs) for Root Cause and Escape Point

Select the best permanent corrective action to remove the root cause. Also select the best permanent corrective action to eliminate escape. Verify that both decisions will be successful when implemented without causing undesirable effects.

D6: Implement and Validate Permanent Corrective Actions (PCAs)

Plan and implement selected permanent corrective actions. Remove the interim containment action. Monitor the long-term results.

D7: Prevent Recurrence

Modify the necessary systems including policies, practices, and procedures to prevent recurrence of this problem and similar ones. Make recommendations for systemic improvements, as necessary.

D8: Recognize Team and Individual Contributions

Complete the team experience, sincerely recognize both team and individual contributions, and celebrate.

24.3.3.2 Accelerated Testing

In many situations, doing life testing of products during development or redesign is infeasible due to the length of time required to perform the test. Manufacturers can compress product usage over time using methods called *accelerated tests,* which may also involve unique statistical models. Lawless (1982), Blischke and Murthy (2000), and Meeker and Escobar (1998) all contain material on accelerated tests.

24.3.3.3 B5 Lifetime

For product reliability purposes, manufacturers evaluate the lifetime of their products. Due to the random nature of lifetime for a population of products, manufacturers specify a lifetime model, usually a random variable that quantifies customer usage. At Ford, when engineers design automotive components, they predict a lifetime value that they expect 95% of the products to exceed, referred to as the *design lifetime*. Ford uses an analogous value, the B5 lifetime, to measure field performance. Warranty and reliability engineers measure field performance, often using warranty data, and develop $\hat{F}(t)$, an estimated cumulative distribution function of actual usage. When an observed value exists for each member in a sample, one can construct a histogram from the data. The histogram serves as a graphical estimate of the probability density function $f(\ )$ and the engineer can use this as an input to model selection. For example, if the histogram possesses the "bell-shaped curve," the modeler of the data should consider using a normal distribution. Blischke and Murthy (2000) provided figures of $f(\ )$ for a variety of distributions (exponential, Weibull, gamma, log-normal, etc.), often fit to lifetime data. In the case of censored data, it is not possible to construct a histogram. In this situation, we suggest graphically evaluating the observed lifetime data with a hazard plot [see Lawless (1982)]. Majeske, Lynch-Caris, and Herrin (1997) used this hazard plotting technique to model warranty data for an automotive audio system. A constant hazard function suggests an exponential distribution, whereas increasing or decreasing shapes may suggest a Weibull distribution. Once the engineer has identified a model, a goodness-of-fit test [see Lawless (1982) or Blischke and Murthy (2000)] to validate model selection should be performed. Using $\hat{F}(t)$—the random variable model fit to the observed data—the warranty or reliability engineer predicts the B5 lifetime as the value they expect 95% of the population to exceed, or

$$B5 = \hat{F}^{-1}(0.05) \tag{24.10}$$

24.3.3.4 Block Diagram

A block diagram [see Blischke and Murthy (2000)] represents a system of assemblies, such as a complete passenger vehicle, or a product subsystem (collection of components), such as the wheel bearing used in this case study. A block diagram contains a block for each component in the assembly. The diagram contains arrows (one-way or two-way) between the blocks to represent relationships between the components. A list of variables (controllable and uncontrollable) that may affect the functionality of the components completes the diagram. As an example, we will develop a block diagram for the automotive wheel bearing. First, we identify the key components for the wheel bearing: anti-lock brake system (ABS) sensor, ABS sensor O-ring, ABS cap, bearing assembly, and hub flange. Next, we list the variables that might affect the functionality of the wheel bearing: dimensional accuracy, outside temperature, thermal aging, internal temperature, weight imbalance, contamination, and road hazards. Lastly, we evaluate the relationships between the components and the variables. Figure 24.2 shows the block diagram for the automotive wheel bearing.

24.3.3.5 Function Tree

A function tree is developed in consort with the block diagram (see above). The function tree has a collection of branches for some or all of the components listed in the block diagram. Starting with a component, the first branch answers the question, "What function does the component perform?" The next branch gives more details on the how; and the last branch answers the question, "Why does this component function?" To demonstrate the technique, we develop a function tree for the automotive wheel bearing, which appears as Figure 24.3.

24.3.3.6 Parameter Diagram

Phadke (1989) presents a parameter diagram (P diagram) as a graphical technique used to organize the various variables that describe and/or affect a product or process. This technique classifies the variables into four categories: response, signal, noise, and control. Response variables are the quality characteristics that describe the product. Signal factors are the variables that are controlled by the customer or user of the product. Noise factors represent variables beyond the control of the customer or manufacturer and can be classified into three groups: external or environmental variables, common cause variation encountered in the manufacturing process, and deterioration attributed to normal usage or wear out. Control fac-

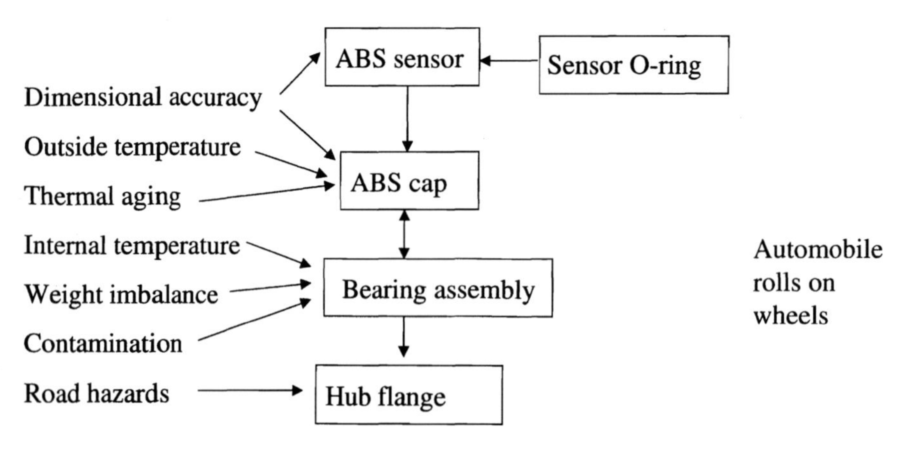

Figure 24.2. Block diagram for automotive wheel bearing.

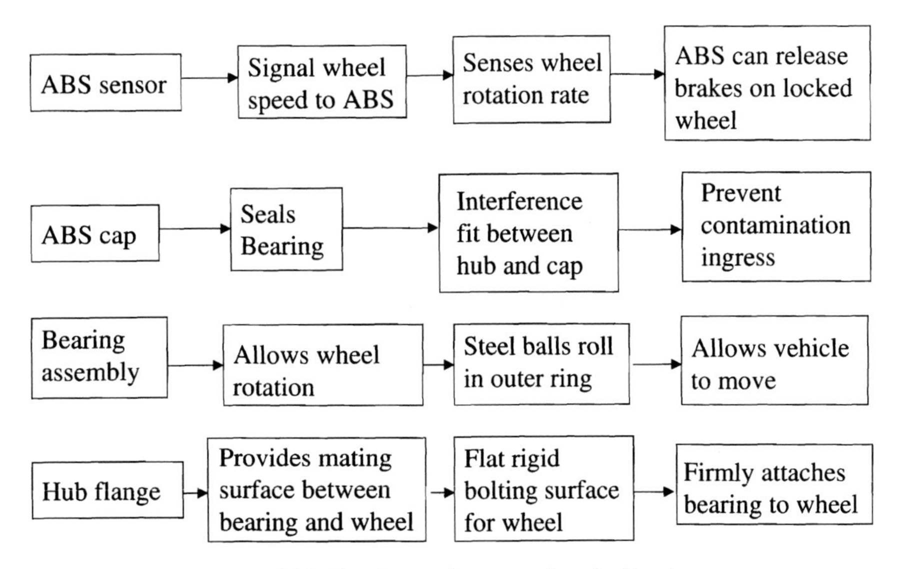

Figure 24.3. Function tree for automotive wheel bearing.

tors are the variables that are defined by the product and or process designers. Figure 24.4 shows the generic structure for a P diagram. In Section 24.4.3 we develop a P diagram for the automotive wheel bearing, which appears as Figure 24.8.

24.3.3.7 Reliability and Robustness Checklist

A great deal of effort has been allocated toward the design of robust products [see Phadke (1989)]. In a statistical sense, robustness is similar to independence. Saying

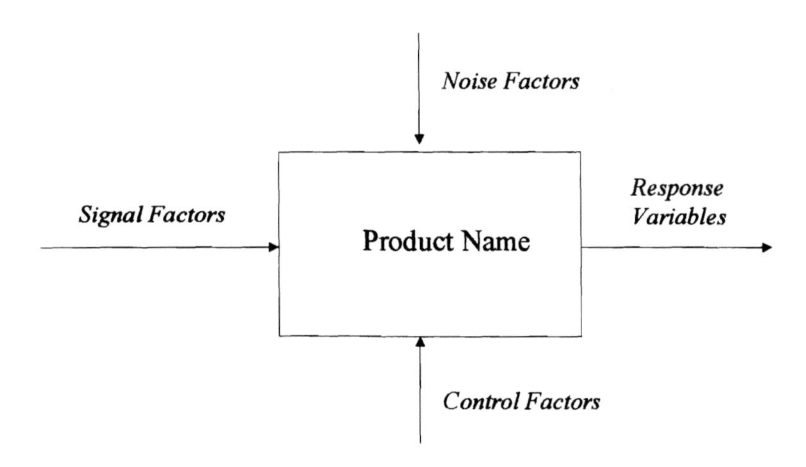

Figure 24.4. Parameter diagram structure.

that a product is robust means that its ability to function or perform is independent of other, usually uncontrollable, variables. When designing robust products, there are two perspectives. One is to design a product that is robust to operating conditions, and the other is to make it robust to uncontrollable variables in the manufacturing and assembly process. The reliability and robustness checklist is just a listing of all uncontrollable variables, both manufacturing and customer, and how the product will remain robust to their existence and variability.

24.3.3.8 Failure Mode and Effects Analysis (FMEA)

This technique assists manufacturers in obtaining robust products by identifying the product's potential failure modes and attempting to prevent them. Pande, Neuman, and Cavanagh (2000) provided a simple five-step approach to performing a product FMEA. Breyfogle (1999) gives a much more thorough treatment of the FMEA, including how to incorporate the technique in the design process. The Automobile Industries Action Group [see AIAG (2001)]—a consortium of Daimler-Chrysler, Ford, and General Motors tasked with developing common procedures for the domestic automobile industry—has documented an FMEA procedure specifically for the automobile industry.

24.4 DATA AND ANALYSIS

This section presents the Ford reliability improvement process as it was applied to a redesign of the automotive wheel bearing that appeared as Figure 24.1. The case study will follow the outline of the Ford methodology discussed in Section 24.2.

24.4.1 Identify Priority Concerns

Ford reliability and product design engineers continually review warranty and customer survey data to identify opportunities for design changes. As part of his regular duties, a reliability engineer identified abnormally high warranty claim rates for a particular wheel bearing. The warranty claims were for the replacement of the bearing, a rather expensive repair, resulting in warranty costs far exceeding target and budget values. The wheel bearing was identified as a candidate for redesign by constructing Table 24.1 with data available at 21 months in service (MIS). Table 24.1 shows data for the case study bearing, two other Ford wheel bearings, and the target or budget values for three quality measures: R(21), repairs per thousand vehicles at 21 MIS, calculated using equation (24.7), average wheel bearing warranty cost per vehicle or unit (CPU) in dollars, and the B5 life of equation (24.10) which represents the predicted useful life. The reliability engineer identified that the case study bearing had warranty and cost rates well in excess of the other two Ford products.

A second technique used in the analysis was to compare actual field performance with the design lifetime of 10 years and 150,000 miles. Based on prior experience

Table 24.1. Warranty Rates and Costs for Various Wheel Bearings

	R(21)	CPU	B5 Life
Target or budget value	7	$1.40	120 MIS
Case study wheel bearing	31.16	$7.49	38 MIS
Wheel bearing 2	8.28	$1.50	79 MIS
Wheel bearing 3	13.73	$2.40	47 MIS

and product knowledge, wheel-bearing design engineers assumed that wheel-bearing lifetime (MIS) followed an exponential distribution, with cumulative distribution function

$$F(t) = 1 - e^{-\lambda t}$$

with $\lambda = 0.0004274$. Thus, the design engineer assumed that the wheel bearing would meet, but not exceed, the useful life criteria, that is, $F(120) = 0.05$. To model actual field performance, the reliability engineer fit random variable models—using the right-censored data maximum likelihood estimation approach outlined in Section 24.3.1—to the warranty claim data. The engineer fit four parametric models to the data (exponential, Weibull, log-normal, and normal) and used a likelihood ratio test [see Meeker and Escobar (1998)] to compare the relative fit of pairs of models. The engineer concluded that a normal distribution, with mean 50 MIS and standard deviation 12 MIS, provided the best fit to the data. Figure 24.5 compares the normal distribution fit to the observed data (predicted lifetime) with the exponential distribution used as the lifetime model during product design (predicted lifetime). It is apparent from Figure 24.5 that the field performance of the wheel bearing falls far short of the designers' intent and suggests that the design engineers' lifetime model (the exponential distribution) did not provide a good forecast of field performance.

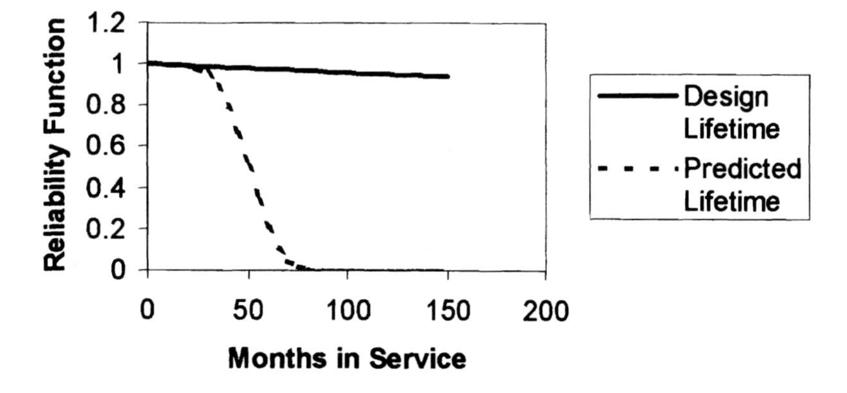

Figure 24.5. Wheel-bearing design lifetime and field performance.

The reliability engineer forwarded the data of Table 24.1 and Figure 24.5 to upper-level management, suggesting that they initiate Ford's reliability improvement process. Based on the high cost and poor field performance, Ford management identified a team champion and began the redesign process for the wheel bearing.

24.4.2 Set-Up Team

Team formation for the wheel-bearing issue was complex and delicate. Ford purchased the wheel bearing from a tier 1 supplier, and the bearing was treated as a "black box" component by Ford. The tier 1 supplier purchased one of the critical components from a tier 2 supplier, thus involving three companies in the design, manufacturing, and assembly of the wheel bearing. The team champion decided that the team should include representatives from both the tier 1 and tier 2 suppliers. Initially, both suppliers were hesitant to be included on the team, fearing they would be tasked with unrealistic assignments without support from Ford. The team champion assured the suppliers that they would be treated as equal partners on the team—sharing the success of the redesign as well as the blame for current failures—and they agreed to join the team. In addition to the champion, the team consisted of eight people: four from Ford, and two from each of the suppliers. The Ford product design engineer took the role of team leader, being assisted by the reliability engineer who first identified the issue.

24.4.3 Detail Failure Modes and Identify Noise

A common technique used by Ford in the problem identification process is to stratify field data by geographic location. Using this technique, initial analysis of the warranty data showed that failures in wheel bearings appeared to vary by state. Figure 24.6 shows the warranty claim rates for six states—the three states with the highest failure rate and the three states with the lowest failure rates—and includes a line showing the average across all 50 states. Notice that the three states with the

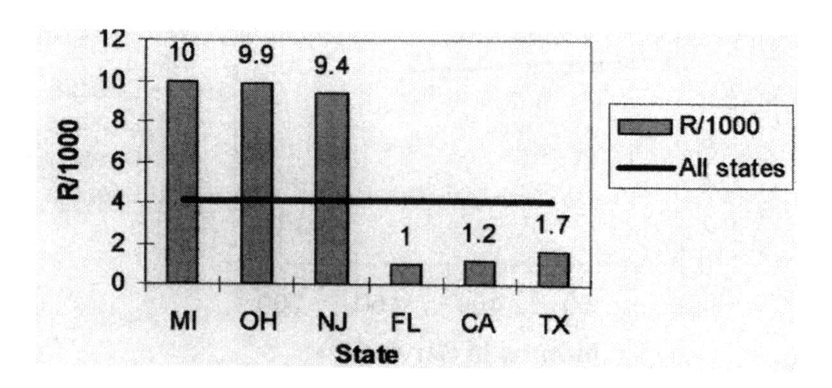

Figure 24.6. Wheel-bearing warranty claim rates by state.

lowest failure rates are warm weather states (California, Florida, and Texas), while the three states with the highest failure rates are cold weather states (Michigan, New Jersey, and Ohio). From this analysis, the team concluded that there was some type of a weather effect in the failure.

The team then began the review of field failures repaired under warranty that had been returned to Ford. While all components repaired under warranty become the property of Ford, due to space constraints Ford does not keep all the individual parts. The team contacted the storage facility and was able to obtain a sample of 18 bearings. They measured and recorded data on 27 different quality characteristics. During this process, the team noted that many of the wheel bearings exhibited a discoloration in the grease, which indicates contaminants entering the wheel bearing. Upon further review, the team noted that 10 of the 18 wheel bearings showed contamination ingress from the inboard side of the wheel bearing. The team then completed page one, the problem description, of the problem-solving worksheet that appears as Figure 24.7.

Next, the team initiated a best in class (BIC) study of wheel bearings. The team contacted the purchasing department for assistance, who completed a benchmark study of seven wheel-bearing suppliers. This study rated suppliers on overall quality and technical capability. After reviewing the purchasing department study, the team identified five wheel bearings to include in the BIC study: two used in Ford products and three used in non-Ford products. The team evaluated the various designs focusing on the ability of the design to prevent contaminants from entering the bearing. The wheel bearing selected as BIC has two major design differences from the current product. First, the BIC bearing has an integrated anti-lock brake system (ABS) sensor. This design eliminates a possible contaminant entry point to the bearing. Second, the BIC bearing has a metal interface in the plastic seal. The team felt that a metal to metal seal could be more effective in preventing ingress than the plastic to metal seal of the current design. The bearing identified as BIC is currently used on some Ford products, which allowed the team to compare failure rates and costs of the BIC bearing to the current design. Table 24.2 adds the BIC bearing to Table 24.1. Notice that the repair rate and costs of the BIC bearing are about one-fiftieth of the case study bearing.

The last step of this phase is to identify the noise factors and error states. The team used a combination of brainstorming sessions, meetings, and informal conversations to develop a list of possible noise factors. The team placed these noise factors into five categories. To organize and present this information, the team used a parameter or P diagram that appears as Figure 24.8.

24.4.4 Analyze Failure Mechanisms

The team then began to identify possible root causes. An analysis of the warranty claim data suggested that customers initially complained about noise in their vehicle. Based on the review of field failures, the replaced wheel bearings had rusted and quite possibly were making noise. The presence of contaminants, specifically water, was indicated by the grease discoloration. The root cause identification then

Problem Statement: Front Wheel Bearing is Noisy (What's wrong with what)				
Problem Description		IS	IS NOT	Get Information
WHAT	1.	Object Front Left & Right Wheel Bearing	Rear Bearing Brake Noise (worn/loose pads) Contact / Rubbing	Team is collecting AWS Data for Rear bearing.
	2.	Defect Noisy	Loss of Pre-Load / Endplay Lack of Grease Fatigue Failure Vibration	Based on data compiled from initial batch of customer returned parts
WHERE	3.	*On Object* Inboard Side Raceways	Outboard Side Raceways	Based on field return noise data
	4.	*First Observed* R-310 Durability for Base Program	Lab Test	
	5.	*Seen since* Seen in all markets predominately in Northeastern States	Predominate in Southern & Western States	Based on AWS Data
WHEN	6.	First Seen Customer Vehicles w/ more than 3000 miles	Customer vehicles with less then 3000 miles	
	7.	*What pattern since* Increasing Failure Rate		
	8.	*How many affected* 13.73 R/1000 at 19 MIS	< 13.73 R/1000	
	9.	*Defects per object* 1	Greater or less then 1	
	10.	*Trend* Increasing	Decreasing, Random, Cyclic	

Figure 24.7. Wheel-bearing problem description worksheet.

focused on potential ingress in two locations. The team concluded that the contaminants could be entering the bearings through the ABS cap to bearing mating surface and/or the ABS sensor to cap interface.

The team initiated the Ford's formal eight-step design change process and completed the first four steps. Having the design change process underway will allow them to quickly implement design changes later in the reliability improvement process. The team then began the process of updating the failure mode and effects analysis (FMEA). Specifically, the team added the noisy failure mode observed in

Table 24.2. Warranty Rates and Costs for Various Wheel Bearings

	R(21)	CPU	B5 Life
Target or budget value	7	$1.40	120 MIS
Best in class (BIC)	**0.62**	**$0.62**	**152 MIS**
Case study wheel bearing	31.16	$7.49	38 MIS
Wheel bearing 2	8.28	$1.50	79 MIS
Wheel bearing 3	13.73	$2.40	47 MIS

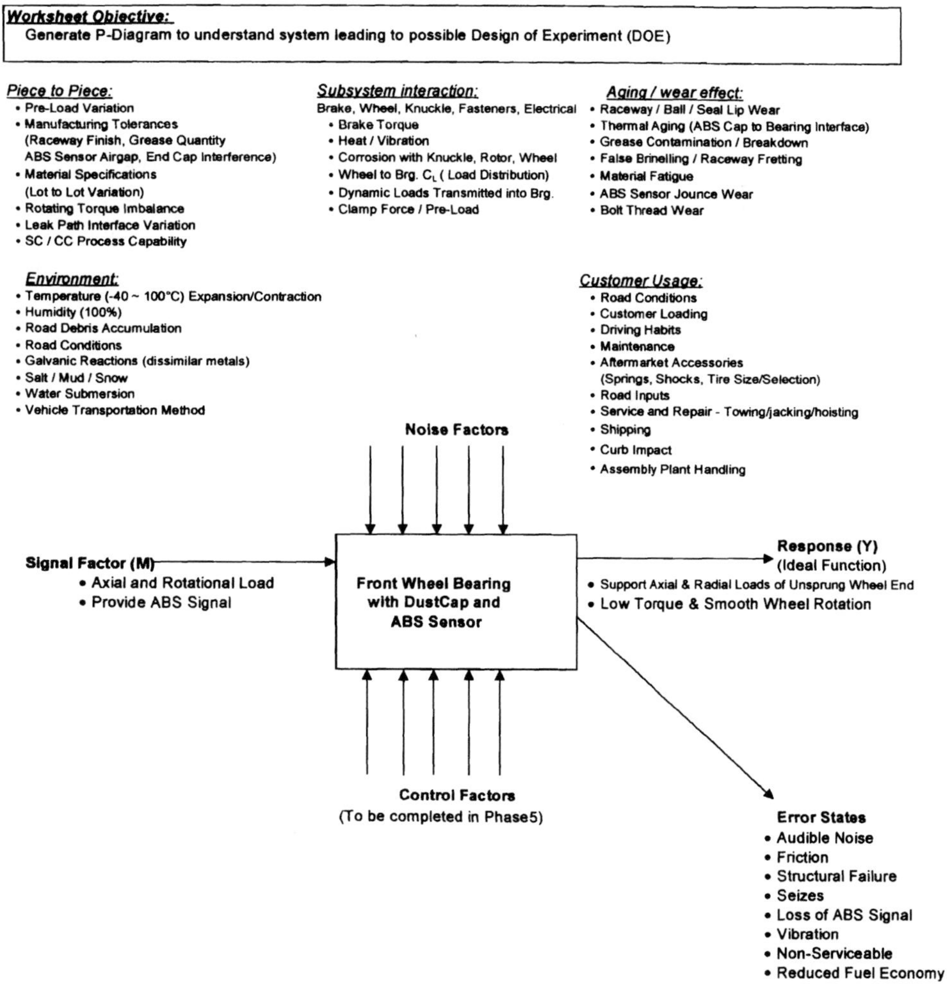

Figure 24.8. Wheel-bearing P diagram.

the field and increased occurrence and detection values on items related to the contamination ingress identified on the failed wheel bearings. The team also completed component-level design FMEAs for the bearing, the ABS cap, and the sensor.

The team performed an analysis of warranty costs and set targets for design changes. Using the data available at 21 MIS, the team fit the log–log $R(t)$ regression model of equation (24.8). Using equation (24.9), the team forecast that bearing would reach 48.3 repairs per thousand ($R/1000$) at 36 MIS, with a predicted cost per unit of $11.63. The team set a goal of a 50% reduction in $R/1000$ and cost per unit. Therefore, before implementing any design changes, the team will need to demonstrate (via bench test data) that they will reduce failures covered under warranty by 50%. Unfortunately, the team did not identify any fast track solutions and moved on to the next phase of the process.

24.4.5 Establish System Design Specification (SDS) and Worldwide Corporate Requirements (WCR) Compliance

By reviewing historical design documentation, the team concluded that the current bearing would satisfy the current bench verification tests. This suggested that the existing apparatus—which tested two bearings at a time—did not induce the failure modes occurring with bearings in the field. The team then tested two production bearings on the bench test, running them 20 days (a simulated 200,000 miles). Neither of these bearings failed, confirming that the original bench test did not induce the failure mode experienced in the field. Rather than develop a new bench test, the team decided to modify the existing test to create failure modes caused by contamination ingress into the bearing.

The team analyzed the current bearing and determined that it met all design standards. The team also noted that to Ford, the wheel bearing is a black-box component. Therefore, the design standards are very broad and generic. At this stage the team concluded that they should leave the standards as written, because from a Ford perspective the details of the design aren't as important as the functionality of the wheel bearing. Ultimately, the bearing must function for the design lifetime in all climates where they operate and must maintain a satisfactory level of customer satisfaction.

24.4.6 Develop Test and Establish Performance of Current Design

The existing bench test for the bearing utilized two factors to replicate field usage: rotation and temperature. Each rotation of the bearing is equivalent to driving a distance equal to the circumference of the tire; therefore, a direct relationship exists between the bench test and driving distance, and varying temperature simulates environmental conditions. Notice that modeling and prediction of field failures occurs in the time domain while the bench test replicates usage in mileage. This seemingly inconsistent strategy is driven by the needs of the various functional groups. Financial managers need to evaluate cash flows in the time domain while the engineering community leverages the ability to accelerate mileage. The team then modified the test to include a contamination component. The team added a pump and movable hoses that could squirt fluids on the bearings while they rotate. The bench test then consisted of rotating the bearings in alternating heat and cooling cycles while exposing the bearing to fluids. During the bench test, bearings are measured once every 24 hours.

The response variable measured on the bearing bench test is in units of Vg, a combined noise–vibration reading. A new production bearing reads in the 10- to 20-Vg range, and the design standard stated that a bearing was considered functional under 200 Vg. However, the team measured some returned field failures and obtained readings in the 50- to 60-Vg range. The team concluded that once a bearing hit 50 Vg it exhibited the field failure and lowered the bench test failure criteria from 200 Vg to 50 Vg. To verify contamination as the root cause, the team purposefully contaminated bearings in an attempt to reproduce the failure mode. The team injected a set of bearings with a muddy water solution. These bearings were placed on the test and at 10 days (100,000 simulated miles) began to deviate from the 10- to 20-Vg range. Because the field failures occurred at much lower mileage, the team determined that muddy water ingress did not cause the failure mode. Next, they injected a pair of bearings with a saltwater solution. After one day on the test, these bearings exceeded the revised bench test criteria of 50 Vg. The team concluded that they would use a saltwater solution (squirted on the bearing with the new pump) to simulate the salt-melted snow encountered in some cold weather states.

The team then conducted a designed experiment (DOE) to study the cap to the bearing interface, the most likely source of the contamination ingress into the bearing. The team identified two factors for the DOE: roundness of the cap and interference of the cap. The team utilized a 2-by-2 full factorial design [see Montgomery (2001)] augmented with current production parts and BIC parts that resulted in six runs. Since the bench test could only accommodate two bearings, and a full test cycle lasted 20 days, each replicate of the experiment could take up to 60 days. The team decided to use four replicates; however, they decided not to randomize the runs. Rather, they prioritized the runs that might allow them to confirm factor effects prior to completing the study. Table 24.3 shows the DOE design with the priorities.

The team obtained a sample of 100 caps and bearings and had them measured. From this sample they identified the 16 cap-bearing pairs that would satisfy the design. After completing testing on nine wheel bearings (one bearing test was considered invalid), the team had observed five failing bearings and four that did not fail

Table 24.3. Runs for Designed Experiment

DOE Run	Roundness	Interference	Description	Priority
1	Best	Best	Optimum condition	B
2	Worst	Best	Roundness test	A
3	Best	Worst	Interference test	D
4	Worst	Worst	Worst case	F
5	N/A	N/A	Case study bearing	C
6	N/A	N/A	BIC bearing	E

Table 24.4. DOE Bench Test Results

Bearing	DOE Run	Roundness	Interference	Bench Test Result
1	2	Worst	Best	Failed at 3 days
2	2	Worst	Best	Failed at 3 days
3	1	Best	Best	Failed at 6 days
4	2	Worst	Best	Failed at 6 days
5	2	Worst	Best	Failed at 4 days
6	1	Best	Best	Did not fail
7	5	Case study bearing		Did not fail
8	3	Best	Worst	Did not fail
9	5	Case study bearing		Did not fail

(see Table 24.4). The team noticed that all four bearings with roundness at the "worst" level had failed, only one of three with "best" roundness had failed, and neither of the two control parts had failed. Due to the long time span of the bench test, the team concluded that they had verified the failure mode and suspended testing, recognizing that they may need to return to this phase at a later date.

24.4.7 Develop Noise Factor Management Strategy

The team evaluated alternative methods to control the noise factors—that is, contamination ingress into the bearing. One approach would be to design a bearing that is robust to contamination that would preclude the need to prevent contaminants from entering the bearing. The team ruled out this approach as cost ineffective. A second approach would be to design a bearing that prevents contamination ingress. The team concludes that they will manage the noise factor (contaminants) by preventing them from entering the bearing.

24.4.8 Develop New Design and Verify Using Reliability Disciplines

The team decided to utilize a two-step approach to the product redesign. In the short-run, they would make changes to the exiting design intended to prevent the water ingress through the cap to bearing interface, and in the long-run they would completely redesign the wheel bearing. Through brainstorming, the team developed a list of 11 possible short-run design changes. For each change, the team identified advantages and disadvantages and then ranked the ideas. The top ranking idea was to add a groove and an O-ring in the cap to bearing interface area. This approach had many advantages (a proven sealing design, short implementation time, and inexpensive) but also had a few disadvantages (an additional assembly step and the inability to detect damaged O-ring). The team implemented the design change and will track the performance of these bearings in the field relative to the old design. For a long-run solution, the team utilized a design similar to the BIC wheel bearing identified in phase 2 of the methodology. This design utilizes an integrated ABS sensor that eliminates one potential ingress point. It also uses a metal-to-plastic seal on the cap. However, it will take approximately two model years for this new bearing to be designed, fully tested, and implemented into production vehicles.

24.4.9 Sign-Off Design and Update Lessons Learned

The team signed off on the long-term redesign strategy and forwarded the documentation to the appropriate design functions. They left the bench test in working order so that product development personnel could utilize it for testing prototypes of the new wheel bearing. The team updated the lessons learned so that in the future, other Ford employees could benefit from their experiences. Lastly, the team was formally recognized for a job well done.

24.5 CONCLUSIONS AND RECOMMENDATIONS

Engineers continually bring quantitative methods to bare on product and process-related problems. This case study demonstrates how engineering and product development functions can bring more structure and quantitative methods to the reengineering process. This case study is not intended to be a "cookbook" approach that can be used for any type of product reengineering. Rather, we show this methodology as an example of how organizations can add more structure to the reengineering process.

We recommend that manufacturers of all products, and possibly even service providers, utilize a structured methodology to continually improve product reliability. We do not intend to intimate that the methodology contained in this case study will work for everyone. Rather, this serves as an example or an outline of how one might develop such an approach for their particular business. To develop a methodology for your company, we recommend that you start with the nine-step approach outlined in Section 24.2. Work through each step and evaluate its applicability to your specific product line or service. If the step is relevant to your situation, modify the content to meet your specific needs. This will include changing wording to match the jargon or terminology of your business, assessing specific analytic techniques to verify that they are consistent with the tools commonly used by your personnel, and verifying that the completion activities will support your business processes. As you work through the nine steps, you may find that some of them can be removed or that you need to add additional steps. As you tailor the general approach, keep in mind that the key elements of success for this methodology are not overconstraining the problem identification criteria and allowing the flexibility to modify the approach to solve specific problems.

ACKNOWLEDGMENTS

The authors would like to thank the Ford Motor Company for allowing them to develop this case study. Publishing this work demonstrates Ford's corporate belief that the engineering community, and society in total, will benefit from the sharing of their business practices. The authors would also like to thank Robert Mince of Ford Motor Company for his support and encouragement during the writing of this case study. The first author would like to thank Al Kammerer of Ford Motor Company for his continued support of my academic research interests. When first approached about writing this case study, Al fully supported the concept, suggested multiple ideas, and provided me with the necessary Ford contacts to pursue the endeavor.

REFERENCES

Automotive Industries Action Group (AIAG) (2001). *Potential Failure Mode and Effects Analysis,* 3rd edition, AIAG, Southfield, MI.

Blischke, W. R., and Murthy, D. N. P. (1994). *Warranty Cost Analysis,* Marcel Dekker, New York.

Blischke, W. R., and Murthy, D. N. P. (2000). *Reliability Modeling, Prediction and Optimization,* Wiley, New York.

Breyfogle, F. W. (1999). *Implementing Six Sigma: Smarter Solutions Using Statistical Methods,* Wiley, New York.

Hu, X. J., and Lawless, J. F. (1996). Estimations of rate and mean functions from truncated recurrent event data, *Journal of the American Statistical Association* **91:**300–310.

Kalbfleisch, J. D., Lawless, J. F., and Robinson, J. A. (1991). Methods for the analysis and prediction of warranty claims, *Technometrics* **33:**273–285.

Lawless, J. F. (1982). *Statistical Models and Methods for Lifetime Data,* Wiley, New York.

Ledolter, J., and Burrill, C. W. (1999). *Statistical Quality Control Strategies and Tools for Continual Improvement,* Wiley, New York.

Majeske, K. D., Lynch-Caris, T., and Herrin, G. D. (1997). Evaluating product and process design changes with warranty data, *International Journal of Production Economics* **50:**79–89.

Meeker, W. Q., and Escobar, L. A. (1998). *Statistical Methods for Reliability Data,* Wiley, New York.

Montgomery, D. C. (2001). *Design and Analysis of Experiments,* Wiley, New York.

Moskowitz, H., and Chun, Y. H. (1994). A Poisson regression model for two-attribute warranty policies, *Naval Research Logistics* **4:**355–376.

Pande, P. S., Neuman, R. P., and Cavanagh, R. R. (2000). *The Six Sigma Way: How GE, Motorola, and Other Top Companies are Honing Their Performance,* McGraw-Hill, New York.

Phadke, M. S. (1989). *Quality Engineering Using Robust Design,* Prentice-Hall PTR, Englewood Cliffs, NJ.

Robinson, J. A., and McDonald, G. C. (1991). Issues Related to Field Reliability and Warranty Data, in *Data Quality Control: Theory and Pragmatics,* pp. 69–90, Marcel Dekker, New York.

Sarawgi, N., and Kurtz, S. K. (1995). A Simple Method for Predicting the Cumulative Failures of Consumer Products during the Warranty Period, in *Proceedings Annual Reliability and Maintainability Symposium,* pp. 384–390.

Singpurwalla N. D., and Wilson, S. (1993). The warranty problem: Its statistical and game theoretic aspects, *SIAM Review* **35:**17–42.

Thompson, W. A. (1981). On the foundations of reliability, *Technometrics* **23**:1–13.

Wasserman, G. S. (1992). An application of dynamic linear models for predicting warranty claims, *Computers in Industrial Engineering* **22**:37–47.

EXERCISES

24.1. Truncated data models represent another method for modeling field failure or warranty data by only including failed parts. How would this affect (a) the definition of a population and (b) the conclusions you would draw from the analysis.

24.2. How would you modify the nine-step approach to apply this methodology to a service organization.

24.3. The failure appears more prevalent in cold weather states. What would be the pros and cons of having location-dependent products.

24.4. This approach used the log–log repairs per thousand vehicles $R(t)$—a nonparametric technique—for modeling lifetime. What limitations do nonparametric techniques place on the analysis of lifetime data?

24.5. The bench test did not incorporate a vehicle. What are the advantages and disadvantages of not using complete products in reliability testing?

24.6. One factor limiting the ability to rapidly implement design changes was determining who should bear the cost for tooling changes. How would you suggest that manufacturers and suppliers resolve these types of issues.

24.7. Develop a block diagram, function tree, and parameter diagram for a product or assembly of your choice.

CHAPTER 25

Reliability of Oil Seal for Transaxle—A Science SQC Approach at Toyota

Kakuro Amasaka and Shunji Osaki

25.1 INTRODUCTION

It is very critical for both vehicle and parts manufacturers worldwide to improve drive train system reliability for ensuring higher customer satisfaction. The reliability of the drive train depends on the reliability of its components. One such component is the oil seal for the transaxle. Failure of this results in oil leaking out; this can have serious implications for the drive train, resulting in high repair cost and high customer dissatisfaction.

Toyota Motor Corporation purchased this component from the NOK Corporation, an external component manufacturer, and developed a new approach to improve the reliability of the oil seal. Because the dynamics of the oil leak were not well understood, the starting point of the joint investigation was to get a better understanding of this. This was done through a "dual total management team" involving both Toyota and NOK using the "science SQC" developed by Toyota.

The "science SQC" involves the "management SQC" and the "SQC technical methods." Toyota and NOK formed the "TDOS-Q5" (teams Q1 to Q5) and"NDOS-Q8"(teams Q1 to Q8) teams to carry out the scientific investigation. Each team had specific goals and objectives. Three management methods, namely "technical management (TM)", "production management (PM)," and "information management (IM)," were adopted to develop new technologies to understand and design a better oil seal for the transaxle.

Note: The first author was TQM Promotion General Manager at Toyota and led the Toyota teams during the course of this case study.

Case Studies in Reliability and Maintenance, Edited by W. R. Blischke and D. N. P. Murthy.
ISBN 0-471-41373-9

This led to two new technologies. The first was a technology to visualize the oil seal leakage dynamic behavior in order to understand the underlying mechanisms, and the second was the use of factor analysis for improving reliability through better design. Use of the "total QA network" method made it possible to build quality into the process, leading to reliability improvement that reduced market claims by 90%.

This case study discusses the approach used by Toyota to improve the reliability of the oil seal for transaxles. The outline is as follows. Section 25.2 deals with the sealing function of the oil seal and earlier approaches to seal design. Section 25.3 describes reliability improvement activities at Toyota using a cooperative approach. Section 25.4 deals with reliability improvement of the oil seal for quality assurance of the transaxle. Finally, in Section 25.5, we conclude with some comments.

25.2 OIL SEAL

25.2.1 Oil Seal Function

An oil seal, shown in Figure 25.1, prevents the oil lubricant within the drive system from leaking from the drive shaft. It is comprised of a rubber lip molded onto a round metal casing. The rubber lip grips the surface of the shaft around its entire circumference, thus creating a physical oil barrier. A garter spring behind the rubber lip increases the grip of the lip on the rubber shaft. As the shaft rotates, a minute quantity of the sealed oil forms a thin lubricating film between the stationary rubber lip and the rotating shaft. This oil film prevents excessive wear of the rubber lip and at the same time reduces frictional loss due to shaft rotation. A properly designed rubber lip rides on this lubricating oil film. On the other hand, an excessively thick

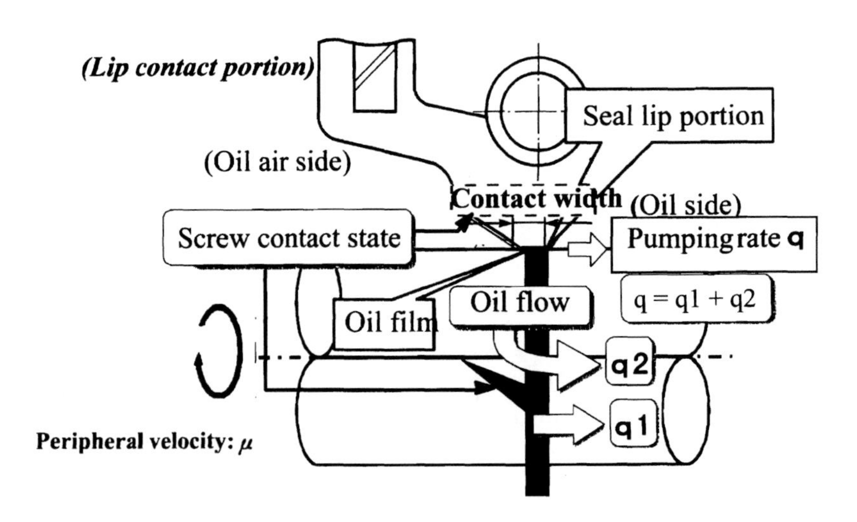

Figure 25.1. Enlarged view of oil seal sealing mechanism.

oil film will itself be a source of leakage. This condition is avoided by precise control of the lubricating oil film between the rubber lip and the rotating shaft.

Extensive experimental and theoretical investigations over the years by NOK have identified several important factors affecting the sealing function of the rubber lip. Of primary importance is the sealing ability of microscopic roughness on the rubber surface. It has been shown that these micro-asperities create shear flows within the lubricating oil film (Nakamura, 1987). The objective of the seal design (Kawahara et al., 1980) is to influence this shear flow so that the net flow of oil is toward the sealed oil side. The pumping ability of the rubber micro-asperities is designated by microscopic pump flow q1. To supplement q1, the rubber lip has also been augmented with helical ribs that function similar to a vane pump, providing macroscopic pumping q2 of leaked oil back to the sealed oil side (Lopez et al., 1997; Sato et al., 1999). Recent efforts by NOK to maximize q2 have resulted in a patented rib shape (Kameike et al., 2000).

The parameters for the sealing condition of the oil film involve not only the design of the seal itself, but also external factors such as shaft surface conditions, shaft eccentricity, and so on (Hirabayashi et al., 1979). Of particular importance is the contamination of the oil by minute particles. Because these are technical issues that involve not only the seal but also the entire drive train of the vehicle, Toyota initiated a corroborative effort with NOK to improve the reliability of the oil seal.

25.2.2 Earlier Approaches to Seal Design

The design quality and the total quality assurance program prior to the recognition of the "transaxle oil seal leakage" problem were mostly centered on treating each part of the item separately. The technical development design staff recovered oil seal units having leakage and analyzed the cause based on proprietary techniques. Corrective measures were then incorporated into the design. Inspection of items with leakage revealed no reason for the leakage and the cause of the oil leak was labeled "unknown," which made it difficult to find a permanent solution to the leaking problem.

To develop an epoch-making quality improvement, it was necessary to study the transaxle as a whole, rather than looking at each part separately by specialists, so as to understand the mechanism leading to seal failure and the effect of the operations during the manufacture process. Two issues of importance in the manufacture of highly reliable transaxle units from a product design viewpoint were identified as follows.

1. It is difficult to characterize theoretically the variation of macroscopic pressure distribution due to shaft eccentricity and the variation of microscopic pressure distribution under the influence of foreign matter in the oil.
2. Design specifications relating to shaft (to mesh with oil seals), on the other hand, were decided by the vehicle manufacturer in the ranges recommended by the component manufacturer based on limited information exchange between the two parties.

For the new approach to yield the optimal design required each party to have the implicit knowledge known to the other, and this was not the case. This highlighted the fact that there were some deficiencies in the total quality assurance program. In order to tackle these issues properly, a new methodology was needed; this is discussed in the next section.

The main objective of the new method was to isolate the true cause of failure. For this reason, it was necessary to implement the "science SQC" approach [see Amasaka (1997) and Amasaka and Osaki (1999)] and carry out a proper study of the mechanism of the sealing performance based on a systematic approach. To achieve this objective, new task management teams comprising of members from related divisions from both the Toyota and NOK organizations were established to improve the reliability of the oil seal.

25.3 RELIABILITY IMPROVEMENT AT TOYOTA: A COOPERATIVE TEAM APPROACH

25.3.1 Dual Total Task Management Team

In the automotive industry (and in other general assembly industries), quality control for parts and units, optimization of adaptation technologies for assembly, and quality assurance are required in all phases of the operations (production, sales, and after-sales service). Effective solution of technical problems requires the formation of teams and an understanding of the essence of problems by the teams as a whole. This allows the bundling of empirical skills of individuals distributed throughout the organization. Solution to technical problems requires harnessing the information (implicit knowledge) among related units of the organization through a cooperative team approach to generate new technologies (explicit knowledge).

If collaborative team activities by the vehicle manufacturer and the parts manufacturer are carried out independently of one other, then the implicit knowledge is often not converted to explicit knowledge due to lack of proper communication between the two. It is necessary to create and conceive new ideas that lead to new technologies by having inputs from the all the parties (component manufacturers, vehicle manufacturer, and end users). This approach was used by Toyota to improve the reliability of the oil seal for the transaxle.

To ensure high reliability of product design and quality assurance, a "total task management team" involving Toyota and NOK personnel was created to transform the implicit knowledge (relating to product and processes in both organizations) into explicit knowledge and to create new technologies of interest to both organizations. The "dual total task management team" named "DOS-Q" (drive-train oil Seal-quality assurance team: T Dos-Q5 and N Dos-Q8) formed between TMC and NOK is shown in Figure 25.2.

Toyota's constituting teams comprise Q1 and Q2 in charge of investigation into the cause of the "oil leakage" and Q3–Q5, which handled manufacturing problems relating to drive shafts, vehicles, and transaxles. Similarly, NOK formed teams Q1

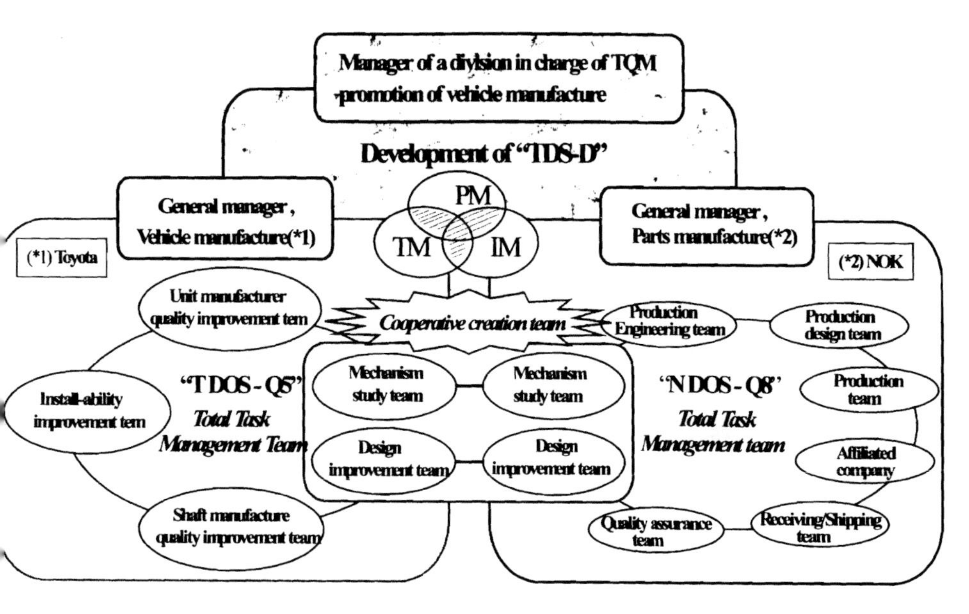

Figure 25.2. Configuration of cooperative creation team dual total task management team.

through Q8 as shown in Figure 25.2. Q1 and Q2 at Toyota interacted closely with their counterparts at NOK to improve the reliability of the oil seals as a single unit; likewise, Q3–Q8 handled the manufacturing problems for quality assurance.

Accordingly, the teams shared their individual knowledge (relating to empirical techniques and other technical information) to apply them to solving the problems under consideration. Each team had a general manager, and the joint team was led by Toyota's TQM Promotion General Manager for the vehicle. The methodology of "TDS-D" (total design system for drive-train development) involving IM (information management), TM (technology management), and PM (production management) was used as indicated in Figure 25.2.

25.3.2 Problem Formulation and Task Setting

According to NOK, oil leaks occurred due to wear. The result of a wear test on oil seals indicates a running distance of 400,000 km (equivalent to 10 years or more vehicle life) is regarded as a sufficiently reliable design. The oil seal leakage from the failure repair history of the Toyota DAS (Dynamic Assurance System [see Sasaki (1972)] can be classified into initial failures that occur during the initial life of new vehicle and failures caused by wear that occur after running for some time. From the investigation by Q1 and Q2 teams, there are cases when oil leaks occur before reaching half of the running distance determined during an earlier test on oil seals alone. Thus, it cannot be said that the design is highly reliable and that the failure mechanism is fully understood.

Judging from the survey and analysis of parts returned from customers for claims, the cause of the failure was identified as being due to the accumulation of foreign matter between the oil seal lip and the contact point of the transaxle shaft, resulting in insufficient sealing. Oil leaks were found not only during running, but also at rest. Thus, it was considered that the cause is poor foreign matter control during the process aiming at an improvement in production quality. It is considered that metal foreign matter in the oil in the transaxle gear box adversely affects the respective contact points, thus accelerating the wear of the oil seal lip during rotation of the axle shaft. However, the permissible limit of the foreign matter particle size that causes oil leaks is unknown for the earlier failure, and the dynamic behavior resulting in oil leaks is not clarified yet for the latter failure. Consequently, the root of the problem is that the oil leak mechanism is not clarified and no quantitative analysis on cause/result correlation has been conducted, which is a true barrier for achieving high design reliability.

Task setting with all the people concerned, discussing the issues of the problems (and not depending on the rule of thumb practices), was done using affinity and/or association diagrams. These diagrams revealed that the essential problem was the lack of complete knowledge about the oil leak mechanism. It also reconfirmed that it was important to scientifically visualize the sealing phenomenon at the fitting contact portion. Consequently, it was decided that the engineering problems need to be solved by (1) clarifying the failure analysis processes through actual investigation of the parts, (2) carrying out a factor analysis of the oil leakage process, and (3) examining the design, manufacturing, and logistics processes involved.

25.4 RELIABILITY IMPROVEMENT OF OIL SEAL

25.4.1 The New Approach Utilizing Science SQC

To clarify the oil seal leakage mechanism for transaxles, "science SQC" [see Amasaka (2000)] was implemented with a mountain-climbing-type problem solution technique and using the "SQC technical methods" [see Amasaka (1998)], as indicated in Figure 25.3. The three elements of "management SQC" [see Amasaka (1997)]—that is, "technology management [TM]," "production management [PM]," and "information management [IM]"—were developed as indicated in each stage of the team activity process.

As the figure illustrates, both teams are linked in the implementation of the "management SQC" by combining the three management methods. In addition, the mountain-climbing method for problem solution utilizing the"SQC technical methods" was used to achieve reliability improvement in terms of both design reliability and production quality. During the first year, the following technical themes were identified and summarized for future study by the teams:

1. Why did oil leak from the oil seal?
2. Had anybody actually seen the phenomenon?

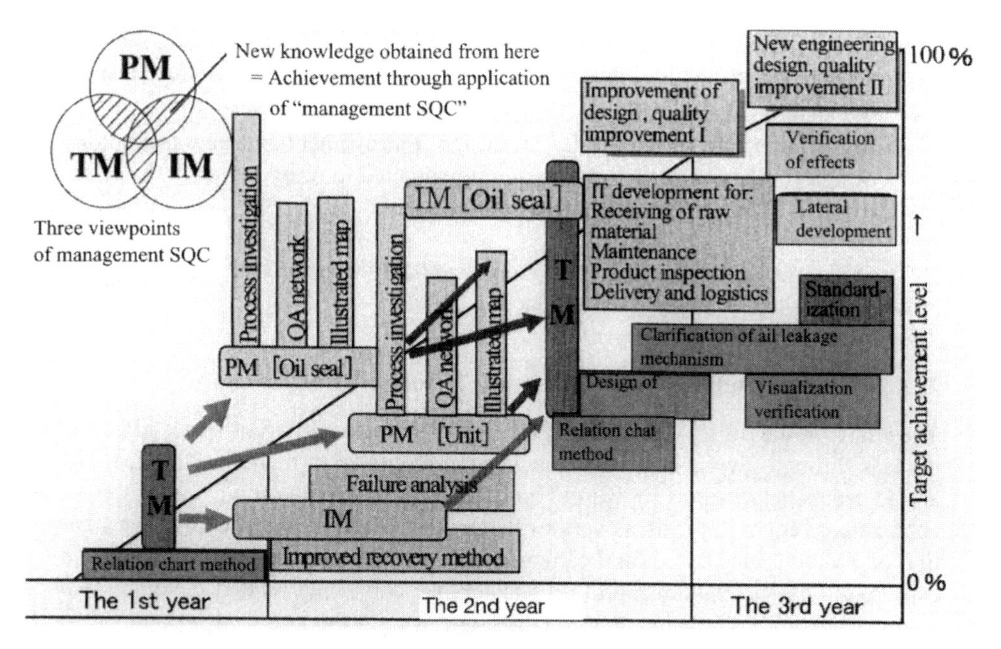

Figure 25.3. Mountain climbing of problem solution utilizing "SQC technical methods."

3. What was the oil seal design concept based on, while the axis is rotating or standing still?
4. Was the oil leak attributable to the oil seal part or unit?
5. Was the quality assurance system complete?

This led to the following actions.

1. The market claim parts recovery method was improved, and proper analysis was carried out for better understanding of the oil leak failure using "information management [IM]" in the second year of the study.
2. A survey of production processes for the oil seal and transaxle unit using "production management [PM]" was done and a correlation analysis between market claims and in process rejects using picture mapping and the "total QA network" [see Amasaka and Osaki (1999)] was carried out for improving process control. Based on the knowledge acquired from these, a cause-and-effect analysis on the oil leak problem was conducted, fully utilizing a science SQC approach, as well as "technology management [TM]," during the latter half of the second year to the first half of the third year. Concurrently, the oil leak dynamic behavior visualization equipment was developed for analyzing and verifying the oil leak mechanism.

3. The information management system was improved for systematic material acceptance inspection, production equipment maintenance, product inspection, shipment, and logistics.
4. In the latter half of the third year, design improvement and new technology design (quality improvement) were implemented in order to verify the effectiveness of the improvement and for horizontal deployment.

The remainder of the section gives more details of these activities that led to improving the design reliability and production quality.

25.4.2 Understanding of the Mechanism Through Visualization

The study on the oil leak mechanism involved looking at different issues affecting the leak through several studies, as indicated in Figure 25.4, involving the TM, PM and IT teams activities as shown in Figure 25.3, in first and second year of the investigation. Figure 25.4 covers various factors that could be acquired by investigation of TM activities based on the knowledge on various factors of PM and IT. To explore the causal relationship of oil leakage, we connected related factors with a (↓) mark using the relation chart method, and we put the knowledge obtained so far into good order. But, because the oil leakage mechanism is unknown, the occurrence route of oil leakage on running and stationary vehicles isn't clear.

As indicated in the figure, for example, although the cause-and-effect relationships in the drive shaft surface roughness were recognized, it was not seen as being the main cause of premature leakage as opposed to wear, since the production

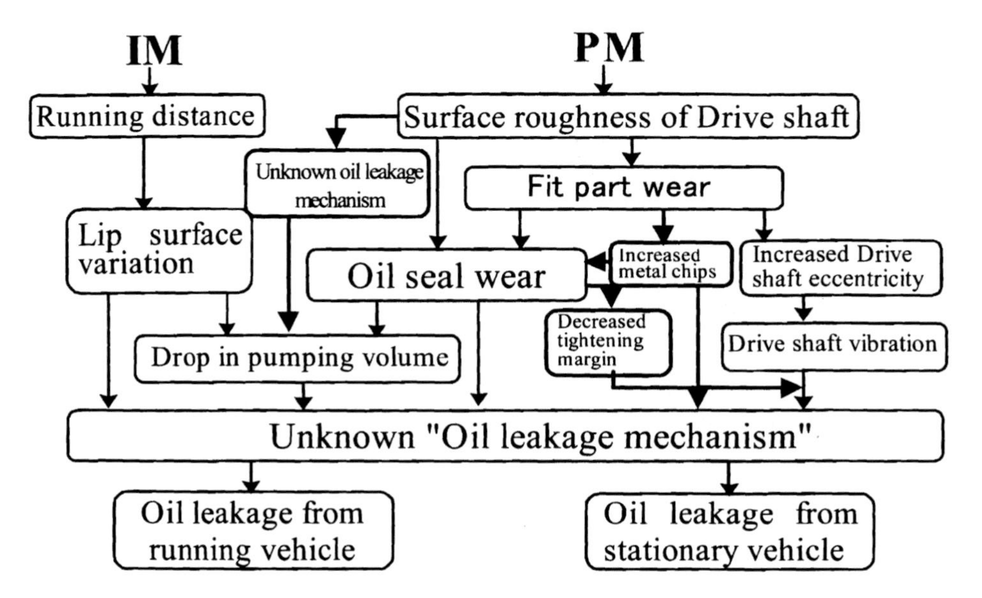

Figure 25.4. Estimation of the oil leakage mechanism.

process capability was ensured. However, this was not certain at that time. With new knowledge, it was deduced that the wear of the transaxle engagement part (differential case) during running increased the fine metal particles in the lubrication oil, which accelerated the wear of the oil seal lip unexpectedly. This, in turn, caused oil leakage due to a decrease in the sealing margin of the oil seal lip, causing the oil seal pump quantity to drop.

The established theory used to be that fine metal particles (of micron order) would not adversely affect the lip sealing effect. When these are combined to produce relatively large particles, however, do they then affect the sealing effect? And what about the effect of alignment between the drive shaft and the oil seal (fixing eccentricity) during assembly? From another aspect, if oil leakage occurs due to foreign matter accumulation to the oil seal lip during transaxle assembly, what is the minimum particle size that causes the problem? These were unknown because the dynamic behavior of oil leakage was not yet visualized, so the true cause had yet to be clarified.

Accordingly, a device was developed to visualize the dynamic behavior of the oil seal lip to turn this "unknown mechanism" into explicit knowledge, as shown in Figure 25.5. As shown in the figure, the oil seal was immersed in the lubrication oil in the same manner as the transaxle, and the drive shaft was changed to a glass shaft that rotated eccentrically via a spindle motor so as to reproduce the operation in an actual vehicle. The sealing effect of the oil seal lip was visualized using an optical fiber.

It was conjectured that in an eccentric seal with one-sided wear, foreign matter becomes entangled at the place where the contact width changes from small to large. Two trial tests were carried out to ascertain if this was true or not. Based on

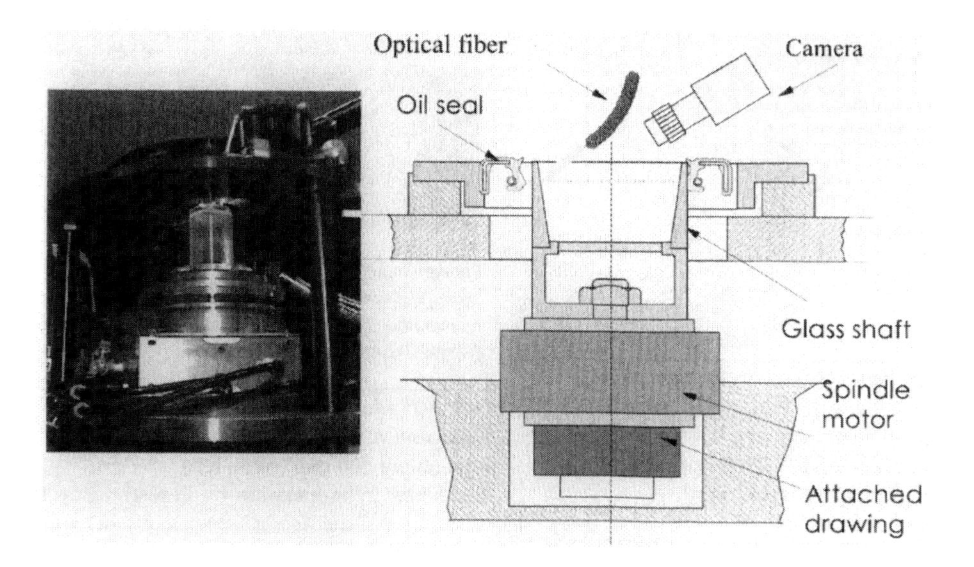

Figure 25.5. Outline of device by visualization.

the observation of the returned parts from the market and the results of the visualization experiment, it was observed that very fine foreign matter (which was previously thought as not impacting the oil leakage) grew at the contact section, as shown in Figure 25.6 (Test 1).

From a result of the component analysis, it was confirmed that the fine foreign matter was the powder produced during gear engagement inside the transaxle gear box. This fine foreign matter on top of microscopic irregularities on the lip sliding surface resulted in microscopic pressure distribution that eventually led to the degrading the sealing performance (Test 2).

Also, the presence of this mechanism was confirmed from a separate observation that foreign matter had cut into the lip sliding surface, causing aeration (cavitations) to be generated in the oil flow on the lip sliding surface, thus deteriorating the sealing performance, as shown in Figure 25.7 (Test 3). The figure indicates that cavitations occur in the vicinity of the foreign matter as the speed of the spindle increases, even when the foreign matter accumulated on the oil seal lip is relatively small.

This confirmed that such a situation leads to a reduction in the oil pump capacity of the oil seal, thus causing oil leakage during running. As the size of the foreign

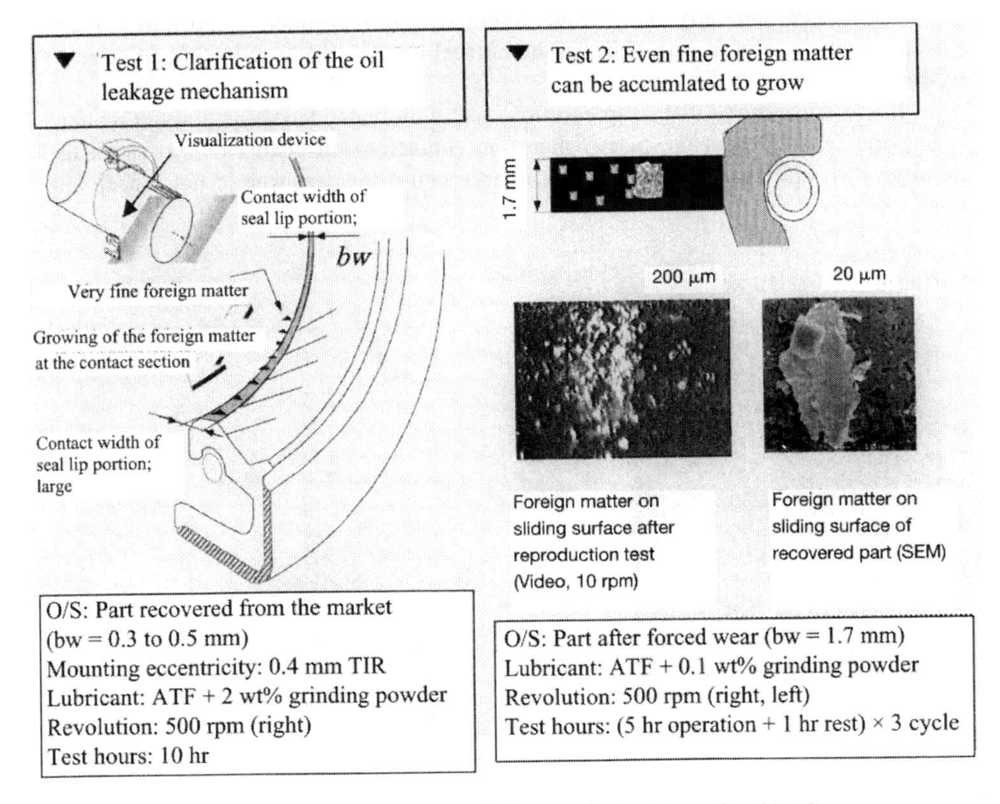

Figure 25.6. Test result of the oil leakage mechanism by test 1 and test 2.

▼Test-3: The presence of cavitations on the lip sliding surface by foreign matters

↓

When foreign matter accumulation occurs, cavitations will cause the oil enclosing balance to move to the atmospheric side, resulting in easier oil leakage.

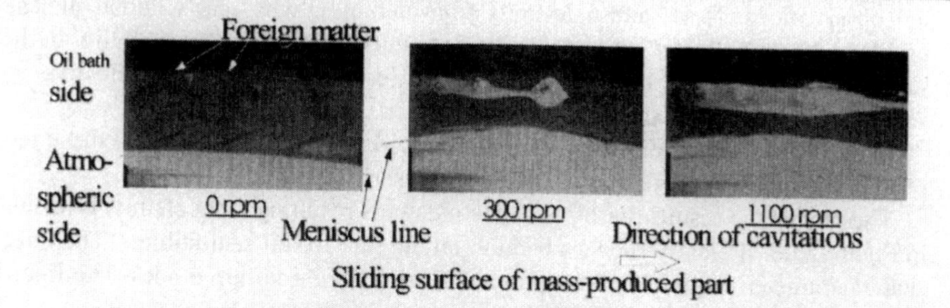

Cavitations (revolution) at near foreign matter
→ Lower pump performance
Oil flow-out toward the atmoshperic side (stopping)
→ The meniscus balance position moves toward the atmospheric side.

OS: Part after forced wear (bw = 0.5 mm)
Mass-produced part and countermeasure screw (packing material)
Lubricant: ATF + grinding powder particle size (#80)

Figure 25.7. Test result of the oil leakage mechanism by test 3.

matter gets bigger, the oil sealing balance position of the oil seal lip moves more toward the atmospheric side and causes oil leaks at low speeds or when at rest. This was unknown prior to the study and hence was not incorporated in the design of oil seals.

25.4.3 Fault and Factorial Analyses

Before studying the mechanism of oil leaks from oil seals described in Section 25.4.2, both NOK and Toyota considered that the wear of leaking oil seal lips would follow a typical pattern. The empirical knowledge based on the results of individual oil seal reliability tests was that the unit axle is highly reliable and would ensure 400,000 km or more in B10 life (less than 10% of the items fail by B10). Because of smooth contact between the oil seal lip and the rotating drive shaft, due to a surface roughness with an oil film in between, it was thought that the oil seal lip should wear gradually.

As a result of the experiment discussed in Section 25.4.2, however, it was found that metal particles generated from gears in the differential case accelerates eccentric wear of the oil seal lip, making the expected design life unobtainable. Since the wear pattern was not simple, it had to be confirmed that faulty oil seals returned under claims would reproduce the oil leak problem.

For this purpose, a survey was conducted along with an experiment as indicated below: First, in addition to defective oil seals, nondefective ones were collected on a regular basis to check for oil leak reproducibility and for comparison through visual observations. Next, transaxle units from vehicles, with and without oil leak problems, were collected on a regular basis to check for leak reproducibility in the same way. Integrating these results of transaxles with and without defective oil seals confirmed the defect reproducibility.

In all of these tests, oil leaks were reproduced as expected. Based on these test results, a Weibull analysis was conducted as described below.

The plot of the results (based on defective items resulting from claims) is shown in Figure 25.8. It clearly shows a bathtub failure rate for oil seal failures. The three shape parameter (m) values correspond to three different failure modes. The figure resulted in the following new knowledge:

1. In the initial period, the failure rate is decreasing (slope (m) < 1), in the middle period it is constant (slope $= 1$) and in the latter period it is increasing (slope > 1), indicating a bathtub failure rate. The failure rate in each of the three section can be modeled by a different Weibull distribution, so that the failures can be modeled by a sectional Weibull model [see Blischke and Murthy (2000)].

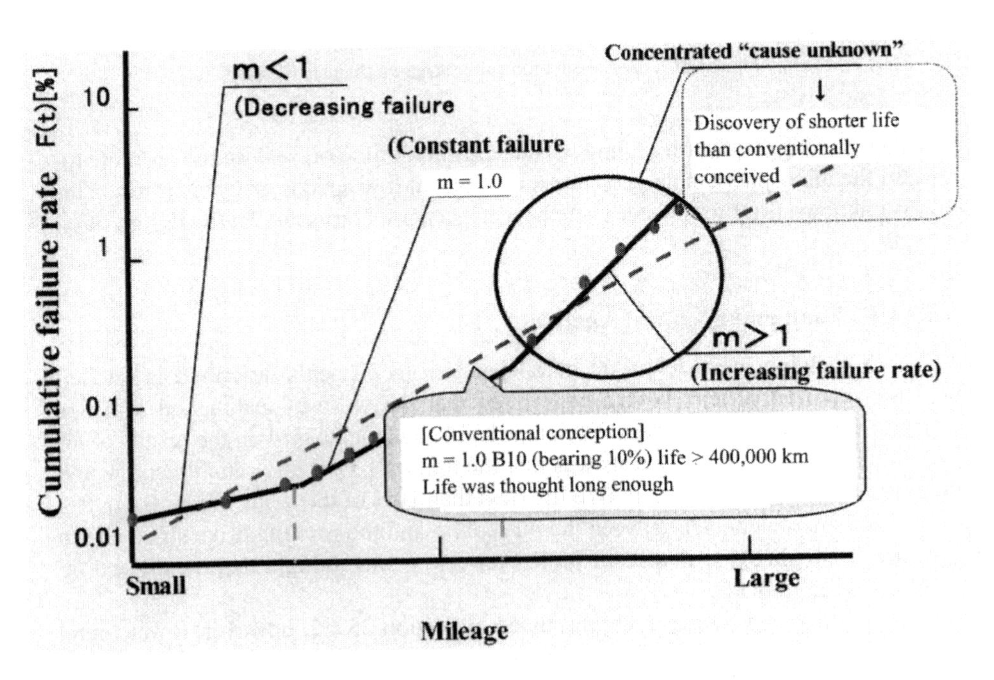

Figure 25.8. Result of Weibull analysis.

2. The initial failures (with failure rate decreasing) occur up to a run distance of 50,000 km. Failures in the intermediate range (with failure rate constant) occur until 120,000 km. Finally, failures occurring above this value (with failure rate increasing) are due to wear.
3. The B10 mode life is approximately 220,000 km, about half the value stated as the design requirement.

To confirm the reliability of these results, the subsequent claims were analyzed using the Toyota DAS system. Within the warranty period (number of years covered by warranty), the total number of claims classified by each month of production (total number of claims from the month of sales to the current month experienced on the vehicles manufactured in the same month) divided by the number of vehicles manufactured in the respective month of production is about twice the design requirement. This agrees with the result of the above reliability analysis.

The influence of five dominating wear-causing factors (period of use, mileage, margin of tightening, hardness of rubber, and lip average wear width) was studied by two-group linear discriminant analysis using oil-leaking and non-oil-leaking parts collected in the past. The result showed high positive discriminant ratios of 92.0% and 91.7% for both group 1 (oil-leaking parts) and group 2 (non-oil-leaking parts). From the partial regression coefficients of the explanatory variables in the linear discriminant function obtained, the most significant influence was found to be the hardness of the rubber at the oil seal lip. The influence ratios of the five factors were obtained by means of an orthogonal experimental design (L27), with three level values, which were thought technically reasonable in consideration of the nonlinear effects, assigned to each of them. [See Steinberg (1996) for a discussion of relevant experimental designs.]

Figure 25.9 shows the influence ratios of each factor contributing to the discrimination. The figure shows that the hardness factor of the rubber is highly influential. This analytical result was also convincing in terms of inherent technologies. To test the validity of this result, the lip rubber hardness and the degree of wear of other collected oil seals was examined further. As a result, it has been confirmed that eccentric wear is more likely to shorten the seal life because the rubber hardness at the lip portion decreases.

The result agrees with experience. Such survey and analysis could not have been carried out with the conventional separate investigation activities of Toyota or NOK. This could only be accomplished through the "dual total task management team" activities between the two companies.

25.4.4 Design Changes for Improving Reliability

Based on the new knowledge acquired (as discussed in Sections 25.4.2 and 25.4.3), several design change plans were created to address the following two reliability problems: (1) The result of the Weibull analysis and visualization tests showed that some items had an unusually short life. It was recognized that it was to necessary to

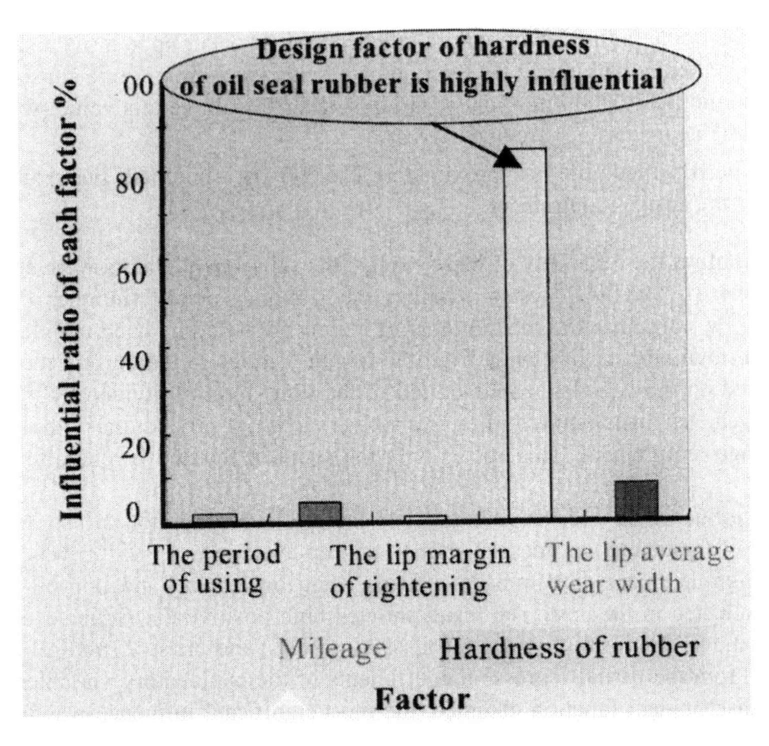

Figure 25.9. Effect of each factor.

prolong the life of items. (2) The study confirmed that there was considerable variation in the oil seal lip hardness, and this had to be controlled.

The following design modifications to improve the transaxle reliability were implemented:

1. Improvement in wear resistance was achieved by increasing the gear surface hardness through changes to the gear material and heat treatment.
2. The mean value of the oil seal lip rubber hardness was increased, and the specification allowance range narrowed. This, in combination with improvements in oil seal lip production technology (including the rubber compound mixing process to suppress deviation between production lots), led to improving the process capability.

Some other issues not discussed here included the following. A design improvement plan was generated which involved a new lip form that prevented the reduction in the pumping capacity even when the sliding surface was worn out. The deposition of foreign matter (particularly when the lip was worn) was reduced by changing the design tolerance for the oil seal lip hardness and by improving the process capability. These together increased the B10 life to over 400,000 km. As a result, the cumulative number of claims by production month decreased by a factor of 10. The final outcome was achieving longer life, as initially planned.

25.4.5 Scientific Process Control

The goal of this activity was to further improve the reliability of oil seals and transaxles, produced in mass, by ensuring process capabilities and proper process management at both NOK and Toyota. This was achieved by controlling the quality variations needed to satisfy the specified design tolerances and a quality assurance management system to maintain the high quality. A new scientific process management, called "inline–online SQC" [see Amasaka and Sakai (1996, 1998)] was used. This is capable of detecting abnormalities in the process that can affect product quality and preventing defective items being released through the use of "science SQC" concepts.

A defect control monitor system was implemented for improving product reliability through intelligent process management and linking it to the IT (Information Technology) used in solving the oil leakage problem. Higher coaxial centers of metal oil seal housings, coil springs and seal lips, their alignment, contact width of oil seal lips, and thread profile identified during the design modification were properly monitored and controlled during the production to ensure high quality for the oil seals.

For transaxles, improvements in roundness and surface smoothness of the drive shaft (resulting in the reduction of metal particles that cause the wear of gears in the differential case) were achieved. Furthermore, the ARIM-BL (availability and reliability information manufacturing system-body line) was introduced, and this contributed to higher reliability in the process.

The quality assurance "QA network" proposed and verified by the author [see Amasaka (1998)] was adopted and the "Toyota's picture maps system" was established to build quality and reliability into production processes from upstream to downstream and improve reliability through prevention of defective items being released. This was combined with the QC process chart, production chart, and process FMEA (failure modes and effects analysis) that Toyota had been using for many years.

The quality failure picture map (called simply "picture map") was posted and used to clearly indicate the correlation between in-process quality failures and initial field failures (resulting in claims) in order to improve the quality awareness of operators at production sites and achieve complete conformance to the specified standards. For controlling foreign matter, the new information that the oil starts leaking with the presence of foreign matters of approximately 75 μm in size (caused by the yarn dust produced from gloves during work, rubbish, powder dust, etc.) was publicized. This required the operators to prevent such material being introduced into the transaxle so as reduce early failures of the oil seal.

Similar "picture maps" were also used for material acceptance inspection and parts logistics as part of the total quality assurance system.

25.5 CONCLUSION

This case study deals with the new type of approach taken jointly by Toyota and NOK to improve the reliability of oil seal for transaxles of vehicles produced by Toyota. A "dual-type total task management team" involving both Toyota and NOK was created to achieve the reliability improvement. This involved the use of "management SQC," "SQC technical methods" and the core technologies of "science SQC."

Toyota and NOK formed "TDOD-Q5" and "NDOS-Q8" teams to improve the reliability technology from technology management (TM), production management (PM), and information management (IM) viewpoints. Their R&D and design teams were closely linked to clarify the oil leakage mechanism, and this led to the visualization technology for understanding the dynamics of the oil leakage. Based on this new knowledge and the Weibull analysis of failure data, the design of the seal was modified and higher reliability was achieved. Through proper process management and "total QA network activities," the claims from oil seal leak were reduced by a factor of 10.

At present, using the knowledge obtained from the studies done, development of the next-generation oil seal is being carried out. It will have little wear and will prevent the leakage even if the lip is worn out.

ACKNOWLEDGMENTS

We are indebted to Dr. D. N. P. Murthy, Professor, Department of Mechanical Engineering, The University of Queensland, and Dr. W. R. Blischke, Professor, Department of Information and Operations Management, University of Southern California, for their support and comments on the original draft that resulted in a much improved article. We would also like to thank the persons concerned at NOK Corporation and Toyota Motor Corporation for their comments and suggestions.

REFERENCES

Amasaka, K. (1997). A study on "science SQC" by utilizing "management SQC"—a demonstrative study on a new SQC concept and procedure in the manufacturing industry, *Journal of Production Economics* **60–61**:591–598.

Amasaka, K. (1998). Application of classification and related methods to SQC renaissance in Toyota motor, *Data Science, Classification, and Related Methods,* Springer, pp. 684–695.

Amasaka, K. (2000). A demonstrative study of a new SQC concept and procedure in the manufacturing industry—establishment of a new technical method for conducting scientific SQC, *The Journal of Mathematical & Computer Modeling* **31**:1–10.

Amasaka, K., and Osaki, S. (1999). The promotion of new statistical quality control internal education in Toyota Motor—A proposal of "Science SQC" for improving the principle of TQM," *The European Journal of Engineering Education on Maintenance, Reliability, Risk Analysis and Safety* **24**(3):259–276.

Amasaka, K., and Sakai, H.(1996). Improving the reliability of body assembly line equipment, *International Journal of Reliability, Quality and Safety Engineering* **3**(1):11–24.

Amasaka, K., and Sakai, H. (1998). Availability and reliability information administration system "ARIM-BL" by methodology in "inline–online SQC" *International Journal of Reliability, Quality and Safety Engineering* **5**(1):55–63.

Blischke, W. R., and Murthy, D. N. P. (2000). *Reliability: Modeling, Prediction, and Optimization,* Wiley, New York.

Hirabayashi, H., Ohtaki, M., Tanoue, H., and Matsushima, A. (1979). Troubles and counter measures on oil seals for automotive applications, *SAE Technical paper series,* 790346.

Kameike, M., Ono, S., and Nakamura, K. (2000). The helical seal: Sealing concept and rib design, *Sealing Technology,* **77**:7–11.

Kawahara, Y., Abe, M., and Hirabayashi, H. (1980). An analysis of sealing characteristics of oil seals, *ASLE Transactions,* **23**(1):93–102.

Lopez, A. M., Nakamura, K., and Seki, K. (1997). A study on the sealing characteristics of lip seals with helical ribs, *Proceedings of the15th International Conference of British Hydromechanics Research Group Ltd 1997 Fluid Sealing,* 1–11.

Nakamura, K. (1987). Sealing mechanism of rotary shaft lip-type seals, *Tribology International* **20**(2):90–101.

Sasaki, S. (1972). Collection and analysis of reliability information on automotive industries, *The 2nd Reliability and Maintainability Symposium, JUSE (Union of Japanese Scientists and Engineers),* 385–405.

Sato,Y., Toda, A., Ono, S., and Nakamura, K. (1999). A study of the sealing mechanism of radial lip seal with helical ribs—Measurement of the lubricant fluid behavior under sealing contact, *SAE Technical Paper Series,* 1999-01-0878.

Steinberg, D. M. (1996). Robust Designs: Experiments for Improving Quality, in *Handbook of Statistics, Vol. 13* (S. Ghosh and C. R. Rao, Editors), Chapter 7, North-Holland, Amsterdam.

EXERCISES

25.1. Discuss the concept of teamwork and its importance in the analysis of product reliability and in the implementation of reliability and quality improvement programs.

25.2. For solving today's engineering problems, it is possible to improve experimental and analysis designs by using N7 (the new seven tools for TQM), the basic SQC method and reliability analysis based on investigation of accumulated technologies.

(a) Describe each of these concepts, using the references given above and/or current quality and reliability texts.

(b) Discuss how these may be applied in a large organization known to you.

25.3. Figure 25.9 above shows the influence of five factors studied in the oil seal leakage problem, with hardness found to account for the vast majority of the variability. Suppose that the result had shown instead that this accounted for only 40% and that mileage also accounted for about 40%. How would this change the way in which you would go about improving the oil seal reliability?

25.4. Discuss the relations that an end user must establish with suppliers to assure product quality and reliability, and describe how this would be done. Suppose that there are a large number of suppliers. Would your approach change if there were only a few suppliers? Explain.

25.5. Discuss the relationship between end user and suppliers from the suppliers point of view, assuming that high reliability of the end product is the goal.

25.6. The network system of "inline–online SQC" was discussed in the case. In order to achieve centralized control of the data concerning availability and defect of equipment by applying various "computer aids," the availability data of equipment groups in domestic and overseas plants are collected and sent via a network. The availability information and the process capability information in signals are transmitted from the site information collecting PC so that each concerned department can check the information on the office monitor in real time and use the information for analysis. Discuss how such a program might be implemented in your own organization or a large company with which you are familiar.

PART F

Cases with Emphasis on Product Warranty

Chapter 26, "Warranty Data Analysis for Assessing Product Reliability"

—Analyzes field data, obtained from warranty claims, on previous generations of CD and DVD products in an attempt to (1) increase reliability by preventing recurrence of quality/reliability problems in current designs and (2) decrease development time in a highly competitive market.

Chapter 27, "Reliability and Warranty Analysis of a Motorcycle Based on Claims Data"

—Deals with analysis of a complex set of two-dimensional (age/usage) warranty data on a motorcycle produced in Indonesia, noting typical data deficiencies encountered in claims data, using two bivariate approaches to analysis, and applies the results in estimating cost models for several warranty policies.

CHAPTER 26

Warranty Data Analysis for Assessing Product Reliability

Peter C. Sander, Luis M. Toscano, Steven Luitjens, Valia T. Petkova, Antoine Huijben, and Aarnout C. Brombacher

26.1 INTRODUCTION

26.1.1 Market Trends

Manufacturers are confronted with several developments that threaten their position—for example, increased competitive intensity (more new products) and the market globalization (more competitors and variants per product). Sander and Brombacher (1999) pointed out another threat that companies must face nowadays: high degree of innovation. Products are, from a sales perspective, obsolete in months. This forces companies to develop and produce their products in a very short period of time (Stalk and Hout, 1991). Because of this, new generations of products are developed before field data about the reliability performance of previous generations are available. In this way it is likely that "old" quality/reliability problems recur in new generations.

26.1.2 Product Development Process

The described time pressure affects the way in which organizations organize their product development process (Wheelwright and Clark, 1992), henceforth denoted as PDP. The traditional functional (Taylor, 1919) organization structures PDPs through milestones and deliverables. The advantage of doing so is that during each individual phase, people concentrate on one aspect. Besides, if the structured checkpoints are effective, this type of organization can be very good at controlling quality and reliability (Meyer, 1998). However, the phases will never be truly independent since deci-

Case Studies in Reliability and Maintenance, Edited by W. R. Blischke and D. N. P. Murthy.
ISBN 0-471-41373-9

sions taken in the first phases, like the design, can seriously affect the following ones. In addition, the introduction of massive milestone procedures increases the throughput time and generates more non-value-adding work as a preparation for the milestones (Minderhoud, 1999). Under time pressure this can result in the skipping or removal of some of the activities—for example, tests, evaluations, and reviews—that might lead to the introduction of (even) more field failures (Lu et al., 1999).

Companies try to reduce their time to market mainly by the introduction of concurrent engineering principles (Minderhoud, 1999). The basic idea behind concurrent processes is that all major decisions in the process are taken in the early phases of the PDP, during the phase of maximum flexibility and minimum costs per change. This means that the traditional principle of separation between phases and disciplines is abandoned; all downstream activities start as early as possible, even when upstream activities have not yet been completed.

Advantages, when successfully implemented, are shorter times to market, lower costs, and better product quality. Since, in concurrent engineering, decisions are taken early in the PDP, this results in long delays between decisions and the effects of these decisions. Therefore concurrent engineering can only be successful if adequate information with respect to downstream phases is available early in the PDP. However, part of the required information is very difficult to get on time, especially in short development processes with long learning cycles. This puts pressure on the quality and the speed of the feedback information flows, especially from the field. Since this information is commonly gathered by the service organization, it will be interesting to see whether this part of the organization is able to meet these added requirements.

26.1.3 Service Centers

Outsourcing service is a common practice (Rothery and Robertson, 1995). An advantage for the outsourcer is an increase in flexibility and a reduction of costs. A disadvantage is that normally the outsourcer loses control over the outsourced activities, and the feedback to the manufacturer about field quality problems is relatively slow and incomplete.

A major problem is that supplying service centers are focused on their own benefit, and therefore all activities that do not contribute to their profit are under pressure. If the outsourcer regards quality improvement as a major goal, the assessment of a service center and the financial compensation must be in line with this. A service center should be assessed on two contributions to quality improvement:

- Short-term customer satisfaction (have the product repaired as soon as possible)
- Long-term customer satisfaction (help the company to improve product quality)

It is an interesting question whether this is an achievable aim, because the outsourcer asks his supplier to provide him with information about the repairs with the aim of improving the reliability of the products, which will have as a consequence, in the long term, less revenue for the service center.

26.1.4 Field Failure Feedback

In order to be able to learn from the past, it is important for manufacturers to collect, analyze, and use information about customer use and field behavior of the products (Brombacher, 1999; Brown and Eisenhardt, 1995; Petkova et al., 1999). From a reliability point of view this serves three objectives:

1. In the very short term (days or weeks), it should demonstrate whether the products fulfill customer needs in a safe and reliable way. If not, necessary actions should be taken, such as design change or changes of the production process, in order to adapt the product to the way it is used. The conclusions should be based on engineering data.
2. In the short to intermediate term (weeks or months), it should demonstrate what design and/or production improvements are most wanted in coming generations or new designs, in order to prevent the recurrence of "old" failure modes in new generations. The main sources of information are engineering data and statistical data about frequency of problems.
3. In the long term (several months), it should make clear whether the product reliability is in line with the predictions and should find out whether the company is learning from the past. The answers to these questions should be based on statistical evidence.

This case study concentrates on objectives 2 and 3.

26.1.5 Case

In 1982 Philips Electronics together with Sony introduced the first CD players on the market. Since then there have been many improvements, but the basic technology remained the same. In 1988 the CD changer was introduced, followed in 1998 by the DVD player. At the moment, companies are working on the next generation using new discs and new (blue) lasers with shorter wavelength. That generation, called digital video recorder (DVR), is still based on the same CD principle, but it has a much larger capacity, namely about 22 GB, compared with the 700 MB for the regular CD. The relation between the different generations of products is given in Table 26.1.

This chapter describes a case study performed by the Storage and Retrieval Group of Philips Research in the Netherlands, in cooperation with Philips Audio. Since the advent of modern, concurrent engineering PDPs require the availability of detailed information early in the PDP, this study was carried out in order to determine whether it is possible to prevent the recurrence of old failure modes in the new generation of DVRs by using field failure information about CD and DVD players.

Table 26.1. Technical Characteristics of Generations of CD-Based Products

	CD	DVD	DVR
Optical wavelength	Infrared	Red	Blue
Data storage capacity	> 650 MB	> 4.7 GB	> 20 GB
Trackpitch	1600 nm	740 nm	320 nm

The purpose of this case study is to address the following question:

- Is it possible to extract sufficient information from warranty data of earlier products to prevent recurrence of failures in new products?

The approach used in this case study is:

- Determine a relevant set of metrics for the purpose of field feedback.
- Determine whether these metrics can be derived from available field data.
- Analyze the results.

26.2 RELIABILITY METRICS

Most textbooks on product reliability present several metrics that can be used to measure product reliability. Basically there are two families of reliability metrics: product-related metrics and logistic or repair-related metrics. Since, especially in high volume consumer products, not all items reach the market at the same moment in time, usually the market will contain a mix of new and slightly older products. Depending on the focus, different organizations will be interested in different aspects. People involved in the product per se will be mainly interested in the number of failures versus the age of the product; people interested in warranty costs or repair logistics will be mainly interested in the number of repair calls in a given (calendar) time interval. The following sections will discuss commonly used reliability metrics together with their main application area. As a next step, the applicability of these metrics in reliability prediction and improvement processes will be discussed.

26.2.1 Reliability Metrics

26.2.1.1 Metrics Based on Call Rates

In this section we discuss three strongly related metrics: the field call rate, the warranty call rate, and the warranty packet method as used commonly in industry.

Field Call Rate

The field call rate (FCR) is used to monitor the fraction number of field failures of a given product. It was developed for logistic purposes, like the estimation of the number of spare parts that will be needed in a given location at a certain point in time. Basically, on the interval (a, b) the FCR is estimated by

$$\text{FCR}(t_1, t_2) = \frac{M(t_2) - M(t_1)}{(t_2 - t_1)N(t_1)} \tag{26.1}$$

where

$M(t)$ = number of failures until time t
$N(t)$ = number of items on the market at time t
t = time since market introduction of the product

Observations

- The expression does not take into account the age of the item.
- Since the failure probability is usually a function of the age of an item, and since usually not all items are sold simultaneously, the FCR has little to do with the expected number of field problems in a given time interval. Therefore the FCR is not very useful for reliability predictions.
- The number of items on the market, $N(t)$, is difficult to determine, for example, by incomplete feedback from retailers.

Warranty Call Rate

Usually companies are only interested in products under warranty. In those cases the FCR is substituted by the warranty call rate (WCR). The only difference between the FCR and the WCR is that the WCR only takes into account the products under warranty. Therefore, this metric uses a kind of moving time-window. Philips Audio uses the WCR instead of the FCR.

Warranty Packet Method

The metric currently used by Philips Audio to estimate the WCR is the warranty packet method (WPM). Suppose the warranty period is 12 months. Then the WPM is defined as follows:

$$\text{WPM} = \frac{\text{Total number of warranty repairs over the last 12 months}}{\text{Average number of units under warranty last year, based on monthly figures}} \times 100 \tag{26.2}$$

The WPM is an improvement over the WCR in the sense that it updates itself every new month. However, it has two major drawbacks:

1. A sudden change in a given month is easily hidden among the figures of the preceding eleven months.
2. The metric is susceptible to changes in sales volume (see Exercise 26.2).

For these reasons, Petkova et al. (2000) advocate the use of the hazard function, although the hazard function has some drawbacks as well (see Section 26.2.1.3).

26.2.1.2 Hazard Function

A second family of reliability metrics does not relate to the logistics aspect of reliability (number of repaired products in the field at a certain moment of time) but to reliability aspects of the product itself (number of products repaired at a certain age of the product). In general the failure probability $F(t)$, or the reliability $R(t) = 1 - F(t)$, gives a good overall impression of the failure behavior of a product. In the case of censored observations, the Kaplan–Meier estimator can be used to estimate $F(t)$, and Greenwood's formula (Meeker and Escobar, 1998) gives a confidence bound for F. Since the goal of our research is to learn from the past, we are not satisfied with functions/graphs that only give the fraction of failures as a function of time. We prefer functions/graphs that help to find the root cause of failures. The hazard function is promising, because it shows whether there is a relation between the age of a product and the instantaneous failure probability. For example, if the hazard function is constant, then, from the point of view of instantaneous failure probability, new products are just as good as old products. This information does not follow from graphs of the failure probability $F(t)$ or the failure density $f(t)$. Unfortunately, the authors were not able to find confidence bounds for the hazard rate in the literature.

The hazard function $r(t)$ represents the instantaneous failure probability and is given by

$$r(t) = \frac{f(t)}{R(t)} \tag{26.3}$$

The behavior of $r(t)$ as a function of time reveals some information about the cause of failure. In the literature, its graphical representation is typically given by the bathtub curve (Figure 26.1) or the roller-coaster curve (Figure 26.2). The bathtub curve distinguishes three phases. First comes the period with relatively many early failures (phase 1). Then a period of random failures follows (phase 2). Finally there is a period with an increasing number of failures that are related to aging of the product (phase 3).

Some authors [e.g., Wong (1988)] distinguish four different classes of failures, not all of which may occur. In this model, two types of early failures are distinguished: hidden 0-hour failures (phase 1a in Figure 26.2), and early wear-out failures (phase 1b). Phases 2 and 3 correspond to phases 2 and 3 of the bathtub curve. A theoretical explanation of this model is given by Brombacher et al. (1992). The first group of failures (1a) is, in this model, attributed to failures already existing during manufacturing but observed only after a certain observation delay in the field. The second group of failures (1b) is due to tolerances in either products or users as a result of, for example, manufacturing spread. Products in this category pass all tests during manufacturing, but these, relatively speaking, weak products show a strongly accelerated degradation behavior.

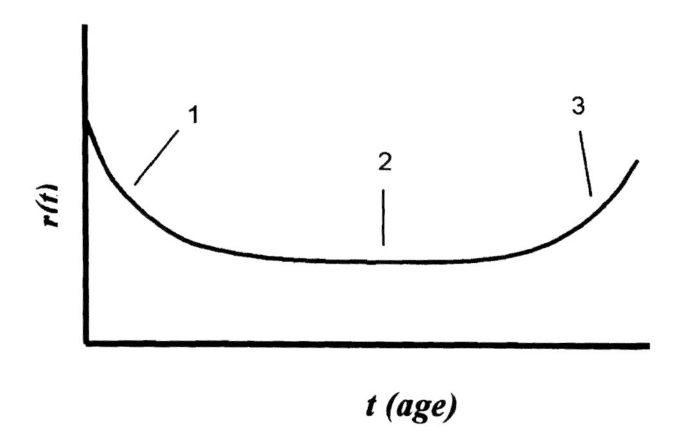

Figure 26.1. Bathtub curve.

The hazard function is defined as the instantaneous probability of failing at time t given that the product is still functioning at time t. From this the following estimate of the hazard function follows [see Petkova et al. (2000)]. Let $h(i)$ be the hazard function of a product at i time units after it has been sold. At time τ_j, the instantaneous failure probability after surviving i time units, say $r_j(i)$, is estimated by

$$\hat{r}_j(i) = \frac{\text{Total number of products that failed immediately after surviving } i \text{ time units}}{\text{Total number of products that survived } i \text{ time units}}$$

$$= \frac{d_{0,i+1} + d_{1,i+2} + \cdots + d_{j-i-1,j}}{n_{0,i} + n_{1,i+1} + \cdots + n_{j-i-1,j-1}} \tag{26.4}$$

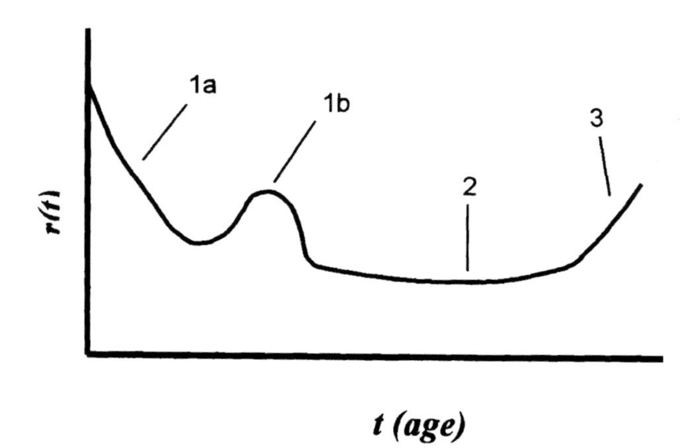

Figure 26.2. Four-phase roller-coaster curve.

where

n_{ji} = number of products functioning at τ_i of all products sold at τ_j
d_{ji} = number of products failing in the interval $[\tau_{i-1}, \tau_i)$ of all products sold at τ_j

Plotting this estimator against i (number of time units, e.g., months), it is possible to compare the graph with the roller-coaster curve as given in Figure 26.2. This might help to understand the failure modes (hidden 0-hour, early wear-out, random or aging failures).

26.2.1.3 Comparing WCR and Hazard Function

The main differences between the FCR/WCR and the hazard function are as follows:

- The FCR and WCR do not take into account the age of the product, while the hazard function does.
- The FCR and WCR use the fraction of repairs over the last year (in case of a 12-month warranty), while the hazard function is able to monitor the "real time" behavior of products.
- If the hazard function is updated when new failure information becomes available, then implicitly it is assumed that the data have been collected in a stable situation. When the product reliability of products depends on the period of production, then separate hazard functions should be used for the different periods.

26.2.1.4 MIS–MOP Diagram

In the theoretical explanation of the roller-coaster model presented in Figure 26.2, it was stated that, according to this model, manufacturing tolerances and tolerances in use play an important role in the actual shape of the curve. Especially in cases where manufacturing is not entirely stable due to, for example, ramp-up effects in early manufacturing, a further refinement would be to measure failures not only against age of the product but also against age of the manufacturing process. In Figure 26.3 it is shown that important systematic problems can be detected by using a simple diagram in which the time to failure (TTF) is given in relation to the month of production, a so-called MIS–MOP diagram (months in service versus month of production). Of course, "time" should be interpreted as number of lifetime units.

The MIS–MOP diagram is very useful to detect the following phenomena:

- Failures that occur mainly during the first production series
- Early failures that occur during the first period of customer use
- Failures that occur during a short production period

With respect to "months in service," the MIS–MOP diagram follows the previously mentioned roller-coaster model. In those cases where manufacturing would not be stable, however, this would show up not in the roller-coaster projection, but in the number of failures against manufacturing time.

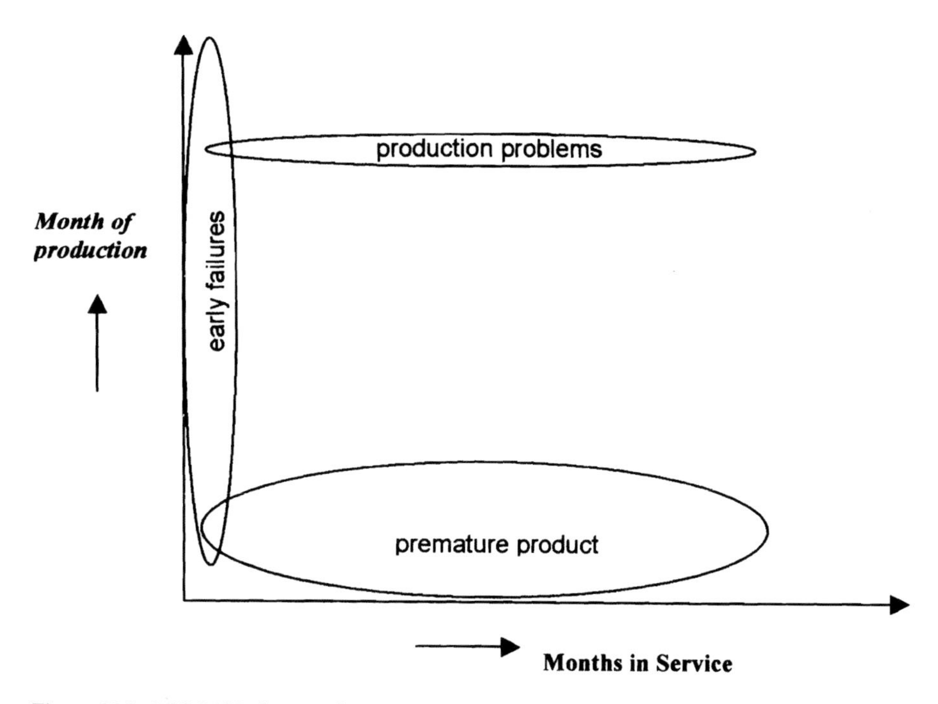

Figure 26.3. MIS–MOP diagram showing the occurrence of systematic problems as a function of the month of production and the number of months in service.

This diagram is an interesting addition to the WCR. The WCR gives the percentage of problems, the MIS–MOP does not, but the MIS–MOP gives the distribution of the TTF for each month of production, which is independent of sales.

26.2.2 Sources of Data

Philips sells the products of concern in a large number of countries. Because of practical reasons, it was decided to concentrate on the Netherlands as sales organization. The reason was that the Netherlands is thought to have the most reliable field data feedback system, and also the one with the "oldest" data. Moreover, it would also be easier to get the data in the limited time that was available for this research. It is widely known that there are big differences in warranty claims between countries, so the concentration on the Netherlands limits the scope of the project.

The data required concerning the sales and repairs for these products in the Netherlands were provided by the Philips service organization in Europe (Euroservice).

26.2.3 Limitation

The aim of the project is to provide a detailed study on the reliability of the CD and DVD products chosen, in order to highlight the most important aspects to be taken into account in the development of the forthcoming third generation. It was the original intention to perform a quantitative and a qualitative analysis of the data. The quantitative analysis should give information about changes in product reliability over generations and about relations between product reliability and the introduction of new technology. For this purpose, the WCR and the hazard function were chosen. It was expected that the hazard function would be more informative than the WCR.

The qualitative study should focus on the root cause analysis in order to be able to prevent the recurrence of old failures in new models. Given the low quality of the available data and the limited time available for the project, this part of the study will not be reported in this chapter.

26.3 FIELD FEEDBACK

26.3.1 Service Centers

Due to a strong "focus on core-business," many companies in the area of consumer electronics outsource their repair activities. The independent service centers are obliged to inform the manufacturer about the repairs. Usually the activities performed during product repair are filled in on a so-called job sheet (either in paper or in electronic form). These job sheets are then analyzed by the manufacturer and used as a basis to pay the costs associated with these warranty repairs.

Philips does not outsource all of its repair activities. In order to maintain a direct link to the field use of the product, it owns three special service centers in Europe; these are called initial workshops or competence centers and are located in Eindhoven (the Netherlands), Paris (France), and Köln (Germany). They are part of Euroservice. These initial workshops are not focused on cost or time, but on collecting quality- and reliability-related information. Normally the most problematic units are taken to these three centers. They usually replace the defective units with new units in order to be able to study the defective ones in detail. The initial workshops are the only ones that inform weekly of their findings; and when a serious flaw is detected, the corresponding factory takes action. Figure 26.4 gives the structure of the field feedback process.

From 1980 up to 2000, Philips used a quality feedback system called the MABRA system. In 2000, a new quality feedback system, called the QFB system (quality feedback system), was implemented. Since the data studied belong nearly totally to the period before 2000, our findings only relate to the old quality feedback system. As mentioned in Section 26.2.5, only field information from the Netherlands has been used in this case study.

26.3.2 The Data

The collected data can be divided into four categories:

- Date of production
- Number of units sold per month (on type number level)

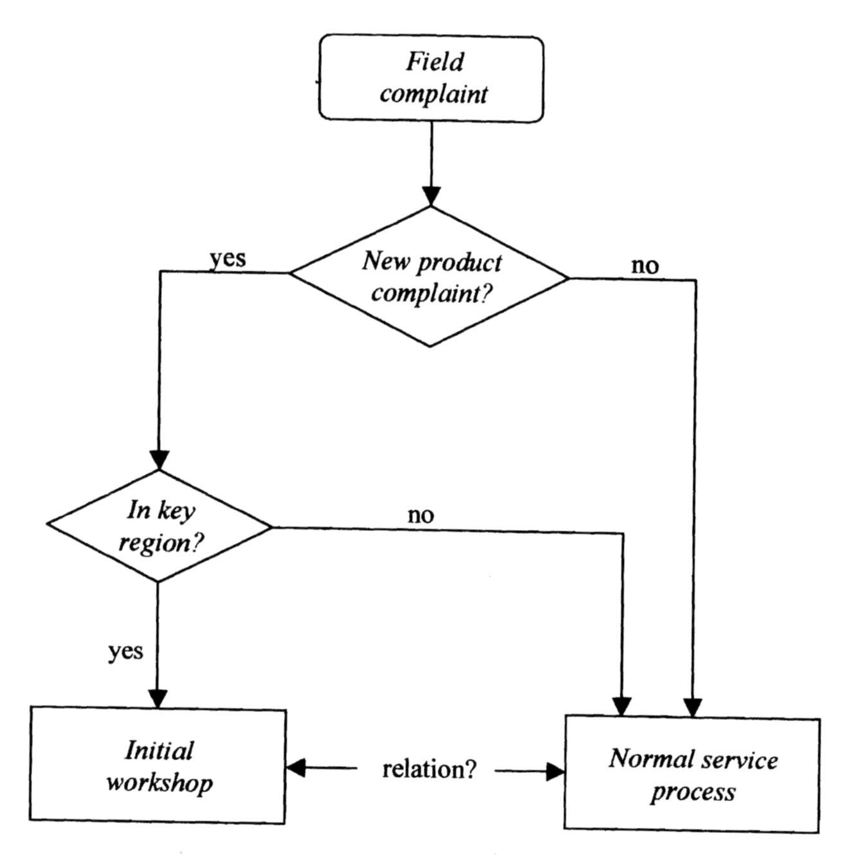

Figure 26.4. Structure of field feedback process.

- Number of warranty repairs
- Repair job sheets (qualitative information)

Since commercial pipelines for these products are usually quite short, it is common practice to take "start of use" as manufacturing date plus a certain offset. In consumer electronics, an often-used assumption for this pipeline is 6 weeks. With respect to the classical, three-phase bathtub curve, this is usually a pragmatic assumption; in the stated period, wear-out can be ignored and early failures take usually much longer than 6 weeks. More recent studies have shown, however, that this assumption creates a considerable amount of noise with respect to phase 1a of the roller-coaster curve. Therefore, currently the manufacturing date and the "start of use" are gathered as different parameters.

Sales per month concerns the number of units sold by the different factories. The real date on which these units are sold to the end user is unknown, since the only available date is the one on which the products are shipped to the dealers. The complete type number of the product is specified.

The involved service centers provide Philips with the number of repairs under warranty (along with the corresponding costs). These figures are checked by the national organizations before being accepted as warranty repairs.

The repair job sheets are the field quality forms that are filled in by the technicians in the service centres. These job sheets are the only source of qualitative information available from the normal service centers (excluding the initial workshops). At present, the qualitative information is mainly recorded in the so-called IRIS codes for brown goods (category of goods that includes consumer electronics products). The acronym IRIS stands for International Repair Information System and was promoted by EACEM (European Association for Consumer Electronics Manufacturers). It is a standardized coding system for the repair data, which must come from dealers and repair workshops to the manufacturers in order to improve the quality of their products. This coding system is seen as a big step toward fully structured, computerized repair administration and was implemented by Philips Audio in 1995. Repair technicians can easily generate these codes by using a PC-Windows freeware program, by just selecting the relevant descriptions from drop-down lists. In this way, simple pointing and clicking can construct one or more standard IRIS data lines.

At present, all activities concerning the quality and reliability of sold products concern the warranty period. Because the warranty period will be extended in the near future, this is a risky situation. European regulations are likely to demand longer warranty periods for consumer electronic products; hence, information about the behavior of the products after the present warranty period ought to be collected as soon as possible.

In general the quality of the database used to store the field information was quite low. The main problems that showed up during the study are the following:

- *Number of items on the market.* An item is considered to be on the market two months after being shipped to the dealer. However, the available data showed that this is not always the case; it can take much shorter, but also much longer, times. Most likely there will be important differences between countries, time of the year, and even different retailers. From the point of view of data analysis, it would make sense to introduce a system in which dealers inform the company of the sales they make every month, even with exact dates of sale. The economic viability of such a system should be studied.
- *Failure date.* It has been observed that the date used for the WCR calculations can differ considerably from the failure date. This will invariably delay the possible reactions of the company.
- *Errors.* Many cells in the available data files contain mistakes. For example, some models that appeared as repaired did not really exist, and many of the cells were empty.

26.3.3 Time Factors

In Section 26.1, it was explained that the speed of the field failure information flow should be increased considerably. Nowadays it is estimated that the time that companies take to learn from the field about the performance of their products is about 7 months for disasters and 1 year for normal-performing cases (Brombacher and Sander, 1999).

For the actual feedback system, some problems related to time are as follows:

- The field quality report. The Philips Audio field quality reports, presenting the results of the quantitative and qualitative analyses, are at present published every 3 months. The report is normally issued 1 month after the end the reported period. This means, for instance, that the report of September 1999, including the WCR for July, August, and September, was issued at the end of October 1999.
- The difference between the day on which the product was taken to the service center and the day on which Euroservice processes the repair into its database varies between 15 days and 4 months, with an average of 1 month and a half. It is to be expected that the delay counting from the moment the failure was observed can be much longer.

26.3.4 Root Cause Analysis

In Philips Audio, there is a structural analysis of the field repairs only in case the WCR of a certain product or component increases sharply. In this case, the failure is analyzed in order to identify the root cause of the failure. By informing development and production, an attempt is made to prevent the recurrence of similar problems in successive products.

The qualitative information received from the service centres concerning the repairs is based on the IRIS codes. (More recent analysis has shown that the IRIS codes are inadequate for a more detailed technical root-cause analysis. Due to the very generic nature of the description, the time-delay between introduction of new technology into the market, and definition/update of IRIS codes, the IRIS code cannot be used to distinguish between different phases of the roller coaster curve via a technical root-cause analysis.) The main IRIS codes used are as follows:

- *Condition code:* Gives information about when the failure occurs (e.g., 1 = constantly, 4 = in a hot environment, etc.).
- *Symptom code:* Describes the problem as perceived by the user (e.g., 110 = power problem or not operating, 620 = irregular mechanical operation, 11X = other "power" problem, etc.).

The customer complaint is supposed to be the information source for these two codes, but often it is not well-specified and is entered or/and corrected by the technician. The remaining codes specify the technician's diagnosis:

- *Section code:* Indicates in which part of the set the intervention was made (e.g., CTR = control panel, SFT = software, etc.).
- *Defect code:* Specifies the type of defect (e.g., N = defective electrical component/module, I = loose/off/stripped, etc.).

- *Repair codes:* Indicates the actions performed by the technician (e.g., A = replacement, G = repaired electrical parts, etc).

As we can see, these descriptions are quite vague and general. Moreover, these cells are sometimes not filled out, which makes it even more difficult to figure out the origin of the problem.

Further available information is about the replaced components.

It is usually not possible to identify the root cause of a failure from the available data, because the real condition under which the failure occurred is often not known. This is especially the case for the so-called no-fault found failures (NFF), where the customer has a complaint but the repair center cannot confirm the problem. In these cases, the problem often seems to be due to an unanticipated mode of use by the customer. In those situations, it is the task of the designer to try to reproduce the failure and to identify the possible root cause(s). This is often quite difficult, partly because not all repair technicians have the right level of education and repair experience, so sometimes components may be replaced which are not defective at all.

26.4 ANALYSIS OF FIELD DATA

26.4.1 Introduction

There are a number of factors that complicate the analysis of field data. We mention several of them in this section. Among them are:

- The environment—for example, humidity
- The usage mode—for example, intensity of use (such as the speed during recording of a CD recorder)
- The characteristics of the CDs used—for example, inferior illegal copies
- The audio system of which the CD player is a part

Because there is no information available about these factors, we cannot take them into account.

Some preliminary meetings were held at the onset of the project with different people from the company to gain insight into the evolution of CD and DVD systems. These discussions concentrated on the CD players, since CD changers and DVD players are quite recent developments and trends are not easily seen.

In one of these meetings, a curve was drawn that, according to the specialists, should give a reasonable estimate of the WCR evolution for the CD player. The curve distinguishes production locations. It starts in the Netherlands, where the first Philips CD player reached the market in 1982, and ends in Singapore in 1999 (Figure 26.5).

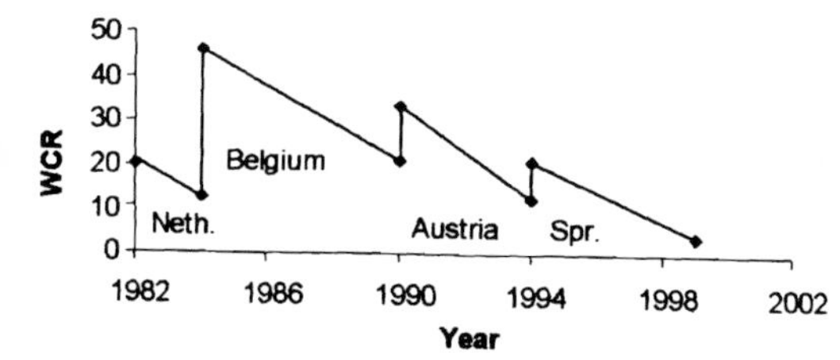

Year	Location	Activities
1977	Netherlands	Development
1982	Netherlands	Dev.+Manuf.(pilot prod.)
1984	Belgium	Dev.+Manuf.
1990	Austria	Dev.+Manuf.(Althofen)
1994	Singapore	Dev.(redesign)+Manuf.
1999	China	Manuf.

Figure 26.5. WCR evolution for CD players.

Notation

For obvious reasons, in all figures of this chapter standardized WCR units have been used instead of the actual WCR values. Standardized WCR units are defined by giving the maximum of all calculated WCRs over all analyzed products the value 100. The same has been done for the hazard function. Thus the different products can be compared with each other and also with the guessed values of Figure 26.5, but the actual field values cannot be traced.

Figure 26.5 shows that according to the feeling of the specialists, the company does learn from the past in the sense that the WCR leaps up when manufacturing moves to a new location. It is interesting to note that experts from the company stated that the top 10 failure mechanisms are solved in a relatively short period, but that these old failure mechanisms sometimes tend to reappear as new employees enter the process.

We wanted to verify this curve with real data and with the real WCR—that is, including the repairs from all service centers. We also wanted to evaluate the effect on the WCR of a change in production location compared to the effect of a change in technology, such as from CD player to CD recorder. However, this turned out to be impossible. The reasons for this are the following:

1. *New Field Feedback System.* There were no electronic field failure data available from before 1994. Contrary to expectations, the field data from the period prior to 1994 had been deleted from the database when a new field failure feedback system was introduced. This was a major inconvenience for the project, since the electronic data available did not comprise more than one production location for each product and, therefore, did not allow us to check out the effect of a change of manufacturing site. We tried to overcome this problem by using the available field quality reports published by Euroservice.
2. *Problems with the Field Quality Reports.* The first report studied dates from March 1990 and includes data from the fourth quarter of 1989, and the last one is from September 1999 and includes data up to that month. With this we

have already lost eight years (1982 to 1990) of CD player information. The lack of homogeneity was another important problem to cope with, because it did not allow for a continued analysis. The difficulties encountered are the following:

- *Change of metric.* The oldest reports do not give the warranty call rate figures but the repair frequency (RF). This means that in the calculations only the repairs executed at Philips own workshops are taken into account. They also give graphs in which the difference between the RF and the total WCR can be visually observed. The trend is similar, but the RF curve is, obviously, lower. From January 1991 on, this difference is not shown any more and the WCR cannot even be estimated. Fortunately, from March 1995 onwards, the reports switch from the RF to the WCR.
- *Change of countries.* The first reports contain data only from the Netherlands. Then, the report from October 1994 includes data from Italy, Austria, and Switzerland. The next report, December 1994, no longer includes Switzerland. Finally, in the March 1996 report, Germany and France were added to the Netherlands, Italy, and Austria (these are referred to as the five key European countries) for the WCR reporting. This has continued to the present. The reports continue to give the trend for the Netherlands for the different article groups up to and including 1997.
- *New and old sets not appearing.* Units younger than 6 months do not have representative WCR figures with the calculation method employed; therefore, only predictions for them are included in the reports. This makes it very difficult to check whether an initial rise of the WCR occurs for the products when starting production in a factory. Also, old units (still under warranty but no longer sold) in which the WCR results are strongly influenced by a decrease in sales do not appear explicitly.
- *Warranty packet not given on type number level.* The warranty packet is detailed for each main type number only from the March 1998 report on, although only the figure for the last reported month is given. Previous reports give, in general, only the WCR percentages.
- *Distinguishing production centers not feasible.* We wanted to separate the figures from different factories, to check whether the first units produced when the manufacturing activities were started in a new location showed a higher WCR, as in Figure 26.5. In most cases this was not possible due to the previously mentioned reasons, lack of data for young units, and lack of warranty packets on type number level. This last reason, provided that the introduction of new production locations is very gradual in the reports, does not allow checking whether the total WCR for a product from a particular production center is low or high, since different models from one factory appear with different WCR levels. The same holds true for the products of older production locations; some of its models show WCRs lower than those of the new centers, and others show higher figures. In other cases, this rise was not observed.

The first preliminary research showed the following:

- Assumptions were to be made concerning the sale dates of the items because these dates were only available for a very limited number of repaired units. We will return to this in the next section.
- There were no data concerning the root causes of the field failures, which, to a large extent, impeded the qualitative analysis. We note that the time constraint also affected this analysis.

In order to check whether there was any indication of production instabilities within the same product group, for each product group a MIS–MOP diagram in which the time to failure was given in relation to the month of production was drawn. We checked the data for the following phenomena:

- Failures that occur mainly during the first production series
- Early failures that occur during the first period of customer use
- Failures that occur during a short production period

However, we found no evidence of any of these phenomena for any of the products we analyzed.

26.4.2 Field Data

Despite all the constraints regarding the available data presented in the previous section, it was decided to continue the analysis for the period for which electronic data (1994–2000) are available. The four products for the analysis were then chosen, namely, CD players, CD changers, CD audio recorders, and DVD sets. Moreover, it was decided to concentrate on the Netherlands as a sales organization. The data studied concern only the warranty period, because the company does not collect any data after this period.

- Because the WCR hardly gives any information concerning reliability aspects, we wanted to estimate the hazard function to see whether this metric could be useful for the company. Furthermore, we wanted to have data about detecting root causes and actions to prevent recurrence of failures in succeeding generations.

Unfortunately, not all the requested information was available. The major differences between the requested and the available data can be summarized as follows:

- The company does not know the exact date on which the sales to the end user take place. Only the dates on which the items are shipped to the dealers are known. It is supposed that it takes two months for the items to reach the end user. Therefore this period is also employed in the analysis, though it introduces inaccuracies (see Exercise 26.3). The purchase date is available only for a variable percentage of the repaired units.

- Information about root causes is not available. The only qualitative information available about repairs is in regard to the IRIS codes and spare parts used. This is a major drawback for the project because with the available data it is not possible to provide insight into the most relevant aspects that should be taken into account for the forthcoming generation of products.
- Mistakes exist in the data received. The excel files containing the available data have many mistakes and empty cells, often due to errors made by technicians when filling out the job sheets in the service centers. Examples of mistakes are wrong dates (e.g., the year 1911), repairs of nonexisting model numbers, model numbers for which the sales start at a particular moment and then the first repair shows up one year later, and so on.

26.4.3 Quantitative Analysis

Before we proceed, the calculation methods of the WCR and the hazard function will be clarified:

1. The following remarks concern the WCR:
   - The number of products under warranty on the market, calculated according to a pipeline of two months may be far from reality in many cases.
   - The date taken as the repair date for the calculations differs from the one used by Philips Audio. We employed the repair receipt date instead of the date on which Euroservice introduces the repair into the database, because the former is closer to the real failure date. All repairs are taken into account.
   - The repair date considered for the calculation is usually at least 1 month later than the date on which the product was taken to the workshop. Apart from the already mentioned effect in delaying reactive and preventive actions, this also affects the WCR calculation itself. For instance, in case of a steep decrease in sales, since the decrease in repairs is likely to arrive late, it will cause a temporary increase of the WCR figure.
   - The number of items under warranty is not in the database, but it is calculated using the following data:
     - Number of units sold per month (per type number)
     - Warranty period in months for the products concerned (currently 12 months for audio and DVD products).
     - Length of the pipeline. The pipeline is the time that elapses from the moment the items are sold to the dealers and the moment these items are actually bought by the consumer. We note that this length may differ from one product to another, from one country to another, and even from one dealer to another. At present, Audio uses a pipeline of two months.

In this way, the date on which a product reaches the market is calculated by adding to the date of sale (to the dealers) the pipeline effect. From this date on, the product is considered to be in the market and under warranty during the next 12 months.

2. Due to the mismatch between the required data and the data received, the calculation method for estimation of the hazard function needs to be clarified. The following steps were taken for this calculation:
   - *Choice of time unit.* One month is the time unit chosen, because the sales were given per month.
   - *Choice of censoring date.* The censoring date is the date in which we "stop" the time for the calculation. Considering the data available, the censoring date is the March 31, 2000. The age of the items at the time of censoring is the difference in months between the date these items were supposed to reach the consumer (2 months after being shipped to the dealers since the actual date is unknown) and the censoring date. Consequently, the fact of not knowing the actual sale date introduces inaccuracies in the results.
   - *Choice of the time to failure.* For the time to failure, we chose the difference in months between the purchase date and the repair receipt date. Therefore, only the repaired units for which the purchase date by the end consumer is available have been taken into account. The ones in which this cell is empty could not be used because in these cases it was not possible to estimate the time to failure. Conversely, for the WCR all the repairs are considered.
   - *Calculation of the hazard function.* The hazard function $h(i)$ is only calculated for the months within the warranty period.

In the next section, the results per product type number are presented. Only the curves for the most representative models are included.

26.4.4 Results

In this section, the results of the quantitative analysis are presented. The report structure for each product contains

- WCR, because this is the metric that is used within the company
- Hazard function, because theoretically this metric should help to identify the cause of field problems

Of course, confidence bounds for these would be informative. However, confidence bounds for the hazard function could not be calculated, because as far as could be determined, formulas for such bounds have not yet been published. It should be noted, however, that since we are dealing with quite large numbers of items in most cases, the hazard functions can be expected to be estimated quite precisely.

Although the definition of the WCR is straightforward, we are also not able to give the precision of these results in terms of confidence intervals. The theory for this has also not been worked out. Given the earlier critical comments, however, the result would be of limited usefulness for our purposes since the calculated WCR is not a clear indicator of a quality/reliability characteristic. As mentioned earlier, the only reason we mention the WCR is that this is the metric that is used by the company.

26.4.4.1 CD Players

We start with the evolution of the field call rate (Figure 26.6). The graph is based on the field quality reports and gives the overall result, in standardized units, for the totality of CD players sold in the Netherlands. It was not possible to discriminate between the factories. Since for some years only data from Philips' own workshops are available, in those years the metric does not really give the WCR but instead the repair frequency (RF). Therefore Figure 26.6 should only be understood as a rough mixture of the WCR and RF.

Although the first introduction of the CD player was in 1982, the first WCR that can be estimated from the graphs in the available field quality reports corresponds to December 1985. It is surprising that this WCR is much higher than the level that was mentioned in the introductory meetings (see Figure 26.5).

Figure 26.6 shows a sharp increase up to 100 units in December 1986, which is the maximum level reached. Then it comes back to the former level in June 1987. Comparing this with Figure 26.5, it seems that the explanation for this rise is not a change in production location. It might be a big volume increase in sales, but unfortunately there are no data available to check this. A new increase shows up in September 1987. This rise also cannot be explained by a change of factory, because the location in Belgium continued to be the production center until 1990.

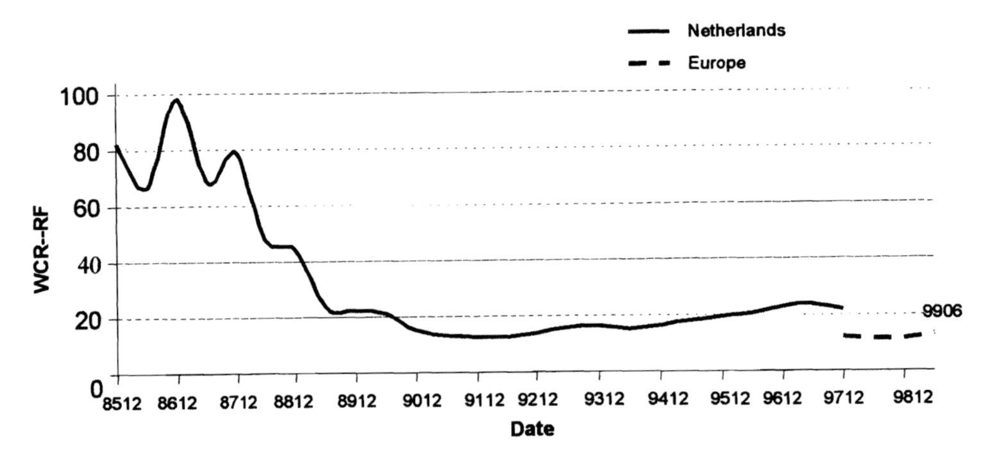

Figure 26.6. Warranty call rate-repair frequency evolution for CD players.

After a big decrease, the level increases again, reaching 24 units at the beginning of 1997. This last increase is the only one that might be related to a factory change, namely the one from Austria to Singapore. However, according to the field quality reports, units produced in Singapore had been sold from 1992 on. From the end of 1997 on, there are no more detailed data for the Netherlands on article group level, and therefore the trend for Europe (Italy, France, Germany, Austria, and the Netherlands) is shown in Figure 26.6 instead.

From the available data the hazard function could not be calculated.

CD Player Model A (Total Sales about 14,400)

We give some facts about one of the oldest models studied; let us call it Model A. For this model, the WCR curve (Figure 26.7), estimated by the WPM, suggests an increase in product reliability by dropping from about 16 units at the beginning to about 8 units at the end. However, this might be the result of a considerable increase in sales. Sales in the second year were 50% higher than those in the first year. Also, the MIS–MOP diagram (Figure 26.8) indicates that the WCR does not give the right impression. Of course this MIS–MOP diagram might be the result of two or more MIS–MOP diagrams for subpopulations. If data for different subpopulations scatter around different MOP values, then the diagram would give a possible explanation for the effect of the factor. Unfortunately, there was no information about meaningful subpopulations available.

The hazard function (Figure 26.9), combining all field data, shows humps at months 2 and 11, but apart from that it is more or less constant. Because of the lack of information, it was not possible to split the data into short intervals and to calculate the hazard function for each of these short intervals. This might have shown that the hazard function is different for different periods, although the MIS–MOP diagram does not support this. Given that this could not be verified, we concentrate on the calculated hazard function. Figure 26.9 has a rough similarity with the roller-coaster curve from Figure 26.2. The major difference is that the last part on the right, corresponding to the *aging failures,* does not show up. This is not surprising, of course, because only the 12 months of the warranty period are reflected. Further-

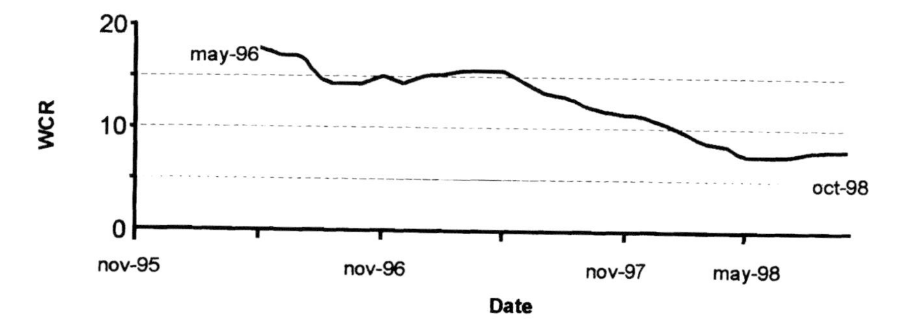

Figure 26.7. Warranty call rate for CD player Model A.

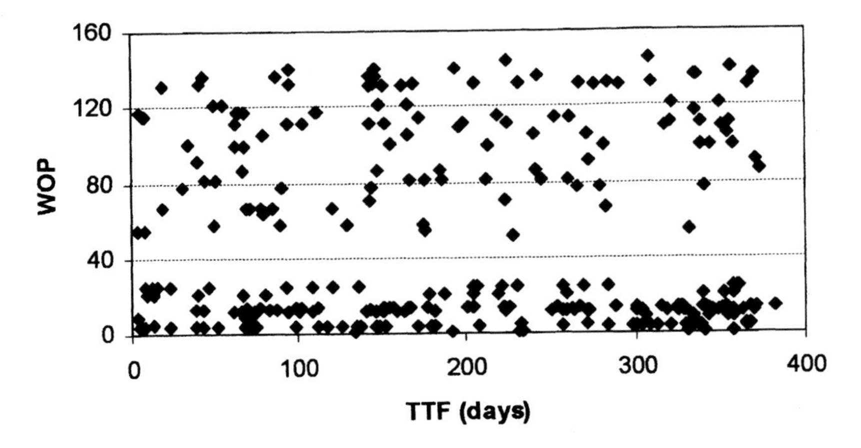

Figure 26.8. MIS–MOP diagram for CD player Model A.

more, the hump linked with the *zero-hour failures* is not present. Finally, more early humps show up, but this may be due by chance; these humps might be just random waves in a flat phase. It makes sense to try to find the root causes behind, in particular, the *early failures*. Some authors state that the first ones (e.g., before the third month) are most likely related to production and that the next humps at older ages may be more related to design. We do not exclude the possibility that heavy users are responsible for part of the early failures.

We mentioned previously that all hazard functions we found show a hump around the eleventh month. Although such a hump might indicate early wear-out, in this situation a nontechnical explanation seems more realistic. It is believed in the company that some retailers, in order to retain their clients, "adapt" the age of a 13- or 14-month old item to let the client benefit from the warranty. Furthermore, it seems realistic that some end users encountering minor problems wait until the end of the warranty period before taking the item to the repair center.

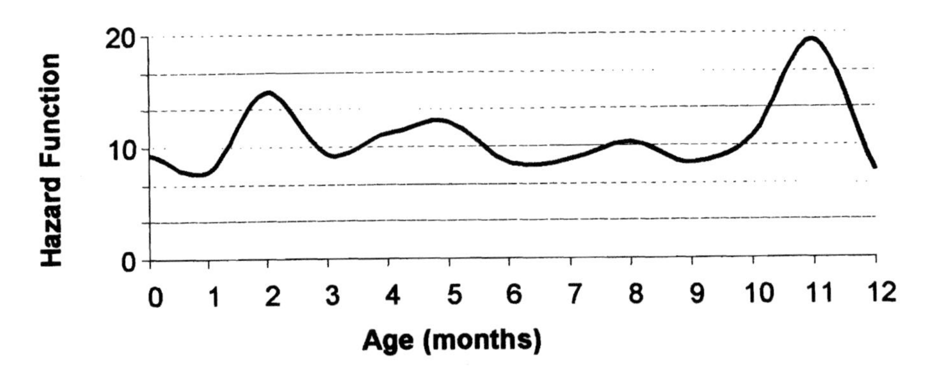

Figure 26.9. Hazard function for CD player Model A.

For the remainder of the models, only special cases or characteristics of the shape and level of the curves will be mentioned, taking the chosen CD player, Model A, as a reference.

26.4.4.2 CD Changers

In 1989, the CD changer (CDC) was introduced. Figure 26.10 gives a rough estimate of the evolution of the WCR-RF for these CDCs. The initial level of the curve in June 1989 (about 70 units) is much higher than that for CD players at that time (about 20 units). (Unfortunately, the initial level for CD players from 1982 is not available in the field quality reports.) The complexity added by the changer, the changer having more mechanical components, and the fact that it was a new product should explain the difference. Comparing Figures 26.6 and 26.10, along with realizing that the first CD player was put on the market in 1982, suggests that the length of the learning process that was necessary by the introduction of the changer (a new technology) was more or less the same as the length of the learning process for the first generation of CD players. For the CD player, as well as for the CD changer, it took about 10 years to reach the level of about 20 units. It is also remarkable that the CD changer starts at a high level of about 70 units. This is another indicator that the learning capability is limited.

CD Changer Model B (Total Sales about 3300)

We give some results for a CD changer that was introduced in the mid-1990s; let us call it Model B. The WCR (Figure 26.11) and the hazard function (Figure 26.12) for Model B have the same shape as the ones for the CD player Model A: The WCR shows a continuous improvement, and the hazard function reveals peaks in months 2 and 11. Notice that the level of both curves for Model B is considerably higher than that for Model A. The new functionality, the changer, obviously caused problems.

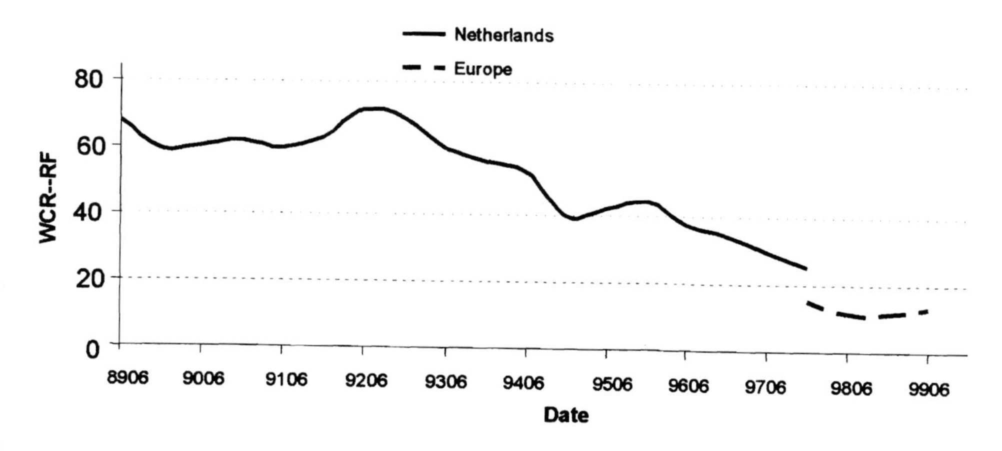

Figure 26.10. Warranty call rate-repair frequency evolution for CD changers.

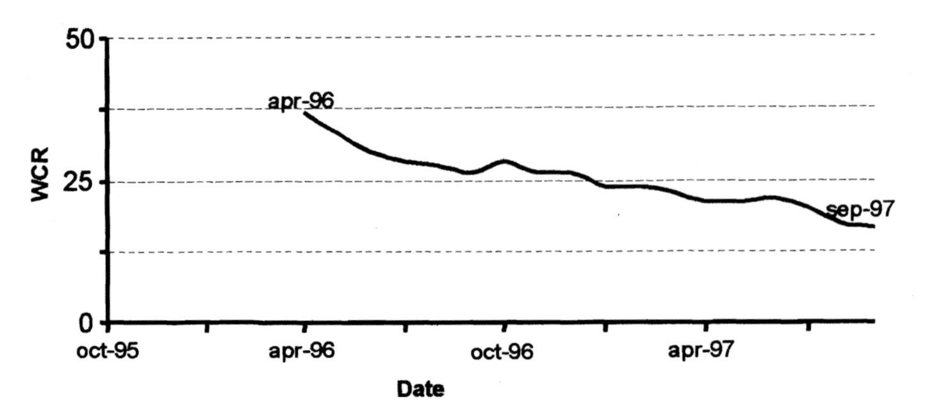

Figure 26.11. Warranty call rate for CD changer Model B.

Again we do not trust the WCR. The decline might be a consequence of the fact that sales during the second year were 44% higher than in the first year, and also the MIS–MOP diagram does not show an improvement. It looks the same as Figure 26.8.

26.4.4.3 CD Audio Recorders

The evolution of the WCR for the CD audio recorders is presented in Figure 26.13. Only the European trend is shown because no detailed data are available in the field quality reports for the Netherlands on type number level. Since it is a very new product, all the numbers are directly copied from the field quality reports (not derived from graphs) and the metric is always the WCR (not the RF). Because the graph is for Europe and not for the Netherlands, the comparisons with the previous graphs should be done with care.

The initial level of the curve, about 40 units, is again much higher than the one of the CD player and CD changer in mid–1998 (less than 20 units), but quite lower than the first results for the CD player and CD changer (about 70 units). There is

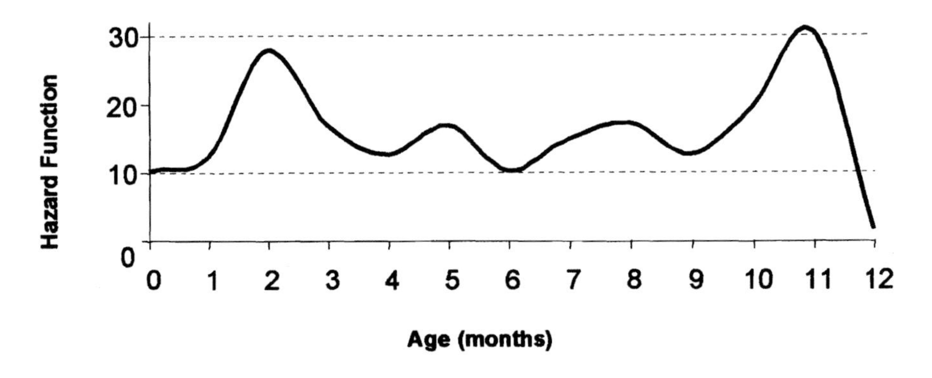

Figure 26.12. CDC 751 hazard function for CD changer Model B.

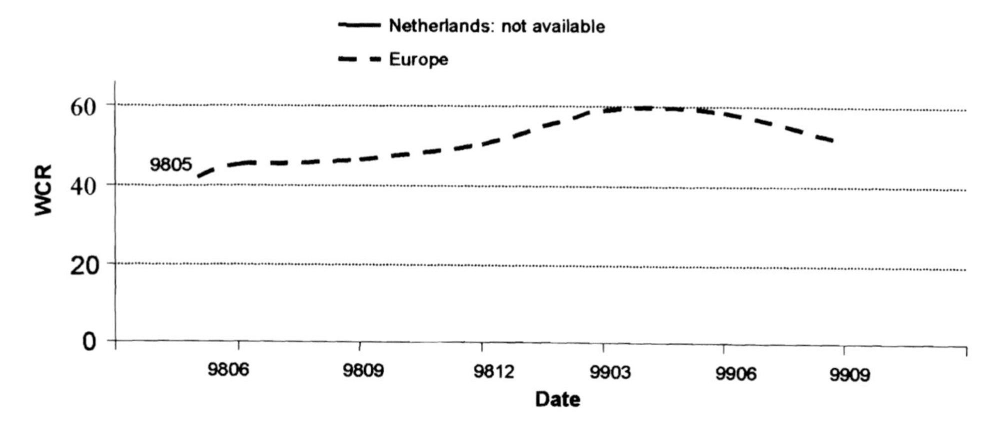

Figure 26.13. Warranty call rate evolution for CD audio recorders in Europe.

hardly 1.5 years of data, but the trend seems worrying, since during the first year the level presents a slow but sustained increase. This increase, however, might be the result of an increase in sales. Given that in the preceding CD products the curve always showed an initial improvement, the technological breakthrough associated with the recording functionality might be the problem.

CD Audio Recorder Model C (Total Sales about 5500)
The CD audio recorder we present, let us call it Model C, has a recorder and a changer. This model confirms what has been said about the CD audio recorder. Although the shape of the WCR is quite similar to those of the previous CD players and changers, the WCR values are much higher (Figure 26.14).

The hazard function (Figure 26.15) does not present a very pronounced initial hump, but we find again the omnipresent problem in the eleventh month. It is re-

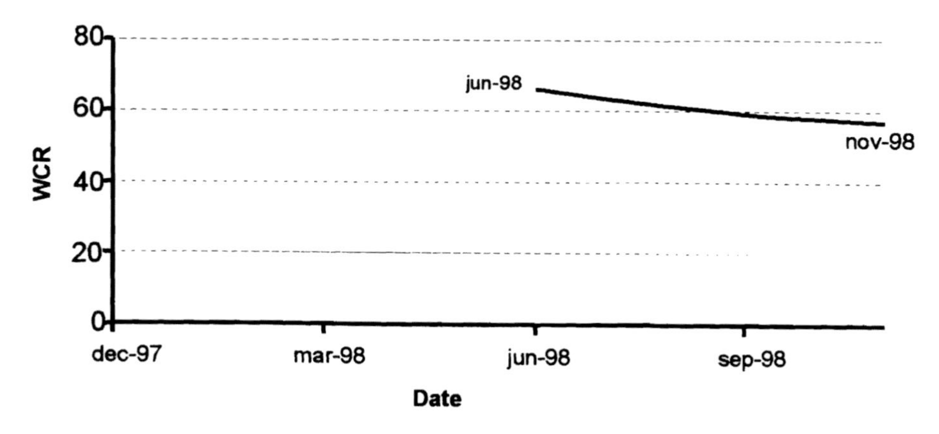

Figure 26.14. Warranty call rate for CD audio recorder Model C.

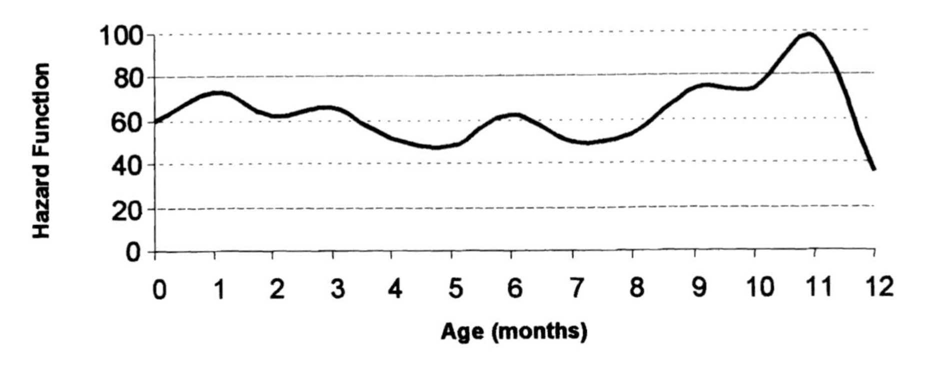

Figure 26.15. Hazard function for CD audio recorder Model C.

markable that for this CD changer the level of the curve is more than three times higher than the ones of the preceding CD-based products.

26.4.4.4 DVD Sets

Figure 26.16 presents the WCR trend for DVD sets. For the DVD, field quality reports data were available for Europe and the Netherlands. The initial difference between both trends is surprisingly big, although the results converge after 1 year. The initial level in the Netherlands is very similar to those of the CD player (Figure 26.6) and the CD changer (Figure 26.10). But it must be taken into account that in this case two changes occur: (1) the increase in the density of information and (2) a new application (video instead of audio). However, and in constrast to the CD recorder case, the evolution in the first year looks promising. Moreover, the current

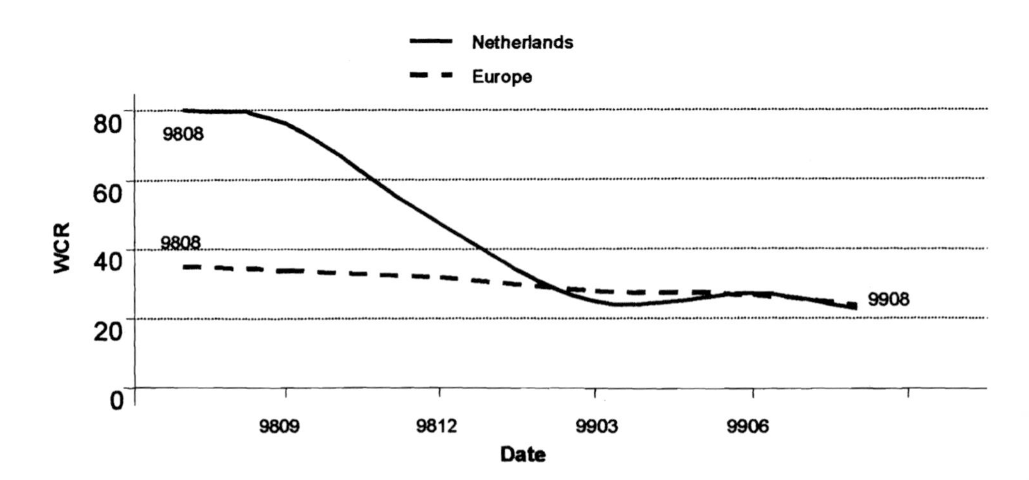

Figure 26.16. Warranty call rate evolution for DVD sets

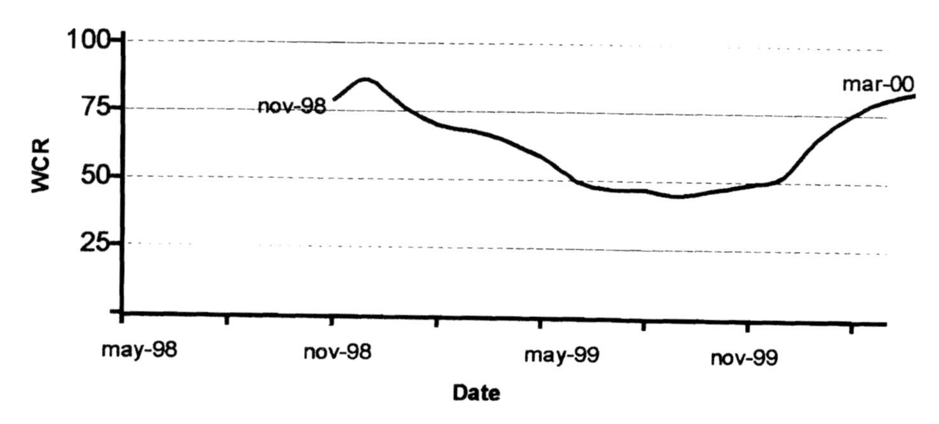

Figure 26.17. Warranty call rate for DVD player Model D.

level in Europe is substantially lower than that of the CD recorder, which confirms the importance of the problems faced by the latter.

DVD Player Model D (Total Sales about 3000)
The WCR trend and the hazard function for a particular DVD model—let us call it Model D—are presented in Figures 26.17 and 26.18. Since this a very new product and the sales volume is not very large up to now, the results may not be very reliable. The figures show that the levels of the WCR and the hazard function are quite high. The WCR shows a big increase from the middle of the curve coming back to the initial level of 75 units. The hazard function is more or less constant over the whole warranty period.

26.5 CONCLUSIONS

The Storage and Retrieval Group within Philips Research Eindhoven desires a better knowledge about the reliability problems affecting optical disc systems in the

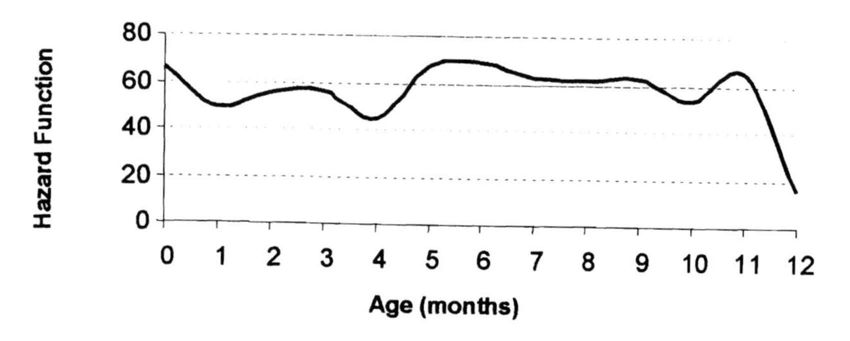

Figure 26.18. Hazard function for DVD player Model D.

field. In this way, special attention can be paid to certain critical aspects during the applied research of the forthcoming third generation. Because this was in line with the research interest of the Product and Process Quality Section of the Eindhoven University of Technology, it was decided to carry out a quantitative and qualitative study of field data. The main conclusions are the following:

a. General Conclusions

- In the evolution CD player–CD changer–CD recorder–DVD player the field information demonstrates that the company did not improve the product reliability. On the contrary, the consecutive hazard functions show that the increasing complexity of the products is accompanied by an increase in product failures.
- Because there is no information about the root causes of problems, it is impossible to determine whether all product failures in the succeeding generations are caused by the introduction of a new technology. Given the increase in product failures over the generations, it is expected that a significant number of the failures are repetitions of earlier problems.
- Because of lack of information about root causes of earlier generations of CD and DVD players/recorders, there is no structural way to prevent the recurrence of "old" problems.
- Given the preceding conclusions, it is to be expected that the field behavior of the new generation of DVRs will not be better than that of the CD recorder.

b. Shortcomings in the Availability of Relevant Data

- In the past, the databases concentrated on information collection for logistics aspects, such as spare parts, and on the financial settlement of warranty claims.
- Many months pass between the act (such as a customer buying a product, a customer reporting a reliability problem) and the availability of this information for reliability analysis.
- The fact that there is no information available about customer use severely limits the data analysis. It makes sense to improve the field information by building in measurement equipment that collects information about customer use that influences product reliability. This information can then be used to discriminate between subclasses of customers.
- The available field quality reports could not compensate for the lack of electronic data from before 1994.
- There is a mismatch between problem needs and collected data. *Example:* unavailability of dates of sale.
- Information on root causes is not collected and stored in any database. The IRIS codes used for the feedback of the repair information are not appropriate for the identification of the root causes. Actually, these root causes are not even known for many of the field failures, because deeper analysis about

these problems is undertaken only in cases where the WCR increases sharply.

c. Warranty Call Rate

- The WCR gives little useful information about product reliability for three reasons:
  1. It does not take into account the time to failure.
  2. It is not up to date because it is based on information that is 1 year old.
  3. It is sensitive to changes in sales volume and therefore might be misleading.
- Concentrating on the initial level and the improvement over time, the WCR evolution for the CD changer shows hardly any improvement over the WCR evolution for the CD player.
- The WCR and the hazard function indicate that adding the recording functionality to the CD player introduced a considerable number of reliability problems. The same holds true for the number of reliability problems when the recording capacity was increased by the introduction of the DVD disc and player. This suggests that a serious rise in the number of reliability problems can be expected when the third generation, digital video recording, is introduced.

c. MIS–MOP Diagram

- The MIS–MOP diagrams do not indicate a relation between the month of production and the time to failure. There was no information that could be used to divide the data set into relevant subclasses with the aim of detecting characteristics that affect product reliability.

d. Hazard Function

- For new products, it is important, particularly during the first few months of production, to have field feedback about reliability. In these situations the hazard function is a perfect metric. Because it is to be expected that immediately after the launch of the product the manufacturer is still improving the reliability of the product, the hazard function should be based on a short period. This makes demands on the speed of field feedback.
- The hazard function has humps around month 2 and month 11. It is important to analyze these early failures. When the root causes and the circumstances in which they show up are known, the right action can be taken.
- It is likely that the hump around the end of the warranty period is caused by customer friendly retailers and by customers who wait until the end of the warranty period before they contact a repair centre.

e. Time Pressure

- In a high-volume consumer industry, the product development process (PDP) is under time pressure. Therefore information on previous products performance must be available as soon as possible. With the current field data feedback system, two main problems are encountered:

1. The repair date used for the calculation was, in many cases, very far in time from the date in which the item was taken to the service centre.
2. There is a delay in reporting.

f. Field Failure Information Only During Warranty Period

- The company does not collect any information once the warranty period has expired. This is most inconvenient since European regulations will require Consumer Electronics manufacturers to increase the length of the warranty period to at least 2 years. Therefore it would be wise to start a pilot study in order to gather information about reliability after the present warranty period of 1 year.

REFERENCES

Brombacher, A. C. (1999). Maturity index on reliability: covering non-technical aspects of IEC61508 reliability certification, *Reliability Engineering and System Safety* **66:**109–120.

Brombacher, A. C., Van Geest, D. C. L., and Herrmann, O. E. (1992). Simulation, a tool for designing-in reliability, *ESREL '92, Schwabisch Gmünd.*

Brombacher, A. C., and Sander, P. C. (1999). Relations between product reliability and quality of business processes: new trends in reliability management. *Proceeding European Safety and Reliability Conference, ESREL '99, München-Garehing.*

Brown, S. L., and Eisenhardt, K. M. (1995). Product development: Past research, present findings, and future directions. *Academy of Management Review,* **20:**343–378.

Lu, Yuan, Loh, H. T., Ibrahim, Y., Sander, P. C., and Brombacher, A. C. (1999). Reliability in a time-driven product development process, *Quality and Reliability Engineering International* **15:**427–430.

Meeker, W. Q., and Escobar, L. A. (1998). *Statistical Methods for Reliability Data,* Wiley, New York.

Meyer, C. (1998). *Relentless Growth,* Free Press, New York.

Minderhoud, S. (1999). Quality and reliability in product creation—extending the traditional approach, *Quality and Reliability Engineering International* **15:**417–425.

Petkova, V. T., Sander, P. C., and Brombacher, A. C. (1999). The role of the service centre in improvement processes, *Quality and Reliability Engineering International* **15:**431–437.

Petkova, V. T., Sander, P. C., and Brombacher, A. C. (2000). The use of quality metrics in service centres, *International Journal of Production Economics* **67:**27–36.

Rothery, B., and Robertson, I. (1995). *The Truth About Outsourcing,* Gower, Aldershot.

Sander, P. C., and Brombacher, A. C. (1999). MIR: The use of reliability information flows as a Maturity Index for quality management, *Quality and Reliability Engineering International* **15:**439–447.

Sander, P. C., and Brombacher, A. C. (2000). Analysis of quality information flows in the product creation process of high-volume consumer products, *International Journal of Production Economics* **67:**37–52.

Stalk, G., Jr., and Hout, T. M. (1991). *Competing Against Time,* Free Press, New York.

Taylor, F. W. (1919). *Shop Management,* Harper and Brothers, New York.

Wheelwright, S. C., and Clark, K. B. (1992). *Revolutionizing Product Development: Quantum Leaps in Speed, Efficiency and Quality,* Free Press, New York.

Wong, K. L. (1988). Off the bathtub onto the roller-coaster curve, *Proceedings Annual Reliability and Maintenance Symposium,* pp. 356–363.

EXERCISES

26.1. There are many reasons why reliability data should be collected. Give five of them, mention the characteristics that should be measured, and propose relevant metrics to solve the problem.

26.2. Demonstrate that, depending on the hazard rate, an increase in sales volume can cause a decline as well as a rise in the WCR, or have no effect at all on the WCR.

26.3. Suppose that the Quality Department calculates the hazard rate based on the difference between the production date and the date a customer finds a defect. This difference, say Z, is the sum of the time in stock, say Y, and the time in service, say X. Suppose that the distributions of the random variables X and Y are known: X has an exponential distribution with expectation $1/\lambda$ and Y has a uniform distribution on the interval $[0, b]$. Calculate the hazard rate of $Z = X + Y$.

26.4. For the majority of the CD players the only available information about the time to failure is the calendar time between production and the end of repair.

a. Discuss the information you would like to have about the time to failure.
b. What other information in relation to a product failure would be most useful?

26.5. A manufacturer of consumer products does control the design, the procurement of (e.g., components and modules), and the assembly. However, the performance of a product is also influenced by the way the customer uses the product. How can this be taken into account in the design phase?

26.6. In this chapter, Figures 26.9, 26.12, 26.15, and 26.18 give the hazard function for a particular CD or DVD player. Discuss the conditions under which these figures give a sound picture.

CHAPTER 27

Reliability and Warranty Analysis of a Motorcycle Based on Claims Data

Bermawi P. Iskandar and Wallace R. Blischke

27.1 INTRODUCTION

Most manufactured goods of all types are sold with warranty. One of the purposes of a warranty is to provide protection to the consumer, in that compensation is provided in the event of premature failure of an item. As a result, the manufacturer incurs costs associated with servicing of the warranty, and these could be considerable under certain circumstances. The costs of warranty depend primarily on the nature and terms of the warranty and the reliability of the item in question. For a detailed discussion of the various types of warranties and associated cost models, see Blischke and Murthy (1994).

A key concern in warranty management is estimation and control of the cost elements associated with warranty (Murthy and Blischke, 2000). A good deal of data are often collected in connection with this effort. These might include marketing information, reliability test results, quality information, responses to consumer surveys, warranty claims data, and relevant data from many other potential sources. Each of these types of data can provide important management information. Even more information can be obtained by analysis of the interrelationships among the many relevant data sets, but this requires a large, composite database and application of the appropriate tools for mining and analysis of this information. Implementation of broad programs of the type required for such analysis is widely underway, particularly in large organizations with extensive product lines and adequate analytical capabilities. For a great many organizations, however, integrated programs of this type are not feasible and a more segmented approach is necessarily taken to analysis of relevant warranty information. Here we look at approaches to analysis of warranty data in the context of such constraints.

Case Studies in Reliability and Maintenance, Edited by W. R. Blischke and D. N. P. Murthy.
ISBN 0-471-41373-9 © 2003 John Wiley and Sons, Inc.

27.1.1 Warranty Data

In this case project, we are concerned with warranty claims data. The specific claims data collected may vary considerably from company to company, but ordinarily the data consist of date of sale of the item and date of claim and, if relevant, usage (these being necessary to verify the validity of the claim), as well as a number of items of identification—for example, product, model number, dealer, identification of part(s) that failed, cost of replacement (usually parts and labor separately), service facility, and so forth. More complete data may also include mode of failure, test results and other evidence of failure, repair time, and so forth, and, for complex items, maintenance records and other relevant information.

Key issues in analysis of claims data include classification and tabulation of failures, analysis of failure frequencies, assessment of cost elements, and other issues that may depend on the product and the intended use of the information. In addition, analysis of a number of more fundamental issues may be desired. These include mean time to failure, failure distribution, root causes of failure, and engineering issues such as materials, design, manufacture, quality programs, and so forth. For most of these purposes (and often even for the simpler issues), claims data are notoriously poor. This is due primarily the fact that the data are provided by consumers, dealers, repair personnel, and others over whom the manufacturer has little or no control. An excellent discussion of data problems and a number of related issues is given by Suzuki et al. (2001).

Some of the principal difficulties frequently encountered in claims data are as follows:

- Inaccurate, incomplete data—missing or incorrect entries; transpositions, and so on
- Delays in reporting—periodic or haphazard reports of claims
- Lags in making claims (particularly for minor failures or failures that do not seriously affect operation of the item)
- Invalid claims—claims after expiration of the warranty, failures due to misuse, or claims on items that did not fail
- Valid claims that are not made—ignorance of warranty terms; compensation deemed not worth the effort of collecting, and so on

As a result of these and other difficulties, warranty data are usually quite messy and a good deal of careful editing may be required. Beyond this, two common and very important characteristics of warranty data that significantly impact the analysis are as follows:

- Very little, if any, information on failure times or usage is obtained on items that fail after the end of the warranty period.
- Data on some important variables are not collected. A common example is total daily sales.

As a consequence of the first item, warranty data are usually very heavily censored (e.g., 95% or more missing observations). Worse, essential censoring information is often not known. (A common example is usage.) In this situation, it may be very difficult to obtain accurate and precise estimates of model parameters. The consequence of missing variables may also be serious, since it may preclude the use of a powerful model.

Because of these shortcomings, many of which are inherent to the process, there is often little done by way of analysis of warranty data other than simple descriptive statistics. It is possible to do a good deal more, however, and a number of techniques have been developed for this purpose. Some of these will be illustrated in this case.

27.1.2 Context and Objectives

The product in question in this case is a motorcycle produced and sold in Indonesia. The manufacturer is a major producer of motorcycles worldwide. The item of interest is a motorcycle sold under a 6-month warranty that covers only first failure of any component. The manufacturer has collected data on warranty claims, including date of purchase, date of claim, identification of failed item, cost of repair, and a number of other variables. The only other information available is monthly sales volume for the product.

The manufacturer, product, and data are described in more detail later in the chapter. Ideally, the goals of the case would be to analyze the data, perform reliability analyses at the component and system levels, model and estimate warranty costs at the component and system levels, and compare several warranty policies. Because of data difficulties, confidentiality, and other constraints, a thorough analysis along these lines is not feasible. We can, however, analyze the data available and begin the study of the broader objectives.

In this context, the more limited objectives of the case are to:

1. Describe and summarize the claims in the context of the warranty terms offered on the product.
2. Characterize the life distribution of the item, insofar as is possible.
3. Analyze the warranty offered and compare it with alternative warranties that might be considered.

Approaches to these problems will involve a number of models, various data analytic techniques, and warranty analysis. The major overall goal of the case is to illustrate the methodology as applied to typically intractable claims data.

27.1.3 Chapter Outline

The organization of the case study is as follows: In Section 27.2, the manufacturer, product, warranty, and data sets used in the study are described in detail. Section

27.3 describes the methods used in analyzing the data and gives a brief description of warranty analysis. Results of preliminary data analyses, including summary statistics and various graphs and charts, are given in Section 27.4. Application of two methods for analyzing bivariate failure data, due to Kalbfleisch and Lawless (1996) and Gertsbakh and Kordonsky (1998), are discussed in Sections 27.5 and 27.6, respectively. In Section 27.7, some results on warranty costing and related issues are discussed. Conclusions of the study and some suggestions for future research are given in Section 27.8.

27.2 PRODUCT, WARRANTY, AND DATA

In this section we discuss the motorcycle and warranty in question, the several sets of claims collected, data editing, and some of the data problems encountered.

27.2.1 The Company, Product, and Warranty

The manufacturer of the motorcycle (hereafter called "The Company") was established in 1971 and is the sole agent, importer, and assembler of this brand of motorcycle in Indonesia. The first motorcycle produced was a well-known model widely used in Asia. At present, The Company produces over 10 models of this brand and has more than 2200 employees at its Indonesian facilities.

International Sales Operations (ISO), a division of the parent company, is responsible for marketing and distribution of motorcycles produced by The Company. ISO has six branches and 777 dealers spread throughout Indonesia. With an aggressive marketing strategy, supported by the wide coverage of ISO's distribution network, the market share of The Company's products in the local market had reached 57.5% in 1998. In addition, ISO has succeeded in marketing the motorcycles in the broader international market, notably in China and Greece.

In order to strengthen its position as a market leader in Indonesia, The Company has implemented modern concepts of management and used high technology in its manufacturing process. High-precision machines, controlled by a computer, and skilled and experienced operators have become the backbone for achieving the current high reliability and overall quality of The Company's products. "Just-in-time" production has been in place for more than 5 years, and, as a result, The Company has been able to increase productivity and efficiency and to manage effectively 63 subcontractors, supplying 80% of local components.

One type of motorcycle, generically called the XF–1 Model, dominates sales of The Company. The volume of sales of XF–1 motorcycles was 151,918 units in 1998. For this reason, we focus on this model in the case study. The XF–1 is a small motorcycle with a 100cc engine. This motorcycle is widely used in Indonesia, because it is considered to be a very efficient vehicle for local transportation. Extensive failure data, at the component level, in the form of warranty claims data, are available and will form the basis of our analysis of failures and warranties.

The Company provides a warranty for each motorcycle sold. The motorcycle is warrantied for 6 months or 6000 km, whichever occurs first. This warranty is given to the original owner and is not transferable on sale of the motorcycle. The warranty covers only the first failure of a given component. Repairs of first failures are made at no cost to the owner, but any subsequent failures of the replaced or repaired component(s) are the responsibility of the owner. The warranty is conditional on normal use as defined in the owner's instruction manual, as well as on proper, regular maintenance in accordance with the service booklet.

Since motorcycles are sold with a warranty, The Company incurs an additional cost due to servicing of the warranty. Although total warranty costs have not been excessive, cost control measures are being implemented companywide, and accurate and precise estimates of all costs are sought. Of particular importance to The Company is an estimate of the warranty cost per unit sale. Realistic estimates of the reliability of the motorcycle and its components are also desired. This information will be useful in improving reliability and overall quality of the motorcycle as well as in investigating the relationship between reliability and warranty costs. Some issues along these lines are addressed in the study.

The motorcycle is a multicomponent system, with many items supplied by vendors. We note that not all of the components are covered by the manufacturer's warranty. Components and materials not covered are those that deteriorate or wear out with usage, such as spark plugs, fuel filter, oil filter, brake pads, light bulbs, tires, and so forth. (Many of these—for example, tires—are covered by warranties provided by the suppliers.) In this context, failure of the motorcycle is defined to be failure of one or more of the components that are covered under The Company's warranty.

27.2.2 Source and Structure of the Data Set

27.2.2.1 Failure Data

Four sets of data on 1998 warranty claims, originally collected by The Company for monitoring claims, were available for analysis. These correspond to files of reports of accepted and rejected claims by owners (denoted C2) and by dealers (C1). The four files are labeled ACC-C1, ACC-C2, REJ-C1, and REJ-C2. Relevant information sought for our purposes is time to failure, measured in both calendar time (age in days) and usage (km). Since claims made by dealers correspond to units not yet sold at the time of claim, both age and km are zero for these reports. As a result, the C1 files provide useful information for our analysis only on mode of failure.

Variables used in the failure analysis are:

Part number: ID number of failed part

Frame number: Serial number of motorcycle

Engine number: Serial number of engine

Description: Type or mode of failure

B = Buy date (date when the unit was sold)

C = Claim date
$X = C - B$ = Age of the unit at failure (days)
Y = Usage at time of failure (km)
C_1 = Cost of parts required in repair
C_2 = Cost of labor
C_r = Total cost of repair or replacement = $C_1 + C_2$

A number of other variables were included in the data set, including dealer, part names, and other types of identification related to the claim, quantity of parts replaced, status of the claim, and so on. Minitab worksheets containing the four data sets are available on the Wiley Internet site.

Note that the data cover all units sold in Indonesia during calendar year 1998. Although all sales are of the same basic Model XF-1, engineering changes are occasionally made. A necessary assumption in our analysis is that any such changes that might have been made do not significantly alter the product reliability. (Note also that improvements in reliability, if any, would lead to reductions in warranty cost, in which case the cost model results would be conservative.)

27.2.2.2 Sales Data

Data on number of units sold were available only on a monthly basis. These results, along with the number of owner claims accepted on units sold that month during calendar year 1998, are given in Table 27.1. Note that the claim counts decrease as the year progresses since many of the claims for these units would be made in 1999 and are not included in the counts. It should also be noted that claims counts are results after editing as discussed in the next section.

Table 27.1. Sales and Warranty Claims by Month, 1998

Month	Number of Units Sold	Total Number of Accepted Claims
1	21,012	844
2	17,804	774
3	12,893	532
4	7,998	526
5	9,909	523
6	12,932	207
7	13,252	118
8	11,266	91
9	16,476	88
10	13,168	33
11	7,853	25
12	7,355	34
Total	151,918	3795

27.2.3 Data Editing

Examination of the data files revealed a number of types of data errors. The nature of the problems and corrective actions taken were as follows:

- *Missing entries.* A relatively small number of records in the original data set were missing essential information. These could not be used in the analysis and were eliminated from the data set. Elimination of these has the effect of biasing the estimated warranty costs downward slightly.
- *Multiple entries.* Warranty claims reports were occasionally entered as several items for the same repair (e.g., one for each part replaced). These results were consolidated into single entries, with cost elements added.
- *Transposed dates.* In a small percent of cases, claim dates prior to date of sale were entered. These were adjusted by reversing the transposed dates.
- *Obvious errors.* In a small number of instances, data were obviously erroneous (e.g., Age $X = 1$ day and $Y = 4000$ km). These cases were eliminated only in analyses where the error is relevant.

Problems such as entries that are missing entirely (i.e., claims made but not entered into the data set) cannot, of course, be accounted for in the editing process. It is hoped that any such problems are relatively minor and will not materially affect the results. The point here is that, as noted, claims data are often of poor quality and care must be taken in editing, analyzing, and interpreting the results.

For some analyses, further editing is required. This is mainly a function of model restrictions and will be discussed later.

27.3 METHODOLOGY

In this section we look at a number of analytical tools that can be applied in analyzing warranty claims and related data, as well as reliability models and warranty cost models.

27.3.1 Approaches to Analysis of Claims Data

Our objective is to estimate/predict future warranty costs and compare different warranty policies. What we have available is a set of edited, but still somewhat dubious, warranty claims data. The approach taken here is as follows: Warranty costs models are based on the warranty structure and the failure distribution of the item in question and some related functions. The model may be estimated by estimating the parameters of the failure distribution. This requires first determining an appropriate distribution for time to failure. The essential first step, then, is a reliability analysis based on the claims data.

A number of approaches have been taken with regard to addressing analysis of claims data and some of the various related issues discussed previously. Many of these are concerned specifically with automobile warranty claims data, which is closely related to the current study, both in product and warranty structure. Of particular note are (a) Kalbfleisch and Lawless (1988) and Robinson and McDonald (1991), who discuss the use of warranty and other data in assessment of field reliability, and (b) Suzuki (1985), Lawless et al. (1995), Hu and Lawless (1996), Hu et al. (1998), and Wang and Suzuki (2001a,b), who dealt with estimation of life distributions and MTTF based on auto warranty data. A different approach to the analysis of claims data is given by Wasserman (1992), Singpurwalla and Wilson (1993), and Chen et al. (1996), who deal with a Bayesian approach to forecasting claims using Kalman filtering in a time series model applied to current and historical data.

The motorcycle warranty is a two-dimensional warranty (age and usage), and data are collected on both variables. It follows that the type of models needed are those that reflect this two-dimensional structure. Several approaches have been taken to this problem. These may be classified as 2D (two-dimensional) approaches, in which basically a bivariate analysis of X and Y is done, and 1D (one-dimensional) approaches, in which an assumed relationship between X and Y is used to reduce the analysis to that of a single variable that is a function of X and Y.

Modeling failures in this context may be done in a number of ways. In the 1D formulation, the failure of items is modeled using a one-dimensional point process. Examples of models of this type are given by Murthy and Wilson (1991), Moskowitz and Chun (1994, 1996), Ahn, Chae and Clark (1998), and Gertsbakh and Kordonsky (1998). The second approach models item failures in term of a two-dimensional point process formulation. Models of this type are dealt with by Iskandar (1993), Murthy et al. (1995), Eliashberg et al. (1997), and Yang et al. (2000). Many additional modeling issues are discussed in the chapters in Jewell et al. (1996). Escobar and Meeker (1999) deal with prediction intervals based on censored data. These and many related data analysis issues are discussed in several excellent texts, including Kalbfleisch and Prentice (1980), Lawless (1982), Nelson (1982), and Meeker and Escobar (1998).

A closely related issue is selection of a time scale on which to characterize and analyze failures. Various approaches have been proposed, including Farewell and Cox (1979), Kordonsky and Gertsbakh (1993, 1995a,b, 1997), and Oakes (1995). These and related issues are discussed in detail by Duchesne and Lawless (2000).

In this chapter, we apply the analyses of Gertsbakh and Kordonsky (GK) and Kalbfleisch and Lawless (KL). These approaches will be discussed in some detail below. First we look briefly at reliability concepts relevant to the analysis.

27.3.2 Reliability Modeling

A motorcycle consists of many hundreds of parts. Modeling of reliability of the motorcycle requires a complex engineering analysis, involving identification of modes and effects of failures—for example, through fault trees and FMEAs—and derivation of models of components, subsystems and systems and their relationships and contributions to product failure. The methodology for reliability analysis is discussed in detail by Blischke and Murthy (2000).

An initial goal of this study was to analyze data at at least the component or subsystem level. Although claims data are, in fact, collected at the part level, the data are not adequate for detailed reliability analysis. In the ensuing, we restrict consideration to simple counts at the part level, and reliability analysis at the system level (i.e., of the motorcycle). A future goal is to break failures down by subsystem (e.g., electrical, engine, frame, other) and then, if possible, by major components of each subsystem. This information may by useful in conjunction with other data (engineering test results, consumer data, and so forth) in a thorough reliability analysis of the product.

27.3.3 The Kalbfleisch–Lawless Approach

27.3.3.1 Models for 2D Claims Data

Lawless and others have developed a number of models for analysis of warranty claims data. These include

- Kalbfleisch, Lawless. and Robinson (1991) and Kalbfleisch and Lawless (1996)—assessment of trends, adjustment for delays in reporting, comparison over time periods, and prediction.
- Lawless (1994)—reporting delays and unreported events.
- Lawless, Hu, and Cao (1995)—modeling and analysis of two-dimensional claims data with supplementary information; estimation of failure distributions
- Lawless (1998)—review of methods of analysis of claims data and uses in prediction, comparisons, estimation of field reliability, and related issues.

See Suzuki et al. (2001) for additional references and discussion.

27.3.3.2 The KL Model

In our analysis, we use the Kalbfleisch–Lawless (1996) model, designated KL. The purpose of the model is to investigate claims patterns (enabling comparisons over different periods, different models, etc.), search for trends, predict future claims, and, to the extent possible, estimate field reliability. The model uses aggregate 2D claims data (typically age and usage at time of claim) and is based on estimating age-specific claims rates. Here we consider estimation of claims rates and of standard errors of the estimators. The model also requires sales data for each time period.

We note at the outset that analysis of the claims data alone is not adequate for most purposes. This is discussed in detail in the references cited above and in Duchesne and Lawless (2000). What is needed is supplementary information on usage patterns. This is usually obtained from other sources—for example, consumer surveys. Information of this type was not available for analysis of the motorcycle claims data. As will be apparent, the consequence is that some of the estimates have relatively large standard errors.

We turn next to the age-specific claims rate model as given by KL (1996). We assume that data for time periods 0, 1, . . . , T are to be analyzed. Claims rates are defined to be

$$\lambda(a) = \text{expected number of claims per unit at age } a \ (0 \leq a \leq T)$$

$$\Lambda(a) = \sum_{i=0}^{a} \lambda(i) = \text{cumulative number of expected claims to age } a$$

Additional notation used is as follows:

$N(t)$ = number of units sold in time period t ($0 \leq t \leq T$)

$N = \sum_{i=0}^{T} N(t)$ = total number of units sold

$n(t, a)$ = number of claims at age a for units sold in time period t ($a, t \geq 0, a + t \leq T$)

$\mu(t, a) = E[n(t, a)]$ = conditional expected number of claims given $N(0), \ldots, N(T)$

It is assumed that claims occur according to a random process. From the above, it follows that the mean function is

$$\mu(t, a) = N(t)\lambda(a) \tag{27.1}$$

Note, incidentally, that $\lambda(a)$ is usually not the expected number of *failures* per unit at age a, because some units may no longer be under warranty coverage at age a, having either exceeded the usage limit or, for the motorcycle warranty, been sold by the original owner.

The claim rates may easily be estimated from the count data defined above. If we assume that the $n(t, a)$ are independent Poisson random variables, the maximum likelihood estimators of the rates are (for $0 \leq a \leq T$)

$$\hat{\lambda}(a) = \frac{n_T(a)}{R_T(a)} \tag{27.2}$$

and

$$\hat{\Lambda}(a) = \sum_{t=0}^{a} \hat{\lambda}(t) \tag{27.3}$$

where

$$n_T(a) = \sum_{t=0}^{T-a} n(t, a) \tag{27.4}$$

and

$$R_T(a) = \sum_{t=0}^{T-a} N(t) \tag{27.5}$$

An estimate of the mean function, $\hat{\mu}(t, a)$, is calculated by substitution of $\hat{\lambda}(a)$ for $\lambda(a)$ in equation (27.1).

We next look at an estimator of the standard error of the estimated rates. Under the assumption of independence of the counts, the estimated variance $\hat{V}(a)$ of $\hat{\Lambda}(a)$, the estimated cumulative claims rate, is

$$\hat{V}(a) = \hat{\sigma}^2 \sum_{t=0}^{a} \frac{\hat{\lambda}(t)}{R_T(t)} \tag{27.6}$$

Here the summation corresponds to the variance assuming independent and identically distributed Poisson counts, and $\hat{\sigma}^2$ estimates "extra-Poisson variation." This may arise because of variability in unit reliability, environmental variability, variation in usage, and so on. The estimate is given by

$$\hat{\sigma}^2 = \frac{1}{\nu} \sum_{t,a} \frac{[n(t, a) - \hat{\mu}(t, a)]^2}{\hat{\mu}(t, a)} \tag{27.7}$$

where $\hat{\mu}(t, a) = N(t)\hat{\lambda}(a)$ as before, and ν = number of terms in the summation minus number of parameters estimated. The result can be used to obtain an approximate confidence interval of the form

$$\hat{\Lambda}(a) \pm z_{\alpha/2}\sqrt{\hat{V}(a)} \tag{27.8}$$

where $z_{\alpha/2}$ is the $(1 - \alpha/2)$-fractile of the standard normal distribution.

These results provide information on the pattern of claims. They may also be used for prediction of future claims assuming that the same basic pattern prevails, as well as for cost estimates of warranty servicing and, under certain conditions, estimation of product field reliability. (See Kalbfleisch and Lawless (1996), Kalbfleisch et al. (1991), and the references cited for further information.) Estimation of product reliability requires additional information. For example, knowledge of the distribution of the ratio $R = Y/X$ can be used for this purpose. Data on R from sources other than the warranty claims data are required for effective estimation. The use of R will be discussed further below.

27.3.4 The Gertsbakh–Kordonsky Approach

The GK approach is described in detail in Gertsbakh and Kordonsky (1998) and Kordonsky and Gertsbakh (1993, 1995a,b, 1997). Duchesne and Lawless (2000) discuss this and several other approaches to scaling in two- and higher-dimensional problems of this type. Applications, extensions, and numerical examples may be found in many of these sources.

27.3.4.1 Basic Approach

As noted previously, the GK approach reduces the 2D warranty problem to a one-dimensional formulation that is essentially equivalent under certain conditions. The

2D warranty provides coverage over a region $\Omega = [0, W) \times [0, U)$, with, in our case, $W = 6$ months and $U = 6000$ km. Thus failures are covered under warranty only if $(X, Y) \in \Omega$ and the failure is the first failure of this component.

The GK approach reduces this to a single variable V, given, for the ith observation, by

$$V_i = (1 - \varepsilon)X_i + \varepsilon Y_i \tag{27.9}$$

where ε is determined so as to minimize the sample coefficient of variation (c.v.) $c = s_V/\bar{v}$, with s_V and $\bar{v}$ being the sample standard deviation and mean of the transformed variable. ε is obtained by a simple search over the region $0 \leq \varepsilon \leq 1$. Given the value of ε, failures are characterized by the univariate distribution function $F_V(v)$, which is often taken to be a Weibull distribution. Alternatively, it may be approximated by fitting selected distributions to the transformed data. Warranty analysis is then done on the V scale.

The time scale of the variable V (called a V scale) is a combination of the age scale and the usage scale. The V scale does not have a physical meaning; it captures the joint effects of age and usage. We note, however, that different scales for age and usage will not change the value of ε, and this in turn will lead to the same essential results regardless of the original units.

27.3.4.2 Parameter Estimation

We next look at estimation for a specified distribution of V, assuming censored data of the type typically found in warranty applications. For this purpose, some commonly used failure distribution such as Weibull, gamma, and log-normal are used to fit the warranty claims data. Parameter estimation for distributions selected is then carried out based on observed failure data and censoring values for unfailed items. In the warranty context, failure data consist of ages and usages at failures (and hence claims) under warranty. Difficulties are encountered in dealing with censored data, since typically age will be known but usage will not. We will return to this in the analysis of the motorcycle data in Section 27.6.

Note that sales data are also required for parameter estimation. As discussed previously, for the motorcycle application, these are available only on a monthly basis.

For the GK analysis, we use the following notation:

k = number of months in period under study

N_j = sales in month j ($j = 1, \ldots, k$)

A_j = number of claims by owners that were accepted on the N_j units sold in month j

B_j = number of claims by owners that were rejected for the N_j units sold in month j

C_j = number of dealer claims (both accepted and rejected) in month j

M_j = number of items sold in month j for which no claims were made = N_j—A_j—$B_j - C_j$

K = number of claims by owners that were accepted in the period = $\Sigma_{j=1}^{k} A_j$
W_j = age at censoring for items sold in month j
U_j = usage at censoring for items sold in month j

Define

$$\Omega_j = (0, W_j) \times (0, U_j), \qquad j = 1, \ldots, k$$

Warranty claims as a result of failures observed are claims that occurred in the region Ω_j—that is, for which $x_i < W_j$ and $y_i < U_j$. Ω_j can be viewed as service ages and usages for unfailed items sold in month j.

We now can specify data available for estimation, consisting of failure data and censored data. In this notation, failure data are observed values $(x_i, y_i) \in \Omega_j$; censored data are unobserved random variables $(X_i, Y_i) \notin \Omega_j$, $j = 1, \ldots, k$. On the V scale, these are observed values v_i ($i = 1, \ldots, K$) and censored values $\tilde{V}_j$, say, where

$$\tilde{V}_j = (1 - \varepsilon)U_j + \varepsilon W_j \tag{27.10}$$

which represents the service times in the V scale of unfailed items at censoring times.

We assume that $V_1, V_2, \ldots, V_n$ are independent random variables of failure times in V scale with distribution and density functions given by $F_V(v; \theta)$ and $f_V(v; \theta)$. The likelihood function is

$$L(\theta) = \prod_{i \in K} f_V(v_i) \prod_{j=1}^{12} \prod_{i \in M_j} [1 - F_V(\tilde{V}_j)] \tag{27.11}$$

Log L is given by

$$\log[L(\theta)] = \prod_{i=1}^{K} \log[f_V(v_i)] + \prod_{j=1}^{12} M_j \log[1 - F_V(\tilde{V}_j)] \tag{27.12}$$

The estimate of θ is obtained by maximizing log L with respect to the unknown parameters of F_V.

Asymptotic normality may be used to obtain approximate confidence interval for the parameters. For distributions selected in our analysis, θ consists of two parameters, say p_1 and p_2. Normal theory requires evaluation of the information matrix given by

$$\begin{pmatrix} \partial^2 \log L(p_1, p_2)/\partial p_1^2 & \partial^2 \log L(p_1, p_2)/\partial p_1 \partial p_2 \\ \partial^2 \log L(p_1, p_2)/\partial p_1 \partial p_2 & \partial^2 \log L(p_1, p_2)/\partial p_2^2 \end{pmatrix} \tag{27.13}$$

The inverse of matrix equation (27.13) is the asymptotic covariance matrix of the parameters. To estimate the standard deviations of p_1 and p_2, we calculate the matrix given in equation (27.13) at the values of $\hat{p}_1$ and $\hat{p}_2$ (the estimates of p_1 and

p_2) and take square roots of the diagonal elements. Approximate confidence interval for the parameters are given by

$$\hat{p}_1 \pm z_{a/2}\sqrt{\hat{V}(\hat{p}_1)}, \qquad \hat{p}_2 \pm z_{a/2}\sqrt{\hat{V}(\hat{p}_2)} \tag{27.14}$$

where $z_{a/2}$ is the fractile of the standard normal distribution corresponding to the desired level of confidence.

27.3.5 Warranty Models

Here we look briefly at warranty cost models. The motorcycle is covered by a 6-month/6000-km warranty that covers only first failure. For simplicity, we assume that these warranty terms cover the system as a whole—that is, that the first failure is rectified by The Company at no cost to the buyer, but that any subsequent failures are not. (In fact, we found only a few instances of multiple failures in the data. An alternate approach would be to treat each component separately.) We model cost to the manufacturer of an item sold under warranty.

27.3.5.1 Alternative Warranties

The warranty described above is one of the simplest warranty structures. Here we look at some alternatives that might be considered by the motorcycle manufacturer. We begin with the one-dimensional case. This is a simpler structure and is appropriate if either age or usage alone is used as the basis of the warranty or if the analysis is done using the GK approach.

The first alternative warranty that is of interest is the nonrenewing free replacement warranty (FRW) with warranty period W. Under the FRW, all failed items (including replacements) are replaced or repaired free of charge to the owner up to time W from time of purchase. By nonrenewing is meant that the warranty period does not begin anew on failure of an item. Thus an item failing at time X, say, from time of purchase is replaced by an item that is covered under FRW for the time remaining in the warranty period, $W - X$.

A second alternative is the renewing pro-rata warranty (PRW). Under this warranty, a failed item is replaced or repaired by the manufacturer at a pro-rated cost to the buyer, and the warranty period begins anew, that is, the replacement is again warrantied for a period W. Suppose that c_b is the selling price (cost to the buyer) of the warrantied item. Under the usual PRW terms, the cost to the buyer for replacement of an item that fails under warranty is $(X/W)c_b$. Thus income to the seller for the replacement item $(1 - X/W)c_b$.

There are a number of ways of extending these concepts to the 2D case. The extension of the FRW is straightforward. A free replacement is provided as long as the failure occurs in the set Ω. (In fact, this extends in an obvious fashion to more than two dimensions.) The two-dimensional PRW is not as easily defined and, in fact, is not often used in practice. For a warranty of this type, the pro-rated cost to the buyer is expressed as a function of both age and usage at the time of failure. One such possibility is the product of age and usage proportions. This gives replacement cost to the buyer of $(XY/UW)c_b$. For other possibilities, see Wilson and Murthy (1996) and Iskandar et al. (1994). We consider only the 2D FRW.

27.3.5.2 Cost Models

The cost of warranty to the manufacturer depends primarily on two factors, the terms of the warranty and the distribution to time to failure of the item. Warranty cost models require characterization of these two factors and evaluation of all costs associated with production, distribution, and marketing of the item as well as all incidental costs. As a result of the many factors involved and their complexity, modeling warranty costs is often difficult, requiring many simplifying assumptions and, even so, challenging mathematical and numerical analysis. For a detailed discussion of various approaches to modeling, the assumptions made in deriving the models, and the many types warranties in use, see Blischke and Murthy (1994).

To model seller's cost, we write c_n = total average cost to seller of providing a new or replacement item. This includes direct costs (materials, production, etc.) as well as all indirect costs (design, marketing, and so forth), amortized over all items sold. For a repairable item, we write c_r = average cost of repair, including parts and labor. To model both repairable and nonrepairable items, we use c_s to be the average cost to the seller, with c_s being c_n or c_r, as appropriate. The cost models given below are average expected cost per unit to the manufacturer for servicing a given warranty.

We begin with the 1D version of the motorcycle warranty, with only first failure covered. Here the expected cost per unit of warranty, denoted $E[C_1(W)]$, is simply c_s times the proportion of items expected to fail within the warranty period, that is,

$$E[C_1(W)] = c_s F(W) \qquad (27.15)$$

Here $F(.)$ is the CDF of the variable defining warranty terms. Depending on the application, this may be age (e.g., for automobile batteries), usage (tires), or, in the GK analysis of the 2D warranty, the composite variable V [with W in equation (27.15) replaced by U, etc., as appropriate].

Similarly, for the actual 2D version of the motorcycle warranty, the expected cost, denoted $E[C_1(W, U)]$, is given by

$$E[C_1(W, U)] = c_s F(W, U) \qquad (27.16)$$

where $F(.,.)$ is the joint distribution of X and Y.

In the ensuing, we use notation as in (27.15) and (27.16) for expected costs, with subscripts 2 and 3 referring to the FRW and PRW, respectively. For the 1D nonrenewing FRW, the expected warranty cost is

$$E[C_2(W)] = c_s M(W) \qquad (27.17)$$

where $M(.)$ is the renewal function associated with the CDF $F(\cdot)$. The renewal function $M(W)$ is defined to be the expected number of replacements in the interval $[0, W)$, assuming independent and identically distributed item lifetimes. For tables and methods of calculation for many life distributions, see Blischke and Murthy (1994, 1996).

The cost model for the 2D version of the FRW requires a two-dimensional renewal function. The result is

$$E[C_2(W, U)] = c_s M(W, U) \tag{27.18}$$

where $M(.,.)$ is the renewal function associated with $F(.,.)$ (see Blischke and Murthy, 1994, Section 8.5.2).

The cost model for the 1D renewing PRW is

$$E[C_3(W)] = c_s + c_b\left[F(W) + \frac{\mu_W}{W}\right] \tag{27.19}$$

where $F(.)$ is as before, and μ_W, the partial expectation, is given by

$$\mu_W = \int_0^W uf(u)\,du \tag{27.20}$$

where $f(.)$ is the density corresponding to $F(.)$.

27.4 PRELIMINARY DATA ANALYSIS

The four sets of claims data that form the basis of our analysis are described briefly in Section 27.2. The data sets include many entries that are incomplete or otherwise unusable. As previously noted, these were edited to produce the final data sets for warranty and reliability analysis. Resulting counts of numbers of accepted and rejected claims are given in Table 27.2.

In this section we look at some charts and summary statistics that describe the failure modes and age and usage at the time of the warranty claims.

27.4.1 Analysis of Failure Modes

A key aspect of failure analysis is determination of the most common modes of failure. In the 1998 composite data set, there were a total of 5119 claims for which the record contained adequate information regarding mode of failure or defect. Over 80 modes were reported in these claims. Figure 27.1 is a Pareto chart of the failure mode data. The top 13 modes charted account for 80.5% of the total cost of warranty.

27.4.2 Preliminary Analysis of Age

Age of the motorcycle at time of claim was calculated as (claim date) – (buy date), after editing as indicated. For purposes of illustration, we look at ACC-C2. A his-

Table 27.2. Sales and Accepted and Rejected Warranty Claims by Month, 1998, Edited Data

Month	Number of Units Sold	C2 ACC	C2 REJ	C1 ACC	C1 REJ	M_j
1	21,012	718	250	411	586	19,047
2	17,804	654	217	247	274	16,412
3	12,893	464	186	230	276	11,737
4	7,998	401	146	127	62	7,262
5	9,909	333	130	89	32	9,325
6	12,932	168	64			12,700
7	13,252	87	21			13,144
8	11,266	68	24			11,174
9	16,476	67	22			16,387
10	13,168	49	10			13,109
11	7,853	50	11			7,792
12	7,355	69	15			7,271
Total	151,918	3,128	1,096	1,104	1,230	145,360

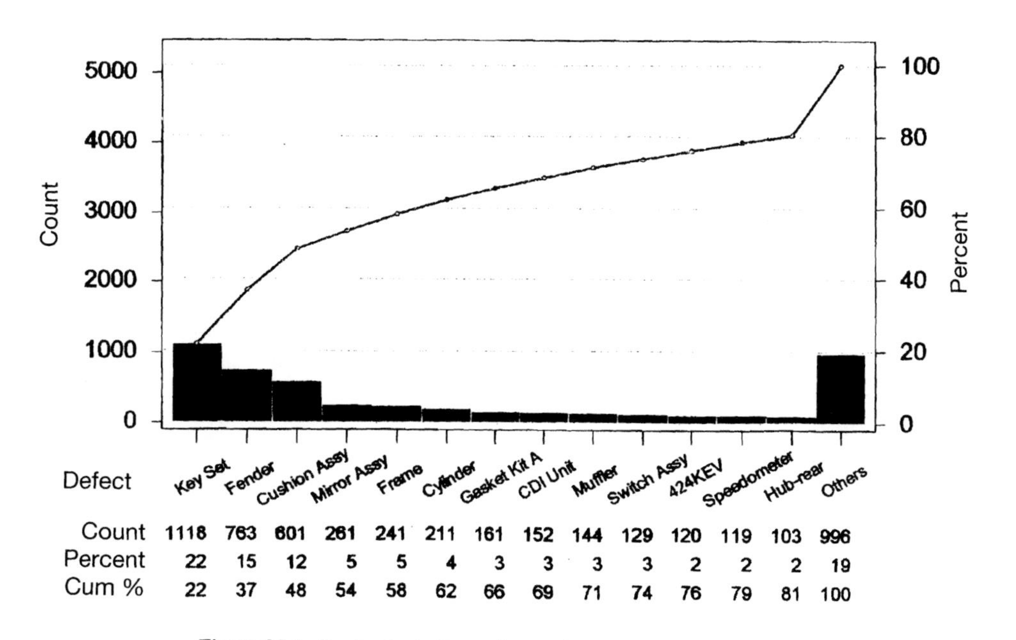

Figure 27.1. Pareto chart of most frequently occurring warranty claims.

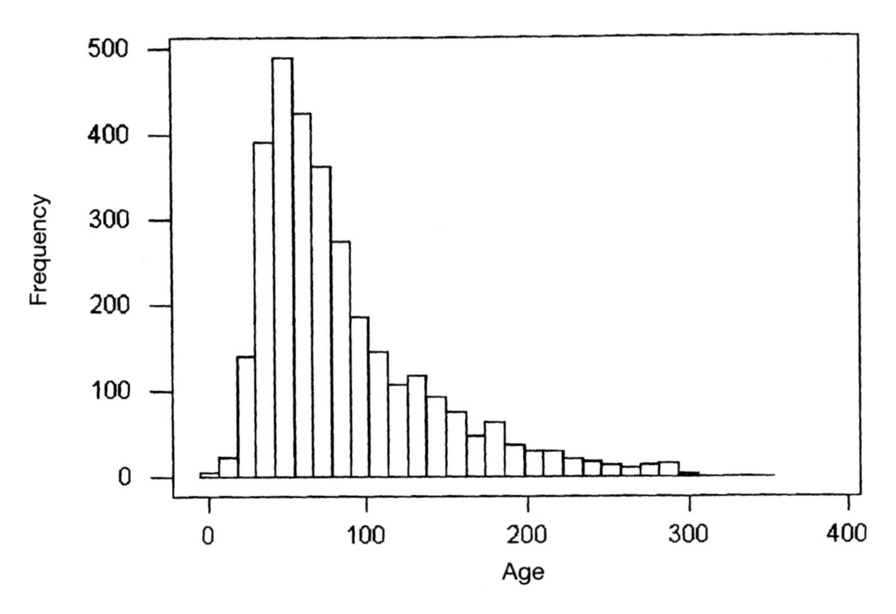

Figure 27.2. Histogram of age, edited data.

togram of the 3128 ages is given in Figure 27.2; summary statistics are given in Table 27.3. The median age for claims is 69.0; the mean is 84.1; the standard deviation is 52.8. These suggest the skewed distribution that is apparent in Figure 27.2. Note that these are conditional results, given a warranty claim, and cannot be interpreted as characteristics of the distribution of time to failure.

Note from the results that many ages exceed 180 days, the nominal age limit of the warranty. Since, as far as we can determine, these were paid claims, we have retained them in the data set. There are a number of possible explanations, including that the claim was legitimate but filing delayed, the claim was honored for good will purposes, and so on. Reporting delays, of course, affect the reliability analysis, though not necessarily the warranty analysis if acceptance of late claims under certain circumstances is, in fact, the policy of The Company.

Table 27.3. Summary Statistics, ACC-C2, Edited (n = 3128)

Variable	Age (days)	Usage (km)
Mean	84.1	1802.6
Median	69.0	1660.5
Standard deviation	52.8	1447.2
Minimum	1.0	1.0
Maximum	305.0	6000.0
First quartile	47.0	500.0
Third quartile	105.0	3000.0

27.4.3 Preliminary Analysis of Usage Data

Summary statistics for usage are also given in Table 27.3 for the edited ACC-C2 data; a histogram of the data is given in Figure 27.3. Average usage is 1802.6 km, again with a large standard deviation. The histogram exhibits a seemingly unusual pattern, with pronounced modes at about 500, 2000, and 4000 km. These, in fact, correspond to required maintenance points, which, if missed, invalidate the warranty. Apparently, owners experiencing noncritical failures sometimes put off making a claim until a trip to the dealer is necessary for maintenance.

This delay in reporting may be modeled by fitting a mixture of distributions to the usage data. The components of the mixture would represent conditional distributions of time to claim, given failure time. This analysis has not been pursued. The effect of the delay on the analysis and interpretation of the results will be discussed further in the next section.

27.5 KALBFLEISCH–LAWLESS ANALYSIS

We begin the further analysis of the claims data by applying the KL analysis. Our objective is to further characterize the claims process. An initial step in the analysis is estimation of claims rates and cumulative claims rates as given in equations (27.2) and (27.3). Although the warranty is nominally for a 6-month period, we complete the computations for the entire year, since the warranty was apparently honored after $W = 6$ and the final claims rate should take this into account for costing purposes. The

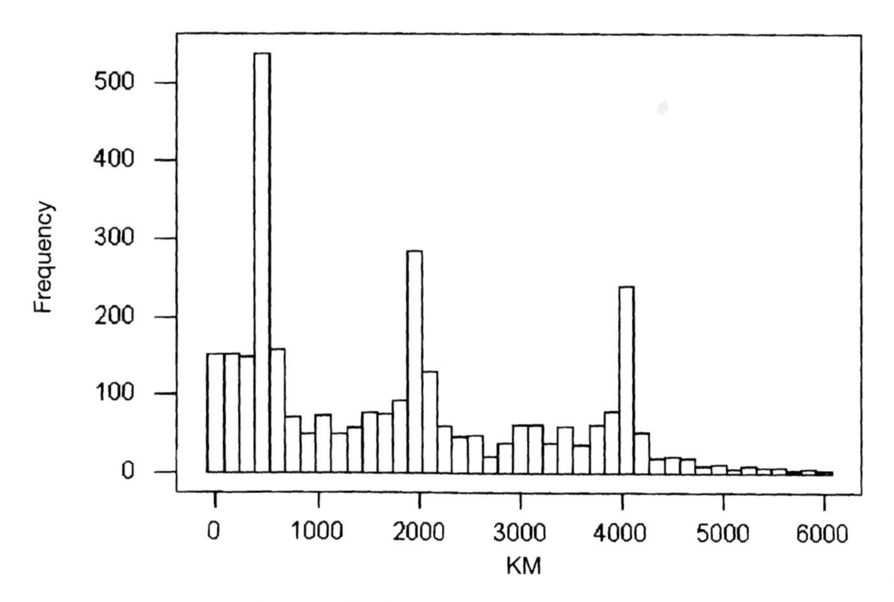

Figure 27.3. Histogram of usage, edited data.

data used are the data on all accepted claims, summarized by total claims per monthly sales in Table 27.1. These data have been edited for errors except for those on usage, since the analysis at this point uses only age at time of claim.

27.5.1 Estimation of Claim Rates

The data needed for the calculations are claims by age. Ideally, since age is given to the nearest day in the data, the analysis should be done at that level. Unfortunately, we have only monthly sales data. One approach to this might be to approximate daily sales (Kalbfleisch and Lawless, 1996). We have chosen instead to illustrate the procedure by aggregating claims to a monthly basis as well. The monthly data are given in Table 27.4 along with cumulative sales $R_T(a)$ for each month. The cumulative claims rate, plotted in Figure 27.4, shows a relatively smooth relationship for the monthly data.

Note that the final value for cumulative claims rate per unit is $\hat{\Lambda}(11) = 0.032$, indicating a claims rate of 3.2%, which is a fairly low rate for an automotive product. This, in addition to the relatively minor nature of many of the failures, is evidence of a reliable product.

A problem not dealt with in this analysis is the apparent delays in reporting discussed previously. Although not evident in the plot of age data (Figure 27.2), these are pronounced in the usage data, and one would assume that, since age and usage are clearly related, the age data are affected as well. A number of techniques have been proposed for dealing with this. [See Kalbfleisch et al. (1991); Kalbfleisch and Lawless (1996), Section 9.3, and Lawless (1998).] A problem is that the solutions require knowledge of the distribution of lags in the population of buyers (e.g., obtained by means of consumer surveys) or knowledge of actual reporting lags. Since no information of this type was available in the study, adjustments for lags was not possible. Note that this does not particularly affect interpretation of the claims patterns, but that it could have a serious impact on any analysis of actual time to failure.

Table 27.4. Estimated Claims Rates

Month (a)	$n_T(a)$	$R_T(a)$	$\hat{\lambda}(a)$	$\hat{\Lambda}(a)$
0	44	151,918	0.000290	0.000290
1	878	144,563	0.006073	0.006363
2	1268	136,710	0.009275	0.015638
3	671	123,542	0.005431	0.021070
4	321	107,066	0.002998	0.024068
5	250	95,800	0.002610	0.026677
6	167	82,548	0.002023	0.028700
7	90	69,616	0.001293	0.029993
8	52	59,707	0.000871	0.030864
9	46	51,709	0.000889	0.031753
10	8	38,816	0.000206	0.031960
11	0	21,012	0.000000	0.031960

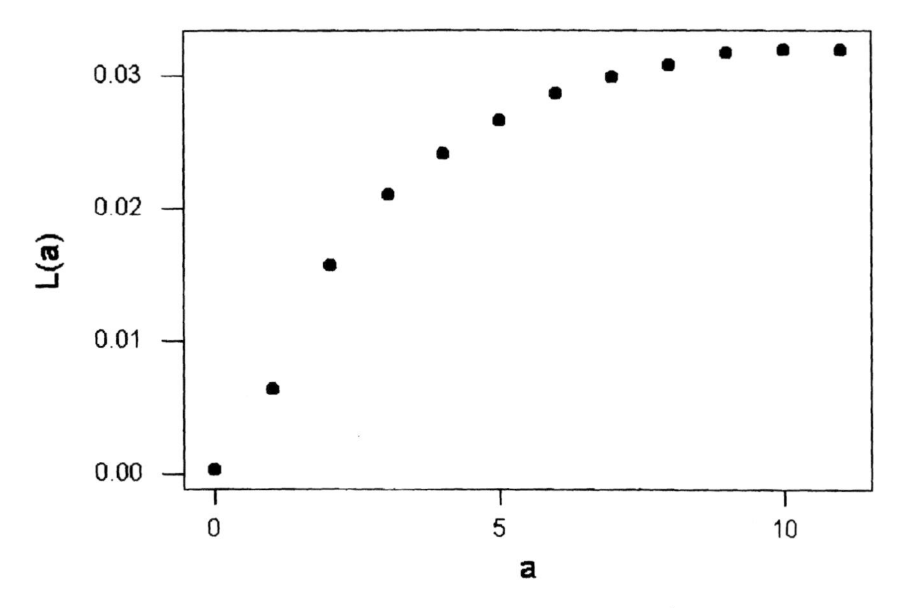

Figure 27.4. Estimated cumulative claims rate $L(a) = \hat{\Lambda}(a)$.

27.5.2 Standard Errors of Estimated Claims Rates

For results to be meaningful, standard errors of the estimates are needed. We assume extra-Poisson variation and estimate this using equations (27.6) and (27.7). Age-specific claims by month, $n(d, a)$, are needed in the calculation. These are given in Table 27.5. Corresponding expected numbers are calculated as indicated, and the estimated variance is found to be

$$\hat{\sigma}^2 = \frac{2357.87}{65} = 32.27$$

This may be used in equation (27.8) to obtain approximate confidence intervals for $\Lambda(a)$. The results for 95% confidence ($z_{a/2} = 1.96$) are given in Table 27.6. The interval is calculated as $\hat{\Lambda}(a) \pm$ (tabulated factor). Note that the claims rates are estimated very imprecisely, especially early on.

The results can also be used to predict future claims. See Kalbfleisch and Lawless (1996) for a discussion of this use of the estimated claims rates.

27.5.3 Estimation of Usage Rates and Field Reliability

Estimation of the distribution of time to failure and of product reliability based on warranty claims data is difficult for a number of reasons. Difficulties include the heavy censoring of the data, other data problems such as lags and errors, and the lack of any control over type of use and a host of other environmental factors that

Table 27.5. Number of Claims, $n(d, a)$, in Month a on Items Sold in Month d

d/a	0	1	2	3	4	5	6	7	8	9	10	Total
0	0	79	414	128	57	58	33	19	15	35	6	844
1	1	277	245	155	43	37	16	25	22	9	2	832
2	17	136	210	60	45	30	38	25	14	2	—	577
3	16	150	138	106	40	45	28	18	1	—	—	542
4	0	133	130	50	66	25	38	3	—	—	—	445
5	1	47	34	95	38	32	14	—	—	—	—	261
6	6	12	45	45	27	23	—	—	—	—	—	158
7	0	16	21	27	5	—	—	—	—	—	—	69
8	2	13	26	5	—	—	—	—	—	—	—	46
9	1	12	5	—	—	—	—	—	—	—	—	18
10	0	3	—	—	—	—	—	—	—	—	—	3
Total	44	878	1268	671	321	250	167	90	52	46	8	3795

typically affect failure rates and related characteristics. In fact, additional information is ordinarily required to estimate the failure distribution and MTTF. This is discussed by Lawless et al. (1995), Hu and Lawless (1996), Lawless (1998), and Hu et al. (1998).

One type of supplementary data that has been used in this regard is data on usage rate, $R = Y/X$. If the distribution of R is known or if it can be estimated from data other than the claims data, improved estimators of $\lambda(t)$ under certain parametric assumptions can be obtained. These lead to improved predictions and can be used to estimate field reliability. It is not uncommon to obtain such information from surveys in the automobile industry, but data of this type have not been collected by the motorcycle manufacturer. We note that information on the distribution of R is also a key element in other approaches to the 2D problem, including the GK analysis discussed in the next section and the approach of Murthy and Wilson (1991).

Table 27.6. Approximate 95% Confidence Intervals for $\Lambda(a)$

a	$\hat{\Lambda}(a)$	Factor
0	0.000290	0.000486
1	0.006363	0.002333
2	0.015638	0.003722
3	0.021070	0.004394
4	0.024068	0.004773
5	0.026677	0.005114
6	0.028700	0.005403
7	0.029993	0.005612
8	0.030864	0.005771
9	0.031753	0.005953
10	0.031960	0.006008

Although inadequate for the purposes under discussion, one can look at the distribution of usage rate in the claims data as a first step. From the buyer data (ACC-C2 and REJ-C2, edited), the average rate was found to be 30.0 km/day; the median rate is 18.5 km/day, indicating a distribution that is highly skewed right. (Note that this is again a conditional distribution, given failure and claim.) Attempts to fit many distributions to the data, including Weibull (often assumed in the analyses cited), gamma, and several other commonly used life distributions, produced no acceptable fits. There are many possible explanations for this, including reporting delays and other problems previously discussed. The use of R in the KL analysis has not been pursued.

27.6 GERTSBAKH–KORDONSKY ANALYSIS

We will model item failures by using the Gertsbakh–Kordonsky approach. We first transform the claims data into V scale and then do parameter estimation for distributions selected on the V scale.

27.6.1 Data Selection and Estimation of ε

We consider warranty claim data for motorcycles sold in 1998 for which claims occurred in 1998. This calendar year is taken as representative of a model year, with analysis done at the end of that year (i.e., after production is completed, but prior to filing of all claims). In practice, an analysis along these lines may be used to predict future claims for similar models.

The censoring structure imposed by the warranty is as follows: For motorcycles that do not fail under warranty, we observe only service age and usage at the time of censoring. Since the warranty is given for 180 days and 6000 km, motor cycles sold prior to July 1, 1998 are censored at $W = 180$ days. For those sold after July 1, the motor cycles are censored at $W < 180$ days. Censoring ages for these motorcycles are determined as follows. Since exact sales dates for these items were not known, we use the midpoint of the month of sale to determine age at censoring. As a result, the censoring age is 165 days for motorcycles sold in July, 135 days for motorcycles sold in August, 105 days for motorcycles sold in September, and so forth.

For all of these cases, we only know ages at censoring times, but not usage. A number of approaches may be taken to approximating usage at censoring. We use three, two of which are based on the usage statistics of drivers with accepted warranty claims. For these, the average usage rate was found to be 0.540 km/day; the median was 0.333 km/day. Pro-ration of these using the midpoint of the sales month provides the approximation of usage. The result based on the mean usage is called Approach I; that based on median of usage is called Approach II. The third alternative considered uses the warranty usage limit, that is, 6000 km (called Approach III).

Note: For convenience, in the remainder of this section, units are changed from days to years and km is changed to 10,000 km. Thus the warranty period becomes

Table 27.7. Ages and Usages at Censoring Times for Approaches I–III

Items sold in month j	Age at Censoring, U_j (in years)	Usage at censoring, W_j (in 10,000 km)		
		Approach I	Approach II	Approach III
1–6	0.5000	0.540	0.333	0.600
7	0.4583	0.495	0.305	0.600
8	0.3750	0.405	0.250	0.600
9	0.2917	0.315	0.194	0.600
10	0.2083	0.225	0.139	0.600
11	0.1250	0.135	0.083	0.600
12	0.0417	0.045	0.028	0.600

$(U, W) = (0.6, 0.5)$. The ages (U_i) and usages (W_i) at time of censoring as indicated above are given in Table 27.7 for $i = 1, \ldots, 12$ and for each of the three approaches. Note that for all approaches, W_j and U_j are constant for $j = 1, \ldots, 6$ (items sold prior to July 1998).

27.6.1.1 Estimation of ε

Using the edited, transformed claims data, we first calculate the estimated value of ε. For this purpose, we use only the 3128 accepted claims in the C2-ACC file. These are buyer data on accepted claims. Data on rejected claims were omitted because many of these are not, in fact, failures and the data set as a whole was considered to be unreliable. Data on C1 are dealer data (rejected and accepted claims) and are not usable for this analysis since all have km = 0, corresponding to problems identified prior to sale.

For the data analyzed, ε was found to be 0.4382. Using this in equation (27.9) transforms the data to the V scale.

27.6.2 Parameter Estimation

The first step in parameter estimation is selection of an appropriate distribution for the variable V. Lacking prior information or any theoretical basis for selection, as is the case here, this is typically done by fitting data to various candidate distributions. Standard statistical goodness-of-fit procedures that may be used for this purpose present two significant difficulties when applied to data of the type with which we are dealing. The first is that we are using estimated parameter values, whereas standard tests such as the Komlogorof–Smirnov, Anderson–Darling, and others require completely specified hypothesized distributions. As a result, standard tabulated values are not valid. This has been addressed for selected distributions by numerical studies, from which critical values can be obtained when parameters are estimated. [See Lawless (1982) and Blischke and Murthy (2000).] The second problem is a result of the censoring. Little is known, either theoretically or through simulation, regarding critical values of the goodness-of-fit statistics when censored data are used, either with specified or estimated parameter values. Furthermore, the test statistics appear to perform very poorly when data are very highly censored, as is the case with the motorcycle data.

As a preliminary check, the complete data (i.e., ignoring all censored values) were fit to various distributions. The gamma distribution appeared to provide the best fit among those tested. We use this as well as Weibull and lognormal distributions in fitting the full set of data. This will illustrate the use of three common failure distributions and will form the basis of a small-scale study of the sensitivity of the results to the distributional assumption.

To fit the three distributions, we need (W_j, U_j), M_j, and $\tilde{V}_j$. These are obtained from Tables 27.7 and 27.2 and equation (27.10). We applied a numerical approach using MATHCAD 2000 to obtain the estimate $\hat{\theta}$ for Weibull, gamma, and lognormal distributions. The results for the three distributions are as follows:

Case 1: Weibull Distribution. The density function is

$$f_V(v) = \frac{\alpha}{\beta}\left(\frac{v}{\beta}\right)^{\alpha-1} e^{-(v/\beta)^{\alpha}} \tag{27.21}$$

In this case, $p_1 = \alpha$ and $p_2 = \beta$. The estimates of α and β for the three approaches, along with estimated standard errors obtained by evaluation of (27.13), are given in Table 27.8, along with the estimated MTTF, calculated from the estimated parameters [see Blischke and Murthy (2000)]. The resulting confidence intervals for α and β, obtained from (27.14), are given in Table 27.9. Note that the confidence intervals are relatively wide, perhaps as a result of the heavy censoring, and that the results depend critically on the censoring approach used. Note also that in each instance, the confidence interval for α (the shape parameter) contains only values exceeding one. This is an indication that the motorcycle as a system has an increasing failure rate.

Table 27.8. Numerical Results for the Weibull Distribution

Approach	$\hat{\alpha}$	$\hat{\beta}$	$M \log L$	S_α	S_β	MTTF
I	1.2047	9.9128	-5.365×10^3	0.0260	0.6812	9.3152
II	1.4095	5.2958	-5.040×10^3	0.0339	0.3493	4.8217
III	1.1511	13.3036	-5.613×10^3	0.0236	0.9165	12.6575

Table 27.9. Confidence Intervals for Weibull Parameters

	95% Confidence Interval for α		95% Confidence Interval for β	
Approach	Lower Bound	Upper Bound	Lower Bound	Upper Bound
I	1.1537	1.2557	8.5776	11.2480
II	1.3431	1.4759	4.6112	5.9804
III	1.1048	1.1974	11.5073	15.0999

Case 2: Gamma Distribution. The density function for the gamma distribution is

$$f_V(v) = \frac{v^{(\alpha-1)}e^{-(v/\beta)}}{\beta^{\alpha}\Gamma(\alpha)} \tag{27.22}$$

Again, $p_1 = \alpha$ and $p_2 = \beta$. The values of estimates $\hat{\alpha}$ and $\hat{\beta}$ for the three approaches are given in Table 27.10. The approximate confidence intervals are given in Table 27.11. As in Case 1, the confidence intervals are relatively wide. Also as in Case 1, the confidence interval for the shape parameter α is indicative of an increasing failure rate.

Case 3: Lognormal Distribution. The density function is

$$f_V(v) = \frac{1}{v\sqrt{2\pi\sigma^2}}\, e^{-(\ln(v)-\mu)^2/2\sigma^2} \tag{27.23}$$

Here $p_1 = \mu$ and $p_2 = \sigma$. The values of estimates $\hat{\mu}$ and $\hat{\sigma}$ for the three approaches are given in Table 27.12. The approximate confidence intervals are given in Table 27.13. In contrast to the first two cases, in this case, the confidence intervals are relatively small. Note, however, that the estimated MTTF is very much larger for the lognormal than for either of the other distributions. This is a consequence of the fact that the fitted lognormal has a much longer tail than either of the others.

Comments

1. The maximum values of the log likelihood, given for the three distributions fitted in Tables 27.8, 27.10, and 27.12, vary somewhat by approach, but very little for the different distributions. On this basis, there is no clear-cut choice

Table 27.10. Numerical Results for the Gamma Distribution

Approach	$\hat{\alpha}$	$\hat{\beta}$	$M \log L$	S_α	S_β	MTTF
I	1.1318	11.6066	-5.359×10^3	0.0304	1.2337	13.1363
II	1.4125	4.3853	-5.036×10^3	0.0384	0.3975	6.1942
III	1.0882	15.7595	-5.603×10^3	0.0260	1.5424	17.1495

Table 27.11. Confidence Intervals for Gamma Parameters

	95% Confidence Interval for α		95% Confidence Interval for β	
Approach	Lower Bound	Upper Bound	Lower Bound	Upper Bound
I	1.0722	1.1914	9.1885	14.0247
II	1.3372	1.4878	3.6062	5.1644
III	1.0372	1.1392	12.7364	18.7826

Table 27.12. Numerical Results for the Lognormal Distribution

Approach	$\hat{\mu}$	$\hat{\sigma}$	$M \log L$	S_α	S_β	MTTF
I	3.6735	2.2543	-5.3×10^3	0.1140	0.0536	499.9131
II	2.7047	1.8616	-4.996×10^3	0.0921	0.0436	84.5613
III	4.3597	2.5083	-5.542×10^3	0.1342	0.0622	1817.986

Table 27.13. Confidence Intervals for Lognormal Parameters

	95% Confidence Interval for μ		95% Confidence Interval for σ	
Approach	Lower Bound	Upper Bound	Lower Bound	Upper Bound
I	3.4501	3.8969	2.1492	2.3594
II	2.5242	2.8852	1.7761	1.9471
III	4.0967	4.6227	2.3864	2.6302

of distribution. For most reliability application, a conservative choice would be desirable. This would argue against the lognormal.

2. Approaches I–III give quite different results for all three distributions. The most conservative choice is Approach II, which uses the median as guessed value for usage. This is also a good choice from a statistical point of view, because it is not influenced by a high degree of skewness or by outlying observations.
3. As noted, standard goodness-of-fit procedures are not applicable to data of the type with which we are dealing. Calculated values of, for example, the Anderson–Darling statistic have been found to be very large for all censored data and all distributions considered, but little is known about the test in this context and there is no way of interpreting the results.
4. For a given distribution, one can calculate confidence bands for the CDF based on censored data. This is discussed in Jeng and Meeker (2001). This has not been pursued in the present study.

In the next section, we look at how these choices affect the results with regard to warranty costing.

27.7 WARRANTY ANALYSIS

The motorcycle in use at the time of this study covered only first failure of a given component for a period of 6 months or usage of 6000 km, whichever occurs first. Cost models for this and several other warranties are given in Section 27.3. In this section we perform a cost analysis of selected warranties, again considering the system as a whole. The primary objective is to analyze alternative warranty terms that might be considered by The Company. The choice of a warranty is influenced by cost considerations as well as by the marketplace. In looking at alternatives, a manufacturer will consider policies that may provide a marketing advantage, so long as the estimated cost is reasonable.

A logical extension of the existing warranty is the nonrenewing FRW, covering the original unit plus all replaced or repaired components up to time W from time on initial purchase. The original warranty and the nonrenewing FRW will be called Warranties 1 and 2, respectively. We translate the warranty terms to the V scale and analyze the warranties on this basis. Cost models for the warranties are given in equations (27.15) and (27.17). The cost analysis will be done on a relative basis, looking at proportional costs of the competing warranties rather than at actual costs, that is, we put $c_s = 1$ in equations (27.15) and (27.17).

In comparing the two warranties, it is also of interest to vary the warranty period. With $\varepsilon = 0.4382$ in the GK approach, the warranty limit of the V scale, say V^*, becomes $V^* = 0.5618(0.6) + 0.4382(0.5) = 0.5562$. We consider Warranties 1 and 2, the estimated distributions corresponding to Cases 1–3, and Approaches I–III. In addition, we look at three values of V^*: the nominal $V_1^* = 0.5562$, $V_2^* = 2(0.5562) = 1.112$, and $V_3^* = 4(0.5562) = 2.225$. These values represent a doubling of the warranty terms to 12,000 km/1 year and a redoubling to 24,000 km/2 years. Terms such as these might be considered if the cost is not excessive, and it is felt that the terms would provide a significant competitive advantage.

Note that evaluation of the cost model for Warranty 1 requires only estimation of the CDF for each case. The results are given in Table 27.14. Note that for the existing warranty terms ($V^* = 0.5562$), the calculation based on the mean usage (Approach II) gives an estimated warranty cost of $P(V < 0.5562) = 0.0294$ for the gamma and lognormal distributions and 0.0306 for the Weibull. These values are very close to the observed proportion of warranty claims of 0.03, which is also the value given by the KL approach. This is as expected, since we are dealing with nearly complete data for the year analyzed.

For Warranty 2, evaluation of the cost model requires calculation of renewal

Table 27.14. Estimated Costs for Warranty 1

		Distribution		
V^*	Approach	Weibull	Gamma	Lognormal
0.556	I	0.0306	0.0294	0.0294
	II	0.0409	0.0401	0.0385
	III	0.0255	0.0248	0.0243
1.112	I	0.0692	0.0629	0.0568
	II	0.1049	0.0994	0.0814
	III	0.0559	0.0518	0.0450
2.225	I	0.1524	0.1312	0.1012
	II	0.2551	0.2296	0.1531
	III	0.1198	0.1062	0.0779

Table 27.15. Estimated Costs for Warranty 2

		Distribution		
V^*	Approach	Weibull	Gamma	Lognormal
0.556	I	0.0310	0.0298	0.0298
	II	0.0414	0.0407	0.0391
	III	0.0261	0.0251	0.0246
1.112	I	0.0712	0.0647	0.0585
	II	0.1088	0.1030	0.0853
	III	0.0572	0.0530	0.0461
2.225	I	0.1629	0.1396	0.1073
	II	0.2806	0.2512	0.1662
	III	0.1265	0.1118	0.0820

functions. This was accomplished by use of a program given in Appendix C_2 of Blischke and Murthy (1994). The results are given in Table 27.15. Note that in every instance Warranty 2 is estimated to lead to higher costs than Warranty 1. This is as expected, since greater coverage is provided. For the 6-month warranty period, however, the difference between Warranties 1 and 2 is quite small. This is due to the fact that multiple failures in this relatively short period are unlikely. As the warranty terms become more generous, the difference is warranty cost between the two policies becomes more significant.

A key consideration in choosing an alternative warranty policy is comparison of estimated warranty costs. As can be seen from Tables 27.14 and 27.15, the most conservative approach in the analysis is to use the Weibull distribution and Approach II. For this choice, the estimated ultimate cost of the existing warranty is 0.041, which can be interpreted as 4.1% of the seller's cost of an item if sold without warranty. The value is only slightly higher for Warranty 2. If the warranty terms are doubled, these costs become 10.5% for Warranty 1 and 10.9% for Warranty 2. These are quite high (more than double the current cost), but may be tolerable if they provide a sufficient increase in competitive advantage. Also, these are conservative estimates and, as such, may be considered upper bounds of a sort on the actual costs. Finally, if more generous warranty terms are desired for marketing purposes, a slight increase in selling price might cover some or all of the warranty cost without unduly affecting sales. Trade-offs among these various alternatives are an important factor in the management of warranties.

27.8 CONCLUSIONS

In this study we have attempted to analyze warranty claims data, with the objectives of characterizing the failure distribution of a motorcycle and evaluating alternative warranties. The motorcycle in question was sold under first-failure warranty coverage with a warranty period of 6 months/6000 km. The two methods of warranty data analysis employed were based on the approaches of Kablfleisch–Lawless (KL) and Gertsbakh–Kordonsky (GK). Our principal findings and recommendations are as follows:

- As noted by many authors, warranty claims data are generally error-prone, incomplete, highly censored, and difficult to analyze. We certainly found this to be the case.
- The claims data exhibited some quite unusual features, including multimodality. This suggests that the probabilistic model representing time to warranty claim in terms of time to failure is quite complex. This complexity was ignored in our analysis.
- The KL approach as used and the GK approach both reduce the analysis of the two-dimensional warranty to a one-dimensional problem. Both approaches provide some useful information regarding the failure distribution of the item. Both, however, would be considerably enhanced by the use of additional data (e.g., from consumer surveys) or other additional information (e.g., the distribution of the ratio $R = Y/X$), none of which were available in the current study.
- In the GK analysis, the estimate of ε was obtained as that value that minimized the c.v. Other possible approaches that can be used are the maximum likelihood estimate and the least-squares estimate of parameter ε [see Duchesne and Lawless (2000) and Ahn et al. (1998)].
- In the KL approach, both distributional assumptions and the method of dealing with censoring can affect the results substantially.
- In this case, some similar useful results were obtained using the KL and GK approaches—for example, the estimated ultimate proportion of claims. Comparing predicted with actual would be a more meaningful exercise, but this was beyond the scope of the project. Some additional useful results for prediction of warranty costs are given by Escobar and Meeker (1999).
- The analysis was done considering the motorcycle as a system. A more thorough analysis would involve breakdown of the system into major components and analysis of each of these individually.
- Based on the results of the cost analysis, the warranty offered could be upgraded to a nonrenewing FRW at very little increase in cost. Extension of the warranty period would lead to increased costs, and at an increasing rate.
- Finally, we note that the problem of characterizing the joint failure distribution (of age and usage to failure) has not been addressed. This requires estimation of a bivariate distribution. Some recent results in this area may be found in Dabrowska et al. (1998), Kooperberg (1998), van der Laan (1996), and Wells and Yeo (1996).

REFERENCES

Ahn, C. W., Chae, K. C. and Clark, G. M. (1998). Estimating parameters of the power law process with two measures of failure rate, *Journal of Quality Technology* **30:** 127–132.

Blischke, W. R., and Murthy, D. N. P. (1994). *Warranty Cost Analysis,* Marcel Dekker, New York.

Blischke, W. R., and Murthy, D. N. P. (1996). *Product Warranty Handbook,* Marcel Dekker, New York.

Blischke, W. R., and Murthy, D. N. P. (2000). *Reliability: Modeling, Prediction, and Optimization,* Wiley, New York.

Chen, J., Lynn, N. J., and Singpurwalla, N. D. (1996). Forecasting Warranty Claims," in *Product Warranty Handbook* (W. R. Blischke and D. N. P. Murthy, Editors), Chapter 31, Marcel Dekker, New York.

Dabrowska, D. M, Duffy, D. L., and Zhang, A. D. (1998). Hazard and density estimation form bivariate censored data, *Nonparametric Statistics* **10:**67–93.

Duchesne, T., and Lawless, J. (2000). Alternative time scales and failure time models, *Lifetime Data Analysis* **6:**157–179.

Eliashberg, J., Singpurwalla, N. D., and Wilson, S. P. (1997). Calculating the reserve for a time and usage indexed warranty, *Management Science* **43:**966–975.

Escobar, L. A., and Meeker, W. Q. (1999). Statistical prediction based on censored lifetime data, *Technometrics* **41:**113–124.

Farewell, V. T., and Cox, D. R. (1979). A note on multiple time scales in life testing, *Applied Statistics* **28:**73–75.

Gertsbakh, I. B., and Kordonsky, K. B. (1998). Parallel time scales and two-dimensional manufacturer and individual customer warranties, *IIE Transactions* **30:**1181–1189.

Hu, X. J., and Lawless, J. F. (1996). Estimation of rate and mean functions from truncated recurrent event data, *Journal of the American Statistical Association* **91:**300–310.

Hu, X. J., Lawless, J. F., and Suzuki, K. (1998). Nonparametric estimation of a lifetime distribution when censoring times are missing, *Technometrics* **40:**3–13.

Iskandar, B. (1993). *Modelling and Analysis of Two Dimensional Warranty Policies,* doctoral dissertation, The University of Queensland, Brisbane.

Iskandar, B. P., Wilson, R. J. and Murthy, D. N. P. (1994). Two-dimensional combination warranty policies, *RAIRO Operational Research* **28:**57–75

Jeng, S.-L., and Meeker, W. Q. (2001). Parametric simultaneous confidence bands for cumulative distributions from censored data, *Technometrics* **43:**450–461.

Jewell, N. P., Kimber, A. C., Ting Lee, M.-L., and Whitmore, G. A., Editors (1996). *Lifetime Data: Models in Reliability and Survival Analysis,* Kluwer, London.

Kalbfleisch, J. D., and Lawless, J. F. (1988). Estimation of reliability in field-performance studies (with discussion), *Technometrics* **30:**365–388.

Kalbfleisch, J. D., and Lawless, J. F. (1996). Statistical Analysis of Warranty Claims Data, in *Product Warranty Handbook* (W. R. Blischke and D. N. P. Murthy, Editors), Chapter 9, Marcel Dekker, New York.

Kalbfleisch, J. D., Lawless, J. F., and Robinson, J. A. (1991). Methods for the analysis and prediction of warranty claims, *Technometrics* **33:**173–285.

Kalbfleisch, J. D., and Prentice, R. L. (1980). *The Statistical Analysis of Failure Time Data,* Wiley, New York.

Kooperberg, C. (1998). Bivariate density estimation with application to survival analysis, *Journal of Computational and Graphical Statistics* **7:**322–341.

Kordonsky, K. B., and Gertsbakh, I. (1993). Choice of best time scale for reliability analysis, *European Journal of Operational Research* **65:**235–246.

Kordonsky, K. B., and Gertsbakh, I. (1995a). System state monitoring and lifetime scales I, *Reliability Engineering and System Safety* **47:**1–14.

Kordonsky, K. B., and Gertsbakh, I. (1995b). System state monitoring and lifetime scales II, *Reliability Engineering and System Safety* **49:**145–154.

Kordonsky, K. B., and Gertsbakh, I. (1997). Multiple time scales and lifetime coefficient of variation: engineering applications, *Lifetime Data Analysis* **2:**139–156.

Lawless, J. F. (1982). *Statistical Models and Methods for Lifetime Data,* Wiley, New York.

Lawless, J. F. (1994). Adjustment for reporting delays and the prediction of occurred but not reported events, *Canadian Journal of Statistics* **22:**15–31.

Lawless, J. F. (1998). Statistical analysis of product warranty data, *International Statistical Review* **66:**40–60.

Lawless, J., Hu, J., and Cao, J. (1995). Methods for estimation of failure distributions and rates from automobile warranty data, *Lifetime Data Analysis* **1:**227–240.

Meeker, W. Q., and Escobar, L. A. (1998). *Statistical Methods for Reliability Data,* Wiley, New York.

Moskowitz, H., and Chun, Y. H. (1994). A Poisson regression model for two-attribute warranty policy, *Naval Research Logistics* **41:**355–376.

Moskowitz, H., and Chun, Y. H. (1996). Two-Dimensional Free-Replacement Warranties, in *Product Warranty Handbook* (W. R. Blischke and D. N. P. Murthy Editors), Chapter 13, Marcel Dekker, New York.

Murthy, D. N. P., and Blischke, W. R. (2000). Strategic warranty management: A life-cycle approach, *IEEE Transactions on Engineering Management* **47:**40–54.

Murthy, D. N. P., Iskandar, B.P, and Wilson, R. J. (1995). Two-dimensional failure free warranties: Two-dimensional point process models, *Operations Research* **43:**356–366.

Murthy, D. N. P., and Wilson, R. J. (1991). Modelling Two-Dimensional Failure Free Warranties, in *Proceedings Fifth Symposium on Applied Stochastic Models and Data Analysis,* pp. 481–492, Granada, Spain.

Nelson, W. (1982). *Applied Life Data Analysis,* Wiley, New York.

Oakes, D. (1995). Multiple time scales in survival analysis, *Lifetime Data Analysis* **1:**7–18.

Robinson, J. A., and McDonald, G. C. (1991). Issues Related to Field Reliability and Warranty Data, in *Data Quality Control: Theory and Pragmatics,* (G. E. Liepens and V. R. R. Uppuluri, Editors), Chapter 7, Marcel Dekker, New York.

Singpurwalla, N. D., and Wilson, S. (1993). The warranty problem: Its statistical and game theoretic aspects, *SIAM Review* **35:**17–42.

Suzuki, K. (1985). Estimation of lifetime parameters from incomplete field data, *Technometrics* **27:**263–271.

Suzuki, K., Karim, Md. R., and Wang, L. (2001). Statistical Analysis of Reliability Warranty Data, in *Handbook of Statistics* (N. Balakrishnan and C. R. Rao, Editors), Chapter 21 Elsevier Science, Amsterdam.

van der Laan, M. J. (1996). Efficient estimation in the bivariate censoring model and repairing NPMLE, *Annals of Statistics* **24:**596–627.

Wang, L., and Suzuki, K. (2001a). Nonparametric estimation of lifetime distribution from warranty data without monthly unit sales information, *Journal of Reliability Engineering Association. of Japan* **23:**145–154.

Wang. L., and Suzuki, K. (2001b). Lifetime estimation based on warranty data without date-of-sale information—Cases where usage time distributions are known, *Journal of the Japanese Society for Quality Control* **31:**148–167.

Wasserman, G. S. (1992). An application of dynamic linear models for predicting warranty claims, *Computers and Industrial Engineering* **22:**37–47.

Wells, M. T., and Yeo, K. P. (1996). Density estimation with bivariate censored data, *Journal of the American Statistical Association* **91:**1566–1574.

Wilson, R. J., and Murthy, D. N. P. (1996). Two-Dimensional Pro-Rata and Combination Warranties, *Product Warranty Handbook* (W. R. Blischke and D. N. P. Murthy, Editors), Chapter 14, Marcel Dekker, New York.

Yang, S.-C., Kobza, J. E., and Nachlas, J. A. (2000). Bivariate Failure Modeling, in *2000 Proceedings Annual Reliability and Maintainability Symposium,* pp. 281–287.

EXERCISES

27.1. Plot the Weibull, gamma, and lognormal distributions obtained in Section 27.6 for Approaches I–III. For each approach, plot the three distributions on the same chart. Comment on the shapes of the distributions and on implications regarding reliability and warranty costs.

27.2. Find formulas for variances of the Weibull, gamma, and lognormal distributions in terms of the parameters of the distribution. Use these results and the estimated parameter values to estimate variances of V for each approach and compare the results.

27.3. Assuming a gamma distribution with parameters estimated by Approach II, evaluate warranty costs for Warranties 1 and 2 for each of the following combinations:

(1) 9 months/9000 km

(ii) 1 year/10,000 km

(iii) at least two other combinations that you feel might be reasonable

Compare the results and give your opinion regarding management's choice of a warranty policy.

27.4. Repeat Exercise 27.3 under Weibull and lognormal assumptions. Do the results change your opinion regarding choice of a warranty policy and terms?

27.5. The results of the preceding two exercises are based on point estimates of the parameters of the distributions. Redo the calculations using upper and lower end-points of the confidence intervals for each of the parameters of each distribution. (Use combinations of upper and/or lower confidence bounds in each case that will provide the most extreme cost estimates.)

27.6. Calculate 99% approximate confidence intervals for the parameters of each of the candidate distributions for V. Use the results of Approach II.

27.7. (Optional) Evaluate the cost of offering a renewing PRW on the motorcycle, with terms of six months/6000 miles. Base the analysis on an appropriate estimate of the distribution of V and compare the results with Warranties 1 and 2.

27.8. Using the KL approach, analyze only the first 6 months of data (given in tables used in Section 27.5). Use the results to predict the remaining periods and compare the predictions with actual, as appropriate.

27.9. Repeat Exercise 27.8 using only the first 3 months' data.

27.10. Find maximum likelihood estimates of ε and θ for each candidate distribution of V, gamma, Weibull and lognormal, using Approach II. Compare the results with those of Section 27.6.

Index

WILEY SERIES IN PROBABILITY AND STATISTICS
ESTABLISHED BY WALTER A. SHEWHART AND SAMUEL S. WILKS

The ***Wiley Series in Probability and Statistics*** is well established and authoritative. It covers many topics of current research interest in both pure and applied statistics and probability theory. Written by leading statisticians and institutions, the titles span both state-of-the-art developments in the field and classical methods.

Reflecting the wide range of current research in statistics, the series encompasses applied, methodological and theoretical statistics, ranging from applications and new techniques made possible by advances in computerized practice to rigorous treatment of theoretical approaches.

This series provides essential and invaluable reading for all statisticians, whether in academia, industry, government, or research.

ABRAHAM and LEDOLTER · Statistical Methods for Forecasting
AGRESTI · Analysis of Ordinal Categorical Data
AGRESTI · An Introduction to Categorical Data Analysis
AGRESTI · Categorical Data Analysis, *Second Edition*
ANDĚL · Mathematics of Chance
ANDERSON · An Introduction to Multivariate Statistical Analysis, *Second Edition*
*ANDERSON · The Statistical Analysis of Time Series
ANDERSON, AUQUIER, HAUCK, OAKES, VANDAELE, and WEISBERG · Statistical Methods for Comparative Studies
ANDERSON and LOYNES · The Teaching of Practical Statistics
ARMITAGE and DAVID (editors) · Advances in Biometry
ARNOLD, BALAKRISHNAN, and NAGARAJA · Records
*ARTHANARI and DODGE · Mathematical Programming in Statistics
*BAILEY · The Elements of Stochastic Processes with Applications to the Natural Sciences
BALAKRISHNAN and KOUTRAS · Runs and Scans with Applications
BARNETT · Comparative Statistical Inference, *Third Edition*
BARNETT and LEWIS · Outliers in Statistical Data, *Third Edition*
BARTOSZYNSKI and NIEWIADOMSKA-BUGAJ · Probability and Statistical Inference
BASILEVSKY · Statistical Factor Analysis and Related Methods: Theory and Applications
BASU and RIGDON · Statistical Methods for the Reliability of Repairable Systems
BATES and WATTS · Nonlinear Regression Analysis and Its Applications
BECHHOFER, SANTNER, and GOLDSMAN · Design and Analysis of Experiments for Statistical Selection, Screening, and Multiple Comparisons
BELSLEY · Conditioning Diagnostics: Collinearity and Weak Data in Regression
BELSLEY, KUH, and WELSCH · Regression Diagnostics: Identifying Influential Data and Sources of Collinearity
BENDAT and PIERSOL · Random Data: Analysis and Measurement Procedures, *Third Edition*

*Now available in a lower priced paperback edition in the Wiley Classics Library.

BERRY, CHALONER, and GEWEKE · Bayesian Analysis in Statistics and Econometrics: Essays in Honor of Arnold Zellner
BERNARDO and SMITH · Bayesian Theory
BHAT and MILLER · Elements of Applied Stochastic Processes, *Third Edition*
BHATTACHARYA and JOHNSON · Statistical Concepts and Methods
BHATTACHARYA and WAYMIRE · Stochastic Processes with Applications
BILLINGSLEY · Convergence of Probability Measures, *Second Edition*
BILLINGSLEY · Probability and Measure, *Third Edition*
BIRKES and DODGE · Alternative Methods of Regression
BLISCHKE AND MURTHY (editors) · Case Studies in Reliability and Maintenance
BLISCHKE AND MURTHY · Reliability: Modeling, Prediction, and Optimization
BLOOMFIELD · Fourier Analysis of Time Series: An Introduction, *Second Edition*
BOLLEN · Structural Equations with Latent Variables
BOROVKOV · Ergodicity and Stability of Stochastic Processes
BOULEAU · Numerical Methods for Stochastic Processes
BOX · Bayesian Inference in Statistical Analysis
BOX · R. A. Fisher, the Life of a Scientist
BOX and DRAPER · Empirical Model-Building and Response Surfaces
*BOX and DRAPER · Evolutionary Operation: A Statistical Method for Process Improvement
BOX, HUNTER, and HUNTER · Statistics for Experimenters: An Introduction to Design, Data Analysis, and Model Building
BOX and LUCEÑO · Statistical Control by Monitoring and Feedback Adjustment
BRANDIMARTE · Numerical Methods in Finance: A MATLAB-Based Introduction
BROWN and HOLLANDER · Statistics: A Biomedical Introduction
BRUNNER, DOMHOF, and LANGER · Nonparametric Analysis of Longitudinal Data in Factorial Experiments
BUCKLEW · Large Deviation Techniques in Decision, Simulation, and Estimation
CAIROLI and DALANG · Sequential Stochastic Optimization
CHAN · Time Series: Applications to Finance
CHATTERJEE and HADI · Sensitivity Analysis in Linear Regression
CHATTERJEE and PRICE · Regression Analysis by Example, *Third Edition*
CHERNICK · Bootstrap Methods: A Practitioner's Guide
CHERNICK and FRIIS · Introductory Biostatistics for the Health Sciences
CHILÈS and DELFINER · Geostatistics: Modeling Spatial Uncertainty
CHOW and LIU · Design and Analysis of Clinical Trials: Concepts and Methodologies
CLARKE and DISNEY · Probability and Random Processes: A First Course with Applications, *Second Edition*
*COCHRAN and COX · Experimental Designs, *Second Edition*
CONGDON · Bayesian Statistical Modelling
CONOVER · Practical Nonparametric Statistics, *Second Edition*
COOK · Regression Graphics
COOK and WEISBERG · Applied Regression Including Computing and Graphics
COOK and WEISBERG · An Introduction to Regression Graphics
CORNELL · Experiments with Mixtures, Designs, Models, and the Analysis of Mixture Data, *Third Edition*
COVER and THOMAS · Elements of Information Theory
COX · A Handbook of Introductory Statistical Methods
*COX · Planning of Experiments
CRESSIE · Statistics for Spatial Data, *Revised Edition*
CSÖRGŐ and HORVÁTH · Limit Theorems in Change Point Analysis
DANIEL · Applications of Statistics to Industrial Experimentation
DANIEL · Biostatistics: A Foundation for Analysis in the Health Sciences, *Sixth Edition*

*Now available in a lower priced paperback edition in the Wiley Classics Library.

*DANIEL · Fitting Equations to Data: Computer Analysis of Multifactor Data, *Second Edition*
DAVID · Order Statistics, *Second Edition*
*DEGROOT, FIENBERG, and KADANE · Statistics and the Law
DEL CASTILLO · Statistical Process Adjustment for Quality Control
DETTE and STUDDEN · The Theory of Canonical Moments with Applications in Statistics, Probability, and Analysis
DEY and MUKERJEE · Fractional Factorial Plans
DILLON and GOLDSTEIN · Multivariate Analysis: Methods and Applications
DODGE · Alternative Methods of Regression
*DODGE and ROMIG · Sampling Inspection Tables, *Second Edition*
*DOOB · Stochastic Processes
DOWDY and WEARDEN · Statistics for Research, *Second Edition*
DRAPER and SMITH · Applied Regression Analysis, *Third Edition*
DRYDEN and MARDIA · Statistical Shape Analysis
DUDEWICZ and MISHRA · Modern Mathematical Statistics
DUNN and CLARK · Applied Statistics: Analysis of Variance and Regression, *Second Edition*
DUNN and CLARK · Basic Statistics: A Primer for the Biomedical Sciences, *Third Edition*
DUPUIS and ELLIS · A Weak Convergence Approach to the Theory of Large Deviations
*ELANDT-JOHNSON and JOHNSON · Survival Models and Data Analysis
ETHIER and KURTZ · Markov Processes: Characterization and Convergence
EVANS, HASTINGS, and PEACOCK · Statistical Distributions, *Third Edition*
FELLER · An Introduction to Probability Theory and Its Applications, Volume I, *Third Edition,* Revised; Volume II, *Second Edition*
FISHER and VAN BELLE · Biostatistics: A Methodology for the Health Sciences
*FLEISS · The Design and Analysis of Clinical Experiments
FLEISS · Statistical Methods for Rates and Proportions, *Second Edition*
FLEMING and HARRINGTON · Counting Processes and Survival Analysis
FULLER · Introduction to Statistical Time Series, *Second Edition*
FULLER · Measurement Error Models
GALLANT · Nonlinear Statistical Models
GHOSH, MUKHOPADHYAY, and SEN · Sequential Estimation
GIFI · Nonlinear Multivariate Analysis
GLASSERMAN and YAO · Monotone Structure in Discrete-Event Systems
GNANADESIKAN · Methods for Statistical Data Analysis of Multivariate Observations, *Second Edition*
GOLDSTEIN and LEWIS · Assessment: Problems, Development, and Statistical Issues
GREENWOOD and NIKULIN · A Guide to Chi-Squared Testing
GROSS and HARRIS · Fundamentals of Queueing Theory, *Third Edition*
*HAHN and SHAPIRO · Statistical Models in Engineering
HAHN and MEEKER · Statistical Intervals: A Guide for Practitioners
HALD · A History of Probability and Statistics and their Applications Before 1750
HALD · A History of Mathematical Statistics from 1750 to 1930
HAMPEL · Robust Statistics: The Approach Based on Influence Functions
HANNAN and DEISTLER · The Statistical Theory of Linear Systems
HEIBERGER · Computation for the Analysis of Designed Experiments
HEDAYAT and SINHA · Design and Inference in Finite Population Sampling
HELLER · MACSYMA for Statisticians
HINKELMAN and KEMPTHORNE: · Design and Analysis of Experiments, Volume 1: Introduction to Experimental Design
HOAGLIN, MOSTELLER, and TUKEY · Exploratory Approach to Analysis of Variance

*Now available in a lower priced paperback edition in the Wiley Classics Library.

HOAGLIN, MOSTELLER, and TUKEY · Exploring Data Tables, Trends and Shapes
*HOAGLIN, MOSTELLER, and TUKEY · Understanding Robust and Exploratory Data Analysis
HOCHBERG and TAMHANE · Multiple Comparison Procedures
HOCKING · Methods and Applications of Linear Models: Regression and the Analysis of Variance, *Second Edition*
HOEL · Introduction to Mathematical Statistics, *Fifth Edition*
HOGG and KLUGMAN · Loss Distributions
HOLLANDER and WOLFE · Nonparametric Statistical Methods, *Second Edition*
HOSMER and LEMESHOW · Applied Logistic Regression, *Second Edition*
HOSMER and LEMESHOW · Applied Survival Analysis: Regression Modeling of Time to Event Data
HØYLAND and RAUSAND · System Reliability Theory: Models and Statistical Methods
HUBER · Robust Statistics
HUBERTY · Applied Discriminant Analysis
HUNT and KENNEDY · Financial Derivatives in Theory and Practice
HUSKOVA, BERAN, and DUPAC · Collected Works of Jaroslav Hajek—with Commentary
IMAN and CONOVER · A Modern Approach to Statistics
JACKSON · A User's Guide to Principle Components
JOHN · Statistical Methods in Engineering and Quality Assurance
JOHNSON · Multivariate Statistical Simulation
JOHNSON and BALAKRISHNAN · Advances in the Theory and Practice of Statistics: A Volume in Honor of Samuel Kotz
JUDGE, GRIFFITHS, HILL, LÜTKEPOHL, and LEE · The Theory and Practice of Econometrics, *Second Edition*
JOHNSON and KOTZ · Distributions in Statistics
JOHNSON and KOTZ (editors) · Leading Personalities in Statistical Sciences: From the Seventeenth Century to the Present
JOHNSON, KOTZ, and BALAKRISHNAN · Continuous Univariate Distributions, Volume 1, *Second Edition*
JOHNSON, KOTZ, and BALAKRISHNAN · Continuous Univariate Distributions, Volume 2, *Second Edition*
JOHNSON, KOTZ, and BALAKRISHNAN · Discrete Multivariate Distributions
JOHNSON, KOTZ, and KEMP · Univariate Discrete Distributions, *Second Edition*
JUREČKOVÁ and SEN · Robust Statistical Procedures: Aymptotics and Interrelations
JUREK and MASON · Operator-Limit Distributions in Probability Theory
KADANE · Bayesian Methods and Ethics in a Clinical Trial Design
KADANE AND SCHUM · A Probabilistic Analysis of the Sacco and Vanzetti Evidence
KALBFLEISCH and PRENTICE · The Statistical Analysis of Failure Time Data, *Second Edition*
KASS and VOS · Geometrical Foundations of Asymptotic Inference
KAUFMAN and ROUSSEEUW · Finding Groups in Data: An Introduction to Cluster Analysis
KEDEM and FOKIANOS · Regression Models for Time Series Analysis
KENDALL, BARDEN, CARNE, and LE · Shape and Shape Theory
KHURI · Advanced Calculus with Applications in Statistics, *Second Edition*
KHURI, MATHEW, and SINHA · Statistical Tests for Mixed Linear Models
KLUGMAN, PANJER, and WILLMOT · Loss Models: From Data to Decisions
KLUGMAN, PANJER, and WILLMOT · Solutions Manual to Accompany Loss Models: From Data to Decisions
KOTZ, BALAKRISHNAN, and JOHNSON · Continuous Multivariate Distributions, Volume 1, *Second Edition*
KOTZ and JOHNSON (editors) · Encyclopedia of Statistical Sciences: Volumes 1 to 9 with Index

*Now available in a lower priced paperback edition in the Wiley Classics Library.

KOTZ and JOHNSON (editors) · Encyclopedia of Statistical Sciences: Supplement Volume
KOTZ, READ, and BANKS (editors) · Encyclopedia of Statistical Sciences: Update Volume 1
KOTZ, READ, and BANKS (editors) · Encyclopedia of Statistical Sciences: Update Volume 2
KOVALENKO, KUZNETZOV, and PEGG · Mathematical Theory of Reliability of Time-Dependent Systems with Practical Applications
LACHIN · Biostatistical Methods: The Assessment of Relative Risks
LAD · Operational Subjective Statistical Methods: A Mathematical, Philosophical, and Historical Introduction
LAMPERTI · Probability: A Survey of the Mathematical Theory, *Second Edition*
LANGE, RYAN, BILLARD, BRILLINGER, CONQUEST, and GREENHOUSE · Case Studies in Biometry
LARSON · Introduction to Probability Theory and Statistical Inference, *Third Edition*
LAWLESS · Statistical Models and Methods for Lifetime Data, *Second Edition*
LAWSON · Statistical Methods in Spatial Epidemiology
LE · Applied Categorical Data Analysis
LE · Applied Survival Analysis
LEE and WANG · Statistical Methods for Survival Data Analysis, *Third Edition*
LEPAGE and BILLARD · Exploring the Limits of Bootstrap
LEYLAND and GOLDSTEIN (editors) · Multilevel Modelling of Health Statistics
LIAO · Statistical Group Comparison
LINDVALL · Lectures on the Coupling Method
LINHART and ZUCCHINI · Model Selection
LITTLE and RUBIN · Statistical Analysis with Missing Data, *Second Edition*
LLOYD · The Statistical Analysis of Categorical Data
MAGNUS and NEUDECKER · Matrix Differential Calculus with Applications in Statistics and Econometrics, *Revised Edition*
MALLER and ZHOU · Survival Analysis with Long Term Survivors
MALLOWS · Design, Data, and Analysis by Some Friends of Cuthbert Daniel
MANN, SCHAFER, and SINGPURWALLA · Methods for Statistical Analysis of Reliability and Life Data
MANTON, WOODBURY, and TOLLEY · Statistical Applications Using Fuzzy Sets
MARDIA and JUPP · Directional Statistics
MASON, GUNST, and HESS · Statistical Design and Analysis of Experiments with Applications to Engineering and Science, *Second Edition*
McCULLOCH and SEARLE · Generalized, Linear, and Mixed Models
McFADDEN · Management of Data in Clinical Trials
McLACHLAN · Discriminant Analysis and Statistical Pattern Recognition
McLACHLAN and KRISHNAN · The EM Algorithm and Extensions
McLACHLAN and PEEL · Finite Mixture Models
McNEIL · Epidemiological Research Methods
MEEKER and ESCOBAR · Statistical Methods for Reliability Data
MEERSCHAERT and SCHEFFLER · Limit Distributions for Sums of Independent Random Vectors: Heavy Tails in Theory and Practice
*MILLER · Survival Analysis, *Second Edition*
MONTGOMERY, PECK, and VINING · Introduction to Linear Regression Analysis, *Third Edition*
MORGENTHALER and TUKEY · Configural Polysampling: A Route to Practical Robustness
MUIRHEAD · Aspects of Multivariate Statistical Theory
MURRAY · X-STAT 2.0 Statistical Experimentation, Design Data Analysis, and Nonlinear Optimization

*Now available in a lower priced paperback edition in the Wiley Classics Library.

MYERS and MONTGOMERY · Response Surface Methodology: Process and Product Optimization Using Designed Experiments, *Second Edition*
MYERS, MONTGOMERY, and VINING · Generalized Linear Models. With Applications in Engineering and the Sciences
NELSON · Accelerated Testing, Statistical Models, Test Plans, and Data Analyses
NELSON · Applied Life Data Analysis
NEWMAN · Biostatistical Methods in Epidemiology
OCHI · Applied Probability and Stochastic Processes in Engineering and Physical Sciences
OKABE, BOOTS, SUGIHARA, and CHIU · Spatial Tesselations: Concepts and Applications of Voronoi Diagrams, *Second Edition*
OLIVER and SMITH · Influence Diagrams, Belief Nets and Decision Analysis
PANKRATZ · Forecasting with Dynamic Regression Models
PANKRATZ · Forecasting with Univariate Box-Jenkins Models: Concepts and Cases
*PARZEN · Modern Probability Theory and Its Applications
PEÑA, TIAO, and TSAY · A Course in Time Series Analysis
PIANTADOSI · Clinical Trials: A Methodologic Perspective
PORT · Theoretical Probability for Applications
POURAHMADI · Foundations of Time Series Analysis and Prediction Theory
PRESS · Bayesian Statistics: Principles, Models, and Applications
PRESS · Subjective and Objective Bayesian Statistics, *Second Edition*
PRESS and TANUR · The Subjectivity of Scientists and the Bayesian Approach
PUKELSHEIM · Optimal Experimental Design
PURI, VILAPLANA, and WERTZ · New Perspectives in Theoretical and Applied Statistics
PUTERMAN · Markov Decision Processes: Discrete Stochastic Dynamic Programming
*RAO · Linear Statistical Inference and Its Applications, *Second Edition*
RENCHER · Linear Models in Statistics
RENCHER · Methods of Multivariate Analysis, *Second Edition*
RENCHER · Multivariate Statistical Inference with Applications
RIPLEY · Spatial Statistics
RIPLEY · Stochastic Simulation
ROBINSON · Practical Strategies for Experimenting
ROHATGI and SALEH · An Introduction to Probability and Statistics, *Second Edition*
ROLSKI, SCHMIDLI, SCHMIDT, and TEUGELS · Stochastic Processes for Insurance and Finance
ROSENBERGER and LACHIN · Randomization in Clinical Trials: Theory and Practice
ROSS · Introduction to Probability and Statistics for Engineers and Scientists
ROUSSEEUW and LEROY · Robust Regression and Outlier Detection
RUBIN · Multiple Imputation for Nonresponse in Surveys
RUBINSTEIN · Simulation and the Monte Carlo Method
RUBINSTEIN and MELAMED · Modern Simulation and Modeling
RYAN · Modern Regression Methods
RYAN · Statistical Methods for Quality Improvement, *Second Edition*
SALTELLI, CHAN, and SCOTT (editors) · Sensitivity Analysis
*SCHEFFE · The Analysis of Variance
SCHIMEK · Smoothing and Regression: Approaches, Computation, and Application
SCHOTT · Matrix Analysis for Statistics
SCHUSS · Theory and Applications of Stochastic Differential Equations
SCOTT · Multivariate Density Estimation: Theory, Practice, and Visualization
*SEARLE · Linear Models
SEARLE · Linear Models for Unbalanced Data
SEARLE · Matrix Algebra Useful for Statistics
SEARLE, CASELLA, and McCULLOCH · Variance Components
SEARLE and WILLETT · Matrix Algebra for Applied Economics

*Now available in a lower priced paperback edition in the Wiley Classics Library.

SEBER · Linear Regression Analysis
SEBER · Multivariate Observations
SEBER and WILD · Nonlinear Regression
SENNOTT · Stochastic Dynamic Programming and the Control of Queueing Systems
*SERFLING · Approximation Theorems of Mathematical Statistics
SHAFER and VOVK · Probability and Finance: It's Only a Game!
SMALL and McLEISH · Hilbert Space Methods in Probability and Statistical Inference
SRIVASTAVA · Methods of Multivariate Statistics
STAPLETON · Linear Statistical Models
STAUDTE and SHEATHER · Robust Estimation and Testing
STOYAN, KENDALL, and MECKE · Stochastic Geometry and Its Applications, *Second Edition*
STOYAN and STOYAN · Fractals, Random Shapes and Point Fields: Methods of Geometrical Statistics
STYAN · The Collected Papers of T. W. Anderson: 1943–1985
SUTTON, ABRAMS, JONES, SHELDON, and SONG · Methods for Meta-Analysis in Medical Research
TANAKA · Time Series Analysis: Nonstationary and Noninvertible Distribution Theory
THOMPSON · Empirical Model Building
THOMPSON · Sampling, *Second Edition*
THOMPSON · Simulation: A Modeler's Approach
THOMPSON and SEBER · Adaptive Sampling
THOMPSON, WILLIAMS, and FINDLAY · Models for Investors in Real World Markets
TIAO, BISGAARD, HILL, PEÑA, and STIGLER (editors) · Box on Quality and Discovery: with Design, Control, and Robustness
TIERNEY · LISP-STAT: An Object-Oriented Environment for Statistical Computing and Dynamic Graphics
TSAY · Analysis of Financial Time Series
UPTON and FINGLETON · Spatial Data Analysis by Example, Volume II: Categorical and Directional Data
VAN BELLE · Statistical Rules of Thumb
VIDAKOVIC · Statistical Modeling by Wavelets
WEISBERG · Applied Linear Regression, *Second Edition*
WELSH · Aspects of Statistical Inference
WESTFALL and YOUNG · Resampling-Based Multiple Testing: Examples and Methods for p-Value Adjustment
WHITTAKER · Graphical Models in Applied Multivariate Statistics
WINKER · Optimization Heuristics in Economics: Applications of Threshold Accepting
WONNACOTT and WONNACOTT · Econometrics, *Second Edition*
WOODING · Planning Pharmaceutical Clinical Trials: Basic Statistical Principles
WOOLSON and CLARKE · Statistical Methods for the Analysis of Biomedical Data, *Second Edition*
WU and HAMADA · Experiments: Planning, Analysis, and Parameter Design Optimization
YANG · The Construction Theory of Denumerable Markov Processes
*ZELLNER · An Introduction to Bayesian Inference in Econometrics
ZHOU, OBUCHOWSKI, and McCLISH · Statistical Methods in Diagnostic Medicine

*Now available in a lower priced paperback edition in the Wiley Classics Library.

Printed in the United States of America/BNB